1.3
1.4
1.5
1.7
Intro to 6
6.1
6.3
6.4
6.5
6.6
458 – 487
Intro to 7
7.1
•7.2
•7.3
•7.4
•7.5
•4.1
•4.2
•4.3
•4.4
•7.6
7.

•10.4   Nuclear Envelope
•11.1   Replication DNA
•5.2   genetic message
•5.3   transcription
•5.4   translation
•5.5   genetic reg. prokaryotes
•5.6   "   "   Eukaryotes
         Histones →

* chapter page 518

_____
     3.5
Final   4.6
covers   5.4
     5.5
    9 review only
    11

CELL BIOLOGY
A Molecular Approach

CELL BIOLOGY
A Molecular Approach

Robert D. Dyson
Oregon State University
　and
*University of Oregon
Health Sciences Center*

SECOND EDITION

Allyn and Bacon, Inc.　　　　　　　*Boston, London, Sydney, Toronto*

Third printing . . . July, 1979

Copyright © 1978, 1974 by Allyn and Bacon, Inc., 470 Atlantic Avenue, Boston, Massachusetts 02210.

All rights reserved. Printed in the United States of America. No part of the material protected by this copyright notice may be reproduced or utilized in any form or by any means, electronic or mechanical, including photocopying, recording, or by any information storage and retrieval system, without written permission from the copyright owner.

Library of Congress Cataloging in Publication Data

Dyson, Robert D
  Cell biology.

  Includes bibliographies and index.
  1. Cytology. I. Title.
QH581.2.D95   1978   574.8'7   77-17813
ISBN 0-205-05942-2
ISBN (international) 0-205-06124-9

To LOUISE
And to LISA, LEANNA, *and* RYAN

Contents

Preface xi
Acknowledgments xiii
User's Guide xv

CHAPTER 1
The Cellular Basis of Life   1

Cell Theory 1 / The Rise of Molecular Biology 4 / The Unity and Diversity of Cells 7 / The Prokaryotic Cell 7 / The Eukaryotic Cell 13 / Cell Cultures 28 / Viruses 30 / The Choice of Experimental Systems in Biological Investigations 41

CHAPTER 2
Chemical Reactions in an Aqueous Environment   46

Bond Energy and Bond Stability 46 / The Biologically Important Weak Bonds 50 / Energy Changes in Chemical Reactions 56 / Coordination Compounds 59 / Acids and Bases 62 / The State of Intracellular Water 65

CHAPTER 3
Molecular Architecture and Biological Function   70

Organic Structure 71 / Lipids 72 / Carbohydrates 78 / Nucleic Acids 84 / Proteins 91 / Macromolecular Assembly 111 / An Introduction to Molecular Specificity: The Antibodies 118

CHAPTER 4
Cellular Homeostasis: Bioenergetics and the Regulation of Enzymes   134

Enzymes as Catalysts 134 / ATP and Coupled Reactions 139 / Glycolysis: A Metabolic Pathway 143 / Biological Oxidations 147 / Enzyme Kinetics and Regulation 150 / Cyclic Nucleotides and Protein Phosphorylation 163

CHAPTER 5
Cellular Homeostasis: Genes and Their Regulation 174

The Genetic Concept 174 / The Genetic Message 182 / Transcription 191 / Translation 197 / Gene Regulation in Prokaryotes 206 / Gene Regulation in Eukaryotes 216

CHAPTER 6
Membrane Structure and Molecular Transport 237

Membrane Structure 238 / Intercellular Junctions 248 / Membrane Permeability 254 / Membrane Transport 258 / Metabolically Coupled Transport 266 / Properties of the Carriers 277

CHAPTER 7
The Oxidative Organelles: Mitochondria, Chloroplasts, and Peroxisomes 287

The Mitochondrion 288 / Electron Transport and Oxidative Phosphorylation 294 / The Metabolic Roles of Mitochondria 302 / The Chloroplast 309 / The Photosynthetic Light Reaction 314 / The Photosynthetic Dark Reaction 324 / The Evolutionary Origin of Mitochondria and Chloroplasts 330 / Peroxisomes and the Evolution of Oxidative Metabolism 335

CHAPTER 8
Excitable Cells 348

The Neuron and the Nerve Impulse 348 / Propagation of Action Potentials 357 / Other Excitable Cells 364

CHAPTER 9
Contractility and Motility 373

Structure of the Muscle Cell 374 / The Mechanism of Contraction 378 / The Metabolism of Muscle 385 / Contraction-Relaxation Control 388 / Contractile Proteins of Non-muscle Cells 395 / Cilia and Flagella 400 / Cell Shape and Cell Movement 408

CHAPTER 10
Bulk Transport and Cellular Recognition 430

The Plasma Membrane 431 / Hormone Receptors 438 / Endocytosis and Lysosomal Digestion 443 / The Nuclear Envelope 452 / Endoplasmic Reticulum 458 / The Golgi Apparatus 466 / Secretion 469 / Synthesis and Turnover of Membrane Components 482

CHAPTER 11
Cell Division 492

Replication of DNA 492 / Mitosis 510 / Meiosis 517 / Spindle Dynamics 523 / Cytokinesis 528 / Cell Cycle Controls 536

CHAPTER 12
Cellular Differentiation 550

Gene Expression in Differentiated Cells 550 / Stability of Differentiation: Transdetermination, Regeneration, and Malignancy 555 / Cellular Senescence 567

APPENDIX
Tools of the Cell Biologist 577

Microscopy 577 / X-Ray Diffraction 589 / Centrifugation and Cell Fractionation 592 / Isotopes 597

Index 603

Preface

The name "molecular biology" . . . implies not so much a technique as an approach, an approach from the viewpoint of the so-called basic sciences with the leading idea of searching below large-scale manifestations of classical biology for the corresponding molecular plan.

                        W. T. ASTBURY
                        Harvey Lecture, 1950

The topic of this book is cellular structure and function, with emphasis on understanding how and why, rather than simple description.

It has been said that a good textbook should tell what is known, what is uncertain about what is known, what is unknown, and how it might come to be known. With that admonition as a guide, I have made it my goal in this work to present a unified introduction to the present state of our knowledge about cell biology, and a framework into which new facts will fit as they become available.

I have been gratified at the acceptance of the first edition, and by the unselfish willingness of many users to tell me not only what they liked but what they would like to see changed. Largely on the basis of those suggestions, and also because of new directions in the thrust of recent research, many changes have been made in this edition. Primarily, they involve (1) a greatly expanded coverage of membrane structure and function, including consideration of cellular recognition, hormone receptors, and more coverage of endocytosis and secretion; (2) much new material on contractility and motility, especially on the contractile systems of non-muscle cells; and (3) a number of clinically relevant medical examples.

It was with considerable pleasure that the last category of changes was made. As a physician as well as a scientist, I have a special interest in the application of cell biology to medicine, and I firmly believe that cell biology is the cornerstone of scientific medicine. I feel that the medically relevant examples will help demonstrate that interaction without diluting the purpose of the text, which is to present basic cell biology—i.e., the study of processes that, for the most part, are common to all eukaryotic cells.

To hold the overall length within bounds, I have abbreviated the discussion of some topics that were covered in the first edition, and dropped others entirely. If I have interpreted the comments of previous users correctly, the missing material will be little mourned.

If, in spite of my attempts to contain the length of the text, it still seems overlong, I take refuge in a remark by Max Delbrück: "Any living cell carries with it the experiences of a billion years of experimentation by its ancestors. You cannot expect to explain so wise an old bird in a few simple words."

Robert D. Dyson

Acknowledgments

Though one name appears as author, many people contributed to making this book a reality. The following scientists were listed in the first edition: Richard Barsotti, Anita Bolinger, Janet Cardenas, Ernst Florey, Dean Fraser, Eric Holtzman, Benjamin Kaminer, Eugene Kennedy, John Menninger, Margot Pearson, and Ralph Quatrano. It is my pleasure again to acknowledge the debt I owe them. Expert advice and counsel specifically on the second edition were obtained from: K. R. Brasch, J. M. Cardenas, P. G. Satir, R. Sinclair, and M. Locke. In addition, scores of comments were received from users of the first edition, in their aggregate providing the basis for most of the changes made in this revision. To all of the above, I am very grateful. No one of these people, however, saw the manuscript in its final form, so it is both safe and fair to attribute any errors that remain solely to me.

There are, in addition, about 350 electron micrographs plus hundreds of other drawings and photographs that should be acknowledged. Where material was obtained from other sources, credit is given in the figure legend. The drawings for the first edition were produced mostly by Jacqueline Atzet. Revisions and new drawings for this edition were done by Vantage Art, Inc., of Massapequa, N.Y.

Last, but by no means least, I am grateful to Allyn and Bacon and its staff for putting together a finished textbook from the bits and pieces furnished them. In particular, Judy Fiske deserves enormous credit for her skill in editing and production, and for guiding the manuscript through a difficult schedule. William Roberts was the Senior Editor whose faith and encouragement made the first edition possible. Harvey Pantzis, whose good humor and understanding are always appreciated, played a similar role for this edition. An author could not hope for more talented or congenial people with which to work!

User's Guide

To me, we have been children, playing on the shore,
Picking up pretty pebbles from the beach . . .
While there beside us lay an Ocean,
Wide with undiscovered Truth.

SIR ISSAC NEWTON

It is the function of this User's Guide to help you obtain as many pebbles as possible from this book. About the ocean of undiscovered Truth, we can do very little; the most important thing is to realize it is there.

Organization. The book begins with an introductory chapter in which the structural features of cells and their organelles are described briefly. The purpose is one of orientation and perspective—it is easy to get so involved with the study of a given cellular component that you forget how this component fits into the overall structural pattern of the cell. You can minimize that problem if you would return from time to time to the descriptions in Chap. 1.

The organization of the rest of the topics is rather traditional: the chemical and biochemical aspects of cell physiology are treated in Chaps. 2–5, followed by an in-depth consideration of the structure and function of the various cellular components that were introduced in Chap. 1. The emphasis throughout is on membranes, their structure, function, and interactions. There is, in addition, a chapter-length appendix that treats physical methodologies used in cell biology. Emphasis, of course, is on microscopy in its various forms.

There has been a conscientious effort to produce a relatively complete index. The purpose was not only to enhance the use of the text as a reference, but also to eliminate the need for a glossary. It is my opinion that scientific terms are best defined in context. Those who prefer the dictionary approach have a number of excellent biological dictionaries from which to choose. A relatively complete and very inexpensive example is the *Dic-*

tionary of Biology by E. B. Steen (New York: Barnes and Noble, 1971).

Chapter Structure. Each chapter is divided into several major sections, with each of the various sections having its own summary and study guide. The objective was flexibility, and the hope that topics could be more easily arranged to suit the design of a particular course or preferences of its instructor.

The summaries are concise presentations of the most relevant points. They are for review. They are not intended to stand on their own and cannot substitute for a careful study of the text and examination of the illustrations.

The study guides are designed to help you determine how much you remember from the associated section. Most of the questions are straightforward, and test only recall. Where appropriate, problems have been included and the answers given. Occasionally, a question demands extrapolation from given material to new situations. That is what science is all about, of course.

References. Each chapter, including the Appendix, has its own group of references, arranged by author within several broad subject-matter categories. The references consist mostly of reviews, but textbooks and reports of individual pieces of work are also listed. The credits in the figure legends should be considered an additional source of references.

When individual research papers are cited, it is either because they were specifically mentioned in the text, or because they can be recommended as classic presentations of a piece of work, or just because they are not adequately discussed in the reviews that are also listed.

In reading individual research papers, it helps to keep in mind that most are constructed with a format that includes: (1) a *summary* or *abstract*, giving the high points of the work; (2) an *introduction*, designed to describe the problem and its relationship to other work in the field; (3) a *materials and methods* section, giving experimental details; (4) a *results* section, containing the outcome of the experiments without much elaboration; and (5) a *discussion* section, in which the meaning of the results is presented, often with speculation about how the results bear on associated problems and what experiments along the same lines are planned for future work.

You may find the following approach to be a useful way for you to get the most out of a research paper without becoming confused by details: Try reading the introduction first, to define the problem; then the summary or abstract to get the outcome; then the discussion to gain the investigators' opinion about the significance of the work. The results section can serve as a reference to get more details about specific aspects of the experiments if you are interested. The materials and methods section is of little use to most students.

Suggestions for Further Study. To facilitate further study, and particularly as an aide in finding papers too recent to be included in the references, the major periodicals that review work on cell biology are listed here, grouped roughly according to their level of difficulty.

Generally easy:
  American Scientist.
  BioScience.
  Hospital Practice. Don't be misled by the name—excellent reviews of cell biology appear regularly.
  New Scientist. A news magazine.
  Perspectives in Biology and Medicine. Strong on history and philosophy.
  The Sciences.
  Science News.
  Scientific American. Always excellent.

Usually of moderate difficulty:
  Essays in Biochemistry. Published yearly.
  The Harvey Lectures. Collected and published yearly.
  MTP International Review of Science. Several volumes a year, each on one topic.
  Nature. Contains reviews plus brief original reports. The "News and Views" section of each weekly issue is an excellent way to keep up.
  Nobel Lectures. Collected and translated by the Elsevier Publishing Co. The lectures on "physiology or medicine" appear annually in *Science.*
  Proceedings of the Royal Society (London). The annual "Croonian lecture" and occasional review are highly recommended. The *Philosophical Transactions of the Royal Society* also publishes reviews.
  Science. Journal of the American Society for the Advancement of Science. Nice reviews plus a weekly section on scientific advances called "Research News."
  Science Progress (Oxford). The reviews are uniformly readable.
  Society of General Physiologists. Society for General Microbiology. Society for Experimental Biology. Each publishes an annual symposium volume with generally readable reviews.
  Trends in Biochemical Sciences. Several short, usually excellent reviews in each monthly issue.

Often more difficult:

Advances in . . . . Many of these. See especially volumes on *Genetics* and on *Morphogenesis*.

Annual Review of . . . . See especially the volumes on biochemistry, on genetics, and on microbology. The volumes on physiology, on plant physiology, and on neuroscience, should also be useful.

Bacteriological Reviews.

Biochimica et Biophysica Acta. Publishes multiple volumes each year, including a volume of reviews on biomembranes, one on bioenergetics, and one on cancer.

Biological Reviews (Cambridge).

Biomedicine. Their reviews stress cancer.

Biomembranes.

Cell. Frequent reviews.

Ciba Foundation Symposia. Each volume a separate topic.

Cold Spring Harbor Symposium on Quantitative Biology. Published yearly, each on a single topic.

CRC Critical Reviews in Biochemistry, and a companion volume in microbiology.

Current Topics in . . . Bioenergetics; Cellular Regulation; Developmental Biology; Membranes and Transport; and others.

Federation Proceedings. Publishes symposia sponsored by the Federation of American Societies for Experimental Biology.

International Journal of Biochemistry. Frequent short reviews.

International Review of Cytology. Published several times a year. Excellent reviews on all aspects of cell biology.

Life Sciences. Most issues carry one or more "Minireviews."

New England Journal of Medicine. Frequent short reviews on biological subjects.

Physiological Reviews.

Progress in . . . Biophysics and Molecular Biology; Molecular and Subcellular Biology; Nucleic Acid Research and Molecular Biology; etc.

Quarterly Review of Biology.

Quarterly Reviews of Biophysics. Sometimes too mathematical.

Reviews of Physiology, Biochemistry, and Pharmacology.

Sub-Cellular Biochemistry.

Periodicals in the first and second categories should be understandable even before the text is read. Periodicals in the third group should probably be consulted only for further detail, after the relevant parts of the text have been mastered.

1

The Cellular Basis of Life

1-1 Cell Theory 1
1-2 The Rise of Molecular Biology 4
1-3 The Unity and Diversity of Cells 7
1-4 The Prokaryotic Cell 7
    Classification
    Structure
1-5 The Eukaryotic Cell 13
    Plasma Membrane
    Nucleus
    Chromosomes
    Nucleolus
    Endoplasmic Reticulum
    Golgi Apparatus
    Lysosomes
    Peroxisomes
    Mitochondria
    Cytoplasmic Filaments
    Microtubules
    Centrioles and Basal Bodies
    Cilia and Flagella
    Vacuoles
    The Plastids and Cell Walls of Plants
    The Origin and Fate of Cytoplasmic Structures
1-6 Cell Cultures 28
1-7 Viruses 30
    Discovery
    The Bacteriophage
    Life Cycle of the Bacteriophage
    The Structure of Animal and Plant Viruses
    Life Cycle of Animal and Plant Viruses
    A Comparison of Cells and Viruses
1-8 The Choice of Experimental Systems in Biological Investigations 41
Summary 42
Study Guide 43
References 44
Notes 45

Long ago it became evident that the key to every biological problem must finally be sought in the cell; for every living organism is, or at some time has been, a cell.

                        E. B. Wilson, 1925[1]

1-1 CELL THEORY

The foundations of cell biology were formed in the seventeenth century, with two of the most important advances of that period coming from the Englishman Robert Hooke (1635–1703) and the Dutch inventor and scientist Antony van Leeuwenhoek (1632–1723).

    In 1665, Hooke published a collection of essays under the title *Micrographia*. One essay described cork as a honeycomb of chambers, or "cells." The chambers are now recognized to be the rigid remains of

CHAPTER 1
The Cellular Basis of Life

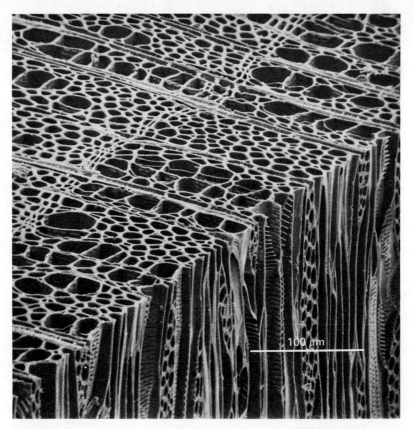

FIGURE 1-1 The Three-dimensional Structure of Wood. Scanning electron micrograph of magnolia shows the honeycomb of chambers left when wood is dry. A similar pattern in cork was described by Hooke as an array of "cells," although each chamber is actually the rigid remains of a dried "cell," as we now use that word. (Courtesy of B. A. Meylan, Physics and Engineering Lab., Dept. of Scientific and Industrial Research, New Zealand.)

dried cells (see Fig. 1-1). Hooke thought of the cells he observed as something similar to the veins and arteries of animals—they were filled with "juices" in living plants—but his microscope did not permit the observation of any intracellular structure, an observation that would have dispelled the notion that cells are merely partitioned channels for the passage of material.

Within a decade after the publication of Hooke's essays, Leeuwenhoek had succeeded in greatly improving the art of polishing lenses of short focal length and had used his lenses to describe a host of "little animals," many of which proved to be single cells (see Fig. 1-2). These little animals were probably protozoa. A typical size for these organisms would be 100 micrometers (100 μm), which is about the resolving power of the human eye.[2] In other words, when two objects get closer than 100 μm they begin to appear as one. Hence, it would take nothing more than a simple magnifying lens to make at least the larger protozoa visible.

Leeuwenhoek was later able to use his microscope to describe for the first time the existence of bacteria, the dimensions of which are typically only about 1 or 2 μm. This incredible feat means that he had an instrument theoretically capable of seeing not only whole animal cells but details of structures within the cells as well. However, in order to take advantage of that capacity, better ways of preparing tissues for microscopy were needed. By the end of the seventeenth century, considerable progress in that area had been made, so that microscopists were able to examine tissues as thin as 10 μm, stained in various ways to bring out different details. (See Appendix A-1, Microscopy.)

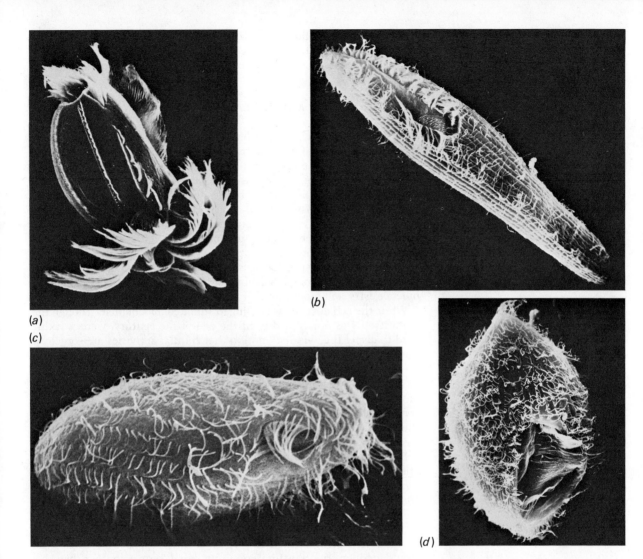

With these improved microscopes and preparation techniques, R. J. H. Dutrochet was able to conclude by 1824 that all animal and plant tissues are actually aggregates of cells of various kinds, and that growth results from an increase in either the size or the number of cells, or both. Though he recognized that intact cells can be separated from each other, he did not realize that each is capable of its own independent existence including, in most cases, the ability to reproduce itself.

The next few decades saw a rapid increase in the understanding and appreciation of cells. In 1838–1839 the German biologists M. J. Schleiden and Theodor Schwann argued convincingly that each cell is capable of maintaining an independent existence, an idea that had a profound influence on the scientific community. Schleiden and Schwann were followed by Rudolf Virchow, who in 1858 published his classical textbook *Cellular Pathology*, which supported the concepts suggested by Schleiden and Schwann and extended them to attribute every cell to a preexisting cell (*omnis cellula e cellula*—every cell from a cell). Virchow's ideas completed what has come to be known as the *cell theory* or

FIGURE 1-2 Leeuwenhoek's "Little Animals"—as he never saw them. These scanning electron micrographs are of protozoa. Each is a single cell. (a) *Uronychia* sp., 100 μm long. (b) *Blepharisma* sp., about 140 μm. (c) *Tetrahymena pyriformis* W. about 25 μm. (d) *Turaniella* sp., 250 μm. [Courtesy of E. B. Small and G. Antipa. (a) and (b) from E. Small, D. Marszalek, and G. Antipa, *Trans. Am. Micros. Soc.*, **90**: 283 (1971).]

cell doctrine, which holds that (1) all living things are composed of one or more units called cells, (2) each cell is capable of maintaining its vitality independent of the rest, and (3) cells can arise only from other cells.

1-2 THE RISE OF MOLECULAR BIOLOGY

Prior to Virchow, the tendency had been to regard the cell as a structural unit rather than as a functional unit of living systems. However, in his book, adapted from a series of lectures given at the Berlin Institute of Pathology, Virchow presented disease as an aberration in normal cellular processes. The properties of a tissue—or of a whole organism—must therefore be due to the properties of its individual cells. Between the 1858 publication of Virchow's book and 1925, when E. B. Wilson wrote the words with which this chapter opened, the cell doctrine as we now understand it became firmly entrenched. During this period also, the influence of the physical sciences on biology was growing ever stronger, complementing the older discipline of physiology with the new fields of biochemistry and biophysics.

While the cell doctrine was being formulated, biochemists (though the word itself was not coined until the end of the century) were working hard to dispel the "vitalist" notion that cellular activities are somehow immune to the laws of chemistry and physics. The first blow came with the laboratory synthesis of urea ($H_2N-CO-NH_2$) by Friedrich Wöhler in 1828, proving that organic compounds are not formed by any mysterious process, but by ordinary chemical reactions. Synthetic organic chemistry flourished thereafter, and with it came the first clues to the nature of the metabolic transformations by which a cell turns its foodstuffs into a vast array of quite different molecules.

By the middle of the nineteenth century, considerable controversy had arisen concerning the nature of one commonly observed metabolic transformation, that of alcoholic fermentation, a process in which glucose is changed to ethanol and carbon dioxide. The German chemist Justus von Liebig (1803–1873) maintained that alcoholic fermentation is catalyzed by nonliving "ferments," later called *enzymes*, that are naturally present in the juices being fermented. He was hotly opposed by those who would attribute the transformation to a "vital" process rather than to a chemical process.

In 1871 a Frenchman, Louis Pasteur (1822–1895), demonstrated that a living organism—yeast—is essential to alcoholic fermentation as it usually occurs. Although vitalists felt that Pasteur's observation proved their point, Eduard Buchner found in 1897 that it is not necessary for the yeast to be alive; extracts from yeast cells can still carry out alcoholic fermentation. The necessary enzymes are made out of simpler organic materials by the yeast cells; however, the function of enzymes does not depend on the vitality of the cells that make them. Thus Liebig's earlier contention that fermentation is a straightforward chemical process was vindicated, even though Liebig did not recognize the source of the enzymes responsible for it.

At the turn of the century, then, the identity and importance of the individual cell were recognized and serious efforts were being made to understand its function. Buchner's work established that individual components such as enzymes could be isolated and studied in the laboratory. Although the success of his approach led many to view the cell as little more than a "bag of enzymes," the interdependence of structure and function gradually became apparent. This recognition fostered an increasing cooperation between those biologists interested in the structure of individual cells (cytologists) or the cellular structure of tissues and organs (histologists), and those whose primary interest is in function (biochemists, biophysicists, and physiologists). As it became more and more difficult to distinguish one discipline from another, interdisciplinary studies, variously called *molecular biology* and *cell biology*, appeared. This pooling of knowledge and experience permitted the rapid expansion of our understanding of cellular processes, some of the early milestones of which are summarized in Table 1-1.

TABLE 1-1. Some Early Milestones in Cell Biology

Year	Milestone
1665	Hooke publishes *Micrographia*, in which he describes and illustrates the cellular structure of cork.
1675	Leeuwenhoek improves microscope lenses, and with them discovers a variety of single-celled life forms, including (in 1683) bacteria.
1824	Dutrochet correctly concludes that all tissues, animal and plant, are composed of smaller units, the cells.
1828	Wöhler synthesizes urea, discrediting the view that organic compounds can only be made by living things, and paving the way for a systematic investigation of cellular reactions.
1830	Meyen suggests that each plant cell is an independent, isolated unit capable of receiving nourishment and building its own internal structures.
1831	Brown reports the existence of nuclei.
1838–9	Schleiden and Schwann argue convincingly for the cell doctrine, holding that all tissues are composed of cells and that the metabolism and development of tissues are the result of cellular activity.
1840	Liebig proposes that alcoholic fermentation is a purely chemical reaction, independent of living cells but catalyzed by substances (ferments, or enzymes) that are naturally present in juices.
1858	Virchow correctly asserts that cells arise only from other cells and that, as the functional units of life, they are also the primary site of disease.
1862	Pasteur disposes of the spontaneous generation theory of microbial appearance.
1871	Pasteur proves that natural alcoholic fermentation always involves the action of yeast.
1897	Buchner finds that alcoholic fermentation requires only the extract of yeast, not the cells themselves.
1907	Harrison finds a satisfactory way of growing isolated animal cells in the laboratory, so that future studies of cellular function can be carried out under controlled conditions.

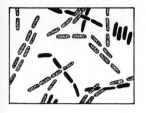
Bacteria: Rods

Flagellated rods

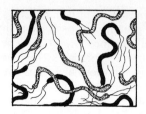
Spirals

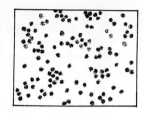
Spheres

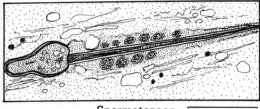
Spermatozoon 5 μm

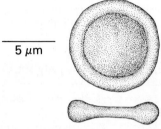
5 μm

Mammalian red blood cells
(side and face views)

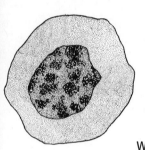
5 μm

White blood cells

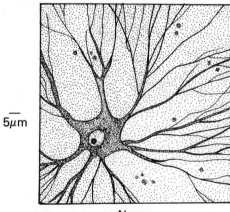
5 μm

Neuron

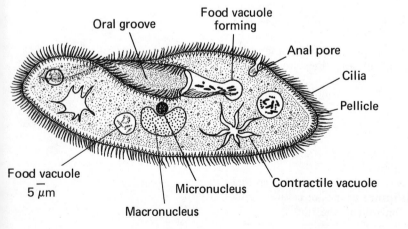
Paramecium

Oral groove, Food vacuole forming, Anal pore, Cilia, Pellicle, Contractile vacuole, Micronucleus, Macronucleus, Food vacuole, 5 μm

5 μm

Multinucleate muscle cells

1-3 THE UNITY AND DIVERSITY OF CELLS

We now recognize that cells are found in an enormous variety of sizes and shapes representing their evolutionary adaptation to different environments or to different specialized functions within a multicellular organism. Cells range in size from the smallest bacteria, only a few tenths of a μm in diameter, to certain marine algae and to various bird eggs with dimensions of centimeters. Between these two extremes, we find that most bacteria have measurements of a couple of μm, human red blood cells have diameters of 6–8 μm and most other human cells fall in the range of about 5–20 μm (see Fig. 1-3).

For all their apparent diversity, however, cells have many characteristics in common, the most basic of which is the potential for an independent existence. That is, cells have the ability to continue living in the absence of any other cell, a capacity that requires first, a metabolic machinery capable of obtaining energy from the environment through capture of light or the degradation of chemical foodstuff; and second, the ability to use this energy to support essential life processes. These processes include the movement of components from one part of the cell to another, the selective transfer of molecules into and out of the cell, and the ability to transform molecules from one chemical configuration to another in order to replace parts as they wear out or to support growth and reproduction. In addition to this metabolic machinery, a cell must have a set of genes to act as blueprints for the synthesis of other components. And finally, a cell must have a physical delimiter, a boundary between it and the rest of the world, called the *cell membrane*.

Most of our attention in the ensuing chapters will be devoted to the similarities among various cells rather than to their differences. Once the discussion begins to focus on the molecular basis for cell structure and function, differences among cells appear as minor variations on a few central themes. Before attacking the subject at that level, however, we shall pause to take a look at some of the more obvious physical features of the typical cell as they appear to the microscopist, and to introduce some of the nomenclature used by cell biologists. This brief overview is presented for purposes of orientation only. Most features discussed briefly here will be presented again in much greater detail. However, it may be useful from time to time to return to this chapter in order to help keep later discussions in perspective.

1-4 THE PROKARYOTIC CELL

It is useful to divide all cells into two broad groups, eukaryotic and prokaryotic, according to whether or not their genes are contained in a well-defined nucleus. (The terms are derived from a prefix *pro*, meaning before; *eu*, meaning true; and the suffix *caryon* or *karyon*, meaning nucleus.)

FIGURE 1-3 The Diversity of Cells. Bacteria (top row) are among the smallest of all cells, with typical lengths of a μm or so. Animal and plant cells are more often 10–20 μm in diameter, although some, such as nerve and muscle cells, may be very long with relatively small diameters. Single-celled animals, like the *Paramecium* (a protozoan) shown here, may be incredibly complex and quite large (over 200 μm).

CHAPTER 1
The Cellular Basis of Life

Only bacteria and the few species of blue-green algae (which are not necessarily either blue or green) belong to the simpler, prokaryotic group. Bacteria are by far the more numerous. Although they have been identified in fossil remains some two billion (2×10^9) years old, they were unknown until 1683, when Leeuwenhoek first reported seeing them with his improved microscope. He noted the major morphological classes of bacteria—spheres, rods, and spirals—and described them in letters to the Royal Society. Leeuwenhoek found that bacteria tend to be associated with decaying organic matter from which, it was generally believed, they arise spontaneously. This notion persisted until the mid-nineteenth century when Pasteur, using newly developed sterile techniques, proved that bacteria do not arise by spontaneous generation, but by contamination. When he sealed boiled organic material (e.g., milk) in its container, no bacterial growth could be detected. However, when the flask was opened to the air for a time, cells invariably appeared. Pasteur correctly concluded, in his 1862 monograph on spontaneous generation, that airborne microbes are the source of the bacterial growth (see Fig. 1-4). Bacteria, it seems, are no exception to the cell doctrine.

Bacteria (*Schizomycetes*), along with the small group of blue-green algae (*Cyanophyta*), have in common a uniquely simple cell structure that sets them apart from all other forms of life. Because of this simplicity, they are often relegated to a kingdom of their own, *Monera* or *Prokaryotae*, rather than to either the plant or animal kingdom. Bacteria and the blue-green algae are, however, true cells. They are bounded by a membrane and contain the genes and metabolic apparatus needed to grow and reproduce, given the proper nutrients and environment.

Classification. Bacteria can be distinguished from blue-green algae in several ways, one of which is that blue-green algae evolve oxygen whereas bacteria do not. Oxygen is released by blue-green algae and higher plants as a by-product of photosynthesis, the process whereby light energy is converted to chemical energy. Although some bacteria are also photosynthetic, they use quite a different mechanism, including a different set of light-capturing pigments. Bacteria, for instance, never evolve oxygen, nor do they have chlorophyll *a*, a green pigment common to all other photosynthetic systems.

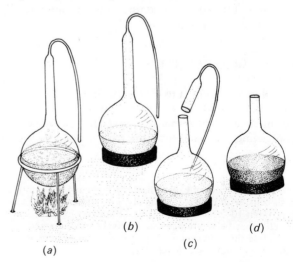

FIGURE 1-4 Pasteur's Experiment. This experiment did more than any other to convince scientists that life does not arise spontaneously from organic matter. Pasteur found (*a*) that if he boiled a solution of organic compounds and left the flask undisturbed, (*b*) no contaminating growth would appear. But when a flask was opened directly to the air (*c*) so that dust could settle into it, a variety of microscopic organisms was soon found flourishing therein (*d*). The long, curved neck of the boiled flask, though unsealed, was an effective trap for dust particles.

Bacteria may be classified according to their shape as *cocci* (sing., *coccus*) for spherical cells, *bacilli* for rods, and *spirilla* for a helical outline. (See Fig. 1-5.) The cocci, especially, are prone to hang together after they are formed by binary fission of the parent cells. They are referred to as diplococci when in pairs, streptococci when in chains or filaments, and staphylococci when found in clusters.

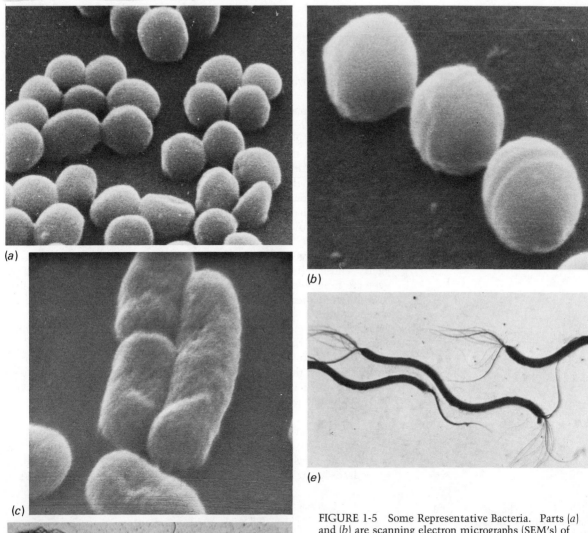

FIGURE 1-5 Some Representative Bacteria. Parts (a) and (b) are scanning electron micrographs (SEM's) of *Staphylococcus aureus* and *Streptococcus pyogenes*, respectively. Both bacteria are about 0.7 μm in diameter. (c) SEM of *Proteus mirabilus*. The larger cell, which is about 1.5 μm long, would normally divide (by binary fission) to yield two smaller cells such as those next to it. (d) Light micrograph of *Salmonella typhosa*, a flagellated rod. The cell bodies are 1–3.5 μm long. (e) *Spirillum volutans*, a 3–5 μm flagellated spiral. [(a), (b), and (c) courtesy of D. Greenwood, St. Bartholomew's Hospital, London. Part (a) from D. Greenwood and F. O'Grady, *Science*, **163**: 1075 (1969) copyright by the American Association for the Advancement of Science (AAAS). (d) and (e) from the Turtox Collection, courtesy of General Biologicals, Inc.]

Bacteria are also broadly classified according to their response to a staining procedure developed in 1884 by the Danish bacteriologist Christian Gram. Those cells that are stained by this method are called gram-positive; the others are gram-negative. The distinction is important because of a fundamental difference in the structure of the walls of the two classes of cells, and a corresponding difference in their sensitivity to various chemical agents other than the stain. For example, lysozyme, a bactericidal enzyme found in egg whites, and in tears and other secretions, is much more effective against gram-positive cells. Penicillin also shows much greater bactericidal action against gram-positive bacteria.

Still another way of distinguishing bacteria is by their nutritional requirements. If a cell requires complex organic molecules of the kind that would normally be supplied only by the destruction of other cells, it is called *heterotrophic* ("feeding on others"). If, however, a cell can utilize carbon dioxide (CO_2) from the air, reducing it to the organic compounds needed, the cell is called an *autotroph* ("self-feeder"). If the energy required for carbon dioxide reduction is obtained from light via photosynthesis, the cell is *photoautotrophic*. If energy is supplied by the oxidation ("burning") of inorganic compounds (e.g., H_2S to S, or H_2 to H_2O), the organism is *chemoautotrophic*.

A fourth useful way of grouping bacteria is by their variable requirement for oxygen. Some are *aerobes*, meaning that they grow only in the presence of oxygen. *Anaerobes*, on the other hand, grow without oxygen. Those cells that can grow in either the presence or absence of oxygen are called *facultative* organisms. Some facultative organisms shift to a different form of metabolism in the presence of oxygen, whereas others do not.

Genus and species names are assigned to bacteria on the basis of a combination of these and other physical and chemical differences, but not without occasional ambiguity. In recent years, attempts to bring greater order to the field of bacterial taxonomy have focused largely on measuring the genetic similarity between various strains of cells using techniques that will be described in later chapters.

Structure. Bacteria and blue-green algae generally have dimensions of a few μm, though the smallest bacteria (mycoplasmas, rickettsiae, and chlamydiae) measure only a few tenths of a μm. Many bacteria have appendages in the form of *flagella* (see Fig. 1-5d and e and Fig. 1-6). They

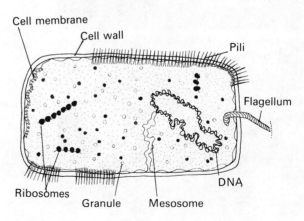

FIGURE 1-6 The Generalized Bacterium. Almost all bacteria are surrounded by a cell wall, which is attached to the cell membrane at relatively few points. Either flagella or pili or both may extend from the interior (*cytoplasm*) through the wall. Ribosomes, granules of various kinds, and DNA are found in the cytoplasm, which may also include an inward extension of the membrane known as a mesosome. In some cases mushroomlike stalks have been reported on the inner surface of the plasma membrane (left side of drawing).

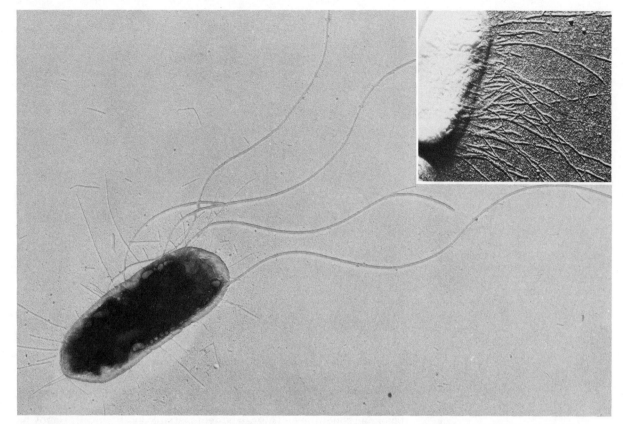

FIGURE 1-7 Pili. Long flagella and shorter pili are seen protruding from the common intestinal bacterium, *Escherichia coli* (*E. coli*). INSET shows pili at higher magnification. [Courtesy of H. C. Berg and G. Fonte, Univ. of Colorado; *inset* courtesy of M. Wilson and A. Hohmann, Univ. of Guelph, Guelph, Ontario, Canada.]

are typically 0.01–0.02 μm in diameter and up to 10 or 11 μm long. Although the bacterial flagellum is a simple structure, consisting of parallel strands of protein wound about each other in a ropelike fashion and anchored within the cell at a *basal body*, it can impart motility to the cell, making it possible for the cell to swim.

In addition to flagella, certain bacteria have numerous finer projections called *pili* ("hair"; sing., *pilus*) or *fimbriae* ("fringe"), with diameters of 0.01 μm or less and lengths up to a μm or two (see Fig. 1-7). Most pili seem to have a construction somewhat similar to that of flagella, including a basal body anchor within the cytoplasm. However, male bacteria of certain strains carry one or two special hollow sex pili that form a bridge with the female cell during mating. In a few pathogenic strains of bacteria (that is, bacteria that cause disease—most do not) pili are responsible for recognizing and attaching to the host.

The surface layer of prokaryotes is very often a gelatinous *capsule* or, in the case of the blue-green algae, a *sheath*. Many pathogenic bacteria can no longer cause an active infection when deprived of their capsule, apparently because of the protection it affords against host defenses.

Inside the sheath or capsule, when one is present, is a rigid *cell wall* (see Fig. 1-8). Cellulose is a common component of this structure in the blue-green algae just as it is in the higher plants; however, bacteria have walls composed of materials (e.g., teichoic acids) not found anywhere else. Gram-positive bacteria have walls varying in thickness from about 0.015–0.08 μm, comprising some 10–25% of the dry weight of the cell. The gram-negative cell wall, though thinner (generally about 0.01 μm),

CHAPTER 1
The Cellular Basis of Life

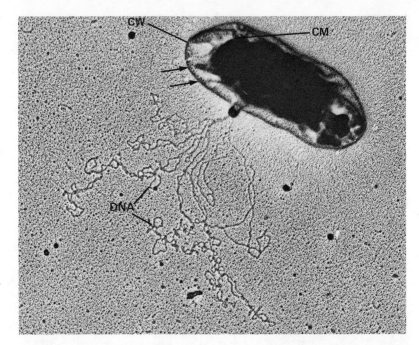

FIGURE 1-8 A Lysed Bacterium. Because of a loss of some cytoplasm, the plasma membrane has collapsed inward, clearly showing that the cell membrane (CM) and cell wall (CW) are separate structures, connected at only a few points (arrows). The extruded fiber is a molecule of deoxyribonucleic acid (DNA), which forms the genetic material of the cell. (Courtesy of M. M. K. Nass, from *Biological Ultrastructure: The Origin of Cell Organelles*, P. J. Harris, ed., Corvallis: Oregon State Univ. Press, 1971.)

is a more complex structure, accounting for the different response to the Gram stain and to certain antibiotics, as mentioned above.

The capsule, sheath, and cell wall are constructed from materials secreted by the cells. They are not essential to viability, at least under favorable conditions. This can be demonstrated by chemically removing them, producing a *protoplast* if removal of wall is complete or a *spheroplast* if removal of the wall material is incomplete. Such cells are still capable of growth and replication. In fact, one genus of bacteria, the tiny *Mycoplasma* (formerly called pleuro-pneumonia-like organisms, or PPLO), is always found without walls.

Although a cell might survive very well without flagella, pili, capsule, or walls, it cannot survive without its membrane. This structure encloses the cytoplasm ("cell plasma") or protoplasm ("chief plasma") of the cell, and hence is also called the *cytoplasmic membrane, plasma membrane,* or *plasmalemma.* The membrane is an essential barrier through which nutrients must pass on their way into the cell and through which waste products or secretions are passed out of the cell. It exercises considerable selectivity in this traffic, a selectivity that is vital to the continuing life of the cell.

The plasma membrane is a deceptively simple structure, physically, only about 7–10 nm thick. Nevertheless, a variety of different molecules are known to be associated with it, some of which act as ferries for the transport of specific materials into and out of the cell (see Chap. 6). In addition, some bacterial membranes appear to have mushroomlike structures pointing toward the interior of the cell (see Fig. 1-6) which are thought to be the major site of oxygen utilization.

While cells from higher organisms have numerous inclusions that are surrounded by their own individual membranes, prokaryotic cells are relatively devoid of such inclusions. There are granules of various sorts that come and go depending on the metabolic condition of the cell, and sometimes these granules are surrounded by a thin 2–4 nm protein membrane, which is much less complex than the cell membrane itself. The poly-β-hydroxybutyrate granules, which accumulate in some cells when carbon and energy sources are in excess, are an example. They may account for greater than 50% of the dry weight of the cell, serving as a reserve of carbon and energy. Spores, too, are found in certain bacterial cells. In addition, the photosynthetic prokaryotes (blue-green algae and green and purple bacteria) have complex layers of membranes (*lamellae*) within their cytoplasm that contain light-capturing pigments and certain other parts of the photosynthetic apparatus. *Mesosomes* are also sometimes seen, but they are really inward folds of the cytoplasmic membrane rather than independent organelles (see Fig. 1-9). In general, the small size of prokaryotes drastically limits the opportunity for internal structure since their longest dimension may be the equivalent of only a few thousand hydrogen diameters.

Ribosomes are the only internal feature aside from the nuclear body (discussed below) that are constant from one prokaryotic cell to the other. Ribosomes are roughly spherical particles about 20 nm in diameter. Some are attached to the plasma membrane, while others are free in the cytoplasm. As we shall see later, ribosomes are the sites of protein synthesis. Hence, cells that are growing more rapidly, and therefore require faster protein synthesis, also have more ribosomes—up to 40% of the cell's total dry weight.

The information needed to guide the construction and continuing functioning of a cell (any cell) is found in its genes. Genes are segments of long molecules of *deoxyribonucleic acid*, or *DNA*. A prokaryotic cell requires only one long double-stranded DNA molecule to contain its entire set of genes, although more than one copy of the DNA molecule may be present. The area of the cell containing DNA is sometimes referred to as a *nucleus*, a *nuclear body*, or a *nucleoid*.

When stretched out, DNA molecules (in both prokaryotes and higher cells) are often a thousand times longer than the cell itself. This great length requires a careful packing if tangles are to be avoided. And tangles must be avoided, because replication of the cell requires a parallel replication and subsequent distribution of its DNA to daughter cells. The distribution problem in prokaryotes appears to be eased by anchoring the DNA to a site on the plasma membrane, often at a mesosome, so that after replication of the DNA the two copies can be separated by growth of the plasma membrane between them.

Many of the prokaryotic structures just described have names that have been adapted from somewhat similar structures found earlier in the cells of higher organisms. Often there is a fundamental difference between the two, however. The nucleus is a case in point, for the nuclear genes of higher cells are enclosed in their own membrane, or nuclear envelope. This is a "true nucleus," and cells that have one are called *eukaryotes*, which is what that name means. The prokaryotes, as that name implies, have a more primitive nuclear arrangement.

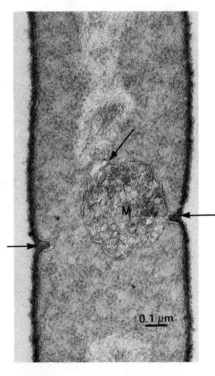

FIGURE 1-9 Bacterial Mesosome. The mesosome (M) represents an inward-folded, convoluted extension of the plasma membrane. As in this micrograph of *Bacillus subtilis*, it is often found at the site of cell division (just beginning at the points marked by outside arrows) and may have DNA attached (inside arrow). [Courtesy of N. Nanninga, *J. Cell Biol.*, **48**: 219 (1971).]

1-5 THE EUKARYOTIC CELL

All the cells from higher animals and those of many microscopic organisms as well are eukaryotic. There is, of course, tremendous diversity among eukaryotic cells, but there are also features common to all and ad-

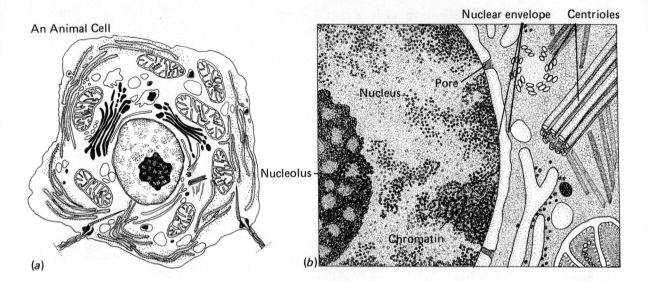

FIGURE 1-10 The Generalized Eukaryotic Cell. (a) An animal cell. (b) A portion of the nucleus, showing the two layers of the nuclear envelope and two centrioles nearby. (c) A cilium (or flagellum) extending from its basal body. (Note that the basal body and centriole have the same construction.) An anchor, in the form of a rootlet, is seen. (d) Detail showing a mitochondrion. On one side is a well-developed Golgi body, on the other side rough endoplasmic reticulum. Note vesicle formation from the endoplasmic reticulum and Golgi membranes. Many filaments are seen just under the cell membrane. (e) A plant cell. It differs from the animal cell primarily in having a rigid cell wall with penetrating plasmodesmata (pores), in the size of the vacuoles, and in having chloroplasts.

ditional features that are not universal but that are very widespread. (See Figs. 1-10 and 1-11.)

The internal structure of eukaryotic cells started to become clear in the 1950s through two different lines of attack: electron microscopy, the usefulness of which required the discovery of new ways of fixing and staining biological material; and cell fractionation by centrifugation, which allowed chemical analysis of isolated components (see Appendix). Many of the pioneering studies were initiated by George Palade, Christian de Duve, and Albert Claude, who shared the 1974 Nobel Prize for this work. The structures about to be discussed were for the most part first described by these investigators and their associates, working at Rockefeller University in New York City.

Plasma Membrane. Plasma (cell) membranes have much the same structure whether found in prokaryotic or eukaryotic cells. They are universal and necessary to every cell type. The function of the plasma membrane was mentioned in the discussion of prokaryotes; it serves the same purpose in eukaryotes—that is, it serves as a boundary to the cell that permits the passage of selected materials in and out at carefully regulated rates. Details of membrane structure and function will occupy a great deal of our attention later, for it is one of the most exciting and rapidly advancing areas of current biological research.

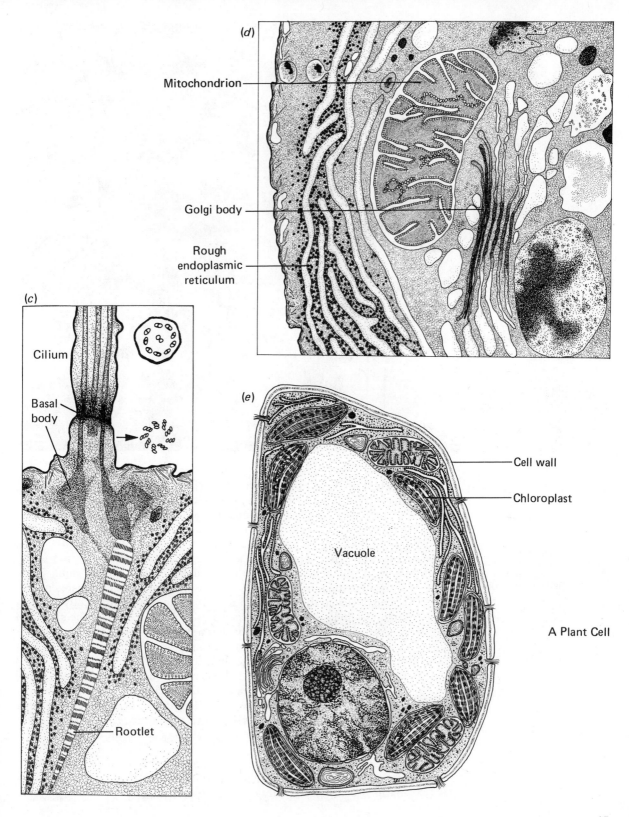

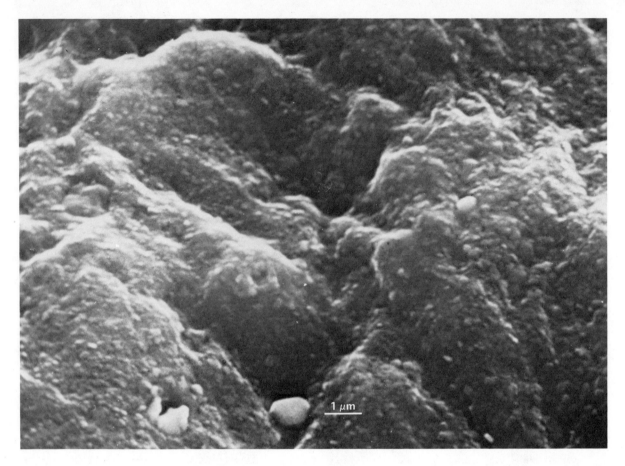

FIGURE 1-11 Bacteria Lying on the Surface of Human Forearm Skin. The crevices are the demarcations between adjacent cells. Note the rod-shaped particle, presumably a bacterium, lying in one of the major crevices. This scanning electron micrograph is presented to emphasize the size differential between a typical prokaryotic and eukaryotic cell. [Courtesy of E. O. Bernstein and C. B. Jones, *Science*, **166**: 252 (1969). Copyright by the AAAS.]

Nucleus. The most prominent internal feature of most cells is the nucleus (Fig. 1-10), discovered in 1883 by Robert Brown. (This is the Brown referred to in the term "Brownian motion.") The nucleus, which is the repository of nearly all of a cell's genetic information, is bounded by the *nuclear envelope* (see Fig. 1-12). The nuclear envelope is usually composed of two membranes, each about 10 nm thick, separated by a space of some 10–15 nm for an overall thickness that is often about 35 nm. Each of the individual membranes is similar in structure to the plasma membrane; however, the two appear to be fused together periodically to form circular windows called *nuclear pores*. (See Fig. 1-13.) The size and spacing of pores vary somewhat, but a diameter of 80–100 nm is quite common. The pores occupy up to about a third of the total surface area of the nuclear envelope and are thought to be responsible for the selective passage of materials into and out of the nucleus.

The prominence and importance of the nucleus are recognized in the nomenclature used to describe the cell contents: everything inside the plasma membrane is collectively called the *protoplasm*, but only that part of the protoplasm outside the nucleus is called the *cytoplasm*.

Chromosomes. The genes of the cell are almost all found in the nucleus, associated with strands of *chromatin* (see Fig. 1-13a). Although chromatin is composed chiefly of DNA, it also contains RNA and protein.

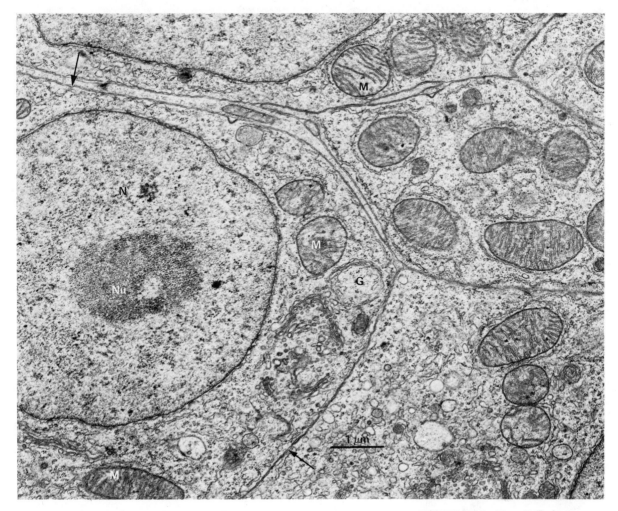

FIGURE 1-12 Liver Cells from a Chick Embryo. This electron micrograph shows portions of several cells. The cell at the lower left (bounded by the plasma membrane running between the two arrows) has a nucleus (N) with a prominent nucleolus (Nu). The nuclear envelope is composed of two membranes, the outer one of which is wavy. Also visible are the Golgi apparatus (G), mitochondria (M), and numerous vesicles within the cytoplasm. [Courtesy of C. A. Benzo and A. M. Nemeth, *J. Cell Biol.*, **48**: 235 (1971).]

Moreover, chromatin may be either dispersed throughout the nucleus or gathered together in discrete, compact bodies called *chromosomes* ("colored bodies"), a name implying that they are dense enough to stain readily. The genes of a human cell, for example, are distributed among its 46 chromosomes. A strand of chromatin varies in thickness from 10–40 nm, giving it a "bumpy" appearance in the electron microscope, but making it too small to be seen with the light microscope. Recent evidence indicates that dispersed chromatin fibers are anchored to the inside of the nuclear envelope.

Nucleolus. The most clearly defined feature within the nucleus is often the nucleolus (the Latin diminutive of nucleus). Nucleolar composition resembles that of chromatin itself except for the presence in the nucleolus of large numbers of granules rich in RNA. In fact these granules are the precursors of ribosomes. Since ribosomes are made there, the size of the nucleolus, and sometimes the number of nucleoli per cell, varies with the requirements for ribosome and protein synthesis. The nucleolus has no membrane of its own, and would not be visible in the light microscope were it not for the relatively high packing density of its

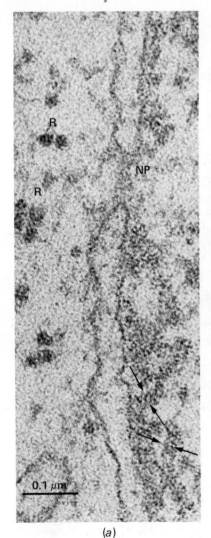

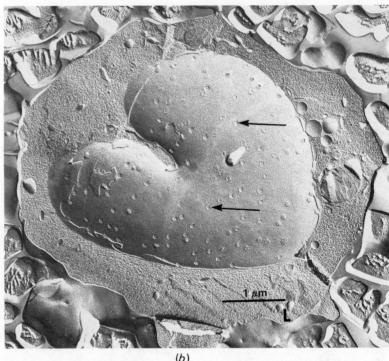

FIGURE 1-13 Nuclear Pores. The two membranes of the nuclear envelope (seen in Fig. 1-12) have periodic openings called pores. (a) Cross section of the envelope, showing a nuclear pore (NP), chromatin fibers (outlined by *arrows*) anchored to the inside surface, and ribosomes (R) out in the cytoplasm. (b) Freeze-fractured and etched view (see Appendix Section A-1) of a human lymphocyte, one of the white blood cells. Note the characteristic indentation of the nucleus and pores (*arrows*) scattered about the envelope. [(a) Courtesy of F. Lampert, *Humangenetik*, **13**: 285 (1971); (b) courtesy of R. Scott and V. Marchesi, *Cellular Immunology*, **3**: 301 (1972), © Academic Press, N.Y.]

fibers and granules. During cellular replication, when the rest of the chromatin is condensed into discrete chromosomes, the material of the nucleolus usually disperses and disappears from view, only to re-form again in the new daughter cells.

Endoplasmic Reticulum. Most eukaryotic cells have a complicated network of cytoplasmic membranes that often appear to be continuous with the nuclear membrane. This network is called the *endoplasmic reticulum* (see Fig. 1-14), a name derived from the fact that in the light microscope it looks like a "net in the cytoplasm." (Eighteenth-century

FIGURE 1-14 Intestinal Absorptive Cell from a Rat. The plasma membranes of adjacent cells (between arrows) form interlocking convolutions. Microvilli (MV) extend into the lumen of the intestine, greatly increasing the surface across which nutrients can pass. The nucleus (N) contains dense chromatin (CH) along its inside edge. A nearby Golgi apparatus (G) has a prominent lipid droplet (black body) in one of its cisternae. Numerous mitochondria (M) and extensive rough endoplasmic reticulum (RER) can be seen, along with some smooth endoplasmic reticulum (SER). Note the well-defined cavities in the endoplasmic reticulum. [Courtesy of H. I. Friedman and R. R. Cardell, Jr., *J. Cell Biol.*, **52**: 15 (1972).]

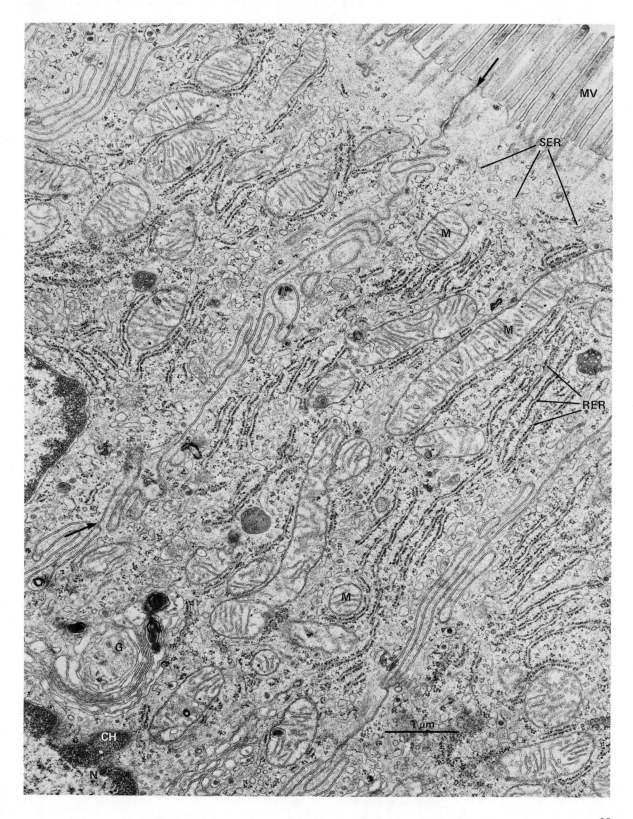

ladies carried purses of netting called reticules.) When a cell is disrupted and its components separated, a *microsomal fraction* is identified that is composed largely of fragments of the endoplasmic reticulum.

The reticulum creates enclosed or semienclosed spaces in the form of tubules or cisternae. (A cisterna is a cavity.) These channels and reservoirs function in intracellular storage and transport and often in protein synthesis, carried out at the ribosomes, which are sometimes found attached to the endoplasmic reticulum. Endoplasmic reticulum with attached ribosomes appears granular in the microscope, and so is known as rough endoplasmic reticulum, or *rough ER*. Without ribosomes, it is *smooth ER*. Proteins destined for use within the cell seem to be produced on ribosomes that are free in the cytoplasm. However, proteins destined for secretion by the cell are usually, if not always, synthesized on ribosomes attached to the endoplasmic reticulum—in other words, the rough ER.

Golgi Apparatus. Almost all eukaryotes have a complex of vesicles and membranes known as the Golgi apparatus or Golgi body, after Camillo Golgi, who first described it at the turn of the century. (He called it an "internal reticular apparatus.") The name *dictyosome* is also used to identify this structure, particularly in plants, where it consists of a stack of flattened saccules, often 20–30 nm apart and surrounded by vesicles at their edges (see Fig. 1-15). Secretion (e.g., of cell wall material in plants or digestive enzymes from pancreatic cells) seems to involve the processing of material by Golgi stacks and packaging of this material within vesicles pinched off from Golgi saccules. The packaged material, now called a *secretory granule* or *secretory vacuole*, fuses with the plasma membrane, opening itself to the exterior in an event that is called *exocytosis*.

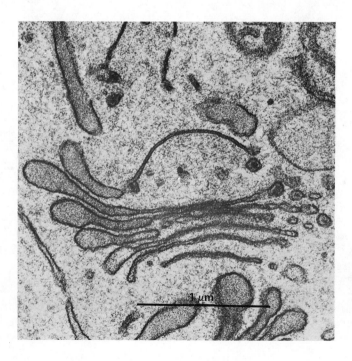

FIGURE 1-15 Dictyosome. The Golgi apparatus of animal cells is seen nicely in both the preceding and following figures. Here we have the Golgi configuration of a higher plant cell, maize root cap, showing the stack of flattened Golgi sacs known as a dictyosome. Note the many small vesicles (membrane-enclosed spheres) derived probably from the Golgi saccules. [Courtesy of H. Mollenhauer, *J. Cell Biol.*, **49**: 212 (1971).]

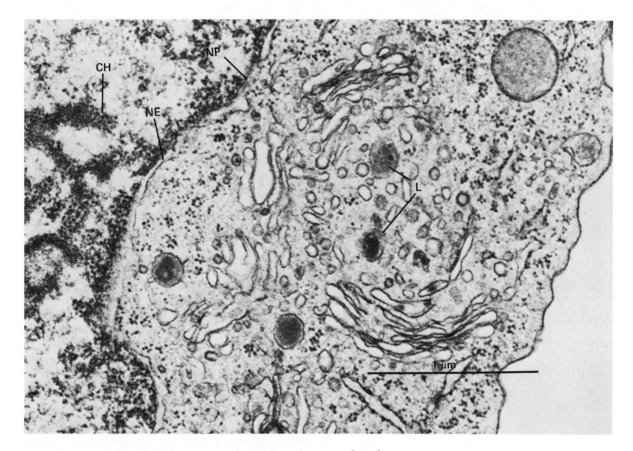

Lysosomes. There are numerous membrane-bound sacs within the cytoplasm for which special functions have been identified. Among them are the lysosomes, discovered by Christian de Duve in 1952 (see Fig. 1-16). Lysosomes, which appear to originate from the Golgi apparatus, serve to isolate digestive enzymes that are capable of destroying a wide variety of substances. Under periods of prolonged fasting, for instance, protein is readily dissolved and reused from our own muscle cells with the help of lysosomes. Material to be digested may also be brought in from outside the cell by enclosing it in membrane (a process called *endocytosis*); this structure may fuse with a lysosome to form a *digestive vacuole*. The fusion exposes ingested material to lysosomal enzymes without exposing the rest of the cytoplasm. Breakdown products apparently leave the vacuole via transport across its membrane. Indigestible material may be eliminated from the digestive vacuole by exocytosis, which opens the vacuole to the extracellular space.

Peroxisomes. Another class of membrane-bound vesicle, the peroxisome (Fig. 1-17), is found in almost all cells. It is similar to a lysosome in size and appearance, though probably a little larger on the average, and was discovered at about the same time. When first identified, it was named a *microbody*. The term peroxisome evolved because microbodies have enzymes responsible for producing and degrading peroxides. The most prevalent peroxisomal enzyme is often catalase, whose function is to degrade hydrogen peroxide (H_2O_2) to water and oxygen. In

FIGURE 1-16 Lysosomes. Lysosomes (L, dense interiors) being formed in the Golgi region of an immature white blood cell (rabbit). Numerous smaller vesicles are also seen. Nuclear envelope (NE). Nuclear pore (NP). Chromatin (CH). [Courtesy of B. Nichols, D. Bainton, and M. Farquhar, *J. Cell Biol.*, **50**: 498 (1971).]

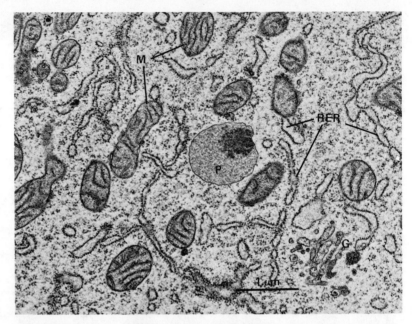

FIGURE 1-17 Peroxisomes. Part of an insect fat body (*Calpodes ethlius*), showing peroxisomes (P) similar to those found in vertebrates. Note the abundance of ribosomes, both free and bound to endoplasmic reticulum (RER). There is a small Golgi complex (G) and numerous mitochondria (M). [Courtesy of M. Locke and J. T. McMahon, *J. Cell Biol.*, **48**: 61 (1971).]

microorganisms and higher plants, peroxisomes may also contain enzymes of the glyoxylate cycle, an essential part of the pathway by which fats and oils are converted to sugars. They are then known as *glyoxysomes*.

Mitochondria. Most eukaryotic cells, both animal and plant, contain mitochondria (see Figs. 1-14 and 1-17). Mitochondria are often long and cylindrically shaped, though round and Y-shaped mitochondria are also seen. They are enclosed by a double membrane, contain ribosomes and some of their own genes, and grow and reproduce by binary fission almost as if they were autonomous organisms instead of an important part of the cell's metabolic apparatus. Mitochondria supply most of the energy needed in a typical nonphotosynthetic cell by oxidizing ("burning") selected foodstuffs, reducing molecular oxygen to water in the process. Mitochondria are the only major site of oxygen consumption in the cell; hence, eukaryotic cells that do not have mitochondria (mature mammalian red blood cells, for instance) cannot use oxygen.

Mitochondria are readily identified by membranes called *cristae* that tend to divide their interior (see Fig. 1-10d). The cristae are actually folds of the inner mitochondrial membrane, and hence are two membrane layers thick with a space (*the intermembrane space*) between the layers. The inner surface of the cristae is studded with mushroomlike projections similar to those found on the inner surface of many aerobic bacteria. These projections are integral parts of the oxidative apparatus.

Cytoplasmic Filaments. The cytoplasm is laced with filaments and tubules of various sizes whose structure and function have only recently started to become clear. There are three readily recognized classes of filaments. The smallest are the *microfilaments*, some 4–6 nm in diameter. They are often found next to the cell membrane where they form a web in the region of cytoplasm known as the *ectoplasm* (see Fig. 1-18).

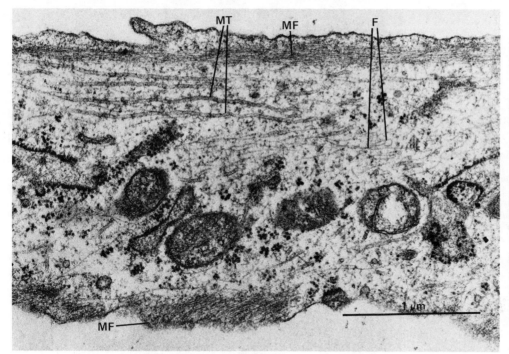

A second class, the *myosin filaments*, was found first in muscle, but the protein of which they are composed has recently been identified in a great many other cell types as well. The contraction of muscle is achieved by sliding filaments of myosin past filaments of *actin*. As it turns out, actin filaments are very similar to, if not identical with, the universal microfilament. This discovery reinforced a long-held suspicion that microfilaments are involved in many kinds of cell movements, including division, the amebalike crawling common to so many cells, and various extensions of the cell surface, such as microvilli (seen in Fig. 1-14), as well as the longer processes seen in Fig. 1-18.

In addition to microfilaments and myosin filaments, a third class of filament is called simply *10 nm* (or *100 Å*) *filaments*. Filaments of this size are seen in many cell types and are given many different names. Whether they are all the same is not clear, nor is their function known for certain. In some cells, however, they appear to be involved in movement, both of the cell itself and of materials within the cell.

FIGURE 1-18 Cytoplasmic Filaments and Microtubules. *RIGHT:* Light micrograph of a cultured hamster kidney fibroblast (BHK-21 strain). Fibroblasts establish collagenous connective tissue, including scars. Note the long processes on these spindle-shaped cells. *LEFT:* An electron micrograph of a longitudinal section through one process, showing the 40–60 Å microfilaments (MF) just inside the plasma membrane oriented longitudinally (top) and at an angle to the cut (bottom). The 100 Å filaments (F) and 250 Å microtubules (MT) are at deeper positions and are usually longitudinally oriented. [Courtesy of R. D. Goldman, *J. Cell Biol.*, **51**: 752 (1971).]

Microtubules. Microtubules are complex structures generally comprised of 13 individual *protofilaments* arranged to form a hollow cylinder about 25 nm in diameter (see Figs. 1-18 and 1-19). Microtubules often seem to form a sort of cellular skeleton and to be responsible for the extension of long cytoplasmic processes as seen in neurons and some other cells. They may also be responsible, at least in part, for movement of cytoplasmic components from one place within the cell to another. In addition to existing as individual structures, microtubules are the basic building blocks of centrioles, basal bodies, cilia, and flagella.

Centrioles and Basal Bodies. Animal cells (but not plant cells except for a few primitive algae) regularly contain near the nucleus at least two

CHAPTER 1
The Cellular Basis of Life

FIGURE 1-19 Microtubules. The large photo shows microtubules from the tail of sea urchin sperm, partially disassembled to reveal individual *protofilaments*. The 13 protofilaments are seen to better advantage in the cross section (INSET), which is a microtubule from *Juniperus*, a common evergreen shrub. [Large micrograph courtesy of P. J. Harris, Oregon State University; inset courtesy of M. C. Ledbetter, *J. Agr. Food Chem.*, **13**: 405 (1965), copyright by the American Chemical Society.]

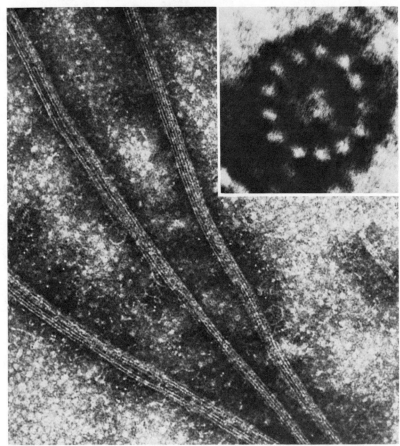

FIGURE 1-20 Centrioles. A pair of centrioles lying at right angles to each other near the nucleus. The centrioles in this organism (*Myrmecaelurus*, an ant) are sometimes very large (up to 8 μm), and are unusual also in that their biosynthesis seems to involve the attached vesicles seen in the micrograph. [Courtesy of M. Friedländer and J. Wahrman, *J. Cell Science*, **1**: 129 (1966).]

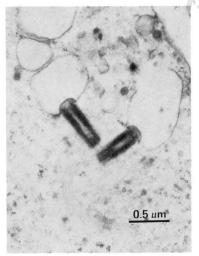

0.5 μm

hollow, cylindrical bodies called centrioles (Fig. 1-20). They are usually about 0.4 μm long by 0.15 μm in diameter, and are frequently found in pairs lying at right angles to one another just outside the nuclear envelope. Their position defines the *centrosome* of the cell. The walls of centrioles and basal bodies appear to be composed of nine sets of microtubules. Generally, each set is a triplet, with the three tubules lying in the same plane and embedded in a dense, granular substance (Fig. 1-10).

Centrioles are responsible for organizing the *spindle apparatus*, which is the collection of microtubules that separates chromosomes into prospective daughter cells at division. In addition, cilia and flagella of both animals and plants project from the cytoplasm at a structure that is identical to a centriole but is called a *basal body* or *basal corpuscle*.

Cilia and Flagella. Cilia are typically 2–10 μm long by about 0.5 μm in diameter, and are numerous on those cells that have any at all (see Fig. 1-21). Flagella are longer (100–200 μm) and have the same diameter and the same structure, but usually there are no more than one or two per cell. Both structures are composed of a microtubular framework enclosed by an extension of the plasma membrane. The tubules are generally found in nine pairs arranged around the circumference, with two single tubules running down the center. This is the so-called "9 + 2" construction, distinct from the "9 + 0" construction of the basal body or centriole (see Fig. 1-22).

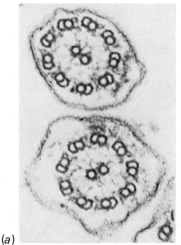
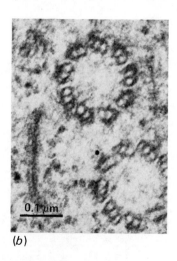

(a) (b)

FIGURE 1-22 Cross Section of the Cilium and Basal Body. From the protozoan *Tokophyra infusionum*. (a) Cilium, showing the "9 + 2" architecture and enclosing membrane. (b) The "9 + 0" structure of basal bodies and centrioles. [Courtesy of Lyndell Millecchia.]

Cilia and flagella are motile appendages. Movement apparently is achieved by sliding microtubules past one another. The cells lining the air passages in the lungs, for example, have cilia that beat in a synchronized, wavelike motion to sweep particles out of the lungs. Flagella are more often used to propel the cell itself—a sperm, for instance, has a single long flagellum for a tail. Note that the cilia, flagella, and basal bodies of eukaryotic cells are much different from the structures of prokaryotes that carry the same name. In fact, many prokaryotic cells would fit quite nicely inside the membrane of the usual eukaryotic cilium.

Vacuoles. The words vacuole and vesicle are sometimes used interchangeably, but in general a vacuole is a large vesicle—i.e., a membrane-enclosed sac. Vacuoles are found in many different cell types, especially in plants, where they generally serve as storage depots of various sorts. In fact, a plant cell often has a single large vacuole occupying as much as 80–90% of its total cell volume. The contents of this large vacuole may be called the *cell sap* (see Fig. 1-23).

The Plastids and Cell Walls of Plants. Many plant cells have outside their plasma membrane a rigid cell wall composed chiefly of cellulose and (especially in woody tissues) lignin (see Fig. 1-23). The hardness of plant tissues, as opposed to the softness of most animal tissue, is due to the presence of walls. Also unique to plants is a series of organelles called *plastids*. Like the nucleus and mitochondria, they are bounded by two membranes, although, at about 6 nm each, the membranes are a little thinner than the typical plasma membrane. A plastid of a particular type may arise by division and differentiation of a smaller precursor, a *proplastid*, or by changes that take place in another mature plastid. Some, the *leucoplasts*, are colorless and are generally concerned with the storage and metabolism of starches and oils. *Chromoplasts*, on the other hand, contain the pigments that give plants their brilliant colors.

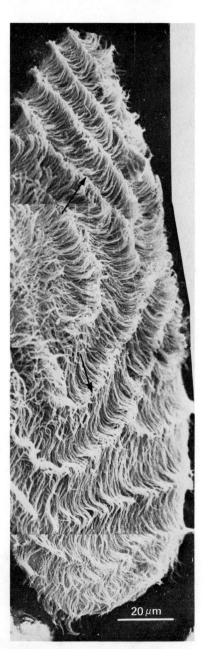

FIGURE 1-21 Coordination of Ciliary Motion. Scanning electron micrographs of the protozoan *Opalina ranarum*. Arrows indicate direction of wave transmission. [Courtesy of G. A. Horridge and S. L. Tamm. *Science*, **163**: 817 (1969). Copyright by the AAAS.]

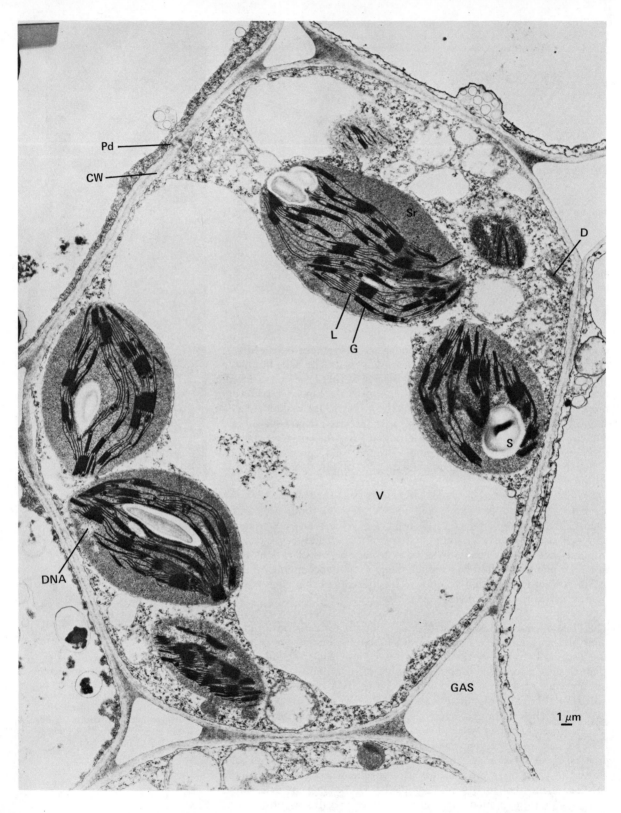

FIGURE 1-23 A Plant Cell. Taken from timothy grass (*Phleum pratense* L.), it has a large central vacuole (V) and several small ones. Five mature chloroplasts can be seen, with numerous grana (G), and interconnecting lamellae (L) traversing the stroma (Sr). The chloroplasts contain starch granules (S) and DNA (light areas). A small dictyosome (D) is present. A plasmodesma (Pd) penetrates the cell wall (CW). Gas-filled cavities (O_2, CO_2, water vapor, etc.) at lower right account for a major portion of the volume of many leaves. [Courtesy of M. C. Ledbetter and K. R. Porter, from *Introduction to the Fine Structure of Plant Cells* (New York: Springer-Verlag, 1970.)]

The most impressive plastid is the chloroplast (Fig. 1-23), which is often lens shaped and 4–10 μm long. Chloroplasts are responsible for the bright green color of leaves, for although they contain pigments of other colors as well, there is generally so much chlorophyll in them that only the green is obvious. When leaves die in the fall or when a tomato ripens, chlorophyll is destroyed first, unmasking the other pigments. Chloroplasts are the site of photosynthesis, using light energy to fix carbon dioxide from the air, reducing it to sugars and other carbohydrates. In recent years, a good deal of attention has been paid to the semiautonomous nature of chloroplasts, including the presence within them of ribosomes and DNA. The DNA contains a portion of the genetic information needed for the synthesis of chloroplasts, making them only partially dependent on nuclear genes. Chloroplasts, like the other organelles just discussed, will be considered in much more detail in later chapters.

The Origin and Fate of Cytoplasmic Structures. Most of the cytoplasmic structures just described are dynamic features that come and go as need dictates. Their formation and fate is more readily understood by dividing the structures into three categories: (1) those comprised largely of protein subunits, such as the various filaments and microtubules; (2) those comprised chiefly of membrane (the main chemical component of which is lipid), including the endoplasmic reticulum and all the assorted vesicles and vacuoles; and (3) the complex organelles, namely the mitochondria and plastids.

Proteins are large molecules that are synthesized from smaller precursors, the amino acids, using genes as blueprints to direct the synthesis. Individual protein molecules may in turn become subunits of larger structures, such as microfilaments, microtubules, ribosomes, and so on. In most cases, assembly of the larger structures occurs spontaneously from the protein subunits. Hence, the key to formation of larger structures is regulation of the genes responsible for producing the subunits.

Membranes, too, may self-assemble from their component parts. However, of the various membranous structures—the endoplasmic reticulum, nuclear envelope, Golgi apparatus, lysosomes, peroxisomes, and so on—only the endoplasmic reticulum seems to grow mainly by the incorporation of molecules of lipid and protein as they become available. The other structures are then derived either directly or indirectly from the endoplasmic reticulum according to the following scheme:

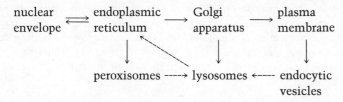

Note that only flow between the nuclear envelope and endoplasmic reticulum is bidirectional. During cell division in all but the most primitive eukaryotes, the nuclear envelope fragments and the pieces are added to the endoplasmic reticulum. New nuclear envelopes are later assembled from the ER and may or may not include some of the original fragments. With this exception, flow of membrane in the preceding scheme is largely through vesicles that bud or are pinched off from the donor structure and then fuse with and become a part of the recipient structure, adding their membrane to the membrane of the latter.

Thus, peroxisomes are derived from the ER by this budding process. Other ER-derived vesicles have quite different contents and coalesce to form saccules of the Golgi apparatus. Golgi membranes, in turn, give rise to lysosomes and to vesicles that fuse with the plasma membrane. Growth of the latter is controlled by a reciprocal return to the cytoplasm of vesicles derived from the plasma membrane by the same budding process, which is known in this case as endocytosis. These vesicles then fuse with lysosomes (dashed lines in the preceding diagram) and are thus exposed to the digestive enzymes contained therein. Other cellular structures, such as the peroxisomes—and, in fact, the ER itself under certain circumstances—are also digested by lysosomes. The digestion products are small molecules, some of which can serve as raw material for more membrane growth at the endoplasmic reticulum, thus completing the cycle from ER to ER.

The plastids and mitochondria are not included in the above scheme. Although they are also subject to lysosomal digestion, their number increases mostly by growth and division of existing organelles. This pattern is more like that of the cell itself—or like that of the prokaryotic cells that may have been the evolutionary ancestors of the mitochondria and plastids, a possibility that will be explored more in Chap. 7.

1-6 CELL CULTURES

Many of the experiments to be described later in this book were carried out *in vitro*, a phrase that means literally "in glass"—i.e., in a laboratory vessel—as opposed to *in vivo*, meaning in a living organism. Cells grown *in vitro* are said to be cultured cells. To avoid digressing at a later time, we shall consider here some of the techniques involved in culturing cells.

Earlier we insisted that all cells have the capacity for an independent existence, meaning that if we can supply them with the proper nutrients they will continue to live for an extended period of time after they have been removed from their original environment. We cannot always make them reproduce, since certain highly specialized cells have lost that capacity (e.g., neurons, most muscle cells, and red blood cells). Specialized cells develop from less differentiated *blast cells* or *stem cells*. Should the specialized cells die (nerve and muscle cells generally do not; red blood cells do), their replacements, if any, come from the less specialized precursor. This of course limits our ability to study some cell types in the laboratory. Others, however, grow quite happily *in vitro*: embryonic cells usually can be made to grow for some time in the laboratory, and cancer cells will grow indefinitely, a fact that is not surprising since uncontrolled growth *in vivo* or *in vitro* is characteristic of cancers.

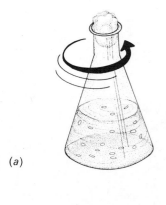
(a)

(b)

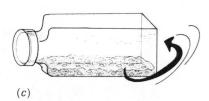
(c)

FIGURE 1-24 Culture Techniques. (a) Most prokaryotic and many eukaryotic cells can be grown in liquid suspension, using shaking or bubbling air to keep them from settling. (b) Agar plates are often used for bacteria. Agar supports the cells on its surface, while the medium in which the agar is dissolved nourishes the cells. (c) Flattened bottles are used for tissue cultures of eukaryotic cells where cell-cell contact is desired.

Single-celled life forms, with only a few exceptions (e.g., the fastidious mycoplasmas) are easy to grow in the laboratory because we can usually provide for them an environment even more hospitable than that to which they were accustomed.

Bacteria and certain animal cell lines are often grown in liquid suspension (Fig. 1-24a). As cells are a few percent denser than water, constant agitation must be employed to keep them from settling to the bottom. Agitation is sometimes provided by bubbling air through the suspension, which also insures an adequate oxygen supply. The liquid solution in which cells are suspended is the *culture medium.* For some bacteria it need contain only a few salts and a source of carbon and energy, often in the form of glucose. This combination, given in Table 1-2, is called a *minimal medium.* Other cells require supplements in the form of vitamins or other nutrients. The more fastidious cells are often grown in blood or serum, or in a medium supplemented with an extract of yeast or meat, which provides a wide variety of organic nutrients.

As an alternative to liquid suspension, cells may be grown on the surface of agar (Fig. 1-24b). Agar is a carbohydrate obtained from seaweeds whose molecules get tangled to form a translucent semisolid at concentrations of about 1% at room temperature. Agar behaves in this respect much like household gelatin which, however, is a protein. Since agar itself cannot be utilized as food by most cells, it must be dissolved in an appropriate culture medium. (The word "agar" is used for the original pure carbohydrate and also, less precisely, for the final gel.) Dissolved agar is sterilized by autoclaving, which means that it is cooked by superheated steam at about 120 °C. It is then poured into shallow, sterile petri dishes, where it is allowed to cool and solidify before being "seeded" with cells. (See Fig. 1-25.)

Cells from the higher animals and plants are often grown as monolayers on a solid surface such as the bottom of a shallow dish or rectangular bottle (Fig. 1-24c). Some will adhere strongly to this surface. A shallow layer of culture medium is added, and the container may then be gently rocked back and forth to aid mixing and aeration. The technique is also suitable for very small quantities of cells, which can be contained between a microscope slide and a domed cover sealed at the edges with paraffin. This technique allows direct microscopic observations of the entire culture without disturbing it.

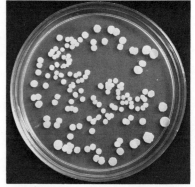

FIGURE 1-25 Growth on Agar. Discrete colonies of a bacterium, *Bacillus subtilis*, are produced by spreading a very dilute suspension of the cells onto agar and then incubating it overnight.

TABLE 1-2. A Simple Minimal Medium for Bacteria

	NH_4Cl	1.0 g
	$MgSO_4$	0.11
	KH_2PO_4	3.0
	Na_2HPO_4	6.0
	glucose	4.0
	water	1000 ml

This medium is designed to supply adequate amounts of available nitrogen, chloride, magnesium, sulfur, potassium, sodium, phosphorus, and carbon. Other metals, such as iron, are required in trace amounts, but are normally provided as contaminants in the salts listed (even when reagent grades are used). Glucose, in this medium, will ordinarily serve as both a carbon and an energy source. The balance between the acidic and basic phosphates is chosen to provide a pH near neutrality.

The rate at which cells grow depends on a number of factors, including the availability of oxygen and nutrients and the concentration of the cells. For example, bacteria cannot be grown readily in suspension to concentrations greater than about 5×10^9 per ml, and many animal cells (but not cancer cells) stop growing when a confluent monolayer is formed in the tissue culture dish. At the other extreme, some animal cells require a critical population density to grow at all, while bacteria and other single-celled forms of life never have any such restriction. Many bacteria are capable of doubling their number every hour *in vitro*, and some every twenty minutes, but animal cells do not usually double oftener *in vitro* than about once a day, even under the best conditions. Since all nutrients must pass into the cell across its plasma membrane, and all waste products must pass out the same way, this difference in growth rates may be due in part to the much higher surface-to-volume ratio of the smaller, bacterial cells.

1-7 VIRUSES

Although viruses are not cells, their study has provided a great deal of information about cells. To understand why, we need to know how viruses, which are cellular parasites, manage to survive.

Discovery. In 1798 an English physician, Edward Jenner, reported that infecting a young boy with a mild disease called cowpox made the child immune to smallpox. Even deliberate attempts to infect him with the much more serious disease failed. This was not as daring an experiment as it sounds, since deliberate exposure to smallpox was then often used in an attempt to produce a mild infection in the hope of protecting the patient from later, more threatening, infections. The term "vaccination," from the Latin for cowpox (*Variolae vaccinae*, meaning small pocks of the cow), was coined to described this procedure. Despite the success of vaccination, the nature of the disease-causing agent, or *pathogen*, remained unknown for another century.

By the 1880s, considerable progress had been made in microbiology, with Pasteur's demolition of the concept of spontaneous generation and the development of culture procedures for growing microorganisms. However, some microorganisms refused to be cultured. The pathogens causing smallpox and rabies fell into that category. Although Pasteur was not able to culture the rabies agent *in vitro*, he was able to pass it from rabbit to rabbit until it had become so adapted to rabbit nervous tissue that it no longer posed a serious threat to humans. He demonstrated this adaptation in 1885 by injecting some of the diseased rabbit tissue into a human who had been bitten by a rabid animal. Instead of causing rabies, Pasteur's vaccine elicited a protective immune response.

Still, the causative agents for smallpox and rabies remained unknown. They were assumed to be particularly fastidious, very small bacteria that could not be grown in the laboratory because of inadequate knowledge of their nutritional requirements. Their true nature as subcellular life forms was not appreciated until a similarly mysterious agent was discovered in plants.

In 1892 a Russian scientist, D. Ivanovsky, studied an infectious agent that causes a destructive mottling of the leaves of tobacco plants. Ivanovsky demonstrated that the agent, now called *tobacco mosaic virus*

(TMV), is small enough to pass a filter of unglazed porcelain (a porcelain candle) that Pasteur and C. Chamberlain had earlier found capable of stopping all known forms of cellular life including bacteria. A few years later, M. W. Beijerinck confirmed Ivanovsky's observations, and in 1899 published a paper suggesting that tobacco mosaic virus is not just a small cell, but something more fundamental: a subcellular form of life that can reproduce only as a parasite in living cells.

The concept of the virus as an obligate cellular parasite was extended to the animal world through experiments performed by F. Löffler and P. Frosch, who in 1898 demonstrated that foot-and-mouth disease in cattle is also caused by an agent that passes a porcelain filter—i.e., the agent is a *filterable virus*. This word *virus*, meaning "poison," was first applied to all pathogens. Later, pathogens were divided into filterable and nonfilterable viruses, using the porcelain filter as a test. As the distinction between cells and what we now simply call "viruses" became clear, however, references to filterability were generally dropped.

It is now known that a host of human ailments, including smallpox, rabies, measles, mumps, polio, and warts, are virus-caused diseases. In addition, it has been known since the second decade of this century that certain viruses cause cancers in birds, and more recently viruses have been implicated in some leukemias and mammalian tumors.

Viruses may be broadly grouped according to their hosts. We have mentioned *animal viruses* and *plant viruses*, but it was the discovery of a third group, *bacterial viruses*, that made possible detailed investigations of viral reproduction. These investigations also revealed much of what is known about the molecular aspects of gene function, as we shall see.

The Bacteriophage. An Englishman, F. W. Twort, published a report in 1915 in which he described a remarkable phenomenon. He watched some bacterial colonies growing on the surface of agar go through a "glassy transformation," becoming watery and transparent. Samples from such colonies could not be used to start new colonies. In fact, when a sample was introduced to a healthy colony, it too went through the same transformation. He diagnosed the situation as a contagious bacterial disease.

Twort observed that even very dilute samples of a "sick" colony could introduce the sickness to healthy cells, and that this process could be repeated as often as he liked, passing material from one colony to the next. Thus, the agent must cause its own reproduction. It proved to be filterable, hence much smaller than the bacteria themselves, and heat sensitive. Twort reasoned that the agent could be an enzyme (also heat labile and very small) or a virus, but his investigations were interrupted by World War I, and he published no more on the subject.

Only two years later, however, a Canadian, Felix d'Herelle, reported a quite independent discovery of a similar nature. He had found a filterable agent capable of causing *lysis* (a clearing of the culture due to destruction of cells) in cultures of a dysentery bacterium. In addition to observations such as those made by Twort, d'Herelle noticed that a very dilute sample from a diseased colony, when mixed with healthy cells and spread on the surface of agar, produced the expected lawn of colonies, but, in addition, a number of round clear *plaques*, or holes (see Fig. 1-26). When sampled, the plaques proved to contain the lytic agent, settling the question of its nature for, according to d'Herelle, "a chemical substance cannot concentrate itself over definite points . . . the antidysentery mi-

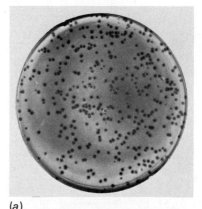

(a)

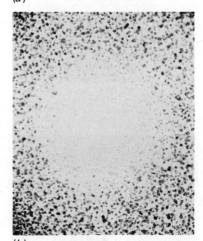

(b)

FIGURE 1-26 Bacteriophage Plaques. (a) A "lawn" of bacterial colonies with bacterial virus plaques. (b) A single plaque, greatly enlarged. Note the spherical colonies of bacteria and how their destruction produces the plaque. (Phage T4, growing on *Escherichia coli*.)

CHAPTER 1
The Cellular Basis of Life

crobe is an obligatory bacteriophage." There was no doubt in his mind that his bacteriophage (or "bacteria eater") was a virus capable of reproducing only as a parasite on bacterial cells.

Since d'Herelle's bacteriophage ("phage" for short) was so effective in destroying the dysentery bacillus *in vitro*, it was natural for him to ask whether it was as efficacious *in vivo*. Although he noted that the recovery of patients suffering from dysentery often coincided with the appearance of phage in their feces, neither he nor an army of other workers was completely successful in developing it into a real weapon in the fight against bacterial disease. There are viruses parasitic to a wide range of (perhaps all) pathogenic bacteria, but bacteria can mutate to virus resistance with a frequency that ensures their survival. Combined with the natural defenses of the body against foreign material, including bacterial viruses, the best efforts of medical microbiologists to use phages to cure human disease were frustrated. Finally, the advent of antibiotics, around 1940, put an end to the study of phages as a medical tool. However, at about the same time, the easily studied phages were recognized as a useful tool in the investigation of gene activity.

Because of the ease with which they are grown, much of the early work with the bacterial viruses was carried out with *coliphages*, which are viruses parasitic to an intestinal bacterium, *Escherichia coli* (*E. coli* for short). We shall have occasion to refer to the members of this group designated as "T" phages (T for "type"), particularly the closely related T-even phages, T2, T4, and T6 (see Fig. 1-27 and Fig. 1-28).

FIGURE 1-27 Bacteriophage T4. The so-called T-even bacteriophages are probably the most complex of all viruses. Each contains about 200 genes. In contrast, simple viruses contain only three or four genes. (Micrograph courtesy of Lee Simon, *New Scientist and Science Journal*, 25 March 1971, p. 670.)

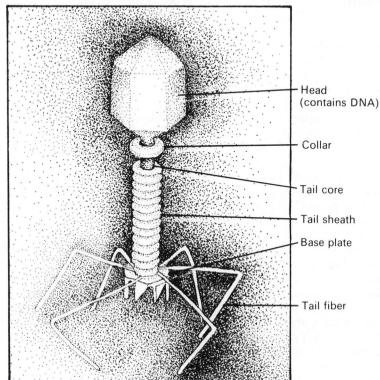

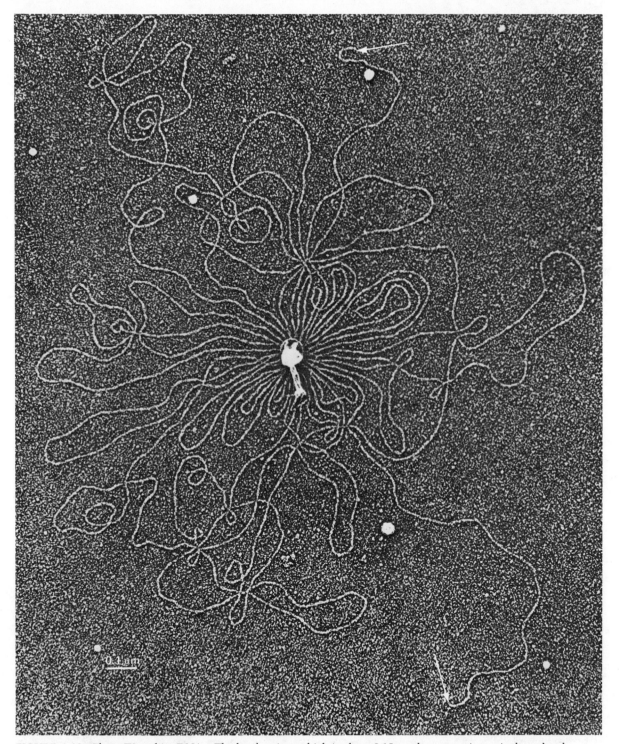

FIGURE 1-28 Phage T2 and its DNA. The head region, which is about 0.15 μm long, contains a single molecule of DNA with a total length of about 50 μm. (Note the two ends, marked by arrows.) During infection, this strand of DNA must pass through the tail of the virus and into the host cell. The preparation has been coated with platinum to improve contrast; hence, the diameter of the extruded DNA has been greatly increased. [Courtesy of A. K. Kleinschmidt et al., Biochim. Biophys. Acta, **61**: 857 (1962).]

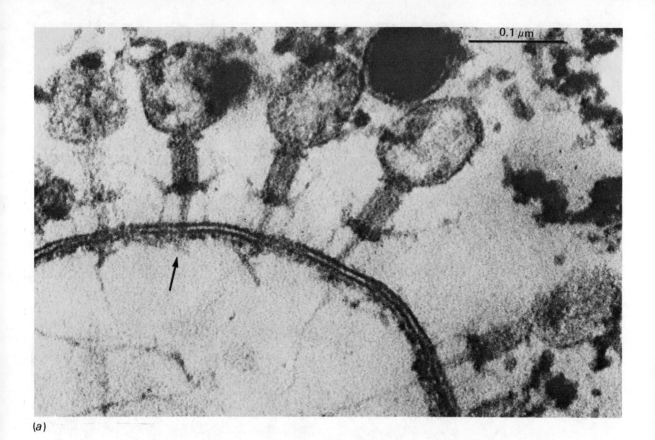

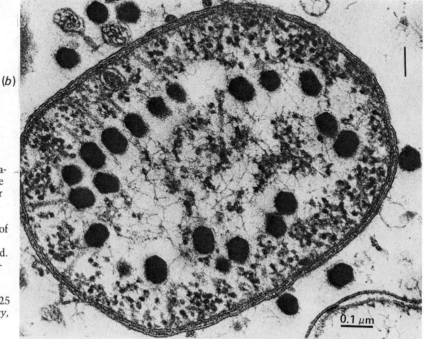

FIGURE 1-29 Infection and Replication of Phage T4. (a) The phages are bound to an *E. coli* cell wall by their tail fibers. The sheaths have contracted and their cores have penetrated into the host (arrow). Fibers of DNA can be seen entering the host cell. (b) New phages are synthesized. Note the regular arrangement of particles: tails, outside; heads, inside. [Courtesy of L. D. Simon, (a) from *New Scientist and Science Journal*, 25 March 1971, p. 671, (b) from *Virology*, **38**: 285 (1969), © Academic Press, N.Y.]

Life Cycle of the Bacteriophage. The life cycle of most bacteriophages is similar to that described by d'Herelle for his antidysentery virus. It consists of (1) adsorption to the host cell, (2) injection of viral genetic material (DNA or RNA), (3) intracellular production of new particles, and finally (4) release of the progeny phage, usually 100 or more.

The *adsorption* of the phage to its host is made possible by a reaction of chemical groups on the two during a random collision. Many phages are shaped a little like tadpoles, with a distinct "head" region and a cylindrical "tail" (see Fig. 1-27). Reactive groups at the end of the tail can join with a complementary set of chemical groups (a *receptor site*) in the cell wall of the bacterium. The T-even phages and a few others also have long fibers extending from the tail which, because of their size, are apt to be the first to contact and attach to the cell. The fibers help to position the phage's tail perpendicularly to the cell wall (see Fig. 1-29). Probably because of them, the T-even phages are unusually efficient at attacking their hosts—calculation of collision frequencies between virus and cell indicate that under some conditions it may take no more than one viral collision to cause adsorption, remarkable when one considers that the collision occurs with a random orientation. Phages without tail fibers seem less efficient in their attacks, though one cannot be absolutely certain that this is the only reason.

Once the phage is attached to its prospective host, *injection* can take place, involving a movement of the genetic material (DNA or RNA) from its position inside the head of the phage through the hollow core of the tail and into the bacterium. Entry is made possible by a hole punched in the cell wall, either by a contraction of the outer sheath of the tail or by the action of enzymes carried by the phage tail, or both (see Fig. 1-29*a*).

Once inside a host cell, phage genes take over the metabolic machinery of the cell and direct it to produce replicas of the infecting virus. In other words, although the cell continues to procure raw material and energy from its environment, the virus genes allow only virus components to be built (see Fig. 1-29*b*). Not only is the normal ability of the host cell's DNA to control the cell lost, but the host's DNA may be destroyed by early products of the viral genes.

Usually the host cell's fate is death by lysis. After anywhere from a few score to several thousand new *virions*, or virus particles, have been manufactured, the cell bursts, permitting the new particles to diffuse away in search of additional hosts.

In the late 1940s, a type of phage was discovered that occasionally adopts a life cycle startlingly different from the usual one (see Fig. 1-30). The genetic material of the phage may be inserted into the chromosome (DNA) of the host. In this condition the host cell may continue to function in the usual way, with no obvious evidence of the viral attack. In other words, the viral genes are seemingly dormant and may remain so for many generations, being replicated along with the rest of the host's chromosome. Thus, a substantial *clone* of cells may develop whose every member carries a copy of the genes of the virus that invaded its single ancestor. A virus with this capacity is called *temperate* to distinguish it from *virulent* viruses that exhibit only the first life cycle described. In its dormant state, the viral genes are said to exist as a *prophage* or *provirus*. The infected cell is said to be *lysogenic* because a prophage may, at any time, become detached from the chromosome of its host and begin directing the synthesis of new virus particles that eventually lyse the cell just as if it had been invaded recently.

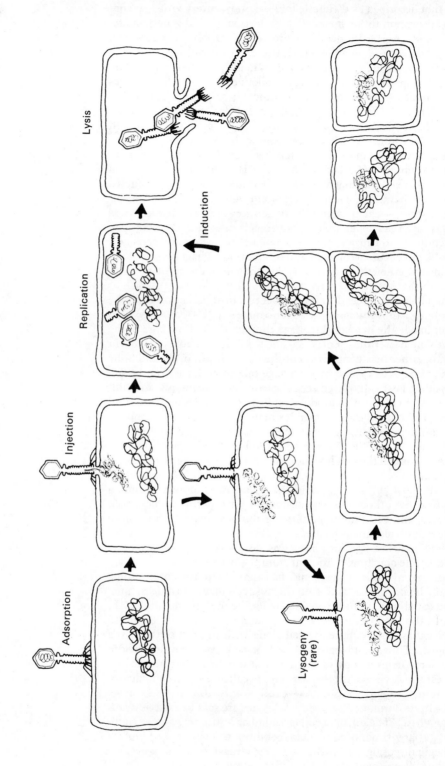

FIGURE 1-30 Life Cycle of Bacterial Viruses. Some bacteriophages (but not the T series) occasionally adopt an alternate life cycle in which injected DNA is incorporated into the host's DNA as a provirus and is replicated for many generations as part of the host. This condition is called *lysogeny*. Occasionally, the virus genes in one of these cells may become detached and cause viral replication just as if the cell were newly infected.

The chances of a spontaneous activation of a provirus may be as low as one in a million per generation; however, if the lysogenic culture is exposed to ultraviolet light or to any one of a number of other physical or chemical agents, it may be induced to begin turning out virus particles *en masse*. This phenomenon, in which an apparently healthy culture of cells can all at once begin producing viruses without any new infection, is called *induction*.

We shall have more to say about phages later, because the study of their interaction with hosts has taught us a great deal, not only about viruses but about how genes behave. This is true because a virus is little more than a set of genes wrapped in a protective coat.

The Structure of Animal and Plant Viruses. Viruses, being cellular parasites, have no metabolism and therefore no repetitive movement and no means for reproduction outside of a host cell. Whether one wishes to call them "alive" is largely a matter of definition, for they occupy the nebulous borderline between living and nonliving. We shall not dwell on these philosophical difficulties, but shall go on to describe more of what is known about viruses because, living or not, they are both interesting and important.

The chemical nature of viruses remained a mystery until Wendell Stanley, who later received the Nobel Prize, purified and crystallized to-

TABLE 1-3. Some Milestones in Virology

1798	Jenner successfully tests his smallpox vaccine.
1885	Pasteur perfects the second viral vaccine, that for rabies.
1892	Ivanovsky finds that tobacco mosaic disease can be transmitted by an agent that readily passes a ceramic filter.
1898	Loeffler and Frosch demonstrate that the agent responsible for foot-and-mouth disease in cattle also passes a ceramic filter, and suggest that animal diseases such as smallpox may be caused by similar agents.
1899	Beijerinck suggests that tobacco mosaic disease is not caused by a cell, but by a cellular parasite, a virus.
1911	Rous finds that a certain kind of cancer in chickens is transmitted by a filterable virus (the Rous sarcoma virus).
1915	Twort reports a bacterial disease characterized by a "glassy transformation" of colonies growing on agar.
1917	d'Herelle independently finds a lytic agent for a dysentery bacterium and concludes that it is a bacterial virus.
1935	Stanley crystallizes TMV, permitting a careful chemical analysis of its composition.
1950	Lwoff proves that lysogenic bacteria carry viral genes without carrying intact viruses. Though the concept had been advanced some years earlier by Sir Macfarlane Burnet, it had been highly controversial.
1952	Hershey and Chase demonstrate experimentally that phage DNA, not its protein, carries genetic information.
1955	Fraenkel-Conrat and Williams show that tobacco mosaic virus can be disassembled and reassembled in the laboratory.
1956	Gierer and Schramm demonstrate that the RNA from TMV is infective even without the protein.

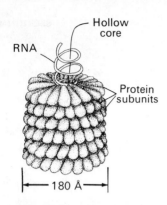

FIGURE 1-31 Partially Disassembled Tobacco Mosaic Virus. Note the strand of RNA coiling out of its protein coat.

FIGURE 1-32 Two Plant Viruses. (a) Tobacco rattle virus. This RNA virus is cylindrical with a hollow core like TMV, but is unusual in that it comes in two different sizes, a long and a short rod. A complete cycle of infection and viral replication requires the presence of both versions. (b) Cucumber mosaic virus, a polyhedral plant virus. Note one empty (i.e., RNA-free) particle into which the stain has penetrated (arrow). (Negatively stained electron micrographs courtesy of T. C. Allen, Jr., Oregon State Univ.)

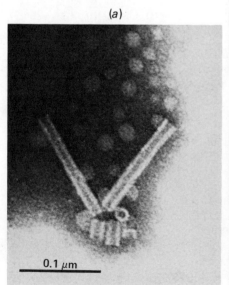

(a)

bacco mosaic virus (TMV) in 1935. (This and other milestones in virology are listed in Table 1-3.) Stanley reported that TMV is extremely simple, consisting almost entirely of protein, although it was soon found that ribonucleic acid (RNA) is also present. Later work established that the average particle is 94.5% protein in the form of about 2100 identical building blocks, or subunits, arranged in a helix some 300 nm long by 18 nm in diameter (see Fig. 1-31). The ribonucleic acid, which contains all the genetic information of the virus, is a chainlike polymer that follows the helix from one end to the other at a distance of about 4 nm from the axis, and is therefore buried in and protected by the protein. This simple structure contrasts sharply with the complex structure of the T-even phages. There are, however, viruslike particles that are even simpler than TMV. These particles, which are responsible for a few types of plant diseases, consist apparently only of ribonucleic acid with little or no protein, and therefore they have no defined three-dimensional structure. They have been termed *viroids*.

Figures 1-32 and 1-33 show examples of other viral structures, including a cylindrical and a "spherical" (actually polygonal) plant virus and two icosahedral (20-sided) animal viruses.

Although most viruses consist almost entirely of protein and one or the other of the nucleic acids, DNA or RNA, some have an outer envelope that has a structure similar to a plasma membrane. In fact, it may actually be derived from the plasma membrane of the host cell, being added to the virus particle as it passes through during release. (See Fig. 1-34.)

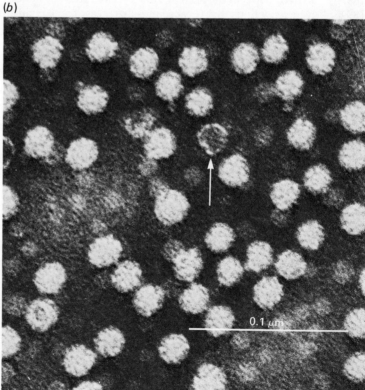

(b)

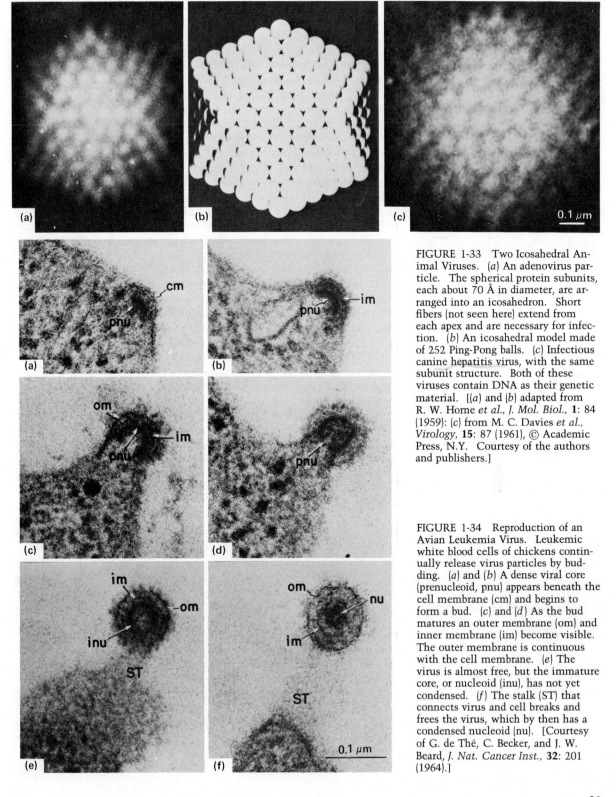

FIGURE 1-33 Two Icosahedral Animal Viruses. (a) An adenovirus particle. The spherical protein subunits, each about 70 Å in diameter, are arranged into an icosahedron. Short fibers (not seen here) extend from each apex and are necessary for infection. (b) An icosahedral model made of 252 Ping-Pong balls. (c) Infectious canine hepatitis virus, with the same subunit structure. Both of these viruses contain DNA as their genetic material. [(a) and (b) adapted from R. W. Horne et al., J. Mol. Biol., **1**: 84 (1959); (c) from M. C. Davies et al., Virology, **15**: 87 (1961), © Academic Press, N.Y. Courtesy of the authors and publishers.]

FIGURE 1-34 Reproduction of an Avian Leukemia Virus. Leukemic white blood cells of chickens continually release virus particles by budding. (a) and (b) A dense viral core (prenucleoid, pnu) appears beneath the cell membrane (cm) and begins to form a bud. (c) and (d) As the bud matures an outer membrane (om) and inner membrane (im) become visible. The outer membrane is continuous with the cell membrane. (e) The virus is almost free, but the immature core, or nucleoid (inu), has not yet condensed. (f) The stalk (ST) that connects virus and cell breaks and frees the virus, which by then has a condensed nucleoid (nu). [Courtesy of G. de Thé, C. Becker, and J. W. Beard, J. Nat. Cancer Inst., **32**: 201 (1964).]

Life Cycle of Animal and Plant Viruses. The mode of replication of animal and plant viruses is not as well understood as that of the phages, partly because the cells themselves are less well understood. Few animal or plant viruses have tails or any other obvious organ of attachment. In fact, both viral protein and viral nucleic acid penetrate into the host. In many animal viruses, uptake by the cell depends upon attachment of the virus to the cell membrane, often at specific receptors, much as a bacterial virus attacks only bacterial cells with appropriate receptors. This requirement for attachment is a partial explanation for why most viruses exhibit a narrow *host range specificity*, meaning that a given virus can reproduce in only a limited number of cell types.

Once inside an appropriate host a virus may be replicated immediately, or its genes may be incorporated into the host's chromosomes for replication in a future generation. The latter situation has generated a good deal of attention, for the incorporated genes are not necessarily entirely dormant. In the bacterial counterpart of this condition, lysogeny, the presence of a virus-introduced gene may be manifested as an ability to utilize some particular food molecule that the host could not use before infection, or by the expression of some other new function. For example, the bacteria responsible for diphtheria in humans only cause the disease if they carry a particular bacterial virus as a prophage. In animal cells, we are more interested in the case where virus-introduced genes cause changes in the normal control of cellular replication, for such cells may become malignant. (Note that it cannot be concluded from this that all malignancies are caused by viruses—see Chap. 12.)

The release of newly formed virus particles may be through the dramatic lysis characteristic of bacteriophage reproduction, or it may occur one at a time while the cell continues to live, grow, and reproduce. Leukemia viruses, as noted, are released singly from infected white blood cells. The latter reproduce uncontrollably, eventually killing the animal. In the case of the avian (bird) leukemia viruses, the serum of an infected animal may have a free virus count exceeding 10^{11} virions per ml before the animal dies from leukemia. The virus count is so high that the serum (clear, cell-free portion of the blood) becomes translucent, resembling thin milk.

A Comparison of Cells and Viruses. We have seen that viruses are much simpler than cells and very different in both structure and life cycle. They range in size from about 10 nm to several hundred nm. The largest viruses, therefore, are at least as large as the smallest bacteria (mycoplasmas, chlamydiae, and rickettsiae), but there is a world of difference between them. In fact, even the tiniest cells may be parasitized by viruses.

Chlamydiae are similar to viruses in being able to reproduce only within living cells. However, chlamydiae are obligate cellular parasites because of stringent nutritional requirements. Viruses are obligate cellular parasites because they are not cells, and only cells can be self-reproducing. Viruses have no metabolism and no cell membrane in the usual sense. To be sure, the envelope that one finds around certain viruses has the same structure as a cell membrane, but it is not a biologically active organ, only the remnant of one. In general, the molecular variety necessary to support an independent existence is simply not present in virus particles. There is no natural way in which their genes can be used to construct new virus particles without the help of a living cell.

The intracellular nature of their life style is also the basis for the difficulty encountered in trying to find chemicals that interfere with virus replication the way that antibiotics interfere with bacterial replication. There are now a handful of such chemicals, however, that are useful in certain restricted situations. Hopefully, more will be discovered.

1-8 THE CHOICE OF EXPERIMENTAL SYSTEMS IN BIOLOGICAL INVESTIGATIONS

I have often had cause to feel that my hands are cleverer than my head. That is a crude way of characterizing the dialectics of experimentation. When it is going well, it is like a quiet conversation with Nature. One asks a question and gets an answer; then one asks the next question and gets the next answer. An experiment is a device to make Nature speak intelligibly. After that one has only to listen.

GEORGE WALD
1967 NOBEL LECTURE

Unlike physical scientists, who usually know where to ask their questions (if not how), biologists have an enormous variety of living things from which to choose. Their success will depend as much on choosing the right system as on asking the right question. The choice of the right system often means finding the simplest form of life that can be handled in the laboratory and yet features the process to be investigated. This consideration may lead to choices that seem, on the surface, to be of limited interest—such as bacterial viruses, squid nerves, or frog muscles. However, simpler systems have fewer variables, and therefore the experimenter has more chance of getting meaningful answers. The information gained can then be used to formulate a model for more complicated systems. The unity of nature being what it is, that model will, more often than not, be correct. In any case, it is much easier to prove or disprove a well-formulated model than it is to start fresh with no concept of how the complicated system might work.

For example, we are interested in how genes function in the cells of higher plants and animals, including human beings. But eukaryotic cells are very complex and relatively difficult to handle in the laboratory. They grow slowly, if they can be made to grow at all, and each of their genes comes in two versions, one an *allele* of the other. Changes in one gene may be masked by the activity of its allele. Bacteria, on the other hand, are much simpler and easier to work with, and grow extremely fast. Although each cell may have multiple copies of the same gene, any changes in the genetic makeup of the cell will be rapidly reflected in the chemical or physical properties of the microbe as replicas of the bad gene are passed on to progeny. We might expect, then, that bacteria would serve as a better experimental system than human cells for studying the basic mechanisms of gene function. And since a virus is nothing more than a set of genes wrapped in a protective coat, we might also expect to learn some fundamental aspects of gene behavior by studying the manner in which viral genes control an infected cell. This is the reasoning used by some of the early molecular geneticists and, of course, their approach has been amply vindicated.

In later chapters we shall be discussing some of the specialized functions of cell types found only in higher organisms. You will find, for example, that progress in neural research was greatly facilitated by the discovery of the squid's "giant axon." This nerve cell is large enough to handle with ease, making it possible to do experiments that simply cannot be done with the tiny nerve cells found in mammals. And there is every reason to believe that the basic mechanisms found in squid nerves are the same as those found in human nerves. Although we readily admit that human systems are frequently more complex than their counterparts in simpler life forms, it is safe to say that because we have studied the lower forms of life, we know a lot more about humans than we would have been likely to learn by direct observation, even with a much greater total effort.

SUMMARY

1-1 The cell theory, or cell doctrine, holds that (1) all animals and plants are composed of cells, (2) each cell is capable of living in the absence of others, and (3) a cell can arise only from another cell.

1-2 The nineteenth century saw the end of vitalism, in which the activities of living things were thought to be somehow immune from the laws of physics and chemistry. We now recognize that biological processes involve completely understandable chemical transformations that can be duplicated in the laboratory, though the efficiency of such reactions in cells is greater due to catalysis by a class of proteins called enzymes.

1-3 All cells have (1) a plasma membrane to control the entry and exit of materials, (2) a metabolic machinery that is capable of converting molecules into cellular components, and (3) a set of genes to guide the synthesis of new cellular components.

1-4 Cells are broadly divided into eukaryotes and prokaryotes, according to whether their genes are enclosed by a nuclear envelope. Eukaryotes have such an envelope whereas prokaryotes do not. Bacteria and blue-green algae are the only prokaryotes.

Prokaryotes have a relatively simple cellular structure. The only cytoplasmic inclusions always present are the ribosomes and a nuclear body in the form of one or more long molecules of DNA. The structures external to the cytoplasmic membrane include (in almost all cases) a wall and, in some bacteria, either flagella, or pili, or both.

1-5 Eukaryotes have all the features of prokaryotes and more, though only the ribosomes and plasma membrane closely resemble their prokaryotic counterparts of the same name. The major eukaryotic structures are as follows:

(1) Units composed of microtubules or structures that look like microtubules include flagella, cilia, basal bodies, and centrioles. Microtubules, with a diameter of about 25 nm, should not be confused with the smaller, solid microfilaments, which are usually less than 10 nm in diameter. All may be involved in movements of various kinds and all are assembled from smaller protein subunits.

(2) Spherical bodies enclosed by a single membrane include lysosomes, containing destructive enzymes; peroxisomes, or microbodies, which contain enzymes involved in peroxide production and destruction (as do glyoxysomes of plants); and vacuoles and vesicles of various kinds.

(3) Bodies enclosed by two membranes are the nucleus, mitochondria (which are sites of oxygen consumption and major transducers of energy), and the plastids of plants, including the chloroplasts where photosynthesis takes place.

(4) The endoplasmic reticulum (ER) and Golgi apparatus are comprised of membranes in the form of flattened sacs and tubes. The ER, which is classified as rough or smooth according to whether or not ribosomes are attached, may by budding give rise to other membranous structures, including the Golgi apparatus (or dictyosome), which is involved in packaging material for secretion.

(5) Chromatin is a complex of protein, RNA, and long molecules of DNA, the latter holding the cell's genetic information. Chromatin may be either dispersed in the nucleus or condensed into discrete bodies, the chromosomes. Nucleoli are also composed largely of a type of chromatin, though they have the more restricted function of producing ribosomal components.

1-6 Cells of all types may be maintained ("cultured") in the laboratory, although not all specialized cells can be made to reproduce. Growth may take place in liquid suspension, or on the surface of agar, or as monolayers on the bottom of shallow dishes.

1-7 Viruses are not cells, but cellular parasites. They are broadly classed, according to their hosts, as bacterial viruses (bacteriophages), animal viruses, or plant viruses. The life cycle of viruses consists of the following steps:

(1) Adsorption of the virus to a prospective host.
(2) Infection. Phages infect a cell by injecting their DNA or RNA, leaving most or all of their protein shell outside. Animal and plant viruses generally gain entry with the virus still intact, though only the nucleic acid is important in the infection.
(3) The infected cell may be caused to produce immediately new virus particles under direction of the viral genes, or the genes of the virus may become incorporated as a provirus into the host's chromosome where they may or may not have an obvious effect on the cell's properties.
(4) Release of new virions. New virus particles may be released *en masse* by lysis of the cell, or slowly over many cell generations by budding from the cell surface.

The most fundamental distinction between viruses and cells is one of movement and change. Cells use energy and nutrients from the environment to maintain their integrity and usually to grow and to reproduce. These activities may be temporarily suspended in some cases (e.g., spores), but the capacity remains. Viruses have genes and a protein coat but must depend on the enzymes and ribosomes of a cell in order to reproduce.

1-8 The main problem faced by biologists is the complexity of the systems with which they work. There are numerous variables in an experiment, not all of which can be controlled. The most fruitful investigations of fundamental life processes, therefore, are usually those that are carried out on the simplest forms of life in which the process is found. Once it is understood there, the information can be used to formulate a model for the study of more complicated organisms.

STUDY GUIDE

1-1 (a) What is meant by the cell theory? (b) What was the contribution of Leeuwenhoek, and why was it important?

1-2 (a) What is vitalism? (b) Liebig viewed alcoholic fermentation as a purely chemical process that occurs in the absence of living things. Pasteur demonstrated (by sterilization) that yeast is always present during fermentation. Yet we can still say that Liebig was essentially correct. How is that possible?

1-3 (a) How many bacteria does it take to equal the volume of a typical human white blood cell? For simplicity, assume that the bacterium is a sphere 1 μm in diameter, and that the white blood cell is a sphere 10 μm in diameter. [Ans: 1000] (b) What features do all cells have in common?

1-4 (a) List and state the function of those structures found outside the plasma membrane of prokaryotes. Do the same for internal structures. (b) A common intestinal bacterium, *Escherichia coli* (*E. coli*), is typically about 2 μm long by about 0.8 μm in diameter. Its genes are contained in a single double-stranded molecule of DNA about 1 mm long by 20 Å in diameter. If there are two such molecules present, what fraction of the cell volume is occupied by them? [Ans: 0.6%] (c) A fully grown culture of *E. coli*, numbering some 5×10^9 cells/ml, will be translucent, like milk. At 10^{-12} ml/cell, what fraction of the culture is actually occupied by cells? [Ans: 0.5%]

1-5 (a) Which of the eukaryotic organelles are also found in prokaryotes? Are their structures the same in the two types of cells? (b) What organisms produce oxygen, and with what cellular process and eukaryotic organelle is oxygen associated? (c) With what process is the Golgi apparatus involved? The lysosomes? (d) What features do mitochondria and chloroplasts have in common? What features are shared by these two organelles and the nucleus as well? (e) List the presumed functions of the microtubules, and the structures that seem to be composed of them. How are microfilaments different? (f) New membrane, assembled as part of the endoplasmic reticulum, may later be found in a variety of other organelles. Outline the routes it may take.

1-6 (a) What are the various ways of growing cells? (b) What types of cells can be cultured readily and which cannot?

1-7 (a) What is a virus? Is it alive? (b) How do viruses compare with cells in size? In structure? (c) What is a typical life cycle of a virus? (d) How do viruses differ in their mode of release from infected cells? (e) What is meant by lysogeny? A provirus? (f) In the late 1950s, TMV was completely disassembled (into RNA and protein subunits) and then reassembled again in the laboratory. What does this suggest about viral assembly *in vivo*? (g) When the RNA from one strain of TMV is reassembled with the protein from a second strain and then used to infect a plant, the viral progeny are all members of the first strain as determined by both their RNA and protein. In addition, purified viral RNA or DNA has been shown to be infective in many cells whereas purified viral protein never is. What does this tell you about the role of the two components? (**h**) Even today one can read in some books that a particular organism is "somewhere between a virus and a bacterium." If it were up to you to decide into which category it falls, how would you go about it?

1-8 An investigator interested in a particular biological process may be able to study that process in any number of different organisms. What kinds of considerations will influence the choice?

REFERENCES

HISTORY AND PHILOSOPHY

BROCK, THOMAS D., ed., *Milestones in Microbiology.* Englewood Cliffs, N.J.: Prentice-Hall, 1961. (Paperback.) The first section, on spontaneous generation, includes writings by Leeuwenhoek, Spallanzani, Schwann, Liebig, Pasteur, Tyndall, and Büchner, plus a commentary on their work.

CLAUDE, A., "The Coming of Age of the Cell." *Science,* **189**: 433 (1975). Nobel Lecture, 1974.

COLLARD, P., *The Development of Microbiology.* New York: Cambridge University Press, 1976.

DELBRÜCK, MAX, "A Physicist Looks at Biology." In *Phage and the Origins of Molecular Biology*, edited by J. Cairns, G. Stent, and J. Watson. Cold Spring Harbor (N.Y.) Lab. of Quant. Biol, 1966.

———, "A Physicist's Renewed Look at Biology: Twenty Years Later." *Science,* **168**: 1312 (1970). 1969 Nobel Lecture.

FRUTON, J. S., "The Emergence of Biochemistry." *Science,* **192**: 327 (1976).

FRUTON, J. S., and E. HIGGINS, *Molecules and Life: Historical Essays on the Interplay of Chemistry and Biology.* New York: John Wiley, 1972. Considers the period 1800–1950.

FULTON, J. F., *Selected Readings in The History of Physiology* (2d ed.). Springfield, Illinois: Charles C. Thomas, 1966.

GABRIEL, M. L., and S. FOGEL, eds., *Great Experiments in Biology.* Englewood Cliffs, N.J.: Prentice-Hall, 1955. (Paperback.) Includes excerpts from papers by Hooke, Brown, Schwann, Leeuwenhoek, Koch, Pasteur, Ivanovsky, Stanley, plus extensive chronologies.

HALL, T. S., *History of General Physiology. 600 B.C. to A.D. 1900.* Chicago: University of Chicago Press, 1975. Two volumes.

HESS, EUGENE L., "Origins of Molecular Biology." *Science,* **168**: 664 (1970).

HOOKE, ROBERT, *Micrographia—or some Physiological Descriptions of Minute Bodies Made by Magnifying Glasses with Observations and Inquiries Thereupon.* New York: Dover, 1961. (Paperback.) Reprint of the first English edition, 1665.

KARLSON, P., "From Vitalism to Intermediary Metabolism." *Trends Biochem. Sci.,* **1**(8): N184 (Aug. 1976).

KOHLER, ROBERT, "The Background to Eduard Büchner's Discovery of Cell-Free Fermentation." *J. Hist. Biol.,* **4**: 35 (1971).

LECHEVALIER, H. A., and M. SOLOTOROVSKY, *Three Centuries of Microbiology.* New York: Dover, 1974. (Paperback.) Reprint of a 1965 history.

LURIA, S. E., "Molecular Biology: Past, Present, Future." *BioScience,* **24**: 1289 (1970).

MILLER, JAMES G., *Living Systems.* New York: John Wiley, 1972. A philosophical discourse on the nature of living systems, with enough biology included to make it readable to the novice. A 200-page excerpt appeared in *Currents in Modern Biology,* **4**: 55 (1971).

MONOD, JACQUES, *Chance and Necessity: An Essay on the Natural Philosophy of Modern Biology* (English ed.). New York: Knopf, 1971. What makes a biologist tick?

NEEDHAM, JOSEPH, ed., *The Chemistry of Life.* Cambridge Univ. Press, 1970. See the essay on the development of microbiology by E. F. Gale, and those on early biochemistry by M. Teich and R. Peters.

PENN, M., and M. DWORKIN, "Robert Koch and Two Visions of Microbiology." *Bacteriol. Rev.,* **40**: 276 (1976).

PORTER, J. R., "Anthony van Leeuwenhoek: Tercentenary of his Discovery of Bacteria." *Bacteriol. Rev.,* **40**: 260 (1976).

QUAGLIARIELLO, E., F. PALMIERI, and T. SINGER, eds., *Horizons in Biochemistry and Biophysics,* Vol. 1. Reading, Mass.: Addison-Wesley, 1974. (Paperback.)

REID, R., *Microbes and Men.* New York: E. P. Dutton, 1975. History of the germ theory.

ROSENBERG, E., *Cell and Molecular Biology: An Appreciation.* New York: Holt, Rinehart & Winston, 1971. (Paperback.) History of cell theory.

SCHAFFNER, KENNETH F., "Chemical Systems and Chemical Evolution: The Philosophy of Molecular Biology." *Am. Scientist,* **57**: 410 (1969).

STENT, GUNTHER S., "That Was the Molecular Biology That Was." *Science,* **160**: 390 (1968). The history and scope of molecular biology.

VIRCHOW, RUDOLF, *Cellular Pathology.* New York: Dover, 1971. (Paperback.) From the 1863 edition.

CELL STRUCTURE

ALTMAN, P., and D. KATZ, *Cell Biology.* Bethesda, Maryland: Federation of American Societies for Experimental Biology, 1976. Biological Handbooks. 1.

BECK, F., and J. B. LLOYD, eds., *The Cell in Medical Science,* Vol. 3. New York: Academic Press, 1976. Considers cells specialized for a variety of organs.

BLOOM, W., and D. W. FAWCETT, *Textbook of Histology* (10th ed.). Philadelphia: W. B. Saunders Co., 1975. Includes some 1200 illustrations, mostly electron micrographs.

BUTTERFIELD, BRIAN, and B. A. MEYLAN, *The Three Dimensional Structure of Wood.* London: Chapman & Hall, and Syracuse Univ. Press, 1972. Beautiful scanning electron micrographs, such as the one shown in Fig. 1-1.

ECHLIN, PATRICK, "The Blue-Green Algae." *Scientific American,* June 1966. (Offprint 1044.)

FAWCETT, D. W., *The Cell: Its Organelles and Inclusions.* Philadelphia: W. B. Saunders, 1966. An excellent atlas.

GUNNING, B., and M. W. STEER, *Plant Cell Biology. An Ultrastructural Approach.* New York: Crane, Russak, 1975. (Paperback.)

JENSON, W. A., and R. B. PARK, *Cell Ultrastructure.* Belmont, Calif.: Wadsworth, 1967. (Paperback.) A low-priced atlas of electron microscopy. Limited in scope but containing many excellent illustrations.

KESSEL, G., and C. Y. SHIH, *Scanning Electron Microscopy in Biology. A Students' Atlas on Biological Organization.* New York: Springer-Verlag, 1974.

LEDBETTER, M. C., and K. R. PORTER, *Introduction to the Fine Structure of Plant Cells.* Berlin: Springer-Verlag, 1970. An atlas, with beautiful electron micrographs.

LENTZ, THOMAS L., *Cell Fine Structure.* Philadelphia: W. B. Saunders Co., 1971. An atlas of drawings of whole cells.

LIMA-DE-FARIA, A., ed., *Handbook of Molecular Cytology.* New York: Wiley-Interscience, and Amsterdam: North-Holland, 1968. Expensive, but a wealth of information.

LOTT, J., *A Scanning Electron Microscope Study of Green Plants.* St. Louis: Mosby, 1976. (Paperback.)

McGEE-RUSSELL, S. M., and K. F. A. FOSS, eds., *Cell Structure and Its Interpretation.* London: Edward Arnold, 1968. A collection of essays, many of which should be easily understood by the beginner.

MOROWITZ, H. J., and M. E. TOURTELLOTTE, "The Smallest Living Cells." *Scientific American,* March 1962. (Offprint 1005.) Mycoplasmas.

PALADE, G. E., "Structure and Function at the Cellular Level." *J. Am. Med. Assoc.,* **198**: 815 (1966).

PORTER, K. R., and M. A. BONNEVILLE, *Fine Structure of Cells and Tissues.* Philadelphia: Lea & Febiger, 1968.

REUSCH, V. M., Jr., and M. M. BURGER, "The Bacterial Mesosome." *Biochim. Biophys. Acta,* **300**: 79(1973). A comprehensive review.

ROBARDS, A. W., ed., *Dynamic Aspects of Plant Ultrastructure,* New York: McGraw-Hill, 1974.

SANDBORN, E. B., *Light and Electron Microscopy of Cells and Tissues.* New York: Academic Press, 1972. An atlas.

SMITH, D. G., "Bacteria with Their Coats Off: Spheroplasts, Protoplasts and L-forms." *Science Progress (Oxford),* **57**: 169 (1969).

VIRUSES

CHAMPE, S., ed., *Phage.* New York: Halsted Press, 1974. Benchmark papers from 1942 to 1972.

DOUGLAS, J., *Bacteriophages.* New York: Halsted Press, 1975. (Paperback.) Concise introduction to bacteriophage biology.

FENNER, F., B. McAUSLAN, C. MIMS, I. SANDBROOK, and D. WHITE, *The Biology of Animal Viruses.* New York: Academic Press, 1974.

FRAENKEL-CONRAT, H., and R. WAGNER, eds., *Comprehensive Virology.* New York: Plenum, 1974 and 1975. Multiple volumes.

FRASER, DEAN, *Viruses and Molecular Biology.* New York: Macmillan, 1967. (Paperback.)

GIBBS, A., and B. HARRISON, *Plant Virology: The Principles.* London: Arnold, 1976. A text.

HAHON, NICHOLAS, ed., *Selected Papers on Virology.* Englewood Cliffs, N.J.: Prentice-Hall, 1964. (Paperback.) Contains many of the most important early papers on animal, plant, and bacterial viruses.

HORNE, R. W., *Virus Structure.* New York: Academic Press, 1974. (Paperback.) Excellent micrographs.

JACOB, F., and E. L. WOLLMAN, "Viruses and Genes." *Scientific American,* June 1961. (Offprint 89.) Lysogeny and viral transformation.

LURIA, S. E., "Phage, Colicins, and Macroregulatory Phenomena." *Science,* **168**: 1166 (1970). Nobel Lecture, 1969.

LWOFF, ANDRÉ, "Interaction Among Virus, Cell, and Organism." *Science,* **152**: 1216 (1966). Nobel Lecture (1965) on lysogeny.

PRIMROSE, S. B., *Introduction to Modern Virology.* New York: Halsted Press, 1975. (Paperback.) Emphasis on the biochemical and genetical aspects.

SMITH, K. M., *Plant Viruses* (5th ed.) London: Chapman and Hall, 1974. (Paperback.)

STANLEY, WENDELL M., "Isolation of a Crystalline Protein Possessing the Properties of Tobacco-Mosaic Virus." *Science,* **81**: 644 (1935). Reprinted in *Selected Papers on Virology,* edited by Nicholas Hahon, and in *Great Experiments in Biology,* edited by M. Gabriel and S. Fogel.

STENT, GUNTHER, *Molecular Biology of Bacterial Viruses.* San Francisco: W. H. Freeman, 1963. Contains an excellent historical introduction.

————, *Papers on Bacterial Viruses* (2d ed.). Boston: Little, Brown, 1965. See Introduction, and reprinted works of d'Herelle, Twort, Bordet, Gratia, Ellis, Delbrück, and Lwoff.

WILKINSON, J. F., ed., *Introduction to Modern Virology.* Oxford: Blackwell Scientific, 1974.

WILLIAMS, R. C., and H. W. FISHER, *An Electron Micrograph Atlas of Viruses.* Springfield, Ill.: Charles C. Thomas, 1974. Animal, plant, insect and bacterial viruses.

NOTES

1. E. B. Wilson, *The Cell in Development and Heredity* (New York: Macmillan Co., 1925.)

2. Dimensions of cellular structures are commonly given in micrometers ($1\,\mu m = 10^{-6}$ meters), nanometers ($1\,nm = 10^{-9}$ m), or angstroms ($1\,\text{Å} = 10^{-8}$ cm $= 10^{-10}$ m). Another name for a micrometer is *micron*, abbreviated μ. It is important to be completely comfortable with interconversion of these units—e.g., $1\,nm = 10\,\text{Å}$; $1\,\mu m = 1000\,nm$.

2

Chemical Reactions in an Aqueous Environment

2-1 Bond Energy and Bond Stability 46
    Boltzmann's Equation
    Bond Energy
2-2 The Biologically Important Weak Bonds 50
    Hydrogen Bonds
    Ionic Bonds
    Hydrophobic Bonds
    Short-Range Bonds
2-3 Energy Changes in Chemical Reactions 56
    Gibbs Free Energy
    Spontaneous Reactions
    Equilibrium Constants
    The Generalized Reaction

2-4 Coordination Compounds 59
    The Dative Bond
    The Stability of Coordination Complexes
2-5 Acids and Bases 62
    The Behavior of Acids in Water
    Bases
    Buffers
2-6 The State of Intracellular Water 65
Summary 66
Study Guide 67
References 68
Notes 69

You were introduced to some of the structural features of cells and their organelles in the preceding chapter. We shall approach their functions by trying to understand the nature of the molecules of which they are composed. To do that, it will first be necessary to have some understanding of the forces responsible for determining the shape of large molecules and for molecular interactions. These forces are properly classed as chemical bonds, but usually they are not covalent bonds. The following discussion is designed to help you appreciate that crucial distinction.

2-1 BOND ENERGY AND BOND STABILITY

When an object falls from a table to the floor, it achieves a lower energy level. The table and the floor are at different elevations and therefore represent two different levels of potential energy in the earth's gravitational field. The amount of energy necessary to lift the object back to the table again is exactly the amount released when it fell to the floor. We could show this if we had a frictionless pulley and string, for the falling object could then be made to raise an identical mass through the same distance. Under these conditions it would be fair to say that the system is in equilibrium because there is movement but no change in net energy, and hence no change in the capacity of the system to perform some outside task—i.e., do "work."

When an object is pushed off the table without such a pulley, one would expect it to fall to the floor and stay there. You might find it a little hard to believe, therefore, if someone told you that the object remained at table height after it was released, or returned there after falling. Of course, we never expect to see that happen, but the existence of thermal (Brownian) movements makes it possible. We can calculate exactly the odds against it; the Austrian physicist Ludwig Boltzmann (1844–1906) showed us how.

Boltzmann's Equation. Boltzmann derived an equation that relates energy to concentration in an equilibrium situation. For our purposes here it can be written in the following terms:

$$\Delta E = -RT \ln \frac{[B]}{[A]} = -2.3RT \log \frac{[B]}{[A]} \qquad (2\text{-}1)$$

where [A] is the number or concentration of particles at energy level A, [B] is the number or concentration at a higher level B, and ΔE is the amount of energy needed to raise a "mole" (Avogadro's number, N_{AV}, or 6×10^{23}) of the particles from level A to level B, or the amount of energy released when they fall from level B back to level A again. The symbol R is the universal gas constant (8.314 joules per mole per degree) and T is the absolute temperature.[1] The term 2.3 converts the equation from natural (base e) logarithms to the customary base 10.

In terms of our previous example, [B] and [A] could be the number of pencils on the table and on the floor, respectively. The difference in energy between the two levels, separated by x cm, would then be calculated as

$$\Delta E = (N_{AV} \times m)g\Delta x \text{ erg/mole} \qquad (2\text{-}2)$$

where m is the mass of each object in grams, $N_{AV} \times m$ is the grams per "mole" (equivalent to molecular weight, M), and g the gravitational constant (980.665 cm/s^2). The calculated number of ergs (an erg is said to be about the amount of energy imparted to a wall when a mosquito flies full tilt into it) must be converted to the standard unit of energy, joules (abbreviated J), by multiplying by 10^{-7}—i.e., there are ten million ergs in a joule. Hence, for an object with a mass of one gram and a table one meter high, there would have to be $10^{10^{18}}$ objects on the floor at 300 K before we could expect to find one floating by itself at table height. (The number of atoms in the earth is about 10^{50}.) Clearly, if you said that you actually saw such an event, there would be every justification for polite doubt.

For particles of molecular size, however, the Boltzmann equation returns very useful information. Consider, for example, a molecule of oxygen (O$_2$) with an M of 32 g/mole. In a gravitational field, the population of oxygen molecules will be about 0.01% less at the height of the one-meter table than at the floor. This is the value predicted by the Boltzmann equation. At 1.6 km (about a mile), the potential energy of oxygen is

$Mg\Delta x = (32)(980)(1.6 \times 10^5 \text{ cm}) = 5 \times 10^9$ ergs/mole
$\phantom{Mg\Delta x = (32)(980)(1.6 \times 10^5 \text{ cm})} = 500$ joules/mole

more than at the surface. At $T = 300$ K Equation 2-1 becomes

$$-\frac{500}{2.3 \times 8.3 \times 300} = -0.087 = \log \frac{[B]}{[A]} \quad \text{or} \quad \frac{[B]}{[A]} = 0.82$$

In other words, the density of oxygen molecules at a constant temperature of 27 °C drops off 18% for every mile of elevation.

From the above, we see that the Boltzmann equation can be used to predict the relative population of two states of a system that differ by a known increment of energy. Conversely, if the population of two states of a system are measured, the equation can be used to predict the amount of energy by which they differ—i.e., the amount of energy gained or lost in moving a particle from one to the other.

These relationships are as applicable to the making and breaking of chemical bonds as they are to movement in the earth's gravitational field. One need only remember that *it always takes energy to break a chemical bond*; otherwise we would all fly apart into atoms. Hence, the formed bond is the low energy state (equivalent to the pencil on the floor or oxygen molecules at the surface of the earth) while a broken bond is the high energy state.

Bond Energy. The Boltzmann equation tells us that no matter how strong the bond between two atoms, a certain fraction of these bonds will always be broken. To put it another way, there is always a finite probability of finding any particular bond ruptured at any moment. The greater the bond energy the more stable the bond, to be sure, but there is no such thing as a permanent chemical bond. However, we shall find it useful to make a distinction between the more stable ("strong") bonds and the less stable ("weak") bonds.

A strong bond, for our purposes, will mean a *covalent bond*. These are, of course, the bonds that are almost universally present in substances, and that are commonly designated by a single solid line, or dash, between the bonded atoms (e.g, H—O—H). They were first described by G. N. Lewis (1875–1946) of the University of California; he attributed their character to the sharing of one or more electron pairs by the two atoms involved. (Two shared electron pairs form a double bond, indicated by =, and so forth.)

Each atom has a particularly stable number of electrons that it tends to achieve by either losing or gaining a few. Hydrogen, for example, has one proton and one electron, but it readily assumes the more stable helium configuration with two electrons. It can do so by accepting an electron from another substance to create the hydride anion, H^-, or by sharing an electron pair with the other substance, each of the atoms donating one electron to the shared pair. Thus, in hydrogen gas, H—H, each proton has, effectively, the more stable two-electron configuration, even with only a single electron pair in the molecule. This need to share the same electron pair bonds the two atoms together. It would take about 400,000 joules (400 kilojoules, abbreviated kJ) to break a "mole" (6×10^{23}) of these H—H bonds, a bond energy that is not unusual for covalent bonds (see Table 2-1).

Other kinds of interactions, which we shall enumerate, are also important in holding atoms and molecules together. They are the weak bonds, so named because they generally have energies of less than 20 kJ per mole in biological systems. The important feature of weak bonds is that

TABLE 2-1. Some Covalent Bond Energies[a]

Single bonds		Double bonds		Triple bonds	
O—H	460 kJ/mole				
H—H	435				
P—O	418	P=O	502		
C—H	414				
N—H	389				
C—O	351	C=O	711		
C—C	347	C=C	611	C≡C	816
S—H	339				
C—N	293	C=N	615	C≡N	887
C—S	259				
N—O	222	N=O	606		
S—S	213				
N—N	159				
O—O	138				

[a] Energies are given as standard state enthalpies. From L. Pauling, *The Nature of the Chemical Bond*, 3d ed. (Ithaca, N.Y.: Cornell, 1960) and T. L. Cottrell, *The Strengths of Chemical Bonds*, 2d ed. (London: Butterworths Scientific Publ., 1958).

they are easily broken in the usual physiological environment. Therefore, structures that are stabilized by weak bonds can be easily rearranged—i.e., they are easy to assemble and disassemble. That simple fact, more than any other, is what makes life possible, for if there is anything that is characteristic of living organisms, it is change.

A bond with an energy of only a few kilojoules per mole has a good chance of being broken at any instant in time. The fraction broken can be calculated from Equation 2-1, which tells us, for example, that a bond energy of 6 kJ/mole results in a ratio of broken to formed bonds of roughly 0.1 at 300 K, an energy of 12 kJ/mole yields a ratio of about 0.01, and so on. The fraction of bonds in the broken state follows by simple algebra.

EXAMPLE. Let the bond energy be 6 kJ/mole and $T = 300$ K. Then

$$\log \frac{[B]}{[A]} = -\frac{6000}{2.3 \times 8.314 \times 300} = -1.046$$

Hence, $[B]/[A] = 0.090$, where $[B]$ is the concentration of bonds broken and $[A]$ the concentration (or relative number) that are intact when the system is at equilibrium.

The fraction of bonds in the broken state is thus

$$\frac{[B]}{[B] + [A]} = \frac{[B]}{[B] + 11.1[B]} = \frac{1}{12}$$

In other words, if a 6000 joule (6 kJ) bond exists between two molecules, then that bond will be broken about 8% of the time.

The above calculation assumes that the two participating groups are clamped into position so that breaking the bond does not permit them to drift apart or change location in such a way that reformation is not pos-

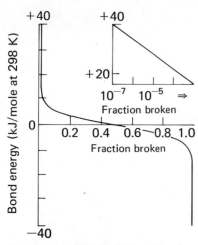

FIGURE 2-1 The Stability of Bonds. The stability of a bond is directly related to its energy. Note that bonds with energies greater than 20–40 kJ/mole are exceedingly stable, with little tendency to rupture spontaneously.

sible. In technical terms, we are neglecting *entropy*, an oversight that will be rectified later.

In contrast to the above example, a covalent bond with an energy of 200 kJ/mole (relatively weak for a covalent bond) will have a broken-to-formed ratio at equilibrium of about 10^{-35}. There would be very little spontaneous rupture in such a case. In fact, at moderate temperatures, it would take more than 10^{13} *moles* of such bonds before we could expect to find even one of the bonds broken. (See Fig. 2-1 for a graphical representation of this point.) If all change in molecular configuration had to depend on the spontaneous making and breaking of covalent bonds, life would be vastly more difficult.

Weak bonds are important to us because they stabilize the shape of many kinds of molecules, and because in many instances they hold together the individual molecules of larger structures as well. The protective coat of most viruses, for example, is a polymer of protein subunits held together by noncovalent (weak) interactions.

2-2 THE BIOLOGICALLY IMPORTANT WEAK BONDS

The weak chemical bonds may be classified as hydrogen bonds, ionic bonds, hydrophobic bonds, and bonds due to short-range forces. Because of their tremendous importance in biological systems, each will be considered in turn.

Hydrogen Bonds. Hydrogen bonds are the result of proton (H^+) sharing. The energy necessary to break a mole of hydrogen bonds in an aqueous environment is about 20 kJ, give or take a little, depending on the circumstances. In biological structures, we usually find the shared proton between nitrogens, between oxygens, or between one of each (as in NH···O), but always closer to the atom to which it really "belongs."

Oxygen and nitrogen atoms tend to hold electrons very tightly, and may retain a partial negative charge even while participating in a covalent bond. We say that such atoms are very *electronegative*. When either oxygen or nitrogen is covalently bonded to hydrogen, the result is a small *dipole*, or separation of unlike charges within the same molecule. In other words, the shared electron pair spends more time on the oxygen or nitrogen than on the hydrogen, leaving the latter with a partial positive charge. The oxygen or nitrogen, of course, carries a partial negative charge of equal magnitude, since the overall structure must be electrically neutral.

Electrical charge originates with the nucleus and produces a field that decreases as the inverse square of distance from the charge. The smaller the atom, the more intense the field will be at the point of closest physical approach to it. Since hydrogen is the smallest atom, consisting of a single proton and a single electron, a given quantity of charge produces an electrostatic field of greater intensity than with other atoms. Hence, a hydrogen atom, covalently bound to an oxygen or nitrogen, still has a significant positive electrostatic field. This raises the possibility of producing an attractive force to an atom with a charge of opposite polarity, such as a second nitrogen or oxygen. Since the charge left on the hydrogen is small, the attraction to the second atom is weak, but it is by no means insignificant. That attraction is the basis for the hydrogen bond.

The positively charged proton, sitting between two negative atoms, shields them from each other. If the proton gets out of line, the negative atoms begin to repel each other again, forcing the proton back. Thus, we find that the three interacting atoms in a hydrogen bond lie very nearly on a straight line, although deviations of up to 30° have been measured in a few cases.

Water is a good hydrogen bonder. Since it consists only of oxygen and hydrogen, there is maximum opportunity for such interactions. The degree of hydrogen bonding increases as the water is cooled until finally it can assume an open but ordered structure—i.e., the water solidifies into ice. Ice is a lattice of water molecules, with each oxygen participating in two hydrogen bonds in addition to two covalent bonds (see Fig. 2-2).

When ice is warmed to the melting point, its three-dimensional lattice begins to break up. Considerable heat energy, called the *heat of fusion*, is necessary to rupture enough hydrogen bonds to turn ice into a liquid. The disruption results in a more compact configuration (greater density), but since only about 15% of the bonds need be broken to melt the ice, there is still a considerable amount of structure left in the water. As the water is warmed to 4 °C, enough additional structure is destroyed to reduce the volume a little further. Above 4 °C water expands as it is warmed, apparently because the natural tendency for molecules to space themselves further apart as their thermal energy increases more than compensates for the additional destruction of lattice structure. Water maintains a considerable amount of structure at all temperatures, however, right up to its boiling point. Then, moving a molecule from the liquid to the vapor state requires the breaking of still more hydrogen bonds, resulting in a high *heat of vaporization*.

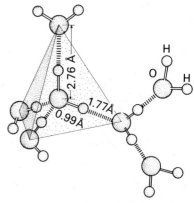

FIGURE 2-2 The Hydrogen Bond in Water. The hydrogen bond is in its most stable position when the three atoms involved (hydrogen and the electronegative atoms on either side) are linearly arranged, since the partial positive charge of hydrogen then shields the two electronegative atoms from one another. This factor, plus the normal bond angles of oxygen, gives ice (and, to a certain degree, liquid water) a tetrahedral arrangement.

The simple hydrogen-oxygen structure of water not only is responsible for its peculiar physical properties—the expansion at very low temperatures, the high heats of fusion and vaporization—but contributes to the fact that water is a very good solvent. Water will dissolve appreciable quantities of almost any molecule that carries a net charge or has *polar groups* (dipoles). That is, any molecule with an asymmetrically distributed electron cloud, or permanent dipole (e.g., O—H), is likely to be soluble in water because of attractive forces between it and the water molecules, which are themselves permanent dipoles. The interactions may be true hydrogen bonds, but other kinds of weak bonds are also important. It is because of such interactions that substances enter aqueous solution. It is a principle of thermodynamics (actually the so-called *second law of thermodynamics*) that any reactive system tends to seek its lowest energy. If more bonding energy can be released by moving a substance into the aqueous phase of a system, thus permitting hydrogen-bonding to water for instance, then that is exactly what will happen.

The hydrogen-bonding capacity of water is also responsible for the "weakness" of the hydrogen bond, since any hydrogen bond that gets broken in an aqueous environment may be replaced with a hydrogen bond to water of nearly equivalent energy. Thus, it is not necessarily true that hydrogen bonds are "weak"—some of them are quite strong—but *it is the ease with which they may be replaced* that makes hydrogen-bonded structures subject to change when water or other potential replacements are about.

Ionic Bonds. Ionic bonds result from the attractive force between completely ionized groups of opposite charge. In a vacuum, the ionic bond is

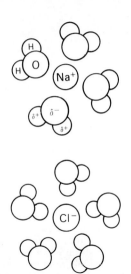

FIGURE 2-3 The Hydration of Ions. Because of their tendency to attract and hold water molecules, each of which behaves as a small dipole, ions in aqueous solution are partially shielded from one another. As a consequence, ionic interactions *in vivo* tend to be weak—in contrast, for example, to the extremely strong ionic bond in a crystal.

very strong, and is actually an extreme example of covalent bonding. The forces that hold Na^+ and Cl^- into a salt crystal are strong ionic bonds, resulting in a melting point (801 °C) that is much higher than the melting point of ice. In solution, however, the polar character of water shields charged groups from each other, and the ionic bond can no longer be considered strong (see Fig. 2-3). In fact, ionic bonds, like hydrogen bonds, have energies of only about 20 kJ/mole in solution.

Water shields an ion because each water molecule is a small dipole. In the liquid state, water molecules are free to orient themselves about an ion, to present their pole of opposite sign to the charged group. This alignment of water has two results: (1) the electrostatic field of the ion is partly neutralized, and (2) some of the water molecules are immobilized, forming a *shell of hydration* about the ion. Typically, an ion of either charge will immobilize from four to six water molecules. For many purposes, the ion is now much larger than it was, since the entire unit—the ion plus its shell of hydration—will move together (see Table 2-2). This *bound water* has properties somewhere between those of free water and ice, and it is the condition in which much of the water in a living cell is found.

TABLE 2-2. The Effect of Hydration on Ionic Size

	Li^+	Na^+	K^+
Anhydrous diameter in a crystal lattice	1.6 Å	2.0 Å	2.3 Å
Hydrated diameter in solution	7.3	5.6	3.8
$\left[\dfrac{\text{hydrated volume}}{\text{anhydrous volume}}\right]$	95	22	4.5

Note from Table 2-2 that the smallest of the three ions listed, lithium (Li^+), actually has the largest hydrated diameter. This difference is easily demonstrated when one measures the ease with which these ions pass through membranes—K^+ behaves as if it were the smallest and Li^+ as if it were the largest, even though we know their actual physical sizes (without the shell of hydration) are the reverse of that. Similarly, their relative velocities when attracted by a negative electrode at 25 °C are 4.01 μm/s for Li^+, 5.19 μm/s for Na^+, and 7.62 μm/s for K^+ in a field of one volt per centimeter. Again, K^+ behaves as if it were the smallest.

The reason for the difference in hydration of the three alkali ions in Table 2-2 is that while each carries the same charge, that charge originates at the nucleus and falls off rapidly as one moves away from the nucleus. In fact, it falls off as $1/r^2$, where r is distance from the nucleus. Because of the small electron cloud of Li^+, the electric field at the point of closest approach is much higher for that ion than for the others. It is thus able to orient a great deal of water about itself.

This capacity of a molecule, demonstrated by water, to orient itself in an electric field is quantitatively expressed as its *dielectric constant*. The dielectric constant is a measure of the effectiveness with which a substance diminishes the electrostatic force between charged groups. The force, F, is given by Coulomb's law,

$$F = \frac{q_1 q_2}{r^2 d} \tag{2-3}$$

where q_1 and q_2 are the magnitude of the charge on the two groups, and r the distance between them. The dielectric constant is given as d. Because it orients itself about charged groups, water has a high dielectric constant, equal to 80 at 20°C, relative to air or a vacuum with $d = 1$. This means that two oppositely charged ions attract each other only 1/80 as strongly when they are in water as when they are in air or a vacuum. Thus, although ionic bonds in solution are weak, in a low-dielectric environment such as a salt crystal, they are very strong.

The dielectric constant, then, is a measure of the polar character of a molecule, and of the ease with which it can orient itself in an electric field. While water has a dielectric constant of 80 at 20 °C, the value is very low (e.g., 1 to 5) for wood, glass, oils, fats, and other nonpolar substances. In contrast, glycerol (CH_2OH—$CHOH$—CH_2OH) and ethanol (CH_3—CH_2OH) have dielectric constants of 42 and 24 at 25 °C, respectively, because the —OH groups of these molecules are all small dipoles. We also find that glycerol and ethanol are soluble in water in all proportions (i.e., they are miscible with water) because of the large number of electrostatic interactions that they may make with water. Molecules with low dielectric constants do not interact with water and are not soluble in it to any appreciable extent. These are the nonpolar molecules involved in our next class of weak interactions, the so-called hydrophobic bonds.

Hydrophobic Bonds. "Hydrophobic bond" is actually a misnomer, because what we really are referring to is the tendency of nonpolar groups to aggregate (associate with each other) when in the presence of water. This hydrophobic, or "water-fearing," behavior minimizes the nonpolar surface exposed to water, and hence minimizes the unfavorable interaction that these groups have with it. The "unfavorable interaction" leads to a structuring of the water through an increase in hydrogen bonding. Water thus forms a lattice about the intruding group—as if the water molecules were clinging together to avoid contact with the nonpolar substance.

If you have been a careful reader, you may be a little puzzled at this point. On the one hand, you were told that the making of new bonds is a desirable situation, leading to a state of lower energy. On the other hand, you are now asked to believe that the increased number of water-water bonds caused by the introduction of hydrophobic groups is an undesirable event. To understand this apparent paradox, we need to introduce a new term, *entropy*.

Entropy, in its simplest sense, is a measure of randomness. The more disordered a system becomes, the greater its entropy content. As you are well aware, it takes energy to bring order out of chaos. When a nonpolar molecule is introduced into water, the energy released by the formation of new hydrogen bonds as water molecules form their "iceberg" about the nonpolar molecule is more than offset by the energy expended in limiting the motion of the water molecules. The result is an unfavorable process, one in which equilibrium will see a small amount of the nonpolar material in solution and a larger amount out of solution. In other words, it is entropy that makes nonpolar substances relatively insoluble in water. This concept will be put on a more quantitative basis later, at which time you will also come to see why hydrophobic bonds are

$$CH_3-(CH_2)_n-\overset{\overset{O}{\|}}{C}-O^-$$

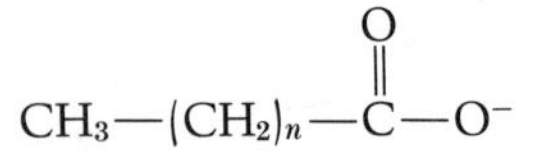

Anion of a fatty acid (soap)

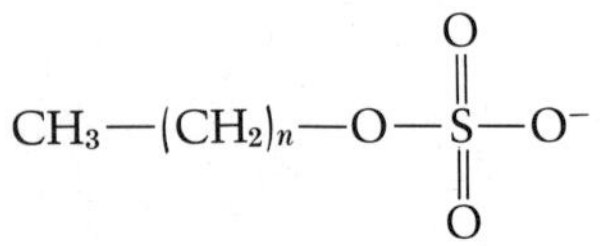

$$CH_3-(CH_2)_n-O-\overset{\overset{O}{\|}}{\underset{\underset{O}{\|}}{S}}-O^-$$

An unbranched synthetic detergent

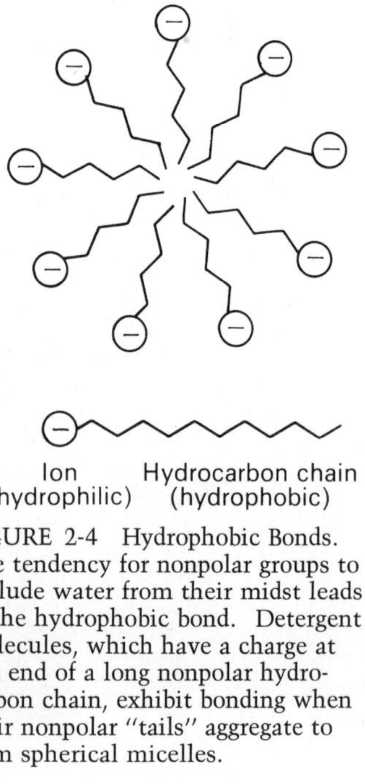

FIGURE 2-4 Hydrophobic Bonds. The tendency for nonpolar groups to exclude water from their midst leads to the hydrophobic bond. Detergent molecules, which have a charge at one end of a long nonpolar hydrocarbon chain, exhibit bonding when their nonpolar "tails" aggregate to form spherical micelles.

unusual in always getting stronger, rather than weaker, as the temperature is increased.

There are many examples of hydrophobic bonding in biological systems, but just to bring the concept closer to home consider an everyday situation. Ordinary detergents, including soap, are long, nonpolar hydrocarbon ($-CH_2-CH_2-CH_2-$) chains, terminated at one end by a charged group, usually an acidic one such as sulfate ($-OSO_3^-$) or carboxylate ($-COO^-$) ion. The charged end is highly water soluble, whereas the hydrocarbon "tail" does not enter readily into an aqueous environment. Although the charged end is soluble enough to drag its tail into the water, a critical concentration may be reached that allows the tails to solve their solubility problem by aggregating. They then form little spheres, or *micelles*, that have a hydrocarbon interior and a surface of charged groups (see Fig. 2-4). Several hundred molecules may participate in a single micelle, effectively excluding water and creating their own nonpolar environment. The micelles, of course, are stabilized by hydrophobic bonds.

If a droplet of fat collides with such a micelle, it will enter the interior where it is soluble. A detergent solution thus presents two solvents, water and a hydrocarbon. The latter substance, confined to the interior of micelles, allows detergent solutions to dissolve fatty material like greases and oils.

Hydrophobic bonds, though important, are not strong. Their strength varies with temperature and with surface area of the aggregating, nonpolar groups, but bond energies are typically less than 10 kJ/mole.

Short-range Bonds. The fourth class of weak interactions includes bonds due to short-range forces (see Fig. 2-5). Such forces are negligible until the potentially bonded groups get very near each other, and they decrease rapidly as the interacting groups are moved apart: The effective bond energy will decrease as $1/r^4$ to $1/r^6$, where r is the distance between the interacting groups. Like hydrophobic bonds, these short-range bonds are individually weak; however, if many occur between two molecules a very stable structure may result. A typical bond energy involving such forces is less than 5 kJ/mole.

We can divide the short-range forces into two groups: those in which an ion interacts with a neutral molecule, called an *ion-dipole interaction;* and those in which two neutral atoms or molecules interact, called a *van der Waals interaction*.

You have already been introduced to the ion-dipole interaction in our discussion of ionic bonds, for it is the ability of ions to form ion-dipole interactions with water (i.e., to structure water, the molecules of which are permanent dipoles) that reduces the force between the ions themselves to the point where we can classify ionic bonds as "weak." However, one need not have a permanent dipole to get an interaction with ions, for a positive ion will attract the electrons of neutral atoms, while a negative ion repels them. In either case, the previously symmetrical charge distribution on the neutral atom is distorted, turning it into an *induced dipole* (Fig. 2-5d). The alignment of the induced dipole is such that attractive forces between it and the ion are created. These forces fall off rapidly with increasing distance, dissipating as $1/r^5$. The effective bond energy between two such groups, then, decreases as $1/r^4$.

The second group of short-range forces, the van der Waals forces, are attractions between neutral atoms or molecules. They include (1) dipole-dipole interactions, (2) dipole-induced dipole interactions, and (3) London dispersion forces, or induced dipole-induced dipole interactions. In these cases, the force between the groups decreases with $1/r^7$ and the bond energy with $1/r^6$. A brief description of each category follows.

(1) Molecules with assymetric charge distributions (dipoles) will attract each other if they are close and oriented with ends of opposite polarity facing each other. But since the charges at the two ends of a dipole are equal and opposite, the attraction between two dipoles falls off rapidly as the intermolecular distance becomes larger. When separated, the dipoles lose alignment, tumbling around in a random way, causing the rapid dissipation of *the dipole-dipole interaction* (see Fig. 2-5b).

(2) P. Debye suggested in 1920 that since a dipole does establish an electrostatic field, it can induce an oppositely oriented dipole in a molecule with a symmetrical charge—i.e., it behaves like an ion, but is less successful. When the negative pole of a dipole is near an uncharged molecule, the electrons of the uncharged molecule will experience a net repulsion and spend more time than usual on the side away from the dipole. Similarly, when the positive pole of a dipole is nearer a neutral atom, the electrons of the atom will be attracted to the side facing the dipole. In either case, the neutral atom or molecule becomes an induced dipole. These *dipole-induced dipole interactions* (Fig. 2-5c) are usually weaker than interactions between permanent dipoles but stronger than the dispersion forces, described next.

(3) In 1930 London extended the concept of induced dipoles to neutral atoms. Although on the average the charge distribution about such atoms is spherically symmetrical, at any one moment in time the distribution will be distorted. You can convince yourself of this by thinking of the hydrogen atom, which has one proton and one electron. Though hydrogen is neutral, with the electron spending equal time on all sides, if you could stop the motion of the electron for an instant, you would have a complete dipole (a "snapshot dipole"). In other words, even a neutral atom represents a fluctuating dipole. If two neutral atoms are brought close together, the transitory dipole nature of one will, for brief instants, induce a dipole of opposite polarity in the other atom, and vice versa. The result is that the two atoms experience a slight mutual attraction. London calculated the magnitude of this attraction from quantum mechanical considerations, and in his honor such forces are sometimes called *London dispersion forces* (Fig. 2-5e).

Short-range forces, especially the dispersion forces, also contribute to the hydrophobic bond, which we presented earlier as a purely entropy-driven interaction. The nonpolar groups that aggregate to avoid water must experience dispersion forces, stabilizing their association and hence increasing the energy of the hydrophobic bond.

The concept of dispersion forces is useful in still another context, for we find that although neutral atoms attract each other with a force that increases rapidly as they approach, a point will be reached when that attraction is offset by overlap in the electron clouds of the two atoms. This overlap produces a repulsive force that becomes extremely strong as they move still closer. At some given separation, the overlap repulsion and the van der Waals attraction will exactly cancel and, in the absence of any disturbing factor, that is the separation we can expect the two atoms to maintain. When two identical atoms are in that position, half the

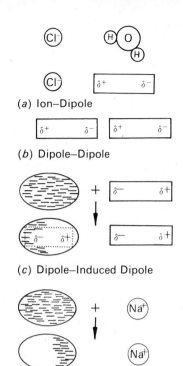

(a) Ion–Dipole

(b) Dipole–Dipole

(c) Dipole–Induced Dipole

(d) Ion–Induced Dipole

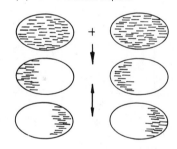

(e) Induced Dipole–Induced Dipole (charge fluctuation forces)

FIGURE 2-5 Short–Range Bonds. (a) The ion-dipole interaction is the kind of behavior exhibited by water (a dipole) in the vicinity of an ion. (See also Fig. 2-3.) (b) Two permanent dipoles tend to orient and attract one another, much as two permanent bar magnets would do. (c) Since a dipole creates an electrostatic field, it can induce a dipole in a neutral atom. The two will then attract one another. (d) An ion can also attract neutral molecules by inducing dipoles in them. (e) The weakest interactions are dispersion forces between two neutral atoms or molecules having generally symmetrical charge distributions.

2-3 ENERGY CHANGES IN CHEMICAL REACTIONS

In order to appreciate fully the later discussions of cellular homeostasis, mitochondrial and chloroplast functions, and certain membrane phenomena, one must have some feeling for the energetics of those processes. The concept of energy change during chemical reactions has already been introduced, albeit indirectly, in the discussion of the various types of chemical bonds. We shall now expand on those ideas to develop a more general description.

Gibbs Free Energy. The kind of energy in which we are most interested here is the *free energy*, or *Gibbs free energy*, designated as G or sometimes F. A change in this parameter, indicated by ΔG, represents the amount of energy available to do work when both the temperature and pressure are held constant. These are the conditions under which almost all biological reactions take place.

Ordinarily, the amount of energy available to do work is less than the total energy content of the system, the rest being tied up in the motion of the molecules. If a change in total energy at constant temperature and pressure is given as ΔH, where H is called the *enthalpy* or heat content, then the amount of energy available to do work is

$$\Delta G = \Delta H - T\Delta S \qquad (2\text{-}4)$$

where T is the absolute temperature and ΔS the change in entropy, or change in randomness, an idea first introduced in Section 2-2. The delta (Δ) always means "final state minus initial state"—i.e., $\Delta G = G_2 - G_1$ in the transition from state one to state two. Thus, for example, if ΔG is negative ($\Delta G < 0$), there is a drop in free energy.

Reactions in which heat is released (i.e., $\Delta H < 0$), are called *exothermic*. They are characterized by a warming of their environment as heat becomes available. On the other hand, reactions that cool their surroundings do so because they are absorbing heat energy. Such reactions are called *endothermic* ($\Delta H > 0$). (See Fig. 2-6.) A similar nomenclature is used for free energy changes: reactions in which there is a release of free energy are *exergonic* ($\Delta G < 0$), while reactions that require energy to be added to the system are *endergonic* ($\Delta G > 0$).

Spontaneous Reactions. There is a natural tendency for all systems to gravitate toward their state of lowest free energy (e.g., the particle falling to the floor). It follows, then, that all spontaneous reactions will have a negative change in free energy ($\Delta G < 0$). From Equation 2-4 it is seen that a negative change in free energy may result from a drop in heat content (enthalpy) or an *increase* in entropy. Both may be true at the same time, or one term may dominate the other in order to drive the reaction and make energy available to do work on another system.

As an example of the importance of entropy, consider the hydrogen bond in the following situation (from Laskowski and Scheraga, 1954: see references).

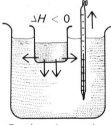

Exothermic reaction

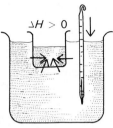

Endothermic reaction

FIGURE 2-6 Heat Transfer. An exothermic reaction is one in which heat energy is released ($\Delta H < 0$), reflected in a warming of the environment. An endothermic reaction absorbs heat energy ($\Delta H > 0$), cooling the reaction mixture, its vessel, and its environment.

Here we have a part of a large protein structure holding two organic side chains in place. (They are portions of the amino acids glutamate and tyrosine.) The groups cannot drift apart when the hydrogen bond (O···H) is broken, but breaking the hydrogen bond does permit rotational freedom about each single bond. When the hydrogen bond is formed, rotation is almost completely stopped. Formation of the hydrogen bond results in a release of about 25 kJ/mole of bond energy ($\Delta H = -25$ kJ/mole), but this is offset by a decrease in entropy because of the loss of rotational freedom, a decrease that requires virtually all of the 25 kJ. Thus, the change in free energy (ΔG) is very close to zero.

The preceding example is not unusual. In general, part of the energy released in the formation of a bond goes into the entropy term instead of being made available to do work (ΔG). In some cases, such as hydrophobic bond formation, the entropy term is actually the dominant one. It is that factor which is responsible (via the negative sign in Equation 2-4) for the unusual temperature dependence of hydrophobic bonds—that is, it explains why they get stronger instead of weaker as the temperature is increased.

Since the terms G, H, and TS each represent a form of energy, Equation 2-4 is really a partial statement of the *first law of thermodynamics*, which says that energy can neither be created nor destroyed, only changed from one form to another. The *second law of thermodynamics* was mentioned in passing in Section 2-2. In its simplest form, it states that all systems tend toward an equilibrium. This concept follows naturally from the statement that systems tend to minimize their free energy content, for when that minimum is reached no further reaction takes place—i.e., $\Delta G = 0$ and the system is in equilibrium.

Equilibrium Constants. The Boltzmann equation as it is written in Equation 2-1 was said to represent a system in equilibrium. That equation can be rewritten as

$$\Delta G° = -2.3RT \log \frac{[B]_{eq}}{[A]_{eq}} \qquad (2\text{-}5)$$

The term $\Delta G°$ is called the *standard state free energy change*. Note that it is *not* the same as ΔG, which is a measure of the energy available to do work. No work can be obtained from a system already in equilibrium; it can be obtained only during the transition toward that condition. The relationship between ΔG and $\Delta G°$ is given as

$$\Delta G = \Delta G° + 2.3RT \log \frac{[B]}{[A]} \qquad (2\text{-}6)$$

The subscripts on A and B in Equation 2-5 were put there to emphasize that equilibrium concentrations were being specified; they are not used here, for ΔG is the energy change when the reaction

$$A \rightarrow B \tag{2-7}$$

is allowed to reach equilibrium after starting with the initial concentrations [A] and [B]. The system is at equilibrium when [A] becomes $[A]_{eq}$ and [B] becomes $[B]_{eq}$. The ratio $[B]_{eq}/[A]_{eq}$ is called the *equilibrium constant*, K_{eq}. At equilibrium, there can be no energy available to do work—i.e., $\Delta G = 0$. Hence, Equation 2-6 becomes

$$\Delta G° = -2.3RT \log K_{eq} \tag{2-8}$$

It is important to realize that $\Delta G°$, which is a term we shall see many times later, describes an equilibrium and thus the end point of a reaction. To determine the amount of energy available to do work (ΔG), we need to know not just the end point but the starting concentrations as well. It is also important to realize that work can be obtained from a reaction when equilibrium is approached from either direction—i.e., when starting with an excess of A or an excess of B. All reactions are in principle reversible, so that Equation 2-7 should be written

$$A \rightleftharpoons B \tag{2-9}$$

By convention, however, ΔG in Equation 2-6 describes the transition $A \rightarrow B$. If the flow is actually in the reverse direction, this will be reflected in a $\Delta G > 0$. In other words, if $A \rightarrow B$ has a ΔG of $+10$kJ/mole the reaction will in fact flow in the other direction, so that $B \rightarrow A$ will have a ΔG of -10 kJ/mole.

The Generalized Reaction. We have up to now limited our discussions of reactions to the simplest possible model, $A \rightleftharpoons B$. Though that reaction can describe a great many situations that arise later—membrane transport, for example—we should also be aware of the more general situation where a moles of substance A, plus b moles of B, react to give the products C and D in amounts c and d, respectively:

$$aA = bB \rightleftharpoons cC + dD \tag{2-10}$$

The free energy change for this reaction is given by

$$\Delta G = \Delta G° + 2.3RT \log \frac{[C]^c[D]^d}{[A]^a[B]^b} \tag{2-11}$$

For example, the reaction of hydrogen and oxygen to give water, $2H_2 + O_2 \rightarrow 2H_2O$, has a free energy change that depends on initial concentrations in the following way:

$$\Delta G = \Delta G° + 2.3RT \log \frac{[H_2O]^2}{[H_2]^2[O_2]}$$

where $\Delta G°$ is the standard state free energy change for the reaction.

The equilibrium constant for the generalized reaction is obtained from Equation 2-11 when $\Delta G = 0$, just as it is for simpler reactions. Since $\Delta G° = -2.3RT \log K_{eq}$, it must be true that

$$K_{eq} = \frac{[C]_{eq}^c[D]_{eq}^d}{[A]_{eq}^a[B]_{eq}^b} \tag{2-12}$$

So far we have talked only about equilibrium, and not about how fast equilibrium might be reached. That subject will be treated in more detail in Chap. 4, but we should point out here that although any system will eventually reach equilibrium, it may take a very long time to get there. Most reactions have an *energy barrier* in the form of some unfavorable step that must be taken before the reaction can proceed to equilibrium. This barrier often involves the formation of an intermediate compound, called an *activated complex*. Any energy required to form such a complex will be returned *in addition* to the normal free energy change once the reaction takes place, because ΔG depends only on the energy content of the initial and final states and not on how you get from one to the other. The rate at which reactants overcome an energy barrier will depend on the fraction of molecules that have an energy content sufficient for activation. As usual, that fraction can be predicted by using the Boltzmann equation.

With this background, let us go on to consider two more classes of bonding of great importance to an understanding of biological structure. These are dative bonds (the basis for chelation), and the bonds responsible for acid-base reactions. Both classes represent covalent bonds of quite typical energy content, but both types can be exchanged for almost equivalent bonds with water molecules. The *effective* bond strength, therefore, is just the difference in energy before and after the water molecules become involved, making these bonds subject to spontaneous rupture and rearrangement much like the weak bonds described earlier.

2-4 COORDINATION COMPOUNDS

As the theory of valence developed during the nineteenth century, it became clear that there are many exceptions to the classical rules. Some of the most common exceptions are due to dative bonds.

The Dative Bond. In addition to the kinds of weak bonds already discussed, nitrogen, oxygen, and a few other elements can participate in a special kind of covalent interaction with certain metals and metal ions. The resulting complex is known as a *coordination compound*; the special bond itself is called a *dative bond* or *coordinate bond*.

The cobalt-ammine ion

$$[Co(NH_3)_6]^{3+}$$

is an example of a coordination compound. It is formed from six molecules of ammonia and the Co^{3+} ion. (Note that cobalt-ammine still retains the 3+ charge of the cobalt ion.) G. N. Lewis suggested in 1916 that this kind of complex results from bonds between the metal ion and nitrogens in which the shared electron pair is donated entirely by the nitrogen in each case. (Such a bond is usually indicated by an arrow drawn from the donor atom.) An ordinary covalent bond is a shared electron pair, of course, but in it one electron actually "belongs" to each participating atom. In the dative bond, both electrons belong to the same atom.

Coordinate or dative bonds, then, represent a class of covalent bonds having energies similar to those of other covalent bonds, often in the neighborhood of 250 kJ/mole. However, the ability of water to form dative bonds makes coordination compounds much less stable than this figure would imply. The situation is parallel to that of the ionic bond, which is very strong in a nonpolar solvent (as in crystals of NaCl) but weak in the presence of a polar solvent such as water.

Molecules containing electron donors are called *ligands*. If there is one donor atom in the molecule, it is an *unidentate ligand*. With two or three donor atoms, it is bidentate and tridentate, respectively, and so on for higher numbers. When there are two or more donor atoms in the same molecule, the ligand is also called a *chelate*.

The dative bond is usually strongest for small metal ions of high charge, so the divalent metal ions, particularly the transition metals (Fe, Co, Ni), are most often involved. The following is a common order of stability for coordination complexes containing divalent metal ions:

$$Hg > Cu > Ni > Co > Pb > Zn > Cd > Fe > Mn > Mg > Ca > Sr > Ba$$

The monovalent metal ions typically form complexes much less stable than those of the above ions. The difference in stability between Mg^{2+} and K^+ chelates, for example, may well be a factor of about 5000.

Of course, the nature of the ligand has much to do with the stability of a coordination compound. Just its physical size compared with that of the metal may be important, for if the metal is exceptionally small or the ligand exceptionally large, the limitations of space dictate that fewer ligands can be accommodated. In general, the divalent metals listed above are capable of being bonded to two, four, or six donor atoms, with four and six being most common.

Like other covalent bonds, dative bonds tend to be directed in space. The most stable complexes form when a single chelate can make a cage about the metal, allowing several bonds to form at their most stable angles. In such cases, the equilibrium constant,

$$K_{eq} = \frac{[\text{free chelate}][\text{free metal ion}]}{[\text{metal-chelate}]} \tag{2-13}$$

may be 10^{-20} or even less, indicating an exceedingly stable structure. The metal ion in such a complex is said to be *sequestered*. Most, if not all, of the divalent metal ions in a cell are sequestered.

Water molecules, when structured about a metal ion, may be considered as ligands in a coordination complex. This is not a contradiction of our former discussion of ionic bonds in which it was pointed out that the dipole nature of water causes it to be attracted to positive or negative charges. Rather, the dative bond is an additional factor, and explains the great tenacity with which water is captured by certain metal ions, even in the crystalline state. Magnesium salts, for instance, are normally obtained as hydrates, as in $MgClO_4 \cdot 6H_2O$. Although the water may be removed by heating, the resulting dry salt will immediately capture new water if it is allowed to stand in moist air.

One should not get the impression that electrostatic attraction plays no part in forming a coordination complex, for all bonding is basically electrostatic. However, the attraction between an ion and a dipole does not by itself explain the dative bond. While the bonds in some coordination complexes may be primarily ionic, in others they are primarily covalent. Most often they are somewhere in between. Lest we be tempted to forget this, we should keep in mind that coordination complexes can sometimes form *even when the metal is not ionized,* though of course the fact that free metal is insoluble limits the opportunity for such complexes in biological systems.

TABLE 2-3. Formation Constants for Copper(II)-Ammine Complexes

Reaction[a]	Equilibrium constant		Free energy change (20 °C)
$M + L \rightleftharpoons ML$	$K_1 =$	$\dfrac{(ML)}{(M)(L)} = 10^{4.0}$	-23 kJ/mole
$ML + L \rightleftharpoons ML_2$	$K_2 =$	$\dfrac{(ML_2)}{(ML)(L)} = 10^{3.2}$	-18 kJ/mole
$ML_2 + L \rightleftharpoons ML_3$	$K_3 =$	$\dfrac{(ML_3)}{(ML_2)(L)} = 10^{2.7}$	-15 kJ/mole
$ML_3 + L \rightleftharpoons ML_4$	$K_4 =$	$\dfrac{(ML_4)}{(ML_3)(L)} = 10^{2.0}$	-11 kJ/mole

[a] $L = NH_3$, $M = Cu^{2+}$.

The Stability of Coordination Complexes. To get an idea of the stability of these complexes in aqueous solution, consider the data in Table 2-3. Note that the first ligand is bound more readily than the others, and that each addition makes the next one harder. This is reasonable, for not only will each addition leave fewer positions to fill, thus reducing the chance that a potential ligand will find the right spot, but the ion is partially shielded by the existing ligands and therefore not as attractive to new ones. In multidentate ligands, on the other hand, the formation of the first bond to the ligand greatly increases the chances for more, because the other donor atoms are forced to remain in the vicinity of the ion. In a typical case, one might find a standard state free energy change for the formation of a metal-chelate to be several kilojoules per mole of dative bond in the complex. Needless to say, these may be very stable compounds.

The standard state free energy change in the formation of a coordination complex is less than one would expect from the strength of the dative bond, reflecting the fact that ligands must displace water which, as we have already noted, also acts as a ligand. Thus, for example, the formation of a copper-ammonia complex in aqueous solution first requires that the copper-water complex be disrupted. The negative ΔG for ammonia binding represents the degree to which the ammonia complex is more stable than the hydrated copper ion; this, in turn, depends on the difference in energy between the two kinds of ion-ligand bonds. Thus, we have yet another class of biologically important reactions whose characteristics depend strongly on the special properties of water.

Coordination complexes are widespread in cells and the dative bond plays many important roles, as we shall see. As an example, oxygen in our lungs is complexed by Fe^{2+} ions of the protein hemoglobin, the major constituent of red blood cells. It is transported in this form to tissues where it is needed. The iron itself is held in hemoglobin as a metal-chelate complex, bonded to the four nitrogens of a ringlike multidentate ligand, the *heme*—see p. 62; the R's represent groups of various composition. The heme structure is drawn with two arrows and two lines to the metal to indicate that two hydrogens are displaced by the ion. Each nitrogen participates in three covalent bonds, but two of the four nitrogens have an additional, dative, bond. We should recognize, however, that because of resonance the four bonds to the Fe^{2+} are really equivalent.

Hemelike structures are also found in other molecules: in myoglobin, which is the oxygen-binding protein of muscle; in the cytochromes, which accept and pass along electrons by alternating between Fe^{2+} and Fe^{3+}; in chlorophyll, which is the green pigment of plants; in vitamin B_{12}; and elsewhere. Hemoglobin, myoglobin, and the cytochromes contain iron; chlorophyll contains Mg^{2+}; and vitamin B_{12} is a complex with Co^{3+}. In addition, many of the simpler organic compounds we shall be discussing are chelates that ordinarily exist in complex with metal ions *in vivo*, making the dative bond in exceedingly common phenomenon in biological systems.

2-5 ACIDS AND BASES

Certain chemical groups have a tendency to lose a hydrogen ion (H^+) in aqueous solution, whereas other groups have a tendency to attract hydrogen ions. The former are called acids and the latter are bases. This is the Brønsted criterion of acids and bases, but it is not the only way of defining them. In 1923 G. N. Lewis suggested a more universal concept: An acid is a substance that can accept a pair of electrons from a donor substance, the base. The process, called *neutralization* in this context, sounds familiar to us because stated in this way it is clear that acid-base reactions are due to the making and breaking of dative bonds. The metal ion, in Lewis's scheme, is the acid; the ligand is the base. Again relatively strong (covalent) bonds are involved, but the reactions take place readily since there is almost always an exchange of one for another, commonly with water as either the displaced or added ligand. Because we are accustomed to thinking in terms of hydrogen ion concentration as a measure of acidity, however, we shall generally use Brønsted's nomenclature.

The Behavior of Acids in Water. An acid has a tendency to transfer a proton (hydrogen ion) to water, creating a hydronium ion, H_3O^+. The reaction is given as

$$HA + H_2O \rightleftharpoons H_3O^+ + A^- \qquad K_{eq} = \frac{[H_3O^+][A^-]}{[HA][H_2O]} \qquad (2\text{-}14)$$

Here HA is the acid (e.g., a carboxylic acid, $-\overset{\overset{\displaystyle O}{\|}}{C}-OH$) and A^- its anionic form $-\overset{\overset{\displaystyle O}{\|}}{C}-O^-$, also called its conjugate base because of the ability of

A^- to accept a proton. For dilute solutions, the water concentration is almost always taken to be 55.55 molar, which is the concentration of pure water, and so it is convenient to combine this value with K_{eq} to give a new constant K_a:

$$K_a = [H_2O]K_{eq} = \frac{[H_3O^+][A^-]}{[HA]} \qquad (2\text{-}15)$$

By taking the logarithm of both sides and rearranging, we can turn Equation 2-15 into the very important relationship known as the *Henderson-Hasselbalch equation*:

$$pH = pK_a + \log \frac{[A^-]}{[HA]} \qquad (2\text{-}16)$$

where, by definition

$$pK_a = -\log K_a \qquad (2\text{-}17)$$

and

$$pH = -\log [H_3O]^+$$

One commonly sees pH defined a little differently, as $-\log [H^+]$, and the acid reaction as $HA \rightarrow H^+ + A^-$, but this is just a shorthand notation for the more correct version given here; its use should cause no confusion if we recognize it as such. The Henderson-Hasselbalch equation tells us that the pH of a solution can be predicted if both the pK_a of an acid and its degree of ionization are known. Conversely, the degree of ionization of an acid can be predicted from the pH and a knowledge of its pK_a. The pK_a, since it describes an equilibrium position, is a constant for a given acid in a given environment, equal to the pH at which half the molecules are charged and half neutral.

Bases. The comparable manipulations for bases begin with the general reaction

$$B + H_2O \rightleftharpoons BH^+ + OH^- \qquad (2\text{-}18)$$

commonly represented in cells by amino groups:

$$R\text{---}NH_2 + H_2O \rightleftharpoons R\text{---}NH_3^+ + OH^-$$

In this situation

$$K_a = \frac{[H_3O^+][B]}{[BH^+]} \qquad (2\text{-}19)$$

and the Henderson-Hasselbalch equation may be written as

$$pH = pK_a + \log \frac{[B]}{[BH^+]} \qquad (2\text{-}20)$$

Thus, bases can be discussed in the same terms used for acids. That is, a base is half ionized when the pH is equal to its pK_a, and so on. The only distinction is that as the pH is lowered, an acid becomes less ionized whereas a base becomes more highly ionized. In both cases the lower pH's correspond to a more highly protonated acid or base.

Note that water itself may act either as an acid by donating a proton, or as a base by accepting a proton. Both conditions are described by the reaction

$$2H_2O \rightleftharpoons H_3O^+ + OH^- \qquad (2\text{-}21)$$

where the equilibrium is such that in absolutely pure water (extremely difficult to prepare because of the tendency of carbon dioxide in the air to form carbonic acid: $CO_2 + H_2O \rightarrow HCO_3^- + H_3O^+$) the pH will be 7. The presence of certain solutes, however, can shift that equilibrium significantly. For example, the coordination complex of water and the mercuric ion (Hg^{2+}) behaves as a fairly strong acid. The reaction

$$H_2O + [Hg \cdot 2H_2O]^{2+} \rightleftharpoons [Hg(H_2O)(OH)]^+ + H_3O^+$$

has a pK_a of 3.7, making it a stronger acid than acetic (pK_a = 4.7) and accounting for the corrosive nature of mercury salts when accidently swallowed.

It must be emphasized that there is nothing peculiar about the covalent bond made to protons in acids or bases. The O—H bond in water, for example, has a bond energy of 460 kJ/mole. Obviously, then, a water molecule in an isolated environment will not ionize. That is, the probability of a spontaneous rupture in such a bond is about 10^{-81}. Yet about 2×10^{-9} of the molecules in pure water will, in fact, be ionized at 25 °C. Clearly, there must be a compensating factor in the form of a new bond to the released proton, leaving the system in a state of energy not so very different from that of undissociated water. The new bond is the H_2O—H^+ bond in the hydronium ion, H_3O^+. As a result, the dissociation of a water molecule has a standard state free energy change of about 100 kJ/mole, rather than the 400 or more one might otherwise expect. Thus acids dissociate readily in biological systems, using water as a proton acceptor.

Buffers. The logarithmic relationship between pH and degree of ionization of an acid or base leads to a *buffering capacity* in the region of the pK_a. In other words, there is a tendency to compensate for added acid or base by shifting the equilibrium in the opposite direction via simple mass action. Thus, when the pH is equal to the pK_a, an acid is half ionized. At one pH unit below the pK_a, the ratio [A^-]/[HA] is 0.1 (i.e., the acid is only about 9% ionized); at one unit above the pK_a, the ratio is 10. Outside that range, buffering capacity falls off very rapidly since the acid is almost entirely in one form or the other (see Fig. 2-7).

Acids that have a pK_a near physiological pH (around pH 7) are of particular interest to us, because they may stabilize the pH of the cell. Phosphoric acid is an example. This acid has three ionizable groups (see Fig. 2-8), with equilibria as follows:

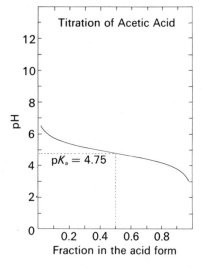

FIGURE 2-7 The Titration Curve of Acetic Acid. The equilibrium constant for the dissociation of acetic acid gives it a single pK_a of 4.75. Since pH is measured on a logarithmic scale, the molecule goes from 91% acid to 9% acid in the range pH 3.75 to 5.75.

HO—P(=O)—OH
|
OH

Inorganic phosphoric acid (P_i)

$$\begin{aligned} H_2O + H_3PO_4 &\rightleftharpoons H_2PO_4^- + H_3O^+ & pK_a &\approx 2 \\ H_2O + H_2PO_4^- &\rightleftharpoons HPO_4^{2-} + H_3O^+ & pK_a &\approx 7 \\ H_2O + HPO_4^{2-} &\rightleftharpoons PO_4^{3-} + H_3O^+ & pK_a &\approx 12 \end{aligned} \quad (2\text{-}22)$$

Since the salts of phosphoric acid and phosphate derivatives are present in the cell in considerable quantities, we can expect them to be important in maintaining physiological pH. The second of the above reactions has a pK_a of 7.2 in very dilute solutions, but about 6.8 under conditions approximating those of the cytoplasm. Thus, the most common intracellular phosphoric acid ion (also called inorganic phosphate or orthophosphate, abbreviated P_i) is HPO_4^{2-}.

2-6 THE STATE OF INTRACELLULAR WATER

Ordinary water is by far the most prevalent compound in a cell, frequently representing about three-quarters of the total mass. Its presence determines the character of the cytoplasm in numerous ways, some of which have been suggested here. We shall see many more examples in later chapters, but it might be wise to summarize what has already been said about the substance.

The electronegative character of oxygen makes water a dipole. This in turn makes it a good dielectric, capable of shielding charged groups by forming structures about them. This shielding causes ionic interactions to be very labile, and hence we classify them as "weak" compared to covalent bonds.

The electronic structure of oxygen allows it to function as an electron pair donor, so that water molecules may participate as ligands in coordination compounds. This ability affects the stability of coordination complexes made with other ligands, because they must replace water or be replaced by it. Ligand replacement accounts for the lower-than-expected energy of formation of coordination complexes which, after all, are stabilized by covalent bonds.

The electronegativity of oxygen also makes water a good hydrogen bonder. It can both donate and accept a shared proton, interacting with other water molecules as well as with other kinds of proton donors and proton acceptors. This capacity has several consequences: (1) it makes the acid-base reaction possible, since the released proton can be bonded to water to form a hydronium (H_3O^+) ion; (2) it causes water to act as a hydrogen bond competitor of other groups; and (3) it gives water a considerable amount of structure, even in the liquid state. The degree of structure is noticeable in many situations, among which is the measurement of ionic mobilities: If one measures the rate at which an ion moves between two oppositely charged electrodes in an aqueous solution, it is found that the hydrogen ion migrates five times as fast as any other positive ion (36.3 μm/sec per volt/cm compared with 7.6 for K^+, 4.0 for Li^+). It appears to do so by combining with a water molecule and "bumping" another H^+ off the other end of the structure:

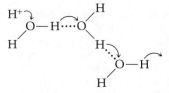

Proton transfer

Hence, hydrogen ions migrate even faster in ice than in liquid water, a mobility that is fostered by the increased structure and in spite of the lower temperature.

The structured character of water may be increased by the presence of other groups. Molecules of water become structured about ionic groups or dipoles due to ion-dipole or dipole-dipole interactions. They may also be structured by forming dative bonds with certain metal ions, and hy-

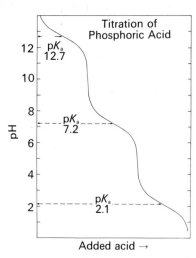

FIGURE 2-8 The Titration Curve of Phosphoric Acid. The biologically important molecule orthophosphoric acid (H_3PO_4, or just P_i) has three titratable hydrogens, and hence three pK_a's. In very dilute solution, the pK_a's are 2.12, 7.21, and 12.67, respectively.

drogen bonds with both proton donors and proton acceptors. We have also seen that the presence of nonpolar groups increases the structure of water in their vicinity. Since the cytoplasm is replete with dipolar, ionic, and nonpolar groups, one wonders whether there is any "free" water in the cell at all. This question has prompted a number of investigations.

By using a technique called "nuclear magnetic resonance," or nmr, one can determine the degree of freedom experienced by hydrogens. It is clear from such studies that water in cells exists in two fractions, neither of which has quite the freedom of ordinary liquid water, and one, which is 10 to 25% of the total in muscle cells, is very nearly icelike in character. These studies strongly support the suggestion, made on the basis of other kinds of evidence, that cytoplasmic water has considerably more structure even than liquid water. Therefore, any conclusions concerning cellular reactions based on studies in free solution should take this factor into account.

SUMMARY

2-1 An isolated system at equilibrium will have some of its members at high energy levels (i.e., with large potential energy) and some at low ones. The Boltzmann equation—(2-1)—predicts that the numbers of elements will decrease exponentially as one examines the distribution at successively higher levels. This relationship applies to gravity or to making and breaking of bonds, the broken bond representing the higher energy level. Thus, the stronger the "bond energy" the less likely spontaneous rupture becomes. The "weak" chemical bonds, however, spend a significant fraction of the time in the broken state, making molecular rearrangements possible.

2-2 Weak bonds can be divided into four groups: (1) Hydrogen bonds, in which a proton (carrying a partial positive charge) is positioned between two electronegative atoms, usually nitrogen or oxygen. (2) Ionic bonds, which result from the mutual attraction of two oppositely charged ions (considered weak in biological systems because of the shielding effect of water). (3) Hydrophobic bonds, where nonpolar groups aggregate to avoid increasing the structure of water. (These are entropy-driven interactions characterized by a bond energy that increases with temperature.) (4) Short-range bonds, due either to ion-dipole interactions (where the dipole is either permanent or induced by the field of the ion) or to van der Waals interactions between neutral atoms or molecules. The van der Waals group consists of interactions between two dipoles, where one or both may be induced. In the case where both are induced, the interaction is also called a London dispersion force.

2-3 Of the heat energy released by a reaction (designated as the change in enthalpy, ΔH), usually only part can do useful work. This available portion of the energy is called Gibbs free energy, G. The remainder of the energy is often offset by a decrease in randomness, or entropy, S. Gibbs free energy is a measure of the energy available to do work at constant temperature and pressure. Any reaction in which there is a drop in free energy

$$\Delta G = (\Delta H - T \Delta S) < 0$$

can be made to do work. Such reactions always occur spontaneously.

The equilibrium constant of a system (K_{eq}) is related only to the standard state free energy change ($\Delta G°$) of the reaction involved. For the reaction

$$aA + bB \rightleftharpoons cC + dD$$

the relationship between ΔG, $\Delta G°$, and K_{eq} is given as

$$\Delta G = \Delta G° + 2.3RT \log \frac{[C]^c[D]^d}{[A]^a[B]^b} \quad (2\text{-}11)$$

Note that when the reaction has reached equilibrium, ΔG is 0; note further that ΔG will be favorable (negative) for progress toward that position from either excess reactant or excess product. This expression also tells us why $\Delta G°$ is such a useful quantity to have, for (1) it provides the equilibrium constant ($\Delta G° = -2.3RT \log K_{eq}$) and (2) it provides the constant necessary to calculate ΔG in a non-equilibrium situation.

2-4 Coordinate bonds, or dative bonds, are a class of covalent bonds in which both electrons of the shared pair are donated by one of the two bonded groups. The donor atom is commonly nitrogen or oxygen; the receptor atom is most often a divalent metal ion such as Ca^{2+}, Mg^{2+}, or Fe^{2+}. Dative bonds have energies similar to those of other covalent bonds, but are relatively weak in biological systems because water itself is a good ligand.

2-5 An acid is a molecule that tends to lose a proton (Brønsted's definition) or to accept an electron pair from a donor group, the base (Lewis's definition). The Lewis concept makes it clear that acid-base reactions are no different fundamentally from reactions involving coordination compounds. As such we recognize them as reactions in which covalent bonds are exchanged, one for another, often with water as a participant. Water can both donate a proton, leaving OH^-, or accept one to create the hydronium ion, H_3O^+. Many substances tend to transfer a proton to water or accept one from it, thereby acting as acids or bases, respectively. An acid or base reaction when it is half complete tends to buffer the solution against changes in pH that would otherwise result from the addition of more acid or base.

2-6 The properties of biological systems depend heavily on the characteristics of water: (1) that it is a dipole and can thus be oriented in electrostatic fields such as those of an ion; (2) that it can act as both a donor and an acceptor molecule in hydrogen bonds; (3) that its oxygen is able to donate an electron pair to a dative bond, making the water molecule a ligand; and (4) that it is able to donate or accept a proton in acid-base reactions. This multiplicity of possible interactions makes water an excellent solvent, but because of them a substantial fraction of protoplasmic water has a structure somewhere between that of ice and that of ordinary liquid water at the same temperature.

STUDY GUIDE

2-1 (a) Find the density of oxygen at the top of Mt. Everest (about 9 km, or 29,000 ft) as a percent of its density at sea level. Assume a constant 27 °C = 300 K. [Ans: 0.32 or 32%] **(b)** Molecules in solution also distribute according to the Boltzmann equation, but their effective mass will be their true mass less the mass of displaced liquid. Thus, we replace M with $M(1 - \bar{v}\rho)$ where $\bar{v}$ (called the partial specific volume) is the ml of liquid displaced per gram of solute and ρ is the density (g/ml) of displaced liquid. If a suspension of virus particles with $\bar{v} = 0.6$ ml/g and $\rho = 1$ g/ml is allowed to stand absolutely undisturbed at 300 K, it may eventually reach an equilibrium distribution in which the concentration at 10 cm from the bottom is 20% of the concentration at the bottom. What is its mass in g/mole? (*Note:* This is the principle behind sedimentation equilibrium; however, gravity is ordinarily replaced with a centrifugal field—see Appendix.) [Ans: about 10^7 g/mole] **(c)** If a bond has an energy of 20 kJ/mole, what percent can we expect to find broken at any one time? (Neglect entropy changes and assume $T = 300$ K.) [Ans: 0.03%]

2-2 (a) Two molecules experience a van der Waals interaction between them equivalent to a 4 kJ/mole weak bond. At $T = 300$ K and with no entropy changes, what is the probability, P, of finding this bond in the broken state? (P is simply the ratio of broken bonds to total potential bonds.) [Ans: 0.16] **(b)** Suppose the two molecules are held together by three interactions like that described in part a. What is the probability that the molecules will be freed from each other at any instant? (*Note:* If the probability of an event is P, then the probability of n such events occurring simultaneously is P^n.) [Ans: 0.004] **(c)** Compare the probability of freeing two molecules by breaking three 4 kJ/mole bonds (Problem 2-2b) with the probability of freeing them by breaking one 12 kJ/mole bond. [Ans: For the 12 kJ/mole bond, $P = 0.008$; note that for higher bond energies the difference in P between n bonds of energy X and one bond of energy nX becomes even smaller.] **(d)** How do the special properties of water contribute to each of the four classes of weak bonds? **(e)** If an ionic bond has a measured energy of 20 kJ/mole in an aqueous environment, what would be the strength of that bond in the interior of a membrane largely composed of hydrocarbon chains with a dielectric constant of about 4? [Ans: 400 kJ/mole]

2-3 (a) Define the terms Gibbs free energy; enthalpy; entropy; exergonic; and exothermic. **(b)** What do we mean by the phrases "entropy-driven reaction" and "enthalpy-driven reaction"? **(c)** Suppose the reaction $A \rightleftharpoons B + C$ has a standard state free energy change ($\Delta G°$) of -33 kJ/mole.
(1) What is the equilibrium constant at 25 °C? [Ans: 6.18×10^5]
(2) What would be the concentration of A at equilibrium if $[B]_{eq} = 5 \times 10^{-3}$ M? (*Note:* The concentration of B equals that of C.) [Ans: 4.0×10^{-11} M]
(3) Using the data from part 2 above, determine the equilibrium ratio $[B]_{eq}/[A]_{eq}$. [Ans: 1.25×10^8]
(4) What is the free energy change, ΔG, for the reaction starting with all species at 2.5×10^{-3} M? [Ans: -48 kJ/mole] (*Note:* The situation and numbers used in this problem could apply to the *in vivo* hydrolysis of ATP to ADP and P_i, discussed in Chap. 4.)

2-4 (a) How do dative bonds differ from other covalent bonds? **(b)** What role does water play in the stability of coordination compounds? **(c)** When Mg^{2+} gathers a shell of hydration about it, are the water molecules held by ion-dipole bonds, by dative bonds, or both?

2-5 (a) Compare the hydrogen bond, ionic bond, dative bond, and the acid-base reaction for similarities and differences. (b) Given a pK_a of 6.8 for the second dissociable group of phosphoric acid, what fraction will exist as HPO_4^{2-} at pH 7.4 in a cell? (Neglect any contribution from other equilibria.) [Ans: The ratio $[HPO_4^{2-}]/[H_2PO_4^-]$ is 4; 80% of the phosphate is HPO_4^{2-}.] (c) If you add 0.01 mole of HCl to one liter of distilled water (pH 7), what will be the final pH? If, instead of distilled water, you have a liter of 0.05 M phosphate (a mixture of the monobasic and dibasic salts) at pH 7, what will be the final pH after adding 0.01 mole of HCl? Assume HCl dissociates completely. Also assume that the pK_a of phosphate at this dilution is 7.0, so that $[HPO_4^{2-}] = [H_2PO_4^-] = 0.025\ M$. (Hint: The answer should be obtained by working it as an equilibrium problem, but a good estimate can be obtained by recognizing that 0.01 mole of HCl will convert almost that much HPO_4^{2-} to $H_2PO_4^-$, and then applying the Henderson-Hasselbalch equation.) [Ans: pH 2; pH 6.63] (d) The solubility of carbon dioxide (CO_2) in water is 1.45 g/liter (0.033 M). Assume complete conversion to carbonic acid (H_2CO_3) and bicarbonate anion (HCO_3^-) with a $pK_a = 6.4$ or $K_a = 4.3 \times 10^{-7}$. What, then, is the pH of a saturated solution? (Hint: Since you start with all acid and a very small initial hydronium ion concentration, let $[HCO_3^-] \approx [H_3O^+]$. Then, from the knowledge that $[HCO_3^-] + [H_2CO_3] = 0.033\ M$, and from the K_a, solve the resulting quadratic equation for $[H_3O^+]$.) [Ans: pH 3.9 by this simplified method; actually pH 3.8] (e) How could a divalent metal ion affect the pH of a solution?

2-6 (a) In what classes of reactions (discussed here) does water participate? (b) What are the special properties of water that allow it to take part in so many different kinds of interactions? (c) We find the following relative velocity of movement toward a negative electrode in a water environment:

$$K^+ > Na^+ > Li^+ << H^+$$

The relative size of the ions is

$$K^+ > Na^+ > Li^+ > H^-$$

We would normally expect size to be inversely proportional to velocity. Explain the discrepancy.

REFERENCES

CHEMICAL BONDS AND THERMODYNAMIC PRINCIPLES

BAILAR, JOHN C., JR., "Some Coordination Compounds in Biochemistry." *Am. Scientist*, **59**: 586 (1971). A review of the role of metal ions in life processes.

BUSCH, DARYLE H., "Metal Ion Control of Chemical Reactions." *Science*, **171**: 241 (1971). The chemistry of metal-ion catalysis.

CHRISTENSEN, H. N. and R. A. CELLARIUS, *Introduction to Bioenergetics*. Philadelphia: Saunders, 1972.

DUNCAN, GEORGE, *Physics for Biologists*. New York: Halsted (Wiley), 1975. (Paperback.)

FREIFELDER, D., *Physical Biochemistry: Applications to Biochemistry and Molecular Biology*. San Francisco: W. H. Freeman, 1976. (Paperback.) A relatively nonmathematical approach.

GUENTHER, W., *Chemical Equilibrium: A Practical Introduction for the Physical and Life Sciences*. New York: Plenum Press, 1976. Acid-base, metal-ion-ligand, solubility equilibria, and redox systems.

JOESTEN, M. D. and L. J. SCHAAD, *Hydrogen Bonding*. New York: Dekker, 1974. Not much about biological systems.

KAUZMANN, W., "Some Factors in the Interpretation of Protein Denaturation." *Advan. Protein Chem.*, **14**: 1 (1959). Includes the first comprehensive discussion of the hydrophobic bond.

KLOTZ, IRVING M., *Energy Changes in Biochemical Reactions*. New York: Academic Press, 1967. (Paperback.)

LASKOWSKI, MICHAEL, JR., and HAROLD A. SCHERAGA, "Thermodynamic Considerations of Protein Reactions: I. Modified Reactivity of Polar Groups." *J. Am. Chem. Soc.*, **76**: 6305 (1954).

LINNETT, J. W., "Chemical Bonds," *Science Progress (Oxford)*, **60**: 1 (1972).

MOROWITZ, H. J., *Entropy for Biologists: An Introduction to Thermodynamics*. New York: Academic Press, 1970. (Paperback.)

ONSAGER, LARS, "The Motion of Ions: Principles and Concepts." *Science*, **166**: 1359 (1969). 1968 Nobel Lecture on irreversible reactions.

PAULING, LINUS, *The Nature of the Chemical Bond* (3d ed.). Ithaca, N.Y.: Cornell Univ. Press, 1960. See Chap. 12, on the hydrogen bond.

SIMMONS, R., and T. L. HILL, "Definitions of Free Energy Levels in Biochemical Reactions." *Nature*, **263**: 615 (1976).

TANFORD, C., *The Hydrophobic Effect*. Somerset, N.J.: Wiley-Interscience, 1973.

WATER

COPE, FREEMAN W., "Nuclear Magnetic Resonance Evidence Using D_2O for Structured Water in Muscle and Brain." *Biophys. J.*, **9**: 303 (1969).

EISENBERG, D., and W. KAUZMANN, *The Structure and Properties of Water*. Oxford Univ. Press, 1969. (Paperback.)

ERLANDER, S. R., "The Structure of Water." *Science J.*, **5**: 60 (1969).

FLETCHER, N. H., "The Freezing of Water." *Science Progress (Oxford)*, **54**: 227 (1966).

FORSLIND, E., "Structure of Water." *Quart. Rev. Biophys.*, **4**: 352 (1972).

FRANK, HENRY S., "The Structure of Ordinary Water." *Science*, **169**: 635 (1970).

GURNEY, R. W., *Ionic Processes in Solution.* New York: Dover, 1953. (Paperback.)

HAZLEWOOD, C. F., ed., "Physiochemical State of Ions and Water in Living Tissues and Model Systems." *Ann. N.Y. Acad. Sci.*, **204** (1973). New work plus reviews.

HAZLEWOOD, C. F., B. L. NICHOLS, and N. F. CHAMBERLAIN, "Evidence for the Existence of a Minimum of Two Phases of Ordered Water in Skeletal Muscle." *Nature*, **222**: 747 (1969).

HORNE, R. A., ed., *Water and Aqueous Solutions: Structure, Thermodynamics and Transport.* New York and London: Wiley Interscience, 1972. Ice, liquid water, ionic solutions, and the hydration of ions and macromolecules.

KUNTZ, I., and A. ZIPP, "Water in Biological Systems." *New Engl. J. Med.*, **297**: 262 (1977).

LING, G. N., and W. NEGENDANK, "The Physical State of Water in Frog Muscles." *Physiol. Chem. Phys.*, **2**: 15 (1970).

SOLOMON, A. K., "The State of Water in Red Cells." *Scientific American*, February 1971. (Offprint 1213.)

SYMONS, M. "Water Structure and Hydration." *Phil. Trans. Roy. Soc. London, Ser. B*, **272**: 13 (1975).

WALTER, J. A., and A. B. HOPE, "Nuclear Magnetic Resonance and the State of Water in Cells." *Progr. Biophys. Mol. Biol.*, **23**: 1 (1971).

WIDOM, B., "Intermolecular Forces and the Nature of the Liquid State." *Science*, **157**: 375 (1967).

NOTE

1. The product $2.3RT$ has a value of about 5700 joules at 25 °C. The joule is now the internationally accepted unit of energy in biological systems, replacing the calorie. The latter, which is the amount of energy needed to raise 1 g of water by 1 °C, was often confused with the Calorie used in nutrition, which is actually 1000 calories. To convert calories to joules (or kcal to kJ), multiply by 4.184.

3

Molecular Architecture and Biological Function

3-1 Organic Structure 71
3-2 Lipids 72
    Fatty Acids
    Glycerides
    Phosphoglycerides
    Sphingolipids
    Steroids
3-3 Carbohydrates 78
    Monosaccharides
    Oligosaccharides
    Polysaccharides
    Mucopolysaccharides
    Glycolipids
3-4 Nucleic Acids 84
    Nucleotides
    The Structure of DNA
    The RNAs
3-5 Proteins 91
    The Amino Acids
    The Peptide Bond
    Primary Structure
    Secondary Structure
    Tertiary Structure
    Quaternary Structure
    Prosthetic Groups
    Lipoproteins
    Glycoproteins
    The Determination of Structure
    Structural Stability
    Solution Properties
3-6 Macromolecular Assembly 111
    Assembly Mechanisms
    Self-Assembly
    The Driving Force for Self-assembly
    Reversible Self-assembly: Microtubules
    Irreversible Self-assembly: Collagen, Elastin, and Fibrin
    Lipid Self-assembly: The Liposome
3-7 An Introduction to Molecular Specificity: The Antibodies 118
    The Antibody Response
    Antibody-Antigen Interaction
    Contributions of the Antibodies to Defense
    Haptens and Antibody Specificity
    Antibody Structure
    Antibody Synthesis

Summary 128
Study Questions 129
References 130
Notes 133

The cell is composed almost entirely of water, assorted inorganic ions, and four classes of organic compounds: lipids, carbohydrates, nucleic acids, and proteins (Table 3-1). Each of these classes has characteristic physical and chemical properties and a set of well-defined biological roles to play in life processes. The basic unit of each is small, almost never over about 300 daltons (one dalton being 1/16 the mass of an oxygen atom or about the mass of a hydrogen atom). However, these small units are held together through a combination of covalent and noncovalent bonds of the type considered in the last chapter, forming in some cases structures of very large relative dimensions called *macromolecules*.

TABLE 3-1. Molecular Composition of Bacteria and Some Mammalian Tissues

	Rat liver	Rat skeletal muscle	E. coli
Water	69%	75%	70%
Protein	16	7	15
Glycogen (a carbohydrate)	3	4	–
Phospholipids	3	2	2
Neutral lipids	2	9	–
RNA } nucleic acids	1	1	6
DNA	0.2	0.3	1

In this chapter, we shall introduce the major biological macromolecules. Some appreciation for their structure and properties is necessary in order to understand the cellular constituents comprised of them.

3-1 ORGANIC STRUCTURE

The biologically important organic molecules are composed of only a few different kinds of atoms. The carbohydrates, for example, consist almost exclusively of carbon, hydrogen, and oxygen. The other three classes contain these elements plus nitrogen (Table 3-2). In addition,

TABLE 3-2. Elementary Composition of Some Macromolecules

	Proteins	Carbohydrates	Nucleic acids
Carbon	50–55%	40%	38%
Oxygen	19–24	53	31
Nitrogen	13–19	–	17
Hydrogen	6–7.3	7	3
Sulfur	0–4	–	–
Phosphorus	–	–	10

phosphorus is found in all nucleic acids but in only certain lipids and a few proteins. Sulfur, on the other hand, is found in almost all proteins but in only a few examples of the other macromolecular types. Although there is an occasional divalent metal ion held by dative bonds, mostly to proteins, these six elements (C, H, O, N, P, and S), out of the more than 100 elements known, account for almost the entire mass of a living organism (Table 3-3).

The structure and properties of organic compounds depend largely on the properties of their major constituent, carbon. Carbon does not participate in any of the weak interactions, except van der Waals attractions, but its electronic configuration is such that an additional four electrons can be accommodated. By participating in four covalent bonds or their equivalent in double and triple bonds, carbon gains the electronic configuration of the nearest inert gas, neon. Similarly, hydrogen normally participates in one bond, oxygen and sulfur in two (in organic molecules), nitrogen in three, and phosphorus in five.

CHAPTER 3
Molecular Architecture and Biological Function

TABLE 3-3. The Elementary Composition of Bacteria and Humans

	The Human Body[a]		E. coli	
	Gross composition	Dry weight	Gross composition	Dry weight
Oxygen	65%	18%	69%	20%
Carbon	18	54	15	50
Hydrogen	10	8	11	10
Nitrogen	3	9	3	10
Phosphorus	1.0	3.0	1.2	4
Sulfur	0.25	0.75	0.3	1

[a] The body also contains appreciable quantities of calcium (1.5% of total mass), potassium (0.35%), sodium (0.15%), chloride (0.15%), and magnesium (0.05%). Iron and zinc each account for 0.004% or less.

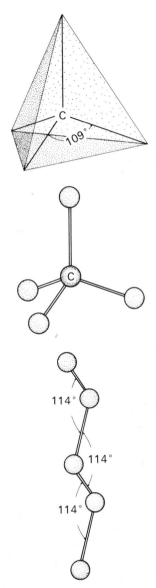

FIGURE 3-1 The Bond Angles of Carbon.

When four single bonds satisfy the electron "deficit" of carbon they will be directed to the four corners of a tetrahedron having the carbon nucleus at its center (Fig. 3-1). (A tetrahedron is a three-sided pyramid whose base and sides are identical equilateral triangles.) When two of the bonds are replaced by a double bond, the remaining two single bonds form the arms of a planar Y with the double bond as its leg. And when two double bonds are made by a single carbon atom, as in carbon dioxide (O=C=O), the three bonded atoms lie in a straight line.

An additional factor that contributes to the spatial configuration of biological molecules is that while free rotation about single bonds is ordinarily allowed, such freedom is not found in double bonds or in partial double bonds. By partial double bonds is meant those bonds which, through resonance or conjugation, are neither single nor double. In benzene, for example, the six bonds are all equivalent, so that the two structures

are identical. (Note that we have also used the convention that omits hydrogens—their number is made clear by remembering that carbon makes four bonds—and assumes a carbon atom to be present at the end of each straight line.) Any structure that can be written in more than one way by interchanging adjacent single and double bonds is a conjugated structure, with the interchanged bonds partially double in character.

These simple principles determine the structure of organic molecules and hence, in large part, the structure of cellular components, and even the structure of cells themselves.

3-2 LIPIDS

Lipids are small organic molecules that tend to be insoluble in water but soluble in organic solvents. In other words, they tend to be hydrophobic or to contain significant hydrophobic regions, a result of their nonpolar nature and low dielectric constant (generally 2 to 4). Lipids form a broad and rather ill-defined class of molecules that includes fats and oils, waxes, and certain of the hormones and vitamins, among other things. Any organic molecule that is not easily classed as a carbohydrate, nucleic acid, protein, or a derivative or breakdown product thereof is apt to be called a *lipid*.

The biological role of lipids is as diverse as their structure. For example, fats and oils form the most important source of energy in the ordinary diet and serve to store energy in the body. But some of the hormones are also lipid. They allow a gland in one part of the body to control the metabolic activity of a tissue in another part. And many of the vitamins, which are essential dietary supplements with a wide range of functions, are lipid. So are the basic structural units of biological membranes.

Lipids can be divided into the major groups shown in Table 3-4 (p. 74). Some of these groups are important enough in later chapters to justify more explanation now.

Fatty Acids. Fatty acids are carboxylic acids (written —COOH or
$$-\overset{O}{\underset{\|}{C}}-OH)$$
that contain from 4 to 20 or more carbons in a chain. The chain may be either saturated (entirely —CH_2—) or unsaturated (containing double bonds). Fatty acids are lipids because the nonpolar character of their "tails" dominates their properties. Their short-chain counterparts, such as acetic acid, CH_3—COOH, are very soluble in water and are not classed as lipids.

The most common fatty acids in animals are the C-16 and C-18 saturated and unsaturated fatty acids given in Table 3-5. (Note that unsaturated fatty acids are not conjugated structures.) Though there are some exceptions, nearly all the naturally occurring fatty acids in animals (95–99%) contain an even number of carbons, a result of being synthesized from acetate, which has two carbons. In general, the fatty acids derived from plant tissues have more double bonds and are more likely to have an odd number of carbons. "Vegetable oil," for example, is largely a mixture of unsaturated fatty acids.

Fatty acids with double bonds in the terminal seven carbons (the ones distal to the carboxyl group) cannot be made by mammals but are required by them. Linoleic and linolenic acids (Table 3-5) are examples of *essential fatty acids*. The essential fatty acids are precursors of *prostaglandins*, a group of 20-carbon fatty acids containing a five-membered ring and various substitutions. Although they were discovered in the 1930s (in prostate secretions, hence their name), the structure and real importance of prostaglandins only became known in the 1960s. They have now been found in small quantities in a variety of different tissues, where they affect a perplexing array of biological processes, including the contraction of smooth muscles. It takes only 10^{-9} g/ml to cause con-

Prostaglandin E_1

TABLE 3-4. Summary of the Lipids

Name	Basic structure	Biological role
Fatty acids	$CH_3-(CH_2)_n-COOH$ straight-chain fatty acid	important source of food energy; constituent of other lipids; precursor of prostaglandins
Glycerol esters	$H_2C-CH-CH_2$ | | | OH OH OH Glycerol	
glycerides	one, two, or three fatty acids esterified to glycerol	source of food energy; fats (triglycerides) serve also as thermal and mechanical insulators
phosphoglycerides	glycerol with one hydroxyl esterified to phosphate and fatty acids at its other hydroxyls	major constituent of all membranes
Sphingolipids	derivatives of sphingosine $CH_3(CH_2)_{12}-CH=CH-CH(OH)-CH(NH_2)-CH_2OH$ Sphingosine	constituent of some membranes
Waxes	fatty acid ester of any alcohol except glycerol	source of energy; the harder waxes (beeswax) may also be structural components
Terpenes	polymers of isoprene $CH_2=C(CH_3)-CH=CH_2$ Isoprene	several vitamins (A, E, K) and some oils are terpenes
Steroids	Steroid nucleus (rings A, B, C, D)	constituent of membranes; some of the hormones (e.g., sex hormones) are steroids, as is vitamin D

traction of uterine smooth muscle, for example, making prostaglandins among the most potent biological effectors found to date. Many of the therapeutic effects of aspirin are attributed to that drug's ability to interfere with the synthesis of prostaglandins.

Thromboxanes and *prostacyclins* are even more potent. These very unstable, short-lived substances are derived from the same precursors as prostaglandins. Their structure is similar to the prostaglandins except that the active forms have an oxygen atom in or bridged to the five-membered, all-carbon ring of the prostaglandins. Investigation of these

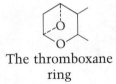

The thromboxane ring

TABLE 3-5. Some Common Fatty Acids

Formula	Common name	Formal name	Structure
$C_{16}H_{32}O_2$	palmitic	n-hexadecanoic	$CH_3(CH_2)_{14}COOH$
$C_{18}H_{36}O_2$	stearic	n-octadecanoic	$CH_3(CH_2)_{16}COOH$
$C_{16}H_{30}O_2$	palmitoleic	9-hexadecenoic	$CH_3(CH_2)_5CH\!=\!CH(CH_2)_7COOH$
$C_{18}H_{34}O_2$	oleic	cis-9-octadecenoic	$CH_3(CH_2)_7CH\!=\!CH(CH_2)_7COOH$
$C_{18}H_{32}O_2$	linoleic	cis,cis-9-12-octadecadienoic	$CH_3(CH_2)_4(CH\!=\!CHCH_2)_2(CH_2)_6COOH$
$C_{18}H_{30}O_2$	linolenic	9,12,15-octadecatrienoic	$CH_3CH_2(CH\!=\!CHCH_2)_3(CH_2)_6COOH$

compounds is just beginning, but it appears that at least some of the biological effects attributed to prostaglandins are in fact due to thromboxanes and prostacyclins.

Glycerides. When glycerol is esterified[1] to one fatty acid, leaving two of the hydroxyls unreacted, the compound is called a *monoglyceride*. If two of the hydroxyls are reacted in this way, the result is a *diglyceride*, and the corresponding triester is a *triglyceride* (see Table 3-4). Mixtures of triglycerides aggregate by hydrophobic bonding to form globules of macromolecular dimensions. Such is the nature of ordinary neutral fat found, for example, in the adipose cells that marble a steak. (It is this association with fat that gave the fatty acids their name.) Neutral fat, or "depot fat," provides the body with insulation and serves as a reservoir of energy, since both the glycerol and fatty acid components may be oxidized by cells in a series of energy-yielding (exergonic) reactions.

Triglycerides may be hydrolyzed back to glycerol and fatty acids by making the solution very alkaline, as in the production of natural soap, where NaOH or KOH is used to hydrolyze the triglyceride. Soap itself is the sodium or potassium salt of the liberated fatty acids. Triglycerides can also be hydrolyzed under mild conditions by certain proteins (enzymes called lipases) at the surface of many cells, permitting the fatty acids to be utilized by cells for energy. This same reaction takes place in the skin of many young adults, when hormonal changes cause an overproduction of sebum, which is almost pure triglyceride. Some of the sebum may become trapped as a blackhead (comedone) allowing irritating free fatty acids to be liberated and thus converting the blackhead into a pimple. The hydrolysis in this case is brought about by enzymes produced by a particular bacterium (*Corynebacterium acnes*), found on all normal skin.

Fat always contains a mixture of fatty acids. In fact, even the three fatty acids of a single triglyceride are not usually the same. Hence, the physical properties of the glycerides, which are dominated by the properties of their fatty acids, vary considerably. However, the glycerides are all quite insoluble (a few mg/liter) because of a total lack of charged groups and a paucity of polar ones. Because of this insolubility, few glycerides (or fatty acids) are found free in the blood, although they can be transported in the form of hydrophobic complexes with serum proteins, especially albumin. Glycerides may also be rendered more soluble by replacing one of the fatty acids with a phosphate or phosphate derivative to form a phospholipid.

Phosphoglycerides. Phosphoglycerides are a major component of membranes of all types (see Chap. 6). The parent compound is *phosphatidic*

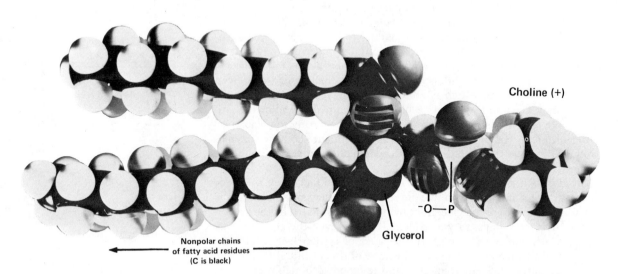

FIGURE 3-2 Amphipathic Lipids. These lipids have both a hydrophilic end and a hydrophobic end—the latter comprised of fatty acids or the hydrocarbon chain of sphingosine. Note that all but the cerebrosides are phospholipids. The structures shown are the major lipids found in membranes. The space-filling model is phosphatidyl choline, also known as lecithin. [Molecular models in this and later figures, except as noted, are from R. C. Bohinski, *Modern Concepts in Biochemistry* (2d ed.), Boston: Allyn and Bacon, 1976.]

acid, which is a diglyceride with the third hydroxyl of glycerol esterified to phosphoric acid (see Fig. 3-2). Phosphoric acid is a strong acid, so the substitution greatly improves water solubility. Hence, phosphatidic acid and its derivatives can be transported readily in the blood of animals. The fatty acid tails remain hydrophobic, however, facilitating micelle and membrane formation. Molecules of this type, hydrophobic at one end and hydrophilic at the other, are called *amphipathic*.

The phosphate of phosphatidic acid is capable of forming a second ester bond. Most commonly this bond is to choline, ethanolamine, serine, or to another glycerol (Fig. 3-2). Alternatively, two phosphatidic acids may be joined by forming phosphate ester bonds to the first and third hydroxyls of a common glycerol to form *cardiolipin*, so named because it was first isolated from heart (cardiac) tissue. As it turns out, cardiolipin is located mostly in mitochondrial membranes, which happen to be in particular abundance in heart muscle cells.

Sphingolipids. Sphingolipids are also common constituents of membranes. Their structure and properties are similar to the phosphoglycerides, though the "backbone" is not glycerol but sphingosine, and the parent molecule of the class not phosphatidic acid but *ceramide*, which is sphingosine plus a fatty acid linked through an amide (—NH—C(=O)—) bond (see Fig. 3-2). When the free hydroxyl of ceramide is esterified to phosphorylcholine, the result is *sphingomyelin*. Sphingomyelin is one of two major groups of sphingolipids; the other is comprised of *cerebrosides*, in which the hydroxyl of sphingosine is linked to a sugar, commonly the monosaccharide galactose. Note that sphingomyelin, like the phosphoglycerides but unlike cerebrosides, is a *phospholipid*.

Steroids. The steroids include the widely distributed lipid, cholesterol, about which we hear so much. Cholesterol is found in the blood, as a part of most animal cell membranes, in nearly all samples of body fat and, unfortunately, as a constituent of the lipid plaque on the arterial walls of patients with atherosclerosis, associated with hardening of the arteries. In spite of its bad reputation, cholesterol is essential to the synthesis of, for example, liver bile, which is needed for digestion of fats; vitamin D, produced in human skin with the help of sunlight and needed for calcium metabolism and hence growth and maintenance of bones; and the sex hormones. The sex hormones include the male *androgens*, of which testosterone is an example, and the female *estrogens*, of which the most active is estradiol (see structures). These hormones, which are produced by the gonads, are responsible for establishing and maintaining the secondary sex characteristics of breast development, body hair, body fat distribution, and so on. A good deal is now known about the specific

H_2C—O—(phosphatidic acid)
HC—OH
H_2C—O—(phosphatidic acid)
Cardiolipin

Testosterone (an androgen)

β-Estradiol (an estrogen)

Cholesterol

Vitamin D

$$R = -\underset{\underset{CH_3}{|}}{CH}-CH_2-CH_2-CH_2-\underset{\underset{CH_3}{|}}{CH}-CH_3$$

3-3 CARBOHYDRATES

Carbohydrates are molecules that generally conform to the empirical formula $(CH_2O)_n$—hence the name "hydrate of carbon," or carbohydrate. The most common carbohydrates are the sugars, or *saccharides*, which are found as simple sugars (monosaccharides), as short chains (oligosaccharides), and as large polymers, the polysaccharides.

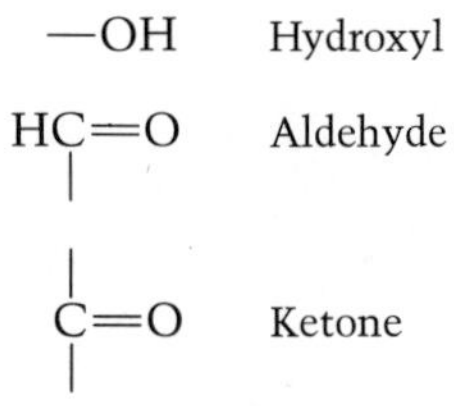

Monosaccharides. The monosaccharides are polyhydroxyl aldehydes or ketones. Typically, each carbon other than the carbonyl (C=O) will have an attached hydroxyl. If the carbonyl is at the end of the carbon chain, the resulting aldehyde is designated an *aldose*. If it is at an interior position, making the molecule a ketone, we refer to the sugar as a *ketose*. Aldoses are much more reactive than ketoses; they are thus sometimes called "reducing sugars." Since hydroxyl groups are very polar (as we saw in Chap. 2), the monosaccharides are quite soluble in water.

Monosaccharides are named by specifying the position of the carbonyl, then the number of carbons, and finally the ending "-ose." Thus, an "aldohexose" is a six-carbon sugar with the carbonyl at the end of the carbon chain, while a "2-ketopentose" has its carbonyl at the second of five carbons.

The aldoses of greatest biological interest are the trioses, pentoses, and hexoses. These have, respectively, three, five, and six carbons. The most common ketoses are the ketotrioses and the pentuloses, hexuloses, and heptuloses, with three, five, six, and seven carbons, respectively. (Note the inserted "ul," indicating the ketose family.)

The monosaccharides usually exist as stereoisomers—two structures that are the mirror images of each other, like your left and right hands (Fig. 3-3). The two isomers are designated as "D" or "L," by analogy to D- and L-glyceraldehyde, which are aldotrioses:

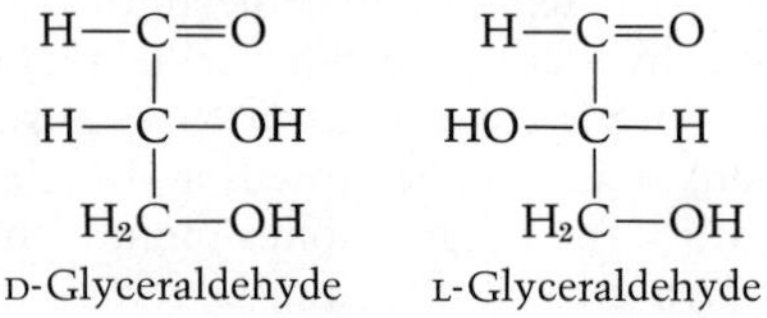

—OH   Hydroxyl

HC=O   Aldehyde

C=O   Ketone

$$\begin{array}{cc} H-C=O & H-C=O \\ | & | \\ H-C-OH & HO-C-H \\ | & | \\ H_2C-OH & H_2C-OH \\ \text{D-Glyceraldehyde} & \text{L-Glyceraldehyde} \end{array}$$

Note that the middle carbon of glyceraldehyde has four different groups attached to it, and hence is asymmetric. That is, because of its tetrahedral bond angles, carbon may be bonded to four different groups in two arrangements that are mirror images of each other. Of the two configurations, D monosaccharides are more common by far.

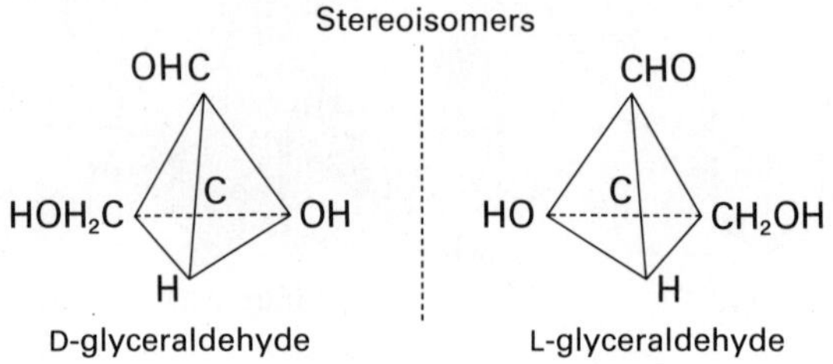

FIGURE 3-3 D- and L-Glyceraldehyde. Note that stereoisomers are mirror images of each other.

Most of the monosaccharides are optically active, meaning that their asymmetric carbon(s) cause the rotation of plane-polarized light. Molecules that rotate the plane of polarization to the right, as you face the light source, are called *dextrorotatory* and are designated *d* or (+), while the opposite case is *levorotation,* designated *l* or (−). It is important to remember that the small capitals D and L refer to structure, whereas the lower case *d* and *l* refer to optical activity, established before the structure could be determined. Thus, one sees reference to D(+)-glucose, also called dextrose, and D(−)-fructose, also called levulose.

For the sake of simplicity, one often writes the sugars in their linear form. In fact, however, the more important configuration is the cyclic one; it is an isomer having an oxygen bridge between two of the carbons, forming structures such as those shown in the following examples:

D-Glucose α-D-Glucose α-D-Glucopyranose

D-Fructose β-D-Fructose β-D-Fructofuranose

D-Ribose α-D-Ribose α-D-Ribofuranose

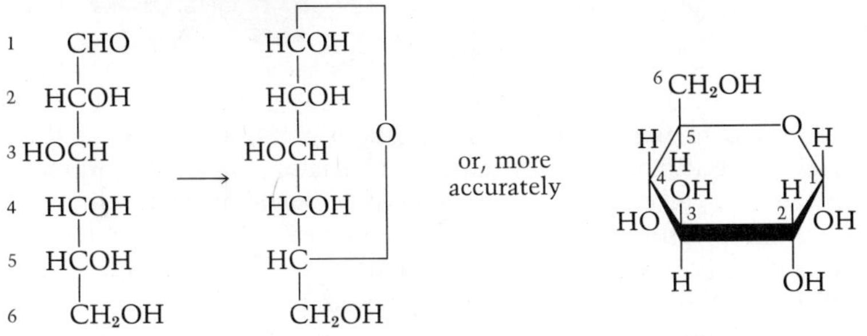

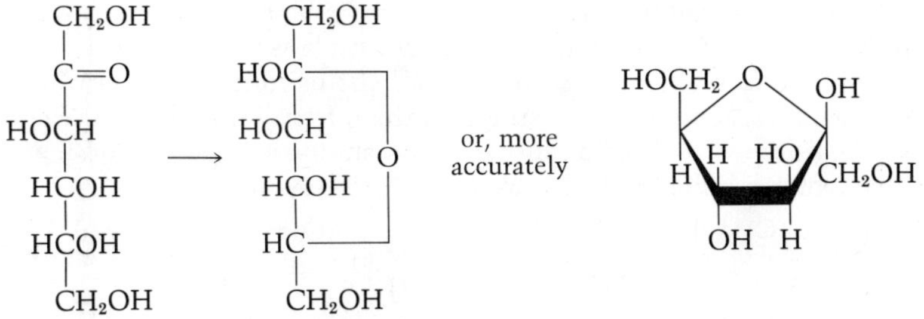

Ring formation introduces a new asymmetric carbon at position one, so the nomenclature must reflect the additional pair of isomers. If the hydroxyl at position one is below the plane of the molecule, it is the α *anomer;* otherwise we have the β anomer. Thus, one can have α-D-glucose or β-D-glucose as well as the two anomers of L-glucose (Fig. 3-4).

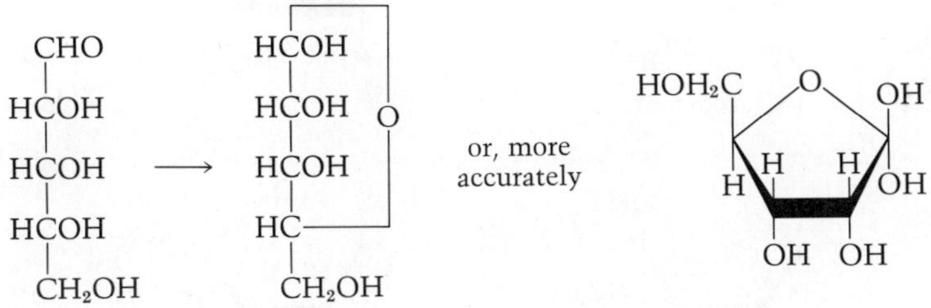

The ring forms of the saccharides are referred to as *furanoses* or *pyranoses* by analogy to furan and pyran, which are five- and six-membered rings, respectively. Thus, one may refer to α-D-glucose and β-D-fructose as, respectively, α-D-glucopyranose and β-D-fructofuranose.

Furan

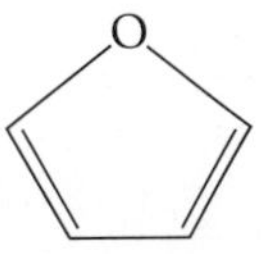

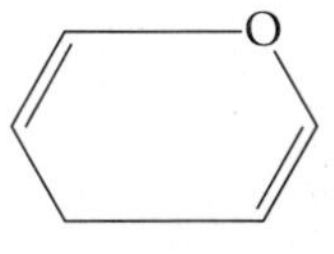

Pyran

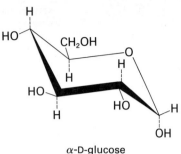

FIGURE 3-4 The Spatial Conformation of D-Glucose. Axial bonds, which are those bonds perpendicular to the plane formed by the center four carbons, are drawn vertically in the diagram.

```
1    O=C ─┐
2   HO─C  │
3   HO─C  O
4      HC ─┘
5   HOCH
6    CH₂OH
```
L-Ascorbic acid
(Vitamin C)

In addition to the monosaccharides themselves, a number of their derivatives are also biologically important. Among the most common is L-ascorbic acid, or vitamin C, an essential carbohydrate that almost all animals except primates can make from glucose. The hydroxyl at the number three position of ascorbic acid ionizes with a $pK_a = 4.2$ to form the ascorbate ion, which is also used as vitamin C.

Oligosaccharides. Monosaccharides may be joined by condensation, removing a hydroxyl from one saccharide and a hydrogen from another to form water. When two are joined, the result is a disaccharide; three form a trisaccharide, and so on. Small chains formed in this way are called *oligosaccharides* to distinguish them from long chains, the *polysaccharides*. The dividing line between an oligosaccharide and a polysaccharide is rather arbitrary, as "oligo-" simply means "few."

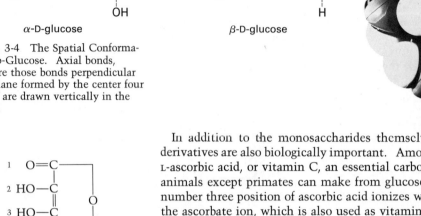

The most familiar disaccharide is ordinary table sugar, or *sucrose*, widely distributed in plants and obtained commercially from sugar cane or sugar beets. It consists of α-D-glucose and β-D-fructose, linked 1–2. That is, the number one carbon of glucose is linked by an oxygen bridge to the number two carbon of fructose as shown. Sucrose may be referred to as an α glucoside by virtue of the fact that some other group has been substituted for the hydrogen of its number one hydroxyl. If, for example, a methyl group were found there instead of fructose, we would have a methyl glucoside, and so on.

Another common disaccharide with which we shall be concerned later is lactose, or milk sugar. Lactose accounts for about 5% of the solid portion of the milk of most mammals. It is a β galactoside and could also be called 4-O-(β-D-galactopyranosyl)-β-D-glucopyranose. Most people would prefer to say "lactose."

β-Lactose

Polysaccharides. The most important polysaccharides are polymers of glucose. These include cellulose, glycogen, and the starches.

More than half of all the carbon in higher plants is in *cellulose*, an insoluble, rigid structural polymer of several hundred to several thousand β-D-glucose units. More than 90% of cotton, for example, is cellulose. The corresponding β-1,4 disaccharide is called *cellobiose*.

Repeating cellobiose unit of cellulose

While cellulose is a linear polymer of glucose, *glycogen* is a branched polymer of glucose. It consists of α-D-glucose units, mostly linked 1–4, but highly branched via frequent 1–6 linkages, each of which starts other α-1,4 chains of some 8–12 glucose units. A complete molecule of glycogen will consist of several thousand glucose monomers that, because of the branching, form a much more compact structure than cellulose.

Glycogen is found mostly in the muscles and livers of animals, and is sometimes called "animal starch." The true starches, however, are plant products and come in two forms, amylose and amylopectin, found together in granules. *Amylose* is an unbranched α-1,4 polymer of glucose. (The disaccharide of the α-1,4 polymers is *maltose*.) The remainder of the starch granule is *amylopectin*, a molecule that has the same structure as glycogen, though it is not so highly branched; each branch is 24–30 units in length. (Amylopectin should not be confused with the pectin of fruits and berries, which is a polymer of a galactose derivative.)

Repeating maltose unit of the starches

(a) Cellulose (β-1,4 linkages)

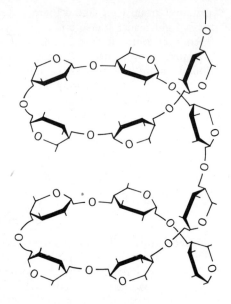

(b) Glycogen and starch
(α-1,4 linkages)

FIGURE 3-5 Cellulose and the Starches. (a) The β-1,4 linkage in cellulose leads to the formation of insoluble fibers. (b) The α-1,4 linkages found in starches (amylose, amylopectin, and glycogen) allow solvent penetration and greater solubility.

Glycogen and the starches serve as depots for glucose. They give the cell a way of storing this important food molecule in a form that is compact, yet accessible to specialized enzymes. Cellulose, on the other hand, is a structural component that is relatively rigid and quite insoluble. The only important difference, however, between cellulose and the starches, plant or animal, is that cellulose is a β-1,4 polymer of glucose whereas the starches are basically α-1,4 polymers. On the surface, this may not seem like an important distinction, but in fact the bond angles of carbon and oxygen are such that α-1,4 polymers are helical, with about six glucose units per turn, while β-1,4 polymers are linear. Hence, cellulose is found in fibers of 100–200 parallel chains (Fig. 3-5a) with few of its hydroxyls exposed to the surrounding solvent. This configuration accounts for both its rigidity and its insolubility. In contrast, the helical α-1,4 polymers (glycogen and starch) provide open spaces for solvent penetration. As a result, glycogen and the starches are more flexible and more soluble (Fig. 3-5b).

The difference between α and β linkages has another biological implication, for while humans have enzymes that can hydrolyze the α-1,4 glucoside bond, they do not have any that will hydrolyze the corresponding β linkage. Therefore, we derive no food value from cellulose, in spite of its being a polymer of nothing more than ordinary β-D-glucose. Certain microorganisms do contain such enzymes. Grazing animals utilize these organisms to break down the cellulose in their diet, and then live by digesting the microorganisms and their metabolic by-products. As carnivores, we take advantage of the food value of cellulose by eating the grazing animals, but since we are only getting it thirdhand, the advantages to be gained from finding a more direct route from plant to man are obvious.

Mucopolysaccharides. There are a large number of polysaccharides, collectively called *mucopolysaccharides* or *glycosaminoglycans,* that contain amino sugars. The term *amino sugar* refers to a monosaccharide in which one —OH group is replaced by an —NH₂ or its derivative. Most of the mucopolysaccharides are found in association with proteins, which may then be called *mucoproteins*. An exception is *chitin*, a polymer of β-linked N-acetyl-D-glucosamine that is similar to cellulose and serves as a structural polysaccharide for fungi and certain other lower plants, and as the exoskeleton of many insects and other invertebrates.

The polysaccharide coat found on most cells and the intercellular material of most animal tissue contains much mucopolysaccharide. Examples include *hyaluronic acid, chondroitin* and *chondroitin sulfates, dermatan sulfate,* and *keratosulfate (keratan sulfate)*. Some of the unusual sugars found in these polymers are shown in the accompanying structures.

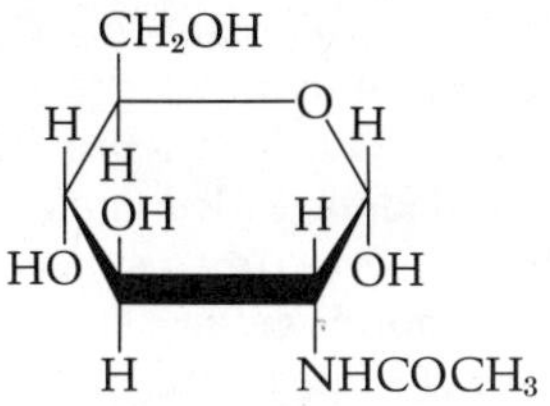

N-Acetyl-D-glucosamine

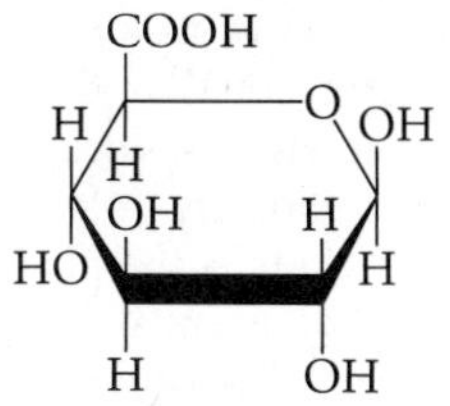

Glucuronic acid

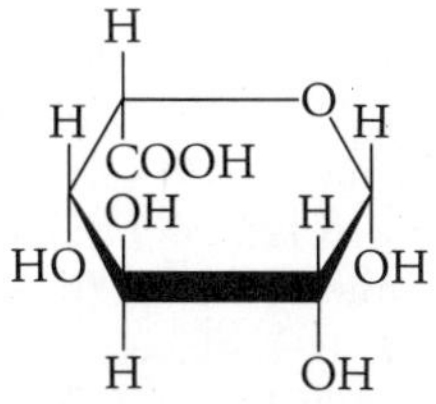

Iduronic Acid

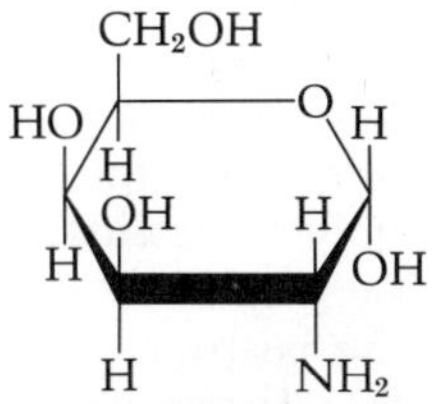

Galactosamine

Heparin is another common animal mucopolysaccharide-protein complex. It is produced by so-called mast cells (Chap. 10) and interferes with the clotting of blood, a purpose for which it is widely employed in biology and medicine.

The inability to properly break down mucopolysaccharides leads to a devastating group of inherited illnesses called *mucopolysaccharidoses*, which result in severe skeletal deformities and frequently mental retardation and an early death. Hurler's syndrome ("gargoylism") is the most common.

Glycolipids. A glycolipid is a combination of lipid and saccharide. The *cerebrosides* have already been mentioned. They contain ceramide, usually linked to galactose. The sulfated derivative of the galactocerebroside is called a *sulfatide*. Both cerebrosides and sulfatides are present in significant quantities in brain and other nervous tissue.

Gangliosides, another important glycolipid of neural tissue, are cerebrosides having a complex oligosaccharide polar end. The oligosaccharide characteristically includes N-acetylneuraminic acid, also called *sialic acid*. The inborn failure to degrade glycolipids (due to a missing lysosomal enzyme) leads to some of the most common and devastating of the lipid storage diseases (lipidoses). Notable among these is Tay-Sachs disease, the recessive gene for which is carried by about 3% of American Jews having Eastern European ancestry.

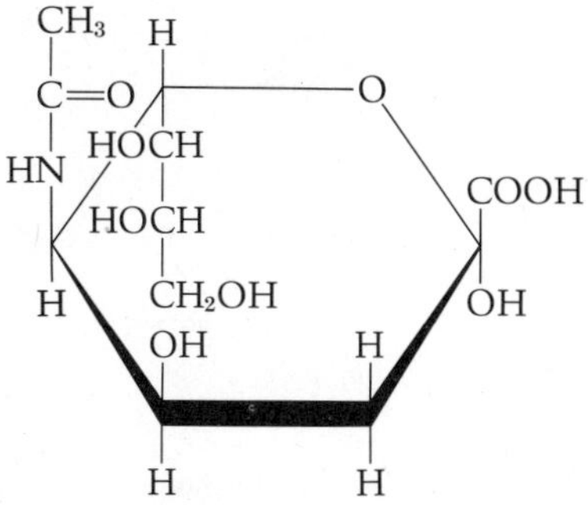

Sialic acid (N-acetylneuraminic acid)

CHAPTER 3
Molecular Architecture and Biological Function

The polysaccharides, like the lipid polymers, do not have precisely defined molecular weights. The biological roles of the materials do not demand it. However, the next two classes of macromolecules do have discrete molecular sizes and, in addition, very definite spatial configurations. These are the proteins and the nucleic acids.

3-4 NUCLEIC ACIDS

Nucleic acids are the informational molecules of the cell. They carry essentially all the information needed to construct a cellular duplicate and, in addition, furnish a portion of the machinery necessary for the construction. Information is coded into deoxyribonucleic acid, DNA, and is retrieved and interpreted with the help of ribonucleic acid, RNA. The main products of this cooperation are *enzymes*, which are specific biological catalysts that determine the kind and amount of the various chemical reactions necessary for the support of life and reproduction.

Nucleic acids are polymers of simpler compounds called *nucleotides*. Each nucleotide is the phosphoric acid ester of a five-carbon sugar that is also linked to an organic base. Because phosphoric acid is a strong acid, and because nucleotide polymers were first associated with the nucleus of the cell, the nucleotide polymers are called "nucleic acids."

Nucleotides. The sugar component of the nucleotides is either ribose or 2-deoxyribose (abbreviated simply deoxyribose). The choice of the sugar alone determines whether the completed polymer is RNA or DNA, for that is the only major difference between the two classes of nucleic acid.

The hydroxyl on the 1′ carbon (read "one prime carbon"—the "prime" identifies the position as being on the sugar) is replaced by a weak organic base, a derivative of either purine or pyrimidine (see Fig. 3-6 and the diagrams that follow). This combination of sugar and base is called a *nucleoside* or, more specifically, a ribonucleoside if the sugar is ribose or deoxyribonucleoside (deoxynucleoside for short) if the sugar is 2-deoxyribose. When phosphate is esterified to the sugar, usually at the 3′ or 5′ position, the nucleoside becomes a *nucleotide*.

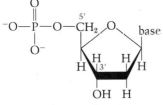
A deoxynucleotide monophosphate

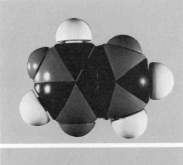

FIGURE 3-6 Purine (left) and Pyrimidine. Note that they are flat (planar) structures.

The organic bases derived from pyrimidine are uracil, thymine, and cytosine, collectively called "the pyrimidines." Likewise there are purines (adenine and guanine), which are simple derivatives of their parent molecule. Normally one finds thymine linked only to deoxyribose, and uracil only to ribose, while the other three bases are found in both kinds of nucleosides and nucleotides.

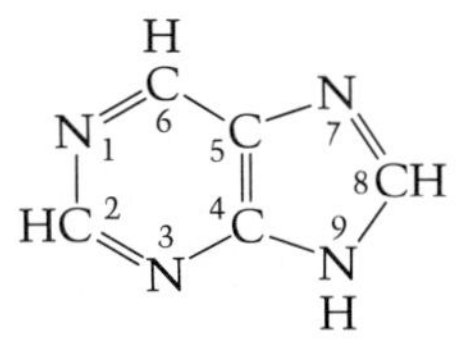

Purine

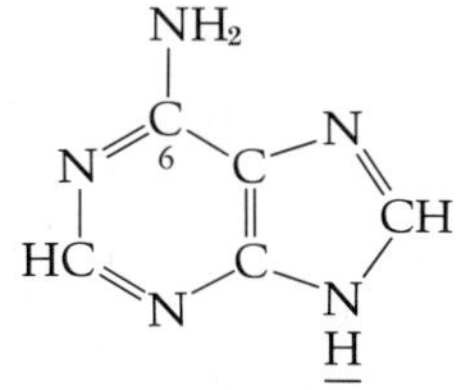

Adenine
(6-aminopurine)

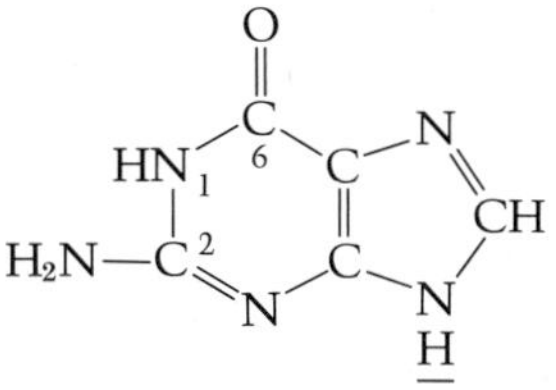

Guanine
(2-amino-6-oxypurine)

The underlined H is replaced by the 1' carbon in nucleosides.

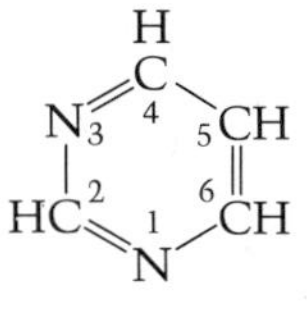

Pyrimidine

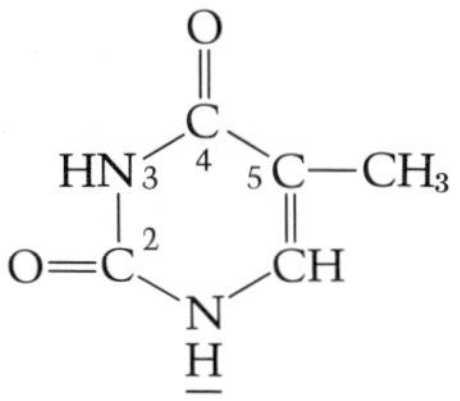

Thymine
(5-methyl-2,4-dioxypyrimidine)

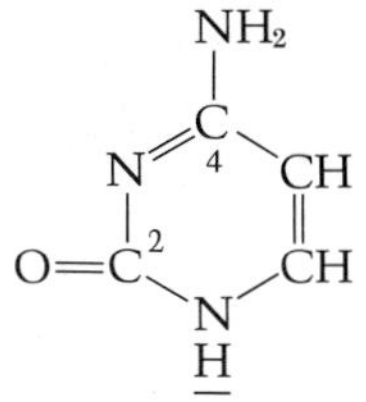

Cytosine
(2-oxy-4-aminopyrimidine)

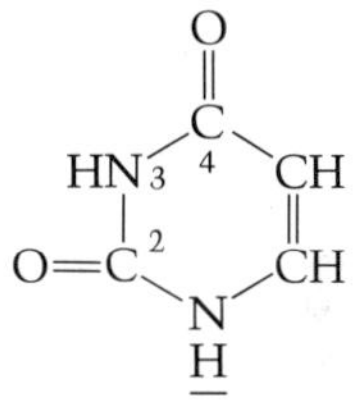

Uracil
(2,4-dioxypyrimidine)

The nomenclature of the nucleosides and nucleotides is summarized in Table 3-6; hereafter, the abbreviations given there will be used. For example, the condensation of adenine with one of the sugars, ribose or deoxyribose, forms adenosine or deoxyadenosine, respectively. When phosphate is esterified to the 5' carbon of one of these nucleosides, it becomes an adenine nucleotide, either adenosine monophosphate (AMP) or deoxyadenosine monophosphate (dAMP). With phosphate at the 3' position, the nucleotide would be 3'-adenosine monophosphate, or 3'-AMP, and so on. (Although it may be found at either or at both positions, phosphate is assumed to be at the 5' position unless otherwise stated.) If the acid portion is pyrophosphate (i.e., phosphoric anhydride) instead of phosphate, the corresponding nucleotide is adenosine diphosphate or ADP. If there is a string of three phosphates at the 5' position, the nucleotide is called adenosine triphosphate, or ATP. The corresponding deoxynucleotides are abbreviated dAMP, dADP, and dATP. As we shall see beginning in the next chapter, the nucleotides by themselves—particularly the adenine nucleotides—play very important biological roles. Our interest in them for the moment, however, is in their polymerization to form macromolecules, the nucleic acids DNA and RNA.

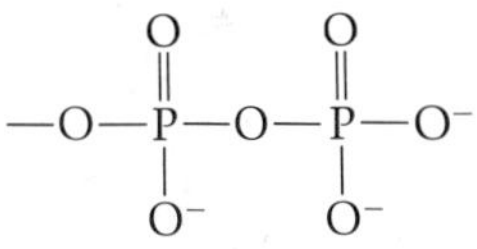

Pyrophosphate

The Structure of DNA. In 1950, E. Chargaff reported that the "base composition" of DNA (the relative amounts of the four bases) shows a peculiar constraint: The number of adenine (A) nucleotides is exactly the same as the number of thymine (T) nucleotides, and the number of guanine (G) nucleotides equals the number of cytosine (C) nucleotides. These constraints apply to DNA from almost every source. However, the ratio (G + C)/(A + T) varies widely from one source to another, clearly indicating that the average composition of DNA is not always the same.

TABLE 3-6. Nomenclature of the Nucleosides and Nucleotides

	Base	Nucleosides[a] (base + sugar)		Nucleotides[b] (base + sugar + phosphate)
Pyrimidines	uracil	$\xrightarrow{\text{+ ribose only}}$ uridine	$\xrightarrow{+P_i}$	UMP uridine monophosphate uridylic acid
	thymine	$\xrightarrow{\text{+ deoxyribose only}}$ deoxythymidine	$\xrightarrow{+P_i}$	dTMP deoxythymidine monophosphate thymidylic acid
	cytosine	$\xrightarrow{\text{+ ribose or deoxyribose}}$ (deoxy)cytidine	$\xrightarrow{+P_i}$	(d)CMP (deoxy)cytidine monophosphate (deoxy)cytidylic acid
Purines	guanine	$\xrightarrow{\text{+ ribose or deoxyribose}}$ (deoxy)guanosine	$\xrightarrow{+P_i}$	(d)GMP (deoxy)guanosine monophosphate (deoxy)guanylic acid
	adenine	$\xrightarrow{\text{+ ribose or deoxyribose}}$ (deoxy)adenosine	$\xrightarrow{+P_i}$	(d)AMP (deoxy)adenosine monophosphate (deoxy)adenylic acid

[a] Note that the names for the pyrimidine nucleosides end with "-dine," whereas the purine nucleosides end with "-sine."
[b] Unless otherwise specified, phosphate is esterified to the 5' hydroxyl of the ribose or deoxyribose. For example, AMP means 5'-AMP; therefore, 3-AMP (where the phosphate is at the 3' instead of 5' hydroxyl) would never be written just AMP.

Certain other information about DNA was also available in 1950. For example, it was known that fibers drawn from highly viscous DNA solutions have some kind of ordered structure, for X rays scattered by the fibers show a nonrandom distribution of reflections. In fact, the pattern of maximum intensities forms an "X," indicating that the individual molecules (which were also known to be very large) have a helical arrangement of atoms (see Fig. 3-7 and the appendix).

These advances stimulated a number of workers to begin the task of actually deciphering the three-dimensional structure of DNA. Among those interested in the problem was a pair of young scientists working at Cambridge, England: J. D. Watson, an American who had recently received his Ph.D.; and F. H. C. Crick, then a graduate student. They had the base ratio data of Chargaff, and they knew from X-ray and other physical studies that the molecule was probably a long, slender helix, with the bases stacked at right angles to the long axis. Starting with this information, they began building "Tinker Toy" models of various possible structures, and calculating what the X-ray diffraction patterns of those structures should look like. They then compared the calculated patterns with recent very high quality scattering diagrams recorded at King's College, London, by M. H. F. Wilkins and Rosalind Franklin. Each discrepancy between predicted and recorded reflections was cause for an adjustment in the model and, after many false starts, they finally found a structure that satisfied all the available evidence.

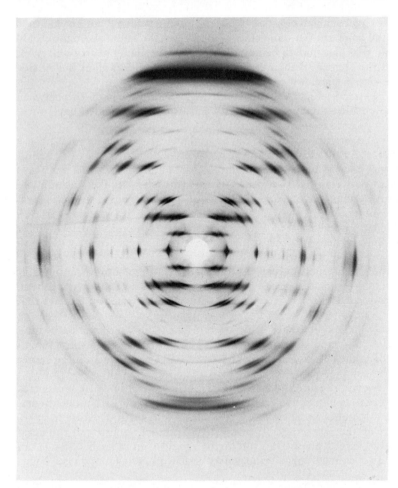

FIGURE 3-7 X-Ray Diffraction Pattern of a DNA Fiber. Note the large X, a characteristic scattering pattern for helices. The fibers are not highly ordered; hence the "fuzziness." (Courtesy of Prof. M. H. F. Wilkins, Medical Research Council Biophysics Unit, Univ. of London, King's College.)

Watson and Crick's model of DNA structure, which was published in 1953, won for them the 1961 Nobel Prize, shared with Wilkins. The model consists of a double-stranded helix, with the two strands lying "antiparallel," meaning that in a given direction one strand repeats the sequence "5' carbon-phosphate-3' carbon" while the other strand has a "3' carbon-phosphate-5' carbon" backbone. The bases of the nucleotides are stacked in pairs at the center of the helix, with the plane of the bases perpendicular to the helix. Chargaff's observation on the distribution of the bases was explained by specifically pairing adenine only with thymine, and guanine only with cytosine.

Adenine   Thymine   Guanine   Cytosine   The DNA "backbone"

Base-pairing in DNA

87

Note that each pair consists of one pyrimidine and one purine, and thus all the pairs are the same size. This allows the sugar-phosphate backbone to form a smooth helix with a constant diameter of about 20 Å. It also provides two regular grooves of 14 and 6 Å widths along the outside (see Figs. 3-8 and 3-9).

The strands of DNA are held together by hydrogen bonds, two between each adenine-thymine pair (A=T) and (it was established later) three between each guanine-cytosine pair (G≡C). The DNA helix in its best known configuration (called the B form) makes one turn for each 10 base pairs, covering a total distance of about 34 Å along the axis. The helical arrangement is stabilized by hydrophobic bonding and by van der Waals forces between the bases that, because of their platelike stacking and 3.4 Å spacing, have a considerable surface area exposed to each other. The importance of these "stacking interactions" to the structure of DNA is clearly demonstrated by the fact that even certain single-stranded polymers of deoxyribonucleotides form helices, relying entirely on a similar pattern of base stacking for stability.

The DNA double helix is quite stable in solution because, although it is held together with weak bonds only, there are a great many of them. This gives the molecule considerable rigidity, reflected in the fact that even dilute solutions (0.01% or less) become very viscous as the long rigid cylinders get in each other's way. This rigidity also makes the molecules vulnerable to breakage; just a vigorous stir of a DNA solution will result in the shearing of phosphodiester bonds, particularly at the center of the helix.

However, because the structure of DNA depends on weak bonds it is subject to *denaturation*, defined as a destruction of its native configuration. Merely dissolving DNA in distilled water will denature it by separating the two strands, a result of the mutual repulsion of the negatively charged phosphates, ordinarily shielded from each other by an excess of cations. In addition, solvents that compete for hydrogen bonds (e.g., urea, $H_2N-\underset{\underset{\|}{O}}{C}-NH_2$) may separate the strands of DNA by interfering with hydrogen bonding between the bases. And, of course, heat will denature the molecule since increasing thermal motion causes the weak bonds to spend less and less of their time in the formed state. However, because G≡C base pairs are held together with three hydrogen bonds whereas A=T pairs share only two protons, DNAs from different sources, and hence of different compositions, will denature or "melt" at different temperatures. In fact, the temperature of melting (T_m) has been correlated with the "G-C ratio," $(G + C)/(A + T)$. This ratio has been used as a way of classifying bacteria and, to some extent, higher organisms as well.

Watson and Crick soon followed their original paper on the structure of DNA with a second, this one entitled "Genetical Implications of the Structure of Deoxyribose Nucleic Acid." In it they point out that because the shape of DNA is so very simple, some other aspect of the molecule must be used for storing genetic information. In fact, the only possibility lies in the sequence of nucleotides—specifically the sequence of bases. In other words, the base sequence of DNA provides a four-letter alphabet (A, T, G, C) with which to record the genetic information of the cell (see Chap. 5).

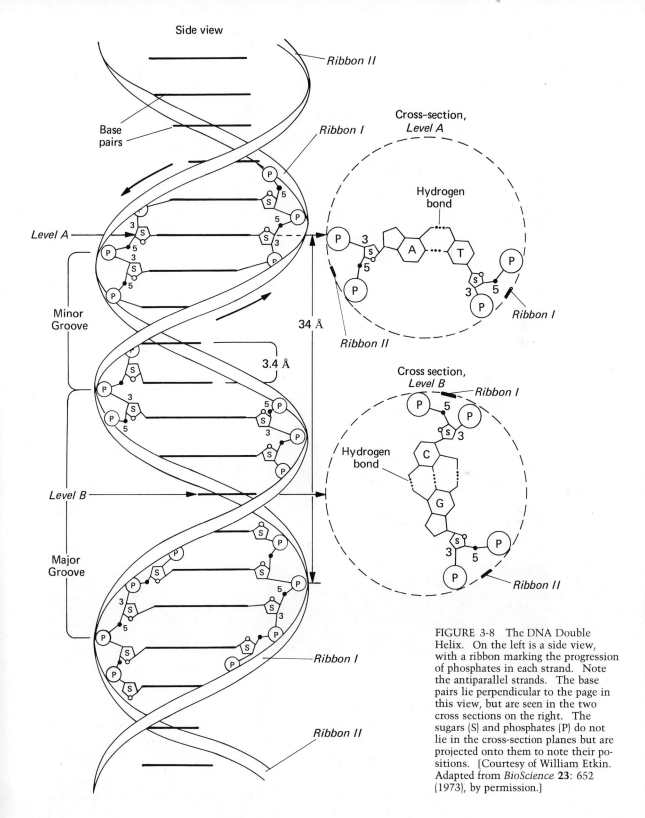

FIGURE 3-8 The DNA Double Helix. On the left is a side view, with a ribbon marking the progression of phosphates in each strand. Note the antiparallel strands. The base pairs lie perpendicular to the page in this view, but are seen in the two cross sections on the right. The sugars (S) and phosphates (P) do not lie in the cross-section planes but are projected onto them to note their positions. [Courtesy of William Etkin. Adapted from *BioScience* **23**: 652 (1973), by permission.]

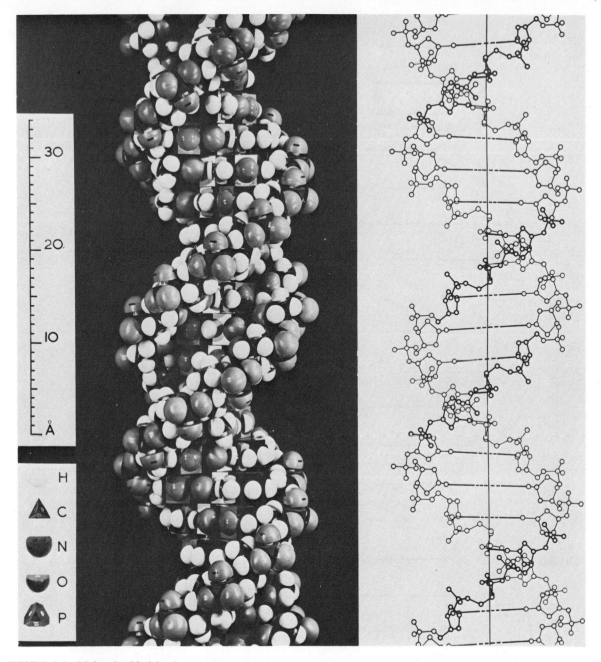

FIGURE 3-9 Molecular Models of the DNA Double Helix. [Courtesy of Prof. M. H. F. Wilkins.]

The RNAs. The history of ribonucleic acid, RNA, has been less dramatic than that of DNA. The RNA of a cell is not considered a part of its genetic material, although it is vital to it and in fact does actually play that role for certain viruses. The main function of RNA, discussed in Chap. 5, is to interpret the information stored in DNA and to translate that information into protein.

There are three classes of cellular RNA: messenger RNA or mRNA; transfer or tRNA, also called soluble RNA or sRNA; and ribosomal or rRNA. All are polymerized by specific base pairing with one strand of DNA, but with uracil replacing thymine. It was in the transfer RNAs that modification of the bases first became evident. While there are only four bases incorporated into RNA when it is formed, these bases can be altered after incorporation to create new chemical entities. Thus, in the 77-nucleotide "alanine tRNA" there are no less than 10 different bases in the final product. Lesser degrees of modication are seen in other nucleic acids as well, including DNA, where there seems to be a species-specificity to the pattern of alterations.

3-5 PROTEINS

The name *protein* is derived from a Greek word meaning "first." This is appropriate, for it could be argued that proteins are of first importance to the continuing functioning of the cell. They provide many of the structural elements of a cell, and they help to bind cells together into tissues. Proteins also catalyze most of the chemical reactions that occur in the cell. Some proteins act as contractile elements, to make movement possible; others control the activity of genes, transport needed material across membranes, and carry certain substances from one part of an animal to another. Proteins, in the form of *antibodies*, protect animals from disease and, in the form of *interferon*, mount an intracellular attack against viruses that have escaped destruction by the antibodies and other defenses. And finally, some hormones are proteins.

It is not surprising that the performance of so many different tasks requires many kinds of proteins, varying markedly in size and shape as well as in function. This diversity is possible because proteins are polymers of 20 or more different kinds of building blocks called *amino acids*—the name is short for "α-amino carboxylic acid." (The α carbon is in the number two position—i.e., adjacent to the carboxyl group.) The amino acids differ from one another in the nature of their side chain, R,

$$H_2N - \underset{R}{\overset{H}{\underset{|}{C}}} - COOH$$

Since a protein may contain anywhere from about 50 to more than 1000 amino acids, an enormous variety is possible. For example, one of the simplest proteins, the hormone insulin, has 51 amino acids. With 20 to choose from at each of these 51 positions, a total of 20^{51}, or about 10^{66} different proteins could be made. As we shall see, it is the sequence of amino acids that determines the shape and biological function of a protein as well as its physical and chemical properties.

The Amino Acids. The amino acids can be broadly classified into two groups, *hydrophobic* and *hydrophilic*, according to the nature of their side chain (see Table 3-7). The hydrophilic amino acids—those whose side chains interact strongly with water—include several that have either a second free amine or a second free carboxyl group. Since the amino and carboxyl groups on the α carbon are not normally available for interaction with a solvent, the solubility characteristics of a protein will depend on the kinds of amino acid side chains on its surface and on their arrangement—i.e., a protein may have areas that are clearly hydrophilic in nature and other areas that are clearly hydrophobic.

TABLE 3-7. The Common Amino Acids[a]

Hydrophilic Amino Acids (polar side chains)

H₂N—CH—COOH | CH₂ | COOH	H₂N—CH—COOH | CH₂ | CH₂ | COOH	H₂N—CH—COOH | CH₂ | SH	H₂N—CH—COOH | (CH₂)₄ | NH₂	H₂N—CH—COOH | (CH₂)₃ | NH | C=NH | NH₂
Aspartic acid asp $pK_a = 3.65$	Glutamic acid glu $pK_a = 4.25$	Cysteine cys $pK_a = 8.18$	Lysine lys $pK_a = 10.53$	Arginine arg $pK_a = 12.48$
Histidine his $pK_a = 6.00$	Asparagine asn	Glutamine gln	Threonine thr	
Tyrosine tyr $pK_a = 10.07$	Serine ser			

Hydrophobic Amino Acids (nonpolar side chains)

Tryptophan try	Phenylalanine phe	Proline pro	Methionine met	Leucine leu
Isoleucine ilu	Valine val	Alanine ala	Glycine gly	

[a] The usual three-letter abbreviation is given, along with the pK_a of side-chain groups that carry significant charge at physiological pH.

All proteins are comprised of the 20 amino acids shown in Table 3-7. There are a number of other amino acids that are present in small quantities and in various organisms. If they are part of a protein, however, they are derived from these 20.

For example, *cystine* is a dimer of cysteine, formed by oxidizing the two sulfhydryl groups into a *disulfide bond*:

$$R_1\text{—SH} + \text{HS—}R_2 \xrightarrow{\frac{1}{2}O_2} R_1\text{—S—S—}R_2 + H_2O$$

Cystine is widely distributed in proteins, as are the methyl derivatives of lysine, arginine, glutamate, and aspartate. (Each methyl group replaces a hydrogen of the side-chain amine or carboxyl group.) In addition, phosphoserine (the phosphate ester of the side-chain hydroxyl of serine) and phosphothreonine occur regularly (in the milk protein, casein, for example) and hydroxyproline and hydroxylysine make up a major fraction of the connective tissue protein, collagen.

Hydroxyproline   Hydroxylysine   Methyllysine

Thyroxine

Other examples of derived amino acids include *thyroxine*, which contains four iodines, and *triiodothyronine*, which contains three. They are synthesized from tyrosines (two each) of *thyroglobulin*, a thyroid protein, and are freed into the blood as hormones referred to as T4 and T3, respectively, when thyroglobulin is degraded. These hormones control the overall metabolic rate of an animal, with even mild overproduction leading to a nervous, "jittery" condition and underproduction leading to a general slowing of physical and mental activity.

Note that one of the common amino acids is, in fact, an *imino* rather than an amino acid. The short hydrocarbon side chain of proline is folded back and joined to the α-amino group. This deviation has important implications for protein structure, as we shall see. Note also that with the exception of cysteine and methionine, which contain one sulfur atom each, the 20 common amino acids are composed entirely of carbon, hydrogen, oxygen, and nitrogen.

The α carbon of each of the amino acids—with the exception of glycine, which has a single hydrogen for a side chain—has four different groups attached to it. Therefore, all the amino acids but glycine exist in two optically active, asymmetric forms that are the mirror images of each other. One form is designated D and the other L, by analogy to the convention used in naming sugars (see Fig. 3-10). As with the sugars, however, there is no correlation between the D and L designation and

CHAPTER 3
Molecular Architecture and Biological Function

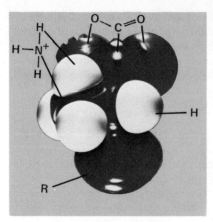
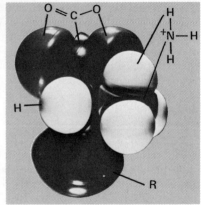

FIGURE 3-10 D- and L-Amino Acids. Note that they are mirror images. The large knobs on the outside represent side chains.

whether the molecule is dextrorotatory (rotates the plane of light to the right) or levorotatory. For example, L-alanine is dextrorotatory in water, while L-tyrosine is levorotatory, leading to the designations L(+)-alanine and L(−)-tyrosine, respectively. The situation is further complicated by the fact that threonine and isoleucine each have a second asymmetric carbon, so four "diastereo" isomers exist for each; these are designated L, D, L-allo and D-allo.

It is important to note that only the L-amino acids occur in proteins. Obviously, the mechanisms whereby amino acids are synthesized and utilized can distinguish between the two isomers. Some D-amino acids are found in microorganisms, however, particularly in the cell walls of bacteria and in several of the antibiotics.

Another important feature of the amino acids is the existence of both a basic and an acidic group at the α carbon. The amino group typically has a pK_a (see Chap. 2) between 9 and 10, while the α-carboxyl group has a pK_a that is usually close to 2 (very low for carboxyls). Thus, at physiological pH, the free amino acids exist largely as *zwitterions*, or double ions, having both a quaternary amine ($-\overset{+}{N}H_3$) and a carboxylate

$$H_3\overset{+}{N}-CH-COO^-$$
$$|$$
$$R$$
Zwitterion

$(-\overset{O}{\underset{\|}{C}}-O^-)$ group. The pH at which the negative and positive charges just balance is called the *isoionic pH*, abbreviated pI. The isoionic pH varies from one amino acid to the next, but in the absence of a charged side chain it is usually about pH 6.

One last feature that will be important to us later is the fact that the nitrogen and oxygens of the amino and carboxyl groups are potential electron donors for the formation of dative bonds, giving each amino acid at least two reactive groups for binding metal ions, and the capacity to act as a chelate.

The Peptide Bond. Amino acids can be linked by a condensation in which an —OH is lost from the carboxyl group of one amino acid along with a hydrogen from the amino group of a second, forming a molecule of water and leaving the two amino acids linked via an amide, called in this case, a *peptide bond* (Fig. 3-11):

$$H_2N-\underset{R_1}{\overset{H}{\underset{|}{C}}}-\overset{O}{\underset{\|}{C}}-OH + HN-\underset{R_2}{\overset{H}{\underset{|}{C}}}-\overset{O}{\underset{\|}{C}}-OH \longrightarrow H_2N-\underset{R_1}{\overset{H}{\underset{|}{C}}}-\overset{O}{\underset{\|}{C}}-\underset{H}{\overset{}{\underset{|}{N}}}-\underset{R_2}{\overset{H}{\underset{|}{C}}}-\overset{O}{\underset{\|}{C}}-OH + H_2O$$

The standard state free energy change, $\Delta G°$, for the reaction is about -2 kJ per mole if the addition is to a large polymer; hence, the reaction is only very slightly favorable under standard conditions. There could be a considerable amount of spontaneous depolymerization of proteins were it not for a high energy barrier that must be penetrated to initiate the reverse reaction, or hydrolysis of the peptide bond.

The peptide group is planar and conjugated, as if it were about 60% like the structure

$$\begin{array}{c} \mathrm{O} \\ \parallel \\ -\mathrm{C}-\mathrm{N}- \\ | \\ \mathrm{H} \end{array}$$

and about 40% like the alternate resonant form

$$\begin{array}{c} \mathrm{O}^- \\ | \\ -\mathrm{C}=\mathrm{N}^+- \\ | \\ \mathrm{H} \end{array}$$

The C—N bond has a length of 1.32 Å, reflecting its approximately 40% double-bond character: The normal carbon-nitrogen single- and double-bond lengths are, respectively, 1.43 and 1.26 Å. The peptide bond holds the four participating atoms rigidly in the same plane, but there is generally free rotation about the bonds on either side of the group—i.e., the bonds to each α carbon. (Proline and hydroxyproline are an exception, since the C—N bond to the α carbon is part of a ring structure.) This flexibility at the α carbon permits chains of amino acids to form a variety of configurations.

Two amino acids joined by a peptide bond form a dipeptide, three form a tripeptide, and so on. Small polymers are termed *oligopeptides* while larger ones are referred to as *polypeptides*. A protein can be defined as one or more polypeptide chains (see Fig. 3-12) folded into a definite spatial configuration. Its spatial configuration, or *conformation*, is the key to its biologicial activity and, as we shall see, is a consequence of the sequence of amino acids in its polypeptide chain(s). It is convenient to describe the conformation of a protein in terms of four levels of organization: primary, secondary, tertiary, and quaternary.

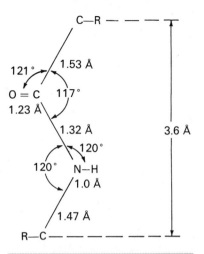

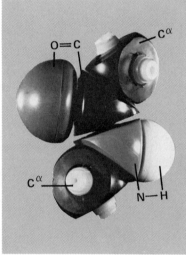

FIGURE 3-11 The Peptide Bond.

FIGURE 3-12 A Polypeptide Chain of L-Amino Acids, Fully Extended. The side chains and hydrogens on the α carbons are omitted.

Primary Structure. The primary structure of a protein is its amino acid sequence. The first protein to have its primary structure determined was insulin, a hormone that regulates glucose metabolism in mammals. Insulin consists of two polypeptide chains of 21 and 30 amino acid residues, called the A and B chains, respectively. (An amino acid residue is that which is left when the elements of water are split out during polymerization.) The sequence of residues within each chain was worked out by F. Sanger and his colleagues at Cambridge in a long and difficult series of experiments completed in 1955. Sanger was awarded the Nobel Prize in 1958 for this accomplishment, which is now done on a routine basis.

Determination of the primary structure of insulin was possible because all insulin molecules isolated from a given animal have exactly the same amino acid sequence. In fact, except for those from occasional exceptional individuals, all the insulin molecules isolated from animals of a given species will be exactly the same. This constancy of primary structure is true for other proteins as well, and is something that was suspected, but not really proved, before Sanger began his studies. (In Chap. 5 we shall learn how the primary structure of a protein is determined by the sequence of deoxynucleotides in a gene.)

Secondary Structure. The next level of protein structure is called secondary and is the result of proton sharing between the oxygen of one peptide group and the nitrogen of another. However, the rigidity of the peptide bond and the linearity of these hydrogen bonds pose certain limitations. In most cases, the first three atoms of the group (NH···O=C) lie in a straight line, with a nitrogen-oxygen distance of 2.79 Å.

In the late 1940s Linus Pauling and R. B. Corey, at the California Institute of Technology, began building models of possible polypeptide chain configurations. With the restrictions just mentioned, a knowledge of the bond angles of carbon, the assumption that the most stable configurations would be those that allowed the maximum number of hydrogen bonds, and some X-ray diffraction patterns of crystallized proteins as a guide, Pauling and Corey found four basic polypeptide configurations that satisfied the known data. Two of these four configurations are *helices*, stabilized by hydrogen bonds from each peptide group to the third peptide group from it in each direction. (In other words, each amino acid is hydrogen bonded to its fourth neighbor on both sides—Fig. 3-13a and b). Both a left-handed and a right-handed helix are allowed, rotating counterclockwise and clockwise, respectively.

The other two configurations that satisfied the known data are called *pleated sheets*; the general form of these was first suggested by W. T. Astbury in the 1930s. The polypeptide chains of pleated sheets are stretched out and lie side by side, either parallel or antiparallel to one another. The bonded groups may be portions of the same chain folded back on itself, or they may be separate chains (Fig. 3-13c).

The helical and pleated sheet arrangements are known as the "alpha" and "beta" structures of proteins, respectively, because they were first identified in α- and β-keratins. Keratin is the fibrous protein of the epidermis (see Fig. 3-14), and of hair, wool, fur, nails, beaks, feathers, hooves, and so on. The fibers of hair and wool, for example, are composed of dead cells containing α-keratin strands (Fig. 3-15). These strands are highly elastic as a result of their ability to assume the more extended β configuration when stretched and a tendency to return to a helix when tension is removed. The return is fostered by interchain disulfide bonds, the rearrangement of which forms the basis for the permanent-wave process used on human hair.

Nature seems to have relied mostly on the right-handed α helix and the antiparallel β sheet for secondary structure, although a number of

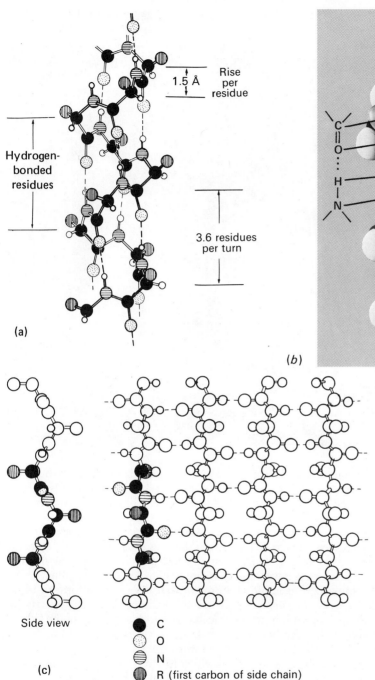
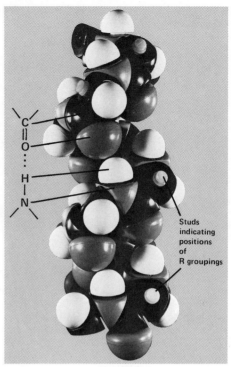

FIGURE 3-13 Secondary Structure. (a) Right-handed α helix. (b) Space-filling model of an α helix, without side chains. (c) An antiparallel β structure.

- ● C
- ⬢ O
- ⊜ N
- ▥ R (first carbon of side chain)
- ○ H

other configurations have also been identified. The α helix, as shown in Fig. 3-13a, has 3.6 residues per turn of 5.4 Å, or 1.5 Å along the axis for each amino acid residue. This spacing allows the carbonyl (C=O) groups to lie parallel to the long axis and to point almost straight at the

97

CHAPTER 3
Molecular Architecture and
Biological Function

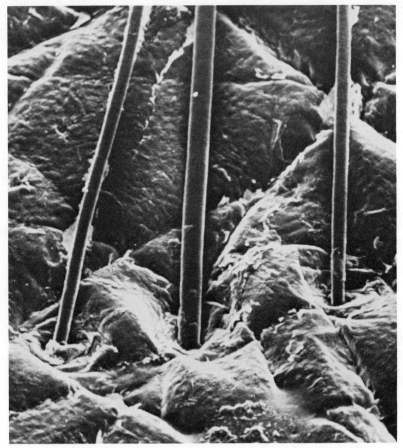

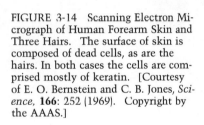

FIGURE 3-14 Scanning Electron Micrograph of Human Forearm Skin and Three Hairs. The surface of skin is composed of dead cells, as are the hairs. In both cases the cells are comprised mostly of keratin. [Courtesy of E. O. Bernstein and C. B. Jones, *Science,* **166**: 252 (1969). Copyright by the AAAS.]

amino groups to which they are hydrogen bonded. A typical β sheet, on the other hand, will have a spacing of about 4.7 Å between polypeptide chains, and one amino acid residue per chain for each 3.5 Å along the axis—a distance per amino acid that is more than double the figure for α helices. Because of this difference, wool can be stretched to about twice its original length as it changes from the α to the β form under tension. In general, the α helix lends elasticity to a structure while the β sheet usually provides rigidity, due to its interchain hydrogen bonding.

One important example of a protein that does not conform to these guidelines is *collagen.* Collagen is the widespread connective tissue fiber that, along with mucopolysaccharides, forms the bulk of the "glue" that holds cells together as tissues in most forms of animal life except arthropods. (The Greek root of the word collagen means "glue.") Collagen also supplies the matrix to which calcium salts are added to form bone and is the substance of scars and tendons, a function to which it is admirably suited by virtue of its enormous tensile strength—a 1-mm fiber can support a 10-kg mass. As much as one-fourth of the protein of an animal may be collagen.

Collagen fibers are comprised of side-by-side aggregates of *tropocollagen,* each a molecule some 280 nm long by 1.5 nm in diameter, with a mass of about 300,000 daltons. Tropocollagen, which is secreted by cells of many types but especially by *fibroblasts,* consists of three poly-

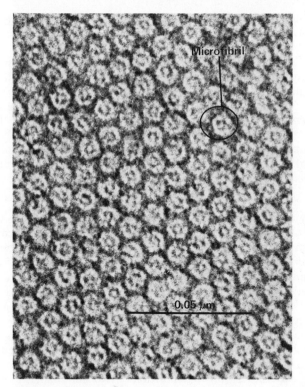

FIGURE 3-15 Cross Section of Merino Wool. Detail of a cell interior, revealing numerous microfibrils, each of which is composed of several α-helical strands of keratin. [Courtesy of G. E. Rogers and B. K. Filshie, *J. Mol. Biol.*, **3**: 784 (1961).]

peptide chains wound together into a three-stranded helix (See Fig. 3-16).

The unusual secondary structure of tropocollagen is the result of its peculiar primary structure: Almost every third amino acid in the chains is glycine, and a large fraction of the remainder is either hydroxyproline

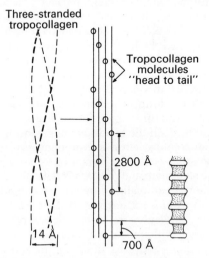

FIGURE 3-16 Collagen. The polypeptide chains form a three-stranded helix in the tropocollagen subunit. Adjacent subunits overlap one another by three-quarters of their length, providing a 700 Å repeat distance to the fibrils, as seen in the electron micrograph. (Micrograph of chromium shadowed human skin collagen courtesy of Dr. Jerome Gross.)

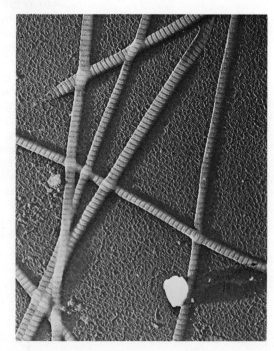

or proline, much of it in the sequence gly-*x*-pro or gly-*x*-hypro, where *x* is any of a number of other amino acids. The geometry of proline and hydroxyproline precludes an α helix—the restricted rotation at the α carbon "points" the nitrogen in the wrong direction. The three strands are interlinked by hydrogen bonds at every opportunity, however, forming a tight unit that is also very insoluble owing to a lack of free polar groups capable of interacting with water. If collagen is boiled, the heat causes denaturation via strand separation. We then call it *gelatin*. The individual strands of gelatin still cannot form compact structures because of the limitations of the proline and hydroxyproline bond angles, but there is a marked tendency to interact nonspecifically through interchain hydrogen bonds, so that a gel may be formed at concentrations of about 1% or more.

Tertiary Structure. A typical protein will consist of sections of secondary structure interspersed with apparently unordered segments called "random coils." The way that regions of secondary structure are oriented with respect to each other is referred to as the *tertiary structure* of the protein. For example, in a protein with a high α-helix content, there will typically be several well-defined helices that could, through rotation about single bonds, be placed in any of several configurations. "Tertiary structure" is a description of their actual arrangement.

One does not expect to find a protein in one continuous helix or pleated sheet for two reasons: Either of the two secondary structures may occasionally result in unfavorable interactions between amino acid side chains—for example, the juxtaposition of two side chains of like charge or two very bulky side chains in a space not adequate for them. In addition, long continuous helices are prevented by the presence of proline, which cannot conform to the required geometry. Its appearance in a helix introduces a kink so that the only position proline can occupy in a helix is the amino end. Proline, then, is a "helix breaker." If considerable proline is present in a protein, the amino acid may even impose a structure of its own, as it does in collagen. But usually a high proline content means only that there will be little opportunity for α helix.

The various regions of helices and pleated sheets (the secondary structures) are held in position by favorable side-chain interactions, including all of the weak bonds mentioned in Chap. 2 and, in addition, an occasional covalent cross-link in the form of a disulfide (—S—S—) bridge between cysteines (Fig. 3-17). The three chains of tropocollagen, for example, are linked at one end by disulfide bonds, complementing the extensive interchain hydrogen bonding. In a few proteins, certain other covalent cross-links are possible—e.g., the formation of an amide between the amine of a lysine side chain and the carboxyl group of aspartic or glutamic acid. It is important to realize, however, that while covalent cross-links may help to *stabilize* a tertiary structure, they do not determine it. This principle is demonstrated by the observation that it is often possible to completely *denature* a protein—i.e., remove all secondary and higher structure—and still find that the polypeptide chains can refold *in vitro* to their original configuration and biological activity (see Fig. 3-18). It is obvious from such experiments that the native structure of a protein can be spontaneously achieved, so that one does not need to propose any kind of directed folding on a template or scaffold to reach a unique tertiary structure. The primary structure, it seems, is enough to determine the higher levels of organization.

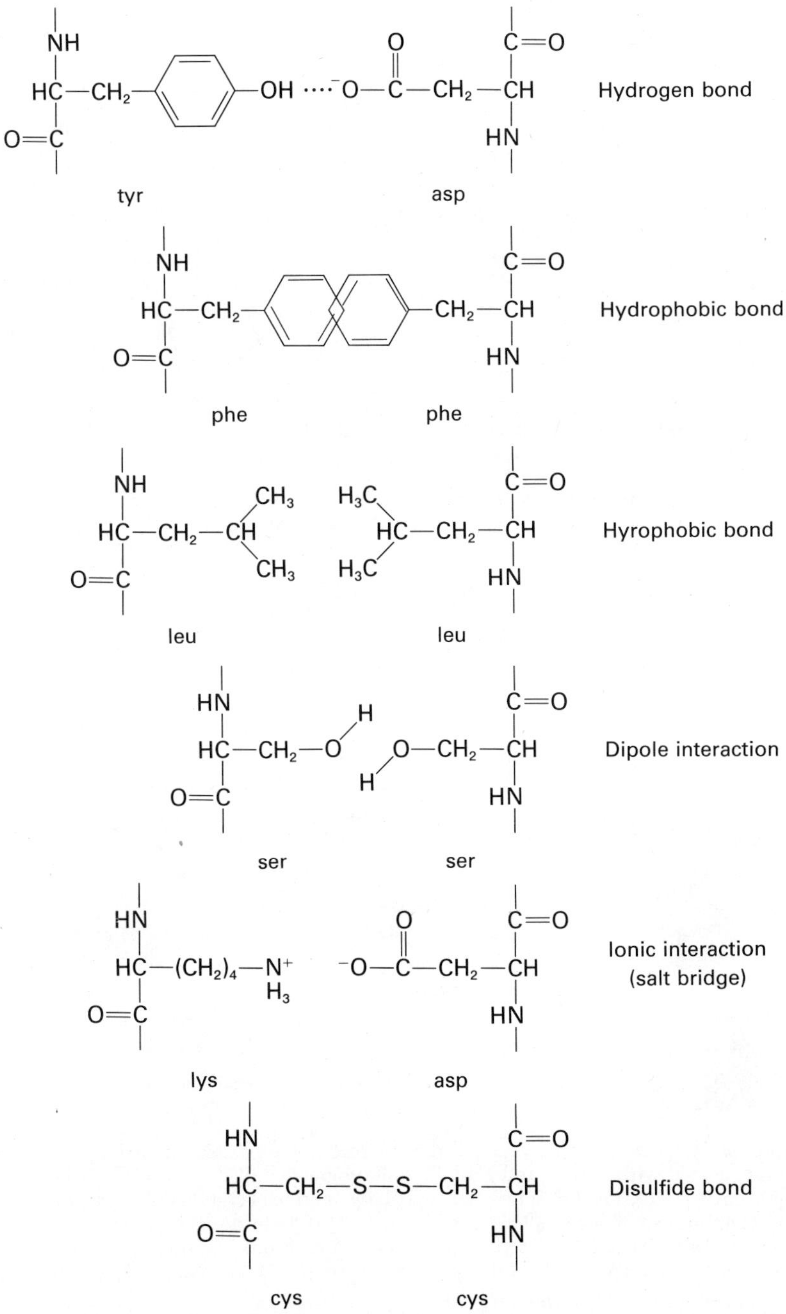

FIGURE 3-17 Bonds That Stabilize Tertiary Structure.

The complete tertiary structure of a protein can only be deduced by a laborious analysis of X-ray scattering patterns from crystals (see Fig. 3-19 and Appendix). The first protein to have its secondary and tertiary structure determined was myoglobin, a 153-amino acid, oxygen-binding protein found primarily in red muscle and largely responsible for the color of that tissue. The work was done at Cambridge under the direction of J. C. Kendrew. The analysis, to a resolution of 2 Å (see Fig. 3-20), was announced in 1961, and was based on about 40,000 reflections made by the

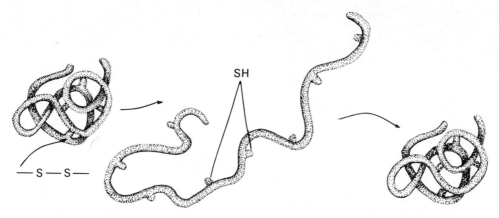

FIGURE 3-18 **The Spontaneous Formation of Tertiary Structure.** Many proteins can be completely denatured and renatured *in vitro*, implying that the native configuration is the most stable. The disulfide bridges in these cases may help to stabilize a structure, but they obviously do not determine it.

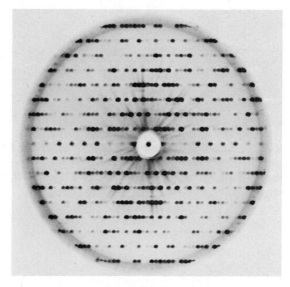

FIGURE 3-19 **The X-ray Crystallography of Proteins.** LEFT: Crystals of the enzyme isocitrate dehydrogenase were chemically fixed with glutaraldehyde, coated with carbon and gold, then photographed in a scanning electron microscope to reveal their shape. RIGHT: In X-ray diffraction, a single, large, unfixed crystal is mounted in an X-ray beam and the scattering pattern examined to deduce molecular structure. This pattern was obtained from a crystal of another enzyme, pork adenylate kinase. [Scanning electron micrograph courtesy of W. Burke, J. Swafford, and H. Reeves, *Science* **181**: 59 (1973) copyright by the AAAS; diffraction pattern courtesy of I. Schirmer, H. Schirmer, and G. E. Schulz, Max Planck Inst. für Medizinische Forschung, Heidelberg, Germany.]

molecule and four chemical derivatives of it. To produce just the final calculation alone took a high-speed (for those days) digital computer about twelve hours. The procedures have been improved a great deal since then, and include the advent of automatic, computer-controlled machines. However, the determination of tertiary structure of proteins still remains a major task.

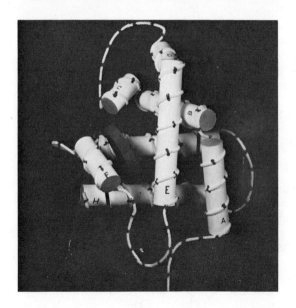 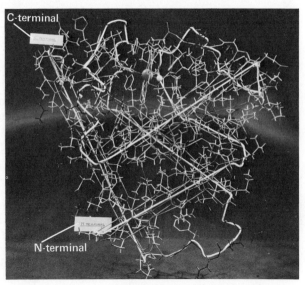

FIGURE 3-20 The Structure of Myoglobin. The pipestem model is designed to show the relationship of the helical segments to each other.

Quaternary Structure. The highest level of protein structure, called quaternary, refers to the *subunit structure* of proteins. Proteins that have more than one polypeptide chain may also have more than one independently folded unit, each of at least one chain. The independently folded units are called *subunits;* the spatial arrangement of these subunits is the *quaternary structure* of the protein.

Whether a protein with two polypeptide chains has one or two subunits depends on the relationship of the chains to each other. If each chain has its own tertiary structure and could maintain that structure in the absence of the other chain, then there are two subunits. On the other hand, if it is impossible to separate the chains without changing their tertiary structure, then they are part of the same subunit. Accordingly, insulin has no quaternary structure; it is a single unit composed of two polypeptide chains. Because there is some ambiguity involved in these definitions, it has been suggested that the smallest unit of independent biological function be called a *protomer.* In most cases, the number of polypeptide chains is also the number of subunits and the number of protomers; however, in other cases it may take two or more subunits to make a protomer.

Subunits may be held together by weak interactions of the kind described in Chap. 2, by covalent linkages, or by both. In general, multisubunit proteins designed for intracellular use rely on weak chemical interactions to hold the subunits together, while covalent cross-links are found in proteins secreted from the cell in which they are made. Most intersubunit covalent cross-links are disulfide bridges. There are special proteins, however, in which other types of linkages are found. For example, lysines and hydroxylysines of collagen can react with the help of a specific enzyme, lysyl oxidase, in order to covalently link tropocollagen molecules into collagen fibrils. A defect in this enzyme is responsible for one form of an inherited condition called Ehlers-Danlos syndrome, which results in the elastic skin and lax joints of an "India rubber man."

$$\begin{array}{c}-CH-\\|\\(CH_2)_3\\|\\CH\\\|\\N\\|\\(CH_2)_4\\|\\-CH-\end{array}$$

Collagen cross-link (formed from two lysines)

In another connective tissue protein, elastin, no less than four lysines are joined also by lysyl oxidase to create a new amino acid, *desmosine* or *isodesmosine*. Four polypeptide chains are thus interlinked, accounting most likely for the ability of elastin to reversibly stretch in any direction—appropriate to its role as part of the elastic layer of major blood vessels.

Still another form of covalent cross-link consists of the *isopeptide bond,* which is an amide link between a side-chain lysine and glutamic acid, the latter starting out as glutamine

$$—(CH_2)_4—NH_2 + H_2N—CO—(CH_2)_2— \longrightarrow —(CH_2)_4—NH—CO—(CH_2)_2— + NH_3$$

Desmosine, an elastin cross-link (formed from four lysines)

Isopeptide bonds are important in several structural proteins, including the keratin of some hair and the fibrin of blood clots.

Frequently, the subunits of a protein are identical to each other and are constructed from information contained in the same gene. This permits large structures to be assembled without an undue commitment of genetic material. The protein portion of tobacco mosaic virus, for example, consists of more than 2000 identical protein subunits, each with a mass of about 17,000 daltons. The use of small subunits in its construction offers the same advantage as building a house of bricks instead of casting it in one piece.

The subunits of a protein are not always identical to one another, however. Hemoglobin offers us an example of a protein with more than one kind of subunit. It is a red protein of 65,000 daltons found in the red blood cells (erythrocytes) where it is responsible for carrying oxygen from the lungs to other tissues. In the human adult, hemoglobin consists of four chains, each folded into a separate subunit. The two "α chains" have 141 amino acid residues each, while the two "β chains" each have 146 amino acid residues. The molecule is thus designated an $\alpha_2\beta_2$ tetramer.

Hemoglobin was the first protein to have the three-dimensional features of its quaternary structure determined by X-ray diffraction. The work was accomplished in the laboratory of M. F. Perutz at Cambridge, England, a project that began some fifteen years earlier than the myoglobin study in Kendrew's laboratory. However, since hemoglobin is four times as large as myoglobin, progress came slowly. In fact, the myoglobin work was of very great value to the hemoglobin project (and vice versa) because, as it turns out, each of the four chains of hemoglobin closely resembles myoglobin in tertiary structure. The two groups of X-ray crystallographers cooperated closely and, as a result, the preliminary three-dimensional model for hemoglobin was presented at about the same time as that of myoglobin. The placement of the subunits in hemoglobin can be seen in Fig. 3-21 and in another oxygen-carrying protein, hemocyanin, in Fig. 3-22. It should be noted that neither myoglobin nor hemoglobin contains covalent cross-links, again emphasizing the importance of the weak bonds in stabilizing protein structure.

Prosthetic Groups. So far, we have considered only the structure of the polypeptide chains of proteins. However, many proteins contain organic or inorganic units that are an integral part of the molecule, but not composed of amino acids. These additions are called *prosthetic groups;* their presence defines a *conjugated protein.* Very often prosthetic groups are simple metal ions, particularly Fe^{2+}, Zn^{2+}, Mn^{2+}, Mg^{2+}, or Cu^{2+}, which

SECTION 3-5
Proteins

FIGURE 3-21 The Quaternary Structure of Oxyhemoglobin. On deoxygenation, the α_1 subunit turns about its axis by 9.4°; the β_1 subunit turns about its axis by 7.4°. Both rotations are clockwise relative to the center. A corresponding change takes place in the α_2 and β_2 subunits. [Courtesy of M. F. Perutz, *Proc. Roy. Soc. (London) Ser. B*, **173**: 113 (1969).]

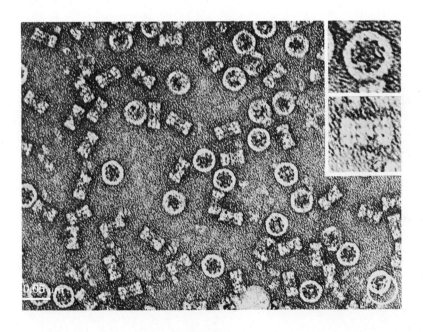

FIGURE 3-22 The Quaternary Structure of a Hemocyanin. These copper-containing molecules transport oxygen in arthropods and molluscs. They are found free in the "blood" rather than packaged into cells as is hemoglobin. The molecules are each composed of a number of subunits arranged in a highly symmetrical fashion, often as cylinders. In this example, from *Octopus vulgaris*, some cylinders are seen end-on and some from the side. (From E. F. J. Van Bruggen and K. E. Van Holde, in *Subunits in Biological Systems*, S. Timasheff and G. D. Fasman, eds. © 1971 by Marcel Dekker, Inc., New York. Reprinted by permission.)

105

are held by dative bonds to oxygens and nitrogens. For example, the enzyme lysyl oxidase, responsible for collagen and elastin cross-linking, contains copper. Hence, severe copper deficiency leads to faulty collagen and elastin, producing a condition similar to the Ehlers-Danlos syndrome already mentioned.

Hemoglobin and myoglobin are additional examples of conjugated proteins. Each subunit of hemoglobin, and each myoglobin molecule, contains a heme group: a ringlike structure with a central iron ion (Fe^{2+}).

Heme
(one of several variations)

Iron makes four dative bonds to heme nitrogens and an additional one to the nitrogen of a histidine. A sixth dative bond is satisfied by a water molecule that can be displaced by oxygen. The heme group is stuck in a hydrophobic crevice of the protein in both hemoglobin and myoglobin, but the lipophilic oxygen molecule is able to penetrate readily into this area and displace the water molecule.

This same prosthetic group, the heme, is found in several other proteins as well. For example, the enzyme catalase, which catalyzes the decomposition of hydrogen peroxide (H_2O_2) into water and oxygen, also contains a heme, the iron of which plays a central role in the catalytic process. The cytochromes are also heme-containing proteins, responsible for transferring electrons by a cyclical change in the valence state of the iron from Fe^{3+} to Fe^{2+} and back again. With the substitution of Mg^{2+} for the iron, plus some changes in the side chains, the heme becomes chlorophyll, which is the green pigment of plants and the key to photosynthesis. These substances and other conjugated proteins will be discussed in more detail later.

Lipoproteins. Some conjugated proteins contain very large quantities of nonprotein material, which may even account for a majority of the mass. Notable examples include the *lipoproteins*, which are complexes of lipid and protein. Hemoglobin and myoglobin, in the strictest sense, are lipoproteins, since the heme group is a lipid. The term, however, is mostly used for components of membranes and for the blood lipoproteins that transport cholesterol, phospholipid, and triglycerides. The basic core of these particles is comprised of protein subunits to which is added the above lipids.

In *very low density lipoproteins* (*VLDL*), protein comprises only about 10% of the mass. The VLD lipoproteins are made primarily by liver

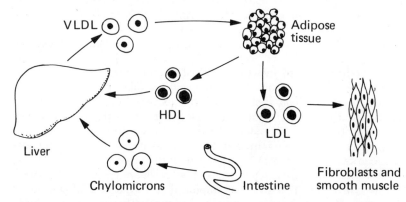

FIGURE 3-23 Lipoproteins. Very Low density lipoprotein (VLDL) is produced by the liver, in part from dietary fat transported as chylomicrons. VLDL donates glycerides to fat cells for storage. In the process, some protein is removed and picked up by HDL (high-density lipoprotein). The VLDL remnant is called LDL (low-density lipoprotein). It can serve as a source of lipids needed for membrane synthesis.

cells, using lipids absorbed by the intestines and transported as fat droplets called *chylomicrons*. The VLD particles are released into the blood stream to be carried to cell types that are capable of binding them and removing glycerides either for storage (as in fat cells) or as a source of energy. In the process of breakdown, a new entity is produced consisting of about half cholesterol and a quarter each protein and phospholipid. This particle, called *low-density lipoprotein* (*LDL*), serves as a source of cholesterol and phospholipid used for membrane synthesis. The removed protein is salvaged by a third species, called *high-density lipoprotein* (*HDL*)—see Fig. 3-23. HDL is about half protein. (It is easy to keep the names straight by remembering that fat is the less dense—fat floats—and protein the more dense.) Chylomicrons are sometimes said to comprise a fourth lipoprotein, but their protein content of only 1–2% hardly justifies the name. It is more useful to think of chylomicrons as fat droplets, rapidly removed from the blood after a fatty meal.

Glycoproteins. *Glycoproteins* (or *proteoglycans*) represent another class of conjugated protein in which the nonprotein part is often a significant fraction of the total mass. Most proteins secreted from the cells in which they are made, as well as most proteins found in membranes and many intracellular proteins (including plant proteins), contain some carbohydrate. Its function is not clear, except that the carbohydrate portion of extracellular and cell surface proteins seems to help make it possible for recognition systems (e.g., antibodies) to tell one protein or cell from another. In the case of mucous secretions, the very large amounts of highly charged mucopolysaccharide seem to foster the mutual repulsion needed to keep the chains from aggregating and thereby increasing the viscosity of these glycoproteins (also called mucoproteins). A high viscosity would be an important feature of proteins designed for lubrication and protection.

The carbohydrate portion of glycoproteins consists typically of short chains of hexoses and pentoses, often terminating in a fucose or sialic acid. (Fucose is a galactose derivative lacking one hydroxyl group.) Acetylglucosamine or galactose, on the other hand, usually start the chain through linkages to asparagine, serine, threonine, or, in those few proteins that have them, hydroxylysine or hydroxyproline. Unlike the proteins themselves, however, the composition and amount of carbohydrate is not highly specific. That is, there may be considerable heterogeneity in the carbohydrate even among molecules of one glycoprotein obtained from a single source.

Collagen is a simple glycoprotein. Anywhere from about 10–80% of the hydroxylysines, depending on the source of the collagen, have galactose attached, and to many of the latter there is attached a glucose.

The Determination of Structure. Even a medium-sized protein of 30,000 daltons will contain about 300 amino acid residues. With free rotation about all single bonds, it should be obvious that a great many different spatial configurations might be possible. Yet very little deviation is observed in the finished product, for the biological role of a protein is a direct consequence of its three-dimensional shape. And that shape must be determined by the primary structure of the protein, at least in those cases where the native structure is spontaneously regained after denaturation *in vitro*. This situation implies that the native configuration is the configuration of lowest energy—i.e., the configuration that allows the maximum number of bonds if all bonds have the same energy.

A great many possible secondary, tertiary, and quaternary configurations are excluded either because they permit too few favorable interactions or because they would require too many unfavorable physical or electrostatic interactions. Thus, the nonpolar amino acids of hemoglobin and myoglobin are found in the interior of those two proteins, where they can form hydrophobic bonds with each other but avoid an aqueous solvent. Conversely, the polar side chains of hemoglobin and myoglobin are largely on the surface, where they can interact with water, lowering the total energy and rendering the proteins soluble. We might expect, then, that proteins in a nonpolar, lipid environment (e.g., membranes) would have a distinctly hydrophobic exterior. And that is, in fact, what we observe.

It is common for a protein to have several conformations of almost equivalent energy, accessible to each other through small energy barriers. Such proteins exist as an equilibrium mixture of several variants, called *conformational isomers*, or *conformers*. The fraction of molecules in each conformation will be a direct result of the energy differences between conformations, and is thus predicted by the Boltzmann equation (Chap. 2). Furthermore, environmental conditions can cause a shift in the equilibrium concentration of the different conformers. Hemoglobin, for example, has at least two stable conformations, with an equilibrium that is shifted toward one or the other according to whether oxygen is bound to the hemes (see Fig. 3-21).

A polypeptide chain is polymerized one amino acid at a time *in vivo* (starting from the amino end, as we shall see). This factor, too, might influence a protein's final configuration. That is, since the amino end is free to find a stable configuration before the carboxyl end is even made, that factor may have some influence on the nature of the structure that eventually emerges.

And finally, one should be aware that a protein may be chemically altered after synthesis. Some proteins, for instance, are synthesized in an inactive form (a *proenzyme* or *zymogen*) and then "turned on" by a specific proteolytic cleavage. Insulin is one of these, for it is synthesized as proinsulin, a protein containing 84 amino acids in a single chain. Proinsulin has no biological activity, but it becomes insulin when a 33-amino acid section is removed by a selective hydrolysis. The 30 residues left at the amino end of the original chain become the B chain of insulin (see Fig. 3-24), while the carboxyl end of 21 amino acid residues yields the A

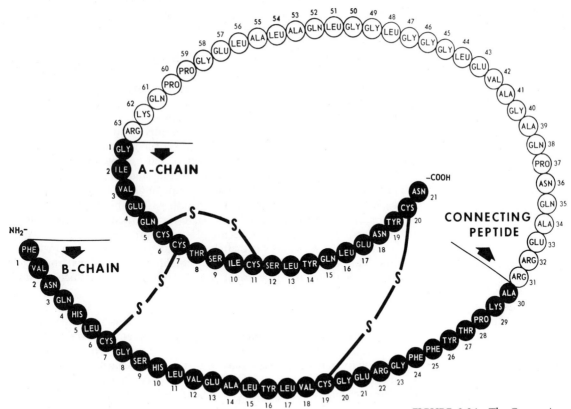

FIGURE 3-24 The Conversion of Proinsulin to Insulin. The molecule is synthesized as a single polypeptide chain of 84 amino acids. It is activated by removing a 33-peptide fragment connecting the A and B chains. [Courtesy of R. E. Chance, *Recent Prog. in Hormone Res.*, **25**: 272 (1969), © Academic Press, N.Y.]

chain. The spatial conformation of insulin, then, is determined by that of proinsulin, since interchain and intrachain disulfide bonds prevent any major changes in shape upon activation.

A similar situation is encountered with collagen, for the protein is synthesized and secreted as *procollagen*. In the extracellular space, enzymes (proteases in this case, for they cleave proteins) remove a piece from both the amino and carboxyl ends of the molecule. This removal alters the molecules in such a way that spontaneous formation of collagen fibrils is possible. This step explains why the fibrils form only outside, and not inside, cells. The importance of this conversion is pointed out by an "experiment of nature" in which certain strains of cattle lack the enzyme necessary to remove the amino terminal piece from procollagen, preventing accurate fibril formation. Such animals are afflicted with the inherited disease called *dermatosporaxis*—"torn skin," a name that tells you what to expect in this condition. A much milder but comparable inherited disease of humans, one of the forms of Ehlers-Danlos syndrome, has a similar defect.[2]

Structural Stability. Protein structure is a rather fragile thing. When a protein loses its biological activity owing to changes in any of the higher levels of structure, we say that it is *denatured*. Denaturation can be accomplished by changes in solvent, pH, or temperature, or often by simple physical abuse. For example, much of the white of an egg is ovalbumin, a single-chain protein of 42,000 daltons. When egg white is cooked, or even beaten hard, ovalbumin and the other proteins lose their compact globular structure owing to the disruption of the weak forces responsible for that structure. In other words, they denature. In an extreme case, the unfolded polypeptide chains may get tangled enough to form a gel, as in gelatin and meringue, or cooked egg white.

Solvent or pH changes can denature proteins by altering side-chain interactions. Changes in pH obviously affect the magnitude of the charge on ionized groups, creating new attractive and/or repulsive forces in the molecule. Because of these forces, most proteins are biologically active only in a relatively narrow pH range around neutrality (e.g., 6–8). Typically there is a well-defined pH optimum within this range, deviations from which result first in a loss of activity and then in an unfolding of the polypeptide chains as extreme pHs are approached.

Solvent changes may have effects similar to those of pH changes. A high concentration of salt tends to shield charged groups from each other, changing the pattern of ionic bonding. Agents such as urea

$(H_2N-\overset{\overset{O}{\|}}{C}-NH_2)$ tend to alter existing hydrogen bonds by substituting for one of the hydrogen-bonded groups. Since urea can hydrogen bond both to water and to groups on the protein, the energetic advantage of maintaining the native structure may be lost in its presence. As a result, the most stable conformation may become an open ("random") coil, which allows a maximum number of interactions with the solvent.

Hydrophobic solvents, on the other hand, may denature proteins by interacting favorably with nonpolar side chains and unfavorably with polar side chains. In water-soluble proteins, nonpolar side chains are usually on the inside, away from the solvent, as they are in myoglobin and hemoglobin, while polar ones are largely on the surface. In this situation, the effect of nonpolar solvents may be literally to turn the protein inside out.

Solution Properties. The question of the solubility of proteins is important to us, for they account for a very large fraction of cytoplasmic solutes. If we cannot predict the behavior of proteins in solution, we can have little hope of understanding the cytoplasm. One sees reference to proteins as colloids, implying that they form aqueous suspensions (i.e., two phases, liquid and solid) instead of solutions. However, S. P. L. Sörenson proved in 1917 that proteins do, in fact, form true solutions, by every thermodynamic criterion.

We can understand Sörenson's conclusions if we remember that the solubility of a substance is determined by the kinds and number of interactions that it makes with a solvent. As noted earlier, many of the amino acid side chains are polar, so that when most of the surface of a protein consists of these polar side chains, water solubility is a natural consequence. In fact, some proteins will form solutions as concentrated as 5–10% by weight, and more. Although very large aggregates of soluble proteins (e.g., viruses) may tend to settle noticeably toward the bottom of a container, settling will stop when the particles have assumed the distribution predicted by Boltzmann's equation. However, if the tendency to concentrate near the bottom causes a protein to reach its solubility limit—the concentration at which unlimited aggregates form—a precipitation will take place.

Even if we concede that proteins can form true solutions, it does not follow that the cytoplasm is a true solution. We know, for instance, that the ionic content of the cytoplasm is quite high, and that proteins tend to precipitate in high concentrations of salt. We also know that a substantial fraction of the water in cytoplasm is bound or structured by the presence of various ionic and hydrophobic groups, and is therefore not available as a solvent in the usual sense (i.e., its "activity" is less than its concentration). So we do have to exercise some caution when trying to predict the cytoplasmic behavior of proteins from studies carried out on dilute solutions *in vitro*. Nevertheless, we have every reason to believe that the behavior of proteins and other cytoplasmic components in relatively dilute solutions in the laboratory accurately reflects, at least qualitatively, their behavior *in vivo*.

3-6 MACROMOLECULAR ASSEMBLY

It may seem like a long way from the individual molecules of the types just surveyed to the construction of a complete cell with all its intricate appearing components. Work done in recent years, however, makes it clear that the supramolecular assemblies observed in the electron microscope in most cases form spontaneously from their macromolecular components. That is, it appears that association into larger structures is a natural consequence of the way in which macromolecules are built. This is a concept to which we shall return many times in future chapters. It is the purpose of this section to introduce the principles of macromolecular assembly.

Assembly Mechanisms. Three routes of assembly are recognized: (1) *Template assembly* involves a master or mold, the structure of which determines the nature of the copies. This mechanism is not found widely in nature, but there are a few exceedingly important examples involving the synthesis of the nucleic acids DNA and RNA. These examples will be discussed in detail in Chap. 5, and hence will not be dealt with here. Suffice it to say that in the replication of DNA or the synthesis of RNA, the order of nucleotides is precisely determined by the strand of DNA being used as a template.

(2) *Enzymatic assembly* is not a good term since enzymes are involved directly or indirectly in all modes of assembly, but we refer here to situations where units are added to a structure according to which enzyme is available. That is, the choice is not due to information specified by a template or due to the properties of the monomer units themselves. We have encountered several examples of these situations, including the synthesis of cellulose, glycogen, and starch. Cellulose, glycogen, and starch, you will recall, are different structures formed from the same basic building block, D-glucose. With one set of enzymes directing the assembly, the insoluble structural polymer, cellulose, will result. With a different set of enzymes, a soluble, branched storage polymer of glycogen or starch will be made. Another example of enzymatic assembly was encountered with the glycoproteins, where it was pointed out that the choice of monosaccharide unit—or whether there would be any at

all—depended on the presence of specific enzymes. Many additional examples of this type of assembly could be cited, and some more will be later. However, it is more important at this point to turn to the third mechanism of construction, that of self-assembly.

Self-assembly. This term refers to the generally spontaneous aggregation of subunits, a phenomenon that is widely found in the assembly of multisubunit enzymes, bacterial flagella, microtubules, microfilaments, membranes, and most of the other structures visible to the microscopist. Protein folding, already discussed, is one example. The assembly of collagen is another, some of the features of which have been mentioned.

In self-assembly, the subunits are constructed in a way that gives them an affinity for one another. This may mean specifically a side-to-side association, an end-to-end association, or something relatively more complex as seen, for example, in the structure of hemoglobin.

Nucleation is a general, but not universal, feature of self-assembly. By this is meant the situation where assembly is slow or haphazard except in the presence of an initial formation to which additional subunits may be added. The formation of organic and inorganic crystals from supersaturated solutions is an example already familiar to you. Another is the growth of bacterial flagella, which are ropelike strands of identical protein subunits extending from a membrane-attached basal body and "hook." Growth proceeds only from the far end of the hook which, since it is outside the cell, means that the individual protein subunits, called *flagellin*, must be extruded by the cell in order for growth of the flagellum to occur. Purified flagellin assembles rapidly onto the ends of isolated hooks *in vitro*, and onto membrane-bound hooks in mutant cells that are unable to make flagellin.

Once self-assembly is initiated in a system, it characteristically proceeds at a rate that depends only on environmental factors such as temperature, pressure, pH, and concentration of available subunits.

The Driving Force for Self-assembly. Self-assembly, in most cases, is a reaction having a positive enthalpy (i.e., it is endothermic, or absorbs heat). Enthalpy change (ΔH) can be determined in the laboratory by noting the small decrease in temperature as the reaction proceeds. In order for such a reaction to occur spontaneously—that is, in order for ΔG to be negative in the equation

$$\Delta G = \Delta H - T\Delta S$$

the entropy term, ΔS, must be positive. However, since the subunits themselves are getting more highly ordered ($\Delta S < 0$) as polymerization proceeds, some other substance must be getting more disordered. That substance, as it turns out, is water bound to the individual subunits and released during polymerization. Self-assembly is, in other words, generally an entropy-driven, hydrophobic interaction among subunits.

Tobacco mosaic virus is an example of a structure held together by hydrophobic bonds. TMV is a helical assembly of about 2100 identical coat protein subunits, each of molecular weight 17,500 daltons. ΔH was found to be about +120 kJ per mole of subunit added and ΔS about +415 joules per deg-mole for the initiated self-assembly reaction under one set of experimental conditions. From these figures, we find that at 37 °C (310 K, human body temperature)

$$\Delta G = \Delta H - T\Delta S = 120{,}000 - (310)(415) = -8.6 \text{ kJ/mole}$$

whereas at 10 °C

$$\Delta G = 120{,}000 - (283)(415) = +2.5 \text{ kJ/mole}$$

In other words, assembly proceeds spontaneously at 37 °C, but the assembled rod dissociates again if the temperature is lowered to 10 °C where ΔG is positive (see Fig. 3-25). The break-even point under these conditions (i.e., where $\Delta G = 0$) is about 16 °C: above that temperature equilibrium favors polymerization; below 16 °C, equilibrium favors depolymerization.

The assembly of TMV is actually a good deal more complicated than indicated here. It involves initiation by preformed discs of subunits, and interactions between protein and an RNA strand that follows the subunits up the helix at a distance of about 40 Å from the axis. The 6400-nucleotide length of the RNA determines the 300 nm length of the completed TMV cylinder. For details, readers are referred to the monograph by Max Lauffer of the University of Pittsburgh, from whose laboratory much of the work on this system has come (see References). Instead of going into more detail on TMV, let us turn instead to a very similar, though more recently elucidated, system of enormous importance to cell biology, that of microtubule assembly.

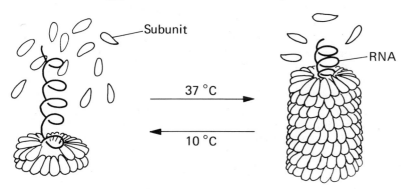

FIGURE 3-25 Assembly of TMV. Tobacco mosaic virus assembles spontaneously from subunits and RNA at 37 °C but dissociates again when the temperature is lowered.

Reversible Self-assembly: Microtubules. Microtubules were introduced in Chap. 1 as 250-Å diameter cylinders having usually 13 "protofilament" strands running longitudinally (12 or 15 in special cases). Microtubules have both structural and motile roles in cells of many types. A bundle of microtubules near the periphery, for example, provides the oval shape of nonmammalian vertebrate red blood cells. Microtubules also run parallel to the axis of nerve processes, where they are involved in both structure and transport; they are also the spindle fibers used to separate chromosomes during nuclear division; and they are responsible for both the structure and motility of eukaryotic (not prokaryotic) cilia and flagella, among many other functions.

Microtubules are comprised basically of two very similar kinds of protein subunits called α and β *tubulin*, each with a molecular weight of about 55,000 daltons. These proteins apparently come in several versions in each cell, which means that not all microtubular structures have interchangeable components. In all cases, however, the α and β subunits have significant affinity for each other so that most higher orders of structure are assemblies of α, β dimers.

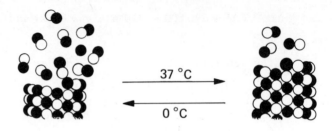

FIGURE 3-26 Assembly of Microtubules. The structure elongates spontaneously at higher temperatures by addition of tubulin dimers. Lowering the temperature causes dissociation again.

Tubulin polymerizes spontaneously at body temperature (37 °C) to form the 13-protofilament structure seen *in vivo*, but depolymerizes again at 0 °C to tubulin dimers (see Fig. 3-26). This reversible process has been used to permit tubulin isolation—centrifugation at 37 °C brings down microtubules, leaving behind small molecular weight debris, while centrifugation at 0 °C brings down large particles, leaving behind tubulin dimers. Repeated cycles result in a preparation containing tubulin and a small amount of *microtubule-associated protein* (MAP).

When the MAPs are removed, purified tubulin still polymerizes at 37 °C but only very slowly. Readdition of a small amount of MAP results in rapid polymerization, implying a nucleation function for the MAPs which, as we shall see in Chap. 9, probably have other functions as well. Other factors required for assembly are the nucleotide GTP, and the divalent ions Mg^{2+} and Ca^{2+}. One molecule of GTP and one Mg^{2+} bind tightly to each dimer and in so doing appear to cause a conformational change in the protein that favors polymerization. The GTP may be hydrolyzed to GDP in the process of polymerization, though the significance is not understood. The role of Ca^{2+} is likewise unclear, although the observation that concentrations above 1 mM (10^{-3} M) cause depolymerization instead of polymerization has led to suggestions that the concentration of free Ca^{2+} regulates assembly *in vivo*.

With the right environment, assembly of microtubules seems to involve a slow initiation step, resulting in a few nucleation centers, followed by rapid elongation. The process is an equilibrium between dimer and polymer, with the tide tipped toward polymer at higher temperatures and toward dimer at lower ones. This feature, of course, is characteristic of hydrophobic, entropy-driven reactions of the type seen with tobacco mosaic virus.

The question of nucleation is of particular importance, for microtubules are not found helter-skelter about the cell, but are organized in specific patterns designed to carry out specific functions. Spontaneous nucleation, as seen *in vitro*, probably does not occur *in vivo*. Rather, initiation of assembly occurs at *microtubule organizing centers* (MTOC). Basal bodies and centrioles are readily recognized examples—indeed, it has been shown that basal bodies isolated from an alga, *Chlamydomonas*, can organize the assembly of tubulin from chick brain, emphasizing the evolutionary conservation of these materials and assembly mechanisms. Other MTOCs exist in membranes, at the poles of mitotic spindles in dividing cells that do not have centrioles, on chromosomes (at a part of the chromosome called a *kinetochore*), and in probably many other places as well. How these organizing centers turn on and off so that assembly occurs only at the right time in the life of the cell is unclear, but may involve changes in the centers themselves, changes in Ca^{2+} concentration, involvement of MAPs, or all three.

Many of the functions involving microtubules have been identified by using drugs that specifically bind to unpolymerized tubulin, shifting the equilibrium toward disassembly *in vivo* and thus halting microtubule-associated processes. The drugs most used are *colchicine* and its derivative *colcemide*; the vinca alkaloids *vincristine* and *vinblastine*; and another plant product called *podophyllotoxin*. There is one binding site on each tubulin dimer capable of reversibly accepting a molecule of colchicine or podophyllotoxin, and another site capable of accepting a vinca alkaloid. When either site is occupied the tubulin dimer is removed from the dimer-polymer equilibrium. The effect is to prevent polymerization or to depolymerize microtubules already formed. Any of these drugs, for example, could be used to prevent cell division by interfering with the spindle apparatus needed to separate chromosomes. In fact, this ability to prevent cell division makes the vinca alkaloids useful in medicine as anticancer drugs. Colchicine, too, is widely used in medicine, but as an anti-inflammatory agent effective in treating gout, again because of its effects on microtubules.

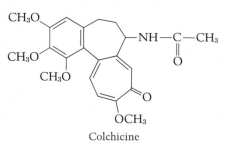

Colchicine

Irreversible Self-assembly: Collagen, Elastin, and Fibrin. Certain of the features of collagen have already been introduced: Collagen is composed of tropocollagen monomers, each a triple helix of three polypeptide chains; tropocollagen is a glycoprotein containing galactose and glucose attached to hydroxylysines; tropocollagen is derived from procollagen by cleavage at both ends of the latter prior to its polymerization; and the tropocollagen subunits are covalently cross-linked by the action of an enzyme, lysyl oxidase.

The actual polymerization of tropocollagen is a hydrophobic, entropy-driven process similar to that responsible for assembly of bacterial flagella, tobacco mosaic virus, or eukaryotic microtubules. Tropocollagen from rabbit skin, for example, is soluble at 4 °C but forms typical collagen fibers at 37 °C, only to disperse when the temperature is lowered again. The staggered side-by-side association that produces the unique appearance of collagen is thus spontaneously formed. It is a structure held together at first by weak chemical bonds and only later stabilized covalently.

Collagen synthesis and assembly illustrate several features not found in the other polymer systems discussed. First, the triple helix of the procollagen molecule is spontaneously formed *in vivo* probably because of the extensions, called "registration peptides," at either end (see Fig. 3-27). These extensions, later removed, appear to nucleate the formation of the triple helix. They serve yet another function in keeping procollagen from polymerizing to form collagen. This is an important feature, for procollagen is an intracellular molecule and collagen an extracellular structural protein.

Once outside the cell, cleavage of the registration peptides takes place (via an enzyme also secreted by the cell) and assembly of collagen fibers occurs. However, the biological role of collagen requires a strong, stable structure. Hence, the last step is stabilization via covalent linkages between tropocollagen molecules. The result is a fiber with the necessary characteristics of strength and resistance to degradation. Its half-life in adult rats is upwards of a year, in contrast to hours or days for most proteins. There is a normal slow turnover of collagen fostered by the presence of a specific destructive enzyme called collagenase, and a more rapid turnover in certain situations—in young animals during growth,

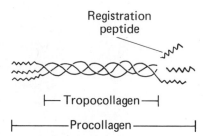

FIGURE 3-27 Tropocollagen and Procollagen. Registration peptides probably aid in spontaneous formation of the triple helix. Their removal converts procollagen to tropocollagen, which is able to assemble spontaneously to collagen. The fibers are then stabilized by covalent cross-links.

wound healing, involution of a uterus following delivery of a child, and also in a few diseases. However, in comparison to other proteins collagen may be considered metabolically inert, and the self-assembly practically irreversible once stablized by cross-links.

An identical series of steps results in the assembly of the connective tissue protein elastin: synthesis as protoelastin, a precursor having a high concentration of repeating peptide units involving glycine and proline; hydroxylation of prolines (but not lysines); secretion from the cell followed by proteolytic cleavage to form tropoelastin; spontaneous self-assembly and covalent cross-links to stabilize the structure. Fewer details are known about elastin than about collagen synthesis, but the major difference appears to be a different type of cross-link (carried out by the same collagen–cross-linking enzyme, lysyl oxidase) and more extensive cross-linking to form a rubberlike polymer instead of the strong fibers of collagen.

A third system that follows this general scheme is fibrin synthesis and assembly. Fibrin is the netlike polymer responsible for blood clots (Fig. 3-28). The protein precursors, *fibrinogen*, circulate in the blood until acted on by the proteolytic enzyme, *thrombin*, whereupon the newly created fibrin monomers polymerize. Stabilization is via isopeptide bonds created by a cross-linking enzyme called factor XIII, itself activated by thrombin. (The key to clotting, then, is the production of active thrombin via mechanisms that will not be considered here.) The major departure of the fibrin system from that of our collagen prototype is in the polymerization, for it is not a result of entropy-driven, hydrophobic interactions. Careful heat measurements reveal that fibrin assembly is strongly exothermic ($\Delta H < 0$). Assembly is presumably due to the formation of hydrogen bonds rather than hydrophobic bonds between monomers.

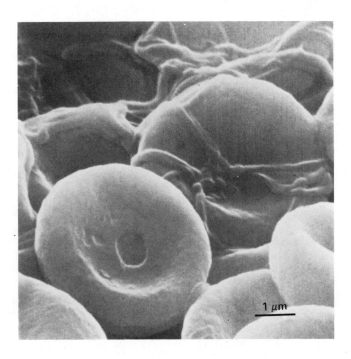

FIGURE 3-28 Fibrin. A clot is formed when the soluble protein fibrinogen is converted into the insoluble, fibrous protein called fibrin. Erythrocytes are seen here, caught in a fibrin mesh. (Scanning electron micrograph courtesy of Thomas L. Hayes and Lawrence-Berkeley Laboratory, Univ. of Calif.)

SECTION 3-6
Macromolecular Assembly

Lipid Self-assembly: The Liposome. The principle of entropy-driven, hydrophobic assembly, though not universal as we have just seen, is widespread and not limited to protein polymers. It describes also the assembly of polar lipids (e.g., phospholipids) into membranes. That phenomenon, an understanding of which is crucial to the study of cell biology, will be introduced here with a synthetic model system, the liposome.

Note that the major membrane lipids (Fig. 3-2) have a polar end and a hydrophobic end. The behavior of such substances in water follows that described for fatty acids in Chap. 2—namely, the hydrophobic ends aggregate to form the interior of a spherical micelle having a polar (water-attracting) surface. Other structures are also possible, however.

In particular, when dried amphipathic lipids are exposed to an excess of water under the proper conditions, they spontaneously arrange in extended bimolecular leaflets with hydrophobic chains on the interior and hydrophilic "heads" on the surface. Such leaflets form vesicles up to 2 μm in diameter. They are typically multiple layered (Fig. 3-29), with each layer a continuous bimolecular sphere separated from the next layer by an aqueous space. The driving force for assembly is the entropy-driven aggregation of the hydrophobic groups, just as it is in the formation of micelles. Such structures, which are called *liposomes*, may be broken up mechanically into single-layered vesicles, some of which are only 25 nm in diameter.

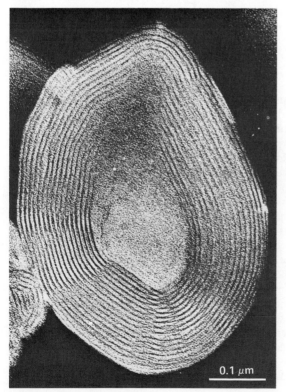
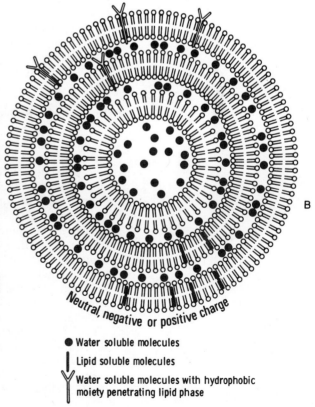

FIGURE 3-29 The Liposome. LEFT: Electron micrograph of a single multilamellar liposome with water-soluble stain between the layers. RIGHT: Schematic of the liposome. These bodies can also be converted to vesicles having a single lipid bilayer wall. [Courtesy of G. Gregoriadis. Reprinted by permission from the *New Engl. J. Med.*, **295**: 704 (1976).]

Liposomes are important to cell biology as models of cells, which are similarly surrounded by bimolecular lipid leaflets. Their usefulness is enhanced by the fact that the aqueous interior is composed of whatever solution was used to disperse the lipid. This solution is cut off from the surrounding solvent by the lipid leaflet. By centrifuging or filtering the liposomes and suspending them in a solution of new composition, the permeability (ease of penetration) of bimolecular lipid leaflets to various solutes can be studied.

Liposomes have also attracted much attention from the medical field in recent years, for the particles are taken up intact by various cells of the body. Thus, the interior of liposomes forms a protected environment into which drugs or other therapeutic substances can be sequestered until transferred directly to a target cell, never to be exposed to degradation by elements of the blood or even, for small liposomes, by digestion. Therapy of metabolic diseases involving single protein deficiencies is a promising application. Administration of insulin to diabetics by mouth instead of by injection, or replacement of faulty enzymes in lipid storage diseases, are two possibilities.

3-7 AN INTRODUCTION TO MOLECULAR SPECIFICITY: THE ANTIBODIES

An important feature of macromolecules is the enormous variety that is possible in a structure comprised of so many smaller units. When these units are arranged in discrete spatial configurations, as they are in proteins, highly specific interactions with other molecules become possible. This idea was implicit in our discussion of protein assembly. What we wish to do in this section is to expand the concept by using as a focal point for discussion those proteins in which the capacity for specific interactions with other molecules was first recognized, namely the antibodies.

The Antibody Response. People have long speculated on the reason why exposure to certain diseases seems to be associated with a prolonged, even lifelong, immunity to reinfection. The observation led to the practice of intentional exposure of children to diseases known to be less severe when contracted young (e.g., smallpox), but it was not until the closing years of the eighteenth century that the phenomenon was exploited by vaccination.

An English physician, Edward Jenner, took note of the fact that milkmaids seldom contracted smallpox (variola), and proposed that immunity to this serious human disease might result from contracting the benign disease, cowpox. Accordingly, in 1796, he inoculated a young boy with material obtained from a cowpox vesicle on a milkmaid's hand. The next year, the boy was inoculated with material that would have normally produced smallpox itself, but he failed to contract the disease.

Jenner's experiments led to his famous publication, "Inquiry into the Cause and Effects of the Variolae Vaccinae," and initiated the practice of vaccination. (The term is derived from *vacca*, the Latin word for cow.) Mass vaccinations against smallpox began in England in 1803, and less than two years later the death rate from smallpox had fallen by two-thirds. Although it was nearly a century before the technique could be repeated with another human disease, rabies, vaccination was thus established as a safe and efficient way of inducing a state of immunity.

Jenner, of course, did not know the mechanism by which he was able to induce immunity, and in fact the existence of pathogenic microbes had not been seriously proposed at that time. But we now know that vaccination results in the production of *antibodies*, which are serum glycoproteins capable of binding specifically to substances present in the vaccine.

Almost any material that is injected will cause the production of antibodies specific for it. Thus, antibodies produced after vaccination with cowpox virus are not the same as those produced after injection of an attenuated rabies, or measles, or polio virus, and so on. Therefore, protection against one of these diseases through vaccination ordinarily has no effect on one's susceptibility to the others.

The production of antibodies, however, is not the only mechanism for establishing immunity. Immunity comes in two forms, *humoral* and *cellular*, distinguished by whether or not circulating antibodies appear in the blood and lymph (i.e., in the "humor"). Either reaction may be triggered by a wide variety of substances described, in this context, as *antigens*. Both reactions are due to special properties of certain white blood cells called *lymphocytes* (Fig. 1-13b).

Specifically, antibodies are produced by cells called *plasma cells* that are differentiated descendants of the so-called type B lymphocyte. The other class of lymphocyte is the T cell. (T stands for thymus, B for bone marrow or, in birds, bursa of Fabricius.) T cells attack antigens directly at close range only; B cells attack antigens by secreting antibodies. Both cells are highly specific in their response.

Antibody-Antigen Interaction. A week or so after an animal becomes infected, or after foreign material is injected into it, antibodies appear in the blood. If serum or plasma from this animal is mixed with additional antigen in a test tube, a visible precipitate often forms.[3] The precipitate from an antibody-antigen reaction carried out in the laboratory may be separated from the plasma by settling or centrifugation. Sometimes the precipitate can then be redissolved in a high concentration of salt or in a solution that is decidedly alkaline or acidic, yielding a mixture of antibodies and the foreign material. The mixture may then be separated by standard laboratory procedures, so that a greatly purified antibody preparation can be examined.

When the above procedure is carried out, one usually finds that the antibody is a protein of the gamma globulin class, with a molecular weight of about 150,000 daltons distributed over four polypeptide chains. Its structure consists of two heavy, or "H" chains, each with a mass of about 50,000 daltons, and two lighter, or "L" chains of about 25,000 daltons each. The chains, which are held together by disulfide (—S—S—) bridges, are folded into a molecule that is shaped like a "Y" or a "T" (Fig. 3-30). At the end of each arm of the Y is an area—the only area on the molecule—that is capable of attaching specifically to the antigen. Because there are two binding sites, two antigens can be cross-linked by a single antibody. If the antigen is also capable of accommodating more than one antibody, an interconnected net of enormous size can be generated, causing precipitation (see Fig. 3-31).

There is nothing mysterious about the antigen-antibody interaction.

SECTION 3-7
An Introduction to Molecular Specificity: The Antibodies

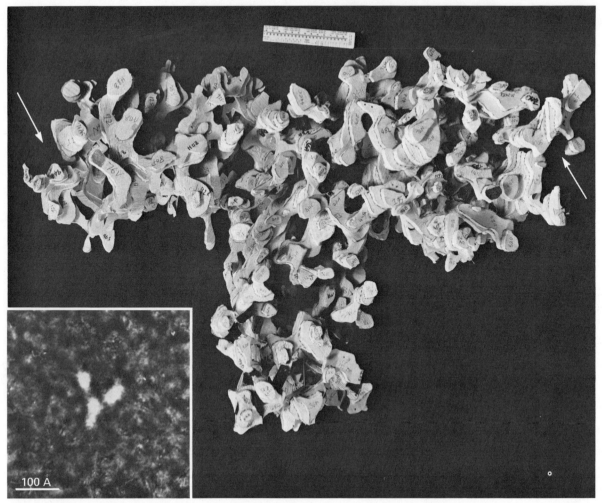

FIGURE 3-30 The Antibody Molecule. A model of a human antibody (class IgG) based on X-ray diffraction studies. Antigen-combining sites (*arrows*). INSET: Electron micrograph of a single antibody of the class known as IgA, isolated from human colostrum (the first milk after birth). [Micrograph courtesy of S-E Svehag. Similar to Svehag and Bloth, *Science*, **168**: 847 (1970). Model courtesy of V. Sarma, E. Silverton, D. Davies, and W. Terry, *J. Biol. Chem.*, **246**: 3753 (1971).]

The types of forces involved are those discussed in Chap. 2. If the antigen is hydrophobic, one will find hydrophobic amino acid residues on the antibody at its binding site. If the antigen is positively charged, the antibody will have negatively charged amino acids at its binding site, and so on. The combination of several weak bonds of this type can produce quite strong interactions, so that equilibrium between unbound and bound antigen lies very far toward the latter. In other words, when antigen and the right antibody are present together in solution, the antigen reacts quickly and completely to form antigen-antibody complexes.

Contributions of the Antibodies to Defense. The antibody-antigen reaction contributes to defense against infection in several ways: (1) The invasiveness of many pathogens, especially viruses, may be effectively neutralized by combination with antibodies. In the case of viruses, the reaction apparently interferes with the mechanism of attachment to host cells. (2) Agglutination or precipitation of an antigen has the obvious advantage of gathering foreign material into clumps, preventing its spread. (3) Antibody-antigen complexes are readily ingested and destroyed by

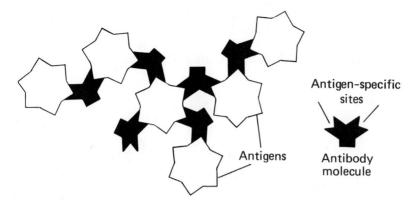

FIGURE 3-31 Antibody-Antigen Reaction. Specific binding between antigen and antibody can lead to extensive cross-linking and subsequent precipitation.

certain scavenger cells. The complex not only presents a larger target, but seems actually to stimulate uptake. (4) Formation of antibody-antigen complexes may trigger other reactions, caused by components of *complement*.

Complement is a collection of 15 or more serum proteins. Although they are globulins, they are not antibodies, nor do their serum concentrations change during the course of an immunization. The various components of complement have different effects when activated. Some destroy the membranes of cellular antigens, causing lysis. Others attract scavenger cells capable of ingesting the antigen-antibody complex. These and other activities are initiated by the binding of certain components of complement to an antigen-antibody complex (specifically, to a portion of the antibody itself) and the removal of a peptide from the complement molecules so bound. This process, called *complement fixation*, leads to activation of the bound complement. These molecules in turn set off a chain reaction that activates other components.

Haptens and Antibody Specificity. When an isolated antibody preparation is subjected to careful physical analysis, it is invariably found to be heterogeneous. That is, the antibody molecules do not all have the same primary structure. Antibody heterogeneity may reflect the fact that the antigen, which is usually a protein or a large polysaccharide, contains a variety of possible antigenic sites, each of which is no more than an area of a given size located on an accessible portion of the molecule. But, as we shall see, there is also a natural heterogeneity to antibodies even when they are all specific for the same antigenic site. This fact became clear as ways were found to elicit antibodies against small, chemically defined antigens, rather than large proteins and polysaccharides.

The ideal antigen for laboratory studies would be a small molecule like benzoic acid. Although small molecules (less than about 1000 daltons) are not antigenic by themselves, we know very well that simple organic molecules often cause allergic reactions of either the cellular or humoral type, and that humoral allergies are caused by the production of antibodies against the organic compound. Investigations by Karl Landsteiner in the 1910s revealed that small molecules can act as allergens when they are attached to larger molecules such as proteins. (The attachment can be to one's own proteins.) It is the complex that is antigenic.

When a small organic molecule is made antigenic by chemically coupling it to a larger molecule, the small molecule is called a *hapten*.

Benzoic acid

(The word "hapten" means, approximately, "to be stuck on.") Studies of the specificity of antigen-antibody reactions often use benzoic acid derivatives as haptens by coupling them to proteins at a tyrosine, histidine, or free amino group in the following kind of reaction:

$$H_2N-C_6H_4-COO^- \xrightarrow{\text{nitrous acid}} {}^+N\equiv N-C_6H_4-COO^- \xrightarrow{\text{protein}} \text{protein}-C_6H_3(OH)-N=N-C_6H_4-COO^-$$

p-Aminobenzoate → p-Benzoate-diazonium salt → p-Azobenzoate protein (tyrosine linkage)

When such a complex is injected into an animal, it is usually found to be antigenic. Some of the antibodies produced will be to the protein carrier, some will recognize the combination of carrier and hapten, and others will recognize hapten alone. When this latter group is isolated, rather good measurements of binding ability and antibody heterogeneity can be made.

By immunizing with one hapten and then testing structurally related molecules for their ability to compete with the hapten and thus inhibit precipitation of the protein-hapten complex, one can get an idea of the size and specificity of the antibody combining site. Some typical results are shown in Table 3-8, where the binding constant of various test haptens is given as K, arbitrarily related to the use of benzoate as a test hapten. The higher the value of K, the tighter the binding. Note, for example, that binding is not simply due to the charge on the hapten, for substitution of one negative group for another makes an enormous difference. The antibody combining site, it appears, rather selectively recognizes the shape of the hapten so that antigen and antibody fit together like a hand and glove. At the same time, however, the binding is relatively tight for quite a variety of chemical groups. Hence, the fit doesn't have to be perfect. It is probable, therefore, that the number of different antibodies needed by an animal is much less than the total number of potential antigens.

The specificity of antibodies for the antigen that induced them has led to the use of antibodies as a way of defining similarities and differences in the structure of biological macromolecules, particularly proteins. Thus, we may see a phrase such as "immunologically distinct," meaning that antibodies made in response to challenge by one antigen do not *cross-react* with the comparison antigen. Conversely, two substances might be described as "antigenically identical" or "immunologically similar" and so on. These last two phrases merely mean, respectively, that antibodies made by the test animal after injection with one antigen react either completely or partially with the other antigen.

Antigenic similarities or differences reflect the arrangement of groups on the surface of the antigen. Antigens that cross-react with the same antibody have similar patterns of surface groups—for example, human and monkey serum albumin are closely related proteins that cross-react strongly as do duck and chicken ovalbumin, and so on. On the other hand, proteins that do not cross-react at all must have very different surface features.

TABLE 3-8 Antibody Specificity

Rabbits were immunized with the *p*-azobenzoate hapten:

protein—N=N—C₆H₄—COO⁻

Test hapten		K (relative to free benzoate)	Test hapten		K (relative to free benzoate)
p-(*p*'-Hydroxyphenylazo)-benzoate	HO–C₆H₄–N=N–C₆H₄–COO⁻	34[a]	*p*-(*p*'-Hydroxyphenylazo)-benzenesulfonate	HO–C₆H₄–N=N–C₆H₄–SO₃⁻	0.12
p-Nitrobenzoate	O₂N–C₆H₄–COO⁻	11.5	Phenylacetate	C₆H₅–CH₂–COO⁻	0.002
Benzoate	C₆H₅–COO⁻	1.0 (arbitrary)	Acetate	CH₃–COO⁻	<0.001
m-Nitrobenzoate	*m*-O₂N–C₆H₄–COO⁻	0.40	Benzenesulfonate	C₆H₅–SO₃⁻	<0.001

Data from various papers by David Pressman, and collected by D. Pressman and A. L. Grossberg in *The Structural Basis of Antibody Specificity*. New York: W. A. Benjamin Inc, 1968.
[a] Note that this hapten most closely resembles the immunizing antigen, since *p*-azobenzoate will usually be coupled to an aromatic amino acid.

Antibody Structure. The heterogeneity of antibodies was long an obstacle to any detailed studies of their structure. Such techniques as amino acid sequencing and X-ray crystallography require a homogeneous preparation of protein. The first breakthrough came in 1959 when R. R. Porter noted that limited digestion of semi-purified antibody with the protein-degrading enzyme papain produced three fragments (see Fig. 3-32). One of these fragments, representing roughly a third of the molecule, would crystallize. Hence this fragment (called *Fc*) is constant in its amino acid composition from molecule to molecule. The other two fragments are identical to each other, but represent the same degree of heterogeneity as the original antibody population. Each of these *Fab fragments* turned out to contain one complete antigen combining site.

The Fab and Fc fragments have been enormously useful in helping to localize function to various parts of the molecule. However, a really detailed analysis, and the X-ray crystallography on which Fig. 3-30 is based, became possible only when it was discovered that a certain disease, *multiple myeloma*, results in the production of large amounts—up to 10% by weight in the serum—of the same antibody molecule.

Multiple myeloma is a malignant condition in which there is uncontrolled proliferation of antibody-producing cells. The cells are malignant, but there is nothing obviously different about the antibody that they produce, even though one does not usually know what antigens are bound by it. In addition, multiple myeloma patients also excrete as much as a gram or two a day of another protein, called *Bence-Jones protein* after the nineteenth-century physician who discovered it. The Bence-Jones protein, as it turns out, is identical with the light chains of the same patient's myeloma proteins.

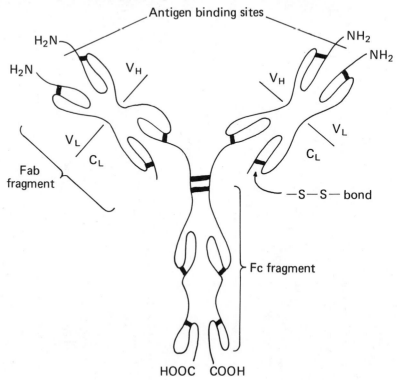

FIGURE 3-32 Antibody Structure. Variable regions of the heavy and light chains are designated by V_H and V_L, respectively; constant region of the light chains by C_L (see p. 125). Domains are indicated by showing the chain folded back on itself, stabilized with disulfide bonds. Papain produces the F_{ab} and F_c fragments.

In examining the amino acid sequence and X-ray diffration pattern of multiple myeloma proteins, the following facts have emerged:

(1) Only about half of each heavy chain and a fourth of each light chain, called the *variable region* (about 110 amino acids in each case), contribute to binding specificity (Fig. 3-32). The rest of the molecule, called the *constant region*, is remarkably the same from one antibody specificity to the next and, with the exception of minor inherited differences in the constant regions called *allotypes*, even from one individual to the next within a species.

(2) Within the variable region, there are *hypervariable segments* representing only a fraction of the amino acids. The variable regions are folded in such a way that these hypervariable segments are brought together to form a groove, cleft, or pocket representing the actual binding site. Thus there is a good deal of similarity between variable regions of different specificity.

(3) Each chain, light or heavy, is folded into several discrete, compact segments called *domains*, which contain much antiparallel β structure and are stabilized by intrachain disulfide (—S—S—) bonds. The domains are joined by exposed *hinge regions* of polypeptide that provide flexibility and account for the selective degradation into Fab and Fc fragments. The variable region of each chain forms one domain. In the type of antibody shown in Fig. 3-32 there are a total of two light-chain and four heavy-chain domains. Carbohydrate attachment (representing 4–18% of the total antibody mass) is limited to the constant region domains found in the Fc fragment.

(4) There are five classes of antibody in humans, differing from one another only in the constant part of the heavy chains. They are designated IgG, IgM, IgA, IgD, and IgE, where Ig stands for *immune globulin* or *immunoglobulin*.[4] The heavy chains that distinguish these classes from one another are referred to by the corresponding Greek letters. The heavy chains are, therefore, gamma (γ), mu (μ), alpha (α), delta (δ) and epsilon (ϵ), respectively. Each heavy chain is paired with a light chain. The light chain, in humans at least, may be one of two types, kappa (κ) or lambda (λ). It appears that any variable-region specificity may be found coupled to any light- or heavy-chain constant region. The major features of the five classes are as follows:

IgG is the workhorse of the antibodies, comprising about 80% of the amount present in serum and thus representing the major part of the humoral response. It also crosses the placental barrier in humans to provide the newborn infant with temporary protection against disease until its own immune system is fully functional several weeks after birth.

IgM often appears before IgG in an infection and is the major antibody elicited by certain very large antigens. Antibodies are found only in vertebrates, and the most primitive vertebrates seem to make only IgM. Hence, it is not surprising that the human fetus is also able to make only this class of antibody. IgM has a unique structure; it is a pentamer of the basic H_2L_2 molecule. The formation of the pentamer is aided by addition of a small polypeptide called a *J chain*.

IgA, like the two classes that follow, is present in only very small quantities in serum. However, IgA is the only antibody found in secretions, including saliva and tears, and secretions from the airways, gastrointestinal tract, prostate, and vagina. The vagina, in a few women, contains IgA specific for their mate's sperm, rendering conception diffi-

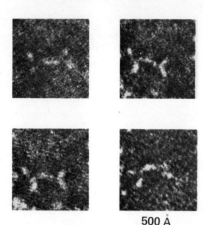

FIGURE 3-33 The IgA Dimer. Human myeloma IgA dimers. Note the "tail-to-tail" configuration in these examples, leaving the four binding sites free. [Courtesy of B. Bloth and S-E Svehag, *J. Exp. Med.*, **133**: 1035 (1971).]

cult. IgA is also found in colostrum and milk, where it helps confer on the infant additional immune protection. IgA is often found as a dimer (Fig. 3-33) containing one J chain. In addition, the IgA in secretions contains another polypeptide called a *secretory component*, the function of which is somehow related to the secretory process.

IgE antibodies are unique in being able to bind, by their Fc portions, to specific receptor sites on the surface of tissue cells called mast cells (see Chap. 10). When the antigen that interacts with these IgE molecules appears, the mast cells secrete *histamine* and other chemicals that may result, for example, in the symptom complex known as *allergic rhinitis* or hay fever.

IgD antibodies were long a class without a known function. It now appears, however, that their major role may be as receptors, charged with recognizing a foreign antigen and triggering the cell proliferation and antibody production needed to eliminate it.

Antibody Synthesis. The generally accepted mechanism of antibody synthesis, known as *clonal selection*, was first proposed by N. K. Jerne and Sir Macfarlane Burnet in the 1950s. In modern terms, clonal selection is now thought to work as follows:

A given B-type lymphocyte has, on its surface, antibodies of a single specificity—that is, all having identical V (variable) regions. These antibodies, which are often IgD but sometimes IgM or another class, are a sample, as it were, of the type of antibody that this cell and its descendants are capable of making. (This idea can be traced back to the turn of the century, as seen in Fig. 3-34.) When an appropriate antigen is encountered, this lymphocyte divides and produces several generations of progeny culminating in a *clone* of *plasma cells*, which are cells specialized for the purpose of turning out antibodies.

A given clone of plasma cells produces antibody of the same specificity as its ancestral B cell. That is, the V region of the antibody will be the same as the V region of the B cell receptor. The C (constant) region, however, may be different. In fact, since most antibody is IgG, it is probably common for a clone to switch from production of the original receptor type (e.g., IgD) to IgM and later to IgG as it proliferates and matures.

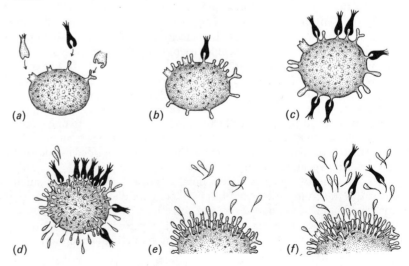

FIGURE 3-34 Antibody Synthesis, 1900. Redrawn from Ehrlich's original paper, this scheme starts with a cell having several receptor sites (*a*). Antigens attach to their specific sites (*b*) and (*c*), causing a proliferation of the receptors. The latter are also discharged as free antibodies (*d*)–(*f*). Ehrlich's proposal differs from clonal selection mainly in showing a variety of sites on a single cell. [Redrawn from P. Ehrlich, *Proc. Royal Soc.* (London), **66**: 424 (1900).]

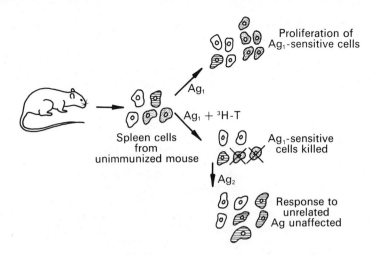

FIGURE 3-35 Clonal Selection. If clonal selection is correct, then removing cells responsive to one antigen Ag_1 should not affect response to an unrelated antigen Ag_2. That was tested in the experiment diagrammed here. Cells responsive to the first antigen were killed by incorporating ^{3}H-thymidine as they began synthesizing DNA. The loss of these cells, however, did not affect response to Ag_2. [Described by R. W. Dutton and R. I. Mishell, *J. Exptl. Med.*, **126**: 443 (1967) and *Cold Spring Harbor Symp. Quant. Biol.*, **32**; 407 (1967).]

The total antibody response, then, is the array of antibodies produced by a number of clones, each making one antibody but not necessarily the same antibody as an adjacent clone. The heterogeneity of the response depends on the number of different B cells that were originally able to respond to the antigen.

Clonal selection predicts that if one could selectively kill only those B cells that recognize a given antigen, response to that antigen would be eliminated without impairing a later response to an unrelated antigen. An experiment that seems to verify this prediction was reported in 1967 by R. W. Dutton and R. I. Mishell of the Scripps Clinic and Research Foundation (Fig. 3-35). They divided lymphocytes from a mouse spleen into two groups, then exposed both groups to sheep erythrocytes, *in vitro*. One group also received a large dose of ^{3}H-thymidine, so that cells undergoing division would incorporate enough radioactivity into their new DNA to be lethal. Only lymphocytes that are stimulated to establish plasma cell clones should be eliminated by this procedure. As predicted, antibody was produced by the control population of spleen cells, but no antibody was formed by the treated sample either initially or upon a second exposure. However, the capacity of the treated sample to respond to an unrelated antigen was not impaired.

According to some estimates, the fraction of lymphocytes that respond to any given antigen represents no more than 0.01% of the total. Destruction of these cells should render the animal *tolerant* (unresponsive) to the antigen in question, a condition that one tries to achieve medically in allergic desensitization or after organ transplantation.

The clonal selection model for antibody synthesis also offers a satisfactory explanation for the *anamnestic* or *secondary response*. These terms refer to the phenomenon where an animal that has once been challenged by an antigen responds faster and with greater output of antibody at a later exposure. (Hence, the rationale for booster shots after an immunization.) The secondary response may be possible because of return to lymphocyte status by some of the cells in an earlier proliferating clone, so that the total number of cells responsive to the antigen at later exposure will be greater than was true at the first exposure. These additional cells are sometimes referred to as *memory cells* (see Fig. 3-36).

Clonal selection is now well established as an explanation for antibody synthesis. We shall return to the topic in later chapters when the origin

FIGURE 3-36 A Model for Immunological Memory. The primary response involves the differentiation and proliferation of "immune-competent" target cells. If some of the differentiated or partially differentiated progeny were to return to the lymphocyte stage as memory cells, the number of cells responsive to the original antigen at a subsequent challenge would be greatly increased.

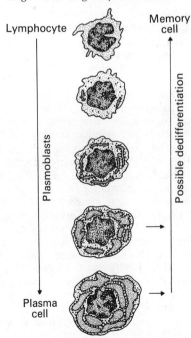

of the genetic diversity of B cells (Chap. 5), the behavior of their surface antibody-receptor (Chap. 10), and the mechanism for triggering cell division (Chap. 11) will be considered. Now, however, we shall turn our attention to a different kind of protein, one that exhibits specificity as great as, or greater than, the antibody. That protein is the enzyme.

SUMMARY

3-1 A cell is typically about three-quarters water. The remainder is largely in the form of macromolecules, either lipid, carbohydrate, nucleic acid, or protein, each of which has a relatively small basic unit that may be joined to others by covalent and/or noncovalent bonds. The structure and properties of the macromolecules are dictated largely by the behavior of carbon.

3-2 Lipids are a diverse group of nonpolar molecules characterized by water insolubility. Four major classes of lipid are (1) Fatty acids and their derivatives, including prostaglandins and thromboxanes (each of which have numerous hormonelike functions); (2) glycerides, including triglyceride (neutral fat) and the phosphoglycerides, all of which are esters of glycerol; (3) sphingolipids, which are derivatives of sphingosine; and (4) steroids. Fat (triglyceride) is important especially as a source of energy. The phosphoglycerides and sphingolipids are particularly important as constituents of membranes, a consequence of their amphipathic nature—hydrophilic on one end, hydrophobic on the other. The steroids include a diverse group of hormones (e.g., the sex hormones) and vitamins, most of which are derived from cholesterol.

3-3 Carbohydrates generally conform to the empirical formula $(CH_2O)_n$. Glucose and its isomers, for example, are all $C_6H_{12}O_6$. Glucose is a simple sugar, or monosaccharide. It can be linked through an oxygen (—O—) bridge to a second monosaccharide, forming a disaccharide such as lactose or sucrose. It can also form higher polymers, the most important of these being the structural polymer cellulose, which is an insoluble, linear β-1,4 polymer, and the starches, which are open, soluble, helical α-1,4 polymers. The starches include the highly branched animal starch called glycogen; the less highly branched plant starch, amylopectin; and an unbranched plant starch called amylose. All serve as temporary storage granules for glucose.

In addition to these pure carbohydrates, there are a large number of carbohydrate derivatives. Mucopolysaccharides are especially common, both on the surface of cells and as intercellular material. They are often associated with protein and by definition contain amino sugars. Glycolipids are another common polysaccharide derivative. Many of them (e.g., cerebrosides, gangliosides) are built with the lipid ceramide as a base.

3-4 Nucleotides and their polymers, the nucleic acids RNA and DNA, are built around one of two five-carbon sugars: ribose for the RNA nucleotides or 2-deoxyribose for the DNA nucleotides. Each nucleotide carries a phosphate (or diphosphate or triphosphate), generally at the 5' carbon of the sugar, and an organic base at the 1' carbon. (Without the phosphates, a nucleotide is called a nucleoside.) The bases are either purine derivatives—commonly adenine or guanine—or pyrimidine derivatives—generally uracil (only in RNA), thymine (only in DNA), or cytosine. The common nucleosides and nucleotides have names derived from their bases, with the ending "-dine" for the pyrimidines and "-sine" for the purines (e.g., adenosine).

The deoxynucleotides may be polymerized through 3' 5' phosphodiester linkages to form DNA, which is normally a double-stranded structure with the bases stacked in pairs lying with their planes at right angles to the long axis. Each pair of bases contains a purine from one strand and a pyrimidine from the other strand: specifically, A with T, G with C, and vice versa. RNA is also a polymer of nucleotides except that it is almost always single stranded, contains the sugar ribose instead of deoxyribose, and has uracil (U) in place of thymine.

3-5 Proteins are polymers of some 20 different amino acids linked by the planar, rigid, peptide bond. The individual amino acids may have either hydrophilic (including ionic) or hydrophobic side chains. The pattern of side chains on the surface of the completed protein, and hence exposed to the solvent, will determine the solution properties of the molecule.

The structure of a protein is discussed in terms of four levels of organization. *Primary structure* refers to the sequence of amino acids in a polypeptide chain. *Secondary structure* is the chain configuration imposed by peptide-to-peptide hydrogen bonds. The most common secondary structures are the α structure, which is a helix, and the β structure, which is a planar pleated sheet. *Tertiary structure* describes the arrangement in space of segments of secondary structures—e.g., the way that several helices within a given molecule are arranged with respect to one another. The fourth level of structure, *quaternary structure*, is the arrangement of individually folded chains or groups of chains called subunits.

The spatial conformation of a protein is determined largely by its pattern of weak interactions. Although the pattern is limited by a protein's primary structure, it can be altered by changes in pH, temperature, solvent, etc. Any of these changes may cause a protein to lose its native configuration and biological activity. This process is called denaturation.

Proteins may contain nonprotein components. Common examples include the lipoproteins and glycoproteins. The carbohydrate portion of the latter is variable and the function unknown, although most secreted proteins and most membrane proteins are glycoproteins.

3-6 Most cellular structures are comprised of subunits. These are sometimes assembled on a template, as in DNA and RNA, and sometimes assembled according to the specificity of whatever enzymes are present (e.g., cellulose, glycogen, and starch), but most often macromolecular structures spontaneously assemble. A typical case (fibrin is an exception) is comparable to the entropy-driven assembly of tobacco mosaic virus and microtubules, where hydrophobic bonds hold the subunits together. This association is highly reversible, so that microtubules, for example, can form and dissolve according to need. Some polymers, however, especially extracellular structural polymers such as collagen, elastin, and fibrin, are stabilized by covalent cross-links. Their assembly prior to cross-linking is reversible. In these three cases, too, the subunits must be shortened by enzymatic cleavage before polymerization can begin, a feature that ensures their assembly at the right time and in the right place. Lipids also self-assemble by hydrophobic bonding. An example is the liposome, which is an enclosed space bounded by a lipid bilayer comparable in structure to a cell membrane.

3-7 When foreign material is injected into an animal, new proteins capable of binding specifically to the foreign material generally appear in the serum. These proteins are antibodies. Since they contain two binding sites for the antigen, and since most antigens can accommodate more than two antibodies, cross-linked meshes can form whose size makes them precipitate readily.

Antibodies have a basic four-chain structure, consisting of two light and two heavy chains or multiples thereof. Each binding site is due to the cooperation of roughly half of one light chain and a fourth of a heavy one. The other half of each light chain, its constant region, may be found in one of two basic amino acid sequences, called κ and λ, plus minor inherited (allotypic) variants. The constant portion of the heavy chain (about three-quarters of the chain) has five basic sequences plus allotypic variants. The sequences, called γ, μ, α, δ, and ϵ, determine the class of the completed antibody, designated IgG, IgM, IgA, IgD, and IgE, respectively.

Antibody synthesis is attributed to clonal selection. That is, precursor cells called B lymphocytes have, on their surface, antibody molecules of a given specificity. Combination with antigen causes proliferation of these cells and their differentiation into plasma cells. A given clone of plasma cells produces antibody of the same specificity (identical V-region sequence) as its ancestral B cell. Some members of the clone may return to B-cell status, enlarging the pool of cells responsive to that same antigen on subsequent exposure and providing thereby the basis for the secondary response.

STUDY GUIDE

3-1 The radioactive isotopes most commonly used in biological research are ^{14}C, ^{3}H (tritium), ^{32}P, and ^{35}S, all of which emit relatively weak β rays (electrons). A cell grown in the presence of any of these isotopes (e.g., with $H^{32}PO_4^{2-}$ or $^{35}SO_4^{2-}$) will incorporate it into cellular constituents just as it would the corresponding nonradioactive element.
(1) If you wished to "label" (make radioactive) the proteins of a cell but avoid labeling the nucleic acids, which isotope would you choose? (2) If you wished to label the nucleic acids without labeling the proteins in any significant way, which would you choose?

3-2 (a) Why are the fatty acids insoluble while their two-carbon counterpart, acetic acid, is highly soluble? (b) Which would be more soluble, a monoglyceride or a triglyceride? A diglyceride or phosphatidic acid? Why? (c) What is an amphipathic lipid? (d) Give a concise definition of a glyceride, sphingolipid, and phospholipid. (e) Draw the general outline of the steroid ring structure. (f) Most of the steroid hormones are derived from what precursor?

3-3 (a) Define the following: aldose, ketose, aldopentose, ketopentose, pentulose, pyranose, and furanose. (b) What is the difference between D and L glucose? Between α- and β-glucose? (c) Why is cellulose insoluble whereas the starches (glycogen, amylose, and amylopectin) are soluble? Since cellulose and the starches are all polymers of glucose, why can we utilize the starches for food but not cellulose? (d) Give two examples of an amino sugar. (e) What is a mucopolysaccharide? (f) Describe the lipid and polysaccharide portions of a common glycolipid.

3-4 (a) What is the difference between a nucleoside and a nucleotide? (b) What is a phosphodiester linkage, and why is it called that? (c) What are the synonyms (including abbreviations and full name) for ATP? (d) What are the forces responsible for the helical structure of DNA? For its double-stranded structure? (e) The melting temperature (T_m) of DNA increases in a quite predictable way with the ratio $(G + C)/(A + T)$. Why? (f) A typical human cell contains about 4×10^{12} daltons $(6 \times 10^{-12}$ g) of DNA. If that amount of DNA were in a single double-stranded helix, how long would it be? (The average mass of a nucleotide in DNA is about 300 daltons.) [Ans: about 7 ft] (g) If the cell in part (f) were a sphere 10 μm in diameter, what fraction of the protoplasm would the DNA occupy, assuming that DNA is a cylinder 20 Å in diameter? [Ans: about 1.3%]

3-5 (a) Define the following: amino acid; peptide bond; primary, secondary, tertiary, and quaternary structure; denaturation of proteins. (b) Some proteins can be denatured *in vitro* but will spontaneously regain their native configuration when their environment is returned to normal. To what do you attribute this? What factors might prevent a protein from behaving in this way? (c) What is a conformational isomer? Why do many (all?) proteins have them? (d) There are three major classes of serum lipoproteins. Name them and describe how they are interrelated. (e) How would you define a glycoprotein?

3-6 (a) Describe the basic features of template assembly, enzymatic assembly, and self-assembly. Give an example of each. (b) What is meant by the term nucleation of self-assembly? (c) Microtubules, tobacco mosaic virus, un-cross-linked collagen and elastin all polymerize at 37 °C and depolymerize again at low temperatures. How do you explain that behavior? (d) Fibrin, before it is cross-linked, does not follow the pattern outlined in the preceeding question. Why? (e) What happens to collagen, elastin, and fibrin to make their assembly essentially irreversible? (f) What is a liposome?

3-7 (a) Describe the structure of a typical antibody molecule. (b) Numerous attempts have been made to determine the evolutionary relationship between species by studying the degree of similarity among certain of their proteins (e.g., serum albumin). Why and how might antibodies be used in such studies? (c) In what ways do antibody-antigen reactions aid in the defense against disease? (d) What is a hapten? (e) How have multiple myeloma patients contributed to our understanding of antibody structure? What is the relationship of this disease to the Bence-Jones proteins? (f) What is an allotype? (g) What are some of the functional and structural differences among the five known classes of antibodies? (h) Explain the clonal selection theory and how it may give rise to immunological memory and the secondary response.

REFERENCES

GENERAL

BANKS, P., W. BARTLEY, and L. BIRT, *The Biochemistry of the Tissues* (2d ed.). New York: John Wiley, 1976. (Paperback.)

BOHINSKI, ROBERT C., *Modern Concepts in Biochemistry* (2d ed.). Boston: Allyn and Bacon, 1976.

CONN, E. E., and P. K. STUMPF, *Outlines of Biochemistry* (3d ed.). New York: John Wiley, 1972.

LEHNINGER, A. L., *Biochemistry: The Molecular Basis of Cell Structure and Function* (2d ed.). New York: Worth, 1975.

MAZUR, A., and B. HARROW, *Textbook of Biochemistry* (10th ed.). Philadelphia, Pa.: W. B. Saunders, 1971.

McGILVERY, R. W., *Biochemical Concepts*. Philadelphia: W. B. Saunders, 1975.

METZLER, DAVID, *Biochemistry: The Chemical Reactions of Living Cells*. New York: Academic Press, 1977. An advanced level text.

ORTEN, J. M., and O. W. NEUHAUS, *Human Biochemistry* (9th ed.). St. Louis: Mosby, 1975.

STRYER, LUBERT, *Biochemistry*. San Francisco: W. H. Freeman, 1975.

WHITE, ABRAHAM, P. HANDLER, and E. L. SMITH, *Principles of Biochemistry* (5th ed.). New York: McGraw-Hill, 1973.

LIPIDS AND CARBOHYDRATES

CURTIS-PRIOR, P. B., *Prostaglandins: An Introduction to their Biochemistry, Physiology, and Pharmacology*. New York: Elsevier/North-Holland, 1976.

FISHMAN, P., and R. BRADY, "Biosynthesis and Function of Gangliosides." *Science*, **194**: 906 (1976).

HORTON, E. W., "The Prostaglandins." *Proc. Roy. Soc. (London) Ser. B*, **182**: 411 (1973). A review.

PIKE, JOHN E., "Prostaglandins." *Scientific American*, November 1971. (Offprint 1235.)

HOUGH, L., and E. TARELLI, "Carbohydrates: Aspects of Structure and Reactivity." *Sci. Progr. (Oxford)*, **64**: 569 (1977).

HUGHES, R. C., "The Complex Carbohydrates of Mammalian Cell Surfaces and their Biological Roles." *Essays Biochem.*, **11**: 1 (1975).

KARIM, S. M., ed., *Prostaglandins*. Baltimore: University Park Press, 1975–76. Three-volume work.

KOLATA, G. B., "Thromboxanes: The Power Behind the Prostaglandins?" *Science*, **190**: 770 (1975).

NOGGLE, G. R., and G. J. FRITZ, *Introductory Plant Physiology*. Englewood Cliffs, N.J.: Prentice-Hall, 1976. Includes discussion of the various polysaccharides.

REES, D., *Polysaccharides*. New York: Halsted Press, 1977. Paperback.

SCHLENK, H., "Odd Numbered and New Essential Fatty Acids." *Fed. Proc.,* **31**: 1430 (1972).

SHARON, N., *Complex Carbohydrates.* Reading, Mass.: Addison-Wesley, 1975.

NUCLEIC ACIDS

CRICK, F. H. C., "The Structure of the Hereditary Material." *Scientific American,* October 1954. (Offprint 5.) Also see Sept. 1957 (offprint 54).

GUSCHLBAUER, W., *Nucleic Acid Structure.* New York: Springer-Verlag, 1976. Heidelberg Science Library, Volume 21. (Paperback.)

HERSHEY, A. D., "Idiosyncrasies of DNA Structure." *Science,* **168**: 1425 (1970). Nobel Lecture, 1969.

MIRSKY, ALFRED E., "The Discovery of DNA." *Scientific American,* June 1968. (Offprint 1109.)

SANGER, F., "Nucleotide Sequences in DNA." *Proc. Roy. Soc.* (*London*), Ser. B., **191**: 317 (1975). Sequencing: how it's done and some early results.

SAYRE, ANNE, *Rosalind Franklin and DNA.* New York: Norton, 1975. To be read after *The Double Helix.*

WATSON, J. D., *The Double Helix.* New York: Atheneum, 1968. A personal—and somewhat controversial—account of the elucidation of DNA structure.

WATSON, J. D., and F. H. C. CRICK, "The Structure of DNA." *Cold Spring Harbor Symp. Quant. Biol.,* **18**: 123 (1953).

———, "Molecular Structure of Nucleic Acids: A Structure for Deoxyribose Nucleic Acid." *Nature,* **171**: 737 (1953). The original presentation.

WILKINS, M. H. F., "The Molecular Configuration of Nucleic Acids." In *Nobel Lectures, Physiology and Medicine 1942–1962.* Amsterdam: Elsevier, 1964, p. 754. Lecture given in 1962.

AMINO ACIDS AND PROTEINS

ANFINSEN, C. B., "The Formation and Stabilization of Protein Structure." *Biochem. J.,* **128**: 24 (1972). A very readable review by this 1972 Nobel Prize winner.

ANFINSEN, C. B., "The Formation of the Tertiary Structure of Proteins." *The Harvey Lectures, 1965–1966* (Series 61). New York: Academic Press, 1967, p. 95.

ANFINSEN, C. B., and H. A. SCHERAGA, "Experimental and Theoretical Aspects of Protein Folding." *Advan. Protein Chem.,* **29**: 205 (1975).

BAILEY, A. J., and S. P. ROBINS, "Current Topics in the Biosynthesis, Structure and Function of Collagen." *Sci. Progr.* (*Oxford*), **63**: 419 (1976).

BELL, E. A., "Uncommon Amino Acids in Plants." *FEBS Letters,* **64**: 29 (1976).

BROMER, WM. W., "Proinsulin." *BioScience,* **20**: 701 (1970).

CORRIGAN, JOHN J., "D-Amino Acids in Animals." *Science,* **164**: 142 (1969).

DICKERSON, R. E., and IRVING GEIS, *The Structure and Action of Proteins.* New York: Harper & Row, 1969. (Paperback.) A beautifully illustrated introduction to the subject.

FRANKS, F., and D. EAGLAND, "The Role of Solvent Interactions in Protein Conformation." *CRC Crit. Rev. Biochem.,* **3**: 165 (1975).

FRASER, R. D. B., "Keratins." *Scientific American,* August 1969. (Offprint 1155.)

GRANT, P. T., and T. L. COOMBS, "Proinsulin, A Biosynthetic Precursor of Insulin." *Essays Biochem.,* **6**: 69 (1970).

HENDRICKSON, W., "The Molecular Architecture of Oxygen-Carrying Proteins." *Trends Biochem. Sci.,* **2**(5): 108 (May, 1977). Hemoglobins, hemocyanins, and hemerythrins.

JACKSON, R., J. MORRISETT, and A. GOTTO, JR., "Lipoprotein Structure and Metabolism." *Physiol. Rev.,* **56**: 259 (1976).

KENDREW, JOHN C., "The Three-Dimensional Structure of a Protein Molecule." *Scientific American,* December 1961. (Offprint 121.) Myoglobin.

LEVITT, M., and C. CHOTHIA, "Structural patterns in globular proteins." *Nature,* **261**: 552 (1976).

MILLER, E. J., "Biochemical Characteristics and Biological Significance of the Genetically Distinct Collagens." *Mol. Cell. Biochem.,* **13**: 165 (1976).

MOORE, S., and W. H. STEIN, "Chemical Structures of Pancreatic Ribonuclease and Deoxyribonuclease." *Science,* **180**: 458 (1973). Nobel lecture, 1972.

MORRISETT, J., R. JACKSON, and A. GOTTO, JR., "Lipoproteins: Structure and Function." *Ann. Rev. Biochem.,* **44**: 183 (1975).

PACE, W. C., "The Stability of Globular Proteins." *CRC Crit. Rev. Biochem.,* **3**: 1 (1975).

PAIK, WOON KI, and SANGDUK KIM, "Protein Methylation." *Science,* **174**: 114 (1972). Proteins may be chemically modified after their synthesis.

PERUTZ, M. F., "The Hemoglobin Molecule." *Proc. Royal Soc.* (*London*) *B,* **173**: 113 (1969). Croonian Lecture, 1969. Also see *Scientific American,* November 1964 (Offprint 196) and *The Harvey Lectures,* 1967–1968, p. 213.

ROBINSON, TREVOR, "D-Amino Acids in Higher Plants." *Life Sci.,* **19**: 1097 (1976).

ROBSON, B., "Protein Folding." *Trends Biochem. Sci.,* **1**(3): 49 (March, 1976).

SHARON, N., "Glycoproteins." *Scientific American,* **230**(5): 78 (May, 1974).

THOMPSON, E. O. P., "The Insulin Molecule." *Scientific American,* May 1955. (Offprint 42.) Determination of primary structure.

WHELAN, W. J., *Biochemistry of Carbohydrates.* Baltimore: University Park Press, 1976. See chapters on glycoproteins (proteoglycans) and polysaccharides.

THE SELF-ASSEMBLY OF MACROMOLECULAR STRUCTURES

ASQUITH, R. S., M. S. OTTERBURN, and W. J. SINCLAIR, "Isopeptide Cross-Links—Their Occurrence and Importance in Protein Structure." *Angew. Chem.,* **13**: 514 (1974).

BAILEY, A. J., and S. P. ROBINS, "Current Topics in the Biosynthesis, Structure and Function of Collagen." *Sci. Progr.* (*Oxford*), **63**: 419 (1976).

BORGERS, M., and M. DE BRABANDER, eds., *Microtubules and Microtubule Inhibitors.* New York: Elsevier/North-Holland, 1975. Proceedings of a symposium.

BOUCK, G., and D. BROWN, "Self-Assembly in Development." *Ann. Rev. Plant Physiol.,* **27**: 71 (1976). Collagen, viruses, bacterial flagella, and microtubules.

BURNSIDE, B., "The Form and Arrangement of Microtubules: An Historical, Primarily Morphological Review." *Ann. N.Y. Acad. Sci.*, **253**: 14 (1975).

COLLEY, C. and B. RYMAN, "The Liposome: From Membrane Model to Therapeutic Agent." *Trends Biochem. Sci.*, **1**(9): 203 (Sept. 1976).

COOK, R. A., and D. E. KOSHLAND, JR., "Specificity in the Assembly of Multisubunit Proteins." *Proc. Nat. Acad. Sci.* (*U.S.*), **64**: 247 (1969).

DOOLITTLE, R. F., "Structural Aspects of Fibrinogen to Fibrin Conversion." *Advan. Protein Chem.*, **27**: 1 (1973).

EDELHOCK, H., and J. C. OSBORNE, JR., "The Thermodynamic Basis of the Stability of Proteins, Nucleic Acids, and Membranes." *Advan. Protein Chem.*, **30**: 183 (1976).

FENDLER, J., and A. ROMERO, "Liposomes as Drug Carriers." *Life Sci.*, **20**: 1109 (1977).

GALLOP, P., and M. PAZ, "Posttranslational Protein Modifications, with Special Attention to Collagen and Elastin." *Physiol. Rev.*, **55**: 418 (1975).

GALLOP, P., M. PAZ, B. PEREYRA, and O. BLOOMFELD, "The Maturation of Connective Tissue Proteins." *Israel J. Chem.*, **12**: 305 (1974). The chemistry of elastin and collagen cross-linking.

GRANT, M., and D. S. JACKSON, "The Biosynthesis of Procollagen." *Essays Biochem.*, **12**: 77 (1976).

GROSS, J., "Collagen Biology." *The Harvey Lectures 1972–73*, p. 351 (1974).

KAPER, J. M., *The Chemical Basis of Virus Structure, Dissociation, and Reassembly.* New York: American Elsevier, 1975.

KLUG, A., "Assembly of Tobacco Mosaic Virus." *Fed. Proc.*, **31**: 30 (1972).

KUSHNER, D. J., "Self Assembly of Biological Structures." *Bacteriol. Rev.*, **33**: 302 (1969). Viruses, bacterial flagellae, microtubules.

LAUFFER, M. A., *Entropy-Driven Processes in Biology (Molecular Biology, Biochemistry, and Biophysics, Vol. 20).* New York: Springer-Verlag, 1975. Mostly on tobacco mosaic virus assembly.

OKADA, Y., "Mechanism of Assembly of Tobacco Mosaic Virus in vitro." *Advan. Biophys.*, **7**: 1 (1975).

OOSAWA, F., and S. ASAKURA, *Thermodynamics of the Polymerization of Protein.* New York: Academic Press, 1976. Mostly on flagellin and actin.

PATEL, H. M., and B. E. RYMAN, "Oral Administration of Insulin by Encapsulation Within Liposomes." *FEBS Letters*, **62**: 60 (1976). It works (at least in rats)!

PERHAM, R., "Self-Assembly of Biological Macromolecules." *Phil. Trans. Roy. Soc. London, Ser. B.*, **272**: 123 (1975).

ROBERTS, K., "Cytoplasmic Microtubules and Their Functions." *Progr. Biophys. Mol. Biol.*, **28**: 371 (1974).

SNELL, W., W. DENTLER, L. HAIMO, L. BINDER, and J. ROSENBAUM, "Assembly of Chick Brain Tubulin onto Isolated Basal Bodies of *Chlamydomona reinhardi*." *Science*, **185**: 357 (1974).

SNYDER, J., and J. MCINTOSH, "Biochemistry and Physiology of Microtubules." *Ann. Rev. Biochem.*, **45**: 699 (1976).

SUTTIE, J. W., and C. M. JACKSON, "Prothrombin Structure, Activation, and Biosynthesis." *Physiol. Rev.*, **57**: 1 (1977).

VEIS, A., and A. G. BROWNELL, "Collagen Biosynthesis." *CRC Crit. Rev. Biochem.*, **2**: 417 (1975).

WILSON, L., ed., "Pharmacological and Biochemical Properties of Microtubule Proteins." *Fed. Proc.*, **33**: 151 (1974). Symposium.

ANTIBODIES

BEALE, D., and A. FEINSTEIN, "Constant Regions of Immunoglobulins." *Quart. Rev. Biophys.*, **9**: 135 (1976).

CAPRA, J., and A. EDMUNDSON, "The Antibody Combining Site." *Scientific American*, **236**(1): 50. (Jan. 1977).

COOPER, M., and A. LAWTON, III, "The Development of the Immune System." *Scientific American*, **231**(5): 58 (Nov. 1974). Offprint 1306.

COOPER, E. L., *Comparative Immunology.* Englewood Cliffs, N.J.: Prentice-Hall, 1976.

EDELMAN, G. M., "The Structure and Function of Antibodies." *Scientific American*, August 1970. (Offprint 1185.)

EDELMAN, G. M., "Antibody Structure and Molecular Immunology." *Science*, **180**: 830 (1973). Nobel lecture, 1972.

EDELMAN, G. M., et al., *Molecular Approaches to Immunology.* New York: Academic Press, 1975.

FUDENBERG, H., D. STITES, J. CALDWELL, and J. WELLS, eds., *Basic and Clinical Immunology.* Los Altos, Calif.: Lange Medical Publications, 1976. A set of brief reviews. (Paperback.)

GLICK, B., "The Bursa of Fabricius and Immunoglobulin Synthesis." *Int. Rev. Cytol.*, **48**: 345 (1977).

HERBERT, W. J., and P. C. WILKINSON, eds., *A Dictionary of Immunology* (2d ed.), Oxford: Blackwell Scientific, 1977. (Paperback.)

HUBER, R., "Antibody Structure." *Trends Biochem. Sci.*, **1**(8): 174 (Aug., 1976)

HUBER, R., J. DEISENHOFER, P. COLMAN, and M. MATSUSHIMA, "Crystallographic Structure Studies of an IgG Molecule and an Fc Fragment." *Nature*, **264**: 415 (1976). Antibody structure to 4 Å resolution.

MILSTEIN, C., "Structure and Evolution of Immunoglobulins." *Prog. Biophys. and Mol. Biol.*, **21**: 209 (1970).

POLJAK, R., L. AMZEL, and R. PHIZACKERLEY, "Studies on the Three-Dimensional Structure of Immunoglobulins." *Progr. Biophys. Mol. Biol.* **31**: 67 (1976).

PORTER, K. R., "Structural Studies of Immunoglobulins." *Science*, **180**: 713 (1973). Nobel lecture, 1972.

ROITT, I. M., *Essential Immunology* (3d ed.). Oxford: Blackwell Scientific, 1977. (Paperback.)

SOLOMON, A., "Bence-Jones Proteins and Light Chains of Immunoglobulins." *New Engl. J. Med.*, **294**: 17 (1976). A two-part series.

WILLIAMSON, A., I. ZITRON, and A. MCMICHAEL, "Clones of B Lymphocytes: Their Natural Selection and Expansion." *Fed. Proc.*, **35**: 2195 (1976).

FOR ADDITIONAL READING:
EVOLUTION OF THE MACROMOLECULES

AYALA, F. J., ed., *Molecular Evolution.* Reading, Mass.: Freeman, 1976. (Paperback.)

BARNICOT, N. A., "Some Biochemical and Serological Aspects of Primate Evolution." *Sci. Progr. (Oxford)*, **57**: 459 (1969).

BUETTNER-JANUSCH, J., and R. L. HILL, "Molecules and Monkeys." *Science*, **147**: 836 (1965). Differences in the primary structure of primate hemoglobins.

CAIRNS-SMITH, A. G., "A Case for an Alien Ancestry." *Proc. Roy. Soc. (London), Ser. B.*, **189**: 249 (1975).

CALVIN, MELVIN, "Chemical Evolution." *Am. Scientist*, **63**: 169 (1975).

DICKERSON, R. E., "The Structure and History of an Ancient Protein." *Scientific American*, April 1972. (Offprint 1245.) Cytochrome c.

DIXON, G. H., "Mechanisms of Protein Evolution." *Essays Biochem.*, **2**: 147 (1966).

DOBZHANSKY, T., M. H. HECHT, and W. C. STEERE, eds., *Evolutionary Biology*. Vol. 8. New York: Plenum, 1975.

FLORKIN, M., and STOTZ, E. H., eds., *Comparative Biology, Molecular Evolution*. New York: American Elsevier, 1975. Vols. 29A and 29B of *Comprehensive Biochemistry*.

FOLSOME, C. E., "Synthetic Organic Microstructures and the Origins of Cellular Life." *Naturwissenschaften*, **63**: 303 (1976).

FOX, S. W., and K. DOSE, *Molecular Evolution and the Origin of Life*. San Francisco: W. H. Freeman, 1972.

KÜPPERS, B., "The General Principles of Selection and Evolution at the Molecular Level." *Progr. Biophys. Mol. Biol.*, **30**: 1 (1975).

LIN, E., A. HACKING, and J. AGUILAR, "Experimental Models of Acquisitive Evolution." *BioScience*, **26**: 548 (1976).

PRICE, CHARLES C., ed., *Synthesis of Life*. New York: Halsted Press, 1974. Chemical evolution and macromolecular systems.

NOTES

1. An ester is the condensation product (meaning water is removed) of an acid and an alcohol, e.g.,

$$R_1-OH + HO-\overset{O}{\underset{\|}{C}}-R_2 \longrightarrow R_1-O-\overset{O}{\underset{\|}{C}}-R_2 + H_2O$$

The reverse reaction, where a compound is split by the addition of water, is called *hydrolysis* ("lysis by water").

2. Faulty tropocollagen (and hence faulty collagen) also results from vitamin C deficiency since that substance is required for the hydroxylation of proline and lysine. Both hydrogen bonding and covalent cross-linking are compromised. The resulting disease, whose symptoms are bleeding, skin lesions, faulty bone growth, and poor wound healing, is called *scurvy*.

3. The liquid part of blood after a clot and cells have been removed is called *serum*. Removal of just the cells and platelets from unclotted blood leaves *plasma*.

4. When serum is subjected to electrophoresis proteins are separated on the basis of electrical charge. At alkaline pHs the fastest moving (toward the positive pole or anode) and most abundant component is albumin. Trailing albumin is a set of more-or-less separated components called, in order from albumin, α, β, and γ (alpha, beta, and gamma) globulins, each of which is a heterogeneous mixture. Most antibodies are in the γ fraction.

4

Cellular Homeostasis: Bioenergetics and the Regulation of Enzymes

4-1 Enzymes as Catalysts   134
   Energy of Activation
   Catalysts
   Enzyme-Substrate Interaction
   Enzyme Nomenclature

4-2 ATP and Coupled Reactions   139
   The Hydrolysis of "High Energy" Phosphates
   Coupled Reactions

4-3 Glycolysis: A Metabolic Pathway   143
   Glucose Catabolism
   Glycolysis
   Stepwise Degradation
   ATP Synthesis

4-4 Biological Oxidations   147
   Examples
   Glycolytic Redox Reactions
   NAD^+ Recycling

4-5 Enzyme Kinetics and Regulation   150
   Feedback Regulation
   Michaelis-Menten Kinetics
   Limitations of the Michaelis-Menten Equation
   Competitive Inhibition
   Noncompetitive Inhibition
   Cooperativity
   Allosteric Regulation
   Control of Glycolysis

4-6 Cyclic Nucleotides and Protein Phosphorylation   163
   Glycogen Synthesis and Degradation
   Hormonal Control of Enzyme Cascades
   Cyclic AMP and Cyclic GMP: The Yin-Yang Hypothesis
   Regulation by Calcium

Summary   169
Study Guide   170
References   171
Notes   173

Cells are able to maintain a favorable internal environment in the face of very wide fluctuations in their external environment. This process of self-regulation is referred to as *homeostasis*. Many different mechanisms are involved, of course, not all of which can be considered here. However, in this and the following chapter we shall consider the two major types of regulatory processes, the regulation of enzymes and the regulation of genes.

4-1 ENZYMES AS CATALYSTS

Every high school chemistry student knows that hydrogen and oxygen gases can react with explosive force to form water:

$$H_2 + \tfrac{1}{2}O_2 \rightarrow H_2O$$

The explosion results from a large negative free energy change for the reaction, which is another way of saying that the equilibrium position of the reaction strongly favors the formation of water. Yet if the two gases are mixed at ordinary temperatures, nothing happens. Actually, water does form at an extremely slow rate, but it is only when a match is dropped into the container that water forms immediately, in a violent reaction.

Energy of Activation. One can explain the drastic change in reaction rate caused by dropping a match into a container of hydrogen and oxygen by saying that the bonds between the hydrogen atoms and those between the oxygen atoms must be broken before new hydrogen-oxygen bonds can form. Although oversimplified, this explanation does introduce the concept of an energy barrier (see Fig. 4-1) because it calls for a certain amount of energy to be put into the system to break up the hydrogen and oxygen molecules before a reaction can take place. This energy is called the *energy of activation*, $\Delta E\ddagger$ or $\Delta G\ddagger$.

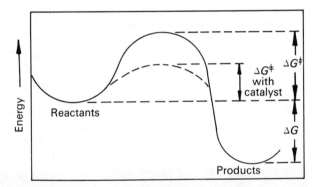

FIGURE 4-1 Energy of Activation. Even a highly favorable reaction may proceed very slowly if there is a significant energy barrier. The input of energy needed to initiate a reaction will be recovered along with the free energy difference between reactants and products. Note that the reaction shown here is essentially irreversible: to proceed in the forward direction, the barrier $\Delta G\ddagger$ must be jumped; to proceed in the reverse direction, the much higher barrier ($\Delta G + \Delta G\ddagger$) must be overcome, where ΔG is the net change in free energy for the reaction.

An energy of activation is supplied through normal thermal motion. Therefore, the chance of reaching it in a collision increases with temperature. Since the standard state enthalpy change ($\Delta H°$) for the formation of water from hydrogen and oxygen is -286 kJ per mole at 298 K, this reaction could itself be a source of considerable heat. At ordinary temperatures, water forms too slowly to have any real effect on the average temperature; however, when a match is introduced, nearby regions may be raised to the critical temperature, causing a reaction that releases more heat. The reaction can thus be propagated throughout the container, leading to an explosion.

Catalysts. Biological systems do not have the option of raising the temperature in order to overcome an energy of activation. Instead, they rely on a set of proteins that are able to speed up reactions by lowering the energy of activation. That process is called *catalysis*. Cells are endowed with a remarkable set of catalysts called enzymes.

All enzymes are proteins, though many of them also contain prosthetic groups of other types of material. Enzymes are true catalysts in that: (1) they are effective in small amounts; (2) they are unchanged by the reaction; and (3) they do not affect the ultimate equilibrium concentrations but only reduce the required activation energy and hence increase the rate of reaction. In addition to these three properties common to all catalysts, enzymes are unique in that each exhibits a very great specificity for a particular reaction. This specificity provides a way of balancing the metabolism of a cell, since the relative rates of each reaction can be controlled by adjusting the amount and/or catalytic efficiency of the enzyme that is specific for it.

The specificity and catalytic properties of enzymes are a result of their structure. This is easy to demonstrate since, like other proteins, enzymes may be denatured by a variety of chemical and physical agents. When their spatial configuration is lost, so is their biological activity.

Enzyme-Substrate Interaction. There is an area on each enzyme, called its *active site*, that is particularly sensitive to chemical or conformational change for it is there that reactants are bound. (See Fig. 4-2.) The interaction between a reactant (also called a *substrate*) and an enzyme is fostered by a complementary arrangement of mutually attractive atoms or groups of atoms—for example, a negatively charged amino acid side chain on the enzyme with a positively charged group on the substrate. Typically, several weak bonds will be formed between enzyme and substrate, creating a relatively stable complex. In addition to van der Waals interactions, hydrogen bonds, ionic interactions, or contact between hydrophobic side chains on the enzyme and hydrophobic areas of the substrate, the binding may also be due to dative bond formation, with a divalent metal ion held between electron donors on the substrate and enzyme. In most cases, the affinity of enzyme for substrate is probably due to a combination of several of these factors.

Once substrates are bound to the active site of an enzyme, catalysis can take place. The mechanism of catalysis varies with the enzyme and type of reaction being catalyzed. In some cases, just orienting the sub-

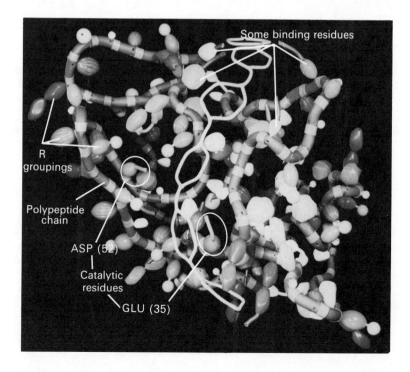

FIGURE 4-2 Lysozyme-Substrate Interaction. Note the complementary fit (in a crevice of the protein) between the enzyme and the carbohydrate chain that forms its substrate. The carbohydrate is hydrolyzed by lysozyme. [Molecular model from R. C. Bohinski, *Modern Concepts in Biochemistry* (2d ed.), Boston: Allyn and Bacon, 1976.]

strates properly and holding them next to each other seems to be explanation enough for the reaction that follows. In other cases, the enzyme is more actively involved and forms temporary covalent complexes with one of the substrates.

For example, the enzyme papain (from papaya fruit) hydrolyzes peptide bonds through formation of an intermediate complex, using the sulfhydryl (—SH) of one of its own amino acids, cysteine:

$$\text{Enz—S} + \underset{\underset{\underset{\underset{\text{COOH}}{|}}{\underset{\text{HC—R}_2}{|}}}{\underset{\text{HN}}{|}}}{\overset{\overset{\overset{\text{NH}_2}{|}}{\text{HC—R}_1}}{\text{C=O}}} \;\rightleftharpoons\; \text{Enz—S—}\underset{\underset{\underset{\underset{\text{COOH}}{|}}{\underset{\text{HCR}_2}{|}}}{\underset{\text{HNH}}{|}}}{\overset{\overset{\overset{\text{NH}_2}{|}}{\text{HCR}_1}}{\text{C=O}}} \;\xrightarrow{H_2O}\; \text{Enz—S} + \underset{\text{OH}}{\overset{\overset{\text{NH}_2}{|}}{\underset{|}{\overset{\text{HCR}_1}{\text{C=O}}}}}$$

The steps must each have a low activation energy if the reaction is to proceed rapidly, but the overall energy change can be no different from what it would have been had the enzyme not participated, for the reactants and products are the same in either case. Many hydrolyses take place by analogous reactions, although not all of them use cysteine. Chymotrypsin and trypsin, for example, catalyze the hydrolysis of peptide bonds through the intercession of a serine hydroxyl, —CH$_2$—OH, instead of the sulfhydryl group.

Many other enzymatic mechanisms have been discovered. We do not know in detail how each enzyme works, but enough has been learned to appreciate that there is nothing mysterious about them. Their mechanisms are straightforward and can be understood using only simple chemical principles.

Enzyme Nomenclature. Except for a few historically important names (chymotrypsin, trypsin, pepsin, papain, etc.), all enzyme names end in the suffix "-ase." Usually, that suffix is added to a stem derived either from the name of one of the substrates or from the name for the type of reaction being catalyzed. The situation is sometimes confusing because many enzymes have two names—a "trivial" name used for its simplicity or because it is historically entrenched, and an "official" name that describes more or less unambiguously the reaction catalyzed by the enzyme.

Another source of confusion is that in many cases more than one enzyme is specific for a reaction. When two or more enzymes have as their primary function catalysis of the same reaction they will have the same name, but will be referred to as *isoenzymes* or *isozymes*. Usually, the isozymes are distributed to different types of cells where special metabolic requirements make one or the other isozyme better suited to carry out the assigned catalytic function. A change from one isozyme to another in a given type of cell is also commonly seen during development and differentiation. For example, an enzyme called pyruvate kinase has only one molecular form in very early mammalian embryos. However, by birth two additional isozymes are added, each specific for the same reaction but limited in distribution to a few highly specialized cell types.

Because we shall have need many times in this and later chapters to refer to specific enzymes, it would be useful to understand from the

CHAPTER 4
Cellular Homeostasis: Bioenergetics and the Regulation of Enzymes

TABLE 4-1. Enzyme Nomenclature

Enzyme category	Type of reaction catalyzed
Oxidase	removal of electrons
Dehydrogenase	removal of electrons plus one or more protons
Kinase	transfer of a phosphate from one substrate (usually ATP) to another
Hydrolase	hydrolysis (breakdown by the addition of water, as in A—B + $H_2O \rightarrow$ A—H + B—OH)
Protease	hydrolysis (see above) of proteins to yield amino acids or oligopeptides
Nuclease	hydrolysis of nucleic acids to nucleotides or oligonucleotides
Phosphatase	removal of phosphates by hydrolysis
Decarboxylase	removal of a carboxyl group, creating CO_2 (A—COOH $\rightarrow$ AH + CO_2)
Carboxylase	addition of CO_2 to create a carboxyl group
Isomerase	interconversion of two isomeric forms

name alone the type of reaction being catalyzed. Hence, Table 4-1 is a brief "dictionary" of enzyme nomenclature, containing the most common enzymatic categories and the reactions catalyzed by them. With only a few exceptions, all the enzymes mentioned later will fall into one of these categories. As an example of how the system works, consider the enzyme lactate dehydrogenase. The name tells you that the reaction involves the removal of electrons and protons (equivalent to hydrogen atoms) from lactate or lactic acid:

$$\underset{\text{Lactate}}{\begin{array}{c} COO^- \\ | \\ HC-OH \\ | \\ CH_3 \end{array}} \longrightarrow \underset{\text{Pyruvate}}{\begin{array}{c} COO^- \\ | \\ C=O \\ | \\ CH_3 \end{array}} + 2e^- + 2H^+ \quad (4\text{-}1)$$

Note that lactate has been *oxidized*, since oxidation is the removal of electrons. Obviously, some other substance must at the same time be reduced—i.e., act as an acceptor for those electrons. As we shall see, that substance is called NAD^+ in the oxidized form and NADH in the reduced form. Lactate dehydrogenase could therefore also be referred to as an NAD^+ *reductase*. However, there are many NAD^+ reductases, hence it is more specific to refer to "lactate dehydrogenase."

As we shall see later, the lactate dehydrogenase reaction as it is written in Equation 4-1 is an "uphill" reaction, thermodynamically. That is, it has a strong positive $\Delta G°$. Nevertheless, it occurs readily because this enzyme, like all enzymes, catalyzes a reaction that may approach equilibrium from either side. Hence, we may speak of "substrates" and "products" of a reaction interchangeably according to a convention that designates substrates as those participants appearing on the left side of the reaction equation and products as those on the right side. In a red blood cell, for example, the lactate dehydrogenase reaction occurs only in the direction pyruvate to lactate—i.e., the reverse of Equation 4-1. Lactate for these cells is a waste product. However, in liver cells the reaction almost always occurs from lactate to pyruvate just as it is written in Equation 4-1, for one of the functions of the liver is to remove lactate from the blood.

The dominant direction of a reaction *in vivo* is dictated by the needs of the cell and is not necessarily consistant at all times with the direction that produces a favorable free energy change nor with the direction most efficiently catalyzed by the enzyme. However, a reaction that would otherwise be extremely unfavorable can be driven by an obligate coupling to a very favorable one. That coupling is most often to the hydrolysis of ATP.

4-2 ATP AND COUPLED REACTIONS

The capture of light or the breakdown of foodstuffs is carried out in a way that makes energy available to the cell. The synthesis of needed organic molecules or the support of movement requires energy. The link between these functions is ATP (see Fig. 4-3) and a few other compounds with similar properties. Energy-yielding reactions typically synthesize ATP from ADP and P_i; energy-requiring reactions hydrolyze ATP back to ADP and P_i again. If the supply of available ATP begins to decline, either the rate of energy-yielding reactions is increased, or the rate of energy-requiring reactions is decreased, or both. The opposite adjustments are made when the supply of ATP rises above its optimum value. The functional parts of the ATP molecule are the two pyrophosphate (or phosphoric acid anhydride, P—O—P) linkages, for the hydrolysis of either is energetically favorable.

The Hydrolysis of "High Energy" Phosphates. It is customary to talk about the two pyrophosphate bonds in ATP as "high energy bonds," a

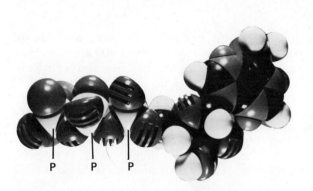

FIGURE 4-3 *ATP*. LEFT: A space-filling model. RIGHT: Diagrammatic representation of adenosine-5'-triphosphate. Hydrolysis to ADP and P_i (i.e., to adenosine diphosphate and inorganic phosphate, $HOPO_3^{2-}$) is an important source of energy for cellular reactions.

TABLE 4-2. Some Standard State Free Energies of Hydrolysis

	$\Delta G°$
phosphoenolpyruvate $\rightarrow$ pyruvate + P_i	-54 kJ/mole
phosphocreatine $\rightarrow$ creatine + P_i	-43
acetylcoenzyme A $\rightarrow$ acetate + CoA	-31
sucrose $\rightarrow$ glucose + fructose	-29
polynucleotide$_n$ $\rightarrow$ polynucleotide$_{n-1}$ + nucleotide	-25
glucose 1-phosphate $\rightarrow$ glucose + P_i	-23
glycogen$_n$ $\rightarrow$ glycogen$_{n-1}$ + glucose	-21
glucose 6-phosphate $\rightarrow$ glucose + P_i	-16
α-glycerol phosphate $\rightarrow$ glycerol + P_i	-10
polypeptide$_n$ $\rightarrow$ 2 polypeptides$_{n/2}$	-2

concept introduced by Fritz Lipmann in 1941. But before you jump to erroneous conclusions concerning the nature of these bonds, we should hasten to point out that the term "high energy" is somewhat misleading. The bonds themselves are not unusual, except that the *hydrolysis* of them is relatively favorable with a $\Delta G°$ of about -33 kJ/mole at pH 7. Whether one wants to call that "high" energy or not is a matter of choice, for it is actually quite an ordinary amount (see Table 4-2). However, its availability makes this energy immensely useful to cells.

In addition to the hydrolysis of ATP to ADP and inorganic phosphate (P_i), a second hydrolysis from ADP to AMP is also used under certain circumstances. As a model for either reaction, let us consider the hydrolysis of simple inorganic pyrophosphoric acid, abbreviated as PP_i.

Since the four hydrogens of pyrophosphoric acid have pK_a's of 0.85, 1.96, 6.54, and 8.44, respectively, we find that PP_i has three negative charges at physiological pHs (around pH 7.4). Since free phosphate will carry two charges at physiological pHs, the hydrolysis of pyrophosphate should be represented as

$$PP_i^{3-} + 2H_2O \rightarrow 2P_i^{2-} + H_3O^+ \qquad (4\text{-}2)$$

or, schematically, as

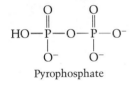
Pyrophosphate

Note that although one P—O bond of pyrophosphate must be broken for the reaction to occur (a process that requires about 400 kJ per mole), a new bond of nearly the same energy is made in the formation of products. As the schematic indicates, water donates the oxygen for the new P—O bond. We drew the reaction this way to emphasize that hydrolysis is a bond replacement or rearrangement rather than a simple cleavage: i.e., there are the same number of covalent bonds in the products as in the reactants. The reaction could not be favorable otherwise.

Because the P—O bond broken during hydrolysis of PP_i is replaced, the only obviously unfavorable part of the hydrolysis is dissociation of water to form the hydronium ion, H_3O^+. The standard state free energy change for that part of the reaction is about $+40$ kJ/mole. However, the overall process of pyrophosphate hydrolysis has a standard state free energy change of -33 kJ/mole. What, then, is there about the pyrophosphate linkage that makes hydrolysis so favorable? That is, why should hydrolysis of PP_i have a net negative $\Delta G°$?

Many factors contribute to making the hydrolysis of PP_i favorable. The most important have to do with the distribution of electrons in the reactant and products. In pyrophosphate, each free P—O bond is a small dipole, with a partial positive charge on the phosphorus (indicated by $\delta+$) and a partial negative charge on the oxygen. The oxygen in the P—O—P bridge, however, carries a partial positive charge. As a result, there are three mutually repulsive positive charges in the core of the molecule (the P—O—P) and around the outside negatively charged oxygens that are also repelling each other. These *charge repulsions* are relieved by hydrolysis.

Note that all the peripheral oxygens in pyrophosphate with the exception of the —OH group are equivalent and carry the same charge. This is a "resonating" structure. The same is true for the hydrolysis product, inorganic phosphate. However, pyrophosphate has five of these partial double bonds (the P—O dipoles) whereas the two products have three each, or a total of six. Since there are the same total number of bonds in reactants (including water) and products, the larger number of partial double bonds in the products means a higher bond energy. Remembering that energy is released when a bond is formed, we recognize that this *resonance stabilization* of products also contributes to the favorable energy change on hydrolysis.

These and other characteristics of the anhydride linkage in pyrophosphate make hydrolysis unusually favorable. Similar features are found also in a few other molecules, making it possible to use them as a medium of exchange. That is, they capture energy in one reaction and deliver it to another. Pyrophosphate itself serves in this capacity for a few reactions. In fact, many bacteria, yeasts, and some other organisms are able to form long polymers of phosphate coupled by anhydride linkages and chelated to Ca^{2+}. This *calcium metaphosphate* may serve to store energy for these cells. It is more common, however, to use relatively complex organic molecules as the link between favorable and unfavorable reactions. We don't know why, but perhaps large molecules are easier to recognize, so that enzymes are less likely to use the wrong ones.

The most widely used of the "high energy" organic compounds, as stated earlier, is ATP. There are two pyrophosphate linkages in this molecule that have roughly equivalent energies of hydrolysis. Most reactions, however, utilize only the terminal phosphate anhydride bond.

$$ATP^{4-} + 2H_2O \rightarrow ADP^{3-} + P_i^{2-} + H_3O^+ \qquad (4\text{-}3)$$

The free energy change for this reaction varies markedly with temperature, pH, and Mg^{2+} availability. (Both ATP and ADP chelate Mg^{2+}, but with different energies.) However, under the usual physiological conditions, the $\Delta G°$ of hydrolysis for the reaction given in Equation 4-3 is in the neighborhood of -30 to -33 kJ/mole. It is not a lot of energy, but it is immensely useful energy.

Coupled Reactions. Suppose that substance A can be transformed to substance B, releasing 50 kJ/mole of free energy. If A is stable—in other words, if there is a large free energy barrier to the reaction—B will not appear at any appreciable rate unless a catalyst is present. That catalyst could be an enzyme constructed in such a way that it is not active unless an ADP and P_i are also bound to its surface. It may also be constructed in such a way that the transformation from A to B simultaneously condenses P_i to ADP, forming ATP. In this way the enzyme, which we shall call E_1, sees to it that in order for A to be changed to B, ATP must also be created. The two reactions are said to be "coupled" by the enzyme as follows:

$$\begin{array}{ll} A \rightarrow B & \Delta G°_1 = -50 \text{ kJ/mole} \\ \underline{ADP + P_i \rightarrow ATP} & \underline{\Delta G°_2 = +30 \text{ kJ/mole}} \\ A + ADP + P_i \rightarrow B + ATP & \Delta G° = \Delta G°_1 + \Delta G°_2 = -20 \text{ kJ/mole} \end{array}$$

(4-4)

One could say that 30 kJ per mole was "captured" by the phosphorylation of ADP. The overall equilibrium for the reactions reflects the total energy change of $-50 + 30 = -20$ kJ/mole. Obviously it is still favorable, but the requirement for an obligate coupling of the two steps has conserved about 60% of the energy that would have been lost to heat and entropy if the first reaction had been allowed to take place alone.

Now suppose there is an essential but energetically unfavorable reaction $C \rightarrow D$. Let $\Delta G°$ for this reaction be $+25$ kJ/mole. No significant amount of C will be converted to D unless the reaction is coupled to a second reaction whose free energy change is negative. Since the hydrolysis of ATP is a favorable reaction, a second enzyme, E_2, might be designed to catalyze the combined reaction:

$$\begin{array}{ll} C \rightarrow D & \Delta G°_3 = +25 \text{ kJ/mole} \\ \underline{ATP \rightarrow ADP + P_i} & \underline{\Delta G°_4 = -30 \text{ kJ/mole}} \\ C + ATP \rightarrow D + ADP + P_i & \Delta G° = \Delta G°_3 + \Delta G°_4 = -5 \text{ kJ/mole} \end{array}$$

(4-5)

Like Equation 4-4, this reaction is also favorable, meaning that the equilibrium will produce more products than reactants.

By comparing Equations 4-4 and 4-5, one can see that part of the energy released by the conversion of A to B was captured in the form of ATP and used to drive the second reaction. The overall reaction, then, is

$$A + C \rightarrow B + D \qquad \Delta G° = -25 \text{ kJ/mole}$$

The intermediate role played by ADP/ATP is indicated by writing the two reactions as follows:

$$\begin{array}{c} A \qquad\qquad D \\ \quad\searrow ADP + P_i \nearrow \\ E_1 \qquad\qquad\qquad E_2 \\ \quad\nearrow ATP \searrow \\ B \qquad\qquad C \end{array}$$

Written in this way it is clear that the energetically unfavorable, or "uphill," reaction $C \rightarrow D$ is driven indirectly by the favorable, or "downhill," reaction $A \rightarrow B$. Actually there is usually more than one step involved in each of the two reactions, often producing a phosphorylated ("activated") intermediate. The result is the same, however.

To couple A → B and C → D directly with one enzyme would be far too rigid. By using ATP as an intermediate energy source, a reaction occurring in one part of the cell can be used to drive a number of different reactions taking place at scattered points elsewhere in the cell. Controlling the flow of energy through the cell then becomes a matter of balancing the production of ATP on the one hand with its utilization on the other. Though maintaining this balance is an intricate task, it is in no way as complicated as direct coupling would have to be, for we need only require that the ATP-yielding reactions work at a variable rate. When the available supply of ATP shrinks, the reactions that create it by breaking down foodstuffs or capturing light are urged on to greater activity. When the ATP supply is plentiful, these ATP-yielding reactions will be slowed.

Even granting the value of an intermediate such as ATP, there is still the need to explain why ATP itself came to fill the role in all cells, from whatever source. We can't say why, of course, except to point out that ATP does have the required characteristics: (1) It has a negative free energy of hydrolysis; (2) it is stable enough to prevent accidentally wasting that energy in a nonenzymatic reaction, and (3) it is complicated enough to be unambiguously recognized by the enzymes that use it. The use of ATP may originally have been accidental, or it may have been because ATP is more readily formed from primordial components. However, once it began to be used, the lack of a competitive advantage and the great difficulty of switching to another substance are probably enough to have assured evolutionary continuity.

4-3 GLYCOLYSIS: A METABOLIC PATHWAY

Reactions that require energy and that result in the biosynthesis of needed compounds are called *anabolic*. Reactions that provide this energy at the expense of the degradation of food molecules are called *catabolic*. *Metabolism* could refer to either or both. We shall discuss one catabolic pathway, glycolysis, and later one anabolic pathway in order to gain some insight into the flow of energy through a cell and the way in which that flow is regulated.

Glucose Catabolism. All our discussions will center about the metabolism of glucose. This preoccupation with glucose metabolism is appropriate for both historical and practical reasons: (1) Glycolysis was among the first metabolic pathways known, for it is nearly identical to alcoholic fermentation, the subject of most early efforts. (2) Humans obtain a substantial fraction of their daily caloric intake from glucose, mostly as starches. (3) Certain animal cells rely very heavily, or even exclusively, on glucose (blood sugar) for their energy. These cells include certain muscle cells, the red blood cells, and nerve cells of all types. (4) And finally, it is easy to tie glucose into the rest of the metabolic scheme, so that (hopefully) we can generate a feeling for the meaning of metabolism without actually going into many of its details.

Glycolysis. When glucose is degraded to lactate, the process is called *glycolysis* (i.e., "glycogen lysis"). The slight variation that provides alcohol as the product is called *alcoholic fermentation*. The individual steps in both pathways have been known since about 1940, with some of

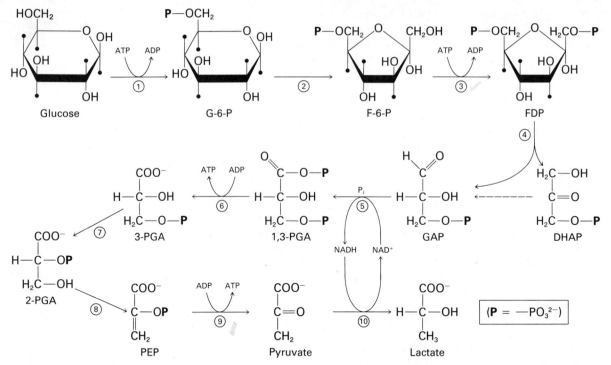

FIGURE 4-4 Glycolysis. See Table 4-3 for a list of the abbreviations and more details on the individual reactions.

$\Delta G° = -200$ kJ/mole

the most important contributions coming from Gustav Embden and Otto Meyerhof. In their honor, the scheme is sometimes called the *Embden-Meyerhof pathway*. For details of the pathway, see Fig. 4-4 and Table 4-3.

The production of two molecules of lactate from one molecule of glucose is highly exergonic, with a standard state free energy change of about -200 kJ per mole of glucose at pH 7.

$$\text{glucose} \rightarrow 2 \text{ lactate}^- + 2H^+ \tag{4-6}$$

A portion of this free energy is conserved in the phosphorylation of two molecules of ADP to form two ATPs. Hence, glycolysis represents a way of deriving useful energy from glucose—the only way this can be done without oxygen. Glycolysis is virtually the sole source of energy for mammalian red blood cells (they make no metabolic use of the oxygen they carry to other tissues), it is an important source of energy to muscles, and it may be utilized by assorted other tissues when oxygen is scarce. The pathway is also used heavily by a few microorganisms (e.g., lactic acid bacteria), but microorganisms generally produce end products other than lactate.

Stepwise Degradation. The transformation from glucose to lactate does not take place in a single step, but in a series of 10 reactions, each catalyzed by its own specific enzyme. This stepwise progression is characteristic of all metabolic pathways, for it allows the cell to make better use of the relatively modest change in free energy that accompanies ATP synthesis or hydrolysis. Hence, the synthesis of a single ATP may conserve a high proportion of the free energy released in a reaction, or the hydrolysis of a single ATP may be used to drive an unfavorable step. In addition, some of the intermediates of one pathway will often serve as

TABLE 4-3. Glycolysis

Step	Reaction[a]	$\Delta G°$, kJ/mole
1	glucose + ATP → G-6-P + ADP + H$^+$	−14
2	G-6-P → F-6-P	+ 2
3	F-6-P + ATP → FDP + ADP + H$^+$	−14
4	FDP → DHAP + GAP	+24
	DHAP → GAP	+ 8
5	GAP + P$_i$ + NAD$^+$ → 1,3-PGA + NADH + H$^+$	+ 6
6	1,3-PGA + H$^+$ + ADP → 3-PGA + ATP	−28
7	3-PGA → 2-PGA	+ 4
8	2-PGA → PEP	− 4
9	PEP + H$^+$ + ADP → pyruvate + ATP	−24
10	pyruvate + H$^+$ + NADH → NAD$^+$ + lactate	−25

[a] G-6-P   glucose 6-phosphate   FDP   fructose 1,6-diphosphate
F-6-P   fructose 6-phosphate   DHAP   dihydroxyacetone phosphate
GAP   glyceraldehyde 3-phosphate   PEP   phosphoenolpyruvate
1,3-PGA   1,3-diphosphoglycerate

convenient starting materials for a second pathway. And, as we shall see, the stepwise pathway, with its several specific enzymes, can be more easily and flexibly controlled.

But aside from these advantages of stepwise degradation, we should admit that probably some multistep mechanism is essential in most cases just because there is no simple chemical reaction that cells could use to effect the transformation in one step under physiological conditions. On the other hand, the pathway as it is presented here is not the only scheme that could be devised to transform glucose to lactate. In fact, minor variations do exist in different cell types, and any good organic chemist could devise other perfectly reasonable alternatives not found in nature, as far as we know.

The energetics of the glycolytic pathway are summarized in Fig. 4-5 and Table 4-3. The vertical axis in Fig. 4-5 is calibrated in kJ per mole, starting from glucose. The rise or fall between any two intermediates represents the standard state free energy change ($\Delta G°$) when the first intermediate is converted to the next. The solid line shows the relationship as it actually exists in the cell, while the dotted line shows what it would be without an obligate coupling to ATP or, in the case of steps 5 and 10, without the coenzyme NAD$^+$. Note that both curves reach the same final position, for both represent the net reaction

$$\text{glucose} + P_i \rightarrow \text{DHAP} + \text{lactate}^- + H^+ \quad (4\text{-}7) \quad \Delta G° = -74 \text{ kJ/mole}$$

In other words, there is no net gain or loss of ATP (two gained and two lost) and no net oxidation or reduction (each occurs once, as we shall see).

It is the utilization of dihydroxyacetone phosphate (DHAP) that accounts for the usefulness of glycolysis, since the conversion of this triose phosphate to its isomer, glyceraldehyde phosphate, permits it to follow the same route to lactate:

$$\text{DHAP} + 2\text{ADP} + P_i + H^+ \rightarrow \text{lactate}^- + 2\text{ATP} \quad (4\text{-}8) \quad \Delta G° = -64 \text{ kJ/mole}$$

The two reactions together, Equations 4-7 and 4-8, are what we call glycolysis:

$$\text{glucose} + 2\text{ADP} + 2P_i \rightarrow 2\text{ lactate}^- + 2\text{ATP} \quad (4\text{-}9)$$
$$\Delta G° = -74 - 64$$
$$= -138 \text{ kJ/mole}$$

Note that $\Delta G°$ for (4-9) is 62 kJ per mole less negative than $\Delta G°$ for Equation 4-6. The difference represents the net synthesis of two moles of ATP from two moles of ADP at $\Delta G° = +31$ kJ per mole each. Hence, glycolysis conserves 62/200 or about one-third of the available standard state free energy.

One of the more interesting features of glycolysis is the requirement that energy be put into the system in the form of 2 ATPs before any recovery is possible. Without this "pump priming," glycolysis would have about a 60 kJ/mole energy barrier—from glucose to GAP—even though

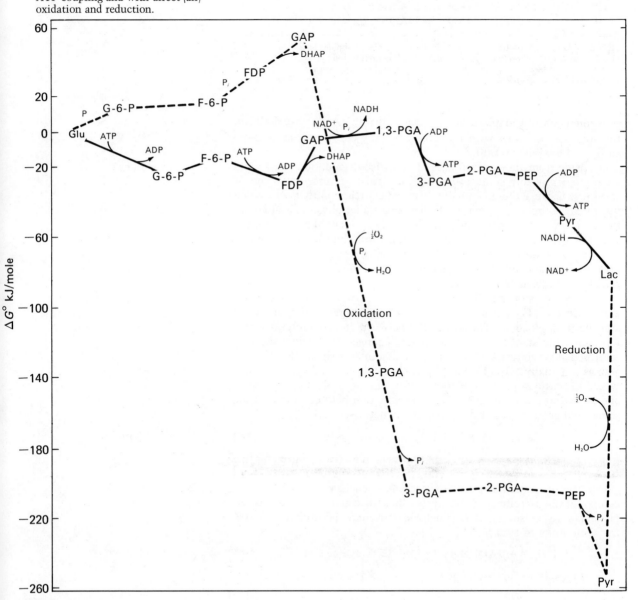

FIGURE 4-5 The Energetics of Glycolysis. The solid line shows the free energy changes as they exist in the coupled reactions. The dashed line shows free energy changes among the main intermediates only, without ATP coupling and with direct (air) oxidation and reduction.

the overall reaction is favorable. With the ATP input, however, GAP lies some 6.5 kJ per mole "below" glucose.

ATP Synthesis. The energy-yielding reactions of glycolysis are at steps 6 and 9 (Table 4-3). Step 6 is the hydrolysis of the phosphoric-carboxylic acid anhydride, 1,3-diphosphoglycerate, which is favorable for the same reasons given earlier for hydrolysis of pyrophosphate, also an acid anhydride. (Hydrolysis of the phosphate ester on the number three carbon is much less favorable.) The cell takes advantage of the large energy drop in the hydrolysis of 1,3-PGA by coupling it to the synthesis of ATP.

The second energy-yielding step in glycolysis is from phosphoenolpyruvate (PEP) to pyruvate. Here, a combination of factors are at work, the most important of which are charge repulsion between the phosphate and nearby carboxylate, and locking of the molecule into an unfavorable, enol configuration. Once hydrolyzed, the immediate product, enolpyruvate, is free to isomerize to the much more stable keto form, a transition that accounts for some 33 kJ per mole of the observed total free energy change.

$$\begin{matrix} COO^- \\ | \\ C-OPO_3^{2-} \\ || \\ CH_2 \end{matrix} \xrightarrow[-P_i]{H_2O} \begin{matrix} COO^- \\ | \\ C-OH \\ || \\ CH_2 \end{matrix} \longrightarrow \begin{matrix} COO^- \\ | \\ C=O \\ | \\ CH_3 \end{matrix}$$

PEP — Enolpyruvate — Pyruvate

(4-10) $\Delta G° = -54$ kJ/mole

$$\begin{bmatrix} C-O \\ || \\ C \end{bmatrix}$$
Enol

Both steps are catalyzed by the same enzyme, pyruvate kinase. Having "kinase" in the name of the enzyme tells you that the reaction *in vivo* is coupled to the synthesis of ATP.

Now direct your attention to two more coupled reactions, steps 5 and 10. The first of these is an oxidation of a glycolytic intermediate, while the second is a reduction.

4-4 BIOLOGICAL OXIDATIONS

An *oxidation* is the removal of an electron from an atom or molecule, while a *reduction* is the addition of an electron. Obviously, the former requires an electron acceptor and the latter an electron donor, for electrons do not just float loosely about in aqueous systems. In other words, a compound can only be oxidized if, at the same time, some other substance is reduced, and vice versa. The combined process is called an oxidation-reduction reaction, or *redox* reaction. The two substances involved are referred to as a "redox pair." In biological systems, many oxidations are accomplished by removing the same number of protons as electrons, thus leaving the charge of the oxidized molecule unaffected. This kind of oxidation is also called a *dehydrogenation*. Conversely, reductions are often hydrogenations, or the addition of neutral hydrogen.

Examples. Four types of oxidations are of particular interest to us here. They are

1. The oxidation of an alcohol to an aldehyde or ketone:

$$-\overset{|}{\underset{|}{C}}-OH \longrightarrow \overset{|}{\underset{|}{C}}=O + 2e^- + 2H^+$$

2. The further oxidation of an aldehyde to a carboxylic acid:

$$HC{=}O + H_2O \longrightarrow COOH + 2e^- + 2H^+$$

3. The creation of an unsaturated bond in a hydrocarbon chain:

$$-CH_2-CH_2- \rightarrow -CH{=}CH- + 2e^- + 2H^+$$

4. The change in oxidation state of a metal ion:

$$Fe^{2+} \rightarrow Fe^{3+} + e^-$$

Note that the first three of these are dehydrogenations in which two protons are removed along with two electrons.

As an example, consider the oxidation of L-ascorbic acid (vitamin C) to produce dehydroascorbic acid, a substance without vitamin activity:

$$\tfrac{1}{2}O_2 + \begin{array}{c} O{=}C \\ | \\ HO{-}C \\ \| \\ HO{-}C \\ | \\ HC \\ | \\ HO{-}CH \\ | \\ CH_2OH \end{array} \longrightarrow \begin{array}{c} O{=}C \\ | \\ O{=}C \\ | \\ O{=}C \\ | \\ HC \\ | \\ HO{-}CH \\ | \\ CH_2OH \end{array} + H_2O$$

L-Ascorbic acid   L-Dehydroascorbic acid

A comparison of the structures reveals two alcohol to ketone dehydrogenations plus the hydrogenation of a double bond. The net result is the removal of two electrons and two protons from ascorbic acid. The electrons and protons are transferred to molecular oxygen in the coupled reduction, forming water. It is this tendency to oxidize in the presence of air that is responsible for loss of vitamin C activity during the storage or processing of foods.

Glycolytic Redox Reactions. Step 5 in glycolysis, the formation of 1,3-diphosphoglycerate from glyceraldehyde phosphate, represents the oxidation of an aldehyde to an acid. The last step in the pathway involves the reduction of a keto group in pyruvate to an alcoholic group in lactate.

The glycolytic pathway, then, contains one oxidation and one reduction, the net effect of which is to transfer two electrons from glyceraldehyde phosphate to pyruvate. The vehicle for this transfer is nicotinamide adenine dinucleotide, or NAD$^+$. By being oxidized in one place and reduced in another, NAD$^+$ is capable of shuffling electrons around the cytoplasm in much the same way that ATP carries phosphate. It is one of several electron carriers with which we shall need to be concerned.

Nicotinamide adenine dinucleotide (NAD$^+$) consists of a nicotinamide mononucleotide, or NMN (the nicotinamide base plus ribose 5' phosphate), linked by a phosphate anhydride (pyrophosphate) bond to AMP. Hence the "dinucleotide" designation. Its functional part is the nicotinamide ring, a substance that humans must have in their diet—it is the vitamin nicotinic acid, or niacin.

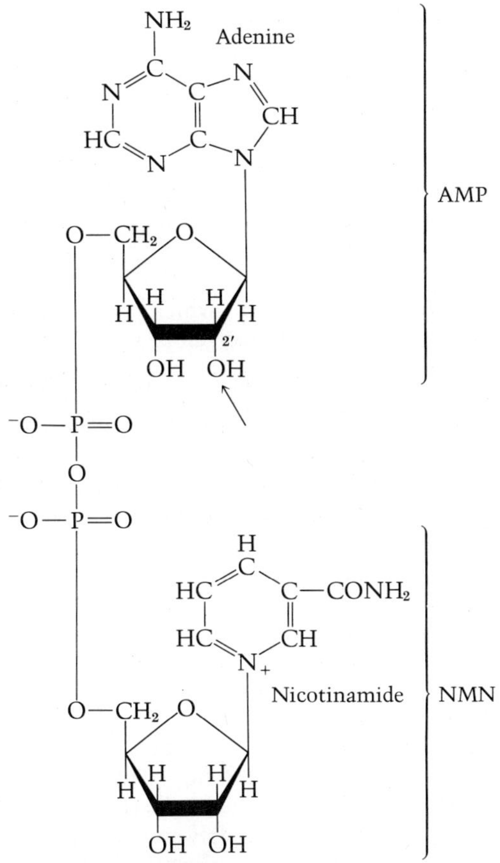

SECTION 4-4
Biological Oxidations

Oxidized nicotinamide adenine dinucleotide (NAD⁺)

[Structure showing AMP (adenine + ribose + phosphate) linked via diphosphate to NMN (nicotinamide + ribose)]

The addition of two electrons and a proton (together a hydride ion, H⁻) to the nicotinamide ring produces the reduced form, NADH.

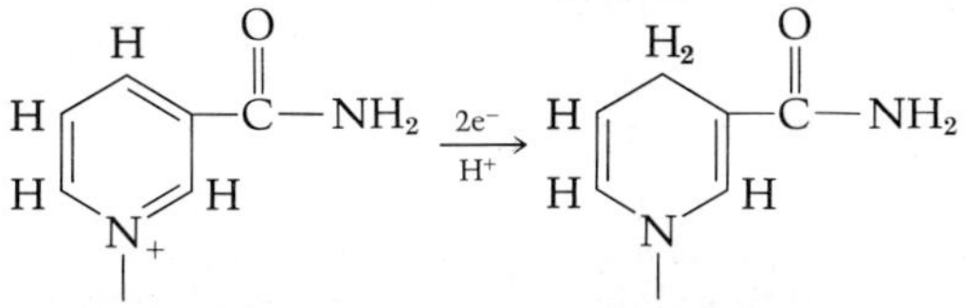

$$\text{Oxidized nicotinamide} \xrightarrow{2e^-, H^+} \text{Reduced nicotinamide}$$

NAD⁺ has a very close relative known as nicotinamide adenine dinucleotide phosphate, or NADP⁺. The only difference between the two is a phosphate ester in NADP⁺ at the 2′ position of the AMP, marked with an arrow in the diagram. The properties of NAD⁺ and NADP⁺ are very similar, but enzymes generally have a strong preference for one or the other.

NAD⁺ Recycling. Since NAD⁺ is present only in tiny amounts in the cell, the function of step 10 in glycolysis, the reduction of pyruvate to lactate, is to regenerate the NAD⁺ needed by step 5 (GAP to 1,3-PGA). Any other mechanism for regenerating NAD⁺ would also serve this purpose, and in various organisms other methods are used. Some microorganisms, for example, accomplish the regeneration by reducing (and decarboxylating) pyruvate to ethanol instead of to lactate.

Pyruvate → (H⁺, −CO₂) → Acetaldehyde → (NADH+H⁺ → NAD⁺) → Ethanol

$$\text{COO}^- \qquad \text{HC}=\text{O} \qquad \text{H}_2\text{C}-\text{OH}$$
$$|\qquad\qquad |\qquad\qquad |$$
$$\text{C}=\text{O} \qquad \text{CH}_3 \qquad\quad \text{CH}_3$$
$$|$$
$$\text{CH}_3$$

The best way of regenerating NAD⁺, however, is to use molecular oxygen as an electron acceptor, for oxygen is readily reduced.

$$\tfrac{1}{2}O_2 + 2e^- + 2H^+ \rightarrow H_2O \tag{4-11}$$

Because of the ease with which oxygen is reduced, the transfer of electrons from NADH to oxygen is a very exergonic process:

$$\Delta G^\circ = -220 \text{ kJ/mole}$$

$$\text{NADH} + H^+ + \tfrac{1}{2}O_2 \rightarrow \text{NAD}^+ + H_2O \tag{4-12}$$

Yeasts, which are unicellular plants, ferment glucose to ethanol under anaerobic conditions, thus producing beer, wine, or other alcoholic beverages. When oxygen is present, however, NAD⁺ is recycled through Equation 4-12 instead—hence, the importance of keeping fermenting beverages tightly sealed.

The size of the free energy change in the oxidative recovery of NAD⁺ is reflected in Fig. 4-5 by the huge difference in ΔG° between the two versions of step 5, in one of which the oxidation is carried out directly by oxygen instead of NAD⁺. We see the same difference in comparing the two versions of step 10. Most cells take advantage of the large negative free energy change of Equation 4-12 by coupling it to the synthesis of several molecules of ATP—not in a single step, but in a chain of reactions called the *electron transport chain*. The chain starts with the oxidation of NADH and ends with the production of water. In between is a series of reactions having manageably small changes in free energy. The electron transport chain, found in the mitochondria of eukaryotes, will occupy a portion of our attention in Chap. 7, where the very great advantage of aerobic (that is, oxygen-utilizing) metabolism will be made clear.

4-5 ENZYME KINETICS AND REGULATION

Glycolysis is an example of a single metabolic pathway out of many that exist in a typical cell, involving often hundreds of individual reactions. The way in which the major pathways interrelate is outlined in the sketchiest way possible in Fig. 4-6. The problem faced by the cell is to coordinate these pathways in order to utilize in an optimum way raw materials and energy available to it. Cells that fail in this task find themselves at a competitive disadvantage and are likely to be replaced in the course of evolution.

All enzymes function at a rate that varies with the availability of substrates and the concentration of products. Many, however, are subject to control by less obvious mechanisms as well. In particular, we find that certain key enzymes are regulated by molecules that are neither substrate nor product, a situation that is often attributable to feedback regulation.

Feedback Regulation. Feedback control means that the intracellular level of a substance regulates the rate of its own synthesis by controlling the amount and/or the activity of one or more enzymes within the

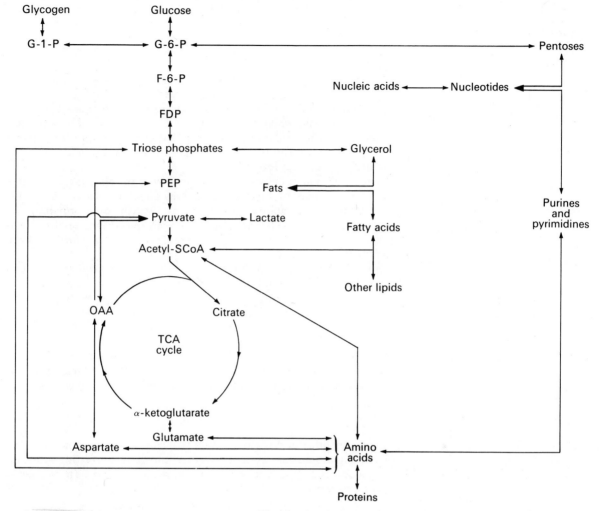

FIGURE 4-6 Outline of Intermediary Metabolism. Only the main mammalian pathways are shown. The TCA cycle (or citric acid cycle—see Chap. 7) is a mitochondrial pathway.

pathway that produces it. The earliest case of feedback regulation was reported in 1940 by Z. Dische, who found that the phosphorylation of glucose is inhibited by phosphoglycerate in red blood cell hemolysates (cell-free suspensions). Since phosphorylation of glucose is the first step in the glycolytic sequence that produces phosphoglycerate as an intermediate, Dische suggested that glycolysis might be regulated by controlling the rate of glucose phosphorylation, perhaps through an inhibition of the enzyme responsible, glucokinase.

The importance of this suggestion went unrecognized until the mid-1950s, when a series of workers began to investigate the biosynthesis of amino acids in microorganisms. Several cases were soon reported in which an exogenous amino acid (that is, an amino acid supplied via the growth medium) was effective in suppressing the cell's ability to synthesize more. Further investigations showed that enzymes needed to produce the amino acid were no longer being made, and that those already present were being inhibited. There are now many similar examples known, where the catalytic rate of enzymes can be inhibited by small molecules that are neither a substrate nor a product of the affected enzyme.

Consider the following hypothetical anabolic (biosynthetic) pathway:

$$A \xrightarrow{E_1} B \xrightarrow{E_2} C \xrightarrow{E_3} D$$

The intracellular level of the product, D, may control its own rate of synthesis by acting on E_1, the enzyme that catalyzes the first step unique to its production. When the level of D rises, its presence causes E_1 to work more slowly; and when the level of D falls, E_1 is allowed to work at a faster rate. The situation is similar to the function of a float valve: a rising liquid level causes the float valve to inhibit the incoming flow, maintaining a preset level of liquid.

The regulation of E_1 by D, the end product of the pathway, is possible only if D can bind to the enzyme. It may do so at the active site, thus interfering with the ability of the site to catalyze the transformation of A to B. Or D may bind elsewhere on the protein and affect the rate of catalysis in more subtle ways. In either case, D is called an *inhibitor* of the enzyme.

Since each reaction in a cell is catalyzed by its own enzyme, the key to maintaining an efficient metabolism is the regulation of either the quantity or activity of enzymes. The regulation of quantity is treated in the next chapter. Here, we deal with the regulation of enzyme activity, which must start with consideration of how enzymes behave before regulatory constraints are placed on them.

Michaelis-Menten Kinetics. Consider the enzymatically catalyzed interconversion of two molecules, one of which we shall refer to as substrate (S) and the other as product (P).

$$S \rightleftharpoons P$$

This or any other enzymatic reaction will consist of at least three steps: (1) binding of substrate to the enzyme, (2) catalysis, converting substrate to product, and (3) release of product from the enzyme. The sequence may be summarized as

$$E + S \xrightarrow{\text{binding}} ES \xrightarrow{\text{catalysis}} EP \xrightarrow{\text{release}} E + P$$

The quantities that can be measured during the course of a reaction are usually limited to the concentrations of substrate and product. The rate at which product appears is commonly referred to as the *velocity* (v) of the reaction. The velocity of an enzymatically catalyzed reaction ordinarily increases with substrate availability up to some level at which the enzyme is saturated with substrate. This *maximum velocity*, V_{max}, is directly proportional to the amount of active enzyme present. For a majority of enzymes these quantities of substrate concentration, velocity, and maximum velocity are simply related through an equation widely known as the *Michaelis-Menten equation*:

$$v = \frac{V_{max}}{1 + K_m/[S]} \qquad (4\text{-}13)$$

The constant, K_m, is called the *Michaelis constant*. It may be defined as

$$K_m = \frac{[E][S]}{[ES]} \qquad (4\text{-}14)$$

Note that in this form, the Michaelis constant appears to be a dissociation (equilibrium) constant. That similarity is only superficial, however, for the indicated concentrations are those measured during the course of the reaction; they are not equilibrium values.

Although it is not a true dissociation constant, K_m is a measure of affinity between enzyme and substrate. Specifically, *the Michaelis constant is numerically equal to the amount of substrate needed to achieve half of the maximum velocity.* (To convince yourself, replace v by $\frac{1}{2}V_{max}$ in Equation 4-13 and then simplify the equation.) Larger values of K_m mean higher substrate concentrations are needed to achieve a given fraction of the maximum potential activity.

The relationship between velocity and substrate concentration predicted by the Michaelis-Menten equation is shown in Fig. 4-7a. An enzyme that behaves in this way is said to exhibit Michaelis-Menten or *hyperbolic* kinetics. Figure 4-7a can be used to obtain values for K_m and V_{max} as shown. With these values, one can calculate using Equation 4-13 the velocity at any given substrate concentration and hence reproduce the entire velocity vs. substrate profile. A more convenient way of determining K_m and V_{max} is shown in Fig. 4-7b. It is called a *Lineweaver-Burk plot*, and is based on the reciprocal of Equation 4-13.

$$\frac{1}{v} = \frac{K_m}{V_{max}} \frac{1}{[S]} + \frac{1}{V_{max}} \quad (4\text{-}15)$$

Equation 4-15 predicts that a double reciprocal plot of $1/v$ against $1/[S]$ will be a straight line with a slope numerically equal to K_m/V_{max} and an intercept on the $1/v$ axis equal to $1/V_{max}$. In addition, the extrapolated intercept on the independent (substrate) axis will have a value equal to $-1/K_m$. (Set $1/v$ equal to zero in Equation 4-15 to establish that relationship.) In this way, both the Michaelis constant and the maximum velocity may be obtained from a relatively small number of observations made under conditions where the maximum velocity is not closely approached.

The two constants K_m and V_{max} tell us a great deal about an enzyme: (1) V_{max} gives us a measure of the catalytic effectiveness of the enzyme; (2)

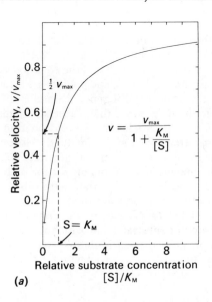

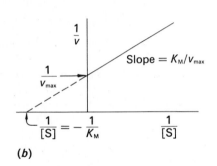

FIGURE 4-7 Michaelis-Menten Kinetics. (a) Hyperbolic velocity plot—see Equation 4-13. (b) Double reciprocal plot. Also called a Lineweaver-Burk plot—see Equation 4-15. It represents a convenient way of obtaining the kinetic constants. K_m and V_{max}.

K_m tells us the concentration of substrate that is necessary to achieve half the maximum velocity (values of 0.001 M or less are common), and (3) V_{max} and K_m can be used as constants in the Michaelis-Menten equation to predict the velocity of the reaction at any other concentration of substrate. Thus, while K_m measures binding affinity, V_{max} is a *turnover number*, or number of molecules of substrate converted to product per minute for each molecule of enzyme (or sometimes for each active site on the enzyme). Turnover numbers of enzymes are often in the neighborhood of a few thousand, but exceed a million or more in some cases. They are clearly related in a simple way to the maximum velocity, which is usually expressed as micromoles of substrate consumed per minute per milligram of enzyme.

Limitations of the Michaelis-Menten Equation. The Michaelis-Menten equation is enormously useful in describing the behavior of enzymes. As we shall see, it also describes the transport of many substances across cell membranes. We shall not go into its derivation here (the derivation is simple, and may be found in any full-length biochemistry book) but it is important to recognize the assumptions made during the derivation and which, therefore, define the conditions where the equation is valid. These assumptions are as follows:

1. Catalysis occurs very rapidly, so that the three steps of binding, catalysis, and release may be reduced to two, each of which is reversible:

$$E + S \rightleftharpoons ES \rightleftharpoons E + P \qquad (4\text{-}16)$$

(The choice of ES instead of EP for the middle term is quite arbitrary.)

2. The rate at which substrate disappears is equal to the rate at which product appears. This *steady-state hypothesis* says only that the concentration of [ES] remains constant during the period of observation, a situation that is almost always true from a few milliseconds after initiating the reaction until equilibrium begins to be approached.

3. The third and most often violated assumption is that the back-reaction—i.e., conversion of product back to substrate—be negligibly small. This condition can usually be met by limiting observation to relatively early times after mixing enzyme and substrate, for if there is no product yet, there can be no back-reaction. If the reaction cannot be carried out in that way, the velocity of reaction is often described approximately by $v_{forward} - v_{backward}$ where $v_{forward}$ is the Michaelis-Menten equation as we have been writing it and $v_{backward}$ the equation as it would be written for the reaction where substrate and product are interchanged. This back-reaction P → S will have a different V_{max} and a different K_m, one that describes product binding rather than substrate binding.

An additional limitation of the Michaelis-Menten equation is that it considers only a single substrate and a single product. Most enzymatically catalyzed reactions involve more than one substrate and/or product. In those cases, the enzyme has a different K_m for each reactant, measured by varying that reactant as a substrate with other substrates held constant at a high (saturating) concentration.

Other deviations from Michaelis-Menten kinetics arise when there are substances present that are not substrates but which nevertheless bind to the active site. The effect of these "inhibitors" will be considered next, followed by discussion of a class of enzymes that do not obey Michaelis-Menten kinetics at all.

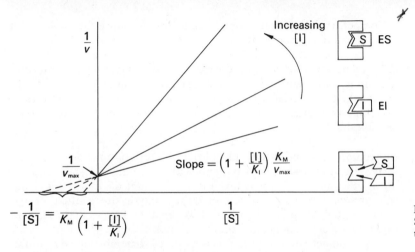

FIGURE 4-8 Competitive Inhibition. Substrate and inhibitor compete for the same site.

Competitive Inhibition. Enzymes gain their specificity from the particular conformation of an active site, a conformation that is designed to recognize and bind to substrate molecules. However, some molecules that merely resemble the substrate also bind to the active site. This results in an inhibition of the enzyme, called *competitive inhibition* because the inhibitor (I) is competing with the substrate (S) for the active site (Fig. 4-8).

In addition to the reaction

$$E + S \rightleftharpoons ES \rightleftharpoons E + P$$

we have the competing reaction

$$E + I \rightleftharpoons EI$$

which may or may not give rise to a product.

This competing reaction is characterized by a dissociation constant given by

$$K_I = \frac{[E][I]}{[EI]} \qquad (4\text{-}17)$$

The overall reaction is described by a double reciprocal equation analogous to the Michaelis-Menten equation:

$$\frac{1}{v} = \left(1 + \frac{[I]}{K_I}\right)\left(\frac{K_m}{V_{max}}\right)\left(\frac{1}{[S]}\right) + \frac{1}{V_{max}} \qquad (4\text{-}18)$$

This equation implies that the only effect of competitive inhibition is to change the slope of the double reciprocal plot. The higher the inhibitor concentration (or the lower the value of K_I), the steeper will be the slope. The maximum velocity, however, will be unchanged; this is not surprising, since V_{max} is defined in terms of an infinite substrate concentration, and under those conditions a finite amount of inhibitor will be swamped and unable to bind. The double reciprocal plots are still linear, so that K_m, V_{max}, and K_I may be obtained from the slope and intercepts of two such plots carried out at different inhibitor concentrations. If we think of K_m as the concentration of substrate needed to achieve half-maximum velocity, K_I becomes the inhibitor concentration that appears to double the value of K_m.

Competitive inhibition is very common *in vivo* since the product of a reaction is often a competitive inhibitor of the substrate, which it may still resemble. Thus, an enzyme-catalyzed reaction may be slowed by the presence of an increasing concentration of product long before equilibrium is approached. This is undoubtedly an important way of maintaining the proper balance between reactant and product for various reactions in which attainment of actual equilibrium concentrations would be undesirable.

Noncompetitive Inhibition. In addition to inhibition by binding at the active site, it is possible for small molecules to inhibit enzymes by binding at other sites. This may result in inhibition if the molecules physically interfere with (sterically hinder) the catalytic site, a situation that is represented schematically in Fig. 4-9. Small molecules that behave in this way are called *noncompetitive inhibitors*.

In noncompetitive inhibition, it is assumed that substrate and inhibitor occupy separate sites, so that both may be bound at the same time. Thus, in addition to the reaction

$$E + S \rightleftharpoons ES \rightleftharpoons E + P$$

where $K_m = [E][S]/[ES]$, there are two other reactions that must be considered:

$$E + I \rightleftharpoons EI$$
and
$$ES + I \rightleftharpoons ESI$$

These reactions are described by the same inhibitor constant

$$K_I = \frac{[E][I]}{[EI]} = \frac{[ES][I]}{[ESI]} \qquad (4\text{-}19)$$

because they really represent the same reaction, namely the binding of inhibitor to its specific site. The velocity in the presence of a noncompetitive inhibitor is given as

$$\frac{1}{v} = \left(1 + \frac{[I]}{K_I}\right)\left[\left(\frac{K_m}{V_{max}}\right)\left(\frac{1}{[S]}\right) + \frac{1}{V_{max}}\right] \qquad (4\text{-}20)$$

Thus, both the slope and the intercept of the double reciprocal plot are a function of inhibitor concentration if the inhibition is noncompetitive.

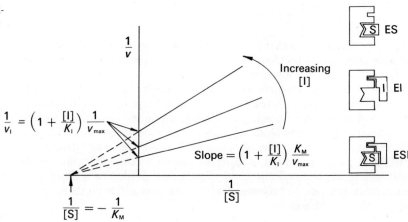

FIGURE 4-9 Noncompetitive Inhibition. Substrate and inhibitor bind to separate sites. When inhibitor is bound, the substrate cannot be released as product. The parameter v_I represents the maximum velocity in the presence of inhibitor.

(Only the slope was changed in competitive inhibition, providing a way to distinguish experimentally between competitive and noncompetitive inhibition.) Note that a noncompetitive inhibitor does not change the intercept on the substrate axis: If the lines are extrapolated back to infinite velocity $(1/v = 0)$, they all meet at the point where the value of $1/[S]$ is numerically equal to $-1/K_m$ (see Fig. 4-9).

Noncompetitive inhibition is characterized by a smoothly decreasing value of the reaction velocity as the inhibitor concentration is increased. In fact, $1/v$ varies linearly with I for both types of inhibition. Thus, the noncompetitive inhibition of an early enzyme in a biosynthetic pathway by the end product of that same pathway could furnish the kind of feedback mechanism needed for efficient regulation. As the end product builds up, the activity of an inhibited enzyme decreases; as the end product disappears, the enzyme is again free to function. Although this is exactly what happens in many pathways, often the enzyme does not exhibit the simple Michaelis-Menten kinetics we have just described. Rather, nature has found a way to sensitize certain enzymes to small changes in inhibitor concentration through cooperative effects. This condition, called *allosteric inhibition*, is a phenomenon of great importance.

Cooperativity Many enzymes have more than one active site. Each site may act independently of the others, just as if they were on different molecules, in which case the enzyme exhibits normal Michaelis-Menten kinetics. On the other hand, the binding of a substrate molecule at one site may affect the catalytic activity of the other sites, even if the other sites are too far away from the first to cause any direct physical interference. Such an interaction, which is attributed to *cooperative effects* among the sites, was first noticed in the oxygen-carrying molecule, hemoglobin.

Hemoglobin, of course, is not an enzyme. But it is considered an "honorary enzyme" because the binding of ligands by hemoglobin is analogous to the binding of substrate by a true enzyme. The function of hemoglobin is to pick up ligands (oxygen in the lungs, carbon dioxide in other tissues) and transport them via the blood to a point where they are discharged. One oxygen molecule can be bound by each of the four heme groups in a reaction that mimics enzyme-substrate binding.

We might expect the oxygen-binding curve of hemoglobin (the percent saturation of the ligand sites vs. the partial pressure of oxygen) to have the same characteristics as the velocity-vs.-substrate curve of an enzyme. However, rather than the hyperbolic curve thus predicted, the actual oxygen-binding curve of hemoglobin is sigmoidal (S-shaped—see Fig. 4-10) and is adequately described, over a limited concentration range, by an empirical relationship known as the *Hill equation*:

$$\frac{y}{1-y} = K(pO_2)^n \tag{4-21}$$

The fraction of binding sites occupied, equivalent to the fractional velocity v/V_{max} of an enzyme, is given the symbol y. The substrate concentration is represented as the partial pressure of oxygen, pO_2, and K is merely a constant of proportionality. The superscript n is called the *Hill coefficient*. Theoretically, the value of n can never be greater than the number of binding sites per molecule. A value of n exactly equal to the

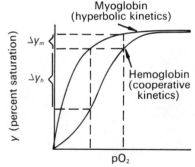

FIGURE 4-10 The Advantages of Cooperativity. The advantages conferred by cooperativity can be seen by comparing hemoglobin with myoglobin. The drop in oxygen tension (change in pO_2) between the lungs and tissues would cause a change, Δy_m, in the binding of oxygen by myoglobin. Hemoglobin, on the other hand, loses Δy_h of its oxygen, a much larger fraction.

number of binding sites would be perfect cooperativity—i.e., the protein could only have n substrate molecules bound to it at any given time, or none at all. Obviously this limiting case cannot be attained, for the ligands will always be added one at a time.

If we rearrange Equation 4-21, we can see that it looks very much like the Michaelis-Menten equation except for n. In fact, when n is one, Equation 4-21 *is* the Michaelis-Menten equation:

$$\frac{1}{v} = \frac{1}{KV_{max}} \frac{1}{[S]^n} + \frac{1}{V_{max}} \qquad (4\text{-}22)$$

Here y has been replaced by v/V_{max} and pO_2 by S to complete the analogy. In the special case where $n = 1$, the constant K is equivalent to $1/K_m$, but otherwise K is a somewhat more complicated term than the Michaelis constant.

The combination of n and K will be such that y (or v) is less at any given substrate concentration than would be expected in the absence of cooperativity. When n is greater than one, the protein is said to exhibit *positive cooperativity*; although it is unusual, an n less than one indicates *negative cooperativity*, and will cause the v or y vs. [S] curve to be a flattened hyperbola rather than a sigmoid.

Clearly, the value of n, the Hill coefficient, is a measure of cooperativity, for the larger its value the sharper the sigmoidicity. The Hill coefficient is measured from plots of $\log(y/(1-y))$ vs. $\log[S]$, which results from recasting Equation 4-21 into the following form:

$$\log\left(\frac{y}{1-y}\right) = n\log[S] + \log K \qquad (4\text{-}23)$$

The slope of such a plot is equal to n.

An examination of Fig. 4-10 shows the value of cooperativity to the biological function of hemoglobin. In the absence of cooperativity, each of its four subunits would bind oxygen very much as does its close cousin myoglobin (upper curve in the graph). But note the steepness of the binding curve of hemoglobin in the region of physiological oxygen tension, a steepness achieved by changing from the hyperbolic to a sigmoidal shape. The degree of oxygen saturation of hemoglobin is very sensitive to the availability of oxygen: it drops a larger fraction of its load in the tissues than it could without cooperativity, while reaching almost the same degree of saturation in the lungs. At oxygen levels characteristic of muscle, myoglobin has a much higher affinity for oxygen than does hemoglobin. The difference fosters an orderly transfer of oxygen from hemoglobin to myoglobin, consistent with the biological role of myoglobin, which is to store oxygen within the cell until it is needed by mitochondria.

This same property—a sensitivity to small changes in substrate concentration—is found in enzymes that exhibit cooperative effects. In addition, as we shall see later, cooperativity can be exploited by molecules other than substrate, in order to achieve a particularly effective kind of control over enzymatic rates.

When cooperative effects were first noticed in hemoglobin in the early part of this century, they were attributed to direct electrostatic interactions among the binding sites. It is now known that direct interaction is unlikely, since the binding sites (the heme groups) are approximately at the corners of a tetrahedron composed of the four roughly spherical subunits that make up the molecule. Because of the physical separation of the binding sites, then, other ideas had to be proposed to explain the apparent interaction among them.

The first real clue to the nature of cooperativity came in the early 1930s, when F. Haurowitz noticed that crystals of deoxyhemoglobin broke when exposed to oxygen. Further studies revealed that oxyhemoglobin and deoxyhemoglobin have different crystal structures, implying that the molecules have different conformations (spatial configurations of the polypeptide chains). X-Ray work by M. F. Perutz and his colleagues confirmed that the binding of oxygen is in fact accompanied by a change in the structure of hemoglobin.

Cooperativity is due to the existence of at least two conformational states: one in which the binding sites have a high affinity for substrate, and another in which they have a low affinity for substrate. Proteins exhibiting cooperative binding effects will have one conformation, on the average, in the presence of substrate and another in its absence. The change from inactive to active occurs before the protein is saturated with substrate, so that unoccupied sites have a maximum affinity for more substrate once the binding process begins (Fig. 4-11).

Enzymes with only a single subunit and only one binding site should always have hyperbolic (Michaelis-Menten) binding curves. Myoglobin, for instance, exhibits hyperbolic oxygen binding with no hint of sigmoid character (see Fig. 4-10). When the hemoglobin tetramer is dissociated, it is found that each isolated subunit by itself not only resembles myoglobin physically, but behaves like myoglobin. Cooperative behavior depends on subunit-subunit interactions, and is lost when the subunits are separated.

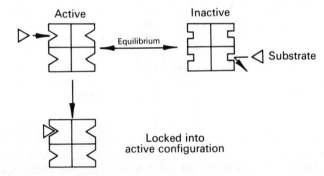

FIGURE 4-11 The Two-state Model for Cooperativity. The enzyme is normally an equilibrium mixture of two configurations. In one configuration, all sites have a high affinity for substrate. In the other configuration, all sites have a low affinity. The presence of a single molecule of bound substrate (in this model) locks the enzyme in the active form, causing the remaining sites to become fully active.

Allosteric Regulation. Suppose now that a second kind of binding site exists, separate from the active site, at which presence of a substance stabilizes the enzyme in one or the other of its possible conformations. If the enzyme is stabilized in the inactive state, then its activity has been inhibited by the substance, which need not be either substrate or product. This kind of noncompetitive inhibition is called *allosteric inhibition*, which means "inhibition without steric hindrance" of the active site. On the other hand, a substance may stabilize an enzyme in its active state, in which case the enzyme will function at a more rapid rate. Ligands that have either of these properties are said to be *allosteric effectors* (see Fig. 4-12). A *positive effector* is one that stabilizes the enzyme in the active state. It is, therefore, an "activator." Conversely, allosteric inhibitors are *negative effectors*. The sites at which allosteric

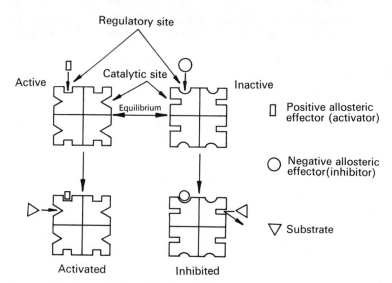

FIGURE 4-12 Allosteric Regulation. When the catalytic sites are active, the regulatory sites will accept a positive allosteric effector (activator). Binding of the activator locks the enzyme into this state and prevents a transition back to the less active state. Conversely, a negative effector (inhibitor) can bind when the enzyme is in its inactive conformation, and in so doing stabilizes that condition.

effectors bind are called *regulatory* or *allosteric* sites, to distinguish them from the catalytic, or active, sites.

Some effectors act by changing K in the Hill equation (4-22 or 4-23); others change the Hill coefficient, n; and some do both. A positive effector often has the property of reducing the Hill coefficient to one, in which case the velocity curve becomes hyperbolic, as we have seen. The enzyme in this condition is indistinguishable from a normal one-subunit enzyme, since it obeys Michaelis-Menten kinetics (see Fig. 4-13).

We shall see other examples of allosteric regulation later in this chapter. However, just as hemoglobin served to introduce the phenomenon of cooperativity, it can serve to illustrate allosteric regulation, for its binding curve can be significantly affected by a number of substances that bind to places other than the active sites—which, for hemoglobin, means to places other than the four hemes. For example, a drop in pH shifts the binding curve to the right, freeing more oxygen from hemoglobin. This phenomenon, known as the *Bohr effect* (Fig. 4-14), is important because both anaerobic and aerobic metabolism can cause a local lowering of the pH. Lactic acid, for instance, is released as a glycolytic by-product in anaerobic metabolism—e.g., in strenuously exercising muscles. One of the products of aerobic metabolism is carbon dioxide, produced largely as a result of decarboxylations in mitochondria. Carbon dioxide increases acidity by combining with water to form carbonic acid, which then dissociates to yield a proton and bicarbonate ($CO_2 + H_2O \rightarrow HCO_3^- + H^+$). Thus, the Bohr effect is a mechanism whereby oxygen delivery can be increased selectively to sites with the greatest need.

The Bohr effect also buffers the blood against wide changes in pH that might otherwise accompany a massive exchange of oxygen and carbon dioxide in the lungs and elsewhere. Oxyhemoglobin is a stronger acid than deoxyhemoglobin. Thus, the following reactions, representing oxygen capture and carbon dioxide release, take place in the lungs.

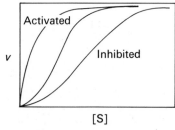

FIGURE 4-13 The Effects of Allosteric Regulators on Reaction Velocity. In the presence of a positive effector (activator), the Hill coefficient in Equation 4-22 may be reduced to unity, whereupon the enzyme becomes indistinguishable from a one-site, Michaelis-Menten type enzyme. On the other hand, the presence of a negative effector (inhibitor) flattens the curve, through changes in both n and K.

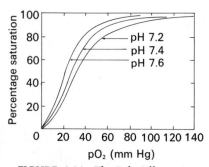

FIGURE 4-14 The Bohr Effect. A lowered pH shifts the oxygen-binding curve to the right, releasing an additional increment of oxygen.

$$\text{Deoxyhemoglobin}^+ + O_2 \rightarrow \text{oxyhemoglobin} + H^+ \qquad (4\text{-}24)$$

$$H^+ + HCO_3^- \rightarrow H_2CO_3 \xrightarrow[\text{anhydrase}]{\text{carbonic}} H_2O + CO_2 \qquad (4\text{-}25)$$

(The proton released in Equation 4-24 is called a "Bohr proton.") In tissues where oxygen is utilized and carbon dioxide is formed, Equations 4-24 and 4-25 should be read from right to left.

It has long been known that the binding curve of hemoglobin for oxygen also undergoes a large shift to the left (smaller oxygen tension) when the protein is released from erythrocytes. The reason for the large difference in the behavior of free and erythrocytic hemoglobin, as it turns out, is mostly presence within the erythrocyte of relatively large quantities (on the order of 0.004 M) of 2,3-diphosphoglycerate (2,3-DPG). The existence of this metabolite was long an enigma to biochemists and physiologists because it is not directly in the pathway of glycolysis, but is formed by the isomerization of the normal intermediate, 1,3-DPG. It is now clear, however, that 2,3-DPG helps to regulate the process of oxygen transfer to and from hemoglobin and provides a mechanism that allows a markedly improved efficiency in times of stress by acting as an allosteric inhibitor. That is, the affinity of hemoglobin for oxygen is significantly reduced in the presence of 2,3-DPG, just as it is when the pH is lowered. The allosteric site for DPG binding has been demonstrated by X-ray crystallography, which shows clearly that it is physically removed from the active site. As noted in Fig. 4-15, the DPG site is the crevice between the N-terminus parts of the β chains. The width of this crevice is greater in the deoxy state and accommodates DPG much more readily. Hence, the presence of DPG shifts the conformational equilibrium in this direction, which is the state of low oxygen affinity.

Regulation of hemoglobin by its allosteric effector, DPG, is thus a mechanism for controlling oxygen-carrying capacity. One of the compensatory changes that take place in a severe anemia (that is, shortage of erythrocytes) or lack of oxygen (resulting from lung disease or high altitudes) is an increase in erythrocyte DPG to permit a greater drop in oxygen at tissue sites.

Control of Glycolysis. Let us apply the regulatory mechanisms just described to the control of our sample metabolic pathway, glycolysis. The two most important regulatory sites in glycolysis are steps 3 and 9 in Fig. 4-4 and Table 4-3:

fructose-6-P + ATP $\xrightarrow{\text{phosphofructo-kinase}}$ fructose-1,6-diphosphate + ADP

and   phosphoenolpyruvate + ADP $\xrightarrow{\text{pyruvate kinase}}$ pyruvate + ATP

The enzyme responsible for the first of these reactions, phosphofructokinase or PFK, is regulated primarily by the level of "high energy" phosphates. At low levels of ATP, the velocity of PFK increases with ATP as one would expect for a cooperative enzyme having ATP as a substrate. At higher levels, however, the velocity begins to decrease as the level of ATP is increased further. In other words, the enzyme is subject to *substrate inhibition* by high levels of ATP. It is also subject to activation by ADP and AMP, which build up during conditions of energy deficit, when stimulation of glycolysis would be desirable.

In addition to regulation by ATP and its hydrolysis products, phosphofructokinase is allosterically inhibited by NADH and citrate. Both are substances that build up during periods of energy surplus, primarily via activity of mitochondria. Since the glycolytic pathway supplies fuel to

$$\begin{array}{c} O \\ \parallel \\ C-OPO_3^{2-} \\ | \\ HC-OH \\ | \\ H_2C-OPO_3^{2-} \end{array}$$
1,3-DPG
(or 1,3-PGA)

$$\begin{array}{c} O \\ \parallel \\ C-O^- \\ | \\ HC-OPO_3^{2-} \\ | \\ H_2C-OPO_3^{2-} \end{array}$$
2,3-DPG

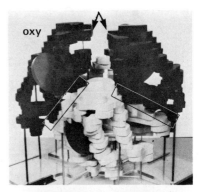

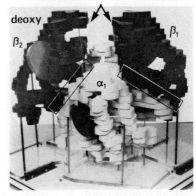

FIGURE 4-15 The Allosteric Site of DPG. DPG binds to hemoglobin in the crevice near the N-terminus of the two β chains (*arrows*). It is accommodated much more readily by the wider spacing found in deoxyhemoglobin, hence it tends to stabilize the molecule in that conformation. [Courtesy of M. F. Perutz, *J. Mol. Biol.*, **28**: 117 (1967).]

mitochondria in the form of pyruvate, this feedback system helps to regulate mitochondrial activity.

The second of the two major regulatory steps in glycolysis is catalyzed by pyruvate kinase. Here, the regulatory system used depends on the cell type being considered, for there are in mammalian systems a total of three important isozymes of pyruvate kinase, each having somewhat different regulatory properties, suitable for different types of cells. Muscle and nerve cells depend on the glycolytic pathway or its aerobic extension via mitochondria as a major source of the energy needed to support contraction and nerve impulse conduction, respectively. The pyruvate kinase isozyme in these cells exhibits Michaelis-Menten kinetics except that it is subject to product inhibition by high levels of ATP. (When, of course, activity of the enzyme is not particularly needed.)

In contrast, the liver cell (specifically, the major parenchymal cell or *hepatocyte*) is a complex cell responsible for many biosynthetic functions including the synthesis of glucose from blood lactate. The pyruvate kinase found in these cells is a protein with significantly different properties from the isozyme found in muscle (Fig. 4-16a). It exhibits sigmoidal rather than hyperbolic kinetics and strong product inhibition by ATP. In addition, fructose-1,6-diphosphate (FDP) is a positive allosteric effector, capable of lowering the Hill coefficient from about 2.4 down to 1—in other words, producing hyperbolic kinetics (Fig. 4-16b).

The activation of pyruvate kinase by FDP makes sense if one examines Fig. 4-5, for FDP is seen to lie at an energy minimum. When activity of the first part of the glycolytic pathway produces substrate at a rate faster than pyruvate kinase can handle it, there will be a backup of intermediates all the way along. The intermediate that accumulates to highest levels, however, is FDP. Hence FDP makes the most sensitive indicator of whether or not there is smooth and unimpeded flow through the pathway and is a logical choice for pyruvate kinase activator, a function which it is able to carry out by binding to an allosteric site on the enzyme.

But why should liver cells need this mechanism when muscle and nerve cells do not? The answer lies in the capacity of liver cells to synthesize large quantities of glucose, a capacity needed in order to help maintain proper blood sugar levels. This function is not shared by muscle or nerve. Glucose synthesis from lactate utilizes most of the steps and most of the enzymes of glycolysis. However, neither of the two control enzymes that we have discussed is used (see Fig. 4-17), for these enzymes catalyze steps with large free energy changes that are favorable for glycolysis but unfavorable for glucose synthesis. Both enzymes are bypassed during glucose synthesis, when more favorable reactions are substituted for them. When these bypass reactions are functioning, both phosphofructokinase and pyruvate kinase must be inhibited, a necessity that is met in part by the presence of very high levels of ATP and very low levels of FDP. High ATP is a prerequisite to glucose synthesis or any other anabolic pathway. Low FDP is ensured by activity of the very favorable bypass enzyme fructose diphosphatase, which is active only during glucose synthesis.

The bypass enzymes utilized for glucose synthesis respond to the same sort of controls already discussed for glycolysis but in the opposite way. That is, when the catabolic enzymes are activated the anabolic enzymes are inhibited, and vice versa.

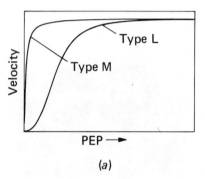

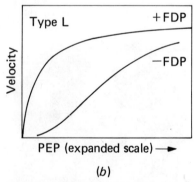

FIGURE 4-16 Isozymes of Pyruvate Kinase. Mammals have three isozymes of this enzyme. One (labeled type M) is found mostly in muscle and nerve cells. Another (type L) is found in liver cells. This latter enzyme is allosterically activated by FDP, whereupon it exhibits hyperbolic kinetics like that of type M. The variable substrate is phosphoenolpyruvate, PEP. (Data from experiments by J. M. Cardenas and R. D. Dyson).

Many pathways are controlled in ways similar to the way in which glycolysis is controlled. There are, however, other types of regulatory processes in addition to the ones just considered. In particular, mechanisms exist for altering the normal homeostatic mechanisms of a cell in response to external stimuli such as hormones. Those processes will be considered next.

4-6 CYCLIC NUCLEOTIDES AND PROTEIN PHOSPHORYLATION

The types of regulatory mechanisms so far discussed provide an efficient way of maintaining the proper intracellular level of a substrate (i.e., its "pool" size) but do not explain how the stable value of various pools might be altered in response to external events. For example, when an animal senses danger its nervous system causes the adrenal medulla (the central part of a gland lying adjacent to the kidneys) to pour into the bloodstream *catecholamines*, specifically *epinephrine* and *norepinephrine*, also known as adrenaline and noradrenaline, respectively. In addition to raising blood pressure and heart rate, catecholamines mobilize fuel reserves by stimulating glycogen breakdown. In muscle, this makes glucose phosphates available to the glycolytic pathway, and in liver free glucose is released, thus raising blood glucose levels. The animal is then ready for "fight or flight." When the danger has passed, excess catecholamines disappear from the blood and glucose handling goes back to normal.

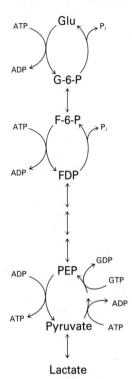

FIGURE 4-17 Gycolysis vs. Gluconeogenesis. The synthesis of glucose from pyruvate, lactate, or similar precursor is called gluconeogenesis. The pathway reverses most of the steps of glycolysis except for those shown, where more favorable reactions are substituted. Reciprocal controls operate so that when the bypass enzymes are activated the original glycolytic enzymes are inhibited and vice versa.

Epinephrine (adrenaline)

Norepinephrine (noradrenaline)

Investigation of the effects of epinephrine on glucose and glycogen during the 1960s opened up an entire new field of cell physiology. The catecholamines, like most other water soluble hormones, are unable to penetrate the lipid barrier posed by the cell membrane. However, the presence of such hormones at the surface is made known by changes in intracellular levels of Ca^{2+} or compounds known as cyclic nucleotides. These changes, in turn, affect the activity of enzymes, often by causing phosphates to be added or removed from specific amino acids.

We shall proceed by describing first how glycogen synthesis and breakdown is normally controlled and then consider the effects of epinephrine on those processes.

Glycogen Synthesis and Degradation. Glycogen is a storage polymer of glucose that is found mostly in liver and skeletal muscle (Fig. 4-18).

CHAPTER 4
Cellular Homeostasis: Bioenergetics and the Regulation of Enzymes

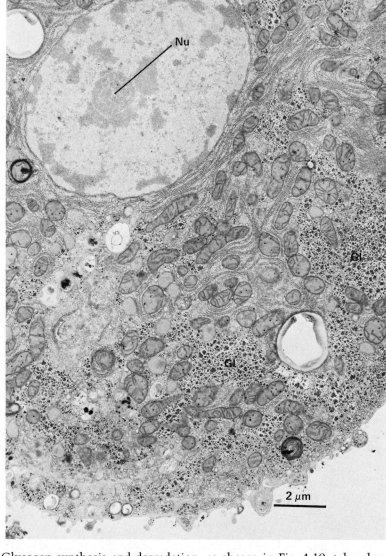

FIGURE 4-18 Glycogen. Portion of a rat liver cell, showing large quantities of glycogen granules (Gl, darkly stained) in the cytoplasm. Glycogen can be broken down and released by these cells as free glucose in order to maintain blood levels. Note the many mitochondria and distinct nucleolus (Nu). [Courtesy of P. Drochmans, J-C. Wanson and R. Mosselmans, *J. Cell Biol.*, **66**: 1 (1975).]

Glycogen synthesis and degradation, as shown in Fig. 4-19, take place by different pathways, each of which has a key regulatory enzyme. The control enzyme in the anabolic pathway is *glycogen synthetase* (or synthase), which catalyzes the elongation of glycogen by transferring glucose from a high energy carrier, UDP-glucose.

$$\text{UDP-glucose} + (\text{glucose})_n \rightarrow (\text{glucose})_{n+1} + \text{UDP} \qquad \textbf{(4-26)}$$

Control of the catabolic pathway is mainly the responsibility of *glycogen phosphorylase*, which removes glucose units from glycogen:

$$(\text{glucose})_{n+1} + P_i \rightarrow (\text{glucose})_n + \text{G-1-P} \qquad \textbf{(4-27)}$$

These two allosteric enzymes respond in reciprocal ways to regulatory signals such as availability of ATP and G-6-P, so that when one of the

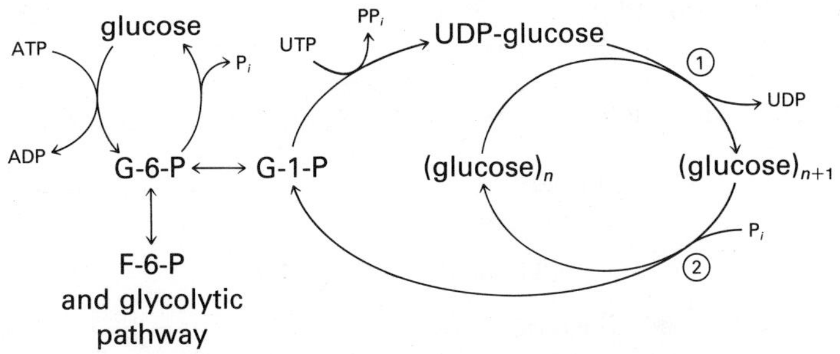

FIGURE 4-19 Glycogen Metabolism. Glycogen is designated $(glucose)_n$ or, when elongated by one more unit $(glucose)_{n+1}$. The synthesis and breakdown, as indicated, occur by different pathways. The key enzymes are (1) glycogen synthetase and (2) glycogen phosphorylase.

two enzymes is activated the other is inhibited. Superimposed on this homeostatic control is another level of regulation, for each of these control enzymes also exists in two molecular forms, one of them phosphorylated and the other not. These two forms have different properties.

The two interconvertible forms of glycogen phosphorylase are called a and b. Conversion to the a form involves the addition of one phosphate to each of the two subunits of the enzyme. The phosphates form esters with serine hydroxyls in a reaction that is catalyzed by a kinase, called *phosphorylase kinase*, using ATP as the phosphate donor. Return to the b form involves hydrolysis of these phosphates through catalysis by a phosphatase known as *phosphorylase phosphatase*.

$$(\text{phosphorylase } b)_2 \xrightleftharpoons[\substack{\text{phosphorylase} \\ \text{phosphatase}}]{\substack{\text{phosphorylase} \\ \text{kinase}}} (\text{phosphorylase } a)_2$$

(2ATP → 2ADP for kinase; 2P$_i$ ← 2H$_2$O for phosphatase)

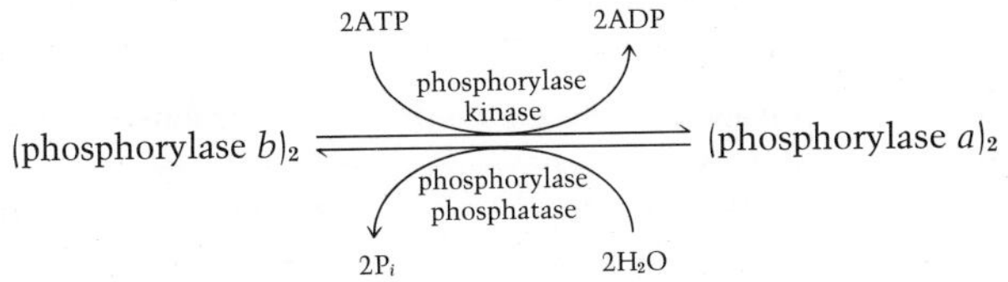

In addition, two molecules of the muscle form of phosphorylase (the a form especially) associate to form a tetramer, or four-chain protein.

$$2(\text{phosphorylase})_2 \rightleftharpoons (\text{phosphorylase})_4$$

The a form (the one carrying phosphates) is a very much more *a*ctive enzyme than the b form.

Glycogen synthetase undergoes a very similar phosphorylation-dephosphorylation cycle. However, in this case phosphorylation produces a less active enzyme, called the *D form* because it *d*epends on the presence of the positive allosteric effector G-6-P for activity. In contrast, the form without phosphates is called the *I form* because its activity is *i*ndependent of G-6-P control.

Phosphorylation of these two enzymes, then, produces a condition that strongly favors glycogen breakdown and associated mobilization of this fuel reserve. Phosphorylation is carried out by enzymes called kinases that can be stimulated by an intracellular increase in free Ca^{2+} (as occurs, for example, in strenuously exercising muscle) or by the presence at the cell membrane of increased levels of certain hormones such as epinephrine and glucagon.

Hormonal Control of Enzyme Cascades. When an animal senses danger, epinephrine is released into the bloodstream. This hormone has many different effects, but among them is the mobilization of glycogen in muscle—this means a stimulation of its breakdown and an inhibition of

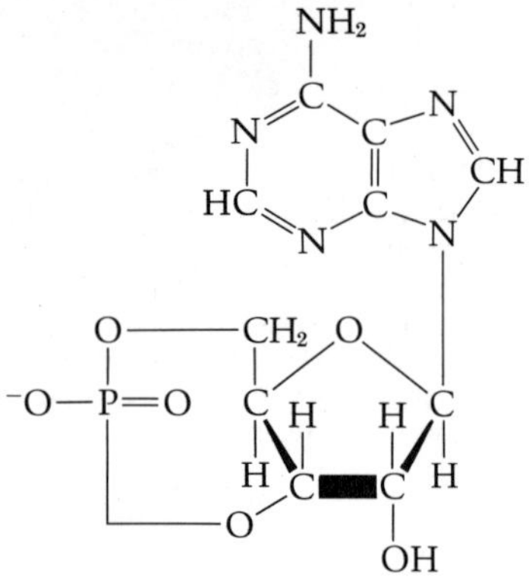

Adenosine 3', 5'-phosphate (cAMP)

its synthesis, both achieved by phosphorylation of a control enzyme. Epinephrine causes these changes by influencing the level of *3',5'-cyclic AMP* or *cAMP*, discovered at Case Western Reserve University in 1956 by Earl W. Sutherland, who received the 1971 Nobel Prize for this contribution.

Cyclic AMP is produced by adenyl cyclase, an enzyme that is activated (in this case) directly by epinephrine.

$$ATP \rightarrow 3',5'\text{-AMP} + PP_i$$

Cyclic AMP is an allosteric activator of several protein kinases. In so doing, it can initiate a *cascade* of activations, with one protein activating the next. The advantage of such a mechanism is the amplification it provides, so that a very small amount of epinephrine at the cell surface can rapidly activate a large quantity of glycogen phosphorylase and just as rapidly inhibit similar quantities of glycogen synthetase scattered throughout the cell. The enzymes involved in the glycogen phosphorylase cascade are shown in the diagram in Fig. 4-20, and the overall effects are summarized in Table 4-4. The inactivation of glycogen synthetase proceeds similarly, catalyzed in the last step by its own kinase. This amplifier effect presumably compensates for the energy expended in these ATP-dependent phosphorylations.

In due course, the stimulus for the cascade will disappear. With a reduced rate of cAMP synthesis and continuing breakdown to AMP via an enzyme called cAMP phosphodiesterase, the intracellular level of cAMP

TABLE 4-4. The Control of Glycogen Synthesis and Breakdown in Muscle

	Resting muscle	Contraction or epinephrine
Intracellular conditions:	low conc. of free Ca^{2+}	high conc. of free Ca^{2+} (during contraction)
	low cAMP	high cAMP (epinephrine stimulation)
	high glycogen	lower glycogen
	high ATP	lower ATP
	low AMP	higher AMP
Glycogen synthetase:	I form (no P) active	D form (phosphorylated) less active
	no effectors	activated by G-6-P
Glycogen phosphorylase:	b form (no P) less active	a form (phosphorylated) active
	activated by AMP inhibited by ATP inhibited by G-6-P	no effectors
Result:	Glycogen synthesis is stimulated and breakdown is inhibited unless the ATP/AMP ratio declines.	Glycogen stores are mobilized as glucose phosphate. Mobilization stops when G-6-P increases, or when the glycogen level is low.

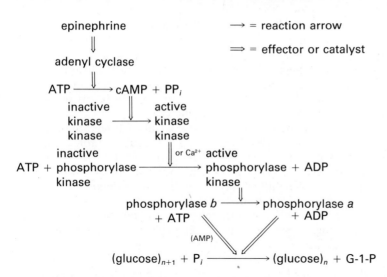

FIGURE 4-20 The Activation of Glycogen Phosphorylase.

soon falls.[1] The phosphorylated enzymes are then returned to their normal, dephosphorylated form by the action of protein phosphatases, the activity of which was probably continuous throughout but no match for the sudden explosive increase in phosphorylating activity stimulated by cAMP.

Cyclic AMP is now known to mediate the effects of many hormones other than epinephrine, most of which are water soluble and hence excluded from the cell. In many cases, these hormones have been shown to bind to specific receptors on the cell membrane (see Chap. 10) and in so doing activate a membrane-bound adenyl cyclase (see Fig. 4-21). Cyclic AMP, then, is said to serve as a "second messenger," carrying the hormonal signal to the interior of the cell.

The effects of cyclic AMP are by no means limited to glycogen synthesis and breakdown. For example, in some species phosphofructokinase is also activated by phosphorylation, overcoming its inhibition by ATP and helping to force an increase in activity of the glycolytic pathway with its resultant yield of ATP. Fatty acid synthesis and breakdown also occur by a set of reactions that are controlled in part by cAMP, as are many other enzymes and certain aspects of gene regulation. This regulator, cAMP, is found in prokaryotes and plants as well as in animal cells.

Most of the effects of cAMP in animal cells are brought about by enzyme phosphorylations catalyzed by a family of protein kinases, not all of which have been studied as yet. One that has, the cAMP-dependent protein kinase of the glycogen phosphorylase cascade, provides a particularly nice example of allosteric regulation. As demonstrated by E. G. Krebs and his associates at the University of California at Davis, the catalytic and regulatory sites are not only different, they reside on different subunits. In the presence of cAMP, the regulatory and catalytic subunits actually separate from each other, freeing the catalytic subunit from the inhibiting effects of its regulatory subunit.

Cyclic AMP is apparently not the only "second messenger" for hormones. There is good evidence now that another cyclic nucleotide, cyclic GMP, plays a somewhat similar role and that many regulatory functions are mediated by altering the availability of free Ca^{2+}, which is itself a sort of second messenger in some systems.

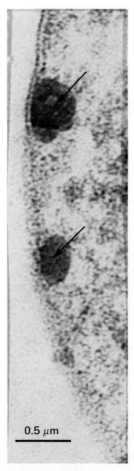

FIGURE 4-21 Membrane-associated Adenyl Cyclase. The location of adenyl cyclase is demonstrated by a technique that reveals its reaction product (arrows), which would normally be cyclic AMP. The product is limited to the cytoplasmic side of the plasma membrane, which in this case belongs to a plant, the slime mold *Dictyostelium discoideum*. [Courtesy of L. Cutler and E. Rossomando, *Exp. Cell Res.*, **95**: 79 (1975), © Academic Press, N.Y.]

Guanosine 3',5'-phosphate
(cGMP)

Cyclic AMP and Cyclic GMP: The Yin-Yang Hypothesis. A great many hormones work in pairs having opposite effects. Thus, while glucagon causes glycogen breakdown and an increase in blood glucose, insulin causes just the opposite changes. In some experiments, insulin seems to lower cAMP levels, which might account for its antagonism to glucagon and epinephrine, but in other experiments insulin seems to raise the level of another cyclic nucleotide, cyclic GMP.

N. D. Goldberg and his associates at the University of Minnesota first presented evidence that cyclic GMP plays a regulatory role comparable to that of cAMP but in many cases having opposite effects. As they put it, cAMP and cGMP may be the Yin and Yang of biology. Experimental verification has been difficult because of the extraordinarily low levels of this cyclic nucleotide—from 10- to 100-fold lower even than cAMP. Nevertheless, a number of hormone pairs may function by stimulating these opposing second messengers. Epinephrine and acetylcholine, for instance, may achieve their opposite effects in this way. In fact, epinephrine itself affects two kinds of receptors, called alpha and beta (α and β), the existence of which has been demonstrated by finding synthetic substances that affect one and not the other. It is the β receptor that has been shown to activate adenyl cyclase. The α receptor seems instead to increase cGMP levels, at least in those few cases so far examined. This may explain why epinephrine causes relaxation of smooth muscle associated with some blood vessels and internal organs (the result of β receptor activity) but causes contraction of others (an α effect).

The cyclic GMP story is too new to give many of its details. It is clear, however, that cGMP is produced from GTP by its own guanyl cyclase and ultimately hydrolyzed to GMP by a phosphodiesterase. Its mode of action, however, is unclear. In at least one system it seems to activate the phosphodiesterase that breaks down cAMP, but in other systems its functions are probably independent of cAMP. There is reason to believe, however, that many of the functions involve the phosphorylation-dephosphorylation cycle of enzymes. A complicating factor in these investigations is that the activity of cGMP, like that of cAMP, is also connected in a complex way with the levels of certain positively charged ions, especially Ca^{2+}.

Regulation by Calcium. In the enzyme cascade leading to glycogen mobilization (Fig. 4-20 and Table 4-4), Ca^{2+} was listed as an activator of a key enzyme, phosphorylase kinase. Actually, both forms of phosphorylase kinase (Fig. 4-20) are activated by Ca^{2+} but the phosphorylated form is more sensitive and can function in the presence of the normal cytoplasmic level of free Ca^{2+}, which is about 10^{-7} molar. A 10-fold higher level of Ca^{2+}, however, activates the unphosphorylated form and hence leads to glycogen mobilization even in the absence of cAMP. Since these higher levels of Ca^{2+} are found in strenuously working muscles, Ca^{2+} thus leads to increased rates of glycogen breakdown in these cells not in anticipation of need (as in the epinephrine-cAMP–induced cascade) but in response to need.

Ca^{2+} in muscle, therefore, has some of the same effects as cAMP, but acts independently. This pattern is repeated in many other cell types and with many other hormone-cAMP responses. In some systems, however, Ca^{2+} mediates cAMP activity instead of working in parallel. For example, cyclic AMP seems to stimulate release of Ca^{2+} from mitochondria of many cells, thus raising the level of free Ca^{2+} in the cytoplasm outside the mitochondria and stimulating enzymatic or other processes that depend on this ion.

There is some evidence for still a third type of system, in which Ca^{2+} increase is the first intracellular event leading to an increase in cGMP, the latter by Ca^{2+}-stimulation of guanyl cyclase. Thus, an external hormone may cause an increase in Ca^{2+} by producing an influx across the cell membrane, which in turn stimulates guanyl cyclase and raises the level of cGMP as a prelude to subsequent events.

In any case, it is clear that levels of free Ca^{2+} fluctuate within the cytoplasm and in doing so bring about numerous changes. We shall encounter many examples of those changes in later chapters.

SUMMARY

4-1 By definition, a catalyst (1) is effective in trace amounts, (2) is unchanged by the reaction, and (3) affects only the rate, not the equilibrium position, of a reaction. Enzymes have all these properties but, in addition, exhibit a very great selectivity in their catalysis, a result of the fact that their active site is designed to bind and affect only molecules with a particular atomic arrangement.

Enzymes increase the rate of reactions by effectively lowering the needed amount of activation energy. Their function depends on a particular spatial configuration, especially at their active sites where substrates are bound.

4-2 Energy-yielding reactions, such as the capture of light or the breakdown of foodstuffs, usually result in the phosphorylation of ADP to form ATP. The object is to store chemical energy in a retrievable form, so that unfavorable reactions can be "driven" by coupling them to ATP hydrolysis. ATP can serve in this capacity because the hydrolysis of its pyrophosphate-type (acid anhydride) linkages is energetically very favorable.

4-3 Metabolic pathways always involve numerous steps, each catalyzed by its own specific enzyme and characterized by a free energy change that is usually less than about 40 kJ/mole. This has the following advantages: (1) It permits an efficient coupling to ATP, to conserve energy from favorable reactions or to drive unfavorable ones; (2) it provides intermediates that may be used by other pathways; (3) it provides for a flexible control by adjusting the concentration or activity of individual enzymes.

Glycolysis is an example of one metabolic (in this case catabolic) pathway. It is the anaerobic transformation of glucose to lactic acid (lactate ion at physiological pH). Glycolysis is the only way cells from higher organisms can obtain useful energy (ATP) from glucose when oxygen is not available or cannot be used.

4-4 An oxidation is the removal of one or more electrons from a substance; a reduction is the addition of electrons. The glycolytic pathway has two oxidation-reduction steps: the first produces NADH from NAD^+; the second restores the NAD^+ again. Other mechanisms can also be used to regenerate NAD^+, however, including the reduction of pyruvate to ethanol (alcoholic fermentation) instead of to lactate. But the most useful way of regenerating NAD^+ is to transfer electrons from NADH to oxygen via mitochondrial reactions, creating water and producing ATP from the energy released by the transfer. NAD^+, then, serves as an electron carrier, using the electrons produced by an oxidation in one place to reduce another substance in another place.

4-5 The simplest kind of enzyme is one that exhibits Michaelis-Menten kinetics (see Equations 4-13 or 4-15). Two useful parameters of such an enzyme are K_m, its Michaelis constant, and V_{max}, its maximum velocity. The maximum velocity represents the greatest rate at which an enzyme could work in a given environment at infinite substrate concentration. The Michaelis constant is the substrate concentration necessary to achieve exactly half of that maximum rate.

Some enzymes with more than one active site have a velocity profile that is sigmoidal rather than hyperbolic. Although the Michaelis-Menten scheme assumes that binding sites are independent of each other, the existence of sigmoidal kinetics is taken as evidence for a cooperative interaction among the sites. Within a certain range of substrate concentration, such enzymes exhibit a very large change in reaction velocity for a given change in substrate. The basis for cooperativity is a conformational change in the enzyme such that the binding of substrate (ligand) to one site increases the affinity of the other sites for substrate.

Both cooperative and Michaelis-Menten type enzymes are subject to inhibition by molecules that compete with substrate for the active site. And both types of enzymes are subject to noncompetitive inhibition due to binding of inhibitors at locations other than the active site. Enzymes exhibiting cooperative kinetics, however, are subject to allosteric regulation, in which the enzyme may either be activated or inhibited according to the way the allosteric effector changes the conformation of the enzyme. Such enzymes are frequently found at critical spots in metabolic pathways, examples of which are the phosphofructokinase and pyruvate kinase reactions of glycolysis.

4-6 In addition to having various homeostatic mechanisms, many cells of higher organisms are subject to external regulatory influences. In the liver and skeletal muscle of animals, for example, glycogen mobilization to produce glucose phosphates or (in liver) free glucose is stimulated by epinephrine and glucagon. These hormones activate membrane-bound adenylate cyclase, producing cAMP and initiating an enzyme cascade that results in phosphorylation of the key enzymes in the glycogen pathway: glycogen phosphorylase, which is thereby activated to break down glycogen more rapidly; and glycogen synthetase, which is inhibited when phosphorylated. When the hormone stimulus ceases, cAMP phosphodiesterase and protein phosphatases return things to normal. Recent evidence indicates that cGMP may be involved in similar regulatory systems, but with actions that are opposite to those of cAMP, and that Ca^{2+} may itself serve a similar role in some cases. In fact, this ion seems to be the sole mediator of some regulatory pathways; in other cases it is a link in the regulatory chain, with levels that either control or are controlled by one of the cyclic nucleotides.

STUDY GUIDE

4-1 (a) What is an "energy of activation"? How does it affect the rate of a reaction, and why? (b) What is a catalyst? Does it affect ΔG or $\Delta G\ddagger$ of a reaction? (c) To what does an enzyme owe its catalytic specificity? (d) What is an isozyme?

4-2 (a) What are the most important reasons given for the favorable free energy of hydrolysis of ATP? (b) What features make ATP a suitable energy-transfer agent? (c) What is a coupled reaction, and why is it useful?

4-3 (a) Define the following: *ana*bolism, *cata*bolism, and *meta*bolism. (b) What advantages are derived by the cell from breaking a metabolic pathway into numerous steps? (c) What would be the energy yield from glycolysis, in terms of net ATP synthesis, in a mutant organism that is unable to change DHAP into GAP? (d) What reasons can be given for the extremely favorable free energy change accompanying hydrolysis of phosphoenolpyruvate to pyruvate?

4-4 (a) Define the terms oxidation, reduction, and dehydrogenation. (b) What is the cellular function of the vitamin niacin? (c) How many oxidations and reductions are there in alcoholic fermentation?

4-5 (a) What is the purpose of feedback regulation? How does it work? (b) Define both K_m and V_{max} formally (in terms of reaction parameters) and in words that give a physical meaning to the values. (c) What distinguishes competitive from noncompetitive inhibition? How is allosteric inhibition different from other noncompetitive inhibitions? (d) Define "positive allosteric effector" and "negative allosteric effector." How do they work? (e) The following table contains three sets of data such as might be obtained from an enzyme-catalyzed reaction using 10 μg of enzyme (MW = 10^5 g/mole) in 1 ml of assay mixture. The first set of velocities is for the enzyme-substrate reaction alone, while the other two represent the reaction in the presence of two different inhibitors. Graph v vs. [S] and $1/v$ vs. $1/[S]$ with all the data presented on each graph. (That is, each graph will have three curves.)

[S] (mM)	1	2	5	10	20
v (μmoles/min, no inhibitor)	2.5	4.0	6.3	7.6	9.0
v (μmoles/min, with inhibitor A)	1.17	2.10	4.00	5.70	7.20
v (μmoles/min, with inhibitor B)	0.77	1.25	2.00	2.50	2.86

(1) For the enzyme in the absence of inhibitor, find the "specific activity" (V_{max}) in international units per mg. (1 IU = 1 micromole of substrate consumed per minute.) What is its turnover number (TON) in molecules of substrate per minute per molecule of enzyme? What is the K_m of the enzyme? (Specify units.) (2) Does either inhibitor compete with substrate for the catalytic site? If so, which, and how do you know? (3) If the inhibitor concentration is 1 mM in each case, what are the values for their respective K_I's? [Ans: (1) 10^3 μmol/min/mg, 10^5 moles/mole/min, 3 mM; (3) 0.66 mM for A, 0.5 mM for B]. (f) The following data might be obtained for

an allosteric enzyme using a 1-ml reaction vessel with 1 μg of enzyme.

[S]	v
0.05 mM	0.017 μmoles/min
0.10	0.041
0.2	0.10
0.4	0.17
0.6	0.21
0.8	0.23
1.0	0.25

Plot the data in three ways: (1) Plot v vs. [S] to illustrate the sigmoidal character of the data and to estimate K_m for PEP. [Your value should be near 0.35 mM.] (2) Plot $1/v$ vs. $1/[S]$ to estimate the maximum velocity. The data will not form a straight line, but you should be able to tell that the activity approaches 300 units (μmol per min) per mg at infinite substrate concentration—i.e., at $1/[S] = 0$. (3) Use Equation 4-23 and the plot indicated by it to estimate the value of the Hill coefficient. [You should find a value near 1.5.] (g) What do we mean by the "feedforward activation" of pyruvate kinase by FDP, and why is it useful?

4-6 (a) Describe the reactions catalyzed by glycogen phosphorylase and glycogen synthetase, and describe the role of these enzymes in the glycogen pathway. (b) How are these two enzymes regulated in the absence and presence of cAMP? (c) What is the advantage of an enzyme cascade? (d) What is the apparent role of cGMP? (e) How does Ca^{2+} affect glycogen phosphorylase? How may Ca^{2+} levels be regulated by cAMP (in cells other than muscle)?

REFERENCES

ENZYMES AND CATALYSIS

BENESCH, R., and R. E. BENESCH, "Homos and Heteros Among the Hemos." *Science*, **185**: 905 (1974). Subunit interactions in hemoglobin.

BERNHARD, SYDNEY A., *The Structure and Function of Enzymes*. Menlo Park, Calif.: W. A. Benjamin, 1968.

BLACKBURN, S., *Enzyme Structure and Function*. New York: Dekker, 1976. Enzymology, Vol. 3.

BLOW, D., and J. SMITH, "Enzyme Substrate and Inhibitor Interactions." *Phil. Trans. Roy. Soc. London, Ser. B*, **272**: 87 (1975).

CHIPMAN, DAVID M., and NATHAN SHARON, "Mechanism of Lysozyme Action." *Science*, **165**: 454 (1969). The first enzyme for which the relationship between structure and function became clear.

CORNFORTH, J., "Asymmetry and Enzyme Action." *Science*, **193**: 121 (1976).

DALZIEL, K., "Dynamic Aspects of Enzyme Specificity." *Phil. Trans. Roy. Soc. London, Ser. B*, **272**: 109 (1975).

EDSALL, J. T., "James Sumner and the Crystallization of Urease." *Trends Biochem. Sci.*, **1**(1): 21 (Jan. 1976). The demonstration that enzymes are proteins.

ENGLE, P. C., *Enzyme Kinetics*. London: Chapman and Hall, and New York: Halsted Press, 1977.

FERDINAND, W., *The Enzyme Molecule*. New York: John Wiley, 1976. (Paperback.)

FERSHT, A., *Enzyme Structure and Mechanism*. San Francisco: W. H. Freeman, 1977. (Paperback.)

GUTFREUND, H. "Kinetics: The Grammar of Enzymology." *FEBS Letters*, **62** (Suppl.): E13 (1976). The history and importance of kinetic measurements.

JENSEN, R. A., "Enzyme Recruitment in Evolution of New Function." *Ann. Rev. Microbiol.*, **30**: 409 (1976). Enzyme evolution.

KOSHLAND, D. E., Jr., "Protein Shape and Biological Control." *Scientific American*, **229**(4): 52 (Oct. 73). (Offprint 1280).

———, "Role of Flexibility in the Specificity, Control and Evolution of Enzymes." *FEBS Letters*, **62** (Suppl.): E47 (1976).

———, "The Evolution of Function in Enzymes." *Fed. Proc.*, **35**: 2104 (1976).

KRAUT, J., "Serine Proteases: Structure and Mechanism of Catalysis." *Ann. Rev. Biochem.*, **46**: 331 (1977).

LIENHARD, G. E., "Enzymatic Catalysis and Transition-State Theory." *Science*, **180**: 149 (1973).

MARKERT, C. L., "Biology of Isozymes." *BioScience*, **25**: 365 (1975).

NEURATH, HANS, KENNETH A. WALSH, and WILLIAM P. WINTER, "Evolution of Structure and Function of Proteases." *Science*, **158**:1639 (1967).

PERHAM, R. N., "The Protein Chemistry of Enzymes." *FEBS Letters*, **62** (Suppl.): E20 (1976). The role of protein chemistry in enzymology.

PHILLIPS, DAVID C., "The Three-Dimensional Structure of an Enzyme Molecule." *Scientific American*, November 1966. (Offprint 1055.) Lysozyme and its catalytic activity.

PLOWMAN, KENT, *Enzyme Kinetics*. New York: McGraw Hill, 1972.

RIGGS, A., "Factors in the Evolution of Hemoglobin Function." *Fed. Proc.*, **35**: 2115 (1976).

SUMNER, JAMES B., "The Isolation and Crystallization of the Enzyme Urease." *J. Biol. Chem.*, **69**: 435 (1926). Reprinted in *Great Experiments in Biology*, edited by M. L. Gabriel and S. Fogel. Englewood Cliffs, N.J.: Prentice-Hall, 1955, p. 50. The first isolation of an enzyme.

TANG, J., "Pepsin and Pepsinogen: Models for Carboxyl (Acid) Preateases and Their Zymogens." *Trends Biochem. Sci.*, **1**(9): 205 (Sept., 1976).

ATP

ALBERTY, ROBERT A., "Effect of pH and Metal Ion Concentration on the Equilibrium Hydrolysis of Adenosine Triphosphate to Adenosine Diphosphate." *J. Biol. Chem.*, **243**: 1337 (1968).

———, "Thermodynamics of the Hydrolysis of Adenosine Triphosphate." *J. Chem. Educ.*, **46**: 713 (1969).

———, "Standard Gibbs Free Energy, Enthalpy, and Entropy Changes as a Function of pH and pMg for Several Reactions Involving Adenosine Phosphates." *J. Biol. Chem.*, **244**: 3290 (1969).

BRODA, E., *The Evolution of the Bioenergetic Processes*. New York: Pergamon, 1975.

DELEY, J., "The Recognition of Bioenergetic Processes." *Proc. Roy. Soc. (London), Ser. B.*, **189**: 235 (1975). "High energy" carriers in cell metabolism.

KALCKAR, HERMAN M., "High Energy Phosphate Bonds: Optional or Obligatory." In *Phage and the Origins of Molecular Biology*, edited by J. Cairns, G. Stent, and J. Watson. Cold Spring Harbor (N.Y.) Lab. of Quant. Biol., 1966, p. 43.

KALCKAR, H. M., "Lipmann and the 'Squiggle.'" In *Current Aspects of Biochemical Energetics*, edited by N. Kaplan and E. Kennedy. New York: Academic Press, 1966, p. 1. On Lipmann's advocacy of a "high energy" phosphate.

LIPMANN, FRITZ, "Metabolic Regulation and Utilization of Phosphate Bond Energy." *Advan. Enzymol.*, **1**: 99 (1941). The original concept of a "high energy phosphate."

PULLMAN, B., and A. PULLMAN, "Electronic Structure of Energy-Rich Phosphates." *Radiation Res. Suppl.*, **2**: 160 (1960).

REEVES, R. E., "How Useful is the Energy in Inorganic Pyrophosphate?" *Trends Biochem. Sci.*, **1**(3): 53 (March 1976). In many systems, it is enormously useful.

SHIKAMA, K., "Standard Free Energy Maps for the Hydrolysis of ATP as a Function of pH, pMg and pCa." *Arch. Biochem. Biophys.*, **147**: 311 (1971).

VAN NIEL, C. B., "Lipmann's Concept of the Metabolic Generation and Utilization of Phosphate Bond Energy: A Historical Appreciation." In *Current Aspects of Biochemical Energetics*, edited by N. O. Kaplan and E. P. Kennedy. New York: Academic Press, 1966, p. 9.

REGULATORY MECHANISMS

ANTONINI, E., and M. BRUNORI, *Hemoglobin and Myoglobin in Their Reactions with Ligand*. Amsterdam: North Holland, 1971. Comprehensive.

ATKINSON, D. E., "Biological Feedback Control at the Molecular Level." *Science*, **150**: 851 (1965).

———, "The Energy Charge of the Adenylate Pool as a Regulatory Parameter: Interactions with Feedback Modifiers." *Biochemistry*, **7**: 4030 (1968).

BALDWIN, J. M., "Structure and Function of Hemoglobin." *Progr. Biophys. Mol. Biol.*, **29**: 225 (1975). A comprehensive review.

BUSBY, S., and G. RADDA, "Regulation of the Glycogen Phosphorylase System." *Current Topics Cell. Reg.*, **10**: 90 (1976).

COHEN, P., "The Regulation of Protein Function by Multisite Phosphorylation." *Trends Biochem. Sci.*, **1**(2): 38 (Feb. 1976).

COHEN, P., *Control of Enzyme Activity*. London: Chapman and Hall, 1976.

DISCHE, Z., "The Discovery of Feedback Inhibition." *Trends Biochem. Sci.*, **1**(12): N269 (Dec. 1976). By the discoverer.

GERHART, J. C., "A Discussion of the Regulatory Properties of Aspartate Transcarbamylase from *Escherichia coli*." *Current Topics Cell. Reg.*, **2**: 276 (1970).

GERHART, J. C., and H. K. SCHACHMAN, "Distinct Subunits for the Regulation and Catalytic Activity of Aspartate Transcarbamylase." *Biochemistry*, **4**: 1054 (1965).

HANSON, R. W., and M. A. MEHLMAN, eds., *Gluconeogenesis: Its Regulation in Mammalian Species*. New York: Wiley-Interscience, 1976.

HOFMANN, E., "The Significance of Phosphofructokinase to the Regulation of Carbohydrate Metabolism." *Rev. Physiol. Biochem. Pharmacol.*, **75**: 1 (1976).

HUIJING, F. "Glycogen Metabolism and Glycogen-Storage Diseases." *Physiol. Rev.*, **55**: 609 (1975).

KATZ, J., and R. ROGNSTAD, "Futile Cycles in the Metabolism of Glucose." *Current Topics Cell. Reg.*, **10**: 238 (1976).

KILLILEA, S., H. BRANDT, and E. LEE, "Modulation of Protein Function by Phosphorylation: The Role of Protein Phosphatases." *Trends Biochem. Sci.*, **1**(2): 30 (Feb. 1976).

KOSHLAND, D. E., Jr., "Conformational Aspects of Enzyme Regulation." *Current Topics Cell. Reg.*, **1**: 1 (1969).

———, "A Molecular Model for the Regulatory Behavior of Enzymes." In *The Harvey Lectures 1969–1970* (Series 65). New York: Academic Press, 1971, p. 33.

KREBS, H. A., "The Role of Equilibria in the Regulation of Metabolism." *Current Topics Cell. Reg.*, **1**: 45 (1969).

LARDY, HENRY A., "Gluconeogenesis: Pathways and Hormonal Regulation." In *The Harvey Lectures, 1964–1965* (Ser. 60). New York: Academic Press, 1966, p. 261.

MADSEN, N., O. AVRAMOVIC-ZIKIC, P. LUE, and K. HONIKEL, "Studies on Allosteric Phenomena in Glycogen Phosphorylase *b*." *Mol. Cell. Biochem.*, **11**: 35 (1976). History and current thoughts.

MONOD, JACQUES, "From Enzymatic Adaptation to Allosteric Transitions." *Science*, **154**: 475 (1966). Nobel Lecture, 1965.

MONOD, J., J.-P. CHANGEUX, and F. JACOB, "Allosteric Proteins and Cellular Control Systems." *J. Mol. Biol.*, **6**: 306 (1963).

NORDLIE, R. C., "Multifunctional Hepatic Glucose-6-Phosphatase and the Tuning of Blood Glucose Levels." *Trends Biochem. Sci.*, **1**(9): 199 (Sept. 1976).

PAIK, W. K., and S. KIM, "Protein Methylation: Chemical, Enzymological, and Biological Significance." *Advan. Enzymol.*, **42**: 227 (1975).

PARDEE, A. B., "Control of Metabolic Reactions by Feedback Inhibition." *The Harvey Lectures, 1969–1970* (Ser. 65). New York: Academic Press, 1971, p. 59.

RAMAIAH, A., "Regulation of Glycolysis in Skeletal Muscle." *Life Sci.*, **19**: 455 (1976). Regulation of phosphofructokinase.

RASMUSSEN, H., "Ions as 'Second Messengers'." In *Cell Membranes*, G. Weissman and R. Claiborne, eds. New York: H. P. Publ., 1975, p. 203. Also see *Hospital Practice*, June, 1974, p. 99.

ROACH, P., and J. LARNER, "Regulation of Glycogen Synthase." *Trends Biochem. Sci.*, **1**(5): 110 (May 1976).

ROACH, P., "Functional Significance of Enzyme Cascade Systems." *Trends Biochem. Sci.*, **2**(4): 87 (April 1977).

RUBIN, C., and O. ROSEN, "Protein Phosphorylation." *Ann. Rev. Biochem.*, **44**: 161 (1975).

TOKUSHIGE, M., ed., *Selected Papers in Biochemistry Vol. 8: Allosteric Regulation.* Baltimore: University Park Press, 1971. A collection of important papers.

TREWAVAS, A., "Post-Translational Modification of Proteins by Phosphorylation." *Ann. Rev. Plant Physiol.*, **27**: 349 (1976).

WYMAN, J., "Regulation in Macromolecules as Illustrated by Haemoglobin." *Quart. Rev. Biophys.*, **1**: 35 (1968).

HORMONES AND CYCLIC AMP

BENTLEY, P. J., *Comparative Vertebrate Endocrinology.* New York: Cambridge University Press, 1976. (Paperback.) An introductory text.

BUTT, W. R., *Hormone Chemistry.* Vol. 1: *Protein, Polypeptide and Peptide Hormones* (2nd ed.). New York: Halsted Press, 1975. The chemistry and biology of the peptide hormones. Not an introduction.

DUMONT, J., and J. Nunez, eds., *Hormones and Cell Regulation.* New York: Elsevier/North Holland, 1977. Symposium.

FLORKIN, M., and E. H. STOTZ, eds., *Comprehensive Biochemistry.* Vol. 25: *Regulatory Functions—Mechanisms of Hormone Action.* New York: Elsevier, 1975.

GOLDBERG, N. D., "Cyclic Nucleotides and Cell Function." In *Cell Membranes*, G. Weissmann and R. Claiborne, eds. New York: H. P. Publ., 1975, p. 185. Also see *Hospital Practice*, May, 1974, p. 127.

GOLDBERG, N. D., and M. K. HADDOX, "Cyclic GMP Metabolism and Involvement in Biological Regulation." *Ann. Rev. Biochem.*, **46**: 823 (1977).

LITWACK, G., ed., *Biochemical Actions of Hormones*, Vol. 3. New York: Academic Press, 1975.

MALKINSON, A. M., *Hormone Action.* New York: Halsted Press, 1976. (Paperback.)

MOSES, V., "Concerning Cyclic AMP." *Sci. Progr. (Oxford)*, **63**: 503 (1976).

PASTAN, IRA, "Cyclic AMP." *Scientific American*, August 1972.

RASMUSSEN, H. "Ions as 'Second Messengers'." In *Cell Membranes*, G. Weissmann and R. Claiborne, eds. New York: H. P. Publ., 1975, p. 203. Also see *Hospital Practice*, Sept. 1974, p. 99.

SEIFERT, G., H. SCHAFER, and A. SCHULZ, "Role of Intracellular Calcium Transport in Cell Function." *Deut. Med. Wochschr.*, **100**: 1854 (1975).

SUTHERLAND, E. W., Jr. "On the Biological Role of Cyclic AMP." *J. Am. Med. Assoc.*, **214**: 1281 (1970).

———, "Studies on the Mechanism of Hormone Action." *Science*, **177**: 401 (1972). Nobel Lecture, 1971.

NOTE

1. Cyclic AMP phosphodiesterase is subject to inhibition by methyl xanthines, a class of compounds that includes caffeine and theophylline, the latter substance a component of tea. Inhibition of phosphodiesterase tends to raise the cellular level of cAMP.

5

Cellular Homeostasis: Genes and Their Regulation

5-1 The Genetic Concept   174
    Mendel's Observations
    Chromosomes and Genes
    One Gene = One Enzyme
    The Basis for Dominance

5-2 The Genetic Message   182
    Chemical Nature of the Gene
    Messenger RNA
    The Genetic Code
    Assigning Code Words
    Universal Nature of the Code

5-3 Transcription   191
    RNA Polymerase
    One Strand or Two?
    Start and Stop Signals
    Posttranscriptional Processing
    HnRNA and the Eukaryotic Message

5-4 Translation   197
    Transfer RNA
    Chain Elongation
    Ribosomes
    Initiation
    Termination
    Bacteriophage φX174: One Gene = Two Polypeptides
    Antibodies: Two Genes = One Polypeptide

5-5 Gene Regulation in Prokaryotes   206
    The Prokaryotic Chromosome
    Induction and Repression
    Operon Theory
    Catabolite Repression
    Other Mechanisms of Transcriptional Control
    Translational Control

5-6 Gene Regulation in Eukaryotes   216
    The Eukaryotic Chromosome
    Gene Expression and Chromatin Condensation
    Histones
    Nonhistone Chromosomal Proteins
    Organization of the Eukaryotic Genome
    Transcriptional Control
    Cytoplasmic Hormone Receptors
    Translational Control and Protein Turnover

Summary   228
Study Guide   230
References   231
Notes   236

5-1   THE GENETIC CONCEPT

The earliest systematic observations of gene behavior date back to the mid-nineteenth century, when an Austrian monk named Gregor Mendel conducted a series of breeding experiments on garden peas. His observations helped define the gene in terms of a separately functioning unit responsible for a particular hereditary characteristic.

Mendel's Observations. Mendel identified characteristics that could be transmitted without change from one generation to the next, but which exist in more than one form. Some plants, for example, had long stems and some short. Generation after generation, inbred long-stemmed plants produced only long-stemmed progeny. The same was true of short-stemmed plants. When, however, Mendel crossed a pure-breeding long-stemmed plant with a pure-breeding short-stemmed one, the offspring in the first generation (F_1, or *first filial generation*) resembled only the long-stemmed parent; and when two of these hybrid plants were bred, they produced a generation (F_2, or second filial generation) of mixed stem length. The ratio of long stems to short in the F_2 generation was about $3:1$. Each of the characteristics behaved the same when plants having alternate forms were crossed.

Mendel concluded that characteristics such as stem length are controlled by hereditary units called *genes*, and that genes behave in the following way:

(1) Genes affecting the same developmental process are always found in pairs in adult plants, with one member derived from each of the plant's two parents. The two genes of a given pair may or may not be identical. When they are different, and hence affect the same process in different ways, they are called *alleles* (or *allelomorphs*) of each other. The organism is then said to be *heterozygous* for the characteristic in question, to distinguish it from a *homozygous* organism in which the two paired genes are identical.

(2) One allele may mask the effect of the other. For example, the allele for long stems is always *dominant,* meaning that it is expressed even in the presence of the allele for short stems, which is *recessive*. Thus, a heterozygote always has a long-stemmed *phenotype* (appearance). However, one cannot tell by looking whether the *genotype,* or gene content, of a long-stemmed plant is homozygous or heterozygous.

(3) A sex cell (gamete, or germ cell) contains only one member of each gene pair. The necessity for allelic pairs to separate during gamete formation is referred to as *Mendel's first law,* or the *law of independent segregation.*

(4) Often, two allelic pairs controlling different characteristics seem to sort themselves at random during gamete formation. Thus, if a plant contains both the dominant and recessive alleles for stem length (represented by L and ℓ, respectively), and the dominant and recessive alleles for seed shape (S for smooth seeds, and s for wrinkled seeds), it may produce gametes of four types: LS, Ls, ℓS, and ℓs. This principle is sometimes called *Mendel's second law,* or the *law of independent assortment.* It results in a reshuffling of dominant and recessive alleles and a corresponding mixing of characteristics in the next generation.

When a pure-breeding (i.e., homozygous) long-stemmed plant is crossed with a pure-breeding short-stemmed plant, as in the first example, the results are easily predicted from the above principles. Schematically, we would represent the cross as follows, keeping in mind that any male gamete usually can combine with any female gamete:

$$
\begin{array}{lll}
\qquad\ ♂\ \text{LL} \times \ell\ell\ ♀ & \text{first cross} & \begin{bmatrix} ♂\ \text{gametes} = \text{L \& L} \\ ♀\ \text{gametes} = \ell\ \&\ \ell \end{bmatrix} \\
F_1 = \overline{\text{L}\ell\quad \text{L}\ell\quad \text{L}\ell\quad \text{L}\ell} & & \\
\qquad\ ♂\ \text{L}\ell \times \text{L}\ell\ ♀ & \text{second cross} & \begin{bmatrix} ♂\ \text{gametes} = \text{L \&}\ \ell \\ ♀\ \text{gametes} = \text{L \&}\ \ell \end{bmatrix} \\
F_2 = \overline{\text{LL}\quad \text{L}\ell\quad \text{L}\ell\quad \ell\ell} & &
\end{array}
$$

Note that F_1 plants are all heterozygotes with a long-stemmed phenotype. In contrast, three-quarters of the plants in the second filial generation have the long-stemmed phenotype, but heterozygotes (Lℓ) outnumber long-stemmed homozygotes (LL) two to one.

The more complicated cross between a homozygous long-stemmed plant with smooth seeds (LLSS) and one with short stems and wrinkled seeds ($\ell\ell$ss) produces gametes that are all LS from one parent and all ℓs from the other. The first filial generation, then, is entirely LℓSs. When two of these hybrids are crossed, however, there are four kinds of male gametes and four kinds of female gametes, yielding sixteen possible combinations containing nine genotypes and four phenotypes. The phenotypes are in a ratio of $9:3:3:1$. This theoretical distribution, which can be predicted from Fig. 5-1, is remarkably close to Mendel's actual observations.

		gametes		
	LS	Ls	ℓS	ℓs
LS	LLSS	LLSs	LℓSS	LℓSs
Ls	LLSs	LLss	LℓSs	Lℓss
ℓS	LℓSS	LℓSs	$\ell\ell$SS	$\ell\ell$Ss
ℓs	LℓSs	Lℓss	$\ell\ell$Ss	$\ell\ell$ss

FIGURE 5-1 The Punnett Square. This scheme can be used to predict the offspring from any cross. In this example, the gametes are derived from two doubly heterozygous parents, each LℓSs. Independent assortment is assumed.

There are, as it turns out, many complications to the principles just outlined. For example, one member of an allelic pair is not necessarily dominant over the other—e.g., a pure-breeding red snapdragon and a pure-breeding white snapdragon produce heterozygotes that are pink. And the law of independent assortment turns out to be violated with great regularity, as determined by the observation that certain genes usually sort together rather than independently. Such genes are said to be *linked*. For example, the genes in humans that confer color blindness and hemophilia are linked to those that determine sex, which is why color blindness and hemophilia are seen mostly in men. Genes that do obey Mendel's second law are said to be unlinked or "in different linkage groups."

In spite of their exceptions, the principles developed by Mendel have been of great value in helping us to understand how hereditary characteristics are transmitted from one generation to the next in both animals and plants. It has also led to some unfortunate oversimplifications, such as the commonly held belief that eye color is controlled by a single gene with the brown allele being dominant. While this does appear to be true most of the time, eye color is actually controlled in a more complicated way, a fact that many blue-eyed parents first learn from their brown-eyed offspring.

Chromosomes and Genes. Mendel's observations were published in the mid-1800s but were largely unappreciated and ignored until the end of the century when they were independently confirmed by several other workers. Then in 1903, Walter S. Sutton, a student of E. B. Wilson's at Columbia University, published a paper entitled "The Chromosomes in Heredity" in which he pointed out the similarity between the behavior of genes and chromosomes. By that time cytologists had learned how to observe chromosomes during cell division, and some had suggested that chromosomes might be responsible for transmitting inherited traits from one generation to the next. It was Sutton's paper in defense of this hypothesis that finally convinced a sizeable segment of the scientific community.

Consider the following parallel between genes and chromosomes:

(1) Genes come in pairs. So do chromosomes.

(2) Although the two members of each gene pair are normally present in the same cell, only one is found in a gamete, or germ cell. Similarly, paired, or *homologous*, chromosomes separate during gamete formation—specifically during a process that is called *meiosis* or *reductive division*. Thus, gametes contain half as many chromosomes as a *somatic* (i.e., nongamete) cell of the same organism. In other words, gametes have a *haploid* chromosome number while somatic cells have a *diploid* number.

(3) Except during gamete formation, genes must be accurately reproduced and transmitted to daughter cells in such a way that each cell contains a set identical to that of the parent. Chromosomes, too, are very carefully replicated and separated during cell division, so that normally all descendants of a cell contain accurate copies of the parental chromosomes. This kind of cell division is called *mitosis*.

(4) Visible changes in chromosomes can often be correlated with genetic changes. For example, the arms of two homologous chromosomes sometimes exchange ends during meiosis. The visible features of this exchange are the joining of the two arms at a discrete point called a *chiasma* (pl. *chiasmata*; from the Greek letter *chi*, or χ, which consists of two crossing arms). The chiasma, first observed in 1909 by a Belgian investigator, F. A. Janssens, was soon found to correlate with a genetic *crossover*, or interchange of alleles between homologous chromosomes. Thus, a plant that would otherwise produce AB and ab gametes might produce Ab and aB gametes after a crossover, where A/a and B/b are dominant/recessive forms of any two linked genes.

These observations and others, some of which are summarized in Table 5-1, led to the conclusion that genes are carried by chromosomes, with the alleles for a given characteristic residing at corresponding points (or *loci*) on the two members of an homologous pair. Genes that are linked are carried by the same chromosome and, except when crossover occurs, such genes will always violate the law of independent assortment or second law (see Fig. 5-2). On the other hand, genes that obey this rule must be a part of nonhomologous chromosomes. The number of linkage groups, then, is the same as the number of homologous pairs or the haploid chromosome number of the organism.[1] The haploid chromosome number is 23 for humans (see Fig. 5-3), seven for the garden pea, and four for the fruit fly, *Drosophila melanogaster*.

Additional evidence for the chromosomal location of genes comes from the observation that some genes behave as if their allele were

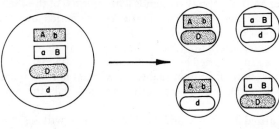

Diploid nucleus   Haploid possibilities

FIGURE 5-2 Linkage. Genes that reside on the same chromosome will be linked. That is, they will sort together during gamete formation in violation of Mendel's second law. If the haploid number of chromosomes is n ($n = 2$ in this example), there will be n linkage groups and 2^n different gametes. Note that the A/a and B/b loci are linked to one another, but both sort independently of the D/d locus.

TABLE 5-1. Historical Development of the Gene Theory

1694	Camerarius publishes *De Sexu Plantarum Epistola*, in which it is demonstrated that plants are sexual.
1779	Lort reports the unusual pattern by which human color blindness is inherited.
ca. 1820	Nasse formulates the laws governing the inheritance of sex-linked characteristics.
1822	Goss and Seton independently note the segregation of recessive characteristics in peas.
1866	Mendel publishes his observations on inheritance in peas.
1875	Chromosomes are described by Strasburger.
1875	Hertwig proves that fertilization involves the fusion of two nuclei, contained in the egg and sperm, respectively.
1879–1885	Flemming publishes his observations of chromosome "splitting" during cell division, with sister chromatids moving to opposite poles.
1883	Van Beneden notes that gametes contain only a haploid chromosome number.
1890	The reestablishment of a diploid chromosome number at fertilization, by the joining of equal sets from the male and female gametes, was noted by Boveri and Guignard.
1902	Sutton points out the parallels between genes and chromosome behavior and suggests a chromosomal location for genes.
1906	Bateson and Punnett present data on linked genes, showing that their assortment during germ cell formation is nonrandom, in violation of Mendel's second law.
1908	Nilsson-Ehle provides a model for the inheritance of continuously variable (nondominant) characteristics.
1911	Morgan advances his gene theory. He proposed that genes are linearly arranged along chromosomes in a definite order. Thus, the number of linkage groups must be equal to the haploid number of chromosomes. Morgan then went on to prove this hypothesis by correlating the movement of chromosomes with the inheritance of sex-linked traits in the fruit fly, *Drosophila*.

present in females and absent in males—i.e., as if males were *hemizygous* for the gene. The basis for this behavior is clear when the chromosomes are examined, for two chromosomes in a diploid set can be without physically homologous counterparts. These are the X and Y chromosomes. At the beginning of meiosis or mitosis in humans and many other animals, the two strands (*chromatids*) of the X chromosome are joined near their center, while the Y chromatids are much smaller and joined near one end, much like chromosomes number 18 in Fig. 5-3. It is this *heteromorphic* (dissimilar) pair that determines sex. In humans, sex genes carried by the Y chromosome are masculine and dominant. Thus, normal human males are XY and normal females are XX (see Fig. 5-4), although in some other animals it is the female that is heteromorphic.[2]

Genes carried by the X and Y chromosomes will be distributed to progeny in a pattern that predictably follows the sex of the offspring—

SECTION 5-1
The Genetic Concept

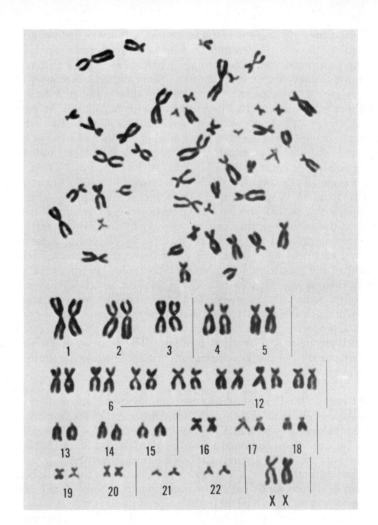

FIGURE 5-3 Human Chromosomes. Ordering of chromosomes for purposes of systematic examination is called *karyotyping* (Greek, *karyon* = nut, nucleus). This is the karyotype of a normal human female. Other staining procedures produce distinctive patterns of bands on the arms, permitting unambiguous identification of each chromosome. [Courtesy of Dr. Jacqueline Whang-Peng. National Cancer Inst.]

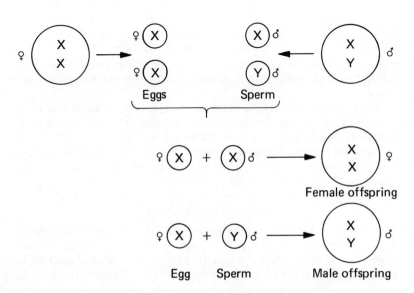

FIGURE 5-4 Determination of Sex. Gender in humans and many other animals is determined by the presence of a heteromorphic pair of chromosomes called the XY pair. A normal female mammal is XX and produces only X gametes (eggs). The male is XY and produces both X and Y sperm. The gender of the offspring is determined by which type of sperm gains entry to the egg.

i.e., the genes are *sex-linked*. The common forms of hemophilia and muscular dystrophy, for example, are caused by recessive genes carried by the X chromosome. The normal allele from a second X chromosome prevents the appearance of symptoms. Since the gene is missing in the abbreviated Y chromosome, all males who inherit the gene from their mothers will be affected while a heterozygous female will be an unaffected carrier. (The only way a female can be a hemophiliac is for her father to be one, and then only if her mother is either a carrier or herself affected with the condition.) Red-green color blindness and several other conditions are transmitted in exactly the same way. These observations are all consistent with the chromosomal location of genes.

The location of genes on chromosomes thus explains several of Mendel's observations, including segregation (the separation of homologous chromosomes during gamete formation) and assortment. Independent assortment occurs when maternally derived and paternally derived members of a given pair are distributed to gametes without regard to the way in which other pairs are split. Independent assortment, then, applies only to unlinked genes—i.e., to genes that reside on nonhomologous chromosomes. To understand the other aspects of gene behavior, such as dominance, we shall have to understand how genes affect the properties of a cell. We know that the properties and activities of a cell are attributed largely to proteins. Therefore we might assume, correctly as it turns out, that genes function by directing the synthesis of proteins.

One Gene = One Enzyme. The connection between genes and proteins was first established experimentally in the early 1940s from work at Stanford University by the 1958 Nobel Prize winners, G. W. Beadle and E. L. Tatum. They examined a number of mutations of the bread mold, *Neurospora crassa*, to discover the biochemical differences between mutant and normal strains. *Neurospora* was chosen for this work because certain of its metabolic pathways were known in some detail, and because the organism is haploid in the stage of its life cycle at which it is ordinarily observed. (Being haploid eliminates the complication caused by dominant and recessive genes, allowing all mutations to be observed.) Furthermore, the fungus can reproduce asexually by spore formation, allowing one to grow large quantities of genetically identical copies of an interesting specimen. And finally, native (wild-type) *Neurospora* can grow on a chemically defined, simple medium containing glucose as the only carbon and energy source, provided that biotin (a B vitamin) is also added.

After exposure to X rays, a number of mutants were found that could no longer grow on this simple medium. By adding various amino acids, nucleotide precursors, vitamins and so on, one at a time, Beadle and Tatum were able to determine the site in the metabolic scheme at which gene damage was manifest. Schematically, let us suppose that a needed substance "D" is manufactured in a three-step process from the starting material "A":

$$A \xrightarrow{E_1} B \xrightarrow{E_2} C \xrightarrow{E_3} D$$

Each step is catalyzed by its own enzyme, so if the cell grows on a simple glucose medium to which C or D is added, but not on the same medium to which A or B is added, the mutation must have affected enzyme E_2. Similarly, if D but not C will support growth, the block would be at E_3.

By this sort of reasoning, Beadle and Tatum concluded that their mutants contained defective enzymes. It was logical to assume, then, that damaged genes were responsible for synthesizing the defective enzymes. Since each of the separately inherited mutations resulted in a defect in a different enzyme, the "one gene = one enzyme hypothesis" was born. In other words, it was supposed that the main (or only) function of genes is to control the synthesis of enzymes on a one-for-one basis.

Actually, the one gene = one enzyme hypothesis is true in only a limited sense. As our knowledge of enzyme structure grew, it became clear that many enzymes are composed of more than one kind of polypeptide chain, and that mutations may affect one type of chain without affecting the others. In addition, proteins other than enzymes may also be affected by genetic mutations. And finally, it is now clear that not all genes make proteins. Thus, the original concept has evolved to the current position, which ascribes each polypeptide chain to a separate gene while conceding that not all genes make polypeptide chains. In fact, the word "gene" no longer has a precise meaning, and so other names with more restricted definitions are used in its stead. In particular, each polypeptide chain is now said to be "coded" for by one *cistron*.

The Basis for Dominance. The reason for dominance can now be understood. The formation of a diploid organism from a mutant and a normal haploid *Neurospora* strain, for instance, would result in the presence of two copies of each gene in the diploid cells of the progeny. They therefore get one good copy and one mutant copy of the gene in question. This mutant gene would usually be recessive to the normal allele, since the presence of the good copy is sufficient, in most cases, to support the synthesis of enough enzyme to permit normal functioning of the cell. However, one functioning gene in a cell ordinarily produces less enzyme than two genes. Hence, where tests for measuring the enzyme are available, it is usually possible to distinguish heterozygotes from homozygotes.

The most clear-cut cases of dominance are those, such as human hemophilia and color blindness, that are sex-linked because the genes involved occupy a site (a *genetic locus*) on the X chromosome that is missing in the abbreviated Y chromosome. A male, being XY, carries only one gene for each of the affected proteins. If that gene is defective, he will have the associated disease. A female (XX) carrying one normal and one mutant allele will be unaffected since the normal gene produces the necessary protein in sufficient quantity to avoid symptoms.

In most cases, however, dominance is a term that is appropriate only because we tend to apply a crude all-or-nothing test. Admittedly, it is adequate for many situations. Thus, most inherited metabolic diseases are clearly recessive in regard to development of symptoms, while most skeletal and other visually apparent abnormalities are clearly dominant. Other conditions are not so simple, leading to descriptions of "dominant with incomplete penetrance" (the dominant gene is there but it doesn't show), "incomplete dominance," and so on. The latter was first recognized in flowers: red snapdragons and white snapdragons produce an F_1 generation that is pink. We now appreciate, however, that in reality nearly all dominance would have to be described as incomplete if examined in the right way.

Attempts to apply dominant and recessive labels to humans are further complicated in that many traits are controlled by more than one genetic locus. In the case of skin, hair, and eye color, for example, a typical offspring is usually about midway between his or her two parents, reflecting a mixture of maternally derived and paternally derived alleles, although there is enough random variation in this mixture to permit a chance deviation of considerable magnitude.

5-2 THE GENETIC MESSAGE

So far we have dealt with genes only in the abstract. To understand them more completely we need to know their chemical structure.

Chemical Nature of the Gene. We know that genes are intimately associated with chromosomes which, in higher organisms, consist of nucleic acids (mostly DNA, but with some RNA) and proteins. Of these components, protein originally seemed the more likely candidate for a genetic role. Proteins contain more than 20 different kinds of amino acids and can assume a wide variety of physical conformations; nucleic acids, on the other hand, are chemically and physically much simpler. This assumption that genes are made of protein survived until the 1940s. The true situation was determined with the help of bacteria, starting with a series of experiments carried out in the 1920s by an English public health officer named F. Griffith.

Griffith was concerned with the epidemiology of pneumococcal pneumonia, a disease caused by a bacterium that is normally surrounded by a polysaccharide capsule. A mutant strain of the bacterium was known to be free of capsule and relatively harmless. Griffith found that nonencapsulated cells, when injected into a mouse along with heat-killed (and therefore harmless) normal cells, caused a fatal infection. Moreover, live encapsulated cells were isolated from the dead mice. Apparently, something from the heat-killed cells was capable of conferring on live cells the ability to make a polysaccharide capsule, and with it the ability to cause pneumonia. The process, which appears to change the genes of living cells, is called *transformation* (see Fig. 5-5).

The chemical nature of the transforming principle was determined in 1944 by O. T. Avery, C. M. MacLeod, and M. McCarty of the Rockefeller Institute. They demonstrated that the only chemical component from one strain that is capable of transforming another strain is DNA (Fig. 5-6). The suggested explanation for transformation was that DNA carries genetic information in a form that can be passed from cell to cell. However, because it ran counter to the popular belief that proteins should carry genetic information, this explanation did not receive universal acceptance. Some argued that the DNA was not pure, or that it was mutagenic rather than informational, or that bacteria are a special case—after all, they do not even have chromosomes in the eukaryotic sense.

Rather than an isolated oddity, bacterial transformation was the first experiment to indicate the true chemical nature of the gene. Though there are still some unanswered questions concerning the process, it is clear that transformation involves the transfer of genetic information from one cell to another via DNA isolated from the donor. A transformed cell incorporates the new genetic information into its own chro-

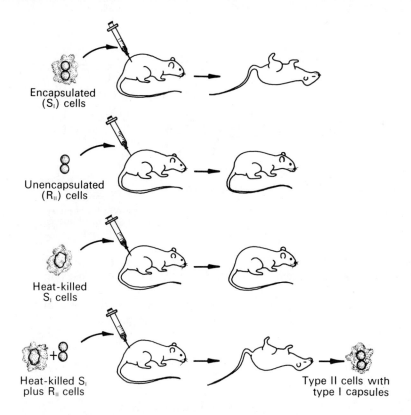

FIGURE 5-5 Bacterial Transformation. The polysaccharide capsule that helps protect normal pneumococcal bacteria from host defenses also gives its colonies on agar a smooth, glistening appearance. These are the S cells. Mutant strains (R cells) lacking capsule have rough-looking colonies. When heat-killed S cells, harmless by themselves, are mixed with R cells (normally also harmless), a fatal infection can occur, yielding live cells having capsules characteristic of the S strain. [Described by F. Griffith in *J. Hyg.*, **27**: 113 (1928).]

mosome, where it is replicated as a normal part of the cell and thus passed along to progeny. However, really wide acceptance of that position came only in 1952, from experiments by A. D. Hershey and M. Chase at the Cold Spring Harbor (N.Y.) laboratory of the Carnegie Institute.

Hershey and Chase were studying the life cycle of bacteriophage T2 (see Chap. 1) with the newly available technique of isotope labeling. Since T2 is composed almost entirely of DNA and protein in nearly equal amounts, adding ^{35}S in the form of radioactive sulfate, or ^{32}P in the form of radioactive phosphate, to a medium with phage-infected *E. coli* results in radioactive virus. But only the protein portion of the phage will be labeled when radioactive sulfur is used, because only protein contains sulfur (in the amino acids cysteine and methionine). Similarly, DNA can be selectively labeled with phosphorus because none of the amino acids contains that element.

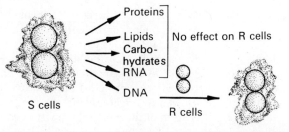

FIGURE 5-6 The Transforming Principle. By fractionating encapsulated pneumococci and performing the transformation experiment *in vitro*, O. T. Avery and his colleagues demonstrated that DNA is the agent responsible for transformation. [Described by O. T. Avery, C. M. MacLeod, and M. McCarty in *J. Exp. Med.*, **79**: 137 (1944).]

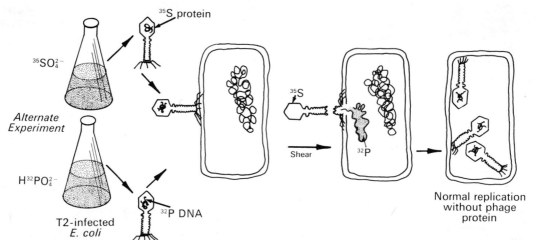

FIGURE 5-7 The Genetic Material of a Bacteriophage. Phage produced in the presence of ^{35}S will have radioactive protein. Phage produced on ^{32}P will have radioactive DNA. Hershey and Chase used this distinction to show that essentially no phage protein enters the host cell at infection. The phage genes, therefore, must be a part of its DNA. [Described by A. D. Hershey and M. Chase in *J. Gen. Physiol.*, **36**: 39 (1952).]

By using radioactive phage, Hershey and Chase were able to determine the fate of the protein and DNA during infection of a host (Fig. 5-7). They found that nearly all the protein remains outside an infected cell and, in fact, can be sheared off by the violent agitation of a kitchen blender without terminating the infection. On the other hand, phage DNA was shown to enter the infected cell. The usual process of phage infection is one in which the genes of the phage commandeer the metabolic machinery of a cell and dedicate it to the task of turning out new phage particles according to information contained in the phage genes. It follows that if phage DNA and not its protein enters a host, it must be the DNA that carries genetic information. Actually, a few percent of the T2 protein does enter with the DNA, but the experiments were repeated with the same results using phages that do not have this DNA-associated "internal protein." And any lingering doubts about the nature of phage genes were clearly dispelled when it was found that naked, purified DNA from certain phages is by itself infective.

While DNA is clearly the genetic substance of some viruses, other viruses contain RNA instead of DNA. An example is tobacco mosaic virus (TMV), isolated and crystallized by Wendell Stanley in 1935 and later found to be 94% protein and 6% RNA. Its RNA carries the genetic information necessary for replication, a fact that was determined in a series of experiments in the mid-1950s, mostly from the laboratories of G. S. Schramm in Germany, and H. Fraenkel-Conrat of the University of California at Berkeley. They found that purified RNA from TMV is infective: When RNA solutions were rubbed onto the surface of tobacco leaves, characteristic lesions developed from which intact TMV was isolated. The control experiment, using TMV protein, produced no infections.

Continuing the investigation of TMV infection, Fraenkel-Conrat showed that protein and RNA components of the virus could be purified and then recombined to give intact, infectious particles. (This feat was "life in a test tube" to the newspapers.) A clear distinction between the roles of RNA and protein was made by experiments in which RNA from one TMV strain was reconstituted with protein from another. When these hybrid particles were allowed to infect a plant, the progeny virus was always a normal representation of the parental type from which the RNA was obtained, not of the one that supplied the protein (see Fig. 5-8).

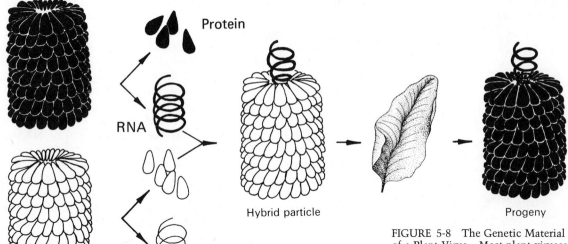

Two strains of TMV

FIGURE 5-8 The Genetic Material of a Plant Virus. Most plant viruses, such as tobacco mosaic virus (TMV), contain RNA instead of DNA. In this experiment, RNA from one strain of TMV was reconstituted with protein from a second strain to produce an infective particle. Plants infected by such particles produced viruses corresponding to the strain that donated its RNA to the hybrid virus, demonstrating that RNA carries genetic information. [Described by H. Fraenkel-Conrat and B. Singer in *Biochim. Biophys. Acta*, **24**: 540 (1957).]

There is little doubt now about the nature of the genetic material: It is DNA in those organisms that contain DNA, and RNA in those that do not. The only ones that do not are viruses. All cells, then, use DNA for their genes. This conclusion was extended to higher organisms first by the following observations: (1) Ultraviolet radiation causes mutations, but only at wavelengths at which nucleic acids absorb. (2) Chemical agents that cause mutations can often be identified as reacting with DNA and not protein. (3) Transformation of some animal cell lines has been achieved with DNA obtained from certain viruses. Later we shall be able to point to similarities in gene function between eukaryotic and prokaryotic cells, leaving no room for doubt about the genetic role of DNA.

The questions that remain are (1) how DNA is capable of storing genetic information, and (2) how that information is utilized to direct the synthesis of proteins. The discovery that an RNA molecule is an intermediary between DNA and protein synthesis helped to solve both of those problems.

Messenger RNA. It has been known since the mid-1950s that protein synthesis takes place in the cytoplasm, not in the nuclei of eukaryotic cells. More specifically, protein synthesis takes place at RNA-protein complexes called *ribosomes*, associated with the microsomal fraction of cell extracts (a fast-sedimenting fraction containing ribosomes bound to membranes). Thus, if one feeds a cell radioactive amino acids or radioactive sulfur, the incorporated label appears first at the ribosomes, and only later in the cytoplasm as a part of free protein molecules.

Clearly, if the genetic information specifying the amino acid sequence of a protein resides in nuclear material, whereas the actual synthesis of the protein takes place somewhere else, there must be a carrier, or "messenger" molecule responsible for the transfer of information. The messenger was found to be a form of RNA called *messenger RNA* or *mRNA*.

CHAPTER 5
Cellular Homeostasis:
Genes and Their Regulation

In 1961 B. D. Hall and S. Spiegelman at the University of Illinois demonstrated by direct means that mRNA reflects the base sequence of DNA. It recently had been discovered that DNA could be denatured by heating in boiling water, a result of "melting" the hydrogen bonds that hold the strands together. If a solution is cooled rapidly, the DNA stays largely single-stranded; however, if the solution is cooled very slowly, there is a specific reformation (*annealing*) of active double-stranded molecules. Hall and Spiegelman reasoned that if slow cooling were carried out in the presence of RNA having a base sequence complementary to some of the DNA, then the RNA might be able to bind to a DNA strand and form a specific DNA-RNA hybrid.

Hall and Spiegelman prepared DNA from stocks of phage T2, and then fractionated the contents of T2-infected cells to obtain some of the hypothetical messenger. Cells infected with T2 were chosen because the only RNA synthesized was known to be a short-lived molecule that associates rapidly and temporarily with ribosomes, and which therefore seemed like a good candidate for mRNA. Radioactive phosphate was added soon after T2 infection so that this RNA would be labeled and distinguishable from other RNA already present.

DNA from T2 was heated, RNA from infected cells was added, and then the mixture was allowed to cool slowly. It was next placed in a centrifuge tube containing CsCl (a dense, high-molecular-weight salt) and spun to allow the salt to form a gradient of concentration, and hence of density. (See Fig. 5-9. Also see the Appendix for more details about this technique, called *density gradient centrifugation*.) After centrifugation, they found the radioactive label in two places: near the bottom of

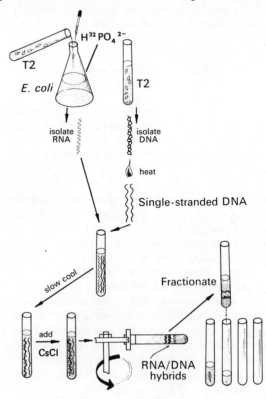

FIGURE 5-9 RNA-DNA Hybrids. Single-stranded DNA can be produced by heating double-stranded DNA. When RNA is added before slow cooling, hybrids will form if the DNA and RNA have complementary base sequences. The hybrids can then be isolated with a CsCl density gradient. (Experiment by B. D. Hall and S. Spiegelman, described in *Scientific American*, May 1964, p. 48.)

the tube, at a density corresponding to that of single-stranded RNA; and at a less dense position, near the band with double-stranded DNA. When the experiments were repeated with DNA from other sources, including *E. coli* DNA, labeled RNA did not associate with the DNA band. It must be that RNA synthesized immediately after phage infection reflects the specific base sequence of phage DNA. It is this RNA that acts as messenger and directs the synthesis of phage protein.

At about this same time, another group announced the purification of an enzyme capable of polymerizing nucleotides into RNA that reflects the base sequence of an added DNA "primer." This step, which is called *transcription*, is catalyzed by a DNA-dependent RNA polymerase that is sometimes called a *transcriptase*. Thus, the flow of genetic information is from the linear sequence of nucleotides (or just "bases," since they distinguish one nucleotide from another) in DNA to the linear sequence of nucleotides in mRNA, and from there to the amino acid sequence of a polypeptide chain:

$$\overset{\curvearrowleft}{DNA} \xrightarrow{transcription} RNA \xrightarrow{translation} protein$$

The circular arrow indicates the need for DNA to replicate, using itself as a template, in order to pass genes along to subsequent generations. That process will be considered in Chap. 11. The last step, *translation* of the genetic message into protein, will be discussed later in this chapter. First, however, we must know more about the way information is stored in nucleic acids.

The Genetic Code. The first indication of how genes might specify proteins came in 1949 from the work of Linus Pauling and his colleagues at the California Institute of Technology. They demonstrated that sickle cell anemia—the recessive gene for which is carried by some 7% of Americans with African ancestry—is due to the presence of a variant protein called hemoglobin S. Pauling was able to distinguish this protein from the usual hemoglobin (Hb A) by its ionic charge, as reflected in its mobility in an electrostatic field.

In 1956 V. M. Ingram of Cambridge showed that normal and sickle cell hemoglobin differ by a single amino acid, with a valine substituting for glutamate. Since glutamate is negatively charged at physiological pH and valine is neutral, the substitution removes one charge and causes a reduction in the solubility of the protein. At low oxygen tension Hb S loses still more charge and may precipitate, causing the red cells to become crescent-shaped and plug fine capillaries (see Fig. 5-10).[3]

Later, other variant hemoglobins were found. They too proved to be single amino acid replacements, as are many mutant proteins obtained from a wide variety of sources. These proteins result from gene alterations, the simplest of which would be the substitution of one nucleotide for another. It turns out that the nucleotide sequences in DNA and in RNA are divided into *code words*, or *codons*, each of which causes a specific amino acid to be incorporated into a growing polypeptide chain. Changing one nucleotide may therefore change the specificity of the codon from one amino acid to another.

Some of the earliest speculations on the nature of the genetic code came from the Nobel Prize-winning physicist George Gamow. He

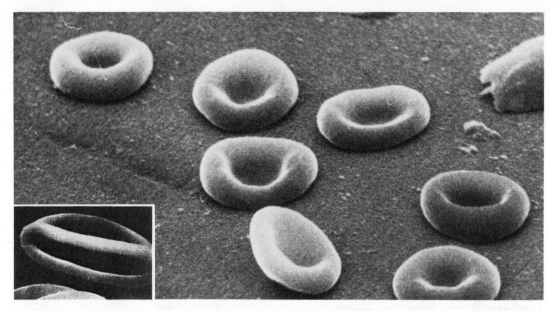

FIGURE 5-10 Normal and Sickle Cell Erythrocytes. Sickle cell anemia is caused by a single amino acid replacement in hemoglobin, resulting in its reduced solubility at low oxygen tensions. The cells may then "sickle" (inset) and plug fine capillaries to cause a sickle cell crisis. [Scanning electron micrograph of normal human erythrocytes courtesy of R. F. Baker, Univ. of Southern California. Inset courtesy of J. A. Clarke and A. J. Salsbury, *Nature*, 215: 402 (1967).]

pointed out that if every two base pairs along the DNA helix determine an amino acid, only 16 different amino acids can be specified. This is not enough, for it implies a certain ambiguity in the sequence of a protein composed of 20 different amino acids. Sequence studies of many proteins rule this out.

If two base pairs are not enough to specify an amino acid, what about three? Now we have gone from too little to too much, for there are 64 possible triplets. (With four bases for each of three positions, there are $4 \times 4 \times 4 = 64$ triplets.) To get a more economical use of DNA, therefore, one might interpret triplets in an overlapping manner. For example, given the base sequence

$$\text{A-T-G-A-G-C-A-T-T}$$

an overlap of one base per triplet would produce the code words ATG, GAG, GCA, and ATT. An overlap of two bases would produce the code words ATG, TGA, GAG, AGC, and so on. In contrast, a simple, nonoverlapping code would produce only the three code words, ATG, AGC, and ATT.

While overlap allows more amino acids to be specified with a given amount of DNA, it exacts a high price: an overlap of one means that only 16 different amino acids can follow any specified amino acid, since the first base in the next code word is fixed by the last base of the current word; an overlap of two bases further reduces the possibilities to one of only four different amino acids. Although the data were not available through which such schemes could be ruled out when they were first presented, as more and more amino acid sequences became known, it also became clear that no such restrictions are present.

Thus, the most efficient way in which DNA can uniquely determine an amino acid sequence without placing prior restraints on the primary structure of the protein, is to use a nonoverlapping sequence of three bases per amino acid. Each group of three bases, then, may define a code word. The validity of that position was demonstrated with the help of synthetic messenger RNA.

Assigning Code Words. The year 1961 was a memorable one for molecular genetics. It saw the appearance of papers describing the *in vivo* characteristics of mRNA, the *in vitro* formation of RNA-DNA hybrids, and the purification of an enzyme (a transcriptase) responsible for DNA-dependent RNA synthesis. As we shall see later, it was also the year in which F. Jacob and J. Monod announced their famous operon model for the genetic regulation of protein synthesis. But one of the biggest bombshells came from the laboratory of Marshall Nirenberg at the National Institutes of Health in Bethesda, Maryland. In a series of papers coauthored with J. H. Matthaei, he reported the development of a cell-free system, obtained from *E. coli*, that is capable of polymerizing amino acids.

The polymerization of amino acids by cell-free extracts of *E. coli* occurs only if high-molecular-weight RNA is added as a replacement for the missing mRNA. (Messenger RNA in bacterial cells is short-lived because of the presence of a destructive enzyme, ribonuclease, in the cytoplasm.) At the International Congress of Biochemistry, held in Moscow in the fall of 1961, Nirenberg announced that synthetic polyuridylic acid (UUUUU···) stimulates cell-free systems to specifically incorporate phenylalanine into a polypeptide chain. This experiment not only defined the first code word (UUU = phe), but presented a system that could be used to define the others (Fig. 5-11).

The first attempts to define the code involved correlating the statistical distribution of bases in a synthetic messenger with the uptake of amino acids by a cell-free system. This statistical approach was necessary because of the difficulty of synthesizing polynucleotide chains with known sequences. Some progress was made with these techniques, but by 1965 H. G. Khorana at the University of Wisconsin's Institute for Enzyme Research had succeeded in producing synthetic messengers with strictly repeating base sequences. Such molecules stimulate a cell-free system to incorporate amino acids also in repeating sequences. For example, alternating uridine and guanine (UGUGUG···) causes the formation of alternating valine and cysteine (val-cys-val-cys···), implying that the polynucleotide is read in groups of odd numbers of bases. A triplet code would produce alternating codons (GUG and UGU in this example), while a code of two or four bases would produce a repeating codon—either UG or GU for two, UGUG or GUGU for four, etc. Thus, Khorana's alternating polymers demonstrated directly that the code is an odd number of bases.

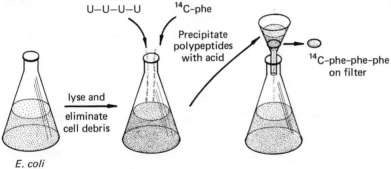

FIGURE 5-11 Breaking the Genetic Code. The first step was the demonstration that polyuridylic acid ("poly-U") causes the specific incorporation of phenylalanine into a polypeptide large enough to precipitate in acid. [Described by M. W. Nirenberg and J. H. Matthaei, in *Proc. Nat. Acad. Sci. U.S.*, **47**: 1588 (1961).]

Using a sophisticated combination of chemical and enzymatic techniques, Khorana was later able to synthesize messengers of repeating triplets and repeating groups of four, the results of which firmly established the triplet nature of the code. Although codons cannot be assigned unambiguously from such polymers because of the difficulty of knowing which nucleotide starts the chain, the possibilities could be narrowed sufficiently to allow assignments when results obtained with several polymers were compared.

Most of the codon assignments, however, came from the ribosomal-binding technique developed by Nirenberg and Leder in 1964. They found that when RNA triplets (synthetic codons) were added to a cell-free extract of *E. coli*, the triplets were bound by ribosomes along with an amino acid and a molecule of tRNA, the function of which will be discussed shortly. By using first one labeled amino acid and then another with a given triplet, they could determine the specificity of the codon. (The complex of amino acid, tRNA, codon, and ribosome sticks to nitrocellulose filters. When the proper radioactive amino acid was used for a given nucleotide triplet, filtering the reaction mixture left the filter "hot.") All possible triplets were tried, and on this basis most of the codons were assigned. The results were not always clear from the binding experiments, but correlation with other available data allowed the codon assignments to be completed (Table 5-2).

TABLE 5-2. The Genetic Code[a]

5' Base	Middle Base				3' Base
	U	C	A	G	
U	UUU } Phe UUC UUA } Leu UUG	UCU } Ser UCC UCA UCG	UAU } Tyr UAC UAA ochre UAG amber	UGU } Cys UGC UGA opal UGG Try	U } pyrimidines C A } purines G
C	CUU } Leu CUC CUA CUG	CCU } Pro CCC CCA CCG	CAU } His CAC CAA } Gln CAG	CGU } Arg CGC CGA CGG	U C A G
A	AUU } Ile AUC AUA AUG* Met	ACU } Thr ACC ACA ACG	AAU } Asn AAC AAA } Lys AAG	AGU } Ser AGC AGA } Arg AGG	U C A G
G	GUU } Val GUC GUA GUG*	GCU } Ala GCC GCA GCG	GAU } Asp GAC GAA } Glu GAG	GGU } Gly GGC GGA GGG	U C A G

[a] Note that in all cases but two (Try, Met), the third position may be occupied by either of the two purines or either of the two pyrimidines without changing the coding specificity. The terminator codons—UAA, UAG, and UGA—stop amino acid incorporation and free the growing polypeptide chain. (The names ochre, amber, and opal refer to the mutant bacterial strains in which the action of these terminators was first studied.) Chain initiation begins with AUG or GUG, marked with an asterisk (*), either of which can code for N-formylmethionine in addition to the amino acid shown for it.

The genetic code is thus known to be a nonoverlapping sequence of nucleotide triplets, read in a 5' to 3' direction. All but three of the 64 codons specify an amino acid. The three unassigned codons are UGA, UAA, and UAG, which serve as chain terminators, stopping the incorporation of amino acids and thus defining the end of a polypeptide chain. Since they do not specify an amino acid, these terminators were formerly called *nonsense codons*.

Examination of the genetic code reveals an interesting pattern of degeneracy. In many cases the terminal (3') nucleotide in a codon can be either of the two pyrimidines, U or C, and still specify the same amino acid. In other cases, the third codon may be either of the two purines, A or G, without changing the specificity. This observation led to the *wobble hypothesis*, in which the greatest importance is assigned to the first two bases, leaving a certain amount of freedom in the third. The result is almost a "$2\frac{1}{2}$" base code. The reason for the wobble lies with the geometry of the ribosome and with the properties of the tRNA molecules, as will be discussed later.

The codon assignments given in Table 5-2 have been verified in the most direct way possible. In 1972, W. Fiers and his associates at the State University of Ghent, Belgium, announced the complete nucleotide sequence of the gene coding for the coat protein of the small RNA phage, MS2. Since the sequence of 129 amino acids in the protein was already known, a direct comparison between amino acid and nucleotide sequences was possible.

Universal Nature of the Code. The genetic code was determined through investigations on *E. coli*. There is every reason to believe, however, that the code is either the same, or very nearly the same, in all forms of life.

For example, tobacco mosaic virus RNA can be used as a messenger to make TMV coat protein in cell-free extracts of *E. coli*. This would not be possible if the message were interpreted in a different way by bacteria and higher plants. In addition, most of the single amino acid replacements in variant human hemoglobins can be explained by single base changes in the *E. coli* code, as Table 5-3 makes clear.

More direct evidence for the universality of the code has come from sequence studies on mRNAs and proteins from various forms of life, including human. So far, the *E. coli* code has always been adequate to explain the relationship between the nucleotide sequence of the mRNA and the amino acid sequence of the protein it specifies. In principle, therefore, a gene from one form of life such as a bacterium (or even a synthetic gene produced in the laboratory) could under the proper conditions become an active gene for cells of another form of life (e.g., human) and be expressed to alter both the characteristics of the recipient cell and the genetics of its offspring. Such experiments are being actively pursued as a possible therapy for certain types of human diseases involving mutant genes.

5-3 TRANSCRIPTION

The two steps in the expression of genetic information are transcription of DNA to produce RNA, and translation of RNA to produce a polypeptide chain. By definition, only messenger RNA (mRNA) gets translated.

TABLE 5-3. Hemoglobin Variants and the Genetic Code

α-Chain position	30	57	58	68
Original amino acid	glutamate⁻	glycine	histidine	asparagine
Possible original codon	GAA	GGU	CAU	AAU
	↓	↓	↓	↓
Possible mutant codon	CAA	GAU	UAU	AAA
Amino acid replacement	glutamine	aspartate⁻	tyrosine	lysine
Name of mutant hemoglobin	Hb G Honolulu	Hb Norfolk	Hb M Boston	Hb G Philadelphia
β-Chain position	6	6	7	63
Original amino acid	glutamate⁻	glutamate⁻	glutamate⁻	histidine
Possible original codon	GAA	GAA	GAA	CAU
	↓	↓	↓	↓
Possible mutant codon	GUA	AAA	GGA	CGU
Amino acid replacement	valine	lysine⁺	glycine	arginine⁺
Name of mutant hemoglobin	Hb S	Hb C	Hb G San José	Hb Zürich

There are two other major classes of RNA, called transfer or *tRNA* and ribosomal or *rRNA*, that do not code for polypeptides but are important nevertheless. Like mRNA, both are produced from DNA (gene) templates by an enzyme called RNA polymerase.

RNA Polymerase. This enzyme is capable of catalyzing both the initiation and elongation of an RNA chain, putting it together from its 3′ end one link at a time. The raw material consists of ribonucleotide triphosphates (see Fig. 5-12). A template is essential. That is, the choice of which ribonucleotide is added depends upon the ability of its base to hydrogen bond to the complementary base in a DNA strand whose genetic message is being transcribed.

The energy for elongating an RNA strand comes from cleavage of the first phosphate anhydride bond of the new nucleotide triphosphate, splitting off inorganic pyrophosphate (PP_i). This exchange of a phosphate anhydride bond for a phosphate ester (ribose-phosphate) is energetically favorable, but the reaction is pulled also by mass action when the PP_i is subsequently hydrolyzed to inorganic phosphate by pyrophosphatase.

The RNA polymerase responsible for this elongation in prokaryotes consists of apparently five subunits of four types. The formula is $\beta\beta'\alpha_2\omega$. To this *core enzyme* may be added still another subunit called a *sigma* (σ) *factor.* It functions only during the initiation of a new RNA chain, then is released and recycled to another RNA polymerase.

While this one RNA polymerase seems to handle all transcription in prokaryotes, eukaryotic cells use a family of RNA polymerases consisting of at least three distinctly different enzymes. *Class A* RNA polymerases are responsible for synthesis of the larger ribosomal RNAs. Class B polymerases produce mRNA and Class C polymerases produce tRNAs. The purpose is presumably to subdivide the task and to provide flexible controls.

One Strand or Two? The transcription of information from double-stranded DNA raises the question of whether one strand of DNA alone serves as a template for the synthesis of RNA, or whether both strands are transcribed. The first experimental evidence that only one DNA strand is transcribed at a given gene was the discovery that the two strands of SP8 DNA (a phage that grows on *Bacillus subtilis*) are different enough in their overall base content to be separable on CsCl gradients. The two strands can thereby be tested independently for hybridizing ability with phage-specific messenger RNA produced after infection. Only one of the two strands of SP8 DNA forms hybrids, however, implying that only one of the two strands serves as a template in the transcription process.

Later work has shown that generally no one strand of DNA has a monopoly on transcription. Rather, some of the genes are transcribed from one strand while others are transcribed from the alternate strand. In other words, the same gene is not ordinarily transcribed in both strands, an event that would lead to two messages bearing the same information in different (but complementary) languages.

Start and Stop Signals. The role of the prokaryotic σ factor is to recognize start signals within the DNA molecule. These regions are called *promoter sites*. The promoter sites are not identical to each other, even though all attract the same RNA polymerase. The difference from one site to another gives some of them a competitive advantage, resulting in more frequent transcription of the associated DNA. Since not all genetic information needs to be turned into protein at the same rate, this difference in promoters presumably serves a useful function.

The recognition (promoter) site for RNA polymerase is not transcribed. Rather, as the polymerase begins to move down the gene it skips the first six or seven bases. Actual RNA initiation then takes place, usually with a purine in prokaryotes. Since this first nucleotide retains its triphosphate, RNAs are either pppApX or pppGpX at their 5' end, where X is the second base in the chain and p a phosphate group. The triphosphate is not necessarily found in the completed molecule, however, since completed RNAs are regularly attacked by nucleases capable of removing some 5' nucleotides.

Once started, RNA polymerase moves down the DNA strand at a more or less constant rate until it comes to a stop signal, at which point it dissociates from the DNA, freeing the newly synthesized RNA strand. The stop signal, like the promoter site, is a sequence of bases. Some of these signals seem to be recognized directly by the polymerase, others appear to require the help of an additional protein called a *rho* (ρ) *factor* whose ancillary function is therefore the counterpart of the σ factor.

The above description is specifically applicable only to prokaryotes. However, although we know less about it, there is no reason to believe that eukaryotic systems are very different.

Posttranscriptional Processing. Both prokaryotes and eukaryotes typically produce RNAs that are shortened and chemically modified after synthesis and before use.

The most extensive chemical modifications are found in the tRNAs. Alanine tRNA of *E. coli*, for example, contains 77 bases, 10 of which are modified by deamination, methylation, hydroxylation, or by other chemical changes after the tRNA is polymerized. The purpose is presumably

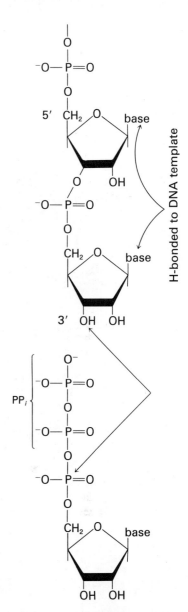

FIGURE 5-12 Nucleotide Addition to RNA. Inorganic pyrophosphate (PP_i) is split off. Note that elongation is only at the 3' end, so that chain growth is 5' to 3'.

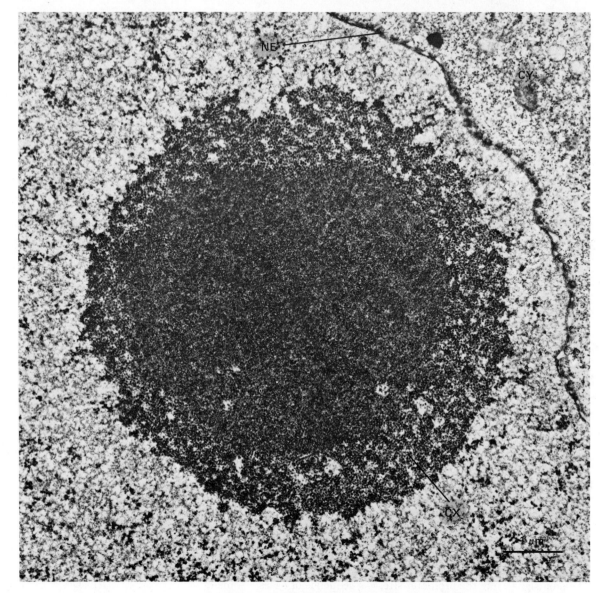

FIGURE 5-13 The Nucleolus. Thin section showing the dense, fibrous core and granular cortex (CX). The granules are ribosome precursors, containing protein and RNA. It is here that rRNA is transcribed and ribosome assembly begins. The nuclear envelope (NE) is visible, with the cytoplasm (CY) beyond. [Courtesy of O. L. Miller, Jr. and B. R. Beatty, *J. Cell Physiol.*, **74,** Suppl. 1: 225 (1969).]

to facilitate the specific interaction between tRNA and ribosome that is needed for protein synthesis. Other classes of RNA generally undergo less extensive modification, but some degree of posttranscriptional chemical change is quite common.

Clipping and trimming of RNA after transcription is also common. Prokaryotic rRNA, for example, comes in three sizes, referred to on the basis of sedimentation velocity in the centrifuge (see the Appendix) as 5S, 16S, and 23S rRNA. The original transcript, however, is a 30S precursor that contains probably one copy of each final RNA plus transcribed but unused spacers on each side. The situation in eukaryotes is similar, with a 5.8S, 18S, and 28S rRNA derived from a single 45S precursor. (There is also a 5S eukaryotic rRNA polymerized at a distant site and by a different polymerase.)

The transcription and processing of eukaryotic rRNA take place in the nucleolus (see Figs. 5-13 and 5-14). Nucleoli are produced at *nucleolus*

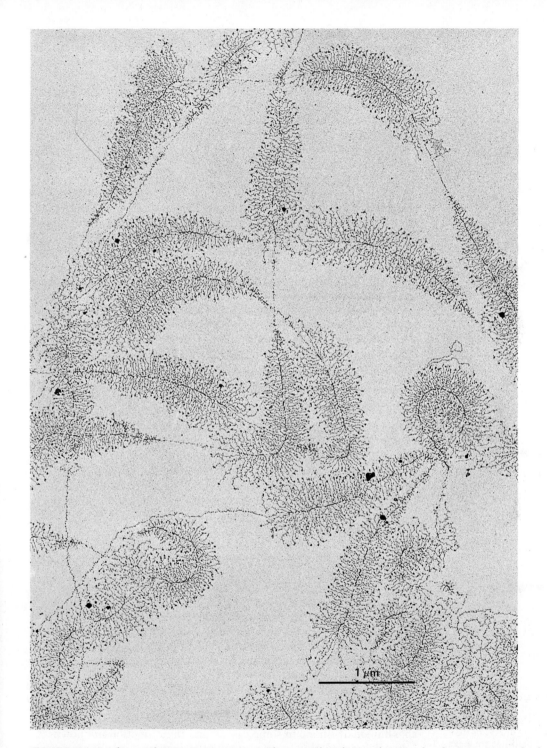

FIGURE 5-14 Ribosomal RNA Transcription. The genes for ribosomal RNA, already present in multiple copies, are amplified by repeated replication. The resulting extrachromosomal DNA forms a nucleolus. Here, one sees some of this DNA in the process of transcription. The gradient of RNA lengths reflects movement of transcriptases from the points of RNA initiation. Newly made rRNA is coated immediately with protein in the first step toward ribosome synthesis. [*Triturus viridescens* oocyte. Courtesy of O. L. Miller, Jr., and B. R. Beatty, Oak Ridge National Lab. From *Science*, **164**: 955 (1969), copyright by the AAAS.]

organizers, which are sites on the chromosome containing multiple copies of the rRNA genes. These repeated units of DNA within the chromosome are reduplicated further outside the chromosome to form the nucleoli.

Messenger RNA is also processed after synthesis, though the situation in prokaryotes is vastly simpler than in eukaryotes. Prokaryotic mRNAs commonly consist of several gene transcripts along with a small amount of untranslated DNA between genes and at the ends. The mRNA usually remains intact during translation—which, because transcription also starts at the 5' end, can begin even before the messenger is completed and released. Not so in eukaryotes.

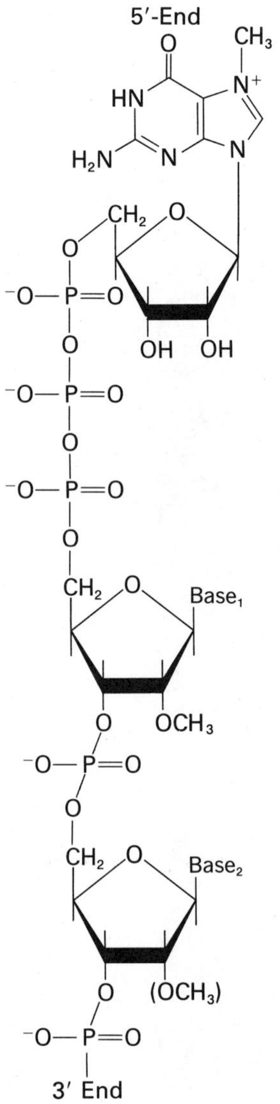

HnRNA and the Eukaryotic Message. When attempts were first made to label eukaryotic mRNA with radioactive isotopes, investigators were puzzled to find that radioactivity could be rapidly incorporated into RNA in the nucleus, but very little of this RNA ever made it to the cytoplasm where it has to go to direct protein synthesis. It turns out that this nuclear RNA is a mixture of very long transcripts, referred to as heterogeneous nuclear RNA or *hnRNA*. HnRNA contains the final messages, but the great majority—perhaps three-fourths or more—of the RNA has an unknown function and is degraded while still in the nucleus.

In addition to being derived from very much larger precursors, most eukaryotic mRNAs undergo three types of further modification: the addition of polyadenylate (poly A) at the 3' end; the addition of "caps" of special nucleotides at the 5' end; and methylation of nucleotides at a few internal sites. The finished product is typically translated into a single polypeptide chain.

The addition of *poly A* to the 3' end was the first modification of eukaryotic mRNA to be discovered. This terminal piece is generally about 200 nucleotides long. It is not a transcript from DNA; rather, it is added to the completed message one nucleotide at a time, using ATP as a precursor. The function is not known with certainty, but in some experimental systems survival time of the mRNA is very much shortened in the absence of a poly-A tail. At least one fraction of mRNA, however, that specifying the chromosomal proteins called histones, does not have these tails.

At the other end of the eukaryotic message, one usually finds a *5' cap*. The cap consists of a methylated guanylate stuck onto the chain "backward" (5' to 5') from the rest of the nucleotides (see Fig. 5-15). This configuration is abbreviated $M^7G(5')ppp(5')X$, where X is the normal 5' terminus. In addition to 7-methylguanosine, cap structures include methylation of the first and sometimes second "normal" nucleotides at their 2' OH. The presence of a cap seems to facilitate translation, though it is obviously not essential in every case since they are absent on some eukaryotic mRNAs.

The third modification of the eukaryotic message is *methylation* of internal adenylate nucleotides at their N-6 position (i.e., at the amine of adenine's number 6 carbon). The number of such methylations is variable but small—none in some cases—and their function unknown.

In addition to this extensive posttranscriptional modification, eukaryotic mRNA differs from its prokaryotic counterpart in leaving the nuclear region prior to translation. This migration seems somehow to involve the nucleolus, for use of UV microbeams to destroy nucleoli rap-

FIGURE 5-15 The 5' Cap. Most eukaryotic mRNAs seem to have a 7-methylguanosine triphosphate attached 5' to 5' to the first nucleotide. In addition, the 2' hydroxyl of that nucleotide and sometimes the next one, too, is methylated. The presence of the methyl group at the second nucleotide distinguishes a "type II cap" from a "type I cap."

idly stops protein synthesis in the cytoplasm. It has been suggested that new mRNA associates with nucleolar-derived ribosomal precursors to facilitate exit from the nucleus. This association, whether or not it takes place in the nucleolus, is the first step in translation of the genetic message.

5-4 TRANSLATION

We have seen that the informational content of DNA lies with its sequence of bases, and that the information may be transferred to messenger RNA by polymerizing the latter with a complementary sequence of bases. The next step is to polymerize the proper sequence of amino acids. That process, called *translation*, involves the ribosomes, several small proteins, messenger RNA, and the additional class of RNA known as transfer RNA, or tRNA.

Transfer RNA. The tRNAs are relatively small—about 80 nucleotides or 25,000 daltons. They are smaller and more soluble in acid than the other RNAs, and hence they were originally referred to as *soluble RNA*, or sRNA.

The first nucleotide sequence of a tRNA was announced in 1964 by R. W. Holley. Since then, it has been possible to determine not only the sequences but the three-dimensional structure of tRNAs (see Fig. 5-16).

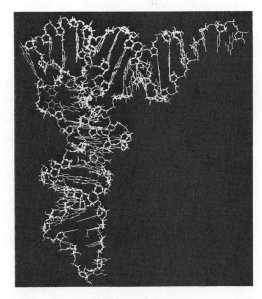

FIGURE 5-16 Transfer RNA. Phenylalanine tRNA from yeast. The *phe* codon in this case is $^{5'}$UUC$^{3'}$, matched by the tRNA anticodon $^{3'}$AAG$^{5'}$ (actually, methylguanosine instead of G). The model was established by X-ray diffraction. The 3' acceptor end (which, like all other tRNA sequences is CCA for the last three nucleotides) is at upper right, the anticodon loop at the bottom. Abbreviations are as follows: m$_y^x$ indicates *y* methyl groups at carbon position *x*; hU, dihydrouridine; ψ, pseudouridine; Y, a purine; T, thymidine. [Courtesy of S. Kim, F. Suddath, G. Quigley, A. McPherson, J. Sussman, A. Wang, N. Seeman, and A. Rich. Diagram from *Nature*, **248**: 20 (1974). Model from *Science*, **185**: 435, copyright 1974 by the AAAS.]

The existence of a specific spatial configuration is important because these molecules must be unambiguously recognized by enzymes and they must bind to specific locations on ribosomes. It is their function to recognize an mRNA codon and to position the corresponding amino acid for incorporation into a growing polypeptide.

Before an amino acid can be incorporated into a growing polypeptide chain, it must first be esterified to a suitable transfer RNA. That event is catalyzed by an enzyme, called an *aminoacyl-tRNA synthetase*, which is specific for the amino acid in question and drives the reaction by coupling it to the hydrolysis of ATP to AMP. The enzyme first attaches the amino acid to AMP, using a high energy anhydride bond between the carboxylate of the amino acid and the phosphate of AMP. This complex remains bound to the enzyme until a second reaction replaces the anhydride bond with a lower energy ester linkage to the 3' terminal hydroxyl of a tRNA (see Fig. 5-17):

$$aa + ATP + E \rightarrow (aa\text{-}AMP\text{-}E) + PP_i$$

$$(aa\text{-}AMP\text{-}E) + tRNA \rightarrow aa\text{-}tRNA + E + AMP$$

Free pyrophosphate gets hydrolyzed to P_i by the enzyme pyrophosphatase, thus helping to drive the reaction to the right.

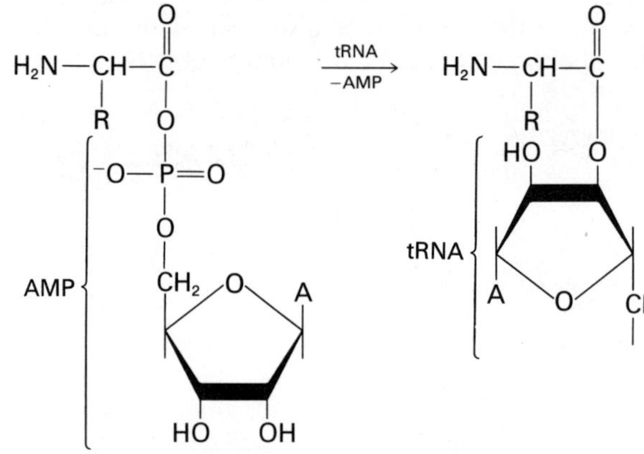

FIGURE 5-17 Formation of Charged tRNA. The amino acid forms an ester bond with the terminal adenosine of the tRNA, replacing an acid anhydride bond in the aa-AMP complex.

The function of the activating enzyme, the aminoacyl-tRNA synthetase, is to recognize a particular amino acid and tRNA and to couple them. It follows, of course, that for each amino acid there is at least one activating enzyme and at least one tRNA.

The amino acid-tRNA complex (called an *aminoacyl-tRNA* or *charged tRNA*) is positioned at an mRNA by matching a three-nucleotide complementary *anticodon* of the tRNA with the corresponding codon of the ribosome-bound mRNA. In this way, the new amino acid is brought into proper position with respect to the growing polypeptide chain.

It is here that the coding ambiguity mentioned earlier as the wobble hypothesis is seen, for according to that concept the third position in the codon is less reliably matched than the others: In particular, either of the two pyrimidines, U or C, might be paired with an anticodon G; and either of the two purines, A or G, might be paired with an anticodon U. (On the other hand, an anticodon C or A in the third position is thought to be rather specific for G or U, respectively, in the mRNA.) Still more flexibility is derived from the appearance of inosine in the third position

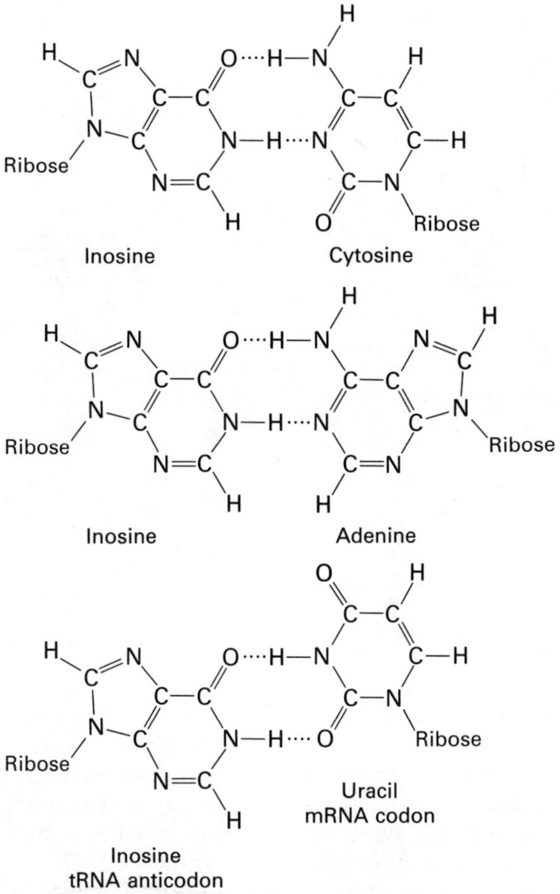

FIGURE 5-18 Wobble Pairing With Inosine. When inosine is in the 5' position of the anticodon, the corresponding 3' base in the codon may be C, A, or U.

of the anticodon, for this base, which is derived from adenine by changing its amino group to a carbonyl (NH_2 to $C=O$), hydrogen bonds to U, A, and to C (see Fig. 5-18).

In spite of this flexibility, there are many different tRNAs, so that in most cases a given amino acid may be picked up and inserted by any of several. Some tRNAs will bind to the codon better than others, however, and presumably more rapidly. Hence, the availability of various tRNAs becomes one way of establishing different rates of synthesis for different proteins.

Chain Elongation. Codon-anticodon pairing brings a new amino acid into the *A*, or *aminoacyl, site* of the ribosome on which the mRNA rests. The existing polypeptide is attached to the *P*, or *peptidyl, site* through an ester bond to its own tRNA residing there (Fig. 5-19a). This ester bond is replaced by a peptide bond, thus transferring the chain, now lengthened by one amino acid, to the newly arrived tRNA. The exchange, which is thermodynamically very favorable, is catalyzed by peptidyl transferase, an enzyme that is an integral part of the ribosome itself (Fig. 5-19b, c).

Lengthening of the polypeptide chain is followed by a translocation that moves the polypeptide and attached tRNA from the A to the P site. Translocation thus frees the A site to make way for the next "charged" (i.e., aminoacyl) tRNA (see Fig. 5-19d). Thus, the messenger RNA is moved past the ribosome three nucleotides at a time, so that a tRNA and

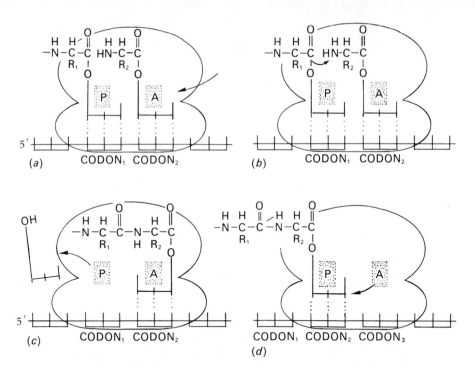

FIGURE 5-19 Polypeptide Chain Elongation. (a) A charged amino acid (tRNA-amino acid) is attracted to the acceptor, or aminoacyl (A) site. (b) The ester bond between the existing chain and the tRNA residing in the peptidyl (P) site is exchanged for a peptide bond to the new amino acid. (c) The tRNA at the P site is now free to leave. (d) Translocation of the ribosome moves the elongated chain from the A to the P site preparatory to starting another cycle.

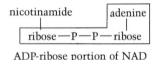

ADP-ribose portion of NAD

mRNA codon occupy each of two ribosomal positions in turn. The whole process of chain elongation occurs very rapidly, with growth rates up to 40 amino acids per second in *E. coli* and perhaps much faster in at least some eukaryotic cells (estimates range up to 600 amino acids per minute for hemoglobin synthesis—therefore much less than a minute per subunit). The cellular rate of protein synthesis is enhanced even further by the presence of multiple ribosomes on the same mRNA, forming a complex that is called a *polysome* (Fig. 5-20).

Even though the equivalent of two ATP hydrolyses (i.e., an ATP to AMP) are needed just to charge the appropriate tRNA with an amino acid, still more energy is needed to complete the chain elongation. The additional input of energy comes in the form of two GTP hydrolyses. One occurs during the binding of a charged tRNA to the ribosome and messenger, and involves at least one nonribosomal protein called *elongation factor T* (EF-T, actually two subunits called Tu and Ts) in prokaryotes and *elongation factor 1* (EF-1) in eukaryotes. The second GTP is hydrolyzed to GDP during translocation, which requires another elongation factor known as EF-G in prokaryotes and EF-2 in eukaryotes.

The complexity of chain elongation makes it a ready target for interfering substances. For example, diphtheria toxin is a protein that is produced by the bacterium *Corynebacterium diphtheriae* when the latter is infected with a virus called phage B. Diphtheria toxin catalyzes transfer of the ADP-ribose portion of NAD to EF-2, thus inactivating EF-2 and eventually killing host cells by preventing protein synthesis. Although the bacteria remain in the throat, spread of the toxin characteristically causes death by heart failure. It has recently been shown that "exotoxin A" of the important unrelated pathogenic bacterium *Pseudomonas aeruginosa* shares this same molecular mode of action, though the toxins produced by the two bacteria are apparently quite different proteins.

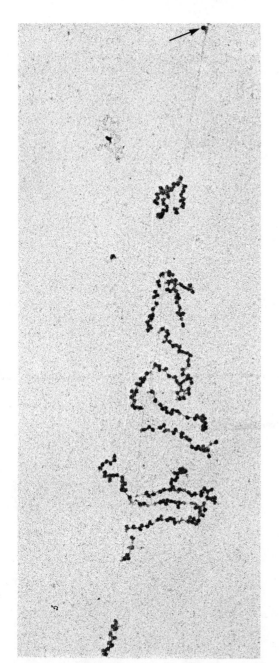
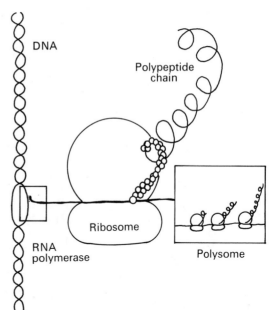

FIGURE 5-20 A Bacterial Gene in Action. *LEFT:* The production of messenger RNA by RNA polymerase (small dots joining side chains to the DNA fiber) is proceeding apace with translation, as evidenced by the numerous ribosomes attached to each mRNA. The longer-chains (called polysomes) are furthest from the point of gene origin. *Arrow* indicates a presumed RNA polymerase molecule at or near the beginning of the gene. *RIGHT:* A schematic interpretation of one polysome showing polypeptide chain elongation. [Micrograph courtesy of O. L. Miller, Jr., B. A. Hamkalo, and C. A. Thomas, Jr., *Science*, **169**: 392 (1970). Copyright by the AAAS.]

Certain antibiotics also act on bacterial cells by interfering with some part of the elongation process. For example, lincomycin, chloramphenicol, and sparsomycin inhibit peptidyl transferase; tetracyclines inhibit binding of charged tRNAs to ribosomes; erythromycin inhibits translocation; and streptomycin and some related aminoglycoside antibiotics inhibit initiation and cause misreading of the codons. Bacteria can usually mutate to become resistant to these antibiotics. In the case of streptomycin, for example, resistance is conferred by a change in one ribosomal protein.

CHAPTER 5
Cellular Homeostasis:
Genes and Their Regulation

The process of chain elongation is also subject to natural errors. The A site may be visited by a number of charged tRNAs before one having the proper anticodon arrives. Ideally, only that one would stay at the site long enough for its amino acid to be transferred to the growing chain. During slow rates of chain growth things happen pretty much that way, but during periods of faster chain growth the wrong tRNA is more apt to get trapped in the A site and the frequency of translation errors increases.

Ribosomes. The ribosome clearly plays an important role in protein synthesis, though the details of that role are by no means all known. The codon-by-codon reading of a messenger RNA implies the existence of a pointer to mark the three bases that constitute a codon. That function falls to the ribosome.

The bacterial ribosome as it exists during translation is roughly spherical, nearly 200 Å across, and has a mass of about 2.4×10^6 daltons. It is referred to as a *70S ribosome* (see Fig. 5-21). Ribosomes of mitochondria and chloroplasts are similar to bacterial ribosomes, but eukaryotic cytosol ribosomes are larger and heavier, reflected in an 80S sedimentation rate. The 70S ribosome is composed of two major subunits, called 30S and 50S (40S and 60S in eukaryotic cytosol ribosomes), with each larger subunit about twice the mass of the smaller. The small 30S subunit may be further broken down into a single 16S RNA and (in *E. coli*) 21 distinct proteins. The 50S *E. coli* ribosomal subunit is composed of 5S and 23S RNA molecules plus about 34 different proteins. In both subunits, the various proteins are present usually in only one copy each, and together constitute about one-third the total mass. In spite of this complexity, dissociation and reassociation studies *in vitro* indicate that ribosomes are self-assembling structures.

Less is known about the 80S eukaryotic ribosome. The larger, 60S subunit contains 5S, 5.8S, and 28S RNAs, while the 40S subunit contains

FIGURE 5-21 (Below left) 70S Ribosomes from *E. coli*. Note that most particles are composed of two unequal subunits, designated 30S and 50S in bacteria. [Courtesy of H. E. Huxley and G. Zubay, *J. Mol. Biol.* **2**: 10 (1960).]

FIGURE 5-22 (Below right) Rat Liver Polysomes. These eukaryotic cytosol ribosomes are 80S. The bar underlines 40S subunits, separated from their 60S subunits and attached to the mRNA. Initiation of transcription begins with this configuration in eukaryotes as in prokaryotes. [Courtesy of Y. Nonomura, G. Blobel, and D. Sabatini, *J. Mol. Biol.* **60**: 303 (1971).]

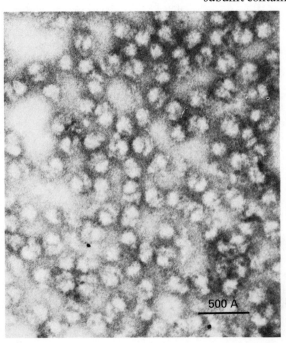
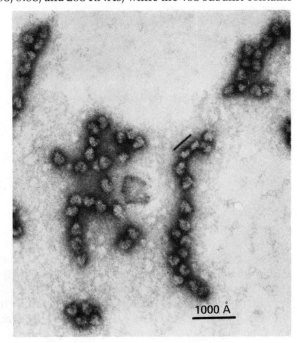

a single 18S RNA. The exact number of proteins remains to be determined. Like its prokaryotic counterpart, however, the RNA is largely structural and, while necessary for assembly, has no messenger function.

Initiation. When they are not involved in translation, ribosomes exist mostly as their free subunits. Every mRNA contains at least one ribosome binding site at which these subunits assemble to reestablish an active ribosome. In bacterial systems, this binding site usually contains the base sequence AGGA at a location roughly 10 nucleotides from the beginning (i.e., 5' end) of the actual message. The 30S subunit binds first, possibly with the help of base pairing between the 3' tail of its 16S RNA and the ribosome binding site (see Fig. 5-22). Other substances must also be present, however, including N-formylmethionyl tRNA and at least three nonribosomal proteins called *initiation factors*. The 50S subunit is added last, a step that results in the hydrolysis of bound GTP (to GDP) and release of the initiation factors.

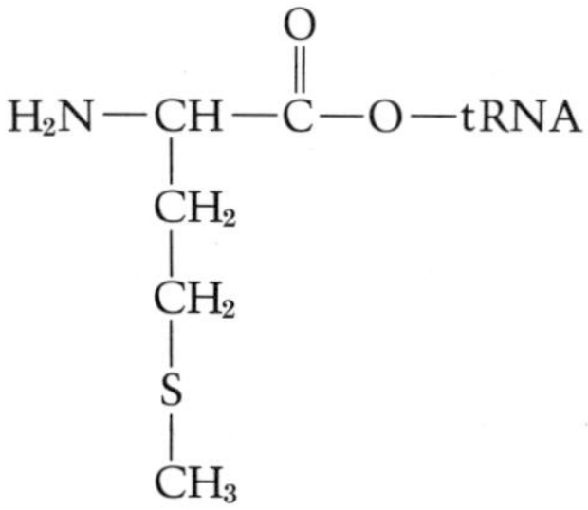

N-Formylmethionyl-tRNA (fmet-tRNA) is a methionine-charged tRNA that has been formylated at the free amino group of methionine. Although there are two methionine transfer RNAs, only one will permit the enzyme-catalyzed formylation of its amino acid. The discovery of fmet-tRNA in 1964 by K. Marker and F. Sanger immediately sparked a great deal of interest, for the previous year J. P. Waller had pointed out the surprising fact that nearly 45% of all proteins in *E. coli* have methionine at their amino ends even though methionine constitutes only about 2.5% of all the amino acids in these same proteins. It was immediately suggested that fmet-tRNA might initiate polypeptide chain synthesis, for the amino end is the only position that formylmethionine could occupy, since its own amino group is blocked.

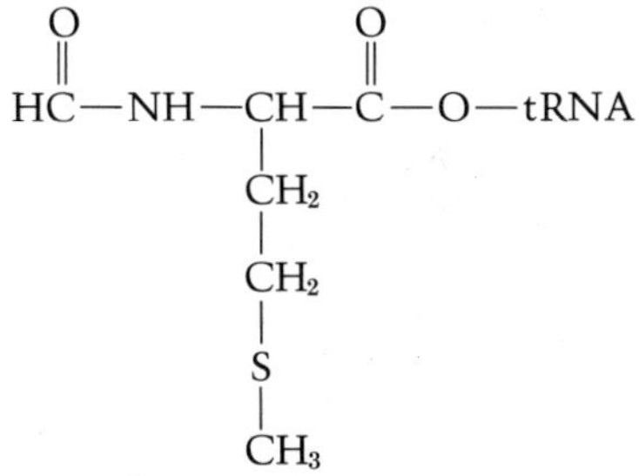

By 1966 several workers had demonstrated the role of fmet-tRNA in the initiation of phage proteins coded by RNA from the small phages f2 and R17. These RNAs act as messengers both *in vivo* and in cell-free extracts of *E. coli* (i.e., *in vitro*). When the phage coat protein made *in vitro* was analyzed, it was found to have N-formylmethionine at its amino end. However, when the same protein is isolated from phages grown *in vivo*, the chains start with alanine. Apparently, the chain is initiated by an N-formylmethionine that is later removed by some peptidase before the protein is incorporated into a maturing phage particle. In other proteins more than one amino acid may be removed or, sometimes, the formylmethionine may merely be deformylated to allow the chain to begin with ordinary methionine.

The role of methionyl tRNA in eukaryotes is somewhat less clear. There are two of them in the cytosol, only one of which can be used for initiation. Although this methionine-charged eukaryotic tRNA can be formylated by bacterial enzymes *in vitro*, formylation may not take place *in vivo*. In addition, cleavage of N-terminal methionine after synthesis appears to be more common in eukaryotic systems.

Formylmethionyl-tRNA binds specifically to the normal methionine codon, AUG. It will also bind less efficiently to GUG, a valine codon. Although AUG seems to be the preferred start signal *in vivo*, proteins have been identified whose message begins instead with GUG. Thus, the same two codons that specify ordinary methionine and valine when they occur in the interior of a polypeptide are also responsible for chain initiation.

Termination. The termination of polypeptide chain growth, like its initiation, takes place at specific codons—in this case UAA, UAG, and UGA. These are the "nonsense" codons that prematurely terminate translation when they appear by mutation. (Such mutants are referred to as "ochre," "amber," or "opal" mutations for the three codons, respectively.) Once a terminator codon is encountered, no further polypeptide chain synthesis occurs until a new initiation codon is reached.

Recognition of the terminator codons comes not from tRNA, but from proteins called *release factors*. At least two of these exist in *E. coli*, one of which is involved in termination at UAA and UAG, and the other at UAA and UGA. When a ribosome encounters these codons in the presence of the appropriate factor, the growing polypeptide chain is freed from its tRNA, a reaction that requires GTP. Hydrolysis of this high energy ATP-like compound thus powers each ribosome-associated step in translation.

In prokaryotic systems, mRNAs frequently contain several cistrons. Termination normally occurs at the end of each cistron, releasing the polypeptide chain to assume spontaneously its final secondary, tertiary, and quaternary structures. Ribosomes may or may not be released and reassembled between cistrons, but each cistron begins with a new initiator codon.

Although eukaryotic systems commonly have monocistronic messengers, there are also many examples where multiple cistrons are translated into one long polypeptide chain, which may later be chopped up into individual proteins. The proteins of polio virus are made in this way, and the hormone ACTH in mammals is also produced by cleavage of a multifunctional polypeptide precursor. Alternatively, eukaryotic enzymes can result from specific folding of only one portion of an intact polypeptide chain that contains the primary structure for several enzymes. The other enzymes in the polypeptide may likewise fold to become active, resulting in a multifunctional complex of covalently linked enzymes.

Errors in the normal termination sequence can be induced with antibiotics, such as *puromycin*, and they also occur as spontaneous mutations. When a mutation converts a normal codon to one of the terminators, the consequence is premature release of the growing polypeptide. Conversely, a mutation may convert a terminator into a codon that can be read as an amino acid, in which case termination fails to take place.

One of the latter type of mutations, terminator to amino acid, occurs in the mutant human hemoglobin called constant spring (Hb CS). Hb CS was found to have an α chain containing the normal 141 amino acids plus an additional 31 amino acid tail at the C-terminus. The tail is derived from that portion of the mRNA beyond the terminator codon. It appears that a UAA terminator codon has mutated to CAA, which is then read as glutamine and followed by the normally untranslated region. The usual function of this untranslated sequence of RNA is not clear, but at least we now know that nothing about it prevents translation except the preceding terminator codon.

The discovery of Hb CS, combined with physical studies, gave us our first look at the microanatomy of a human mRNA. Starting from the 5' end, the α-chain mRNA consists of (1) about 50 nucleotides that includes the ribosome binding site and initiator codon; (2) the α-chain cistron of 423 nucleotides plus terminator; (3) about 108 nucleotides that include the Hb CS tail; and (4) about 70 nucleotides of poly A at the 3' end.

Bacteriophage φX174: One Gene = Two Polypeptides. If a single base is either added or deleted within a cistron, from that point on the reading frame is out of phase and incorrect amino acids will be inserted one after the other until translation stops. By shifting the reading frame, a new message is created which in almost every case is worthless. An interesting exception has been identified in the small bacteriophage called φX174. Here the gene for "protein E" is contained entirely within the gene for the larger and completely unrelated protein D. The start signal for E is about halfway through the D cistron, but it is recognized by shifting the reading frame by one base. Thus, one stretch of DNA and its RNA transcript contains not one but two cistrons (see Fig. 5-23).

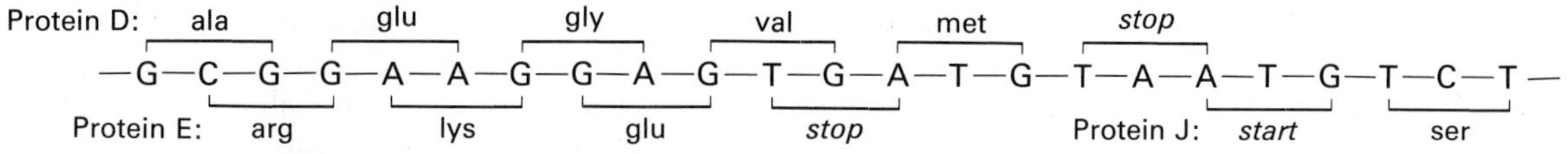

Protein D: ala   glu   gly   val   met   stop
         —G—C—G—G—A—A—G—G—A—G—T—G—A—T—G—T—A—A—T—G—T—C—T—
Protein E: arg   lys   glu   stop   Protein J: start   ser

FIGURE 5-23 One Gene, Two Proteins. The gene for protein E of phage φX174 is contained within the gene specifying protein D. The start signal for E defines a reading frame that is out of phase with that used for D. Hence, the two amino acid sequences are entirely different. Note that the nucleotides given are those of the gene, not an mRNA. (Replace T with U to get the usual codons). Note also that the last base of protein D's stop codon is the first base of protein J's start signal, an extraordinarily thrifty use of DNA. [Described by D. Barrell, G. Air, and C. Hutchinson, *Nature*, **264**: 34 (1976).]

This unusual situation is also an interesting lesson in gene evolution. Protein E is used only to help lyse the host cell, releasing progeny phage sooner than would be the case without this gene. It is probable, therefore, that protein D existed first and that a base substitution resulted in the chance occurrence of the phase-shifted start signal. The protein produced, because it facilitates lysis and therefore reduces generation time, would give the mutant phage a competitive advantage. Clearly, the probability of such an overlapping gene being useful would be small, but there are two such examples within φX174 and it seems likely that others will eventually be found elsewhere.

Antibodies: Two Genes = One Polypeptide. The structure of antibodies was discussed in Chap. 3. You will recall that the typical molecule is comprised of two types of polypeptide chain. Each of these chains has a variable region (V region) of some 110 or so amino acids that differs from one antibody specificity to another, plus a constant region (C region) that is the same for that particular chain type in antibodies of all specificity. Each chain, as it turns out, is coded for not by one but by two genes. Thus, there need be only seven constant-region genes (one gene for the constant region of each of the five heavy-chain classes plus one gene for each of the two light-chain constant regions). To these constant regions can be added polypeptides drawn from a pool of variable region genes. The amount of DNA committed to antibodies is therefore very much less than would otherwise be the case.

The problems that remain are explaining how the V-region and C-region gene products get linked together, and establishing just how many V-region genes are present. The latter problem stems from data indicating a far smaller number of V-region genes in the germ line (that is, genes passed from generation to generation) than is needed to account for the wide variety of antibodies that lymphocytes can produce. It is thought that some sort of rearrangement of sequences within the V regions during maturation of the immune system may add to the available pool of V-region specificities.

Rearrangement of DNA also appears to be the explanation for the participation of two genes in the synthesis of each antibody chain. In embryonic cells, the C region and V region genes seem to be clearly separated. In antibody-producing cells, however, data now available indicate that the specific C-region and V-region genes being used are adjacent.

CHAPTER 5
Cellular Homeostasis:
Genes and Their Regulation

Hence the determination of specificity for a lymphocyte line may be established by physical rearrangement of DNA to allow these two genes to be transcribed together and later translated without a stop signal to yield a single polypeptide.

5-5 GENE REGULATION IN PROKARYOTES

The mechanics of gene expression—that is, transcription and translation—are very much the same in eukaryotes and prokaryotes. Whether regulation of genetic expression shares the same degree of homology is not so certain. First, we know much less about gene regulation in eukaryotes. Second, there are some rather obvious and important differences in the way genes are arranged in the two classes of cells requiring, perhaps, different types of control. We shall proceed by summarizing some of the better understood aspects of gene regulation in prokaryotes, then examine eukaryotic systems to make comparisons and contrasts.

The Prokaryotic Chromosome. Eukaryotes and prokaryotes are, of course, very different in the way in which their genetic information is stored. Whereas humans, for example, have 46 separate chromosomes, each a complex mixture of DNA, RNA, and protein, the commonly studied bacterium *E. coli* has its 4000 or more genes spaced along a single double-stranded molecule of DNA. The mass of this DNA is over

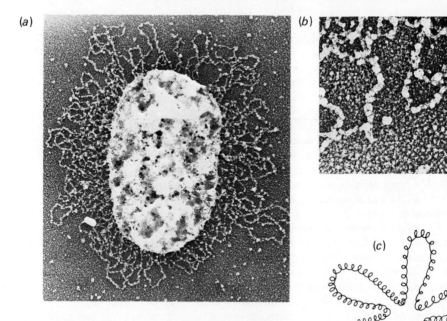

FIGURE 5-24 The Prokaryotic Chromosome. (a) Gently lysed *E. coli* showing loops of "chromatin." (b) Detail of the chromatin, showing beaded structure. (c) Model for the nucleoid structure, interpreting the loops as supercoiled segments of DNA. (d) [Opposite page] *E. coli* with its DNA spread onto a surface. Note retention of the loops, now unfolded. Bar = 1 μm [(a) and (b) courtesy of Jack D. Griffith, *Proc. Nat. Acad. U.S.*, **373**: 563 (1976); (d) courtesy of R. Kavenoff, B. Bowen, and O. Ryder. From *Chromosoma*, **55**: 13 (1976), by permission.]

2×10^9 daltons and, when it is fully spread out, the contour length is around 1100 μm (1.1 mm). *In vivo*, this enormous molecule is packed to become a nuclear body (or *nucleoid*) only a μm in diameter.

To get so much DNA in so small a space requires a high degree of order. We are largely ignorant of the details, but when a cell is gently disrupted loops of DNA are extruded (see Fig. 5-24). The fibers comprising these loops have a beaded appearance with a diameter of about 120 Å (DNA itself has a diameter of 20 Å) and consist of about 60% DNA, 30% RNA, and 5–10% protein, most of which is RNA polymerase. One Å of this fiber appears to contain about 7 Å of DNA. There is some evidence to indicate that this packing density is achieved by coiling ("supercoiling") the DNA strand (see Fig. 5-24c). There may be 100 or more of these supercoiled loops per nucleoid.

We don't know what factors maintain the beaded or supercoiled configuration *in vivo*, although divalent cations such as Mg^{2+} and highly basic (amine-containing) small molecules and peptides are indirectly implicated. Certainly, the positive charges carried by these ions would be expected to interact strongly with negatively charged phosphates along the DNA backbone. There is some evidence that a partial unfolding of these loops can occur *in vivo*, perhaps to support transcription. It is

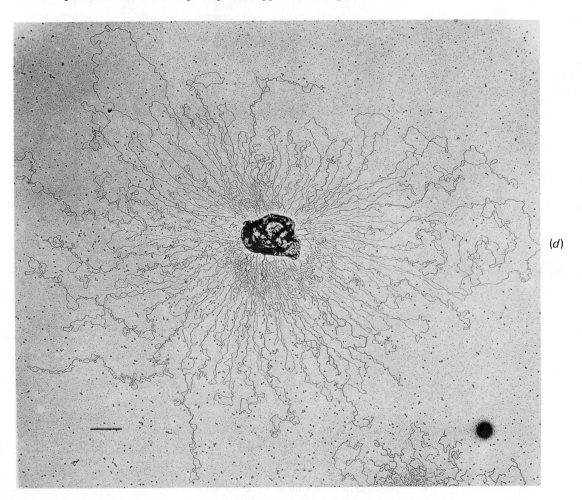

(d)

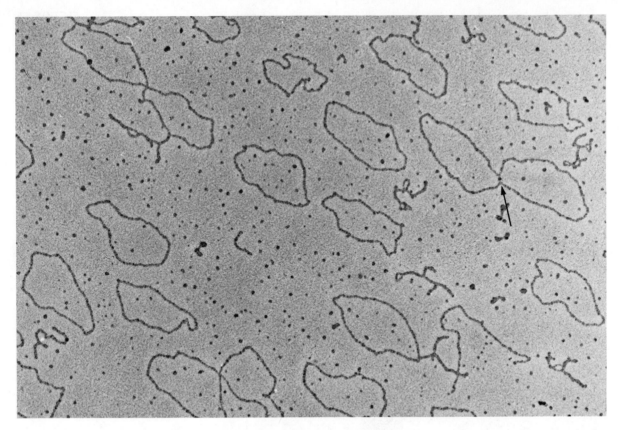

FIGURE 5-25 Plasmids. From *Neisseria gonorrhoeae*, causative agent of gonorrhea. Contour length = 1.5 μm. Note the presence of occasional oligomers (*arrow*), presumably due to incomplete replication. [Courtesy of S. Palchaudhuri, E. Bell, and M. R. J. Salton. From *Inf. and Immunity*, **11**: 1141 (1975).]

even possible that supercoiling may be one parameter that controls the rate of gene expression, since a tightly folded or coiled section of DNA might be relatively less accessible to RNA polymerase or admit fewer ribosomes for polysome formation.

Many bacteria have some additional DNA in the form of autonomously replicating *plasmids* (Fig. 5-25). These small circles of DNA, almost like a virus without a protective coat, may contain anywhere from 2 to over 200 genes. Some plasmids can reversibly insert themselves into the main chromosome, in which case they are called *episomes*.

The first plasmids (actually, episomes) to be discovered were the *F factors*, or fertility factors. The presence of the F factor defines a male bacterium. During mating, chromosomal DNA passes through a special hollow pilus on the male cell and into the recipient. The semiautonomous nature of these factors is apparent from the observation that they alone may be transferred to a recipient cell, which is thereby converted from F^- to F^+—in other words, from female to male. (The male gender in bacteria is thus a contagious condition, a situation that presumably causes them less confusion than it would cause us.)

Plasmids are of extraordinarily practical importance, for they can carry genes conferring antibiotic resistance. These drug-resistance factors, or *R factors*, may recombine to produce plasmids specifying resistance to many drugs at once. R factors not only spread drug resistance among cells of a given species but in some cases can cross the species barrier. They are now so prevalent that certain species of common bacteria are almost always found to have them.[4]

SECTION 5-5
Gene Regulation in Prokaryotes

Since plasmids replicate independently of the main chromosome, they may be present in many copies per cell. And since a gene that is present in more copies can be expected to produce more protein, plasmid replication may be a crude form of genetic regulation, complementing the much more sophisticated mechanisms of induction and repression, to be described.

Induction and Repression. When *E. coli* is grown on the sugar lactose as a carbon and energy source each cell contains several thousand copies of the enzyme, β-galactosidase, that hydrolyzes lactose to glucose and galactose. Sister cells, genetically identical to the first culture but growing on glucose instead of lactose, will have only a few copies of this enzyme per cell. If lactose is added to this second growth medium, it will remain unused until all the glucose is consumed. There will then be a pause in the growth of the culture while cells begin making β-galactosidase and two companion proteins, galactoside permease (to facilitate lactose transport across the cell membrane) and a galactoside acetylase. That having been accomplished, growth resumes. By the use of radioactive amino acids, one can readily show that new enzyme activity appearing during this metabolic readjustment is due to the complete synthesis (i.e., *de novo* synthesis) of the enzymes in question, rather than to the activation of already existing proteins (see Fig. 5-26).

Enzymes like β-galactosidase that do not appear until they are needed are called *inducible*. Inducible enzymes are synthesized in the presence of an appropriate substance, called an *inducer*, that is often a substrate of the enzyme itself, a substrate of another enzyme in the same pathway, or a chemical relative of one or the other.

In addition to inducible enzymes, there are *repressible* enzymes, the synthesis of which decreases, rather than increases, in response to the appearance of certain metabolites. A metabolite that lessens their rate of synthesis is called a *corepressor*. (The origin of the "co-" will be made clear in a moment.) For example, the enzymes of the histidine pathway are repressible: When histidine is presented to a cell, the cell has less need for the enzymes that make histidine.

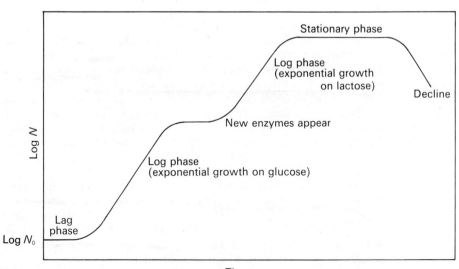

FIGURE 5-26 Biphasic Growth. A bacterial culture started with N_0 cells generally exhibits an initial lag phase during which metabolic adjustments are made, followed by an exponential phase terminating in a plateau, or stationary phase. In the example shown, both glucose and lactose were available. Thus, during the first plateau, cells are adjusting for the exhaustion of glucose by making the additional enzymes needed for lactose catabolism. This type of biphasic growth is also called *diauxie*.

To get rid of an enzyme after its synthesis is stopped, bacteria must ordinarily depend on diluting the enzyme to progeny cells. For example, if lactose is removed from an *E. coli* culture that has been adapted for growth on it, or if glucose is added to the culture, synthesis of β-galactosidase stops. The total amount of β-galactosidase in the culture then remains nearly constant while the amount per cell decreases as the number of cells increases.

A third group of enzymes, whose rate of synthesis is neither increased nor decreased by reasonable changes in the environment or food supply, are called *constitutive*. Even though their rate of synthesis does not change in response to external stimuli, there is always just the right amount of each. Depending on the job the enzyme must do and on the efficiency with which it can do it, there may be a few copies per cell or many thousands of copies per cell. Whatever the number, it must be maintained in the face of loss due to degradation or dilution by growth. There must be a way, therefore, of synthesizing each protein at its own optimum rate. And in the case of inducible and repressible enzymes, that rate must be variable.

Operon Theory. Once synthesis of a particular mRNA begins, it seems to proceed at roughly the same rate for any gene in a given cell (about 55 nucleotides/sec in *E. coli* at 37 °C) and to do so until the polymerase terminator sequence is reached. Hence regulating the amount of mRNA usually means controlling the rate of initiation of RNA polymerase—that is to say, *transcriptional control*.

Our knowledge of transcriptional control began in the 1950s with studies, especially by François Jacob and Jacques Monod of the Institut Pasteur in Paris, on the induction of β-galactosidase. As mentioned above, the appearance of this enzyme is coupled with the appearance of two others, galactoside permease and galactoside transacetylase. Experiments revealed that genes carrying the message for these three proteins, called *structural genes* z, y, and a, are controlled by two others called the *regulator* and *operator*, or i and o genes, respectively. The four genes o, z, y, and a (operator, galactosidase, permease, and acetylase—referred to collectively as the *lac operon*) were found to be next to each other and in that order. The i gene proved to be a short distance away from the others, on the operator side (see Fig. 5-27).

The functions of the two control genes, the regulator and the operator, were revealed by the properties of their mutants. Mutation in either of them (the mutants are called i^- and o^c, respectively) results in a constitutive synthesis of the three normally inducible proteins. The mutant cells continue to synthesize all three proteins regardless of the presence or absence of lactose. When partially diploid cells—containing both mutant and normal i genes, or mutant and normal o genes—were examined, the i^+ and o^c genes appeared to be dominant over their alleles, i^- and o. (The superscript +, added for emphasis, always means the normal, or "wild-type" gene.) A cell containing both a wild-type and a mutant regulator gene continues to behave like a normal cell, but when both a normal and an o^c operator are present one gets constitutive synthesis of the three proteins (see Fig. 5-28). In addition, a second kind of mutation in the i gene was found, called i^s. The i^s mutation is dominant over the wild-type regulator and results in a cell that cannot be induced at all, in contrast to the i^- mutant, which is constitutive.

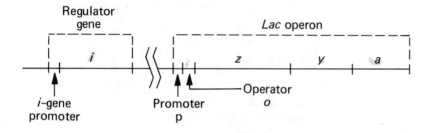
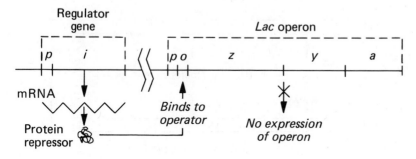
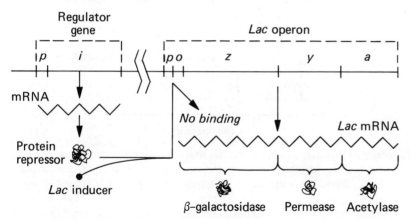

FIGURE 5-27 The *Lac* Operon. The operon consists of three adjacent structural genes coding for proteins, plus an operator and promoter. (The promoter is the site to which RNA polymerase first binds.) The presence of the i-gene product prevents transcription of the operon unless an inducer is also present. The inducer is *allolactose*, which is produced from lactose by the enzyme β-galactosidase, the other function of which is to hydrolyze lactose to galactose and glucose.

In 1961 Jacob and Monod proposed a model that offered a logical explanation for these observations. Their paper had an enormous impact on molecular biology and helped to win for them the 1965 Nobel Prize, shared with André Lwoff. Jacob and Monod suggested that the i gene makes a molecule, called a *repressor*, that can bind to the operator gene. When it does so, synthesis of mRNA from the three structural genes is blocked. The presence of lactose (or a chemical derivative of it) blocks the repressor-operator interaction and results in induction. The i^- regulator gene, then, is recessive because the repressor from a wild-type allele in a partially diploid cell can bind to both operators. An o^c mutant is dominant because the structural genes that it controls will be actively transcribed regardless of the existence of repressors. And finally, the i^s mutant appears to make a repressor that always binds to a normal operator in spite of the presence of inducer.

In support of their model, Jacob and Monod offered the results of an experiment in which an i^+z^+ donor stain (male) was mated to an i^-z^-

CHAPTER 5
Cellular Homeostasis: Genes and Their Regulation

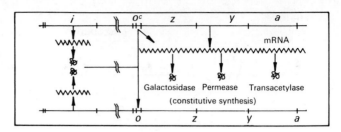

o^c is dominant over o^+

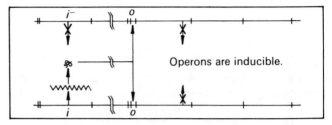

i^+ is dominant over i^-

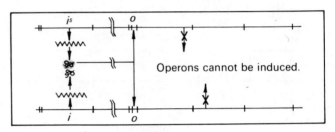

i^s is dominant over i^+

FIGURE 5-28 *Lac* Operon Mutations. The model of the *lac* operon, presented in Fig. 5-27, was deduced from observations made on three mutant strains, o^c, i^-, and i^s. Their behavior was observed in partially diploid cells containing both a normal operon and a mutant version.

recipient (female) strain. Bacterial mating, or *conjugation*, involves the linear transfer of DNA from the donor to the recipient. Transfer always starts at the same point on the chromosome (for a given strain) and proceeds at a more or less constant rate with time. It may thus be interrupted at any point by a vigorous stirring in a blender, producing partially diploid cells of the type just mentioned. In the present experiment, however, the recipient began making β-galactosidase as soon as it received the z^+ gene, even in the absence of lactose. About an hour after the i^+ gene was transferred, however, synthesis of β-galactosidase required the presence of lactose or other inducer. This response is precisely what one would expect if the i gene were releasing to the cytoplasm some product that is capable of inhibiting the activity of the structural genes.

When the operon theory was first proposed in 1961 it was not clear how many genes were included in the lactose (*lac*) operon, since the presence of galactoside acetylase had not yet been confirmed. The nature of the regulator molecule, the repressor, was also uncertain. But in 1966, Walter Gilbert and B. Müller-Hill of Harvard succeeded in isolating the repressor for the *lac* operon and demonstrated that it is a protein of four subunits (about 37,000 daltons each), present in only a few copies per cell and quite ordinary in most respects. The repressor acts by binding directly to DNA at the operator site unless the appropriate inducer is also present. In that case, the inducer binds to the repressor and destroys its capacity to recognize the sequence of nucleotides defining the operator, presumably through a conformational change in the repressor protein.

The *lac* operator consists of apparently 21 nucleotides, including the start codon for operon transcription. The binding site for RNA polymerase (the promoter) overlaps the operator site slightly and is on the side away from the structural gene. Thus, the presence of a repressor physically blocks a portion of the promoter and prevents the binding of RNA polymerase.

Though it has been modified in several ways since its original inception, Jacob and Monod's operon theory has served as a useful model for the study of enzyme induction in both prokaryotes and eukaryotes. (As we shall see, however, the situation in eukaryotes is decidedly more complicated.) The operon theory also explains the ability of certain viruses to remain in a provirus state. In fact, a repressor for phage λ was isolated by Mark Ptashne of Harvard at about the same time as the *lac* repressor, and proved to control the λ provirus genes in much the same way as the *lac* repressor controls the *lac* operon. While preventing transcription of a prophage, the λ repressor also prevents replication of superinfecting λ particles, making the cell appear to be immune to them.

Among the variations of the original operon model is one proposed by E. Englesberg and coworkers to explain the induction of the arabinose operon. Like lactose, the sugar arabinose cannot be metabolized without the help of certain enzymes that are made only on demand. These enzymes are the products of an operon that is under the control of a regulator gene, called the C gene. However, the control seems to be *positive* rather than negative. That is, a complex of regulator protein and inducer (arabinose) must bind to an operatorlike site at the edge of the operon in order for transcription to take place. Hence, mutants with deletions in the C gene are uninducible, as are mutations that prevent binding of the complex of arabinose and regulator protein. In contrast, both classes of mutants are constitutive in the *lac* system, which is under *negative* control by its repressor.

Actually, control of the arabinose operon is a bit more complicated, for there are two adjacent binding sites for the regulator protein. The protein in the absence of arabinose binds to one site and shuts down the operon. In the presence of arabinose it binds to the adjacent site and turns on the operon. The overall effect in the normal cell, however, is one of positive control.

Although we have mentioned only inducible systems, it should be clear that repressible enzymes could be controlled in much the same way. And, in fact, they are. That is, the operon may be under either negative or positive control of a regulator molecule. The difference is that small molecules produced by the pathway cause the operons to be turned off instead of on.

The operon concept is now firmly established, though there are many variations. For example, there is no reason why a given regulator could not control more than one operon. If several operons at different sites were to share the same operator sequence, they would be coordinately controlled by the same inducers and repressors. This type of system has been termed a *regulon* (Fig. 5-29). Nor is it required that the regulator protein be the product of a distant gene or have as its only function operon control. Cases have been described in which the regulator pro-

CHAPTER 5
Cellular Homeostasis:
Genes and Their Regulation

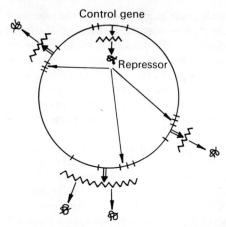

FIGURE 5-29 The Regulon. A single regulator gene controls several operons at scattered sites around the chromosome. The arginine pathway of *E. coli*, for example, involves structural genes located at four distinct sites, all under the control of the same repressor.

tein is made by one of the structural genes within the operon itself, a situation known as *autogenous* regulation. Control is then established by feedback, modulated by small molecules. Indeed, this type of regulator may even double as one of the enzymes in the pathway established by the operon.

Catabolite Repression. Another example of transcriptional control through modification of a polymerase-promoter interaction is the phenomenon of catabolite repression. Catabolite repression was described earlier as the ability of *E. coli* to exhaust the supply of glucose in its growth medium before allowing induction of the *lac* operon. It is more general than that, however, for the catabolism of a number of other sugars is similarly repressed in the presence of glucose. The mechanism involves the intracellular concentration of cyclic AMP (cAMP), which falls to low levels when glucose is being catabolized but not when the cell is living on one of the other sugars. The effect of cAMP is mediated through a specific receptor protein which, in the presence of cAMP, can increase the affinity of RNA polymerase for the promoter regions of the various sugar operons. The cAMP receptor protein, also called *catabolite gene activator protein* (CAP), acts by binding directly to promoter sites. In the case of the *lac* operon, the CAP binding site is in the half of the 80-base-pair promoter region away from the structural gene. Just how the presence of the cAMP-CAP complex facilitates polymerase-promoter interaction and transcription is not clear.

Other Mechanisms of Transcriptional Control. While operon theory satisfactorily explains the regulation of certain genes, it is a relatively "expensive" mechanism. It requires, for each operon, an operator gene and a regulator gene, though the latter may be shared with other operons if their coordinate expression is desirable. The regulator gene must be transcribed and translated into a protein; hence, this mechanism not only presents a considerable genetic burden but, if widely used, would clutter the cell with numerous proteins whose only function is to control genes instead of carrying on the immediate task of metabolism. In addition, not all genes need to be turned on and off regularly, even though all do require a controlled rate of synthesis. It should not be surprising, then, that other mechanisms exist to control transcription. Most have to do with the affinity between polymerase and promoter.

Not all promoters have the same nucleotide sequence and not all promoter sequences are equally good at permitting polymerase to bind and transcription to begin. Thus, for example, proteins that need to be present in only small amounts would be expected to have a low-affinity promoter. On the other hand, it is possible to modify RNA polymerase so that it interacts with greater or lesser affinity with the same promoter. One mechanism has already been introduced in the form of the σ factor, required for normal initiation. It was originally expected that a family of σ factors would be found, the properties of which would allow for variable rates of transcription under different conditions. So far, that does not seem to be the case. However, the idea is incorporated by certain bacteriophages to control the sequence in which their genes are expressed, one example of which follows.

Only some of phage T4's genes can be transcribed by the host polymerase and σ factor. Among the products of these genes is one that neutralizes the host σ factor, thus stopping cellular as well as early T4-specific RNA synthesis, and another that binds with the core polymerase to permit recognition of additional ("late") phage T4 genes. These late genes code, among other things, for coat protein, which is obviously not needed until the infection is well advanced. The polymerase-modifying factors thus provide a mechanism for sequentially transcribing blocks of genes.

A rather special case of polymerase modification is that used to control the rate of ribosome synthesis. Under conditions of rapid growth, rRNA synthesis can account for up to 40% of the total RNA being made, yet the rate can be quickly decreased with changing conditions. It turns out that when an uncharged tRNA occupies a ribosomal site, as would happen during a shortage of amino acids or ATP, an idling reaction occurs that produces a guanine nucleotide having diphosphate groups at both the 5' and 3' positions. This molecule, abbreviated ppGpp, appears to bind directly to RNA polymerase, decreasing its ability to bind to rRNA promoters but not necessarily to other promoter sites.

Translational Control. The amount of a specific protein can be controlled in three ways: (1) by regulating the rate of mRNA synthesis; (2) by regulating the number of times its mRNA is translated; and (3) by regulating the lifetime of the protein. The latter is not much used in bacterial cells (though it is in eukaryotes), and we have already discussed regulation of mRNA synthesis. We are left, then, with the second option, translational control, which can be achieved by varying the lifetime of the mRNA or the rate at which proteins are made from it.

It is obvious that the amount of protein made by an mRNA will depend on how long the mRNA survives before it is degraded. (Degradation starts at the 5' end, hence promoters are lost first.) The average lifetime of mRNA in *E. coli* is only about 2 minutes, but there is a good deal of variation. It appears that much of the variation is due to differences in mRNA structure and hence is genetically determined. A protein requiring especially rapid fluctuations in level might therefore be coded by a very short-lived mRNA so that effects of gene repression could more quickly be felt. In addition to these fixed differences in lifetime, it appears that survival of some mRNAs changes with conditions. Thus, when arginine synthesis is repressed in *E. coli*, existing mRNA seems to be more rapidly degraded.

The more widely used mechanisms of translational control regulate the rate at which proteins are made. This rate varies with the secondary structure of the mRNA, which may fold back onto itself to form hairpin turns; it varies with the choice of codons specifying abundant vs. less abundant tRNAs; and it varies with the efficiency of the ribosomal binding sites. The presence of tryptophan, for example, not only reduces the rate of transcription from the tryptophan operon but stops translation of existing mRNAs. And even within the same mRNA, different cistrons may be translated at different rates. Thus, the *lac* mRNA produces β-galactosidase, permease, and acetylase in a ratio of $1:\frac{1}{2}:\frac{1}{5}$, respectively.

The extent to which these translational controls actually regulate the metabolism of bacterial cells is unclear. Certainly the primary site of regulation is at transcription. Most of these mechanisms, however, have also been identified in eukaryotic systems, where translational control seems to be much more common.

5-6 GENE REGULATION IN EUKARYOTES

Eukaryotic cells share with prokaryotes a need to adjust constantly the concentration of enzymes in response to changing conditions. In addition, there is a more basic level of control that establishes the differentiated properties of cells in multicellular organisms. As we shall see in Chap. 12, the very great difference between, say, a neuron and a liver cell is due to which genes are expressed, not to any difference in the DNA itself.

In discussing gene regulation in higher eukaryotes, then, we are looking for both short-term and long-term controls. Many of the prokaryotic mechanisms already introduced can be extrapolated to eukaryotes, but only with a good deal of caution for there are important differences in the two classes of organisms. As noted in the last section, the most obvious difference is chromosomal organization.

The Eukaryotic Chromosome. Eukaryotic DNA is found associated with proteins in fibers called *chromatin*. Chromatin is usually dispersed within the nucleus (Fig. 5-30) except just prior to and during cell division when it is gathered together into discrete bodies, the chromosomes (Fig. 5-31). Though the appearance is quite different, chromatin composition in these two conditions is much the same. DNA forms the core of the fiber which contains as well small amounts of RNA and large amounts of two kinds of protein, the *histone* and *nonhistone chromosomal proteins*.

Histones are basic proteins because of a high proportion of the amino acids arginine and lysine, both of which have positively charged amino groups ($-\overset{+}{N}H_4$) at neutral pH. Histones have molecular weights in the range of 11,000 to 21,000, and together comprise a mass that is always about equal to that of the DNA itself. In most eukaryotic nuclei there are only five principal histone types, called H1, H2b, H2a, H3, and H4. (An older system of nomenclature, still used, refers to them as f1, f2b, f2a2, f3, and f2a1, respectively.) In the order given, they vary in composition from "very lysine rich" (29% lysine, 2% arginine in calf thymus H1), to "arginine rich" (11% lysine, 14% arginine). Avian erythrocytes, which retain a nonfunctioning nucleus in the mature cell, have an addi-

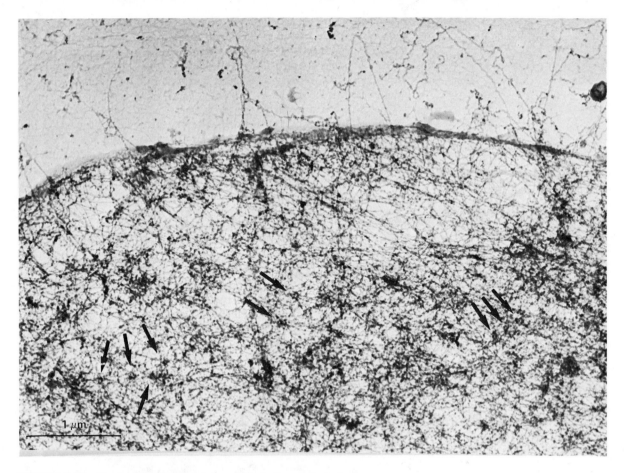

FIGURE 5-30 Chromatin. Dispersed chromatin in the nucleus of a human lymphocyte, isolated and spread onto the surface of water. *Arrows* point to fragments of nuclear envelope to which some of the fibers are still attached. [Courtesy of F. Lampert, *Humangenetik*, **13**: 285 (1971).]

FIGURE 5-31 A Human Chromosome. From a dividing lymphocyte. Note the complex folding of the chromatin fiber. [Courtesy of G. F. Bahr, Reprinted from *Fed. Proc.*, **34**: 2209 (1975).]

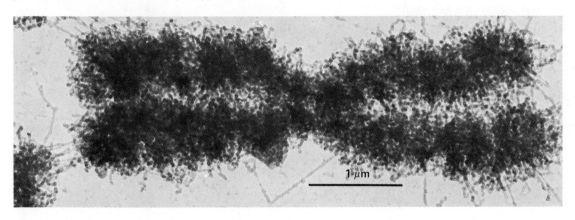

tional histone called H5 or f2c. Except for H1, histones of the same class from different eukaryotes have very similar amino acid sequences. In particular, H4 has been rigidly conserved through evolution, varying by two amino acids out of 102 between calf thymus and pea seedlings. This strict conservation implies an important function, the interference with which is apt to be lethal.

One of the most important recent advances in the study of eukaryotic nuclei is the discovery that histones interact specifically with each other and with DNA to form chromatin subunits called ν (nu) bodies, PS particles, or *nucleosomes* (see Fig. 5-32). The composition and size of these subunits seems to vary somewhat, but in general they appear to be 70–100 Å in diameter and to have about 140 base pairs of DNA within each particle and perhaps another 30–60 base pairs of DNA as a flexible strand between particles. With a repeat distance for the nucleosomes of about 200 base pairs, a typical gene will have three or four.[5]

Nucleosomes probably do not contain histone H1. The other four histones are typically present, however, and in many cases at exactly two copies each per particle. The best guess at present is that H1 is distributed largely between nucleosomes and has as its major function the imposition of higher orders of structure.

But what is the structure of the nucleosome, and what are these higher orders of structure? As to the nucleosome, we can say only that its DNA behaves as if it were wound about the *outside* of the histone complex. Not only is the DNA accessible to chemical reagents of various types,

FIGURE 5-32 Nucleosomes. Chromatin particles such as these, formed from a specific interaction of DNA and histones, have been called ν (nu) bodies, PS particles, or nucleosomes. TOP: from rat thymus. BOTTOM: from chicken erythrocytes. [Courtesy of A. L. and D. E. Olins, from *Science*, **183**: 330, copyright 1974 by the AAAS.]

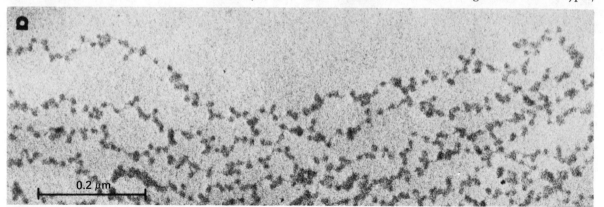

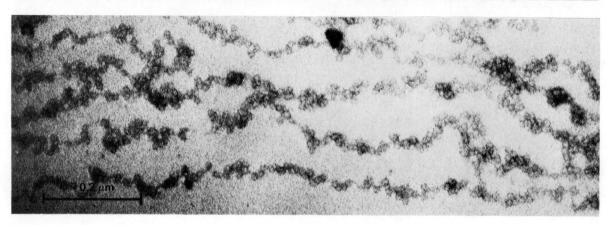

but it can be transcribed while still in the nucleosome. Observations of this type are not consistent with models in which DNA is buried within and therefore protected by the histones of the particles.

The situation regarding higher orders of structure is also in a state of flux. It is clear that some higher order is needed. The DNA in an average human chromosome, for example, has a mass of about 6×10^{10} daltons. If this DNA were to exist as a single strand (and there is evidence to indicate it might) the total length when extended would be about 3 cm. Yet a chromosome like that of Fig. 5-31 is only about 5 μm (0.0005 cm) in its longest dimension. Packing DNA into nucleosomes helps reduce the length, of course, but only by a factor of 6 or 7 (Fig. 5-33). In other words, the nucleosome string for a human chromosome would still be about 0.5 cm (5,000 μm) long.

Various models have been proposed for higher orders of structure, both to bring the total length down to manageable levels and to account for the fact that chromatin fibers seen by electron microscopy are often much thicker than the 100-Å nucleosome. Based on available data, it seems probable that the string of beads is itself coiled to form a fiber roughly 300 Å in diameter. The ratio of lengths, DNA to chromatin, might thus be on the order of 40 to 1. At this value each chromosome would have something like 700 μm of supercoiled chromatin—still a lot, but certainly a more reasonable number.

Gene Expression and Chromatin Condensation. Dispersed chromatin is referred to as *euchromatin*. Condensed, compact chromatin is referred to as *heterochromatin*. When these names were coined, euchromatin and heterochromatin were assumed to be different entities, rather than different configurations of the same material. It is now clear, however, that most chromatin can undergo the transition between dispersed ("euchromatic") and condensed ("heterochromatic") configurations.

Prior to cell division, most chromatin is in the condensed form. Electron microscopy of the resulting chromosomes often reveals fibers 250–350 Å in diameter. Between cell divisions, the bulk of the chromatin in most cells is well dispersed within the nucleus (Fig. 5-34). The fiber diameter at this time is much more variable, with some of it measuring about 100 Å in diameter and some of it having thicker dimensions. It has long been felt that the more compact (supercoiled?) heterochromatin is probably transcriptionally silent, and that loosening of chromatin structure may permit transcription and perhaps even regulate it.

For example, a typical eukaryotic cell ceases to make most RNA at about the time when chromatin condenses into chromosomes—i.e., just prior to cell division. And chromatin that does not get dispersed at all between cell divisions may never be transcribed into RNA. One of the two X chromosomes in female mammals is known to fall into that category. It forms a noticeable clump of heterochromatin, called a *Barr body*, in the nucleus between cell divisions (see Fig. 5-35). Genes on this heterochromatic X chromosome are not expressed, although their alleles on the euchromatic X are expressed normally. (Note that male and female cells are thus alike in having one active X chromosome. The male cell, in addition, has the abbreviated Y chromosome on which only a few active genes have been identified.)

The silent X chromosome is clearly demonstrated in women who are

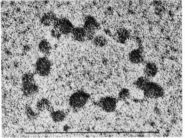
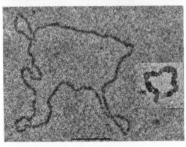

FIGURE 5-33 The SV40 Minichromosome. SV40 is a DNA virus capable of causing tumors in certain mammals. TOP: native chromatin. MIDDLE: after dilution with distilled water, 10-nm nucleosomes became visible. BOTTOM: protein removed to reveal the DNA strand. INSET: a native chromosome at the same magnification to compare size. (Contour ratio = 7 to 1). Bars = 100 nm. [Courtesy of J. D. Griffith, *Science*, **187**: 1202, 1975, copyright by the AAAS.]

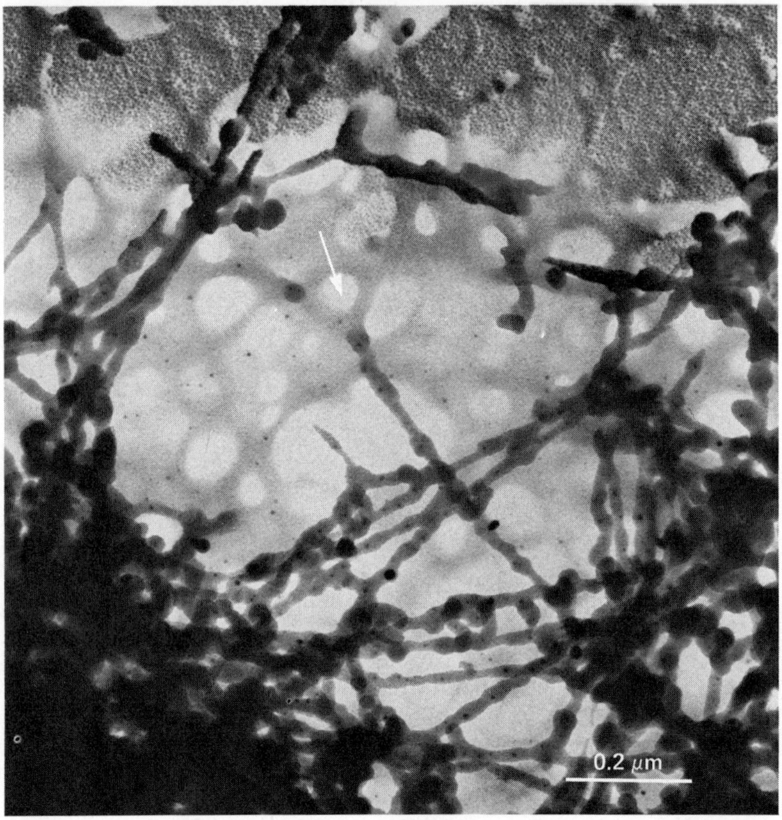

CHAPTER 5
Cellular Homeostasis:
Genes and Their Regulation

FIGURE 5-34 Honeybee Chromatin. From inside a honeybee nucleus looking toward the nuclear envelope. Note the attachment between chromatin and envelope (arrow) and the uneven diameter of the fibers. [Courtesy of E. J. DuPraw, *Proc. Nat. Acad. Sci. U.S.*, **53**: 161 (1965).]

heterozygous for two isozymes of glucose 6-phosphate dehydrogenase (G6PDH), the gene for which occupies a locus on the X chromosome. When individual cells are examined in heterozygotes, one or the other isozyme is found, never both. The original choice as to whether the maternal or paternal X chromosome will be expressed in cells of most female mammals is apparently made randomly at an early stage in embryonic development—a process that is sometimes called *Lyonization* after Mary Lyon, who is credited with being the first to realize what was happening. Once a given X chromosome is inactivated, the same choice will be made in all of a cell's descendents. Females are therefore genetic mosaics for traits carried by the X chromosome.

Another example where chromatin unfolding accompanies transcription is found in the lampbrush chromosome (Fig. 5-36). In the oöcytes (developing eggs) of some species, especially amphibia, chromosomes temporarily develop lateral loops extending from a central axis of two DNA helices. It can be demonstrated that transcription is taking place only in the loops, and that each homologous pair of chromosomes has its own highly specific pattern of looping, reflecting the activation of genes. After meiosis, these chromosomes return to a normal configuration.

One of the earliest correlations between unfolding and transcription was noted in the salivary gland of the fruit fly, *Drosophila*. Like certain other insect tissues, these cells are very large, with correspondingly large nuclei and large chromosomes. The number of chromosomes is normal for the species (four pair) but the amount of DNA per chromosome may

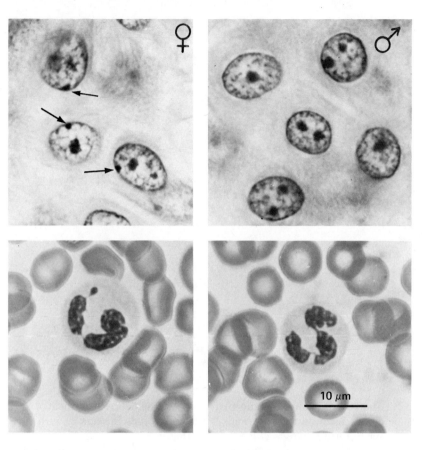

FIGURE 5-35 The Heterochromatic X Chromosome. In female mammals, one of the two X chromosomes is transcriptionally inactive heterochromatin. TOP: "Barr bodies" in the female cell (*arrows*) correspond to the inactivated X chromosome. BOTTOM: Human neutrophils (white blood cells) showing the heterochromatic X as a "drumstick" off one of the lobes of the female nucleus. [Courtesy of Murray L. Barr, Univ. of Western Ontario.]

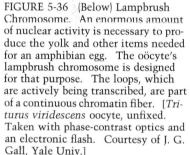

FIGURE 5-36 (Below) Lampbrush Chromosome. An enormous amount of nuclear activity is necessary to produce the yolk and other items needed for an amphibian egg. The oöcyte's lampbrush chromosome is designed for that purpose. The loops, which are actively being transcribed, are part of a continuous chromatin fiber. [*Triturus viridescens* oocyte, unfixed. Taken with phase-contrast optics and an electronic flash. Courtesy of J. G. Gall, Yale Univ.]

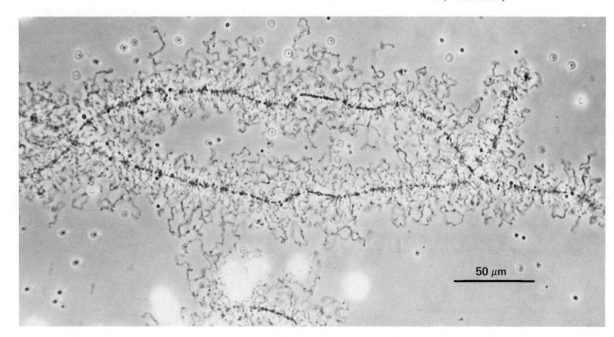

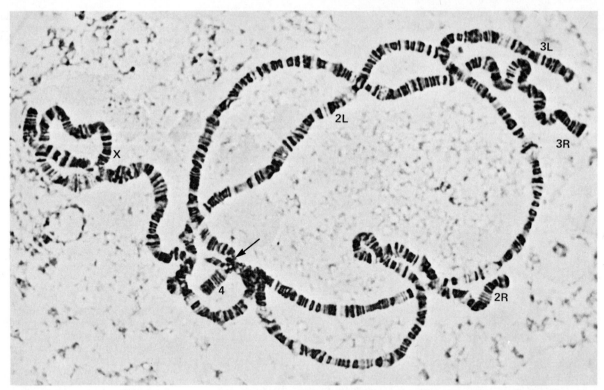

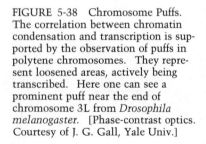

FIGURE 5-37 The Polytene Chromosome. The salivary gland chromosomes of *Drosophila* larvae, showing the four polytene chromosomes. Radiating from the chromocenter (arrow) are five long arms and a short one. Four of the long arms represent the "left" and "right" ends of two autosomes (chromosomes 2 and 3). The fifth arm is the sex (X) chromosome, or chromosome 1. The short arm (about 10 μm) is the complete chromosome 4. [Courtesy of Paul A. Roberts, Oregon State Univ.]

be 1000 times greater than normal. The original DNA is duplicated over and over again, with the strands lying side by side in register with each other to form *polytene chromosomes* (Fig. 5-37). Note in Fig. 5-37 the striking pattern of transverse bands. There are a total of about 5000 of these bands, called *chromomeres*, which may represent individual genetic loci. The pattern of banding is identical in the two members of each homologous pair, but different from one pair to the next. The banding pattern is also the same in each of the tissues in which these chromosomes are visible. However, the chromosomes differ from one tissue to the next in the pattern of localized uncoiling called *puffs* (Fig. 5-38). Since puffs appear to represent areas of active gene transcription we would expect the puffing pattern, but not the banding pattern, to vary with the tissue type.

These examples all suggest that tightly coiled heterochromatin might be inaccessible to RNA polymerase. The question is what causes the

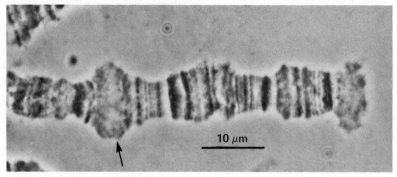

FIGURE 5-38 Chromosome Puffs. The correlation between chromatin condensation and transcription is supported by the observation of puffs in polytene chromosomes. They represent loosened areas, actively being transcribed. Here one can see a prominent puff near the end of chromosome 3L from *Drosophila melanogaster*. [Phase-contrast optics. Courtesy of J. G. Gall, Yale Univ.]

transition back and forth between heterochromatin and euchromatin, and whether this transition is a cause of gene activation or merely a result.

Histones and Gene Expression. Histone H1 seems to be the major factor in determining higher orders of chromatin structure. It is absent or much reduced in chromatin that is uncoiled and actively being transcribed, whereas the other histones seem to be present in the usual concentration. Conversely, untranscribed heterochromatin has much H1. A special case is found in chicken erythrocyte nuclei, where most of the H1 is replaced by a unique histone, H5, that is in many ways similar to H1. As H5 accumulates in the maturing erythrocyte, transcription is progressively inhibited.

All of the histones are capable of being reversibly modified by the addition and removal of phosphate or acetate groups, or in some cases by oxidation and reduction of sulfhydryl (—SH) groups. These changes may be responsible for altering the interaction between histone and DNA and hence altering transcriptional ability. Phosphorylation is of particular interest because several cyclic AMP (cAMP) dependent histone kinases have been identified. Thus, changes in histones—for example, the phosphorylation of H1 associated with chromatin condensation into chromosomes (discussed in Chap. 11)—can be tied to other regulatory processes through cAMP.

There is good evidence that histone modification is able to turn blocks of genes off and on. Histones, however, do not have the diversity necessary to be regulators of individual genes the way the *lac* repressor controls the *lac* operon in *E. coli*. For that role we need molecules capable of more specificity. Most attention in this regard is currently centered on the nonhistone chromosomal proteins.

Nonhistone Chromosomal Proteins. While the ratio of histone to DNA is relatively constant in cells of many types, the amount of nonhistone chromosomal protein (NHC protein) varies widely. In general, cells with more active genes tend to have more NHC protein, indicating a role in gene activation. Consistent with this view is the observation that NHC proteins also differ from histones in being polydisperse—that is, there are a very large number of individual species.

The first demonstration that NHC proteins can actually control individual genes was by J. Paul and his colleagues in Glasgow, Scotland. Normally, mRNA for globin (the protein of hemoglobin) is produced by erythrocyte precursors but not by brain cells. However, when brain chromatin was dissociated *in vitro* and reassembled using erythroid NHC proteins, the reconstituted brain chromatin was able to produce globin mRNA. When brain NHC protein was used for the reconstitution, globin mRNA was not made. These and other experiments leave no doubt that components of NHC proteins are capable of regulating the expression of individual genes.

It appears that NHC proteins have the capacity to bind to regulator sites of specific genes and thus make the genes available for transcription—perhaps by displacing and/or chemically altering histone H1, which would otherwise mask the gene in question. To go from this finding to an integrated concept of eukaryotic gene control is not so very hard, but to do so we have to take into account some peculiarities in the way the collection of genes (i.e., the genome) is arranged in eukaryotes.

Organization of the Eukaryotic Genome. With very few exceptions, each protein has only one structural gene present in a haploid set of chromosomes. Nevertheless, the haploid genome size ranges from about 4×10^8 to 8×10^{10} base pairs among vertebrates, even though there is no reason to believe that a comparable 200-fold ratio exists in the number of proteins made by the different species. Nor is there any clear correlation between our estimate of the complexity of the animal and its genome size—humans, for example, are about in the middle of the vertebrate range. There must be some fundamental difference in the organization of eukaryotic and prokaryotic genomes, for in prokaryotes more DNA in the chromosome means more genes and greater metabolic complexity.

There are a great many unanswered questions concerning the arrangement of genes in eukaryotic chromosomes, but the following important observations have been made:

(1) Between 50 and 75% of the DNA consists of gene-sized (i.e., 1000 base pairs or so) segments that are present only once in each haploid set of chromosomes. Exceptions to this rule include the genes for ribosomal RNA, which may be tandemly repeated hundreds of times in the nucleolus organizer region of the chromosome, and the genes for tRNAs and histones, which are also present in multiple copies. The five histone genes, for example, are present side by side in tandemly repeated sequences that are transcribed as single units.

(2) The great majority of these "single-copy" genes never make an mRNA and hence do not code for existing proteins. They may be transcribed to hnRNA, but their RNA does not leave the nucleus.

(3) Most of the rest of the DNA consists of small (about 300 base pair) segments. There are a number of different sequences in these short segments, but copies of most sequences are found at many different places in the genome. Much of this *repetitive DNA* is transcribed, but again very little of the RNA ever reaches the cytoplasm.

(4) Most, if not all, of the single-copy DNA has adjacent repetitive sequences. The repetitive sequences, in other words, are typically interspersed among the single-copy sequences.

In addition to the above, it has been observed that a small fraction of the genome consists of short (e.g., 20 base pair) simple sequences repeated many times. Most of this *highly repetitive DNA* is found in the centromere region, where two chromatids join. (This DNA frequently is different enough from the bulk of the DNA in its base composition to be separable on CsCl gradients as *satellite DNA*.)

Still another small fraction of the DNA is found in 300–1200 nucleotide *palindromes* scattered throughout the chromosome. A palindrome is a sequence of DNA that would be read the same in either direction, for example

$$\begin{array}{c} \longrightarrow \\ \text{C A T A T G} \\ \text{G T A T A C} \\ \longleftarrow \end{array}$$

No matter which strand is transcribed (5' to 3') the resulting RNA is the same. However, the RNA will be internally complementary and form hydrogen-bonded hairpin loops:

$$\begin{array}{c} -\text{C}-\text{A}-\text{U} \\ -\text{G}-\text{U}-\text{A} \end{array}\Big)$$

We have no good idea why most single-copy DNA of gene size does not make messenger RNA. Nor do we know the function of satellite DNA or palindromes. The logic behind tandemly repeated segments seems clear, however, as this pattern provides increased template in those few cases where especially frequent transcription is needed. Of more interest is the existence of repetitive sequences interspersed among single-copy genes, for it is probable that these repeated sequences have a role in gene control.

Transcriptional Control. The first comprehensive report of repetitive genes was made in 1968 by R. J. Britten and D. E. Kohne, then both at the Carnegie Institute in Washington. The following year Britten and E. H. Davidson proposed a theory which, though modified by others since then, provides a satisfying if not wholly proven explanation for the existence of repetitive DNA.

Most or all structural genes are assumed to have associated with them multiple control genes. Each cell type in a multicellular organism would produce a set of gene regulators characteristic of that cell type. These regulator molecules would be capable of turning on those genes needed to carry out the differentiated functions of that cell. If a particular structural gene has a control element corresponding to one of the regulators, it will be turned on; otherwise it will be silent, perhaps due to associated histones.

In a hypothetical case, a structural gene might have adjacent controlling genes recognized by tissue-specific activators A, B, E, and F, but not by C and D:

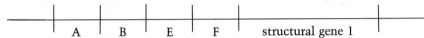

The four control elements in this model permit the gene to be turned on in those tissues having any of the four activators. A second structural gene might have control elements A, C, and D:

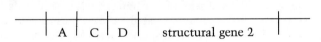

Gene 2 would thus be on in those cell types producing one of these three activators. Thus, the two structural genes are recognized by an overlapping but nonidentical set of activators. In this way, any combination of structural genes could be selected by a much smaller number of different control genes and their products, which are likely to be NHC proteins.

The above scheme, while speculative, does satisfy most of the observations. It is also consistent with data from the California Institute of Technology, published by Davidson and W. H. Klein, in which it was demonstrated that only 10–20% of the 5000 or so families of repetitive gene sequences in sea urchin are adjacent to the structural genes actually expressed at one particular stage in early development. It appears that particular repetitive sequences are associated with specific structural genes, and may well be involved in their control.

To this tissue-specific pattern of gene control must be added mechanisms for induction or repression of individual genes as conditions change. It seems likely that nonhistone chromosomal proteins provide both levels of control. Tissue specificity could result from the selection of NHC proteins, which may then undergo chemical or conformational changes due to the influence of small molecules, thus fine-tuning the metabolism. These changes probably include phosphorylation by cAMP-dependent protein kinases. Through cAMP, genes may thus be responsive to external hormones. Other hormones, however, penetrate cells and seem to bind directly to NHC proteins.

Cytoplasmic Hormone Receptors. Water-soluble hormones generally act on their target cells via membrane receptors and cytoplasmic mediators such as cAMP. Lipid-soluble hormones, on the other hand, typically penetrate cells to bind to specific receptors found within the cytoplasm. Hormones having known cytoplasmic receptors include indoleacetic acid (an *auxin*) of plants; the sex steroids, cortisol, and activated vitamin D of mammals; and ecdysone, the molting-hormone of insects.

One of the clearest examples of the mode of action of lipid hormones is seen with the effect of ecdysone on polytene chromosomes of the fruit fly, *Drosophila*. Within 5–10 minutes after ecdysone addition, six major new "puffs" are seen in these chromosomes, indicating the transcription of previously silent genes. (Recall that a puff represents activation of a chromomere containing—in a great excess of DNA—possibly a single structural gene.) Some of these early puffs, however, apparently produce not enzymes but regulator molecules that are capable of turning on an additional 100 or so "late puffs" to produce an amplified secondary response to the hormone. In addition, one or more of the proteins produced by the early puffs acts to terminate RNA synthesis in the early puffs. As a result, the original effect of the hormone is transient.

The original, early puffs of the ecdysone-stimulated polytene chromosome presumably result from a direct effect of the hormone and its receptor protein on nuclear genes. The actual steps involved have been more clearly defined for stimulation of mammalian cells by sex steroids. The sequence appears to be as follows: The steroid penetrates the cell membrane and binds specifically to a protein receptor in the cytoplasm. This receptor, which is absent in hormone-unresponsive cells (i.e., in other than target cells), is thus activated and becomes capable of penetrating the nucleus to interact with specific acceptor sites on chromatin. Through mechanisms not yet understood, the genes associated with these acceptor sites then become available for transcription (Fig. 5-39).

Receptors for estrogens (female hormones), progesterone, androgens (male hormones), and cortisol have been isolated in varying degrees of purity. The estrogen receptor, for example, is a protein of about 80,000 daltons that binds estrogens but not molecules having similar structures but no estrogenic activity. It is activated by binding estrogen in the presence of a second protein, with which it forms a complex.

Various mutations have permitted direct demonstration of the importance of receptor proteins to steroid function. For instance, there is an X-linked recessive condition in humans (as well as other mammals) in which the cytoplasmic receptor for the male hormone dihydrotestosterone is defective. The defective receptor leads to a condition known as "testicular feminization" in which the genotype is normal male (46, XY) but the external phenotype is normal female.[6]

The above model for steroid hormone action, in spite of the gaps in our knowledge, is presently the best understood example of transcriptional control in eukaryotes. Transcriptional control, however, is not the only way of regulating protein synthesis.

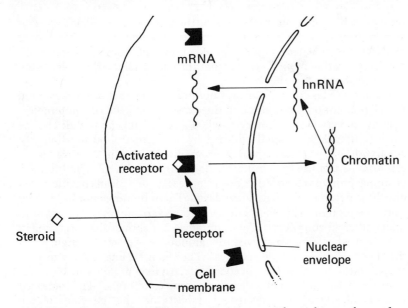

FIGURE 5-39 Steroid Hormone Receptors. Steroids penetrate cell membranes and bind to specific cytoplasmic receptor proteins. The complex then interacts with acceptor sites on the chromatin to activate genes.

Translational Control and Protein Turnover. The subject of translational control in prokaryotes was discussed in the preceding section. The mechanisms mentioned there are as applicable to eukaryotes. The importance of translational control in eukaryotes appears to be much greater, however, for mRNA in eukaryotes has typically a much longer lifetime.

The advantage of translational control is the speed with which protein synthesis can be turned on and off. Its disadvantage lies in the waste involved in making mRNA that is not used before it is degraded. The relative weight given to these two criteria obviously depends on how important immediate control of enzyme synthesis is in a given situation, keeping in mind that the regulation of already functioning enzymes, as discussed in Chap. 4, is very rapid.

The longer an mRNA can survive intact, the more important it becomes to be able to control its rate of translation. And the more rapidly mRNA turns over, the more wasteful it becomes to avoid translating it without a feedback signal to also control its synthesis. The lifetimes of mRNA in both eukaryotes and prokaryotes are highly variable, but usually in a logical way. That is, enzymes whose quantity must be increased or decreased in response to external signals (e.g., diet) have shorter mRNA lifetimes than do enzymes whose synthesis is constitutive. The range of half-lives of most mRNAs in eukaryotes is from roughly an hour to several days. The extreme case of mRNA stability is found in unfertilized eggs, for their mRNA may remain untranslated for very long periods, even years, until fertilization occurs. The analogous situation in plants is the formation of seeds and spores, both of which also contain stable messenger RNA.

One way of controlling the amount of translation of a given message, then, is to regulate the lifetime of the mRNA. Unfortunately, it is not at all clear what makes some mRNAs more stable than others. One possibility is that lifetime is determined by secondary structure. Complementary sequences within the same mRNA molecule result in a specific folding, consisting of loops and hairpins. While these local regions of hydrogen bonding must come apart as the ribosomes pass over them, they probably make an mRNA less attractive to the ribonuclease that will eventually degrade it. An alternative scheme suggests that some mRNAs are translated a predetermined number of times. A possibly relevant observation in this regard is that the poly A tail on the 3' end often seems to shorten with time.

The level of an enzyme depends not only on the rate of synthesis and translation but on the lifetime of the enzyme. In bacteria, protein degradation is very slow. However, when a bacterial gene is shut off, the enzyme it makes is rapidly diluted by cell growth and division. In eukaryotic cells, some of which never divide, protein turnover is an essential part of the system for regulating protein level. The lifetimes of animal proteins range from a few minutes (about 5 for an enzyme called ornithine decarboxylase) to a few days. These lifetimes not only differ from one protein to another, but seem to be variable for a given protein under different conditions. For example, R. T. Schimke demonstrated in 1966 that when rats are starved, the amount of the enzyme arginase in their liver doubles in about six days. The change is due entirely to a decreased rate of degradation. In contrast, most liver proteins are degraded more rapidly than normal under these same conditions. The increased stability of arginase and decreased stability of other enzymes in the starving rat may be related to the availability of substrates, for it is known that the *in vitro* digestion of enzymes by proteases is hindered by the presence of substrate.

When two proteins differ in their rate of degradation, their level will also respond differently to gene regulators. Thus, Schimke reported that cortisone stimulates the production of liver tryptophan pyrrolase and arginase to the same extent, yet the administration of cortisone has a much more profound effect on the level of tryptophan pyrrolase. The reason for this difference is that tryptophan pyrrolase is degraded more rapidly, providing it with a lower basal level and greater sensitivity to changes in its rate of synthesis.

The degradation of proteins is, for the most part, the function of cytoplasmic organelles called lysosomes, to be considered in more detail in Chap. 10. To do that subject justice, however, we shall first have to understand much more about the structure and function of some other organelles and especially of the membranes that surround them. That task starts with the next chapter.

SUMMARY

5-1 The earliest systematic observations on the nature of genes were made by Gregor Mendel: (1) Genes for a given characteristic come in pairs, one member of each pair being derived from the mother and one member from the father; (2) when the two genes of a pair are alternate alleles, one allele may be dominant; (3) the two genes of a pair are segregated during gamete formation—i.e., they end up in separate germ cells; and (4) the genes for some characteristics sort independently during gamete formation.

Observations 1, 3, and 4 above result from the fact that genes are carried by chromosomes and that chromosomes come in homologous pairs, each parent furnishing one member of each pair. Segregation (observation 3) is the result of separating the two members of a homologous pair during germ cell formation. The maternal and paternal members of each pair are segregated randomly, leading to the independent assortment (observation 4) of unlinked characters—namely of characters carried on nonhomologous chromosomes.

The second of Mendel's observations, the phenomenon of dominance, exists because genes produce proteins. A recessive gene may produce an inactive or less active protein, or none at all. This condition is sometimes masked in a heterozygote by the presence in the cell of a gene producing the normal protein in sufficient quantity.

5-2 Genes consist of sequences of deoxyribonucleotides (DNA) except in the case of the RNA viruses. The genetic function of DNA was demonstrated first in bacterial transformation experiments and later in experiments showing that only viral nucleic acid, not viral protein, is needed to produce a normal infection.

The major task of genes is to make proteins. The nucleotide sequence of DNA is first transcribed to a complementary sequence of RNA nucleotides, preserving the informational content of DNA in the nucleotide sequence of mRNA. That information consists of an uninterrupted sequence of nucleotides, functionally grouped into code words or codons, most of which specify an amino acid. Each codon consists of three bases (nucleotides). By the use of synthetic RNA polymers of three or more nucleotides, 61 of the 64 possible codons have been assigned to amino acids. The remaining three are terminator (or nonsense) codons, used to stop the incorporation of amino acids and to release the polypeptide chain.

5-3 The flow of information from gene to protein may be represented as

$$\text{DNA} \xrightarrow{\text{transcription}} \text{RNA} \xrightarrow{\text{translation}} \text{protein}$$

The first step, transcription, is catalyzed by an RNA polymerase. This enzyme is a multisubunit protein that begins its function at a sequence of bases on the DNA called a promoter. In prokaryotes, at least, recognition of the promoter requires the help of another protein added to the polymerase as a subunit and called a σ factor. The polymerase then moves down the DNA using one of the two strands as a template to lace up the complementary RNA strand from nucleotide triphosphate precursors. The RNA grows 5' to 3' (that is, with new nucleotides at the 3' end). The new RNA chain is released when a specific stop sequence in the DNA is reached.

The new RNA often contains more nucleotides than are needed in the final product. Both tRNAs and rRNAs, for example, are produced by cleaving and trimming longer precursors. (Ribosomal RNAs are processed in the nucleolar region, where ribosomal precursors are assembled.) After trimming, chemical modification of the bases may occur. This modification is especially extensive in tRNAs.

Messenger RNA also contains sequences other than those actually specifying polypeptides. These sequences are relatively short in prokaryotes, where multiple-gene messengers are common. In eukaryotes, however, single-polypeptide messengers are derived from much longer hnRNA sequences, most of which is degraded. The mRNAs of eukaryotes are also typically modified by having a string of adenylates (poly A) attached to the 3' end and a methylguanosine at the other end, attached 5' to 5' with a triphosphate separating the guanosine from the first base. The poly A (missing in histone mRNA) seems to confer stability while the 5' cap facilitates later translation.

5-4 Translation includes the following steps:

(1) Formation of an initiation complex consisting of the smaller of the two ribosomal subunits plus proteins called initiation factors and a special tRNA containing (in prokaryotes) N-formylmethionine. The large ribosomal subunit is added last. The process has a total energy requirement of one GTP to GDP hydrolysis plus one ATP → AMP used in charging the tRNA.

(2) An initiator codon, AUG or GUG, is recognized by the N-formylmethionyl-tRNA. Formylmethionine thus becomes the original N-terminal amino acid. The formyl group (which may never be present in eukaryotes) will later be removed from the finished peptide, often with one or more of the N-terminal amino acids.

(3) Elongation takes place by (a) esterification of an amino acid to a particular tRNA, catalyzed by an aminoacyl synthetase and driven by an ATP to AMP hydrolysis; (b) hydrogen bonding of the charged tRNA to its mRNA codon at the acceptor or aminoacyl (A) site on the ribosome; (c) exchange of the ester bond between the amino acid and tRNA for a peptide bond between the amino acid and the C-terminal residue of the existing polypeptide; (d) translocation, during which the elongated chain is shifted from the aminoacyl site of the ribosome to the peptidyl (P) site. Steps (b) and (d) each require proteins called elongation factors and each step requires the hydrolysis of a GTP to GDP.

(4) Termination of the polypeptide occurs when at least one of three terminator codons is encountered along with an appropriate release factor, and it requires another GTP hydrolysis.

5-5 The prokaryotic chromosome consists of a single long DNA molecule folded and coiled into a compact nucleoid. (In addition, extrachromosomal rings of DNA called plasmids are frequently found.) The genetic information contained in these structures can be expressed at rates that vary with (1) the rate of mRNA synthesis (transcription), and (2) the rate of mRNA translation. The best understood regulatory mechanism is operon control of transcription.

The operon is a set of adjacent genes whose coordinate expression is under the control of a repressor (or regulator) protein, itself the product of a regulator gene. When the regulator protein attaches to the operator site of the operon, transcription may be either inhibited (negative control) or made possible (positive control). The ability of a regulator protein to bind to the operator may be influenced by small molecules acting as either corepressors or activators.

The affinity of RNA polymerase for a promoter can also be affected in other ways. An example is catabolite repression of *E. coli*, where, in the presence of cyclic AMP, a cAMP receptor protein increases the normally low affinity of RNA polymerase for a variety of sugar operons. Since cAMP is at a low level during glucose catabolism, glucose will be utilized in preference to these other sugars. Other examples include differences in the design of the promoter itself, so that some constitutive proteins will be transcribed more often than others, and modification of the RNA polymerase, as during phage T4 infection.

Translational control may be achieved through variations in the lifetime of the mRNA, through variations in the rate of ribosome binding or initiation, or through factors (e.g., tRNA availability) that affect translational rate.

5-6 Eukaryotic nuclear DNA is found complexed with large amounts of two types of protein called histone and nonhistone chromosomal protein. The resulting fiber, or chromatin, seems to have a repeating subunit structure formed by clusters, called *nucleosomes*, of histones and associated DNA. The chromatin may be further coiled or folded to provide still more compact fibers. This higher degree of structure seems to prevent transcription, and since histones appear to be primarily responsible they thus become capable of preventing translation of large blocks of DNA. Histones do not, however, have the diversity (there are only five histone types in most cells) or specificity needed for regulation of the many individual genes present in a cell. That function is thought instead to be the province of the nonhistone chromosomal proteins. Unlike histones, the NHC proteins are highly heterogeneous and are present in variable amounts, being more common on actively transcribed sections of DNA.

Although the organization of the eukaryotic genome is not well understood, it appears in most cases that short regulatory elements of DNA are interspersed among the single-copy genes, some of which code for proteins. Nonhistone proteins may well interact with these regulatory segments to turn genes on (or off). The best studied example of this sort of interaction is the transcriptional stimulation of specific genes by steroid hormones. These hormones penetrate the cell membrane, bind to specific receptor proteins in the cytoplasm, and then as a protein-steroid complex interact with chromatin to turn genes on.

As with prokaryotes, the regulation of transcription is complemented by translational controls. Such controls are more important in eukaryotes because of the longer relative lifetime of eukaryotic mRNAs. The lifetime of the mRNA, the rate of its translation, and the lifetime of the protein itself obviously interact with transcriptional rates to establish the final level of a protein. All seem to be variable in eukaryotic systems.

STUDY GUIDE

5-1 (**a**) What are the four principles of Mendelian genetics, as summarized here? (**b**) If one were to follow *n* human genes through several generations, what is the largest value of *n* that could be used without necessarily breaking the law of independent assortment? (**c**) Define these terms: haploid, diploid, homologous chromosomes, hemizygous, somatic cell, sex-linked. (**d**) What parallels exist between chromosome and gene behavior that would lead one to suspect a chromosomal location for genes? (**e**) What was the experimental foundation for the one gene = one enzyme hypothesis? (**f**) Hemoglobin is a protein of four subunits, two of one variety and two of another. What is the minimum number of cistrons dedicated to its manufacture?

5-2 (**a**) Describe the first transformation experiments as they were carried out both *in vivo* and *in vitro*. (**b**) How do we know that DNA carries genetic information: (1) in bacteria? (2) in viruses of all types? (3) in eukaryotic organisms? (**c**) What is messenger RNA, and how was it discovered? (**d**) One of the more popular early proposals for the genetic code had three bases per codon, but with each codon overlapping the next by one base. How do the results obtained with alternating RNA copolymers (e.g., UGUGUG···) eliminate that scheme? What does protein sequencing have to tell us about it? (**e**) Another early proposal separated the codons from each other by "commas," consisting of one or more bases. Can you design experiments with synthetic RNA messengers to test (and eliminate) that scheme? (**f**) What observations indicate that the genetic code is identical in probably all forms of life?

5-3 (**a**) Describe the reaction catalyzed by RNA polymerase. (**b**) What experiment first indicated that only one DNA strand in a gene region is transcribed? (**c**) Describe the RNA polymerase start and stop processes in prokaryotes. (**d**) What is meant by "posttranscriptional processing?" What kind of processing is received by tRNAs and rRNAs in prokaryotes and in eukaryotes? (**e**) Contrast the prokaryotic and eukaryotic messengers with respect to gene content, relative precursor size, and 5' and 3' modifications.

5-4 (**a**) What are the three classes of RNA, and what

role does each play in translation? (b) Briefly summarize the shape and major features of tRNA and ribosomes. What is meant by a 70S and 80S ribosome? (c) What is the basis for the "wobble hypothesis"? (d) At which steps in translation is GTP involved? (e) What are the functions of the A and P ribosomal sites? (f) In terms of ATP to ADP equivalents, what is the energy required for each amino acid inserted? (g) Summarize the events of initiation and termination.

5-5 (a) Describe the structure of the prokaryotic chromosome. (b) What is a plasmid? Give examples of plasmid types. (c) What distinguishes inducible, repressible, and constitutive genes from one another? (d) Describe the structure of the *lac* operon, and how it is controlled. (e) What is the difference between negative and positive control of an operon? What experiments can be performed to distinguish one from the other? (f) What is meant by catabolite repression? How does *E. coli* respond to catabolite repression by glucose? (g) What other mechanisms help to establish different transcriptional rates for different genes? (h) What are the various possibilities for translational control?

5-6 (a) Describe the basic structure of chromatin as we understand it. What is the role of histones in this structure? (b) What is the difference in structure and activity of heterochromatin and euchromatin? Give an example of transcriptionally silent heterochromatin. (c) Contrast the two classes of chromosomal proteins, histone and nonhistone. (d) What is meant by "tandemly reiterated," "repetitive," and "single-copy" DNA? What are the presumed roles of each? (e) Describe the mode of action of steroid hormones. (f) What factors besides transcriptional rate might regulate the cytoplasmic level of an enzyme?

REFERENCES

GENERAL

BEADLE, GEORGE W., "The Genes of Men and Molds." *Scientific American*, September 1948. (Offprint 1.)

———, "Genes and Chemical Reactions in Neurospora." In *Nobel Lectures, Physiology or Medicine, 1942–1962.* Amsterdam: Elsevier, 1964, p. 587. Nobel Lecture, 1958.

BODMER, W. F., and L. L. CAVALLI-SFORZA, *Genetics, Evolution and Man.* San Francisco: Freeman, 1976.

CARLBERG, D. M., *Essentials of Bacterial and Viral Genetics.* Springfield, Illinois: Charles C Thomas, 1976.

CARLSON, ELOF AXEL, *The Gene: A Critical History.* Philadelphia: W. B. Saunders, 1966.

COHEN, S. N., "Gene Manipulation." *New Engl. J. Med.*, **294**: 883 (1976). Genetic Engineering. Also see *Scientific American* July, 1975, p. 24. (Offprint 1324).

COLEMAN, WILLIAM, "Cell, Nucleus and Inheritance: An Historical Study." *Proc. Am. Phil. Soc.*, **109**: 125 (1965).

DAVIDSON, J. N., *The Biochemistry of the Nucleic Acids* (8th ed.). London: Chapman and Hall, 1976. (Paperback.) Replication and function.

ERBE, R. W., "Principles of Medical Genetics." *New Engl. J. Med.*, **294**: 381 and 480 (1976). Chromosomal disorders.

GUSCHLBAUER, W., *Nucleic Acid Structure.* New York: Springer-Verlag, 1976. (Paperback.)

JOHN, B., and K. R. LEWIS, *Chromosome Hierarchy: An Introduction to the Biology of the Chromosome.* New York: Oxford University Press, 1976. (Paperback.) Covers architecture, replication, and movement plus some discussion of function.

KING, R. C., ed., *A Dictionary of Genetics.* New York: Oxford University Press, 1974. (Paperback.)

KING, R. C., *Molecular Genetics* (*Handbook of Genetics*, Vol. 5). New York: Plenum Press, 1976.

LEVINE, LOUIS, *Biology of the Gene* (2d ed.). St. Louis, Mo.: C. V. Mosby, 1973.

LEWIN, BENJAMIN M., *The Molecular Basis of Gene Expression.* London: Wiley-Interscience, 1970.

LURIA, S. E., *36 Lectures in Biology.* Cambridge, Mass.: MIT Press, 1975. (Paperback.) Genes and their evolution.

MACLEAN, N., *Control of Gene Expression.* New York: Academic Press, 1976. Considers both prokaryotes and eukaryotes.

MOORE, D. M., *Plant Cytogenetics.* London: Chapman and Hall, 1976. (Paperback.)

NEWLON, C., C. GUSSIN, and B. LEWIN, "Molecular Information in Developmental Genetics." *Cell*, **5**: 213 (1975). Recent advances in transcription, translation and gene control.

NIEDERMAN, R. A., ed., *Molecular Biology and Protein Synthesis.* New York: Halsted Press, 1976.

OLBY, R., *The Path to the Double Helix.* Seattle: University of Washington Press, 1976. Discovery of the Molecular structure of DNA.

ROTHWELL, N. V., *Understanding Genetics.* Baltimore: Williams & Wilkens, 1976.

STURTEVANT, A. H., *A History of Genetics.* New York: Harper & Row, 1965.

TATUM, E. L., "A Case History in Biological Research." In *Nobel Lectures, Physiology or Medicine, 1942–1962.* Amsterdam: Elsevier, 1964, p. 602. Nobel Lecture, 1958.

WATSON, J. D., *Molecular Biology of the Gene* (3d ed.). Menlo Park, Calif.: W. A. Benjamin Co., 1976.

WATSON, J. D., and F. H. C. CRICK, "Genetical Implications of the Structure of Deoxyribonucleic Acid." *Nature*, **171**: 964 (1953).

WOODWARD, D., and V. WOODWARD, *Concepts of Molecular Genetics.* New York: McGraw-Hill, 1977.

REPRINT COLLECTIONS

Featuring original papers or excerpts thereof.

ADELBERG, E. A., *Papers on Bacterial Genetics* (2d ed.). Boston: Little, Brown, 1966. The introduction contains a clear discussion of mutation, transformation, and other processes.

CARLSON, ELOF A., *Gene Theory.* Belmont, Calif.: Dickenson Publ. Co., 1967. Excerpts from important papers, with annotation.

CARPENTER, BRUCE H., *Molecular and Cell Biology.* Belmont, Calif.: Dickenson, 1967. A collection of readings, consisting of excerpts from important papers.

GABRIEL, M. L., and S. FOGEL, eds., *Great Experiments in Biology.* Englewood Cliffs, N.J.: Prentice-Hall, 1955. Includes writings of Mendel, Sutton, Wilson, Morgan.

HAHON, NICHOLAS, ed., *Selected Papers on Virology.* Englewood Cliffs, N.J.: Prentice-Hall, 1964.

LEVINE, LOUIS, *Papers on Genetics: A Book of Readings.* St. Louis, Mo.: C. V. Mosby, 1971.

LOOMIS, W. L., ed., *Papers on Regulation of Gene Activity During Development.* New York: Harper & Row, 1968.

PETERS, JAMES A., *Classic Papers in Genetics.* Englewood Cliffs, N.J.: Prentice-Hall, 1959. Reprints many of the earlier papers, with annotation. Included are works by Mendel, Sutton, Morgan, Beadle, and others.

STENT, GUNTHER S., *Papers on Bacterial Viruses* (2d ed.). Boston: Little, Brown, 1965. The introduction presents a clear overview of the subject matter.

STERN, CURT, and E. R. SHERWOOD, eds., *The Origin of Genetics: A Mendel Source Book.* San Francisco: W. H. Freeman, 1966. Writings of Gregor Mendel in translation and with annotation.

TAYLOR, J. HERBERT, *Selected Papers on Molecular Genetics.* New York: Academic Press, 1965.

VOELLER, BRUCE R., *The Chromosome Theory of Inheritance: Classic Papers in Development and Heredity.* New York: Appleton Century Crofts, 1968. Excerpts from historically important works.

WAGNER, R. P., *Genes and Proteins.* New York: Halsted Press, 1975. Benchmark papers in Genetics, Vol. 2.

ZUBAY, GEOFFREY L., *Papers in Biochemical Genetics* (2d ed.). New York: Holt, Rinehart & Winston, 1973.

TRANSCRIPTION AND TRANSLATION

BERNFIELD, MERTON R., and MARSHALL W. NIRENBERG, "RNA Codewords and Protein Synthesis." *Science,* **147**: 479 (1965). Describes the trinucleotide binding technique developed in Nirenberg's lab.

BRIMACOMBE, R., "Analysing the Structure of the Ribosome." *Trends Biochem. Sci.,* **1**(8): 181 (Aug. 1976).

BRIMACOMBE, R., K. H. NIERHAUS, R. A. GARRETT, and H. G. WITTMANN., "The Ribosome of *Escherichia coli.*" *Progr. Nucleic Acid Res. Mol. Biol.,* **18**:1 (1976).

BUSCH, H., "The Function of the 5′ Cap of mRNA and Nuclear RNA Species." *Perspectives Biol. Med.,* **19**: 549 (1976).

CHAMBON, P., "Eukaryotic Nuclear RNA polymerases." *Ann. Rev. Biochem.,* **44**: 613 (1975).

CHAN, L., S. HARRIS, J. ROSEN, A. MEANS, and B. W. O'MALLEY, "Processing of Nuclear Heterogeneous RNAs." *Life Sci.,* **20**: 1 (1977).

CRICK, F. H. C., "On the Genetic Code." In *Nobel Lectures, Physiology or Medicine, 1942–1962.* Amsterdam: Elsevier, 1964, p. 811. Nobel Lecture, 1962.

———, "The Genetic Code." *Scientific American,* October 1962. (Offprint 123.) Also see Oct. 1966 (offprint 1052).

DUNN, J. J., ed., *Processing of RNA.* Upton, New York: Brookhaven National Laboratory, 1975. Brookhaven Symposium in Biology No. 26.

ENGELMAN, D., and P. MOORE, "Neutron-Scattering Studies of the Ribosomes." *Scientific American,* **235**(10): 44 (Oct. 1976).

FURUICHI, Y., A. LAFIANDRA, and A. SHATKIN, "5′-Terminal Structure and mRNA Stability." *Nature,* **266**: 235 (1977). The 5′ cap protects against exonuclease digestion.

GOIDL, J. A., "Intranuclear Messenger-Ribosome Complexes and Protein Synthesis." *Trends Biochem. Sci.,* **1**(4): 76 (April 1976).

GOLDBERG, A., and A. ST. JOHN, "Intracellular Protein Degradation in Mammalian and Bacterial Systems." *Ann. Rev. Biochem.,* **45**: 721 (1976).

GREEENBERG, J. R., "Messenger RNA Metabolism." *J. Cell Biol.,* **64**: 269 (1975). A review.

GRIFFEN, B., "Eukaryotic mRNA: Trouble at the 5′-end." *Nature,* **263**: 188 (1976). A short review of 5′ "caps."

HADJIOLOV, A., "Patterns of Ribosome Biogenesis in Eukaryotes." *Trends Biochem. Sci.,* **2**(4): 84 (April 1977).

HAYAISHI, O., "Poly ADP-ribose and ADP-ribosylation of Proteins." *Trends Biochem. Sci.,* **1**(1): 9 (Jan. 1976).

HOLLEY, R. W., "The Nucleotide Sequence of a Nucleic Acid." *Scientific American,* February 1966. (Offprint 1033.)

HURWITZ, J., and J. J. FURTH, "Messenger RNA." *Scientific American,* February 1962. (Offprint 119.)

INGRAM, V., "How Do Genes Act?" *Scientific American,* January 1958. (Offprint 104.) Sickle-cell hemoglobin.

JIMÉNEZ, A., "Inhibitors of Translation." *Trends Biochem. Sci.,* **1**(2): 28 (Feb. 1976).

JOU, W. M., G. HAEGEMAN, M. YSEBAERT, and W. FIERS, "Nucleotide Sequence of the Gene Coding for the Bacteriophage MS2 Coat Protein," *Nature,* **237**: 82 (1972). The first nucleotide sequence of a complete gene. The whole MS2 genome sequence is given in *Nature,* **260**: 500 (1976).

KAZIRO, Y., ed., *Selected Papers in Biochemistry V. 7 Protein Synthesis.* Baltimore, Md.: Univ. Park Press, 1971.

KHORANA, H. G., "Polynucleotide Synthesis and the Genetic Code." In *The Harvey Lectures, 1966–1967* (Ser. 62). New York: Academic Press, 1968, p. 79.

———, "Synthetic Nucleic Acids and the Genetic Code." *J. Am. Med. Assoc.,* **206**: 1978 (1968). Lasker Award, 1968.

KOLATA, G., "Intracellular Bacterial Toxins." *Science,* **190**: 969 (1975). A short review.

KURLAND, C., "Structure and Function of the Bacterial Ribosome." *Ann. Rev. Biochem.,* **46**: 173 (1977).

LANE, C., "Rabbit Hemoglobin from Frog Eggs." *Scientific American,* **235**(8): 60 (Aug. 1976). The universal code.

Lewin, B., "Units of Transcription and Translation." *Cell,* **4**: 11 and 77 (1975). Two papers on the relationship between hnRNA, and mRNA, the first a review.

Lodish, H., "Translational Control of Protein Synthesis." *Ann. Rev. Biochem.,* **45**: 11 (1976).

Losick, R., and M. Chamberlin, eds., *RNA Polymerase.* Cold Spring Harbor (N.Y.) Laboratory, 1976. The first third of the book is devoted to reviews.

Maden, B., "Ribosomal Precursor RNA and Ribosome Formation in Eucaryotes." *Trends Biochem. Sci.,* **1**(9): 197 (Sept. 1976).

Maugh, T. H., II., "Ribosomes." *Science,* **190**: 136 and 258 (1975). A brief, two-part review of structure.

Miller, O. L., Jr., "The Visualization of Genes in Action." *Scientific American,* March 1973.

Miller, O. L., Jr., and B. A. Hamkalo, "Visualization of RNA Synthesis on Chromosomes." *Int. Rev. Cytology,* **33**: 1 (1972).

Molloy, G., and L. Puckett., "The Metabolism of Heterogeneous Nuclear RNA and the Formation of Cytoplasmic Messenger RNA in Animal Cells." *Progr. Biophys. Mol. Biol.,* **31**: 1 (1976).

Nirenberg, M. W., "The Genetic Code: II." *Scientific American,* March 1963. (Offprint 153.)

Nomura, M., "Ribosomes." *Scientific American.* October 1969. (Offprint 1157.)

Nomura, M., A. Tissieres, and P. Lengyel, eds., *Ribosomes.* Cold Spring Harbor (N.Y.) Laboratory, 1974.

Ochoa, Severo, "Enzymatic Synthesis of Ribonucleic Acid." In *Nobel Lectures, Physiology or Medicine, 1942–1962.* Amsterdam: Elsevier, 1964, p. 645. Transcriptase. Nobel Lecture, 1959.

Ochoa, S., "Initiation of Protein Synthesis." *Naturwissenschaften,* **63**: 347 (1976). A short review.

Perry, R., "Processing of RNA." *Ann. Rev. Biochem.,* **45**: 605 (1976).

Richter, D., and K. Isono, "The Mechanism of Protein Synthesis." *Curr. Topics Micro. Immunol.,* **76**: 83 (1977).

Roscoe, D., "Recognition of Specific Nucleic Acid Sequences by Proteins." *Sci. Progr.* (Oxford), **64**: 447 (1977).

Rottman, F., A. Shatkin, and R. Perry., "Sequences Containing Methylated Nucleotides at the 5' Termini of Messenger RNAs: Possible Implications for Processing." *Cell,* **3**: 197 (1974). Perhaps the most influential of the early papers on mRNA "caps."

Sebastián, J., "Structure and Function of the Yeast RNA Polymerases." *Trends Biochem. Sci.* **2**(5): 102 (May 1977). The eukaryotic transcriptases.

Shatkin, A., A. Banerjee, G. Both, Y. Furuichi, and S. Muthukrishnan, "Dependence of Translation on 5'-Terminal Methylation of mRNA." *Fed. Proc.* **35**: 2214 (1976).

Smith, J. D., "Transcription and Processing of Transfer RNA Precursors." *Progr. Nucleic Acid Res. Mol. Biol.,* **16**: 25 (1976).

Spiegelman, S., "Hybrid Nucleic Acids." *Scientific American,* May 1964. (Offprint 183.)

Stark, G., "Multifunctional Proteins: One Gene—More than One Enzyme." *Trends Biochem. Sci.,* **2**(3): 64 (March 1977). Some eukaryotic enzymes are linked covalently, apparently as different parts of the same polypeptide.

Travers, A., "RNA Polymerase Specificity and the Control of Growth." *Nature,* **263**: 641 (1976). A review.

van Heyningen, S., "The Structure of Bacterial Toxins." *Trends Biochem. Sci.,* **1**(5): 114 (May 1976). Diphtheria toxin and protein synthesis.

Vazquez, D., "Inhibitors of Protein Synthesis." *FEBS Letters,* **40**(Suppl.): S63 (1974). A review of the site of action of various inhibitors.

Watson, J. D., "The Involvement of RNA in the Synthesis of Proteins." In *Nobel Lectures, Physiology or Medicine, 1942–1962.* Amsterdam: Elsevier, 1964, p. 785. Nobel Lecture, 1962.

Weatherall, D. J., and J. B. Clegg, "Molecular Genetics of Human Hemoglobin." *Ann. Rev. Genet.,* **10**: 157 (1976). Human hemoglobins reveal an enormous variety of genetic errors.

Yanofsky, Charles, "Gene Structure and Protein Structure." In *The Harvey Lectures, 1965–1966* (Ser. 61). New York: Academic Press, 1967, p. 145. Also see *Scientific American,* May 1967. (Offprint 1074.)

GENE ORGANIZATION AND REGULATION IN PROKARYOTES

Adler, K., K. Beyreuther, E. Fanning, N. Geisler, B. Gronenborn, A. Klemm, B. Müller-Hill, M. Pfahl, and A. Schmitz, "How Lac Repressor Binds to DNA." *Nature,* **237**: 322 (1972).

Altman, S., "Biosynthesis of Transfer RNA in *Escherichia coli.*" *Cell,* **4**: 21 (1975).

Barrell, B., G. Air, and C. Hutchinson III, "Overlapping Genes in Bacteriophage φX174." *Nature,* **264**: 34 (1976). One gene = two proteins.

Bertran, K., C. Yanofsky, et al., "New Features of the Regulation of the Tryptophan Operon." *Science,* **189**: 22 (1975).

Bourgeois, S., and M. Pfahl, "Repressors." *Advan. Prot. Chem.,* **30**: 1 (1976).

Clarke, Patricia H., "Positive and Negative Control of Bacterial Gene Expression." *Sci. Progr.* (Oxford), **60**: 245 (1972).

Gilbert, Walter, and B. Müller-Hill, "Isolation of the Lac Repressor." *Proc. Nat. Acad. Sci. U.S.,* **56**: 1891 (1966).

Goldberger, R., R. Deeley, and K. Mullinix, "Regulation of Gene Expression in Prokaryotic Organisms." *Advan. Genet.,* **18**: 1 (1976).

Goldberger, R., "Autogenous Regulation of Gene Expression." *Science,* **183**: 810 (1974).

Helinski, D. R., "Plasmids." *Fed. Proc.,* **35**: 2024 (1976).

Jacob, François, "Genetics of the Bacterial Cell." *Science,* **152**: 1470 (1966). Nobel Lecture, 1965.

Jaskunas, S., and M. Nomura, "Organization of Ribosome Genes in *E. coli.*" *Trends Biochem. Sci.,* **1**(7): 159 (July 1976).

Lewin, B., "Interaction of Regulator Proteins with Recognition Sequences of DNA." *Cell,* **2**: 1 (1974). A review.

LEWIN, B., *Gene Expression.* Vol. 1: *Bacterial Genomes.* New York: Wiley-Interscience, 1974. (Paperback.)

MANIATIS, T., and M. PTASHNI, "A DNA Operator-Repressor System." *Scientific American,* **234**(1): 64 (Jan. 1976). Phage lambda.

MOSES, V., "Concerning Cyclic AMP." *Sci. Progr.* (Oxford), **63**: 503 (1976).

MÜLLER-HILL, B., "Lac Repressor and Lac Operator." *Progr. Biophys. Mol. Biol.,* **30**: 227 (1975).

NOMURA, M., "Organization of Bacterial Genes for Ribosomal Components." *Cell,* **9**: 633 (1976).

PASTAN, IRA, "Cyclic AMP." *Scientific American,* August 1972. (Offprint 1256.)

PASTAN, IRA, and S. ADHYA, "Cyclic AMP in *E. coli.*" *Bacteriol. Rev.,* **40**: 527 (1976).

PETERKOFSKY, A., "Regulation of *E. coli* Adenylate Cyclase by Phosphorylation-Dephosphorylation." *Trends Biochem. Sci.,* **2**(1): 12 (Jan. 1977). The mechanism of catabolite repression.

PETTIJOHN, D. E., "Prokaryotic DNA in Nucleoid Structure." *CRC Crit. Rev. Biochem.,* **4**: 175 (1976).

Pirrotta, V., "The λ Repressor and its Action." *Current Topics Microbiol. Immunol.,* **74**: 21 (1976).

PTASHNE, M., and W. GILBERT, "Genetic Repressors." *Scientific American,* June 1970. (Offprint 1179.)

SANGER, F., et al., "Nucleotide Sequence of Bacteriophage ɸX174." *Nature,* **265**: 687 (1977). A series of four articles detailing the peculiarities of this tiny phage.

WITMER, H. J., "Regulation of Bacteriophage T4 Gene Expression." *Progr. Mol. Subcell. Biol.,* **4**: 17 (1976).

GENE ORGANIZATION AND REGULATION IN EUKARYOTES

ALLFREY, V. G., et al., eds., *Organization and Expression of Chromosomes* (*Life Sci. Res. Reports,* Vol. 4). Berlin: Dahlem Konferenzen, 1976.

BACK, F., "The Variable Condition of Euchromatin and Heterochromatin." *Int. Rev. Cytol.,* **45**: 25 (1976).

BARR, M. L., "The Significance of the Sex Chromatin." *Int. Rev. Cytol.,* **19**: 35 (1966). On the "Barr body."

BASERGA, R., *Multiplication and Division in Mammalian Cells.* New York: Dekker, 1976. See Chap. 7 on chromosomal proteins.

BASERGA, R., and C. NICOLINI, "Chromatin Structure and Function in Proliferating Cells." *Biochim. Biophys. Acta,* **458**: 109 (1976).

BEREZNEY, R., and D. S. COFFEY, "The Nuclear Protein Matrix." *Advan. Enzyme Reg.,* **14**: 63 (1976).

BISWAS, B., A. GANGULY, A. DAS, and P. ROY, "Action of Indoleacetic Acid: A Plant Growth Hormone on Transcription." In *Regulation of Growth and Differentiated Function in Eukaryote Cells,* G. Talwar, ed. New York: Raven Press, 1975, p. 461.

BRADBURY, E. M., "Current Ideas on the Structure of Chromatin." *Trends Biochem. Sci.,* **1**(1): 7 (Jan. 1976).

BRITTEN, R. J., and E. H. DAVIDSON, "Gene Regulation for Higher Cells: A Theory." *Science,* **165**: 349 (1969).

———, "Repetitive and Non-Repetitive DNA Sequences and a Speculation of the Origins of Evolutionary Novelty." *Quart. Rev. Biol.,* **46**: 111 (1971). Includes a summary of genome sizes and redundancy.

———, "DNA Sequence Arrangement and Preliminary Evidence on its Evolution." *Fed. Proc.,* **35**: 2151 (1976).

BRITTEN, R. J., and D. E. KOHNE, "Repeated Sequences in DNA." *Science,* **161**: 529 (1968).

———, "Repeated Segments of DNA." *Scientific American,* April 1970. (Offprint 1173.)

BROWN, D. D. and I. B. DAWID, "Specific Gene Amplification in Oocytes." *Science,* **160**: 272 (1968).

BUSCH, H. *The Cell Nucleus.* New York: Academic Press, 1977.

CIBA FOUNDATION, *The Structure and Function of Chromatin* (Symposium No. 28). Amsterdam: Elsevier, 1975. Many fine reviews. See especially the articles by Bradbury, Paul, and Bonner.

COX, R. P., and J. C. KING, "Gene Expression in Cultured Mammalian Cells." *Int. Rev. Cytol.,* **43**: 282 (1975).

CRICK, F. H. C., and A. KLUG, "Kinky Helix." *Nature,* **255**: 530 (1975). Model of chromatin structure.

DANEHOLT, B., "Transcription in Polytene Chromosomes." *Cell,* **4**: 1 (1975). A review.

DAVIDSON, E., B. HOUGH, W. KLEIN, and R. BRITTEN, "Structural Genes Adjacent to Interspersed Repetitive DNA Sequences." *Cell,* **4**: 217 (1975).

EDSTRÖM, J. E., and B. LAMBERT, "Gene and Information Diversity in Eukaryotes." *Progr. Biophys. Mol. Biol.,* **30**: 57 (1975).

ELGIN, S., and H. WEINTRAUB, "Chromosomal Proteins and Chromatin Structure." *Ann. Rev. Biochem.,* **44**: 725 (1975).

FINCH, J. T., and A. KLUG, "Solenoidal Model for Superstructure in Chromatin." *Proc. Nat. Acad. Sci. U.S.,* **73**: 1897 (1976).

GALL, JOSEPH G., "Differential Synthesis of the Genes for Ribosomal RNA During Amphibian Oogenesis." *Proc. Nat. Acad. Sci. U.S.,* **60**: 553 (1968). Reprinted in WLL.

GARTLER, S. M., "X-Chromosome Inactivation and Selection in Somatic Cells." *Fed. Proc.,* **35**: 2191 (1976).

GELEHRTER, T. G., "Enzyme Induction." *New Engl. J. Med.,* **294**: 522, 589, & 646 (1976). A three-part review, stressing clinical applications.

GILMOUR, R., and J. PAUL, "Role of Nonhistone Components in Determining Organ Specificity of Rabbit Chromatins." *FEBS Letters,* **9**: 242 (1970). A classic experiment.

HARRIS, H., *Nucleus and Cytoplasm.* Oxford: Clarendon Press, 1974.

HNILICA, L. S., *The Structure and Biological Function of Histones and Other Nuclear Proteins* (2d ed.) Cleveland: CRC Press, 1976.

KANDUTSCH, A., et al., "Symposium on Gene Regulation in Mammals." *J. Cell Physiol.,* **85** (Suppl. 1): 341 (1975).

KEDES, L. H., "Histone Messengers and Histone Genes." *Cell,* **8**: 321 (1976).

KORNBERG, R. D., "Structure of Chromatin." *Ann. Rev. Biochem.* **46**: 931 (1977).

LATT, S. A., "Optical Studies of Metaphase Chromosome Organization." *Ann. Rev. Biophys. Bioeng.,* **5**: 1 (1976).

LEWIN, B., *Gene Expression*, Vol. II: *Eucaryotic Chromosomes*. New York: Wiley-Interscience, 1974. (Paperback.)

LI, H. J., "Chromatin Structure." *Int. J. Biochem.*, **7**: 181 (1976).

LI, H. J., and R. ECKHARDT, eds., *Chromatin and Chromosome Structure*. New York: Academic Press, 1977.

LYON, MARY F., "The Activity of the Sex Chromosomes in Mammals." *Sci. Progr. (Oxford)*, **58**: 117 (1970).

———, "X-Chromosome Inactivation and Developmental Patterns in Mammals." *Biol. Rev. (Cambridge)*, **47**: 1 (1972).

MACGREGOR, H. C., "The Nucleolus and Its Genes in Amphibian Oogenesis." *Biol. Rev.*, **47**: 177 (1972).

MACLEAN, N., and V. HILDER, "Mechanisms of Chromatin Activation and Repression." *Int. Rev. Cytol.*, **48**: 1 (1977).

MARX, J., "Nitrogen Fixation: Prospects for Genetic Manipulation." *Science*, **196**: 638 (1977). Practical application of modern cell biology.

McKUSICK, V., and F. RUDDLE, "The Status of the Gene Map of the Human Chromosomes." *Science*, **196**: 390 (1977). Chromosomal location of human genes.

MITTWOCH, URSULA, "Sex Differences in Cells." *Scientific American*, July 1963. (Offprint 161.) The Barr body.

NAGL, W., "Nuclear Organization." *Ann. Rev. Plant Physiol.*, **27**: 39 (1976).

ORD, M. G., L. A. STOCKEN, and S. THROWER, "Histone Phosphorylations and their Disparate Roles in Interphase." *Sub-Cellular Biochemistry*, **4**; 147 (1975).

PAUL, J., R. GILMOUR, N. AFFARA, G. BIRNIE, and P. HARRISON, "The Globin Gene: Structure and Expression." *Cold Spring Harbor Symp. Quant. Biol.*, **38**: 885 (1973).

ROBERTS, T. M., G. D. LAUER, and L. C. KLOTZ, "Physical Studies on DNA from Primitive Eucaryotes." *CRC Crit. Rev. Biochem.*, **3**: 349 (1975).

SALSER, W., et al., "Investigation of the Organization of Mammalian Chromosomes at the DNA Sequence Level." *Fed. Proc.*, **35**: 23 (1976). Satellite DNA.

SCHWARZACHER, H. G., *Chromosomes in Mitosis and Interphase*. New York: Springer-Verlag, 1976. The structure and organization of the chromosomes of man.

SLUYER, M., "The H1 Histones." *Trends Biochem. Sci.*, **2**(9): 203 (Sept. 1977).

SMITH, D. W. E., "Reticulocyte Transfer RNA and Hemoglobin Synthesis." *Science*, **190**: 529 (1975).

STEIN, G. S., and L. KLEINSMITH, eds., *Chromosomal Proteins and Their Role in the Regulation of Gene Expression*. New York: Academic Press, 1975.

STEIN, G., and J. STEIN, "Chromosomal Proteins: Their Role in the Regulation of Gene Expression." *BioScience*, **26**: 488 (1976).

STEIN, G., J. STEIN, and L. KLEINSMITH, "Chromosomal Proteins and Gene Regulation." *Scientific American*, **232**(2): 46 (Feb. 1975).

SZARSKI, H., "Cell Size and Nuclear DNA Content in Vertebrates." *Int. Rev. Cytol.*, **44**: 93 (1975).

TARTOF, K. D., "Redundant Genes." *Ann. Rev. Genet.*, **9**: 355 (1975). tRNA, rRNA, and histones.

TREWAVAS, A., "Post-Translational Modification of Proteins by Phosphorylation." *Ann. Rev. Plant Physiol.*, **27**: 349 (1976). Histones and NHC proteins.

VAN HOLDE, K. E., and I. ISENBERG, "Histone Interactions and Chromatin Structure." *Accounts Chem. Res.*, **8**: 327 (1975).

WELLS, R. D., et al., "The Role of DNA Structure in Genetic Regulation." *CRC Crit. Rev. Biochem.*, **4**: 305 (1977).

WEATHERALL, D., and J. CLEGG, "The α-Chain Termination Mutants and their Relation to the α-Thalassaemias." *Phil. Trans. Roy. Soc. London, Ser. B.*, **271**: 411 (1975).

WEINTRAUB, H., and M. GROUDINE, "Chromosomal Subunits in Active Genes Have an Altered Conformation." *Science*, **193**: 848 (1976).

VAGIL, G., "Quantitative Aspects of Protein Induction." *Current Topics Cell. Reg.*, **9**: 183 (1975).

THE STEROID HORMONES

ASHBURNER, M., C. CHIHARA, P. MELTZER, and G. RICHARDS, "Temporal Control of Puffing Activity in Polytene Chromosomes." *Cold Spring Harbor Symp. Quant. Biol.*, **38**: 655 (1973). A classic example of steroid hormone effects.

ATTARDI, B., "Genetic Analysis of Steroid Hormone Action." *Trends Biochem. Sci.*, **1**(10): 241 (Nov. 1976).

CHAN, L., and B. W. O'MALLEY, "Mechanism of Action of the Sex Steroid Hormones." *New Engl. J. Med.*, **294**: 1322, 1372, & 1430 (1976). A three-part review.

DUMONT, J., and J. NUNEZ, eds., *Hormones and Cell Regulation*. Amsterdam: Elsevier/North Holland, 1977.

GERALD, P. S., "Sex Chromosome Disorders." *New Engl. J. Med.*, **294**: 706 (1976).

GORSKI, J., and F. GANNON, "Current Models of Steroid Hormone Action." *Ann. Rev. Physiol.*, **38**: 425 (1976).

KING, R. J. B., "Intracellular Reception of Steroid Hormones." *Essays Biochem.*, **12**: 41 (1976).

LEAKE, R., "Current Views on Oestrogen Receptors." *Trends Biochem. Sci.*, **1**(6): 137 (June 1976).

MAKIN, H., ed., *Biochemistry of Steroid Hormones*. London: Blackwell Scientific, 1975.

MEANS, A., S. WOO, S. HARRIS, and B. W. O'MALLEY, "Estrogen Induction of Ovalbumin mRNA: Evidence for Transcription Control." *Mol. Cell. Biochem.*, **7**: 33 (1975).

O'MALLEY, B. W., and W. T. SCHRADER, "The Receptors of Steroid Hormones." *Scientific American*, **234**(2): 32 (Feb. 1976).

PALMITER, R. D., "Quantitation of Parameters that Determine the Rate of Ovalbumin Synthesis." *Cell*, **4**: 189 (1975).

ROUSSEAU, G. G., "Interaction of Steroids with Hepatoma Cells: Molecular Mechanisms of Glucocorticoid Hormone Action." *J. Steroid Biochem.*, **6**: 75 (1975).

SCHIMKE, R. T., et al., "Hormonal Regulation of Ovalbumin Synthesis in Chick Oviduct." *Rec. Progr. Hormone Res.*, **31**: 175 (1975).

SCHIMKE, R. T., D. J. SHAPIRO, and G. S. McKNIGHT, "Ovalbumin mRNA and Ovalbumin DNA and the Molecular Biology of Steroid Hormone Action." In *Control Mecha-*

nisms in Development, R. Meints and E. Davies. New York: Plenum Press, 1975.

SCHULSTER, D., S. BURSTEIN, and B. COOKE, Molecular Endocrinology of the Steroid Hormones. New York: John Wiley, 1976. (Paperback.)

VILLEE, C. A., and G. M. LORING, "Estrogenic Control of Uterine Enzymes." Advan. Enzyme Reg., 13: 137 (1975).

YAMAMOTO, K., and B. ALBERTS, "Steroid Receptors." Ann. Rev. Biochem., 45: 721 (1976).

YAMAMOTO, K., and B. ALBERTS, "The Interaction of Estradiol-receptor Protein with the Genome: An Argument for the Existence of Undetected Specific Sites." Cell, 4: 301 (1975). A review.

GENETIC BASIS FOR ANTIBODY DIVERSITY

ASKONAS, B. A., "Immunoglobulin Synthesis and its Induction in B-Lymphoid Cells." Acta Endocrinol., 78 (Suppl. 194): 117 (1975). A review.

COLD SPRING HARBOR SYMPOSIUM, "Origins of Lymphocyte Diversity. (Symposium No. 41)" Cold Spring Harbor (N.Y.) Laboratory, 1977.

EICHMANN, K., "Genetic Control of Antibody Specificity." Immunogenetics, 2: 491 (1975).

HOOD, L., J. CAMPBELL, and S. ELGIN, "The Organization, Expression, and Evaluation of Antibody Genes and Other Multigene Families." Ann. Rev. Genet., 9: 305 (1975).

JERNE, N. K., "The Immune System: A Web of V Domains." The Harvey Lectures, Ser. 70 p. 93 (1976).

HOZUMI, N., and S. TONEGAWA, "Evidence for Somatic Rearrangement of Immunoglobulin Genes Coding for Variable and Constant Regions." Proc. Nat. Acad. Sci. U.S., 73: 3628 (1976).

PADLAN, E. A., "Structural Basis for the Specificity of Antibody-Antigen Reactions and Structural Mechanisms for the Diversification of Antigen-Binding Specificities." Quart. Rev. Biophys., 10: 35 (1977).

WILLIAMSON, A., "The Biological Origin of Antibody Diversity." Ann. Rev. Biochem., 45: 46 (1976).

WILLIAMSON, A., "Control of Antibody Diversity." Trends Biochem. Sci., 2(1): N8 (Jan. 1977). A short review.

NOTES

1. There is an interesting but unexplained exception: Some species of plants and animals have, in addition to their usual chromosomal complement, so-called B chromosomes. These are chromosomes that are present in some members of a species but not others, are not necessary for survival, and have little or no effect on phenotype. Their function is unknown.

2. Aberrant individuals with XXY and XYY configurations are males, while those with an unpaired X (called XO) are female. This sort of distribution error is not without its consequences, though less devastating than autosomal (i.e., non-XY) errors, the most common of which is trisomy 21 (three number 21 chromosomes), which leads to Down's syndrome.

3. People who are heterozygous for Hbg S have both normal and abnormal hemoglobin in their erythrocytes and usually escape serious harm. Homozygotes, on the other hand, often suffer a sickle cell crisis and die while still in their childhood. It should be noted that the evolutionary survival of this gene is apparently related to the ability of carriers to withstand malaria, a disease that is caused by a red cell parasite that is prevalent in the same geographical areas where sickle cell anemia is most prevalent.

4. It takes energy to produce plasmids. Hence, if the selective agent—i.e., the antibiotic—were to disappear, strains without R factors would eventually take over. That is always our "ace in the hole." So far, however, new antibiotics have become available fast enough to keep ahead of the bacteria and it is still possible to cure almost every bacterial infection in an otherwise healthy person. In addition, some common bacteria fortunately seem unable to harbor R factors, for reasons we don't understand. Plasmids also have the potential for being very useful; the insertion into bacteria of artificial plasmids could turn the cells into factories capable of producing any number of useful proteins—human growth hormone, insulin, etc.

5. If an average protein has a molecular weight of 25,000 daltons, it will be composed of about 250 amino acids and be coded for by 750 nucleotide pairs of DNA.

6. To become male, a fetus needs two factors, both provided by the embryonic male gonad: a "Mullerian duct inhibiting factor" that prevents formation of the uterus and tubes, and testosterone to cause development of the male internal and external sex organs. Without testosterone, or without cytoplasmic receptors for it, the baby will be externally female. At puberty enough estrogen will be available to result in a normal looking woman. The lack of uterus is usually discovered when she seeks medical attention for absent menstruation or because she can't get pregnant.

6

Membrane Structure and Molecular Transport

6-1 Membrane Structure 238
    The Lipid Bilayer
    The Danielli-Davson Model
    The Unit Membrane
    The Fluid Mosaic Model
    Membrane Stability
    Membrane Asymmetry
6-2 Intercellular Junctions 248
    Nomenclature
    Desmosomes
    Tight Junctions
    Gap Junctions
6-3 Membrane Permeability 254
    Plasmolysis
    Oil/Water Partition Coefficients
    Ion Trapping
    Membrane Pores
6-4 Membrane Transport 258
    Transport Proteins
    Transport Kinetics
    Competition for Transport
    Counter Transport
    The Genetics of Transport
6-5 Metabolically Coupled Transport 266
    Donnan Equilibrium
    Osmotic Pressure
    The Sodium Pump
    Membrane Potentials
    Criteria for Active Transport
    Na^+-Coupled Transport
    Group Translocation
6-6 Properties of the Carriers 277
    The Isolation of Transport Proteins
    The Advantages of Protein Carriers
    Nonprotein Carriers
Summary 281
Study Guide 282
References 283
Notes 286

The cell membrane divides this universe into two parts, the "inside" and the "outside." By this the membrane becomes the most important organ of the cell for it is here where the two worlds, the inside and the outside, meet.

                    A. Szent-Gyorgyi

The membranes of a cell are an integral part of its metabolic machinery. The plasma membrane, for instance, helps to determine the composition of the cytoplasm by controlling which materials get in and out and the rate at which they do so. It thus permits the selective uptake of nutrients and secretion of waste products. Membranes found elsewhere in the cell, such as those around the nucleus, mitochondria, chloroplasts, various vacuoles, sacs, and so forth (e.g., lysosomes, peroxisomes) also

CHAPTER 6
Membrane Structure and Molecular Transport

represent highly selective barriers to the passage of materials. Though most membranes are quite similar in structure, their properties vary considerably. We shall proceed by first outlining those structural features that seem to be common to most membranes. Then we shall point to experiments that indicate a need for modifications of that basic structure, and go on to suggest the probable nature of those modifications.

6-1 MEMBRANE STRUCTURE

When cells are grown together in culture, there will be an occasional merging of two cells to form one. The frequency with which this occurs can be greatly increased with inactivated *Sendai virus* (named after a city in Japan), which adheres strongly to the cell surface. Its "sticky" nature glues cells together, providing more opportunity for fusion between neighboring cells (Figs. 6-1 and 6-2). With cells of different strains, a *somatic hybrid* or *heterokaryon* forms. (A somatic cell is any cell other than a reproductive cell. The name heterokaryon refers to the two unlike nuclei.)

In 1970 L. D. Frye and M. Edidin of Johns Hopkins University reported an experiment in which specific surface features of two cells were followed after fusion. They labeled cultured mouse cells with antibodies having a green fluorescent dye attached and labeled cultured human cells with an antibody–red-dye combination. Immediately after fusion, half of the heterokaryon was red and the other half green, but within 40 min there was total mixing of the two surface markers (Fig. 6-3). It appears from such experiments that cell membranes behave more like fluids than solids.

One gets the same impression of a fluid structure after puncturing a cell with a fine needle, for the membrane seems to flow over the hole

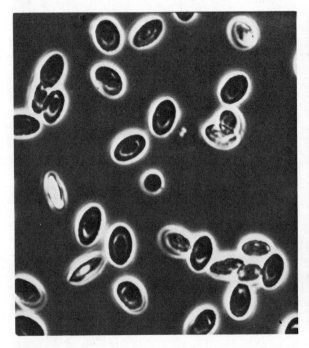
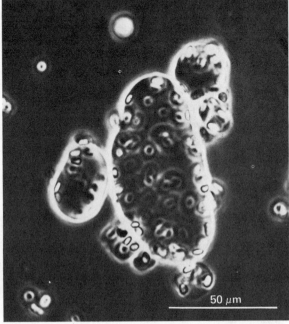

FIGURE 6-1 Cell Fusion. The fusion of cell membranes is a common event. Here the process is studied by observing the fusion of separate cells, induced by the presence of inactivated Sendai virus, a parainfluenza myxovirus. These phase-contrast photomicrographs are of chicken erythrocytes. *LEFT:* Before fusion. *RIGHT:* After fusion. Note the many nuclei in the fused cell. [Courtesy of Z. Toister and A. Loyter. *Biochem. Biophys. Res. Commun.* **41**: 1523 (1970) © Academic Press, N.Y.]

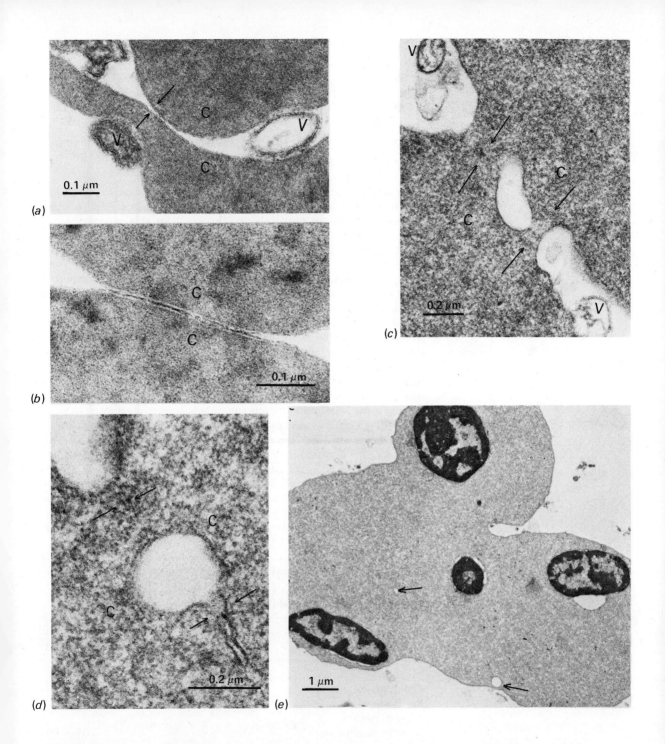

FIGURE 6-2 The Mechanism of Cell Fusion. Electron micrograph study of Sendai virus-induced fusion of chicken erythrocytes: (a) two cells (C) held in close proximity by attachment to a common virus particle (V); (b) The initiation of membrane fusion; (c) the formation of cytoplasmic bridges (arrows) between adjacent cells; (d) a closer view of the bridges, showing disorganization of the membranes; and (e) a fused cell with four nuclei. Note the small vacuoles (arrows), presumably the residue of those formed in the fusion process. [Courtesy of Z. Toister and A. Loyter. *J. Biol. Chem.*, **248**: 422 (1973).]

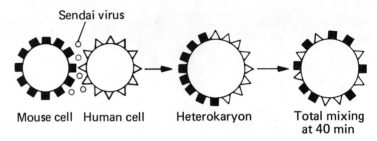

FIGURE 6-3 Mixing of Surface Antigens after Cell Fusion. Antibodies tagged with a green dye were reacted with surface antigens of cultured mouse cells; antibodies tagged with a red dye were similarly attached to cultured human cells. After Sendai-virus–induced fusion, total mixing was seen in 40 minutes. (Described by L. D. Frye and M. Edidin, *J. Cell Sci.*, **7**: 319 (1970).)

after withdrawal, preventing the escape of cytoplasm. These sorts of observations lead us to conclude that a typical membrane is not a rigid structure composed of stiffly connected subunits, but a fluid in which the components can move about. The basis for this structure is apparently a lipid film containing considerable protein (see Table 6-1) and small amounts of carbohydrate.

The Lipid Bilayer. The first indication that lipids might be an important constituent of biological membranes came in the last years of the nineteenth century, primarily from the work of E. Overton. In 1895 he published a report on the permeability properties of various membranes in which he noted how easily lipid-soluble substances penetrate them compared with their relative impenetrability to hydrophilic substances. On the basis of "like dissolves like," Overton concluded that the surface permeability barrier of cells must be predominantly lipid. As such, it would readily be penetrated only by lipid-soluble substances.

The most common membrane lipids are the phospholipids—molecules of glycerol containing a negatively charged phosphate ester at one hydroxyl and long-chain fatty acid esters at the other two hydroxyls (see Chap. 3). Phospholipids have both a hydrophilic end (the phosphate, often with other polar residues attached to it) and a hydrophobic end (the fatty acid "tails"). The physical chemist I. Langmuir demonstrated that molecules of this nature arrange themselves into monomolecular films at air-water interfaces. He proposed that the molecules in such films are all oriented with their polar ends into the water and their hydrophobic ends away from the water. Furthermore, Langmuir devised a physical technique for measuring the expanse of such films by gathering the molecules together with a thread until a resistance is met, indicating close-packing.

TABLE 6-1. The Composition of Some Membranes[a]

	Lipid	Protein	[Cholesterol] / [Polar lipid]
Myelin	80%	20%	0.7–1.2
Chloroplast lamellae	50	50	0
Erythrocytes	20–40	60–80	1
Mitochondrion			
Outer membrane	45	55	0.03–0.09
Inner membrane	25	75	0.02–0.04
Endoplasmic reticulum	64	36	0.07
Bacteria	20–30	70–80	0

[a] Most of the data taken from E. D. Korn, *Annual Reviews of Biochemistry*, **38**: 263 (1969).

Capitalizing on Langmuir's work, the Dutch scientists E. Gorter and F. Grendel of the University of Leiden (1925) set out to measure the total film size produced by the lipids extracted from human erythrocytes (red blood cells). Knowledge of the physical area covered by the lipids, they reasoned, might enable them to produce a model for the way in which the lipids are arranged in the membrane. Human erythrocytes were a logical choice for these experiments, since the cells are easy to obtain and were known to be extremely simple. Since erythrocytes have little or no internal structure, and therefore little or no internal membrane, one might presume that extracted lipids all come from the surface.

Gorter and Grendel extracted erythrocytes with acetone and measured the monomolecular film formed from the extraction product with Langmuir's device, called a *Langmuir trough*. They found a total film area of about 200 μm^2 per cell, about twice their estimate for the surface area of the erythrocyte. Thus they concluded that there is just enough lipid in each cell to cover it twice, and that the lipids are associated in a bilayer (Fig. 6-4a). As it turns out, both area measurements were in error: the extraction process had only removed a portion of the lipid, and the erythrocyte area is closer to 150 μm^2. (Their measurements were based on light microscopy and the assumption that erythrocytes are disks when, in fact, they are biconcave.) However, their conclusion that the membrane is composed of a lipid bilayer was quite correct.[1]

This didn't solve the problem completely, however, for Gorter and Grendel's measurements did not lead directly to a complete model for membrane structure. Erythrocytes do not always behave as if they have a lipid exterior. For one thing, their surface tension, which is a measure of the tenacity with which molecules at the surface of a liquid cling to their own kind, is far too low. Fats and oils in an aqueous environment have very large surface tensions because of hydrophobic bonding; lipids extracted from erythrocytes do not.

The Danielli-Davson Model. J. F. Danielli and E. N. Harvey suggested that the anomalously low surface tension of erythrocyte lipids is the result of contamination by protein, which would naturally seek the surface of a lipid droplet and thereby change its character. The hydrophilic behavior of intact cells could also be explained by this assumption, which led Danielli and Hugh Davson to propose the first complete membrane model in 1935.

Danielli, then at Princeton University, envisioned a membrane with a lipid center, coated on either side with protein (Fig. 6-4b). The phospholipids in his model are oriented in two monomolecular layers, with their hydrophobic tails toward the inside of the structure and their hydrophilic phosphates on the surface, contacting the layers of protein. The fundamentals of this structure are still accepted today, though there have been numerous proposals for modification.

The Unit Membrane. In the middle 1950s J. D. Robertson of University College, London, and others found stains that resolve most membranes into two distinguishable lines on micrographs, whereas both electron and light microscopists had hitherto seen only single lines (Fig. 6-5). It was later demonstrated that partial extraction of membrane lipids with acetone leaves these double lines intact. They were assumed to represent protein layers on the two surfaces of the membrane, as in Fig. 6-4c. The

SECTION 6-1
Membrane Structure

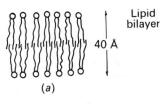
(a)

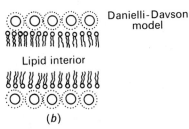
(b)

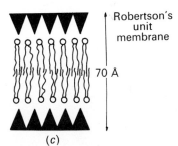
(c)

FIGURE 6-4 The Early Membrane Models. (a) Gorter and Grendel suggested a simple lipid bilayer. (b) Later, Danielli and Davson extended the model to include a coating of protein. The interior in their model was an unspecified "lipoid" region. (c) The unit membrane, first proposed by Robertson in the 1950s, explained the trilaminate appearance of many membranes in the electron microscope by including an extended layer of protein on either side of the bilayer.

241

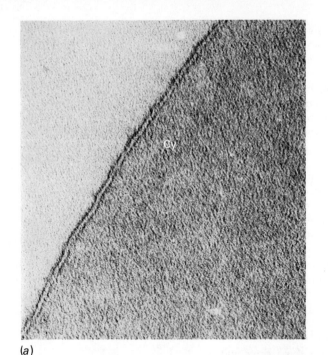 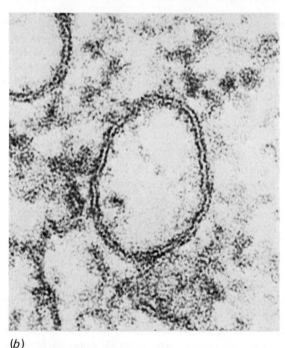

FIGURE 6-5 The Electron Microscopy of Membranes. Note the trilaminate appearance of these stained preparations. (a) Red blood cell membrane. Cytoplasm (Cy). (b) A vacuole from a mussel cell. [(a) Courtesy of J. D. Robertson, Duke Univ. (b) Courtesy of N. B. Gilula and P. Satir, *J. Cell. Biol.*, **51**: 869 (1971).]

stability of membranes to acetone suggests a structural role for these protein layers. This represented a departure from the concepts presented in Danielli and Davson's paper, as they had dismissed the protein portions rather lightly and assumed that the important part of the membrane, both structurally and biologically, is the lipid barrier.

Robertson's membrane model had rather thin protein layers on the two surfaces, more consistent with polypeptides in a flat β structure than with the compact globular proteins suggested by Danielli. Furthermore, Robertson's model is not necessarily symmetrical—whereas the internal surface in the model is coated with protein, the outer surface could be either a mucoprotein or a mucopolysaccharide. These features, together with the single lipid bilayer leaflet 40 to 65 Å thick, are known as the *unit membrane* model, a name that implies a homogeneity in structure from one source to another. Though it adequately explains the appearance of many membranes in electron micrographs, the reliance of the model on electron microscopy for support is also the basis for its vulnerability to criticism. This vulnerability is because electron micrographs can be made only after elaborate preparation of the sample, usually consisting of fixing, staining, drying, and embedding in a hard plastic. (See the appendix for a discussion of electron microscopy and sample preparation for it.) If these fixing and staining procedures do not preserve the original structure, then one may observe only artifacts in the microscope. (The literature on electron microscopy abounds with such artifacts.) Therefore, a number of other physical techniques have also been used to investigate membrane structure.

One alternative to electron microscopy involves the use of X-ray diffraction, which can be carried out on untreated samples. Whenever there is a repeating sequence of molecules, X-ray scattering diagrams will be nonrandom and can be interpreted (albeit often with great difficulty) to reveal the atomic dimensions and spacing of the repeating units.

SECTION 6-1
Membrane Structure

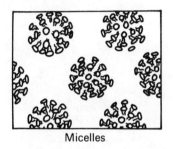
Micelles

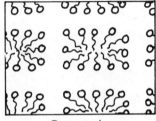
Rectangular

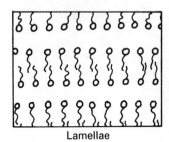
Lamellae

Complex hexagonal

FIGURE 6-6 Orientation in Lipid-Water Systems. X-ray studies by Luzzati and Husson revealed a number of stable "liquid-crystalline" structures in aqueous solutions of a fatty acid (potassium palmitate): Micelles; lamellae; rectangular (note "pores"); complex hexagonal. [Redrawn from V. Luzzati and F. Husson, *J. Cell. Biol.*, **12**: 207 (1962).]

Many X-ray scattering studies on membranes and membrane models have been published. V. Luzzati and F. Husson, for example, demonstrated in 1962 that various lipid-water mixtures reveal several kinds of structure, depending on environmental conditions of concentration, temperature, and so forth (see Fig. 6-6). One of these structures is the bilayer leaflet upon which the unit membrane concept is built. The bilayer leaflet is not, however, the only stable structure revealed by the scattering patterns. Among the alternatives are the micelle and a nearly rectangular arrangement formed by adding molecules to the edge of a rather irregular bilayer. In an actual membrane, this arrangement would provide a less-than-perfect alignment of lipids in the basic bilayer and periodic interruptions that could take the form of hydrophilic channels, or pores, extending through the lipid layer. The significance of this model in terms of membrane permeability, especially to water, will be evident later.

The Fluid Mosaic Model. Current concepts of membrane structure were developed in large part from observations made on membranes after *freeze-fracture* and *freeze-etching* (Fig. 6-7). These names refer to a preparation procedure for electron microscopy in which a solidly frozen specimen is broken by a sharp blow. Part of the surface ice is then sublimed from the fracture faces in a vacuum, "etching" the surface and thus exposing part of the specimen. The etched surface is next coated with a thin film of platinum and carbon (see Appendix) after which the organic matter is dissolved away with acid, leaving the metal replica to be examined in the electron microscope. Results obtained with this technique changed completely our concept of membrane structure.

The importance of freeze-fracture and freeze-etching to the study of biological membranes became clear in 1966 when Daniel Branton, of the University of California at Berkeley, demonstrated that the fracture line frequently follows the *middle* of a membrane, separating the two layers of the lipid leaflet from one another. This separation revealed not a

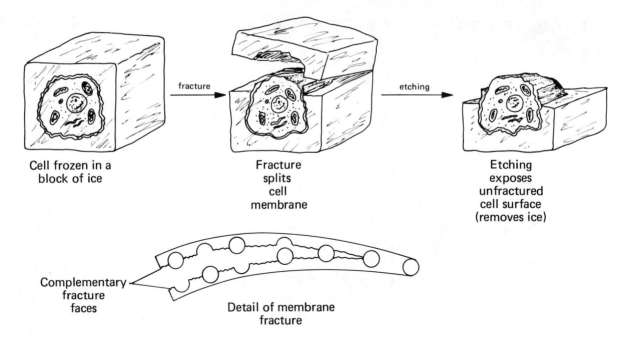

FIGURE 6-7 Freeze-Fracture and Freeze-Etching. The frozen cell is fractured with a sharp blow. The fracture line very often runs through the hydrophobic interior of membranes, separating the two leaflets of the lipid bilayer and revealing membrane proteins otherwise buried therein. Etching, or sublimation of surface ice, exposes unfractured membrane.

smooth lipid surface, as expected, but a surface that looked like cobblestones (Fig. 6-8). The "stones" are membrane-associated proteins, sitting not on the surface but within the lipid leaflets. In contrast, the true inner and outer surfaces of the membrane are usually rather smooth, similar to the outer surface of the erythrocyte seen in Figure 6-8.

This intimate association between lipid and protein in membranes explains a number of features that would otherwise be puzzling. For example, isolated membrane proteins tend to be relatively compact structures often 40 Å or more in diameter, not the flat sheets needed for some earlier membrane models. Membrane proteins also tend to be very insoluble in aqueous solution under mild conditions unless detergents are present, indicating the existence of a hydrophobic exterior unable to interact with the surrounding water. These features would be expected for proteins that are designed to be embedded in a lipid membrane rather than situated on its surfaces.

Many of the pebbles seen in the fractured erythrocyte membrane of Fig. 6-8 contain molecules of *glycophorin*. This glycoprotein has a molecular weight of about 50,000 daltons, more than half of which is carbohydrate, mostly N-acetylgalactosamine, galactose, and sialic acid in chains of from 4 to 12 monosaccharides each (see Chap. 3). Structural studies by V. T. Marchesi and his colleagues at Yale University revealed that the carbohydrate is found attached only to asparagine, threonine, and serine residues in the first half of the glycophorin chain, starting from the amino end. Immediately following this segment is a stretch of about 25 hydrophobic amino acids; the remainder, or carboxyl end of the chain, contains hydrophilic amino acids without carbohydrate. This sequence correlates very nicely with observations on the placement of glycophorin within the membrane—it contributes carbohydrate to the outside surface of the cell, where its amino terminus is found, and appears to extend through to the interior surface, exposing to the inside its hydrophilic, nonglycosylated carboxyl end.

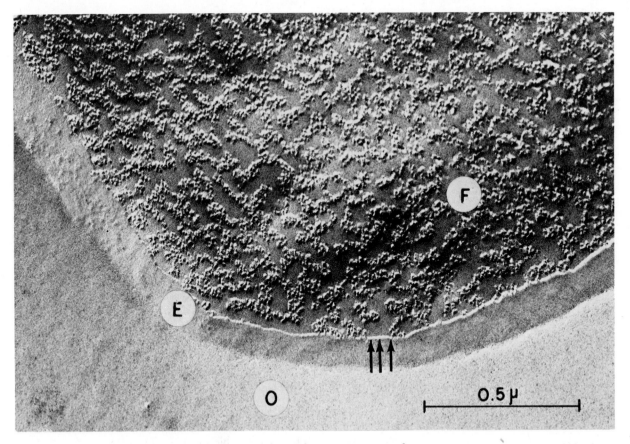

Attachment of antibodies to surface-exposed membrane proteins such as glycophorin (in experiments like the one described in Fig. 6-3) led to the realization that membrane proteins are not fixed within the lipid bilayer but are free to move laterally, like icebergs floating submerged in a sea of lipid (Fig. 6-9). This picture inspired the name *fluid mosaic*

FIGURE 6-8 A fractured red blood cell. The fractured surface (F) reveals many particles, presumably membrane proteins. The edge of the fracture is marked by triple arrows. Adjacent to this edge is a thin region (E) where the true exterior surface is exposed by etching (sublimation of surface ice after fracture). The region marked O was still covered with ice at the time of metal shadowing. [Courtesy of P. Pinto da Silva and D. Branton, *J. Cell Biol.*, **45**: 598 (1970).]

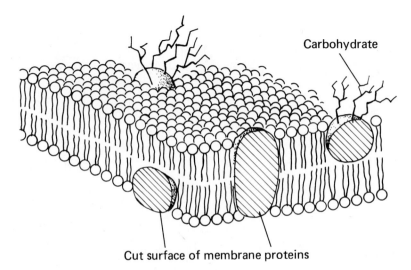

FIGURE 6-9 Fluid Mosaic Model. Proteins floating in a sea of lipid. Note carbohydrate on exterior surface only. Some proteins span the lipid bilayer, others are exposed only to one surface or the other.

model, coined by S. J. Singer and G. L. Nicolson. If this analogy of icebergs floating in a sea of lipid is valid, then it should be possible to freeze the proteins into place by solidifying the lipid sea in which they float. Indeed, that is readily accomplished, for as one lowers the temperature a point is reached at which a *phase transition* in the lipid sea occurs from a relatively liquid (called *"liquid crystal"*) to a relatively solid (crystalline) state. At this point, mobility of the membrane proteins is greatly restricted.

If this temperature (or phase) transition is really due to the freezing and melting of membrane lipids, then the temperature at which it occurs should vary with the fatty acid composition. Specifically, the transition temperature should increase with chain length and decrease with degree of unsaturation. (Double bonds put kinks in the hydrocarbon chains, as in Fig. 6-10a, making crystalline close-packing harder to achieve.) Since the fatty acid composition of both bacterial and eukaryotic membranes reflects to some degree the availability of fatty acids from outside sources (diet in the case of animals), it has been possible to verify the correlation between phase transition temperature and fatty acid composition. In fact, one of the mechanisms used by bacteria and fish to adapt to a very cold environment is alteration of their membrane lipids by using shorter, more unsaturated fatty acid chains.

Cholesterol drastically lowers or abolishes the apparent transition temperature by imposing a structure of its own on membrane lipids (Fig. 6-10c). It is because of their cholesterol content (which is not rigidly fixed, but varies somewhat with availability) that animal cell membranes maintain a liquid crystalline state at body temperature in spite of having relatively long-chain (16 and 18 carbon), saturated fatty acids. The membranes of higher plants and bacteria, most of which do not contain cholesterol, maintain a liquid crystalline state by using shorter-chain fatty acids with more double bonds.

The fluid mosaic model of membrane structure is now almost universally accepted. Though most of the work has been on plasma membranes, the model is presumed to apply to membranes of all types, in spite of their varying and characteristic differences in lipid and protein composition.

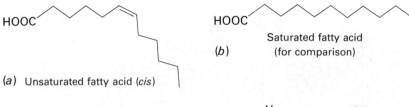

FIGURE 6-10 Structures that Disrupt Lipid Close-Packing. Cholesterol and fatty acids having double bonds do not pack well in membrane bilayers because of the "kinks" in their configuration. The effect is to inhibit crystalline close-packing and hence to lower the phase transition temperature. In fact, cholesterol may abolish it altogether.

(a) Unsaturated fatty acid (cis)

(b) Saturated fatty acid (for comparison)

(c) Cholesterol

Membrane Stability. The fluid nature of membranes is responsible for their resistance to damage. Small puncture wounds, as noted earlier, self-seal as membrane flows in to cover the defect. Localized chemical damage is also apt to be repaired, although more extensive damage by enzymatic attack from certain insect, snake, and bacterial toxins (especially species of *Staphylococcus* and *Streptococcus* bacteria), or from components of complement or *lymphotoxins* produced by T-type lymphocytes—the so-called killer lymphocytes—during an immune response is apt to lead to membrane destruction and cell death.

The integrity of cell membranes is also threatened in an unknown way by oxidations, molecular oxygen itself being the most common destructive agent. Oxidations are repaired by cytoplasmic reducing substances that include an —SH-containing tripeptide called glutathione and, at least in mammals, probably also by ascorbic acid and tocopherol (vitamins C and E, respectively). These agents are kept in their active (i.e., reduced) forms by reducing power obtained through the phosphogluconate pathway (see Chap. 7), which involves the cytoplasmic oxidation of glucose. Most of the glucose consumed by mammalian erythrocytes is utilized via this pathway.

The importance of the phosphogluconate pathway to maintenance of cell membranes is illustrated by humans who are deficient in glucose-6 phosphate dehydrogenase (G6PDH, the product of an X-linked gene), which is an early enzyme in the phosphogluconate pathway. Without normal levels, erythrocytes in particular are fragile and short-lived. Any added oxidative stress (e.g., from antimalarial drugs or sulfa-type antibiotics) is apt to result in massive destruction of erythrocytes and subsequent anemia.

Membrane Asymmetry. To describe biological membranes as fluid or liquid crystalline does not imply that all components of a given membrane are free to move about without constraints. Though lateral mobility (that is, in the plane of the membrane) is well established for both lipids and at least some proteins, movement is not necessarily random. As will be explained in Chap. 9, some membrane proteins are connected to underlying networks of microtubules and microfilaments, thus restricting and controlling their movement. This restriction may lead to marked differences in the protein content from one part of a membrane to another.

There is also marked asymmetry across membranes. Carbohydrate, which is primarily contributed by glycoproteins that are specific for cell and species type but also by glycopeptides and glycolipids, is found almost exclusively on the outer surface of cell membranes. In cells lining the intestine, for example, carbohydrate produces a 0.1–0.5 μm thick branching network of fine filaments extending from the outer leaflet of the membrane (Fig. 6-11). The inner surface of plasma membranes, in addition to its lack of carbohydrate, has typically a quite distinct set of antigenic determinants. Some of this asymmetry is due to proteins that extend from surface-to-surface, such as erythrocyte glycophorin; some is also due to proteins that do not span the lipid bilayer but are confined to one side or the other, available therefore to chemical reagents or antibodies introduced from one side only.

Of course, finding a different set of proteins on the two sides of a membrane does not by itself rule out the possibility of movement from one

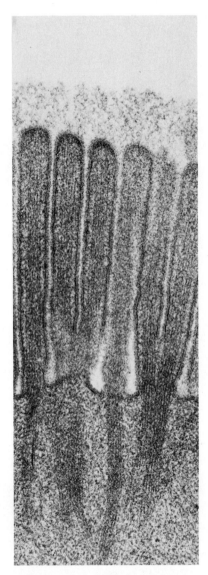

FIGURE 6-11 Carbohydrate Coat on an Intestinal Brush Border Cell. Cells of the intestinal lining have a lush carbohydrate coat with branching chains of carbohydrate extending into the intestinal lumen. (The diameter of each fingerlike microvillus is about 0.1 μm.) [Courtesy of S. Ito, *Phil. Trans. Roy. Soc. London Ser. B*, **268**: 55 (1974).]

side to the other. However, when proteins are "tagged" in various ways and then reexamined later, they are characteristically still found on their original side. Although lateral movement of some membrane proteins is allowed, it appears that there is very little tendency for them to flip, rotate, or translocate to the other side. This observation has important implications in the study of transport through membranes, as we shall see.

Lipids, too, are asymmetrically arranged across those membranes amenable to study. By subjecting erythrocytes to various lipid degrading enzymes, it has been established that phosphatidyl choline and sphingomyelin are confined mostly to the outer surface, while phosphatidyl ethanolamine and phosphatidyl serine are found on the inner surface. There is some transfer back and forth, but it is very slow, requiring hours in most experiments.

Lipids can, however, move in and out of membranes. Lipid exchange between cytoplasmic lipoproteins and the inner surface of the cell membrane is readily demonstrated. A similar exchange between serum lipoproteins (see Chap. 3) and the outer surface is also common and is demonstrated with particular ease in erythrocytes since the normal turnover of plasma membrane (Chap. 10) is not possible in these simple cells. The cholesterol content of erythrocyte membranes, for example, varies rapidly with changes in serum cholesterol levels, which in humans is carried by serum lipoproteins.

Membrane asymmetry is important to membrane function. The two sides of a membrane typically participate in very different activities. There is, for instance, a constant traffic of small molecules across the cell membrane, in some cases made possible by highly directional transport systems contributed by membrane proteins. The plasma membrane surface also has the task of recognizing extracellular regulators such as hormones, and in many cases recognizing other cells as well. Intercellular recognition during the formation of solid tissues is almost always followed by the development of specialized areas of contact furnished by membrane proteins in the two adjacent cells. These areas of specialized contact are known as *intercellular junctions*.

6-2 INTERCELLULAR JUNCTIONS

In multicellular animals and plants, individual cells are found grouped together into *tissues*. Fat, bone, and muscle are examples. The cells within a given tissue are not necessarily identical, but they cooperate to carry out a common task. The plasma membranes of these cells are usually separated from one another by a 20 nm space filled with mucopolysaccharide secreted by the cells and found either free in the intercellular space or as extensions of membrane proteins. In some cases, this pattern is interrupted by areas in which the membrane proteins of adjacent cells interact with each other in a highly specific way to provide an intercellular junction.

Some junctions are devices that glue cells together, others are barriers that prevent fluids from passing between the cells (e.g., between cells lining fluid-filled cavities such as the gut), and some junctions serve primarily to establish channels of communication from one cell to another. (See Chap. 8 for a discussion of a quite different type of intercellular communication, the neuronal and neuromuscular synapse.)

Nomenclature. Cell junctions are given names that describe their shape and the relative closeness of the cooperating membranes. If the junction is a strip that passes completely around the cell, it is called a *zonule*; if it is a spot, the appropriate term is *macule*. If the outer surfaces of the two membranes are very close together, in some cases even fused, the specialization is said to be an *occludens* (i.e., occluding) junction; a wider separation, filled with an indefinite arrangement of dense material forms an *adhaerens* (adhering) junction. Both adhering and occluding type junctions may be found as either macules or zonules.

A great variety of specialized junctions have been described. We shall consider in detail only the three types most common in higher animals; they are also representative of the three main functional classes: the *desmosome*, a form of macula adhaerens, helps glue cells together; the *tight junction*, a zonula occludens, provides a barrier to the flow of fluids between cells; and the *gap junction*, which is a unique type of macula occludens, provides intercellular communication by making possible direct passage of small molecules and ions between cells.

Desmosomes. These examples of macula adhaerens junctions are found in nearly all epithelia (skin, inside lining of the intestinal tract, etc.) and in many other tissues where they apparently serve mainly to glue the cells together, providing structural strength. (See Fig. 6-12a.) They have been likened to spot welds in a piece of machinery. The opposing membranes have the usual 20–30 nm spacing, but there is a dense fibrous material between the cells. In addition, there are dense clusters of filaments in the cytoplasm of both cells looping in and out of the region of the desmosome. (These are *tonofilaments*, possibly related to the 10 nm or 100 Å filaments seen in other cellular locations.) Freeze-fracture of either of the membranes in this region reveals a cluster of granules approximately 8–10 nm in diameter, presumed to be the specialized membrane proteins that provide the basis for the desmosome structure (Fig. 6-12b).

Tight junctions. Tight junctions, the common form of which is a *zonula occludens*, serve to seal the space between cells (Fig. 6-12d). In the intestines, tight junctions form a ring (or zone) completely around each cell, sealing it to its neighbors and ensuring that intestinal contents pass through the cells rather than between them, presenting the opportunity for careful regulation of that flow by mechanisms to be considered shortly. A comparable arrangement is found in the cells lining other cavities (glands, gall bladder, etc.) where intercellular leakage is undesirable. These junctions also presumably contribute to tissue integrity. If the desmosome is a spot weld, tight junctions are seam welds.

In the electron microscope, tight junctions appear to involve fusion of the outer leaflets of adjacent membranes so that the total cytoplasm-to-cytoplasm distance is less than twice the usual membrane width. When the membranes are freeze-fractured, their proteins in this region are seen to consist of long ridges which, on careful analysis, appear as a double row of particles, 3–4 nm in diameter, one row contributed by each of the cooperating membranes. The ridges seen in Fig. 6-12b, and removed from the complementary fracture face in Fig. 6-12c, are the structures that actually span the junction, thus sealing the cells together.

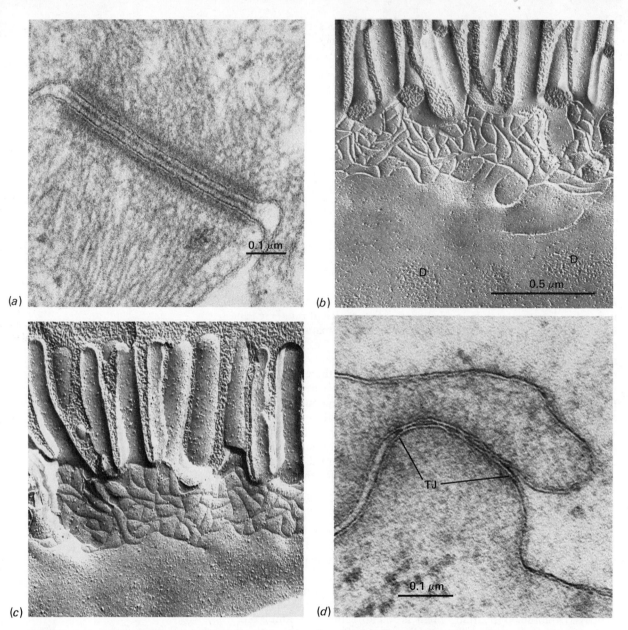

FIGURE 6-12 The Desmosome and Tight Junction. (a) A desmosome from frog skin. Note the intermembrane filling and cytoplasmic filaments looping into the desmosome area. (b) Freeze-fractured view of an intestinal cell. Note the membrane particles identifying the desmosome (D). Nearer the top is one face of a tight junction. The ridges form from a double row of membrane particles, one row donated by each membrane. (c) Another tight junction, showing grooves left in the complementary fracture face. (d) Thin section of one tight junction (TJ). The points of actual fusion are where one finds the membrane particles seen in the previous two views. [(a) and (d) courtesy of M. G. Farquhar, G. E. Palade, *J. Cell Biol.*; a from **26**: 263 (1965), b from **17**: 375 (1963). (b) and (c) courtesy of L. A. Staehelin, b from *Int. Rev. Cytol.* **39**: 191 (1974) © Academic Press, N.Y.]

Gap Junctions. The gap junction, also called a *nexus*, is a form of macula occludens. Examples are seen in Fig. 6-13 along with a specialized junction called a *septate desmosome*, found mostly in invertebrates.

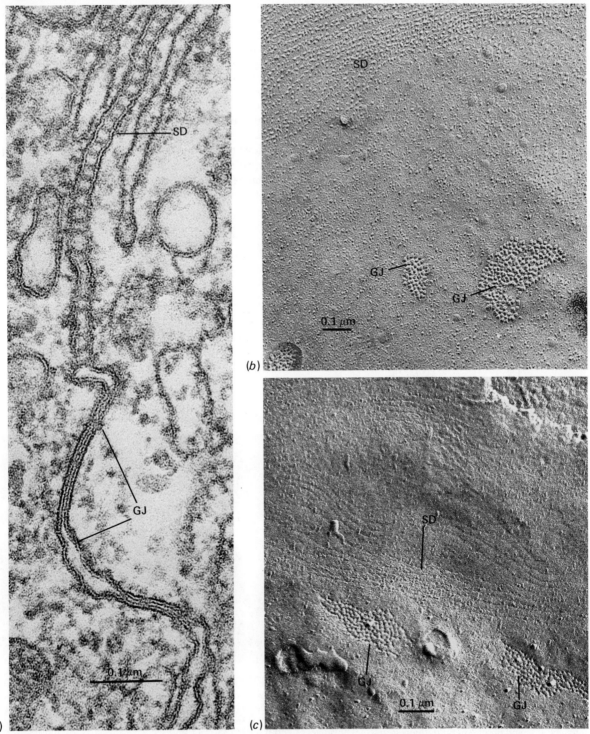

FIGURE 6-13 Gap Junction and Septate Desmosome. (a) Thin section, stained with osmium tetroxide. Note that the space between membranes at the gap junction (GJ) is filled with closely packed material that admits some stain. The intermembrane space at the septate desmosome (SD) has a ladderlike arrangement of material. (b) and (c) are complementary fracture faces of one participating membrane. Note that the particles on one face leave depressions in the other. [Courtesy of N. B. Gilula and P. Satir, *J. Cell Biol.*, **51**, 869 (1971).]

In thin sections, the gap junction appears as a 20 Å (2 nm) "gap" between adjacent membranes that is filled with a highly ordered array of particles. These particles are associated with (or are extensions of) a hexagonal arrangement of proteins in the two adjacent membranes. The arrangement appears to provide open channels through the hexagon extending directly from the cytoplasm of one cell to the cytoplasm of the adjacent cell. This property of gap junctions was discovered by W. R. Loewenstein during the investigation of electrical conductivity (ion flow) across membranes.

Low-resistance junctions between cells can be demonstrated by using electrodes—in the form of tiny micropipets filled with a salt solution—to pierce the membranes of cells and thereby measure the flow of current (ions) between them (Fig. 6-14a). The electrical resistance be-

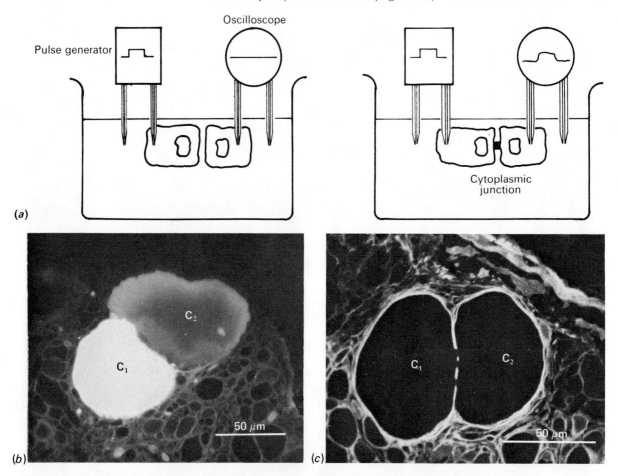

FIGURE 6-14 Gap Junctions and Ionic Channels. (a) The existence of ion-permeable channels between cells can be demonstrated with a pulse generator (here producing a square wave), using an oscilloscope as a detector. (b)–(c) Certain nerve cell processes (axons) of the crayfish (*Procambarus*) exhibit "electrotonic" coupling via gap junctions, detectable as described in (a). (b) A dye, Procion Yellow M4RS (about 500 daltons) is injected into one cell (C_1) and observed by its fluorescence. The adjacent cell (C_2) also shows fluorescence, especially around its edges. (c) The same dye is added to the extracellular space, where it stains the fatty covering (Schwann sheath) around a similar set of two cells, C_1 and C_2, penetrating neither and eliminating diffusion into and out of cells as an explanation for figure b. (Note the discontinuous nature of the covering between the cells. The cell membranes are intact in these regions, but not stained.) [(b)–(c) Courtesy of B. W. Payton, M. V. L. Bennett and G. D. Pappas, *Science*, **166**: 1641 (1969). Copyright by the AAAS.]

tween a single cell and its surrounding solution is very high, reflecting the integrity of the cell membrane and its relative impermeability to ions. Thus, if two cells are impaled by electrodes, the insulation provided by the intervening membranes should keep any current from passing between the cells. But when the two cells come together, an electrical pulse introduced to one can sometimes be seen in the second, indicating a low-resistance contact between the adjacent cells.

In order to find out more about the nature of low-resistance intercellular contacts, Loewenstein injected fluorescein dye into one of the cells engaged in such contact. Within a few minutes, the use of an ultraviolet lamp revealed a telltale green fluorescence in a number of cells grouped together with the injected cell, results similar to those seen in Fig. 6-14b. These intercellular contacts permit the transfer not only of simple ions, but of organic molecules at least as large as fluorescein, which has a molecular weight of 332 daltons. Subsequent work in Loewenstein's laboratory has made it clear that the junctions through which fluorescein passes are gap junctions, and that gap junctions permit intercellular movement of a variety of small molecules, probably up to a mass of 1000 daltons or so.

Gap junctions are found in a wide variety of embryonic and adult tissues, and even in cultured cells *in vitro*. Their functions in some cases (certain muscle and invertebrate nerves) are apparently understood and will be discussed in later chapters. In many cases, however, we can only speculate on the kind of communication that is carried from cell to cell by these junctions.

Gap junctions, as noted, are formed by a regular array of special plasma membrane proteins. Because of our view of the membrane as a fluid, we should not be surprised to learn that gap junctions are not permanent fixtures, but can form and dissolve in response to appropriate stimuli. Their formation at sites of contact between appropriate cells takes only a few minutes. When one of the cells is damaged or killed, closing of the channel comes even more quickly. This property is an important one, for it means that destroying a few cells in liver or skin, for example, does not cause leakage from adjacent viable cells, a situation that would endanger them all.

The integrity of gap junction communication appears to depend on the cytoplasmic level of free Ca^{2+}, which is normally on the order of 10^{-6} M or less. When it rises substantially above this level, intercellular channels quickly close. Since the fluid bathing most animal cells contains about 10^{-3} M Ca^{2+} or more, an unsealed puncture of the cell membrane will cause Ca^{2+} influx. In most cells, mitochondria sequester Ca^{2+} and thus control the level of free ion in the cytoplasm. If the cell is poisoned, leakage of Ca^{2+} from mitochondria causes gap junctions to close even before the cell is irreversibly killed, an obvious protective feature for adjacent cells.

The widespread intercellular communication offered by gap junctions should not be confused with the larger and relatively more permanent channels found, for example, between the cytoplasm of cells in many plants (where the channels are called *plasmodesmata*), between developing sperm, and between surrounding nurse cells and a developing egg. These are instances of true windows between two plasma membranes. Gap junctions yield a more restricted flow of materials and are clearly the result of a careful arrangement of specialized proteins that are integral parts of intact cell membranes.

6-3 MEMBRANE PERMEABILITY

The cytoplasm of the cell has a composition very different from that of the fluid surrounding it. This fact is *prima facie* evidence for a selectively permeable membrane. Though it is clear that molecules do get into and out of a cell, it is just as clear that the passage of substances across the cell membrane is a regulated—and regulatory—function. The first experimental verification of this came at about the turn of the century, primarily through the work of several plant physiologists including W. Pfeffer, E. Overton, and others.

Plasmolysis. Early experiments on membrane permeability utilized the phenomenon of *plasmolysis* to demonstrate the selective permeability of membranes. A cell is plasmolyzed by suspending it in a medium with a higher total solute concentration than the cytoplasm. Water then leaves the cell in an attempt to equalize its concentration on the two sides. If water is the only molecular species capable of making that transfer, the volume of cytoplasm shrinks. With a microscope, one can watch such a cell quickly shrivel down to a small fraction of its former size. In the case of plant and bacterial cells, which have both a cell wall and a membrane, one can see the membrane and wall separate as the volume of the cytoplasm gets smaller (Fig. 6-15). It was through such experiments that the existence of two structures, a wall and a separate membrane, first became clear.

Overton found that cells of certain plant root-hairs could be plasmolyzed by 7.1% sucrose solutions, although 7.0% sucrose (0.21 M) had no such effect. With the latter amount of sucrose, the total concentration of dissolved molecules or ions is the same as that of the cytoplasm. At a sucrose concentration of 7.5%, a complete and uniform plasmolysis was achieved in 10 seconds or less, and the cells remained in their shrunken condition for 24 hours or more. Apparently the cells are not even slightly permeable to sucrose; otherwise they would slowly regain their former volume (*deplasmolyze*) as sucrose enters them.

When Overton used solutions containing 7% sucrose plus about 3% methyl or ethyl alcohol, no plasmolysis was observed. Overton correctly decided that plasmolysis did not occur because alcohol distributes itself equally on the two sides of a plasma membrane in less time than plasmolysis requires—in other words, in something less than 10 seconds. The membrane, so impermeable to sucrose, must be extremely permeable to these simple alcohols.

In an attempt to define better the permeability characteristics of plasma membranes, Overton repeated his plasmolysis experiments with a wide range of solutes, taking care to use relatively nontoxic materials in order to keep the cells from dying. He found that a number of substances penetrated cells readily. Others, such as sucrose, could not get in at all. A third group was found to penetrate cells only at a very slow rate: glycerol, the hexoses, and amino acids, for example, caused a rapid plasmolysis followed by a slow water regain, indicating a slow penetration. When the chemical nature of the various solutes was examined, it became clear that in most cases there was a direct correlation between ease of entry and the lipid solubility of the substance. Though there are some notable exceptions to this rule, water itself being one of them, the cell seems to present a lipid barrier to most substances. This, of course,

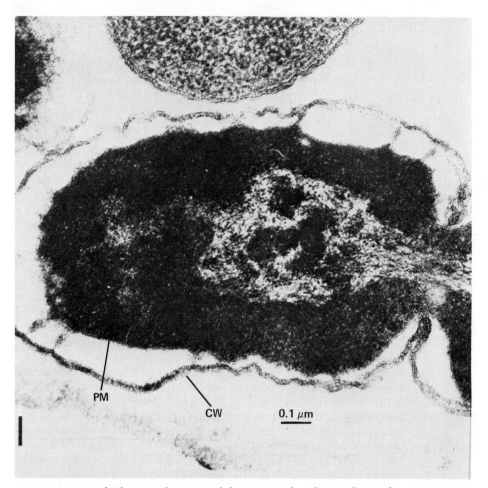

FIGURE 6-15 A Plasmolysed Cell. An *E. coli* cell plasmolysed with 20% sucrose. (Almost all cells are impermeable to sucrose.) Note connections between the plasma membrane (PM) and the cell wall (CW). A portion of a normal, unplasmolysed cell is seen at the top. The light area in the plasmolysed cytoplasm is DNA. [Courtesy of M. E. Bayer and Cambridge Univ. Press. From *J. Gen. Microbiol.*, **53**: 395 (1968). Reprinted by permission.]

is consistent with the membrane models presented earlier and was the basis for proposing them.

Oil/Water Partition Coefficients. To put the plasmolysis experiments on a more quantitative basis, later investigators measured and compared the *permeability coefficients* of various solutes with the *oil/water partition coefficients* of the same materials. Permeability coefficients are easily measured by determining the rate of entry of radioactively labeled solute into the cytoplasm at various external concentrations. An oil/water partition coefficient, on the other hand, is measured by shaking the solute in an oil-water mixture and then letting the phases separate. The partition coefficient is the concentration found in the oil phase at equilibrium divided by the concentration found in the aqueous phase. The relationship between these two coefficients is shown in Fig. 6-16.

It is clear from these experiments that many molecules are able to enter a cell by simple diffusion. However, diffusion represents a movement from a region of higher concentration to a region of lower concentration. It is possible for diffusion to maintain a constant flow into a cell if the molecules are chemically altered immediately upon entry, thus keeping their internal concentration lower than their external concentration. Similarly, waste products may leave the cell by simple diffusion as

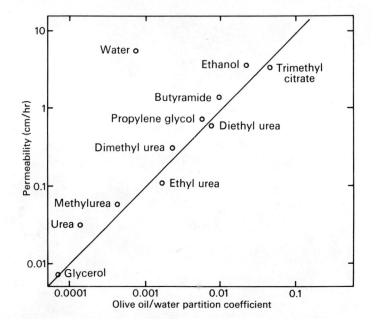

FIGURE 6-16 Permeability to Solutes as a Function of Oil/Water Partition Coefficient. For many substances, there is a direct relationship between cell permeability and lipid solubility. For others, however, the relationship fails completely. Note that water is in the latter category. [Data replotted from R. Collander, *Physiol. Plantarum*, **2**: 300 (1949).]

long as they are washed away from the cell's surface at a rate fast enough to maintain a favorable concentration gradient for efflux—in other words, as long as the external concentration of the waste product remains less than its cytoplasmic concentration.

Ion Trapping. Diffusion is an attractive mechanism for getting things into and out of cells because it does not require any specialized apparatus, nor does it require any expenditure of energy by the cell. In fact, free diffusion is quantitatively the most important way of getting things into and out of cells, for lipid-soluble molecules readily dissolve in and hence penetrate the plasma membrane. Oxygen, carbon dioxide, and steroid hormones are examples of such lipid-soluble materials. There are also many substances that, because they are weak acids or weak bases, exist in an equilibrium mixture of charged and uncharged forms at physiological pH. These substances are typically lipid soluble in their uncharged form, and in this form can therefore diffuse across cell membranes.

The proportion of a weak acid or base that is present in the lipid-soluble, uncharged form can be predicted from the pK_a of the substance and the existing pH. That fraction will penetrate membranes until an equilibrium is reached at which the uncharged form will be in the same concentration on the two sides of the membrane. If the pH is the same on the two sides, then the total concentration of the substance (charged plus uncharged) will also be the same. But if the pH is different, the side having the largest fraction of charged (ionized) acid or base will also have the largest total concentration of the substance.

Consider, for example, acetylsalicylic acid, which is ordinary aspirin. It is a convenient choice because its pK_a at body temperature is close to 3.4, a nice round four pH units below the physiological pH of 7.4 that is characteristic of most body fluids of animals. Aspirin, at pH 7.4, exists predominantly as acetylsalicylate—in fact, only 10^{-4} is in the acid, or uncharged form, a number that you can get from the Henderson-Haselbalch

aspirin
(acetylsalicylic acid)

equation (2-16) or just by remembering that the ratio of ion to acid changes by a factor of 10 for each pH unit away from the pK_a. The acid form of aspirin is very soluble in lipids—1 gram dissolves in only 10 ml of ether as opposed to 300 ml of water. Hence, this form is able to penetrate the cell membrane, eventually coming into equilibrium with nearly equal concentrations inside and outside the cell.

Now consider absorption of aspirin from the stomach into surrounding cells. The stomach pH is about 1.4, or two units below the pK_a of aspirin. At this acid pH, most of the drug in the stomach (99%) is in the lipid-soluble, uncharged form. At equilibrium, which is reached very quickly, the uncharged molecules are equally distributed on the two sides of the cell membrane. The ratio of acetysalicylate ion to acid in the stomach is 0.01 to 1 while the same ratio in the cytoplasm at pH 7.4 is 10,000 to 1. Hence, if the concentration of uncharged (acid form) of the drug were 1 mM in both places, the total concentration of the drug in the stomach (ion plus acid) would be 1.01 mM while the total concentration in the cytoplasm at equilibrium would be 10,001 mM. The ionized form, being relatively insoluble in the lipid membrane, is "trapped" in the cytoplasm and does not contribute to the equilibrium. Thus, a large concentration gradient of the drug is created as a result of the pH gradient.

This principle, *ion trapping*, is of great importance in understanding the biological distribution of acids and bases. In fact, since the cytoplasmic pH is commonly a few tenths of a pH unit below that of the surrounding fluid in animals, diffusible, weak acids have somewhat lower cytoplasmic concentrations than in the surrounding fluid at equilibrium. Weak bases, on the other hand, tend to be accumulated by cells for the same reason. Ammonia, for example, has a pK_a of 8.9. At physiological pH, it exists mostly as the ion ($NH_3 + H^+ \rightarrow NH_4^+$), but because of the NH_3 form it nevertheless gets into and out of cells readily by free diffusion and at equilibrium will be almost twice as concentrated in a cytoplasm at pH 7.1 as in the surrounding fluid at pH 7.4.

Ion trapping is often exploited for medical purposes—e.g., urine pH is raised with bicarbonate to hasten the elimination of aspirin after accidental poisoning. The concept of ion trapping is also important in physiology, especially when discussing gastrointestinal absorption and kidney function. It provides a mechanism for the cellular accumulation of substances that, because of their charge, would otherwise be excluded. However, many substances (sugars and simple ions such as Na^+ and K^+, for instance) that are nearly insoluble in lipids are neither acids nor bases but still get into and out of cells at relatively rapid rates. The earliest explanation for this behavior, still thought to be true in a limited way, was that membranes contained small holes, or pores.

Membrane Pores. Pores or channels of some sort have long been postulated for most if not all biological membranes. Their presence is inferred from measurements on the rate at which certain very small molecules, especially water, pass through membranes. Since they would need to be only transient rearrangements in the lipid bilayer, 4–10 Å in diameter, one would not expect to see them in the electron microscope. Hence, direct evidence for their existence is lacking and, in fact, recent calculations made with artificial lipid bilayer membranes (formed like a soap bubble across small openings) casts some doubt on whether they are needed even to explain the permeability of most membranes to water.

When one begins to add substances to the basic bilayer, however, evidence for the existence of pores becomes greater. The antibiotics nystatin and amphotericin B, for example, are used in medicine for the treatment of fungal infections because they incorporate into the fungal membranes and produce pores. These long hydrocarbons are thought to arrange themselves like the staves of a barrel to produce a pore some 8 Å in diameter, about the cutoff size for glucose. If enough of these pores are created, the fungal cell may become so leaky that it fails to survive. The bacteria that produce these antibiotics are immune, because pore formation by the antibiotics also requires cholesterol or related sterols which bacteria do not have but which fungi have in abundance. (So do animal cell membranes, requiring careful control of the dosage of antibiotic to avoid human toxicity.)

There is good reason to believe in the existence of natural analogs of these antibiotic-produced pores. In most cases, proteins are involved. A protein that spans the membrane could well have a channel down its middle adequate to permit the passage of water, urea, or other small substances. X-Ray diffraction studies on hemoglobin reveal a 10-Å channel through its structure, so the precedent is at least known. Evidence for the involvement of proteins comes from experiments in which it has been found that treatment of erythrocytes with sulfhydryl reagents inhibits water flow. These reagents react largely with the —SH group of cysteines in membrane proteins. The voltage-sensitive Na^+ and K^+ channels of electrically excitable membranes (e.g., nerves and muscles) are also thought to be provided by proteins whose conformation is changed by electrical fields (see Chap. 8).

Water probably gets into cells lining the ducts and tubules of kidneys through pores. When there is need to conserve fluid, a pituitary peptide called *antidiuretic hormone* (or in higher animals *vasopressin*) causes greatly increased water permeability of cells lining the distal tubules and collecting ducts. In terms of the pore model, the calculated channel size in these cells is 20–30 Å when maximally stimulated. The receptors for the hormone are on the side of the cell accessible to the blood and lymph—that is, on the side opposite the lumen of the duct or tubule where water absorption must first occur. Hence, the effect of antidiuretic hormone on water permeability of kidney cells must be mediated through some intracellular change that, in turn, affects membrane proteins. The mediator is thought to be cyclic AMP.

Thus, pores or channels have their place in understanding the passage of substances through membranes. They are not, however, the whole answer. Membranes possess a selectivity that demands further explanation.

COOH
|
Cys ─┐
| │
Tyr │
| │ S
Phe │ │
| │ │
Gln │ │
| │ S
Asn │ │
| │ │
Cys ─┘
|
Pro
|
Arg
|
Gly
|
NH₂

Vasopressin

6-4 MEMBRANE TRANSPORT

Cells of all types share a common problem in having to choose from among the many molecules and ions present in their environment exactly those substances needed to maintain life. These substances, to the exclusion of others, must be taken up by cells at rates fast enough to maintain growth, reproduction, or specialized functions. For example, in the kidneys of animals blood is filtered so that virtually all substances below about 30,000 daltons pass into the urine. Some of these substances—various salts, glucose, amino acids, and so on—are then reclaimed, mostly by selective uptake that leaves behind numerous different waste products, drugs, and other things that need to be eliminated from the animal.

Specific transport mechanisms are required to account for the degree of selectivity exercised by these and other cells, and to allow the passage of essential lipid-insoluble molecules (sugars, for example) at a sufficiently rapid rate to maintain life. It is now well established that many substances gain entry to the cytoplasm, or are removed from it, by a reversible combination with specific molecules designed to assist translocation across the membrane. Once on the other side, the transported material leaves its *carrier*, which is then free to assist in the passage of the next molecule much as a ferry assists automobiles and passengers to cross a river.

Transport Proteins. Carrier-assisted transport exhibits a very great selectivity, so that one molecule may enter readily whereas a nearly identical molecule is totally excluded. This ability to make extremely fine distinctions between closely related molecules is usually associated with the binding sites of enzymes and certain other proteins, because only proteins have the molecular variety and structural flexibility necessary to provide a wide range of specificities. Accordingly, the transport of certain molecules is attributed to a class of *transport proteins*, which in bacterial systems have also been called *permeases*. The ending "-ase," normally associated with enzymes, was picked to emphasize certain analogies between transport proteins and true enzymes: (1) transport proteins accelerate the reaction being considered, namely membrane translocation; (2) they provide great selectivity; and (3) they are themselves unchanged in the process, being recycled after each assisted entry or exit. However, the analogy between transport proteins and enzymes breaks down at a very important point: It is the function of an enzyme to bring about some chemical change in the substrate, but that process is not intrinsic to membrane transport.

We would expect that entry of a substance, whether assisted or not, should be possible only so long as its external concentration is greater than its internal concentration. But, in fact, many substances may be accumulated to internal levels many times greater than their concentration in the surrounding medium even when the substance cannot be metabolized or chemically altered in any way. Thus, some transport systems can move a molecule across a membrane only in response to a favorable concentration gradient, while others may do so in spite of an unfavorable one.

We must recognize, then, that material may get into and out of a cell by at least two carrier-mediated transport mechanisms in addition to any flow due to free diffusion. If transport is driven by a diffusion gradient, the process is called *facilitated diffusion*. The name indicates that the carrier improves the cell's permeability to a substance without altering the direction in which the substance would tend to move on its own. In other instances, transport is thermodynamically unfavorable, and can be achieved only with the expenditure of energy supplied by the cell. Such translocations are referred to as *active transport*. For example, glucose uptake by microorganisms and by cells lining the intestines and kidney tubules of animals is typically active; in most other animal cells, it is achieved by facilitated diffusion.

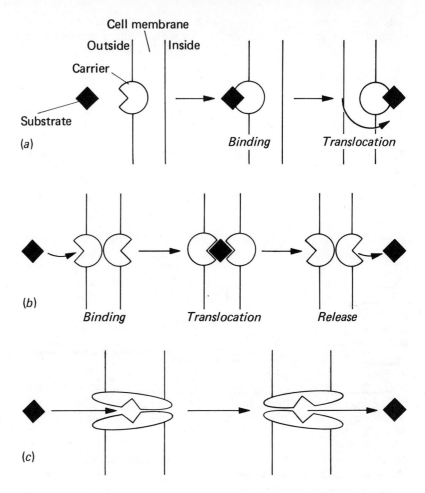

FIGURE 6-17 Facilitated Diffusion: Three Models. (a) Rotation and translocation of a carrier is the simplest model, but the reluctance of membrane proteins to flip or move between bilayers makes it seem unlikely. (b) The problem would be less if substrate were passed from one subunit to another of a complex carrier. (c) An alternate approach is to think of a channel that somehow has the desired specificity. It could be opened and closed by conformational changes in the protein.

Figure 6-17 shows schematically three ways in which facilitated diffusion might be accomplished. The rotating carrier model is perhaps the easiest to picture mentally, and the easiest to adapt to active transport, but requires that the carrier rotate, translocate, or both, something that we said earlier large membrane proteins are reluctant to do. To circumvent this problem, one can imagine the substrate being passed from one subunit to another of a multisubunit carrier, or through a channel created between subunits by conformational change of the protein, as pictured. Either of these modifications may be closer to reality, but until the real situation is defined we'll stick in the future to drawing the rotating carrier model and let your imagination, rather than the artwork, supply alternatives.

The existence and properties of a transport system can be established by studying the kinetics of uptake, including competition among closely related substrates; by looking for induced efflux of accumulated material when a second substrate using the same carrier is added (called *counter transport*); by looking for induction or repression of transport systems; or by examining mutants lacking the ability to transport specific substances. These approaches, which will now be explained in more detail, do not distinguish active transport from facilitated diffusion. That problem will be considered later.

Transport Kinetics. The kinetics of membrane transport obey precisely the same rules as do the kinetics of enzymatic catalysis, developed in Chap. 4. One does not, however, need to presume that an enzyme or any other kind of protein is needed for the process, which may be represented as the following reaction:

$$S_{ext} + A \rightleftharpoons AS \rightleftharpoons A + S_{int} \qquad (6\text{-}1)$$

Here S_{ext} and S_{int} are the substrate outside the cell (external) and inside the cell (internal), respectively, and A is a transport receptor site within the cell membrane. Hence the complex AS exists only when substrate is in contact with the membrane.

If we now assume (1) that the concentration of AS is constant (i.e., the rate of binding to the receptor on the outside is the same as the rate of release inside), and (2) that the internal concentration of S is low enough to prevent any significant efflux, then the Michaelis-Menten equation for the rate (velocity, v) of transport may be written as

$$v = \frac{V_{max}}{1 + K_m/[S]_{ext}} \qquad (6\text{-}2)$$

or

$$\frac{1}{v} = \frac{K_m}{V_{max}}\left(\frac{1}{[S]_{ext}}\right) + \frac{1}{V_{max}} \qquad (6\text{-}3)$$

The maximum rate of entry is V_{max}, and K_m is the Michaelis constant for substrate binding on the outside. In other words, K_m represents the external concentration of free substrate necessary to half-saturate the receptor sites, or the concentration necessary to achieve half of the maximum rate of entry.

It is often difficult in practice to collect data sufficiently precise to establish definitely Michaelis-Menten kinetics. However, if it can be done, we have strong evidence for carrier-assisted entry.

A simplified test that is often employed to distinguish Michaelis-Menten kinetics from free diffusion is to look for *saturation*. At very high levels of substrate, Michaelis-Menten transport will level off so that further increases in substrate do not result in appreciably greater rates of uptake. The reason, of course, is because all the carriers are busy and V_{max} is approached. Simple diffusion, on the other hand, takes place at a rate that is always proportional to the difference in concentration between the two sides:

$$v = K_D([S]_{ext} - [S]_{int}) \qquad (6\text{-}4)$$

Here, K_D is a constant equal to the diffusion coefficient (characteristic for a given molecule in a specified environment) times the membrane thickness. Although the rate of free diffusion as indicated by Equation 6-4 does not ordinarily saturate, it can do so in a few special cases when pores are involved, making saturation an unreliable test. A much more sophisticated approach is to look for competition among transported substances.

Competition for Transport. When molecules at moderate concentrations enter a cell by free diffusion, one does not expect them to compete with each other. However, if carrier-mediated transport obeys Michaelis-Menten kinetics, as suggested above, then a limited number of receptor sites should lead to competition between molecules transported by the same carrier. As an example, glucose and 2-deoxyglucose compete with each other, although neither affects the transport of galactose. Since galactose and glucose are isomers, differing only in the position of the hydroxyl at carbon 4, there must be a highly specific receptor for glucose and a different one for galactose. This, of course, is consistent with the concept of a transport protein. It is also the kind of behavior actually observed.

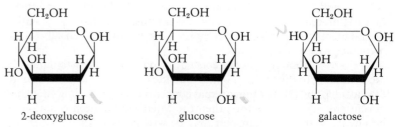

Molecules compete for transport sites in the same way that they compete for the active sites of enzymes. That is, if S is the substance for which transport is being measured, then any other substance I that competes for the same site becomes a competitive inhibitor of S. The velocity of transport of S will be given by

$$\frac{1}{v} = \left(1 + \frac{[I]_{ext}}{K_I}\right)\left(\frac{K_m}{V_{max}}\right)\left(\frac{1}{[S]_{ext}}\right) + \frac{1}{V_{max}} \quad (6\text{-}5)$$

where K_I is the Michaelis constant for the inhibitor. In other words, K_I is the concentration of inhibitor that seems to double the value of K_m so that twice the usual level of substrate is required to reach half-maximum velocity.

If a substrate and inhibitor compete equally well for the same site, then K_m for the substrate should be numerically equal to K_I for the inhibitor. This relationship holds because K_I and K_m are defined in the same terms (see Chap. 4). In the case of transport, the choice as to which substance is labeled "substrate" and which "inhibitor" is often quite arbitrary, for both may be transported equally well. In other cases, the choice is made for very good reasons. For example, it is difficult to measure the uptake of readily metabolized sugars such as glucose and mannose. On the other hand, the nonmetabolizable analog 2-deoxy-D-glucose is readily transported and the rate of uptake, and hence K_m, is easily determined. By adding increasing amounts of glucose or mannose to the external medium during uptake of 2-deoxyglucose by cultured animal cells, competitive inhibition was demonstrated with a K_I for glucose or mannose equal to what the K_m for those substances would have been had they been measured directly (see Fig. 6-18). These three sugars, it appears, are transported by the same system.

There are, of course, instances where one substrate inhibits another's transport without itself being taken up by the cell. In addition, transport is also subject to noncompetitive inhibition, or "poisoning," with kinetics that are predicted by the equations derived for the noncompetitive inhibition of enzymes. In this case, the presence of the inhibitor blocks binding or release of the transportable substrate. Many chemicals that

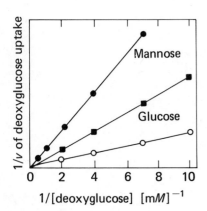

FIGURE 6-18 Competitive Inhibition of 2-Deoxyglucose Transport in Rat Tumor Cells (Novikoff Hepatoma). Deoxyglucose uptake (O) was competitively inhibited by glucose (K_I = 1.3 mM) and by mannose (K_I = 2.8). These values for K_I are about what the corresponding K_m's for glucose and mannose would be if they could be measured, but metabolism of transported glucose and mannose makes that experiment difficult. [Data obtained from E. Renner, P. Plagemann, and R. Bernlohr, *J. Biol. Chem.*, 247: 5765 (1972).]

are capable of altering proteins also act as noncompetitive inhibitors, again leading one to suspect that proteins may be acting as carriers for transported substances.

In general, when two similar but different substances compete for entry to a cell but neither is affected by a third transported substance, there is strong (albeit not conclusive) evidence that two quite different carriers are present. For example, when erythrocytes are studied for their ability to transport amino acids, it becomes clear that several systems are operative. One system favors basic amino acids (those carrying a positive charge on the side chain), one favors acidic amino acids, and there are at least two more systems for neutral amino acids. Although competition can be clearly demonstrated within these groups, there is also overlap, so that each amino acid is transported by two or more systems with one system generally favored under ordinary conditions. In contrast, glucose seems to have only one significant transport system in animal cells.

An overlap among transport systems, as seen with amino acid uptake by red cells, in no way violates our concept of selective transport carriers—consider, for comparison, the range of binding specificities noted for antibodies in Chap. 3. Even in instances of transport overlap, some seemingly insignificant change may render the substance virtually untransportable. Examples would be a D vs. an L amino acid, or galactose vs. its isomer, glucose. This sort of selectivity leads one naturally to consider the existence of transport proteins.

Counter Transport. Competition for a common binding site on a transport protein, in addition to altering the kinetics of uptake produces a phenomenon called *counter transport*, the existence of which is probably the most specific criterion for a carrier mechanism. The same test has also been referred to as *exchange diffusion*, though that term is applied to other situations as well, and hence may lead to some confusion.

Suppose we let a cell accumulate a substance S until equilibrium is reached. Then if a second substance R is added that competes for the same transport system, uptake of R in response to its greater concentration outside the cell fosters a net efflux of S against its own concentration gradient. Initially, the carriers were busy transporting S in and out at the same rate in a dynamic equilibrium. When R is added some of the inward carriers become occupied by it, leaving the outward carriers to continue expelling S, now incompletely replaced by uptake. If S is radioactively labeled, the net drop in internal concentration is readily detected.

If the above argument is not intuitively clear, it may help to put the problem in quantitative terms. We have initially a situation that violates Michaelis-Menten criteria in that the internal concentration of S is significant. In such situations, however, uptake is approximately given as the difference between inward and outward velocities, where both are expressed by the Michaelis-Menten equation (6-2). Thus, net uptake is

$$v = \left\{\frac{V_{max}}{1 + K_m/[S]_{ext}}\right\} - \left\{\frac{V_{max}}{1 + K_m/[S]_{int}}\right\} \quad (6\text{-}6)$$

At equilibrium, there will be no net transport (i.e., $v = 0$) so that the two terms in { } must be equal. (I have avoided saying that $[S]_{ext} = [S]_{int}$ at equilibrium, because that will be true only if K_m for influx and efflux are identical. They will be for facilitated diffusion, but in active transport they need not be, as we shall see later.) Now add the competitive substrate, R, which will reduce the inward velocity. The reduced value is given by Equation 6-5, which may be rewritten as

$$v_{in} = \frac{V_{max}}{1 + K_m/[S]_{ext}(1 + [R]_{ext}/K_R)} \tag{6-7}$$

On the other hand, the outward velocity remains initially unchanged because $[R]_{int} = 0$. Hence, a net outward flow occurs, and will continue until the internal concentration of R builds up to establish a new equilibrium.

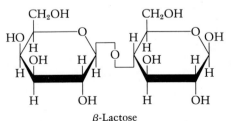

β-Lactose
(a β-galactoside with R = glucose)

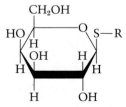

A thio-β-D-galactoside

R = —CH₃
Thiomethyl-β-D-galactoside (TMG)

R = —⌬

Thiophenyl-β-D-galactoside (TPG)

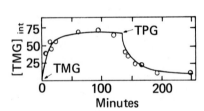

FIGURE 6-19 Counter Transport. *E. coli* was allowed to accumulate TMG at 0 °C. (Concentration expressed in micromoles per gram of cells.) Then TPG was added. Competition for the same binding sites caused TMG efflux. [Adapted from H. V. Rickenberg, G. N. Cohen, G. Buttin, and J. Monod. *Ann. Inst. Pasteur*, **91**: 829 (1956). By permission.]

An application of counter transport is seen in Fig. 6-19. Here, the bacterium *E. coli* was permitted to accumulate thiomethyl-β-D-galactoside (TMG), a nonmetabolized (because of the sulfur linkage instead of oxygen) substrate that is transported by the same system as lactose. The transport protein is a product of the *lac* operon—see Chap. 5. At equilibrium, thiophenyl-β-D-galactoside (TPG) is added causing, by competition, efflux of TMG. This situation, it should be noted, is not as straightforward as it first appears, since galactosides are actively transported, being accumulated by *E. coli* to internal concentrations hundreds of times greater than the external concentration. The demonstration of counter transport, however, works the same.

Thus, counter transport is a powerful tool. If it can be demonstrated, then free diffusion, through pores or otherwise, can be ruled out as a mechanism of uptake. Though there are many variations of this and other tests that could be presented (for an excellent discussion, see the reference by Wilbrandt), let us turn instead to a quite different approach to the study of transport systems, one that uses genetics instead of kinetics.

The Genetics of Transport. Further support for the existence of specific transport systems has come from genetic studies in which mutants have been discovered that are incapable of transporting a given class of substances. For example, mutants of *E. coli* have been found that are fully capable of metabolizing lactose once the sugar is inside the cell, but that cannot transport it inward fast enough to sustain life. These transport-defective cells are called *cryptic mutants*. Such mutants result from mistakes in a single gene. The transport of a family of galactosides may be equally affected by one mutation, while the selective transport of other substances is left unchanged. In general, any class of substances that compete with each other for entry, and that are equally affected by a single genetic mutation, are presumed to be transported with the aid of the same carrier. The carrier would be, at least in part, the direct or indirect product of the gene in question.

Metabolism is regulated at two levels, by controlling the biological activity of proteins such as enzymes, and by controlling their amount via induction or repression (see Chaps. 4 and 5). So it is with transport systems. *E. coli*, for example, has only a few molecules per cell of the galactoside transport protein when no galactosides are present in the growth medium. When lactose or another β-galactoside is added, the cells respond by producing several thousand copies of the appropriate transport protein, each presumably serving as a separate channel for galactoside uptake. This inducibility is, of course, good evidence for the existence of specific carrier mechanisms. In fact, most sugar transport systems in bacteria are inducible, glucose transport being a common exception.

Comparable situations exist in animal cells, though the changes are usually not so dramatic. In addition to induction by substrate, animal cells may also alter transport in response to certain hormones. Insulin, for instance, has a wide variety of metabolic effects, among which is the promotion of glucose uptake. In mouse embryo cells in confluent culture (nongrowing), insulin increases the V_{max} of glucose uptake as much as three- to fivefold without changing the K_m. This effect, which lasts for some hours after washing away the insulin again, indicates an increase in the number of functional transport sites. The effect is not seen in actively growing cells, which already transport glucose at the more rapid rate.

Another example of induction is the observation that vitamin D increases the amount of a protein needed for Ca^{2+} transport in intestines and kidneys of animals and in the eggshell glands of birds. This finding helps explain why vitamin D deficiency leads to poor calcium absorption and increased loss of calcium in the urine, accompanied by the soft, misshapen bones of the child with rickets or the adult with osteomalacia.

Inborn errors of metabolism in humans have also helped to establish the nature of mammalian transport systems. Cystinuria is the best known example. This autosomal recessive disease affects about one person in 7000 and leads to failure to reabsorb from urine the four amino acids cystine, lysine, arginine, and ornithine (an amino acid formed during protein breakdown). Cystine, because it is relatively insoluble, tends to precipitate and form "stones" in the kidney or lower urinary tract, the pain from which brings patients to their doctors. These four amino acids are affected by a single gene, suggesting a common transport system. (The affected system is probably between the kidney cell and blood, rather than between urine and cell as one would expect—see the reference by Segal.) Additional support for a common system comes from the observation that loading the urine of a normal person with lysine results in simultaneous loss of cystine, arginine, and ornithine from his or her urine, presumably because of competitive inhibition. And finally, the intestines of cystinuric patients fail to absorb at normal rates only these four amino acids, pointing again to a common transport system.

These sorts of studies make us more comfortable with the concept of specific transport proteins and help define the degree of their specificity, even though we don't always know the details of how they carry out their function.

6-5 METABOLICALLY COUPLED TRANSPORT

Whenever a cell accumulates a substance to an internal level that is substantially greater than its external level, we are tempted to call the process active transport. We would not always be right, however. When free diffusion was discussed in Section 6-3, it was noted that a cell might accumulate a weak acid or weak base by ion trapping as long as the cytoplasmic pH were higher or lower, respectively, than the external pH. But since differences in pH depend on metabolic activity, ion trapping might be a special case of active transport, or at least of metabolically coupled transport. There is obviously some ambiguity in our terminology, which we shall have to clear up. Suffice it to say for now that ion trapping, while metabolically dependent, is not active transport in the usual meaning of that word, which applies to instances where a substrate moves against an electrochemical gradient, as will be explained.

If a substrate were chemically altered after transport, making it unacceptable to its membrane carrier, facilitated diffusion alone could account for its apparent accumulation. Glycerol is accumulated by *E. coli* in this way, since phosphorylation by the cytoplasmic enzyme glycerol kinase is very efficient. Glucose is taken up by most animal cells in a comparable manner.

Other ways in which accumulation can be accomplished include binding or precipitation of the substance to remove it from solution. The mechanism probably contributes to the uptake of many substrates, particularly some of the inorganic ions, though the importance in most cases is controversial. In skeletal muscle cells, however, accumulation of Ca^{2+} by endoplasmic reticulum is clearly due in part to the formation of a calcium phosphate gel, though active transport also is at work.

It is common, in fact, to find an unequal distribution of ions across any membrane that is semipermeable, which includes all biological membranes. The explanation in many cases is not active transport but a Donnan (or Gibbs-Donnan) equilibrium, described by an English chemist, Frederick Donnan, in 1927.

Donnan Equilibrium. Consider what would happen if a solution of protein and nucleic acids were contained in a cellulose bag with pores large enough to pass salt ions but too small to pass any of the macromolecules. Since nucleic acids and most proteins carry a net negative charge at neutral pH, the solution placed into the bag will always contain enough counter ion, say K^+, to balance the macromolecular charge. If the bag were now dipped into a dilute solution of KCl, eventually an equilibrium would be reached in which the products of diffusible ion concentrations inside would equal the products of the diffusible ion concentrations outside:

$$[K^+]_{in} [Cl^-]_{in} = [K^+]_{out} [Cl^-]_{out} \qquad (6\text{-}8)$$

However, the trapped macromolecules will have a tendency to retain K^+ inside to balance their charge. Hence, $[K^+]_{in} > [K^+]_{out}$. An opposite gradient of chloride must be established to balance that potassium gradient and satisfy the equilibrium condition—i.e., $[Cl^-]_{in}$ must be less than $[Cl^-]_{out}$.

To put this relationship on a quantitative basis, let K^+ and Cl^- each have an initial concentration outside the bag given by C_{out} while the concentration of K^+ and macromolecular anion (call it Pr^-) inside is initially C_{in} each. At equilibrium, electrical neutrality requires that

$$\text{inside:} \quad K^+ = Cl^- + Pr^-$$
$$\text{outside:} \quad K^+ = Cl^-$$

If the final concentration of chloride inside is given by X, and the volume inside and outside is assumed to be equal, Equation 6-8 becomes

$$(C_{in} + X)(X) = (C_{out} - X)(C_{out} - X)$$

which is easily solved for X:

$$X = \frac{C_{out}^2}{C_{in} + 2C_{out}} \tag{6-9}$$

As an example, let KPr be initially present at 10 mM and the outside solution of KCl be 5 mM. At equilibrium, the inside concentrations will be:

$$[Cl]_{in} = X = 1.25; \quad [Pr^-]_{in} = 10; \quad \text{therefore } [K^+]_{in} = 11.25 \text{ m}M$$

The outside concentrations are easily calculated since you now know that 1.25 millimoles of Cl^- per liter and an equal amount of K^+ crossed from outside to inside. The value of $[Cl]_{out}$ must be 3.75 and $[K]_{out}$ the same. Note that the total concentration of dissolved species inside at equilibrium is 22.5 mM while outside it is only 7.5 mM. The significance of that difference will become clear shortly.

This relationship, the Donnan equilibrium, affects the cytoplasmic content of virtually every living cell. When one analyzes the ion distribution across plasma membranes, one frequently reaches the conclusion that Cl^- is distributed according to the Donnan relationship and that the concentration of K^+ is often close to the ratio predicted. In other words, there is usually more K^+ and less Cl^- inside cells than outside. In fact, the ratio across the cell membrane is often 10 to 1 or even more for these ions.

Osmotic Pressure. An important point to note about the Donnan equilibrium is that the *total* molar concentration of solutes inside the bag at equilibrium will be greater than the total molar concentration outside. This imbalance is a natural consequence of the trapping of macromolecules by the semipermeable membrane. Because of this greater concentration of solute inside, water is partially displaced. But since water passes the membrane freely, it will enter the bag as if it were trying to equalize its concentration on the two sides; in the process the internal pressure is increased. At equilibrium the pressure due to dissolved solutes, called the *osmotic pressure*, inside the bag will always be greater than the pressure outside the bag.

The osmotic pressure of a dilute solution is given approximately by the van't Hoff equation (after the Dutch chemist Jacobus van't Hoff). The expression may be written as

$$\pi = RT \,([A] + [B] + \cdots) \tag{6-10}$$

where pi (π) is the common symbol for osmotic pressure; and A, B, etc. are solutes (individual ions or molecules) given usually in molar (more precisely in molal) concentration. This equation has the general form of the ideal gas law, which makes it easy to remember. In applying the

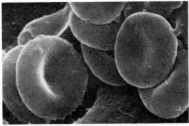
Normal

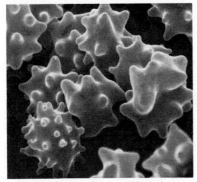
Hypertonic → crenated

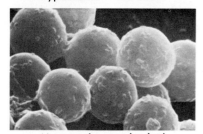
Hypotonic → spherical

FIGURE 6-20 The Effects of Hypertonic and Hypotonic Solutions on Erythrocytes. Hypertonic solutions (solutions of high osmolarity) cause water to leave erythrocytes, resulting in shrunken cells with a plasma membrane thrown into folds (*crenated*). Hypotonic solutions cause water influx and result in spherical cells that may eventually burst (*hemolyse*). The micrographs of erythrocytes illustrate these shapes, though they were achieved in a different way. [Micrographs courtesy of M. Sheetz, R. Painter, and S. J. Singer, *J. Cell Biol.* **70**: 193 (1976).]

van't Hoff equation, keep in mind that it is the total number of solute particles per liter that is important. Thus, for example, one mole of NaCl gives two *osmoles* of solute, $MgCl_2$ gives three, but glucose gives one for one because it does not dissociate.

Two solutions having the same osmotic pressure are said to be *isoosmotic* or *isosmotic*. The cytoplasm of a human red blood cell, for example, is isoosmotic with an external solution of 0.154 M NaCl or 0.33 M glycerol. (One value is not exactly twice the other because we are not dealing with ideal solutions.) The same concentration of NaCl is also *isotonic* with erythrocytes, which means that the cells neither shrink nor swell when placed in that concentration of salt. However, 0.33 M glycerol is not isotonic with human erythrocytes. In fact, it causes the cells to swell and eventually to lyse or undergo a *hemolysis*, the opposite of the unfortunately named plasmolysis. This same concentration of glycerol is, however, isotonic for rabbit or bovine erythrocytes. The difference is that human red cells transport glycerol by facilitated diffusion; the others do not. When glycerol gets into the human red cell, adding its osmotic pressure to that already existing, cellular swelling is the consequence (see Fig. 6-20). Thus, osmolarity is a calculated quantity, but tonicity (whether a solution is "hypertonic"—i.e. too high—"isotonic," or "hypotonic") is an observed property that depends on permeability of the cell membrane.

We said earlier that a Donnan equilibrium typically exists across most cell membranes. If allowed to take its natural course, this factor would result in a higher cytoplasmic concentration of solute molecules and a consequent increase in internal osmotic pressure. This situation cannot be tolerated by animal cells since they have no restraining wall. A hypertonic cytoplasm would lead to cellular swelling and eventual lysis. The same would be true, of course, for those plant and bacterial cells that lack rigid walls.

Cells typically control their internal osmotic pressure, hence their volume, by controlling the cytoplasm's concentration of small ions, especially sodium. In most cases this means the active extrusion of sodium in an energy-linked process, carried out by a membrane protein called the *sodium pump*.

The Sodium Pump. The situation as we have developed it so far is as follows: Because of the trapping of macromolecules inside a cell, there is a tendency for increased internal osmotic pressure and water uptake that, if unopposed, would lead to swelling and lysis of those cells not protected by a rigid cell wall. To counteract that pressure a mechanism has evolved for lowering the sodium concentration and hence the osmotic pressure of the cytoplasm. That mechanism is a membrane protein (or complex of proteins) known as the *sodium pump*. It is readily demonstrated that the sodium pump requires metabolic energy and has the effect on volume regulation just described, for when the cell is poisoned with cyanide or dinitrophenol, water influx takes place and swelling and lysis follow. The pump can also be selectively inhibited by a class of substances known as cardiac glycosides, of which ouabain and the digitalis preparations are examples, with the same result—swelling and probably lysis.[2]

SECTION 6-5
Metabolically Coupled
Transport

Ouabain
(strophanthin G)

The sodium pump was discovered by A. L. Hodgkin and R. D. Keynes in 1955 and associated with ATP hydrolysis (its source of energy) *in vitro* by J. C. Skou in 1957. It was the first documented instance of active transport. The details of its mechanism are not known, but it is clear that it functions by coupling the hydrolysis of ATP to the extrusion of Na^+ from the cytoplasm (Fig. 6-21).

Erythrocytes have been a popular cell type in which to study the sodium pump. Not only are they readily available, but the cytoplasm can be exchanged for solutions of different composition, allowing the effects of various internal conditions to be examined. If erythrocytes are hemolyzed in a small volume of water, the result is a "ghost" cell containing a leaky membrane capable of allowing almost unrestricted pas-

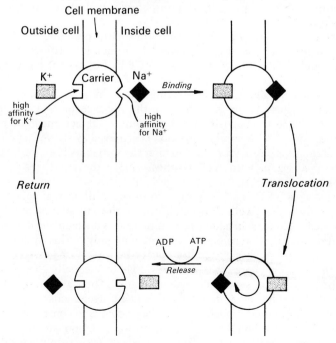

FIGURE 6-21 The Sodium Pump. In this hypothetical model, the carrier (which in this case moves K^+ as well as Na^+) can assume two configurations, one with high affinities for the ions and one with low affinities. The transition is fostered by the hydrolysis of ATP, and lasts until the carrier has returned to its initial position—i.e., with its K^+ binding site on the outside and its Na^+ site on the inside. As a result, Na^+ is pumped out and K^+ is pumped in.

sage of solutions in and out and hence replacement of a portion of the cytoplasm by some other solution. If isotonicity is restored within 15 minutes or so, the membrane reseals and the process called *reverse hemolysis* is complete.

Erythrocytes, after reverse hemolysis, resume energy-requiring membrane activities such as sodium pumping if a source of energy is available in the artificial cytoplasm. ATP and substances that yield ATP are the obvious and very effective choices.

Reverse hemolysis has been used to define many properties of the erythrocyte sodium pump. For example, it is now clear that the erythrocyte sodium pump is actually a *sodium-potassium exchange pump.* For every ATP hydrolyzed three Na^+ are transported out and two K^+ are transported in. The coupling is an obligate one—unless there is external K^+ available for uptake, no Na^+ is transported out. The activity of the pump is stimulated by an increase in internal Na^+ or by an increase in external K^+. But even without stimulation, there is a high level of normal activity, presumably due to a constant leakage of Na^+ inward that must be countered by the pump. The flux amounts to roughly 100 K^+ pumped per minute for each of the 200–300 pumps on a human erythrocyte (estimated from the number of digitalis binding sites).

The vectorial nature of the ion translocations (Na^+ out, K^+ in), and the inside-vs.-outside differences in response to ions, indicate that the *Na^+/K^+-stimulated ATPase,* or sodium pump, spans the erythrocyte membrane with binding sites exposed to both surfaces. The examples isolated so far are consistent with that model, for they are relatively large protein complexes (about 250,000 daltons), easily big enough to span a membrane. Although details of how they work remain unclear, we do know that pumping is a two-step process: (1) the presence of Na^+ fosters phosphoryl transfer from ATP to a membrane protein associated with the pump; (2) if K^+ is present, the phosphoryl is hydrolyzed and transport takes place.

In spite of our ignorance of the details of sodium pumping, many clever experiments have soundly established the principles. For example, the tight coupling between transport and ATP hydrolysis has been demonstrated by establishing erythrocyte ghosts with a high K^+, low Na^+ artificial cytoplasm and then suspending them in a high Na^+, K^+-free medium. These ionic gradients, in the absence of ATP, force the pump to run in the opposite direction from usual, accompanied by the *synthesis* of ATP. A similar situation is achieved by starved (energy depleted) intact erythrocytes suspended in the same medium, for K^+ efflux is observed, again with ATP synthesis. (See the references by I. M. Glynn.)

The Na^+/K^+-ATPase is common to most or all animal cells and to many plant cells as well. It is not, however, the only sodium pump to be identified. There is, in addition, a pump that is ouabain and digitalis insensitive and that translocates sodium ions only. This system is referred to as an *electrogenic* sodium pump because of the greater separation in electrical charge that it fosters. There is equally good evidence for the existence of several other ion pumps. The Ca^{2+} pump is one of these that will be encountered in subsequent discussions.

The sodium pump probably evolved as a device to control water flow across animal cell membranes (i.e., osmotic regulation) and it seems capable of satisfying that need. To see the advantage of osmotic regulation via salt control as opposed to the evolution of a water-impermeable membrane and a molecule-by-molecule membrane transport of water, consider that isotonic saline is approximately 55 M H_2O and only 0.154 M salt. Thus the transport of 0.154 moles of Na^+ causes 55 moles of water to follow passively, a ratio of about 360 water molecules per sodium ion transported. Note that Cl^- penetrates most membranes relatively easily and thus is able to follow Na^+, although in some cases, apparently including the red cell, most of the permeability is due to proteins that exchange chloride for bicarbonate ($Cl^- \rightarrow HCO_3^-$).

Even in organs that are concerned especially with water flow, that flow is fostered by ion pumps. In the kidney, for instance, the amount of water in the urine (i.e., urine concentration or osmolarity) is controlled in part by the permeability of cell membranes responding to antidiuretic hormone as noted earlier. The actual movement of water, however, is in response to local osmolarity gradients and ion flows maintained by ion pumps.

While osmotic regulation is thus an established function of the sodium pump, the activity of the pump creates conditions on which other functions are known to depend. These conditions are low internal sodium, high internal potassium, and an electrical gradient across the cell membrane: High internal K^+ concentration is needed for protein synthesis, as can be demonstrated by experiments *in vitro*; low internal sodium is used to allow the entry of certain other substances (glucose, some amino acids) coupled to sodium influx; and the combination of sodium and electrical gradients created by ion pumps is used to power still other transport processes and to make possible the membrane excitability found in nerves and muscles. (See Chap. 8.) To understand those processes we shall first have to understand membrane potentials.

Membrane Potentials. The Donnan equilibrium originates with the trapping of nondiffusible macromolecules within a semipermeable (or, in the case of living cells, *selectively permeable*) membrane. There are two consequences of this distribution: an increased osmotic pressure inside the membrane, the results of which have already been considered; and the establishment of an *equilibrium potential*, or voltage gradient, across the membrane. The sodium pump corrects the former, so that cells do not swell and lyse, but at the same time the equilibrium potential may be accentuated further.

In the case of living cells, the trapped macromolecules, chiefly proteins and nucleic acids, typically carry a net negative charge—i.e., they are *anions*. As a consequence, the resulting voltage across the membrane is negative inside with respect to the outside. Sodium pumping, by causing a net extrusion of *cations* (positive charges) can only make this electrical gradient steeper. Typical values for living cells would be -60 to -90 mV (millivolts), the negative sign by convention indicating negative inside with respect to the outside surface.

One must appreciate that the existence of a charge separation (voltage gradient) across a membrane does not imply any significant violation of the electroneutrality principle used in developing the Donnan equilibrium. The bulk of the solution inside the cell still has effectively equal numbers of negative and positive charges, and the same is true for the solution outside. Separation of charge exists only in the immediate region of the membrane and involves but a small quantity of ions. Even so, these voltage gradients are capable of drastically altering membrane transport, as we shall see.

The magnitude of the equilibrium potential is easily derived from the principles introduced in Chap. 2. We learned there that free energy (Gibbs free energy) is logarithmically related to concentration so that flow from a concentration $[S]_{ext}$ to $[S]_{int}$ will be given by

$$\Delta G = -2.3RT \log \frac{[S]_{ext}}{[S]_{int}} \quad (6\text{-}11)$$

where R is the gas constant and T the absolute temperature. ($\Delta G°$ is omitted. It is zero where equilibrium results in an equal distribution on the two sides.) When $[S]_{ext} > [S]_{int}$, the free energy change for entry will be favorable ($\Delta G < 0$) as expected. At 25 °C (298 K) the relationship is approximately

$$\Delta G_{entry} = -5700 \log \frac{[S]_{ext}}{[S]_{int}} \text{ joule/mole} \quad (6\text{-}12)$$

Now we must add the complicating factor of voltage gradients and the effects they have on transport of a charged substance. The relationship between electrical energy and Gibbs free energy is given as

$$\Delta G_{entry} = zFE \text{ joule/mole of ion} \quad (6\text{-}13)$$

where z is the ionic charge ($+1$ for Na^+, -1 for Cl^-), F is the Faraday constant (96,500 coulombs, which is the charge on one mole—i.e., Avogadro's number—of electrons) and E is the electrical potential energy given in volts. Note that a negative membrane potential would result in a negative ΔG for entry of a cation, hence spontaneous influx.

An ion at equilibrium across a membrane must, by definition, have a ΔG of zero. In that case, the contribution to free energy from the concentration gradient (Equation 6-12) must be equal and opposite to the contribution from the membrane potential (Equation 6-13). Combining these two terms leads directly to the *Nernst equation* which, for purposes of routine calculations, may be written as

$$E = \frac{60}{z} \log \frac{[S]_{ext}}{[S]_{int}} \text{ mV} \quad (6\text{-}14)$$

Thus, there is a 60 mV change for each 10-fold change in concentration. To take K^+ as an example, a gradient $[S]_{ext}/[S]_{int}$ of 0.1 would be in equilibrium with a -60 mV membrane potential. For chloride, the ratio would have to be $[S]_{ext}/[S]_{int} = 10$.

It is possible to check the validity of the Nernst equation as it applies to living cells by impaling the cells with a tiny electrode to measure the membrane potential, and then determining how that potential changes with changing external concentration of some ion that is present at a known concentration inside the cell. The conclusion from such experiments is that the Nernst equation comes very close to, but does not quite, explain the distribution of those ions whose distribution is not actively disturbed by such things as pumping. Hence, another factor must be at work to account for the discrepancy. It appears that the missing ingredient is the *diffusion potential*.

Suppose NaCl and KCl solutions of equal molar concentrations, initially separated by a vertical partition, are allowed to diffuse into one another. The KCl solution would be negative with respect to the NaCl solution because K⁺ diffuses about half again as fast as Na⁺ (see Chap. 2). Thus, K⁺ leaves one side faster than it is replaced by Na⁺ diffusing in from the other side. The voltage that is generated is called a diffusion potential, and the fact that membranes are more permeable to K⁺ than to Na⁺ (73 times more permeable in the case of frog muscle) leads to the same phenomenon in cells.

A diffusion potential decays with time as equilibrium is approached. Equilibrium potentials, as predicted from the Nernst equation, do not. But we already know that in the living cell there is a continuous transport of ions across the membrane and that equilibrium is prevented by metabolically driven ion pumps. Hence, the diffusion potential is maintained as long as the cell has energy to drive the pumps. Should they fail, eventually a Donnan equilibrium and Nernst potential would be reached—if the cell didn't lyse from increased internal osmotic pressure in the meantime.

For most purposes the diffusion potential, as it applies to membranes, is adequately included in an expression derived in 1943 by D. E. Goldman and known as the *Goldman equation.* Of all the ions that are distributed across the typical cell membrane, only K⁺, Na⁺, and Cl⁻ are quantitatively important, leading to the following form of this equation:

$$E = \left(\frac{2.3\,RT}{F}\right) \log \frac{P_K[K^+]_{ext} + P_{Na}[Na^+]_{ext} + P_{Cl}[Cl^-]_{int}}{P_K[K^+]_{int} + P_{Na}[Na^+]_{int} + P_{Cl}[Cl^-]_{ext}} \qquad (6\text{-}15)$$

Here $(2.3RT/F)$ at 25 °C is approximately 60 mV as before. The quantities denoted by P are *permeability coefficients,* which can be measured by noting the rate at which radioactive ions get into or out of cells at equilibrium. Conversely, if one assumes the Goldman equation to be correct, it can be used to calculate permeabilities. By experimentally replacing Cl⁻ with SO_4^{2-}, the permeability coefficient of which is near zero for most cells, the Goldman equation reduces to

$$E = 60 \log \frac{[K^+]_{ext} + b[Na^+]_{ext}}{[K^+]_{int} + b[Na^+]_{int}} \text{ mV} \qquad (6\text{-}16)$$

where b is the ratio of permeabilities, sodium to potassium. For frog muscle, b has a value of about 0.013. Note that if Na⁺ were also removed from consideration, Equation 6-15 or 6-16 would become identical to the Nernst equation. In other words, the membrane potential would be determined by the K⁺ equilibrium potential alone. The fact that it isn't can be attributed largely to the presence of a sodium flux, maintained by the sodium pump. In fact, in any situation in which the permeability or concentration of one ion greatly exceeds all others, the Goldman equation reduces to the Nernst equation. Hence, the use of the Nernst equation is often justified in practice, even though a cell does not have a true Donnan equilibrium, which is the situation to which it theoretically applies.

Criteria for Active Transport. We have described the sodium pump as an example of active transport. Evidence that sodium translocation is active includes its sensitivity to metabolic poisons, the measured stoichiometry between transport and ATP hydrolysis, and the ATPase properties of the isolated pump enzyme. For most membrane translocations, however, things are not so clear cut and it is not always obvious whether

TABLE 6-2. Ionic Concentrations and Measured Potentials across Frog Skeletal Muscle and Human Red Cell Plasma Membranes

		Na^+	K^+	Cl^-	E
Frog skeletal muscle fibers:	int	13	140	3	-90 mV
	ext	110	2.5	90	
Human red blood cells:	int	19	136	78	-10 mV
	ext	145	5	110	

transport is active or passive. Hence, many criteria have been developed to aid in making that decision.

The clearest indication that transport is active—i.e., depends on a source of energy—is the demonstration that it occurs against an *electrochemical gradient*. In other words, it must be shown that the sum of ΔG due to concentration (Equation 6-12) plus that due to the electrical potential (Equation 6-13) is positive, hence unfavorable.

In the case of the ionic distribution across frog skeletal muscles, the following relationships are calculated from the figures given in Table 6-2 (the subscripts c and e indicate contributions from concentration and electrical gradients, respectively):

$$\Delta G_{entry} = \Delta G_c + \Delta G_e$$
for Na^+ entry: $\Delta G = -5.3 - 8.7 = -14.0$ kJ/mole
for K^+ entry: $\Delta G = +10.0 - 8.7 = +1.3$
for Cl^- entry: $\Delta G = -8.4 + 8.7 = +0.3$

In other words, Na^+ is a very long way from electrochemical equilibrium, with both concentration gradient and electrical gradient favoring entry. Transport of Na^+ out of the cell, then, is clearly against an electrochemical gradient and must require a considerable amount of energy, namely $+14$ kJ/mole. Potassium is slightly out of equilibrium, with efflux favored, and Cl^- is very nearly in electrochemical equilibrium. Of the 3 ions, then, the calculations provide good evidence only for sodium pumping.

There are other criteria that could be applied to establish that a given translocation is active instead of passive. For example, H. H. Ussing, one of the pioneers in the study of active transport, demonstrated Na^+ transport across frog skin when the solutions on the two sides of the membrane were identical in composition and electrically short-circuited. Without a concentration or an electrical gradient, transport must clearly depend on other—i.e., metabolic—sources of energy.[3]

Still another indication for the existence of active transport is the finding of asymmetric Michaelis constants for a translocation. If the apparent Michaelis constant for exit is different from the Michaelis constant for entry, then steady-state conditions (apparent equilibrium) can only be reached when the internal concentration is different from the external. From the Michaelis-Menten equation, we have, as in Equation 6-6,

$$v_{total} = 0 = \frac{V_{max}}{1 + K_{m(ext)}/[S]_{ext}} - \frac{V_{max}}{1 + K_{m(int)}/[S]_{int}} \qquad (6\text{-}17)$$

If the maximum velocity is the same in the two directions (it should be since the same carriers, and hence same number of carriers, are responsible), then $[S]_{ext}/[S]_{int} = K_{m(ext)}/K_{m(int)}$. Although a change in conformation of the carrier protein could account for the difference in K_m, if that change is going to cause work to be done, then it must take energy to make the change. Looking at it this way, the coupling of ATP hydrolysis to active transport is easy to understand: Phosphorylation by ATP could cause a conformational change affecting K_m that is restored when the phosphate is hydrolyzed off again at the end of the pumping cycle.

Na^+-Coupled Transport. The above criteria for active transport help one to decide whether metabolic energy is required for a given translocation, but the consumption of that energy need not be directly and tightly coupled to the translocation.

For example, if a piece of small intestine is removed from an animal, turned inside out, and tied at both ends, the resulting sack will accumulate glucose from the surrounding medium to internal levels so high that there can be no doubt about the existence of some form of active transport. However, as demonstrated by R. K. Crane and others, glucose uptake by the "brush border" cells lining the intestine (as the first step in its passage to the other side—into the everted sack, for instance) is a process that utilizes the sodium gradient as a direct source of energy.

If one were to use a rotating carrier as a model for Na^+-coupled transport—easy to diagram but probably incorrect in detail for reasons discussed earlier—then one would imagine a protein having either two binding sites, one for glucose and one for Na^+, or a single binding site that the two substrates cooperate in filling (see Fig. 6-22). At the surface of the membrane, where the Na^+ concentration is high relative to its

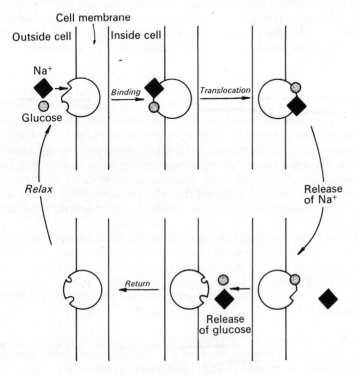

FIGURE 6-22 Sodium-coupled Transport. The favorable electrochemical gradient for Na^+ entry can be utilized to transport a second substance against the latter's own gradient. In this model, binding Na^+ increases the affinity of the carrier for glucose. Both substances are then translocated and released together inside the cell.

concentration inside the cell, the carrier accepts a Na^+ ion that causes it to assume a conformation with high affinity for glucose. When both substrates are bound, the carrier rotates or translates to expose its binding sites to the cytoplasm. Because of the low internal Na^+ concentration, the transported ion is released. This causes the protein to relax to its conformation having a low affinity for glucose, thus releasing that substrate also and freeing the carrier to return for another load.

This model for Na^+-coupled transport results from the observations that glucose transport is Na^+ dependent and that the rate of transport increases with an increasing external concentration of Na^+. Furthermore, a number of monovalent cations, including K^+, compete for the sodium binding site and inhibit the transport of glucose. Competition from K^+ could help force the internal release of glucose, for the cytoplasmic K^+ concentration is relatively high compared to its external concentration (the opposite of the Na^+ gradient). Such a model also predicts that if the transmembrane gradient of Na^+ were reversed, the cell should be able to transport glucose out against a concentration gradient as easily as it normally transports it inward. Crane performed this experiment and reported that the mechanism is, indeed, reversible.

Many other substances (e.g., some amino acids) are not known to be transported by a Na^+-coupled mechanism, and the phenomenon has been observed in many cell types. The process depends on a carrier with variable and controllable affinity for the substance that must be moved against its concentration gradient. Cooperative behavior of this type is normally associated only with proteins.

Sodium-coupled transport fits our criteria for active transport as long as we apply the criteria only to the substance that is cotransported with Na^+. Even though there is no immediate input of metabolic energy, the process would soon come to a halt if sodium were not forcibly extruded from the cytoplasm again by the sodium pump. Thus, in a steady-state situation every three glucose molecules transported, if accompanied by Na^+ on a one-for-one basis, will require one ATP hydrolysis via the sodium pump.

Other kinds of gradients could serve the same purpose and, in fact, H^+-coupled transport appears to take place across bacterial, mitochondrial, and chloroplast membranes. For example, galactoside transport in *E. coli* seems to take place this way. As noted earlier, *E. coli* is capable of accumulating galactosides such as lactose to very high internal concentrations. The influx appears to be obligately coupled to the uptake of H^+ in response to a negative membrane potential and relatively alkaline cytoplasmic pH. If the pH and electrical gradient across the plasma membrane are neutralized, or if the cells are poisoned so that gradients cannot be maintained, the active transport system reverts to facilitated diffusion.

Other sugars in *E. coli* are transported by different systems. The facilitated diffusion of glycerol has already been mentioned, and galactose itself is transported primarily by a system powered directly by ATP hydrolysis. Most of the sugars examined, however, at least in *E. coli* and other facultative organisms, are transported by a system known as *group translocation*.

Group Translocation. In group translocation, phosphorylation of the sugar accompanies transport. However, the phosphate donor is not ATP but phosphoenolpyruvate, PEP. The overall reaction is

$$\text{sugar} + \text{PEP} \rightarrow \text{sugar—P} + \text{pyruvate} \qquad (6\text{-}18)$$

The standard state free energy change for this reaction is extremely favorable, about -40 kJ/mole, because of the properties of PEP outlined in Chap. 4. The needed PEP can come directly from glycolysis or it can be regenerated from pyruvate by an enzyme unique to bacteria and plants called phosphoenolpyruvate synthetase. The regeneration reaction

$$\text{pyruvate} + P_i + \text{ATP} \rightarrow \text{PEP} + \text{AMP} + PP_i \qquad (6\text{-}19)$$

is followed by PP_i hydrolysis to pull the equilibrium to the right. This reaction is specifically mentioned here because we will see in the next chapter that it also plays a very important role in photosynthesis.

Group translocation can be broken down into two steps

$$\text{PEP} + HP_r \xrightarrow{\text{Enzyme I}} \text{pyruvate} + P\text{—}HP_r \qquad (6\text{-}20)$$

$$P\text{—}HP_r + \text{sugar} \xrightarrow{\text{Enzyme II}} \text{sugar—P} + HP_r \qquad (6\text{-}21)$$

HP_r is a small, 9500-dalton protein, a histidine of which gets phosphorylated by PEP. The first reaction is catalyzed by the cytoplasmic protein, Enzyme I. Transfer of phosphate from HP_r to a sugar is catalyzed by Enzyme II, which is a complex, membrane-bound protein. Mutant cells lacking HP_r or Enzyme I fail to transport actively any of the sugars handled by this system, but Enzyme II mutations typically affect the uptake of only one sugar. Hence, Enzyme II appears to be a family of proteins, each specific for a different sugar—one for glucose, one for fructose, and so on. In fact, there is reason to believe that Enzyme II is the carrier protein itself.

To understand the mechanism of group translocation or any other transport system, we should know something about the transport proteins involved. Fortunately, some progress in this task has been made, examples of which follow.

6-6 PROPERTIES OF THE CARRIERS

To better understand the mechanism of carrier-mediated transport, we should like to be able to examine the responsible carriers *in vitro*, and to use them to reconstruct model transport systems.

The Isolation of Transport Proteins. A number of molecules that apparently serve as transport proteins have now been isolated, beginning with the galactoside carrier in the mid-1960s. One of the first successes was achieved by C. F. Fox and E. P. Kennedy, who took advantage of the fact that, in the absence of galactoside, *E. coli* does not manufacture any significant quantity of the three proteins required specifically and solely for lactose metabolism. When a galactoside is added to the culture, however, synthesis of the required proteins begins again. Galactoside permease, the presumed galactoside carrier, is one of the proteins. Its synthesis is inducible—i.e., it is made only in response to need.

Inducibility of the galactoside transport system permitted Fox and Kennedy to look for proteins that were present in induced cultures but absent in uninduced cultures. In addition to the other products of the *lac* operon, they were thus able to identify and purify a membrane-associated protein, which they called the M protein, that may be the sought-after carrier. The M protein was present at about 10,000 copies per cell in the induced state, had a molecular weight of around 30,000 daltons, and in the original cells was apparently exposed to the outside surface as judged by its ability to react with externally applied reagents.

Since then, a number of other proteins have been isolated that probably serve as carriers for membrane transport, but it is difficult to prove that they actually have this function *in vivo* because the appropriate assay (measure of biological activity) does not exist *in vitro*. One can measure their capacity to bind the substance to be transported, but of course a protein does not have to be a membrane carrier to have a specific affinity for a small molecule. Any enzyme, and a host of other proteins, also fit that description.

However, some of these isolated proteins have been shown to be necessary for transport, even if it should turn out that they themselves are not carriers. For example, several proteins have been isolated from osmotically shocked bacterial cells. When bacteria are plasmolyzed in sucrose and then quickly diluted with water, their rapid reexpansion causes some proteins to be released and some transport capacity to be lost. That some of the released proteins participate in transport was inferred from the observations that: (1) mutant cells lacking the capacity to transport a particular substance (cryptic mutants) also fail to release the corresponding binding protein when osmotically shocked; (2) in cases where transport can be induced, uninduced cells fail to release the binding protein; (3) the binding constants between the free proteins and transportable substances have been shown to be virtually identical with the binding constants for transport of these substances in normal cells; and (4) there are reports that incubation of the shocked cells with solutions of some of the released binding proteins restores at least part of a lost transport capacity.

This group of osmotically released proteins includes individual species capable of binding calcium, sulfate, phosphate, various amino acids, galactose, and other substances. They are sometimes called *periplasmic binding proteins* because they appear to reside in the periplasmic space between the cell membrane and the outer membrane that is part of the cell wall in gram-negative bacteria (Fig. 6-23). Their function is apparently not transport as such, but to present substrates to the true carriers. Presumably, the hydrophobic nature of the true carriers prevents them from being so easily removed from the cell. Why bacteria have evolved this extra layer of complexity in so many of their transport systems is not certain, but it probably has to do with the difficulty of getting material through their protective but complex cell walls.

There have also been a number of successes in isolating and characterizing intrinsic membrane proteins that apparently are true carriers. The properties of the Na^+/K^+-activated ATPase have already been mentioned. The function of this protein has been proved by incorporating it into artificial membranes and demonstrating ATP-dependent ion translocation. Similar results have been obtained with Ca^{2+}-activated ATPase from the endoplasmic reticulum of muscle, with a H^+-translocating ATPase from mitochondria, and so on. It is only a matter of time, therefore, before the mechanism of translocation is understood in more detail.

The Advantages of Protein Carriers. From the foregoing observations and arguments, we must conclude that carrier-mediated transport across

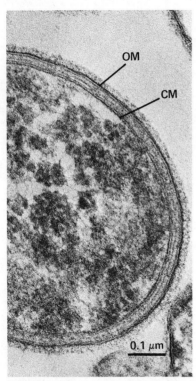

FIGURE 6-23 Periplasmic Space. Gram-negative bacteria have a lipid bilayer membrane within their complex cell wall, as readily seen in this micrograph of *Bacteroides fragilis*, a common intestinal anaerobe. The periplasmic binding proteins may reside between this outer membrane (OM) and the true cell membrane (CM). [Courtesy of D. L. Kasper, Harvard Medical School. See *J. Infec. Dis.* 133: 79 (1976).]

cell membranes exists and that at least some of the carriers involved are proteins. One might, however, wonder why proteins should be chosen for this role, since they are relatively large and seemingly "expensive" to make. There are, of course, very good reasons for that choice:

(1) Proteins are specific in their capacity to bind other molecules. They may be given exactly the right shape, with precisely the right distribution of positive and negative charges, of hydrophobic and hydrophilic regions, and of reactive groups of various sorts to allow them to make fine distinctions between those molecules that will be bound and those that will not. This specificity can be of great importance to the cell, for a close regulation of metabolic efficiency can only be achieved if the entry and exit of molecules are effectively controlled. Thus, it is possible for the cell to provide a mechanism for a particular molecule to get in and out while in the presence of much larger quantities of other kinds of molecules. Furthermore, the rate of transport is easily controlled by regulating the number of specific carriers irrespective of the rate of transport of other molecules gaining entry by their own mechanisms.

(2) The capacity of proteins to alter their binding affinity for substrate has some important applications when this function is present in a transport molecule. In Na^+-coupled transport, for instance, we have a requirement for cooperativity; specifically, affinity of the carrier for glucose at its "active" site is dependent on the presence of sodium ion at its "regulatory" site.

(3) Proteins, as catalysts, also make active transport possible by allowing an obligate coupling between two reactions—energetically unfavorable transport with an energetically favorable reaction such as the hydrolysis of ATP. The sodium pump, for example, depends on this kind of coupling.

And finally, it should be pointed out that the use of proteins for transport is actually a conservative choice, for they are the direct products of genes. When small organic molecules are chosen as carriers, their synthesis is apt to require not just one, but several proteins in the form of enzymes. That is, the production of a protein carrier may require only one gene, whereas establishing a biosynthetic pathway to produce any other kind of carrier might require several enzymes, and therefore several genes.

Nonprotein Carriers. The suitability and economy of using proteins as carriers does not preclude the use of other organic molecules, for we may easily imagine transport tasks for which proteins are not ideal.

Only a few small molecular weight carriers have been isolated. This is not surprising, as they would be present in only tiny quantities in the cell and would therefore be difficult to identify. Much of the work on small organic transport molecules has centered on the ionophorous (ion-carrying) antibiotics. These are generally macrocyclic (ringlike) compounds, produced by microorganisms and capable of sequestering inorganic ions. By virtue of its lipid solubility, the ion-antibiotic complex can diffuse through a lipid barrier that would be quite impermeable to the ion itself.

One of the more widely studied ionophorous antibiotics is *valinomycin* (Fig. 6-24). It is a 36-atom ring of alternating hydroxy and amino acids, consisting of the sequence [D-valine-D-hydroxyisovalerate-L-valine-L-lactate] repeated three times.

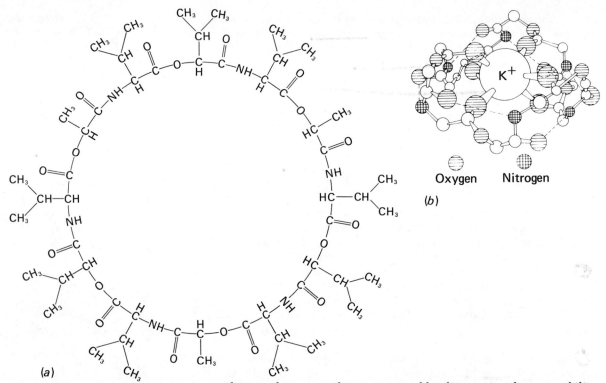

FIGURE 6-24 Valinomycin. (a) Chemical structure of valinomycin. (b) Three-dimensional representation based on X-ray crystallography. Dashed lines indicate possible hydrogen bonds. The side chains, which would point either straight up or straight down, are omitted. Note that K^+ is held in the 4.5 Å diameter pore by bonds to six oxygens. [Adapted from M. Pinkerton, L. K. Steinrauf, and P. Dawkins, (*Biochem. Biophys. Res. Commun.*, **35**: 512 (1969).]

The ionophorous antibiotics are capable of increasing the permeability of both natural and artificial lipid membranes to small inorganic ions. They often exhibit remarkable selectivity in this role. For instance, valinomycin has about a 10,000 to 1 preference for K^+ over Na^+, an observation that is not easily explained on the basis of ionic size.

The biological action of the ionophorous antibiotics has been studied with model systems such as simple chloroform barriers in which ordinary salts (e.g., KCl) are quite insoluble. In one such experiment, chloroform was placed in the bottom of a U-tube with a KCl solution above it in one arm and water above it in the other (Fig. 6-25). Valinomycin, which is capable of transporting a cation-anion pair, was added to the chloroform. Although potassium is the cation favored by valinomycin, K^+Cl^- is not a suitable ion pair for the antibiotic. However, picrate, Pc^- or 2,4,6-trinitrophenol, was found to be transportable with potassium. Thus, when potassium picrate was added to KCl solution on one side, K^+Pc^- ion pairs were transported across the chloroform by diffusion and concentrated on the other side until a Donnan equilibrium was reached:

$$[K^+]^L[Pc^-]^L = [K^+]^R[Pc^-]^R \qquad (6\text{-}22)$$

Here Pc^- is the picrate anion, and L and R stand for the left and right arms of the U-tube, respectively. This equilibrium relationship was obeyed regardless of the starting concentrations of potassium picrate and potassium chloride on the two sides. The Cl^- is not picked up by valinomycin and is thereby effectively ignored when the equilibrium is established.

Because valinomycin itself is uncharged, it must move both a cation and an anion in order to maintain neutrality. (Charged species, of course, are generally less soluble in lipids.) Other ionophorous antibiotics are charged and may therefore transport a single ion, promoting an ion exchange (e.g., Na^+ for K^+, H^+ for K^+, etc.) or creating a voltage gradient, in which case the equilibrium position will follow the Nernst equation (6-14). The principle is the same, however: The antibiotic alone, or the complex of antibiotic and ion(s), is soluble and can therefore diffuse across the lipid barrier. An ion or ion pair is picked up on one side, according to the equilibrium constant between it and the antibiotic, and dropped on the other side according to the equilibrium prevailing there.

Most of the ionophorous antibiotics are products secreted by microorganisms to protect themselves against other microorganisms. They kill by disrupting normal, essential ionic gradients. There are also a few nonprotein carriers that may be secreted but are designed for transport within the cell from which they arise. The most closely studied is a family of bacterial iron-binding substances known collectively as *siderochromes* or *siderophores*. They are typically secreted from the cell, bind iron, and in that state diffuse back across the bacterial membrane. This may seem inefficient, since obviously many of the carrier molecules will be lost, but in fact bacteria have an absolute requirement for iron, which is typically scarce and for which there is competition from the host. In vertebrates, for instance, iron in the blood is transported tightly bound to a protein called transferrin and therefore is not readily available. Thus, the bacteria are fighting for survival and produce siderochromes at a rate proportional to their need.

We shall have occasion in several future chapters to refer again to the ionophorous antibiotics and their relatives, for they have served as useful tools in elucidating various membrane-dependent bioenergetic processes.

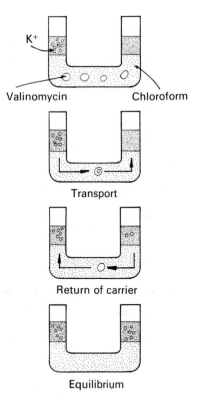

FIGURE 6-25 A Model System for Facilitated Diffusion. Valinomycin can transport a K^+/anion pair through the chloroform barrier until an equilibrium described by Equation 6-22 is attained.

SUMMARY

6-1 Membranes behave more like a fluid than a layer of rigidly connected subunits. We see this in the redistribution of surface features when two cells fuse and in the ability of most membranes to seal themselves after a small puncture. The basis for this behavior seems to be a bilayer of phospholipid in which "float" proteins of various sorts; this is the fluid mosaic model of membrane structure. The fluid nature of membranes results (at temperatures above the phase transition) in lateral movement of both lipid and protein. There is, however, very little interchange between bilayers, leading to marked differences in the composition of the two surfaces.

6-2 Adjacent cell membranes sometimes cooperate to form junctions, which are ordered arrays of specialized membrane proteins. There are three common types of junctions (and many less common ones) called desmosomes, tight junctions, and gap junctions. Desmosomes, with their cytoplasmic filaments and intercellular "glue," add strength to tissues. Tight junctions have a characteristic fusing of membranes, hence seal the intercellular space against leakage. Gap junctions are unique in that cooperating membrane proteins are arranged to provide cytoplasm-to-cytoplasm channels between cells, permitting intercellular communication via ions or small molecules.

6-3 Membranes present a lipid barrier to the passage of most substances. That is, substances penetrate cells with a facility determined in large part by their lipid solubility. In the case of weak acids and bases, which exist as equilibrium mixtures of an ionized (lipid insoluble) and nonionized (lipid soluble) form, only the nonionized form is able to pass through membranes. Hence, only this form will come to equilibrium with equal concentrations on both sides. The ionized form is trapped, leading in some cases to enormous concentration gradients. An apparent exception to the lipid barrier is made for some substances. One explanation for these exceptions is the existence of pores furnished (probably) by membrane proteins.

6-4 Many substances get in and out of cells at rates much faster than can be accounted for by free diffusion or by diffusion through small pores. In addition, cells typically exercise great selectivity in the passage of water-soluble substances and discriminate on the basis of relatively trivial differences in molecular structure. These observations are explained by the existence of carrier-mediated transport, usually involving membrane proteins. Carrier-mediated transport exhibits Michaelis-Menten kinetics, including saturability as V_{max} is approached, and both competitive and noncompetitive inhibition. It also exhibits counter transport, where efflux of one substrate is driven by entry of a second as long as the same carrier is used. And finally, carrier-mediated transport systems are in many cases inducible, indicating genetic regulation of the responsible proteins.

6-5 Cells have selectively permeable membranes that trap macromolecules inside. Most of these trapped molecules carry a net negative charge and so require a counter ion such as K^+ in order to maintain electrical neutrality. Since the counter ion is diffusible, permitting the system to come to equilibrium would result in (1) a Donnan equilibrium having a net excess of solute particles inside the cell, hence an increased internal osmotic pressure; and (2) a negative membrane potential described by the Nernst equation. To control the internal osmotic pressure, ion pumps actively extrude sodium. In the process, however, the resulting ion flow produces a diffusion potential that contributes further to the membrane potential. The ion pumps also indirectly power other translocations, including the Na^+-coupled transport of glucose into intestinal cells. This system takes advantage of the low internal sodium concentration and highly favorable electrochemical gradient for Na^+ entry in order to cotransport glucose against its own gradient. Sodium-coupled transport thus relies on metabolically supplied energy, but only indirectly, in contrast to the direct utilization in transport coupled to ATP hydrolysis or, as in group translocation, the hydrolysis of another high energy compound such as phosphoenolpyruvate.

6-6 The existence of transport proteins has been demonstrated directly by their isolation and incorporation into reconstituted or artificial membranes with resulting resumption of function. The properties of proteins make them uniquely suitable choices for transport functions. These properties include their specificity; the ease with which their activity can be regulated; their adaptability to complex functions such as cotransport of two substrates; and the fact that they are direct products of genes so that special metabolic pathways need not be set up for their synthesis. Nevertheless nonprotein carriers also exist. The ionophorous antibiotics —carriers capable of transporting ions across membranes, with sometimes fatal results for the cell— have been the most closely studied. Interest in them stems from their usefulness in experimental systems where one wishes to affect ionic gradients across membranes.

STUDY GUIDE

6-1 (a) Define the terms somatic hybrid and heterokaryon. (b) What is meant by a "lipid bilayer"? (c) Describe the fluid mosaic model and some of the more important observations that led to it. (d) Given what you know now about the structure and properties of membrane proteins and lipids, can you explain why there should be little or no movement of these components from one bilayer to the other? (Hint: Consider hydrophilic and hydrophobic features and what that means for solubility.) (e) How and why do cholesterol and unsaturated fatty acids affect the phase transition temperature of a membrane?

6-2 (a) Describe the appearance and characteristic arrangement of membrane proteins in desmosomes, tight junctions, and gap junctions. (b) What is the presumed function of each of these junctions? (c) When a cell is damaged, what mechanisms protect adjacent cells by keeping them from leaking cytoplasmic contents through gap junctions into the damaged cell?

6-3 (a) What is plasmolysis and how can it be used to define the permeability of a membrane? (b) What is the usual relationship between oil/water partition coefficients and membrane permeability? What is the reason for that relationship? (c) A weak acid has a pK_a of 5.4. If the cytoplasmic pH is 7.4 and the extracellular pH 8.4, will that acid penetrate cells faster or slower than a second weak acid of similar molecular weight but with a pK_a of 6.4? (d) Using again the weak acid having pK_a of 5.4 determine the concentration ratios, inside to outside, of the acid form and the ion form of the substance at equilibrium. [Ans: 1:1; 1:10] (e) What evidence exists for pores in erythrocyte membranes, and how are they probably constructed?

6-4 (a) What properties of proteins lead us to expect their involvement in carrier-mediated transport? (b) Demonstrate that at very low concentrations of substrate (i.e., when $K_m/[S] \gg 1$) Equation 6-2 becomes equivalent to the equation for free diffusion. (c) In trying to decide whether a given uptake was carrier-assisted, what properties would you look for? (**d**) The following data were obtained for lactose uptake by *E. coli*:

Velocity of uptake μmol/g of cells per min	[S] millimolar (mM)
20	0.01
35	0.02
57	0.04
84	0.08
110	0.16

Find the K_m and V_{max} for uptake. [Ans: 0.07 mM, 160 μmol/g/min] (**e**) Lactose accumulation is subject to competitive inhibition by thiomethylgalactoside (TMG). The measured value of K_I for TMG is 0.5 mM. At 0.1 mM lactose, what will be the rate of lactose accumulation with and without 1.0 mM TMG? (Use the K_m and V_{max} for lactose given in part (d).) [Ans: 52 and 94 μmol/g/min]

6-5 (**a**) What is a Donnan equilibrium and how does it arise? (**b**) When a cell is poisoned, so that it cannot make ATP, it usually swells and breaks. Why? (**c**) What is the difference between a resting potential (Nernst potential) and a diffusion potential (Goldman potential)? Under what conditions are they the same? (**d**) To determine if a given translocation is active transport, what criteria would you apply? (**e**) Describe transport via group translocation. (**f**) What is an "electrogenic" sodium pump? (**g**) Lactose can be accumulated by *E. coli* to an internal concentration. That is about 2000 times its external concentration. What is the *standard state* free energy change for this process at 25 °C? (Note that accumulation can be treated like any other reaction.) [Ans.: -19 kJ/mole] (**h**) The acidity of your stomach, which is about pH 2, is a result of proton pumps in the surrounding cells. Assuming that the pH of cell protoplasm is 7, what is the standard state free energy change for this translocation? Is one ATP/H$^+$ transported likely to be adequate? (Neglect voltage gradients and assume normal body temperature of 37 °C.) [Ans: 30 kJ/mole] (**i**) What membrane potential would the H$^+$ gradient used in part (h) balance, neglecting all other factors? [Ans: about $+300$ mV] (**j**) Assuming that a cell maintains the same Na$^+$ gradient as frog muscle (Table 6-2), what would be the maximum equilibrium concentration ratio (inside/outside) of a substance that is accumulated by sodium-coupled transport? [Ans: about 8.5:1]

6-6 (**a**) What is a periplasmic binding protein? (**b**) What properties of proteins make them uniquely suited for roles as membrane carriers? (**c**) What is a siderochrome and why do bacteria bother making them? (**d**) What is an ionophorous antibiotic, and why are they of interest in the study of transport? (**e**) A 0.05 M KCl solution was put in the left arm of a U-tube (Fig. 6-25), and the same volume of 0.10 M KCl on the right. Then enough potassium picrate (KPc) was added to the right side to make its final concentration $2.1 \times 10^{-5} M$. With a few micromoles of valinomycin in the chloroform phase equilibrium was reached in about 5 days. The observed picrate concentration was then 14 μM on the left and 6.8 μM on the right. Theoretically, what should it have been? (Note that the measured and theoretical values agree closely, and that the picrate concentration is too low to make any significant change in the K$^+$ concentration.) (**f**) If the carrier in part (e) were to transport K$^+$ without a picrate counter ion, would the total amount of K$^+$ moved be greater or less? Why?

REFERENCES

MEMBRANES: GENERAL REFERENCES

See also more recent reviews in *Biomembranes, Current Topics in Membranes and Transport,* and *Biochimica et Biophysica Acta's Reviews on Biomembranes.*

BAKER, D. A., and J. L. HILL, eds., *Ion Transport in Plant Cells and Tissues.* New York: American Elsevier, 1975. See particularly Chap. 1 on ion transport and Chap. 2 on membrane structure.

BRETSCHER, M., and M. RAFF, "Mammalian Plasma Membranes." *Nature,* **258**: 43 (1975).

FENOGLIO, C., C. BOREK, and D. W. KING, eds., *Cell Membranes. Structure, Receptors, and Transport.* New York: Stratton, 1975. (Paperback.) Advances in Pathobiology, Vol. 1.

FINEAN, J. B., R. COLEMAN, and R. H. MITCHELL, *Membranes and Their Cellular Functions.* New York: Wiley, 1975. A small, easily read introduction.

FOX, C. F., ed., *Biochemistry of Cell Walls and Membranes.* Baltimore: Univ. Park Press, 1975. MTP International Review of Science Series.

HARRISON, R., and G. LUNT, *Biological Membranes: Their Structure and Function.* New York: Halsted Press, 1976. (Paperback.)

JAMIESON, G., and D. ROBINSON, eds., *Mammalian Cell Membranes.* Woburn, Mass.: Butterworth, 1976 and 1977. Five volumes. General reviews in Vol. I.

NYSTROM, R. A., *Membrane Physiology.* Englewood Cliffs, N.J.: Prentice-Hall, 1973. The structure, transport, and excitability of biological membranes.

PARSONS, D. S., ed., *Biological Membranes: Twelve Essays on Their Organization, Properties, and Functions.*

London: Clarendon (Oxford University Press), 1975. An intermediate level introduction.
QUINN, P. J., *The Molecular Biology of Cell Membranes.* London: Macmillan Press, 1976. Emphasis on model systems.
SAIER, J. H., Jr., and C. D. STILES, *Molecular Dynamics in Biological Membranes.* New York: Springer-Verlag, 1975. (Paperback.)
WALLACH, D., *The Plasma Membrane: Dynamic Perspectives, Genetics, and Pathology.* (Heidelberg Science Library, Vol. 18.) New York: Springer-Verlag, 1973. (Paperback.)
WEISSMANN, G., and R. CLAIBORNE, eds., *Cell Membranes: Biochemistry, Cell Biology, and Pathology.* New York: H. P. Publ., 1975. Excellent reviews reprinted from *Hospital Practice.* Highly recommended.

MEMBRANE STRUCTURE

BAR, R. S., D. DEAMER, and D. CORNWELL, "Surface Area of Human Erythrocyte Lipids: Reinvestigation of Experiments on Plasma Membrane." *Science,* **153**: 1010 (1966).
BERNHEIMER, A. W., "Interactions Between Membranes and Cytolytic Bacterial Toxins." *Biochim. Biophys. Acta,* **344**: 27 (1974). A review.
BRANTON, D., "Fracture Faces of Frozen Membranes." *Proc. Nat. Acad. Sci. U.S.,* **55**: 1048 (1966). Classic paper on freeze-fracture.
BRANTON, D., and D. DEAMER, *Membrane Structure.* New York: Springer-Verlag, 1972.
BRANTON, D. and R. PARK, *Papers on Biological Membrane Structure.* Boston: Little, Brown & Co., 1968. (Paperback.) An excellent introduction, plus reprinted papers by earlier workers in the field.
BULLIVANT, S., "Freeze-etching Techniques Applied to Biological Membranes." *Phil. Trans. Roy. Soc. London, Ser. B.,* **268**: 1 (1974).
CAPALDI, R., "A Dynamic Model of Cell Membranes." *Scientific American,* **230**(3): 26 (March 1974). (Offprint 1292).
CHAPMAN, D., "Fluidity and Phase Transitions of Cell Membranes." *Biomembranes,* **7**: 1 (1975). Also see *Quart. Rev. Biophys.,* **8**: 185 (1975).
CHERRY, R. J., "Protein Mobility in Membranes." *FEBS Letters,* **55**: 1 (1975).
DEMEL, R., and B. DE KRUYFF, "The Function of Sterols in Membranes." *Biochim. Biophys. Acta.,* **457**: 109 (1976). A review.
EDIDIN, M., Y. ZAGYANSKY, and T. LARDNER, "Measurement of Membrane Protein Lateral Diffusion in Single Cells." *Science,* **191**: 466 (1976).
EISENBERG, M., and S. MCLAUGHLIN, "Lipid Bilayers as Models of Biological Membranes." *BioScience,* **26**: 436 (1976). Good discussion of structure and noncarrier mediated permeability.
EISENMAN, G., ed., *Membranes. A Series of Advances.* Vol. 3. *Lipid Bilayers and Biological Membranes: Dynamic Properties.* New York: Dekker, 1975.
EPHRUSSI, B., and M. C. WEISS, "Hybrid Somatic Cells." *Scientific American,* April 1969. (Offprint 1137.)
FOX, C. F., and A. KEITH, eds., *Membrane Molecular Biology.* Stamford, Conn.: Sinauer Associates, Inc., 1972. Composition, isolation, properties, and reassembly.

FOX, C. F., "The Structure of Cell Membranes." *Scientific American,* February 1972. (Offprint 1241.)
GORDESKY, S., "Phospholipid Asymmetry in the Human Erythrocyte Membrane." *Trends Biochem. Sci.* **1**(9): 208 (Sept. 1976).
GULIK-KRZYWICKI, T., "Structural Studies of the Associations Between Biological Membrane Components." *Biochim. Biophys. Acta,* **415**: 1 (1975). Lipid and lipid-protein arrangement.
ITO, S., "Structure and Function of the Glycocalyx." *Fed. Proc.,* **28**: 12 (1969). Also see *J. Cell Biol.,* **27**: 475 (1965).
JAIN, M. K., "Role of Cholesterol in Biomembranes and Related Systems." *Current Topics in Membranes and Transport,* **6**: 1 (1975).
LENAZ, G., G. CURATOLA, and L. MASOTTI, "Perturbation of Membrane Fluidity." *J. Bioenergetics,* **7**: 223 (1975). An extensive review of the effects of temperature, ionophores, and anesthetics.
LUCY, J., "Cell Fusion." *Trends Biochem. Sci.* **2**(1): 17 (Jan. 1977).
LUFTIG, R., E. WEHRLI, and P. MCMILLAN, "The Unit Membrane Image: A Re-Evaluation." *Life Sci.* **21**: 285 (1977).
MADDY, A. H., ed., *Biochemical Analysis of Membranes.* London: Chapman and Hall; New York: Halsted, 1976.
MARCHESI, V., H. FURTHMAYR, and M. TOMITA, "The Red Cell Membrane." *Ann. Rev. Biochem.,* **45**: 667 (1976).
MELCHIOR, D. L., and J. M. STEIN, "Thermotropic Transitions in Biomembranes." *Ann. Rev. Biophys. Bioeng.,* **5**: 205 (1976). Phase transitions.
OVERATH, P., L. THILO, and H. TRAUBLE, "Lipid Phase Transitions and Membrane Function." *Trends Biochem. Sci.,* **1**(8): 186 (Aug. 1976).
POSTE, G., "Mechanisms of Virus-Induced Cell Fusion." *Int. Rev. Cytol.* **33**: 157 (1972).
POSTE, G., and G. NICOLSON, eds., *Dynamic Aspects of Cell Surface Organization (Cell Surface Reviews,* Vol. 3). New York: Elsevier/North Holland, 1977.
ROSEMAN, S., "Sugars of the Cell Membrane." *Hospital Practice,* **10**(1): 61 (Jan. 1975).
ROTHMAN, J., and J. LENARD, "Membrane Asymmetry." *Science,* **195**: 743 (1977).
SALTON, M., and P. OWEN, "Bacterial Membrane Structure." *Ann. Rev. Micro.,* **30**: 451 (1976).
SINGER, S. J., and G. L. NICOLSON, "The Fluid Mosaic Model of the Structure of Cell Membranes." *Science,* **175**: 720 (1972). The now classic paper.
VERKLEIJ, A., and P. VERVERGAERT, "The Architecture of Biological and Artificial Membranes as Visualized by Freeze Etching." *Ann. Rev. Phys. Chem.,* **26**: 101 (1975).
ZINGSHEIM, H. P., "Membrane Structure and Electron Microscopy." *Biochim. Biophys. Acta,* **MR1**: 339 (1972).
ZWAAL, R., et al., "The Lipid Bilayer Concept of Cell Membranes." *Trends Biochem. Sci.,* **1**(5): 112 (May 1976). Historical review.

INTERCELLULAR JUNCTIONS

FRIEND, D. S., and N. B. GILULA, "Variations in Tight and Gap Junctions in Mammalian Tissues." *J. Cell Biol.,* **53**: 758 (1972).

GILULA, N. B., "Junctions Between Cells." In *Cell Communication*, ed. R. P. Cox. New York: John Wiley, 1974.

GOODENOUGH, D. A., "The Structure and Permeability of Isolated Hepatocyte Gap Junctions." *Cold Spring Harbor Symp. Quant. Biol.*, **40**: 37 (1976).

LOEWENSTEIN, W. R., "Permeable Junctions." *Cold Spring Harbor Symp. Quant. Biol.*, **40**: 49 (1976).

PAPPAS, G., "Junctions Between Cells." In *Cell Membranes*, G. Weissmann and R. Claiborne, eds. New York: H.P. Publ., 1975, p. 87. Also see *Hospital Practice*. Aug. 1973, p. 39.

PERACCHIA, C., "Gap Junction Structure and Function." *Trends Biochem. Sci.*, **2**(2): 26 (Feb. 1977).

ROSE, B., and W. R. LOEWENSTEIN, "Permeability of Cell Junctions Depends on Local Cytoplasmic Calcium Activity." *Nature*, **254**: 250 (1975).

STAEHELIN, L. A., "Structure and Function of Intercellular Junctions." *Int. Rev. Cytol.*, **39**: 191 (1974).

WEINSTEIN, R. S., F. B. MERK, and J. ALROY, "The Structure and Function of Intercellular Junctions in Cancer." *Advan. Cancer Res.*, **23**: 23 (1976). Largely devoted to reviewing the various types of normal junctions.

MOLECULAR TRANSPORT

ANDREWS, K., and E. LIN, "Selective Advantages of Various Bacterial Carbohydrate Transport Mechanisms." *Fed. Proc.*, **35**: 2185 (1976).

BAKER, A., and J. L. HALL, eds., *Ion Transport in Plant Cells and Tissues*. Amsterdam: North-Holland; New York: Elsevier, 1975.

BRADLEY, W. E., and L. A. CULP, "Stimulation of 2-Deoxyglucose Uptake in Growth-Inhibited BALB/C 3T3 and Revertant SV40-Transformed 3T3 Cells." *Exp. Cell Res.*, **84**: 335 (1974). Insulin effect on glucose transport.

BRONK, J. R., ed., *Membrane Adenosine Triphosphatases and Transport Processes*. London: The Biochemical Society, 1974.

CHRISTENSEN, H. N., "Recognition Sites for Material Transport and Information Transfer." *Current Topics in Membranes and Transport*, **6**: 227 (1975). Selectivity of transport systems.

CHRISTENSEN, H. N., *Biological Transport*. (2d ed.) Reading, Mass.: W. A. Benjamin, 1975.

CLAUSEN, T., "The Effect of Insulin on Glucose Transport in Muscle Cells." *Current Topics in Membranes and Transport*, **6**: 169 (1975).

CORDARO, C., "Genetics of the Bacterial Phosphoenolpyruvate: Glycose Phosphotransferase System." *Ann. Rev. Genet.*, **10**: 341 (1976).

CRANE, R. K., "Structural and Functional Organization of an Epithelial Cell Brush Border." In *Intracellular Transport (Int. Symp. Cell Biol.*, **5**), edited by K. B. Warren. New York: Academic Press, 1966, p. 71.

DIAMOND, J., and W. BOSSERT, "Standing-Gradient Osmotic Flow: A Mechanism for Coupling of Water and Solute Transport in Epithelia." *J. Gen. Physiol.*, **50**: 2061 (1967). Where salt goes, water follows.

DIETZ, G. W., Jr., "The Hexose Phosphate Transport System of *Escherichia coli*." *Advan. Enzymol.*, **44**: 237 (1976). Sugar phosphates *are* transported, contrary to early assumptions.

EDELMAN, G. M., ed., *Molecular Machinery of the Membrane*. Cambridge, Mass.: MIT Press, 1975.

ELBRINK, J., and I. BIHLER, "Membrane Transport: Its Relation to Cellular Metabolic Rates." *Science*, **188**: 1177 (1975).

GARRAHAN, P. J., and I. M. GLYNN, "The Incorporation of Inorganic Phosphate into Adenosine Triphosphate by Reversal of the Sodium Pump." *J. Physiol. (London)*, **192**: 237 (1967). A classic experiment.

GLYNN, I. M., and S. KARLISH, "The Sodium Pump." *Ann. Rev. Physiol.*, **37**: 13 (1975).

GLYNN, I. M., and V. L. LEW, "Synthesis of Adenosine Triphosphate at the Expense of Downhill Cation Movements in Intact Human Red Cells." *J. Physiol. (London)*, **207**: 393 (1970).

GUIDOTTI, G., "The Structure of Membrane Transport Systems." *Trends Biochem. Sci.*, **1**(1): 11 (Jan. 1976).

HARRIS, E. J., ed., *Transport and Accumulation in Biological Systems* (3d ed.). London: Butterworth, 1972.

HEPPEL, L. A., "Selective Release of Enzymes from Bacteria." *Science*, **156**: 1451 (1967). By osmotic shock.

HILL, A. E., "Solute-Solvent Coupling in Epithelia: An Electro-Osmotic Theory of Fluid Transfer." *Proc. Roy. Soc. (London), Ser. B.*, **190**: 115 & 537 (1975). Alternative to the standing gradient model.

JAIN, M. K., A. STRICKHOLM, and E. H. CORDES, "Reconstitution of an ATP-mediated Active Transport System Across Black Lipid Membranes." *Nature*, **222**: 871 (1969).

KIMMICH, G. A., "Coupling Between Sodium and Sugar Transport in Small Intestine." *Biochim. Biophys. Acta*, **300**: 31 (1973). Review, including contradictory findings.

KINNE, R., "Properties of the Glucose Transport System in the Renal Brush Border Membrane." *Current Topics in Membranes and Transport*, **8**: 209 (1976). The kidney uses Na^+-coupled transport to remove sugar from the urine and facilitated diffusion on the other side of the cell to transfer it to lymph.

KOLATA, G., "Water Structure and Ion Binding." *Science*, **192**: 1220 (1976). Controversial aspects of ion pumps.

KORENBROT, J., "Ion Transport in Membranes: Incorporation of Biological Ion-Translocating Proteins in Model Membrane Systems." *Ann. Rev. Physiol.*, **39**: 19 (1977).

KORNBERG, H. L., "The Nature and Control of Carbohydrate Uptake by *E. coli*." *FEBS Letters*, **63**: 1 (1976).

KOTYK, A., and K. JANAČEK, *Membrane Transport (Biomembranes*, Vol. 9). New York: Plenum Press, 1977.

LEFEVRE, P. G., "The Present State of the Carrier Hypothesis." *Current Topics in Membranes and Transport*, **7**: 109 (1975).

MACROBBIE, E., "Ion Transport in Plant Cells." *Current Topics in Membranes and Transport*, **7**: 1 (1975).

MARTONOSI, A., ed., *The Enzymes of Biological Membranes*, Vol. 3: *Membrane Transport*. New York: John Wiley, 1976.

MCLAUGHLIN, S., and M. EISENBERG, "Antibiotics and Membrane Biology." *Ann. Rev. Biophys. Bioeng.*, **4**: 335 (1975). Ion carriers and pore formers.

NAKAO, M., ed., *Selected Papers in Biochemistry:* Vol. 9, *Active Transport*. Baltimore, Md.: Univ. Park Press, 1972.

NEAME, K. and T. RICHARDS, *Elementary Kinetics of Mem-*

brane Carrier Transport. New York: John Wiley, 1972. (Paperback.)

Oxender, D., and S. Quay, "Binding Proteins and Membrane Transport." *Ann. N.Y. Acad. Sci.,* **264**: 358 (1975).

Pietras, R., and C. Szego, "Specific Binding Sites for Estrogen at the Outer Surfaces of Isolated Endometrial Cells." *Nature,* **265**: 69 (1977). Contrary to the traditional view, some steroids seem to penetrate membranes via carriers.

Racker, E., "Structure and Function of ATP-Driven Ion Pumps." *Trends Biochem. Sci.,* **1**(10): 244 (Nov. 1976).

Rothstein, A., Z. Cabantchik, and P. Knauf, "Mechanism of Anion Transport in Red Blood Cells: Role of Membrane Proteins." *Fed. Proc.,* **35**: 3 (1976).

Satir, B., C. Schooley, and P. Satir, "Membrane Reorganization during Secretion in Tetrahymena." *Nature,* **235**: 53 (1972).

Schatzmann, H., "Active Calcium Transport and Ca^{2+}-Activated ATPase in Human Red Cells." *Current Topics in Membranes and Transport,* **6**: 126 (1975). Explanation for the low cytoplasmic levels.

Schuldiner, S., G. Rudnick, R. Weil, and H. Kaback, "Mechanism of β-Galactoside Transport in *E. coli* Membrane Vesicles." *Trends Biochem. Sci.,* **1**(2): 41 (Feb. 1976).

Segal, S., "Disorders of Renal Amino Acid Transport." *New Engl. J. Med.,* **294**: 1044 (1976).

Selkurt, E. E., ed., *Physiology* (4th ed.). Boston: Little, Brown, 1976. See Chap. 1 by W. M. Armstrong for a clear discussion of membranes and transport.

Shamoo, A., ed., *Carriers and Channels in Biological Systems.* Ann. N.Y. Acad. Sci., Vol. 264, 1975. Symposium.

Silbernagl, S., and E. C., Foulkes, "Renal Transport of Amino Acids." *Rev. Physiol. Biochem. Pharmacol.,* **74**: 105 (1975).

Skou, J. C., "The (Na^+-K^+)-Activated Enzyme System." In *Perspectives in Membrane Biology.* S. Estrada-O and C. Gitler, eds. New York: Academic Press, 1974, p. 263. A nice review by a pioneer worker in this field.

Whittam, R., and A. R. Chipperfield, "The Reaction Mechanism of the Sodium Pump." *Biochim. Biophys. Acta,* **415**: 149 (1975).

Wilbrandt, W., "Criteria in Carrier Transport." *Biomembranes,* **7**: 11 (1975). An easily read review of transport kinetics.

Wilson, T. H., and P. Maloney," Speculations on the Evolution of Ion Transport Mechanisms." *Fed. Proc.,* **35**: 2174 (1976).

NOTES

1. In fact, the lipids are sufficient to cover only about two-thirds of the surface; the rest of the surface is protein.

2. Digitalis is commonly used to increase the muscle strength of failing hearts. Its inhibition of the sodium pump alters the internal Na^+ and K^+ concentration which, in turn, probably makes more Ca^{2+} available to the contractile elements, the significance of which is discussed in Chap. 9.

3. Frog skin normally is positive inside with respect to outside, and uses a sodium pump in the epithelial cells to actively transport Na^+ from the surrounding solution into the animal. This transport plays an important role in electrolyte regulation in these animals. The situation is analogous to the transport of Na^+ from urine back into the bloodstream in kidneys, or the absorption of Na^+ from the intestines, and is an easily manipulated model for those processes.

7

The Oxidative Organelles: Mitochondria, Chloroplasts, and Peroxisomes

7-1 The Mitochondrion 288
   Discovery
   Distribution
   Structure
   Structure-Function Relationships

7-2 Electron Transport and Oxidative Phosphorylation 294
   The Electron Transport Chain
   Oxidative Phosphorylation
   Chemical Coupling
   Conformational Coupling
   Electrochemical Coupling

7-3 The Metabolic Roles of Mitochondria 302
   The Citric Acid Cycle
   The Advantages of Aerobic Catabolism
   Heat Production by Mitochondria
   Biosynthetic Activities of Mitochondria
   Transport Properties

7-4 The Chloroplast 309
   Plastids
   Chloroplast Structure
   Development and Distribution of Chloroplasts

7-5 The Photosynthetic Light Reaction 314
   The Photoelectric Effect
   Light Capture in Chloroplasts
   The Dual Pigment System
   Photophosphorylation

   Cyclic Photophosphorylation
   The Stoichiometry and Efficiency of Photosynthesis
   Bioluminescence

7-6 The Photosynthetic Dark Reaction 324
   Carbon Dioxide Fixation
   The Calvin Cycle
   Alternate Pathways of Carbon Fixation
   The Phosphogluconate Pathway of Animal Cells

7-7 The Evolutionary Origin of Mitochondria and Chloroplasts 330
   Photosynthesis in Prokaryotes
   Oxidative Metabolism in Prokaryotes
   Protein Synthesis in Mitochondria and Chloroplasts
   Endosymbiotic Evolution of Mitochondria and Chloroplasts

7-8 Peroxisomes and the Evolution of Oxidative Metabolism 335
   Peroxisomal Reactions
   The Glyoxylate Cycle (Glyoxysomes)
   Photorespiration
   Evolution of Oxidative Metabolism

Summary 341
Study Guide 343
References 344
Notes 346

The typical eukaryotic cell depends on mitochondria or chloroplasts to supply most of its energy. Chloroplasts, found only in plant cells, capture light energy and transform it to chemical energy in the form of ATP and reduced coenzymes, a process called *photosynthesis*. Mitochondria, which used to be known as *chondriosomes* or, in muscle, *sarcosomes*, are found in both animal and plant cells. They oxidize reduced coenzymes and certain other organic molecules, using oxygen from the air as an electron acceptor. Oxidation releases energy, a portion of which is captured by mitochondria and used to form ATP.

Many of the reactions associated with mitochondria and chloroplasts are found also in prokaryotes, where some are catalyzed by elements associated with the plasma membrane. Most prokaroytes can oxidize organic material, and the blue-green algae and a few species of bacteria are capable of photosynthesis. On the other hand, there are a few specialized eukaryotic cells that can do neither. Mature red blood cells, for example, can obtain energy only from glycolysis.

In spite of a few exceptions, mitochondria and chloroplasts are the "powerhouses" of most eukaryotic cells. We shall now take a closer look at their structure and function. In addition, we shall consider a third oxidative organelle, the *microbody* or *peroxisome*, the function of which is closely associated with that of both mitochondria and chloroplasts.

7-1 THE MITOCHONDRION

Mitochondria are found in all eukaryotic cells capable of utilizing oxygen. They are thus found in aerobically growing yeast, in protozoa, and in virtually every cell of higher plants and animals. (Red blood cells are a notable exception.) These important organelles have been intensively studied for over a century (see Table 7-1), yet only in the past few years have we begun to understand how they actually function. And, of course, not all questions have been answered even now.

Discovery. Perhaps the first description of mitochondria was by A. Kölliker, who in 1857 described them as the "sarcosomes" of muscle. Much later he was able to show that the mitochondria of muscle are individual entities and not connected directly to other parts of the cell. However, only after the application of appropriate staining procedures by Altmann in the latter years of the nineteenth century did it become possible to make detailed descriptions of mitochondrial distribution. From such studies, Altmann concluded that mitochondria are autonomous organisms living within the cytoplasm of a host cell. Many doubted this theory and held to the modern notion that they are integral parts of the cell itself. Nevertheless, there is now the suspicion that mitochondria are, in fact, the highly evolved descendants of bacteria that once lived symbiotically in higher cells.

The function of these organelles was much debated during the early decades of this century. In 1912, B. F. Kingsbury proposed that they might be the sites of cellular respiration—that is, the sites of oxygen utilization. He was quite right, of course, though direct proof had to await the development, primarily by George Palade and his associates at Rockefeller University, of improved methods of isolating cellular components. By 1950, however, it was recognized that mitochondria are not only the sites of cellular respiration, but also the major source of ATP production in aerobic animal cells.

Distribution. In many cells, mitochondria are distributed rather uniformly about the cytoplasm. Where there is an area of particularly

TABLE 7-1. Some Early Milestones in Respiratory Research

ca 1500	Leonardo de Vinci likens animal nutrition to the burning of a candle.
1770–74	Priestly shows that oxygen is consumed by animals.
1780	Lavoisier and Laplace conclude that the respiration of animals is an oxidation ("burning").
1857	Kölliker discovers mitochondria in muscle.
1872	Pluger finds that oxygen absorbed by the lungs is actually consumed in part by all tissues.
1886	MacMunn discovers the cytochromes (originally "histohematins").
1888	Mitochondria are isolated by Kölliker.
1890	Altmann finds a specific stain for mitochondria and suggests that they are autonomous organelles.
1912	Warburg demonstrates that iron is essential to respiration.
1923	Keilin shows that cytochromes have an altered oxidation state during respiratory activity.
1928–33	Warburg determines the basic structure of heme.
1931	Engelhardt shows that phosphorylation is coupled to oxygen consumption.
1933	Keilin succeeds in the partial reconstitution of an electron transport chain.
1937	Krebs formulates the citric acid cycle.
1937–41	Kalckar and Belitser each devise ways of quantitatively studying oxidative phosphorylation.
1939–41	Lipmann proposes a central metabolic role for ATP.
1947–50	Lipmann and Kaplan determine the structure of coenzyme A.
1948–50	Kennedy and Lehninger show that the citric acid cycle, oxidative phosphorylation, and fatty acid oxidation take place in the mitochondria.
1951	Lehninger shows that oxidative phosphorylation requires electron transport.
1954	Palade describes the *cristae mitochondriales*, or crista membranes.

heavy utilization of ATP, however, one typically finds an unusually high concentration of mitochondria. Thus, they are often found in rows adjacent to the contractile elements of muscle, wrapped about the base of a sperm flagellum, and adjacent to lipid droplets of fat cells.

The number of mitochondria per cell also varies widely. Cells with large ATP requirements tend to have more—up to a thousand or so in a liver cell, for instance. Since they replicate autonomously, the number is not fixed and may fluctuate with need.

Structure. Mitochondria are about the size of many common bacteria—that is, they are 0.5–1 μm in diameter and several μm long. The size and shape are highly variable, however. Some are round, some are branched, and some are long and filamentous (see Fig. 7-1) with lengths up to about 40 μm in extreme cases.

Mitochondria are bounded by a 60-Å outer membrane, separated from an inner membrane of the same thickness by the *intermembrane space*. This space is usually about 80 Å wide, though in some spots it disappears entirely and the inner and outer membranes appear fused.

CHAPTER 7
The Oxidative Organelles

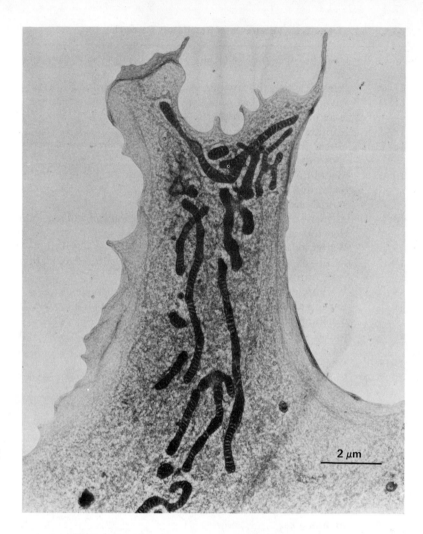

FIGURE 7-1 Mitochondria. High voltage (10^6 volt) electron micrograph of a human fibroblast showing highly elongated, cylindrical mitochondria. The longest one visible here is about 7 μm. [Courtesy of J. Wolosewick, Univ. of Colorado.]

The inner membrane is thrown into folds called *cristae* that project into the interior (*matrix*) of the organelle (see Fig. 7-2). These folds divide the matrix into compartments that are only partially separated from one another.

Fortuitously cut thin sections of inner mitochondrial membranes reveal the usual trilaminate structure. However, when isolated fragments of the inner membrane are negatively stained, they are seen to be studded with mushroomlike projections pointing toward what would be the matrix (Fig. 7-3). Each projection is part of a *respiratory assembly* consisting of a cluster of membrane proteins, a 80–100 Å head piece (or *inner membrane sphere*, seen in Fig. 7-3a), and a stalk some 30–40 Å in diameter. The stalk is not ordinarily seen in thin-sections, suggesting that it may be withdrawn into the membrane *in vivo* leaving the spheres sitting on the surface of the membrane. (Even the spheres are not seen in routine sections, since special fixatives must be used to preserve them.)

The inner membranes also contain phospholipids, the composition of which is unusual, consisting of about 20% *cardiolipin*. Altogether, the inner membrane, at least in liver cell mitochondria, is remarkable in

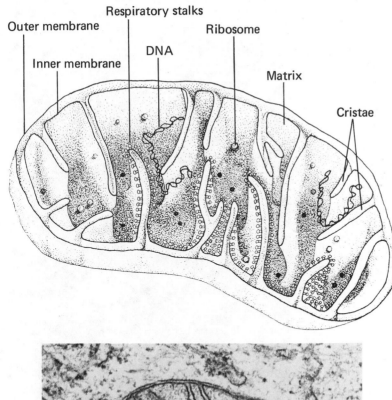

FIGURE 7-2 Mitochondrial Structure. Note that cristae are formed from folds in the inner membrane. [Micrograph courtesy of E. Gordon and J. Bernstein, *Biochim. Biophys. Acta*, **205**: 464 (1970).]

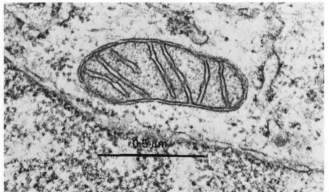

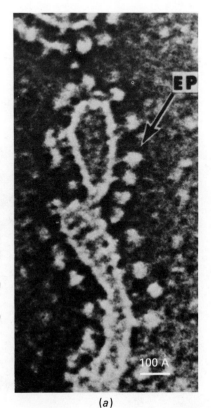

being about 80% protein (Fig. 7-3b, c). Most of the protein is associated with respiratory assemblies, though ill-defined structural proteins may also be present.

The outer membrane is much more usual in its composition. The protein content (about 50%) and phospholipid composition is much like that of the surrounding endoplasmic reticulum, although mitochondrial membranes in general tend to have more unsaturated fatty acids and less cholesterol. The outer membrane is also relatively permeable to a variety of substances. The inner membrane, in contrast, is the major permeability barrier and is endowed with many highly selective pumps that maintain the unique composition of the matrix, which at almost 50% protein is nearly a gel.

FIGURE 7-3 Mitochondrial Cristae. (a) Negatively stained side view of an inner membrane fragment. Elementary particle (EP) is an older name for the respiratory sphere. (Parts b and c appear on p. 292.) [(a) Courtesy of H. Fernández-Morán, T. Oda, P. Blair, and D. E. Green, *J. Cell. Biol.*, **22**: 63 (1964).]

(a)

FIGURE 7-3 Mitochondrial Membranes. (b) Fracture face of an inner mitochondrial membrane. The exposed membrane particles, which consist of clusters of proteins, are often arranged in rows (*arrow* at right). Circled arrow indicates direction of metal shadowing. (c) Freeze fractured and etched view of two mitochondria. Note numerous cristae at left, the relatively particle-free, fractured outer membrane (*double arrow*) and the particle-studded fracture face of the inner membrane (*single arrow*). [(b), (c) Courtesy of J. Wrigglesworth, L. Packer, and D. Branton, *Biochim. Biophys. Acta*, **205:** 125 (1970).]

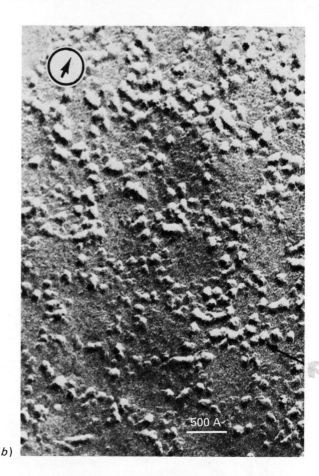

(b)

(c)

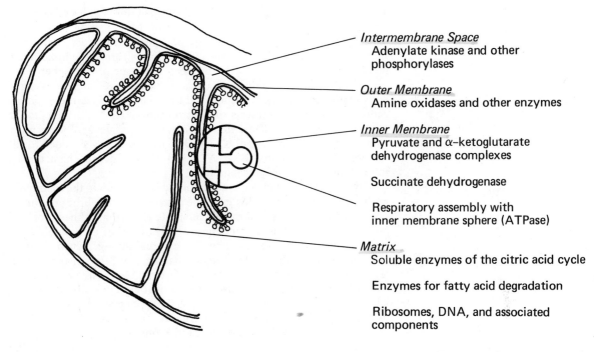

FIGURE 7-4 Localization of Mitochondrial Reactions.

Structure-Function Relationships. Each of the membranes and compartments (that is, matrix and intermembrane space) contributes in its own way to mitochondrial function (see Fig. 7-4). The main part of that function is to oxidize pyruvate, fatty acids, and certain of the amino acids, relying largely on an enzyme sequence known as the *citric acid cycle.* Electrons removed in this process or brought into the mitochondria from oxidative steps in the cytosol are transferred ultimately to oxygen. The overall process produces carbon dioxide and water, and is coupled to the production of ATP.

The outer membrane can be removed from the mitochondrion by certain detergents or phospholipid hydrolases, leaving the rest of the organelle intact. Under these circumstances, cristae disappear as the matrix swells to stretch the inner membrane. In addition, soluble enzymes from the outer space are released, allowing their identification. Among them are some nucleoside phosphorylases, including adenylate kinase, which catalyzes an interconversion of the adenine nucleotides.

$$2ADP \rightleftharpoons ATP + AMP$$

Fractionation also allows the outer membrane itself to be examined, revealing several associated enzymes, including an amine oxidase. The latter is an enzyme that removes amino groups from various molecules and is often used as a "marker" for the outer membrane.

Most of the metabolic activity of mitochondria takes place in the inner compartment, catalyzed by enzymes associated either with the inner membrane or with the matrix. The majority of oxidative steps are associated with membrane-bound enzymes, clusters of which work together as an *electron transport chain*. The function of this chain is to pass electrons from coenzymes such as NADH to oxygen, creating water. The usual objective is to produce ATP by phosphorylating ADP, a function that is actually carried out by an ATPase that is part of the inner membrane spheres. The association between inner membrane spheres and ATP synthesis was established by Efraim Racker of Cornell University, who demonstrated that mitochondrial fragments that lack these spheres cannot couple ATP synthesis to electron transport. A purified protein complex capable of ATPase activity *in vitro* was prepared from mitochondria. When this complex, called factor one (F_1), was added to the membrane preparations, both ATP synthesis and the characteristic inner membrane spheres reappeared.

Why the ATPase of mitochondria should be stuck on to the membrane in this way is not clear. However, the enzyme is hydrophilic and thus is not suited to be an integral membrane protein. Perhaps the hydrophilic properties are important in binding and activating the highly soluble substrates ADP and P_i. In any case this external arrangement is seen in other places as well: Aerobic bacteria having electron transport chains feature similar spheres on the inner surface of their plasma membranes; and chloroplasts have a comparable spherical complex, also associated with ATP synthesis, on the surface of their photosynthetically active membranes. This similarity in structure among the various ATP-synthesizing devices will be explored in more detail later. It is all the more impressive when one understands that not only the structure but the mechanism of these complexes is basically the same in each case.

7-2 ELECTRON TRANSPORT AND OXIDATIVE PHOSPHORYLATION

Many cells can utilize glucose under either aerobic or anaerobic conditions. However, when oxygen becomes available to a cell that is capable of using it but is growing anaerobically, the amount of glucose consumed for a given increase in cell mass falls dramatically. Simultaneously, anaerobic end products such as ethanol or lactate are no longer made. These changes in cell metabolism, called the *Pasteur effect*, are possible only in those cells capable of *oxidative phosphorylation*. This term refers to the production of ATP by a series of oxidative reactions carried on (in eukaryotes) by components of the mitochondrial inner membrane. The sequence of oxidative steps is referred to as the *electron transport chain* or *respiratory chain*.

The Electron Transport Chain. It was pointed out in Chap. 4 that oxidation could be a source of considerable energy but that glycolysis and fermentation fail to take advantage of that fact. The oxidative step in glycolysis removes two electrons from glyceraldehyde phosphate, transferring them to NAD^+. The latter substance is thus reduced to NADH. Since only small amounts of NAD^+ are present, there must be a rapid recycling of NADH if metabolism is to continue. Under anaerobic conditions, recycling is accomplished by using NADH to produce the reduced end products, lactate or ethanol. If oxygen and mitochondria are available, however, a much more attractive possibility exists.

Mitochondria are able to catalyze the reaction

$$NADH + H^+ + \tfrac{1}{2}O_2 \rightarrow NAD^+ + H_2O \qquad (7\text{-}1)$$

The standard state free energy change of this reaction is about -220

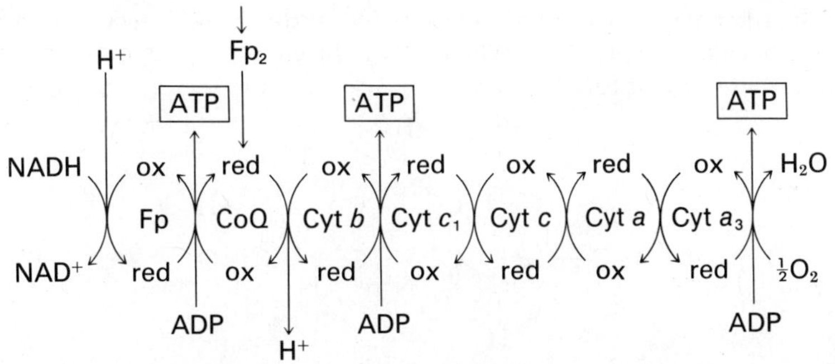

SECTION 7-2
Electron Transport and
Oxidative Phosphorylation

FIGURE 7-5 The Electron Transport Chain. In sequence: flavoprotein, coenzyme Q, and the cytochromes. Fp_2 indicates nonchain flavoproteins that transport electrons from fatty acid oxidation or the citric acid cycle. Note that entry at Fp_2 results in one less ATP.

kJ/mole. The sequence of steps by which it is accomplished is outlined in Fig. 7-5. Note that electrons removed from NADH are passed from one component to another until they are used to reduce molecular oxygen to water. Note also that three steps are coupled to the phosphorylation of ADP, thus producing ATP. Each molecule of NADH can therefore be used to generate three molecules of ATP under aerobic conditions. That potential is lost when NAD^+ is regenerated anaerobically in the production of lactate or ethanol from pyruvate.

Pyruvate itself can also be oxidized by mitochondria to produce electrons for the electron transport chain, some of which are carried by NADH and some of which enter the chain from reduced *flavoproteins* (designated Fp_2 in Fig. 7-5). The stoichiometry of these reactions will be presented, but first let us define the components of the chain, for most of them will be encountered again later in other locations.

The flavins. The immediate acceptor for electrons removed from NADH in the respiratory chain is a *flavoprotein* (*Fp*). Flavoproteins contain a prosthetic group built about vitamin B_2, also called *riboflavin*. There are two common derivatives of riboflavin that act as electron-transporting coenzymes: *flavin mononucleotide* or FMN, and *flavin-adenine dinucleotide* or FAD.

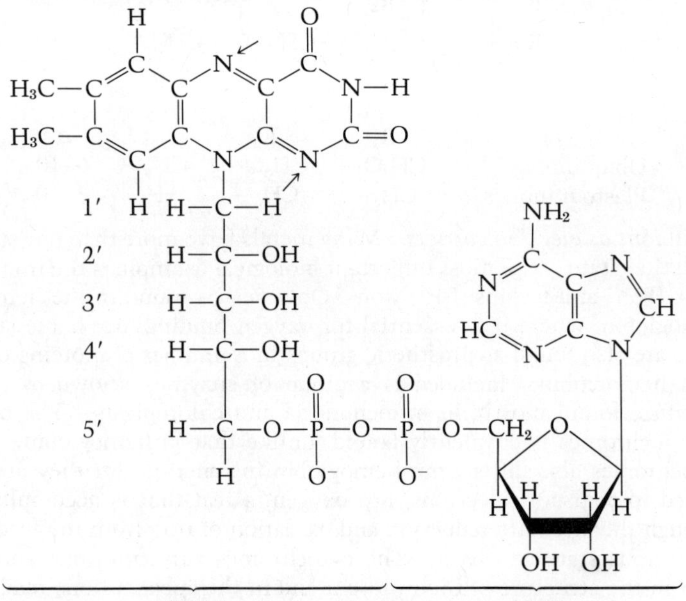

Oxidized flavin adenine dinucleotide (FAD)

295

The riboflavin ring can be reduced twice (at the arrows), accepting two electrons and two protons. When fully reduced, the coenzymes are designated FMNH$_2$ or FADH$_2$.

Fully reduced flavin ring

The quinones. Derivatives of ordinary quinone can act as as electron carriers by being reduced to semiquinones, and then further reduced to dihydroquinones:

Quinone   Semiquinone   Dihydroquinone
            (free radical)

Of the many types of quinones found in a cell, those of most interest here are the *ubiquinones* (also called "coenzymes Q") and the *plastoquinones*. The former are mitochondrial components and the latter are found in chloroplasts. Both are derived from the closely related vitamin K group.

	R_1	R_2	R_3	n
Ubiquinones	CH_3O-	CH_3O-	CH_3-	6–10
Plastoquinones	CH_3-	CH_3-	$H-$	6–10

Metal ions as electron carriers. Many metals have more than one stable oxidation state. The most important biological example is the iron pair ferric (Fe^{3+}) and ferrous (Fe^{2+}) iron. Only Fe^{2+} is found in the heme of hemoglobin, where it is essential for oxygen binding, but heme structures are also found as prosthetic groups in a number of proteins other than hemoglobin. Included is a group of enzymes known as *cytochromes,* found mostly in mitochondria and chloroplasts. The heme in cytochromes is covalently bound, unlike that in hemoglobin. The cytochromes also differ from hemoglobin in function, for they are designed to transport electrons, not oxygen, a feat that is accomplished through the alternate reduction and oxidation of iron from the ferric to the ferrous state and back. The cytochromes vary from one another both in the structure of their protein and in the substitutions made on the basic heme (porphyrin) ring. The cytochromes as a group are remarkably similar from one species to the next, however, and even between plants and animals.

Iron-containing proteins other than cytochromes have also been identified as electron carriers. The most common are called *iron-sulfur proteins* or *nonheme iron proteins*. They have been found in bacteria, in plants, and in various animal respiratory systems. Their location in the electron transport chain is not firmly established, however.

Cytochrome oxidase. The cytochromes are divided into three classes, a, b, and c, according to their absorption spectra. The mitochondrial type a cytochromes are further subdivided on the same basis into cytochromes a and a_3. These two exist together in a complex known as *cytochrome oxidase*, which has a total of two hemes (a plus a_3), two copper ions, and seven protein subunits. Since copper as well as iron can be reduced, $(Cu^{2+} + e^- \rightleftharpoons Cu^+)$, fully reduced cytochrome oxidase has a total of four extra electrons, which is the exact number needed for the complete reduction of one molecule of oxygen (O_2) to water.

The mechanism by which the oxygen atoms in an oxygen molecule are reduced by cytochrome oxidase is still under investigation, but the actual transfer of electrons to oxygen seems to be a function reserved for cytochrome a_3. It is this step that is specifically sensitive to cyanide, accounting for the toxicity of that substance.

Oxidative Phosphorylation. The function of the electron transport chain is to use oxygen as an electron acceptor in the regeneration of flavins or NAD^+, and in the process to make energy available for ATP synthesis. In intact mitochondria, oxygen is not consumed unless ADP and inorganic phosphate are present. This dependence represents an obvious control mechanism, for it means that the rate of respiration is limited by the amount of ATP required, a need that is reflected in the level of its hydrolysis product, ADP. More than that, it indicates that electron transport and mitochondrial ATP synthesis are tightly coupled processes. They are, however, separate processes, as we can demonstrate with certain poisons like dinitrophenol that allow electron transport to continue without ATP production—i.e., poisons that *uncouple* the two processes.[1]

The three sites at which ATP synthesis is coupled to electron transport were defined by the use of poisons that block specific reactions in the chain and by analysis of the energetics of the individual steps (determined by measuring oxidation-reduction potentials). The three sites indicated certainly have energy changes sufficient to account for ADP phosphorylation. How that energy is utilized in the actual production of ATP, however, remains controversial.

There are three basic theories that attempt to explain how electron transport is coupled to ADP phosphorylation. In chronological order, these theories attribute phosphorylation to the following mechanisms: (1) a direct chemical coupling to certain redox reactions in the chain; (2) the collapse of a voltage and/or pH gradient established through electron transport; and (3) a conformational change in macromolecular components of the mitochondrial membranes, the unstable (high energy) state of which was achieved as a result of electron transport.

Chemical Coupling. The oldest theory of oxidative phosphorylation dates back to the work of E. C. Slater and others in the early 1950s. It proposes only direct enzymatic coupling between particular redox steps and phosphorylations. There is reason to believe that the respiratory redox reactions would have to be coupled to the formation of a phosphorylated intermediate other than ATP itself. The phosphate would then be passed to ADP, regenerating the intermediate. In spite of an intense search, this postulated intermediate has yet to be isolated, a fact that does not discourage proponents of the chemical coupling theory because only trace amounts would be required. Until the intermediate is found, however, there can be no experimental verification of chemical coupling. (This elusive intermediate is thought to be required by the other mechanisms also, but it is less central to those theories.)

The theory of direct chemical coupling has several advantages over its competitors. Not the least of these is that it does not propose any new mechanisms, but only new applications of reactions for which there are several well-characterized examples. In effect, it proposes nothing more than an enzymatically coupled oxidation and phosphorylation not so very different from the combined phosphorylation and oxidation of glyceraldehyde phosphate to form 1,3-diphosphoglycerate, a reaction from the glycolytic pathway.

Conformational Coupling. The most recent theory of oxidative phosphorylation was proposed in the mid-1960s by P. D. Boyer of the University of California, Los Angeles, and in a modified version by David E. Green of the University of Wisconsin. This theory, conformational coupling, assumes that macromolecular components within the mitochondrion have at least two conformational configurations with a significant difference in energy between them. The high energy configuration results from electron transport. The free energy change in dropping from a configuration of higher energy to a configuration of lower energy must be sufficient to force the phosphorylation of ADP by an ATPase that is an integral part of the conformational unit. Since we discussed earlier (Chap. 4) how conformational changes might play important roles in enzymatic catalysis, this postulated mechanism should not seem completely foreign to us.

Those in favor of the conformational theory can point to the observation that respiring and resting mitochondria have different light-scattering properties, that electron micrographs show a correlation between respiratory activity and the configuration of the crista membranes (Fig. 7-6), and that the conformation of some cytochromes is dependent on the oxidation state of their associated iron. These changes, which reflect electron transport, are consistent with the idea of energy capture via conformational isomerization.

Electrochemical Coupling. This proposal is both the most interesting and probably the most popular at the present time. It was originally described in 1961 by an English scientist, Peter Mitchell, who termed it the *chemiosmotic hypothesis*. The chemiosmotic hypothesis takes advantage of the fact that whereas electrons are passed smoothly from one component to the next in the respiratory chain, protons are not. Some of the steps require protons and some release them. We can see this in the chain as it was drawn earlier, but additional proton transport might also be achieved by slightly rearranging the order of intermediates or introducing new steps at strategic points.

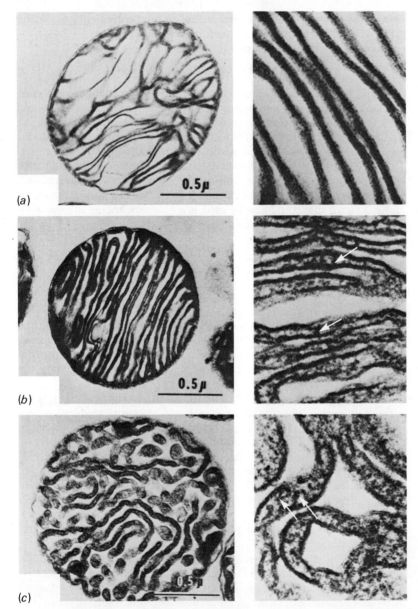

FIGURE 7-6 Conformational Changes in Mitochondrial Cristae. (a) The isolated beef-heart mitochondrion is inactive, with swollen cristae that press tightly against each other. (The light area is the intracristal and outer space.) (b) Electron transport causes the stalks of the respiratory assemblies (arrows) to push adjacent cristae apart. (c) The addition of phosphate to active mitochondria causes an "energized twisted" configuration in which the stalks and head pieces are clearly evident (arrow). [Courtesy of D. E. Green and J. H. Young, *American Scientist*, **59**: 92 (1971). High resolution micrographs were taken by T. Wakahayashi.]

Mitchell suggested that proton-requiring and proton-releasing reactions are associated with membranes in such a way that proton fixation occurs always on one side while release occurs always on the other. If these membranes (the cristae) are otherwise impermeable to hydrogen ions, then the effect of respiratory chain activity will be to pump protons from one side to the other, creating a proton (pH) gradient across the membrane along with a charge imbalance or voltage gradient.

Voltage and pH gradients are each a potential source of energy. That is, it takes energy to create them, so one should be able to capture almost the same amount from their collapse. In order to utilize this energy, Mitchell suggested that an ATPase must be arranged through a membrane in such a way that an electrical and/or proton gradient can tilt the ATP/ADP equilibrium in the direction of ATP synthesis.

Since ATP hydrolysis releases protons at physiological pH

$$ATP^{4-} + H_2O \rightleftharpoons ADP^{3-} + P_i^{2-} + H^+ \tag{7-2}$$

it should be quite clear that if a membrane were erected between reactants and products, then an increase in hydrogen ions on the right would shift the equilibrium toward the left. Alternatively, one could imagine that the enzyme is placed within a membrane so that the active site for protons is by itself on one side, while the phosphates are all on the other. Then the synthesis of ATP would reduce the proton gradient and hence be driven by it, just as the hydrolysis of ATP could create the gradient. In addition, the reaction causes a change in the voltage across the membrane, a factor that could also be used as a source of energy to force ATP synthesis, as in the following schematic:

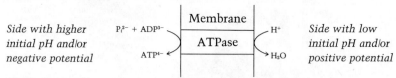

The reaction as it is shown here will make the left side less negative and the right side less positive, equivalent to transferring one electron from left to right. This reduction of charge difference could be the force that drives the reaction. In addition, a low pH on the right would have the same effect.

The chemiosmotic theory, then, assumes that activity of the respiratory chain causes the formation of pH and electrical gradients across membranes, and that those gradients can provide the energy needed to generate ATP. These two assumptions can be tested independently.

First, does electron transport create an electrochemical gradient? Careful measurements indicate that mitochondrial respiration is accompanied by acidification of the external medium, which indicates an outward flow of hydrogen ions. There is an efflux of six or more H^+ per oxygen atom used—i.e., per electron pair transported. Such translocations would result in a proton (pH) gradient and in a corresponding membrane potential.

There is also direct evidence for mitochondrial membrane potentials. Though measurements with electrodes are apt to be unreliable because of the small size of the organelle, lipid-soluble ionized compounds such as picrate penetrate the inner membrane of actively respiring mitochondria. Negatively charged ions move outward while positive ions move inward, a situation that is consistent with a net negative charge on the inside. Inhibiting electron transport, for example with rotenone, blocks these ion translocations, presumably by preventing formation of a chemiosmotic gradient.

It appears, then, that electron transport results in pH and electrical gradients. When these gradients are disturbed with valinomycin or other ionophores (see Chap. 6), ATP synthesis is compromised. The antibiotic valinomycin uncouples oxygen consumption from ATP synthesis by changing the permeability properties of mitochondria. Electron transport is normal in the presence of the compound, while ATP synthesis is apparently prevented by an increased potassium permeability of the inner membranes. By allowing K^+ to cross the membranes, charge differentials are reduced or eliminated. The rate of ATP synthesis at constant oxygen consumption has been shown to be proportional to the remaining electrical gradient, consistent with Mitchell's proposals.

Work with isolated systems further supports the coupling of electrochemical gradients and ATP synthesis. For example, Efraim Racker and his associates at Cornell University have succeeded in achieving electron transport and ATP synthesis in vesicles produced from purified lipid plus proteins from inner mitochondrial membranes. ATP synthesis depends on the integrity of these vesicles, and fails to take place when they are leaky.

The most elegant demonstration of electrochemically driven ATP synthesis, however, was due to a collaboration between Racker and Walther Stoeckenius of the University of California at San Francisco. Stoeckenius had been studying a peculiar bacterium called *Halobacterium halobium* that lives in concentrated brine—it grows best at 4.3 M NaCl (blood is 0.14 M, sea water 0.6 M). These cells have large purple patches on their membrane, the composition of which is mostly lipid plus a protein with properties remarkably like those of *rhodopsin*. (Rhodopsin is found in light-sensitive cells of animal eyes.) This *bacteriorhodopsin*, like animal rhodopsin, can be bleached by strong light and has as its light-absorbing pigment retinal, a vitamin A derivative.

Stoeckenius and his associates performed experiments with *Halobacterium* that revealed the presence of a unique energy transducing system. It appears that the purple membrane can use light to power a proton pump, thus creating a pH and electrical gradient across the cell membrane—the same kind of gradient that, according to the chemiosmotic hypothesis, can be used to phosphorylate ADP. To test the impression that this gradient is a source of energy for the cell, Stoeckenius and Racker produced artificial vesicles of purified lipids containing purple membrane plus an ATPase preparation from beef-heart mitochondria. When these vesicles were illuminated, they synthesized ATP from ADP and P_i (see Fig. 7-7).

Thus both questions concerning the chemiosmotic hypothesis have been answered in the affirmative: Electron transport can produce electrochemical gradients; and such gradients *in vitro* can drive an ATPase "backwards" to produce ATP. It remains only to show that the combined electrical and pH gradients *in vivo* are sufficient to do the same thing. Using the principles developed in Chaps. 2 and 6, we can estimate that at 10 mM phosphate and 300 K *either* a membrane potential in excess of 210 mV (negative inside) *or* a proton gradient over 3.5 pH units (acid outside) is probably sufficient to tip the ADP/ATP reaction toward ATP synthesis. Either value alone would be physiologically unreasonable, but since the two sources of energy are additive, adequate combinations of the two can be constructed that are near experimental estimates of their true values.

But if available evidence supports the validity of the chemiosmotic hypothesis, where does that leave the alternate theories of oxidative phosphorylation? It does not disprove them. In fact, aspects of the conformational hypothesis are applicable to either of the other two theories since conformational changes could serve as intermediate sources of energy in either chemically coupled or electrochemically coupled reactions. At the present time, however, the chemiosmotic hypothesis is

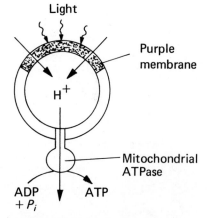

FIGURE 7-7 Light-driven ATP Synthesis. The drawing represents reconstituted vesicles containing purple membrane from *Halobacterium halobium* plus mitochondrial ATPase. Bacteriorhodopsin in the purple membrane catalyzes light-induced proton uptake. The resulting electrochemical gradient is used by mitochondrial ATPase to phosphorylate ADP. [Described by E. Racker and W. Stoeckenius, *J. Biol. Chem.*, **249**: 662 (1974).]

rather widely accepted as the basis not only for oxidative phosphorylation in mitochondria, but also for ATP production by chloroplasts and by at least some bacteria.

7-3 THE METABOLIC ROLES OF MITOCHONDRIA

Mitochondrial enzymes are responsible for the oxidation of NADH, pyruvate, fatty acids, and certain of the amino acids. The energy released is utilized by electron transport to produce ATP. Mitochondria also participate in a few biosynthetic activities including, in some cases, certain steps in the synthesis of heme and of steroid hormones. The only one of these pathways to be considered in any detail here is the oxidation of pyruvate, obtained mostly from carbohydrates, for it illustrates the truly great advantage of aerobic catabolism and hence the real importance of mitochondria.

The Citric Acid Cycle. Most of the reduced coenzymes used in oxidative phosphorylation originate with the mitochondrial pathway known as the *citric acid cycle* or, since citrate is a tricarboxylic acid, the *tricarboxylic acid (TCA) cycle*. It is also known as *Krebs cycle* in honor of H. A. Krebs, who first proposed it in 1937.

The cycle requires as a starting material acetate that has been "activated" through a thiol ester ($CH_3-\overset{\overset{O}{\|}}{C}-S-R$) link to *coenzyme A*, forming acetylcoenzyme A (acetyl-SCoA). Coenzyme A itself is an adenosine derivative that also contains pantothenic acid (a vitamin) and β-mercaptoethylamine:

Coenzyme A (HSCoA)

Acetyl-SCoA can be produced by the degradation of carbohydrates (via pyruvate), fatty acids, and amino acids. The free energy of hydrolysis of this and other thiol esters is even more favorable than the hydrolysis of ATP.

Pyruvate is transported into mitochondria, then converted to acetyl-SCoA by the action of an enzyme system called the *pyruvate dehydrogenase complex* (Fig. 7-8). This complex, which is associated with the inner membrane, contains multiple copies of three different enzymes. (One of them, pyruvate decarboxylase, uses as a prosthetic group thiamine pyrophosphate, derived from thiamine or vitamin B_1.) The reaction catalyzed by the pyruvate dehydrogenase complex results in the *oxidative decarboxylation* of pyruvate and in its coupling to coenzyme A:

$$\underset{\text{Pyruvate}}{\begin{matrix} COO^- \\ | \\ C=O \\ | \\ CH_3 \end{matrix}} + HSCoA + NAD^+ \longrightarrow \underset{\text{Acetyl-SCoA}}{\begin{matrix} SCoA \\ | \\ C=O \\ | \\ CH_3 \end{matrix}} + CO_2 + NADH \quad (7\text{-}3) \quad \Delta G° = -39 \text{ kJ/mole}$$

The extremely large drop in free energy helps make this reaction virtually irreversible. Note that the keto group of pyruvate becomes an aldehyde (acetaldehyde) by decarboxylation then an acid by oxidation:

$$\underset{\text{Pyruvate}}{\begin{matrix} COO^- \\ | \\ C=O \\ | \\ CH_3 \end{matrix}} \xrightarrow[-CO_2]{H^+} \underset{\text{Acetaldehyde}}{\begin{matrix} HC=O \\ | \\ CH_3 \end{matrix}} \xrightarrow[H_2O]{NAD^+ \quad NADH} \underset{\text{Acetate}}{\begin{matrix} COO^- \\ | \\ CH_3 \end{matrix}} + 2H^+$$

Fatty acids are prepared for entry into the citric acid cycle by the removal of two carbons at a time, each pair in the form of activated acetate (i.e., acetyl-SCoA). After an initial input of energy in the form of an ATP → AMP hydrolysis, which is used to couple coenzyme A to the free carboxyl group of the fatty acid, each acetyl-SCoA cleaved from the chain is accompanied by the production also of one reduced flavoprotein and one NADH from their oxidized counterparts. Note that flavoproteins can enter the electron transport chain at the site marked Fp_2 in Fig. 7-5. As indicated in that figure, however, a very favorable step at the beginning of the chain is bypassed and thus so is the opportunity to make the first ATP. Still, two ATP are produced from each reduced flavin.

When the fatty acid chain has only two or three carbons left (depending on whether the original fatty acid had an even or odd number of carbons) the remnant is also fed into the citric acid cycle as its coenzyme A derivative but without production of the extra NADH and reduced flavoprotein.

Amino acids, after removal of the amine group (which in vertebrates is excreted in the form of urea), can for the most part be metabolized to glycolytic or citric acid cycle intermediates, or directly to acetyl-SCoA. Thus, excess amino acids—or even essential amino acids during starvation—can be made available, together with carbohydrates and fatty acids, to supply cells with ATP via the citric acid cycle and oxidative phosphorylation.

The *reactions of the citric acid cycle* begin with the condensation of acetate, in the form of acetyl-SCoA obtained as above, to oxaloacetate (also spelled oxalacetate) forming citrate:

$$\underset{\text{Acetate}}{\begin{matrix} CO-SCoA \\ | \\ CH_3 \end{matrix}} \Big\{ \underset{\text{Citrate}}{\begin{matrix} COO^- \\ | \\ CH_2 \\ | \\ ^-OOC-C-OH \\ | \\ CH_2 \\ | \\ COO^- \end{matrix}} \Big\} \underset{\text{Oxaloacetate}}{\begin{matrix} COO^- \\ | \\ C=O \\ | \\ CH_2 \\ | \\ COO^- \end{matrix}}$$

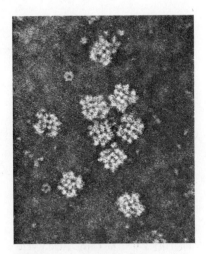

FIGURE 7-8 Pyruvate Dehydrogenase Complex. Negatively stained preparation of the pyruvate dehydrogenase complex from *E. coli*. Diameter about 300 Å. [Courtesy of L. J. Reed, from *The Enzymes*, Vol. 1, 3d ed., P. D. Boyer, ed., p. 213. © 1970 by Academic Press, N.Y.]

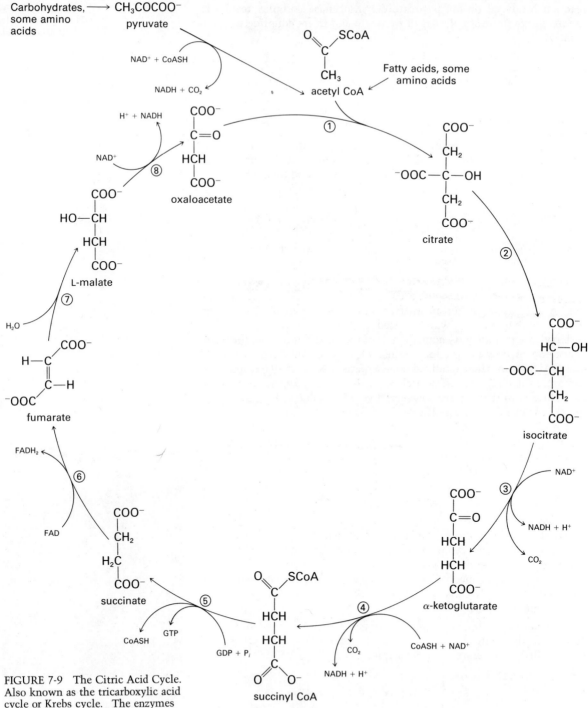

FIGURE 7-9 The Citric Acid Cycle. Also known as the tricarboxylic acid cycle or Krebs cycle. The enzymes are (1) citrate synthase; (2) aconitase; (3) isocitrate dehydrogenase; (4) α-ketoglutarate dehydrogenase; (5) succinyl-SCoA synthetase; (6) succinate dehydrogenase; (7) fumarase; (8) malate dehydrogenase.

A total of four oxidations and two decarboxylations follow, regenerating oxaloacetate ready to start a new round.

The reactions are shown in Fig. 7-9. Together they have a $\Delta G°$ of -54 kJ/mole and may be represented by the summary equation

$$3H_2O + \text{acetyl-SCoA} \rightarrow 2CO_2 + HSCoA + 8H^+ + 8e^- \quad (7\text{-}4)$$

$$\text{coupled to} \begin{cases} 3NAD^+ \to 3NADH \\ FAD \to FADH_2 \text{ (as a reduced flavoprotein)} \\ GDP + P_i \to GTP \end{cases}$$

Note that the effect is to oxidize acetate completely to carbon dioxide and water, producing NADH and reduced flavoproteins for entry into the electron transport chain. Each turn of the cycle also produces one "substrate level phosphorylation" in the form of a GTP. The latter may be used directly in a variety of reactions (see Chap. 5) or to phosphorylate ADP via a mixed-nucleotide kinase

$$GTP + ADP \rightleftharpoons ATP + GDP \qquad (7\text{-}5)$$

Note that since electron transport and oxidative phosphorylation produce 3 ATP from each NADH and 2 ATP from each reduced flavoprotein that gets recycled, the net yield per turn of the citric acid cycle is 12 ATP equivalents. In addition, by the time the cycle comes full circle, oxaloacetate to oxaloacetate, two carbon atoms will have been added in the form of acetate and two carbon atoms lost (not the same two) in the form of carbon dioxide.

The Advantages of Aerobic Catabolism. We have now followed two catabolic pathways of glucose: (1) glycolysis, and (2) its aerobic extension in the mitochondrion. Let us compare the efficiency of these processes. Glycolysis, the anaerobic degradation of glucose, was summarized in Chap. 5 as

$$glucose + 2ADP + 2P_i \to 2 \text{ lactate} + 2ATP \qquad (7\text{-}6)$$

The aerobic oxidation of glucose, on the other hand, consists of the following reactions:

(1) The cytoplasmic oxidation and degradation of glucose to pyruvate:

$$glucose + 2ADP + 2P_i + 2NAD \to$$
$$2 \text{ pyruvate} + 2ATP + 2NADH \qquad (7\text{-}7)$$

(2) The oxidative decarboxylation of pyruvate by the pyruvate dehydrogenase complex in the mitochondrion:

$$2 \text{ pyruvate} + 2NAD + 2CoASH \to$$
$$2 \text{ acetyl-SCoA} + 2NADH + 2CO_2 \qquad (7\text{-}8)$$

(3) Degradation of acetyl-SCoA via the citric acid cycle, also in the mitochondrion:

$$2 \text{ acetyl-SCoA} + 2GDP + 2P_i + 2FAD + 6NAD \to$$
$$4CO_2 + 2CoASH + 2GTP + 2FADH_2 + 6NADH \qquad (7\text{-}9)$$

In summary, the aerobic oxidation of glucose, Equations 7-7 through 7-9, combined with the respiratory chain oxidation of the coenzymes gives (using the chemical formula for glucose)

$$C_6H_{12}O_6 + 6O_2 \to 6CO_2 + 6H_2O$$

This reaction is accompanied by the production of 2 ATP, 2 GTP, 2 $FADH_2$, and 10 NADH. In terms of net ATP equivalent, using the stoichiometry factors introduced previously, glucose oxidation has an energy yield of 38 ATP. Since the combustion of glucose has a $\Delta G°$ of -2870 kJ/mole, the production of 38 ATP with a recoverable energy of about 33 kJ/mole each makes the overall process about 45% efficient.

When we compare the 38 ATP available from the aerobic oxidation of glucose with the 2 ATP available from glycolysis, the advantages of aerobic growth become obvious. It must be remembered, however, that the difference between aerobic and anaerobic oxidation of glucose has a somewhat different meaning in animals than in microorganisms. Although it is true that lactate, as the end product of glycolysis, is normally excreted from the cell as waste, in multicellular organisms this does not automatically mean that it is passed out of the animal. In higher animals, for example, lactate produced in red blood cells or vigorously exercising skeletal muscle is carried by the bloodstream to other tissues, such as the liver, where it may be oxidized to pyruvate. From there it can enter the citric acid cycle or be converted back to glucose. The difference in energy between glycolysis and the aerobic oxidation of glucose is therefore not wasted as far as the animal is concerned: It is only the individual cell that fails to derive as much energy from the carbohydrate as it might. This kind of cooperation among various cell types is, of course, one of the advantages of being multicellular.

The same analysis that we applied to the aerobic catabolism of glucose could be applied to fatty acids and to amino acids as well. If we did so, we would find that while the production of 38 ATP from glucose is impressive, the yield from fatty acid degradation is even more impressive. In fact, the yield in terms of ATP per gram of fat is over twice that of either carbohydrate or protein.[2] Since it is ATP that supports biosynthesis of fat it should be clear that, contrary to the opinion of some fad dieters, how much you eat does make a difference.

Heat Production by Mitochondria. If only 45% of the energy released during the oxidation of glucose is captured in the form of ATP, then over half must be lost as heat. Whether that really is a "loss," however, depends on the cell type and on environmental conditions. Warm-blooded animals require an internal source of heat to maintain body temperature in cool surroundings. Metabolism must be that source.

Metabolic heat production has been studied most thoroughly in mammals during adaptation to extreme cold and during hibernation. An important source of heat in these instances seems to be *brown fat*. This tissue is found to some extent in all mammals at one time or another in their lives. Humans, for example, have some when born, presumably to help protect the infant against the sudden drop in temperature at birth. Although brown fat normally disappears during the first year of life in humans, it is abundant in those species and under those conditions where maintenance of body temperature poses an unusual problem.

The color of brown fat comes from its high concentration of mitochondria, which are sparse in ordinary fat cells. These mitochondria appear to catalyze electron transport in the usual way but are much less efficient at producing ATP. Hence, a higher-than-usual fraction of the oxidatively released energy is converted directly to heat (called *nonshivering thermiogenesis*). Measurements made on isolated mitochondria from these cells reveal a relatively leaky inner membrane. Thus much of the electrochemical gradient produced by electron transport is dissipated without generating ATP—the mitochondria, in other words, are uncoupled. The degree of uncoupling may well be variable, thus providing heat as needed.

Biosynthetic Activities of Mitochondria. Although the major activity of mitochondria is catabolism and the subsequent production of ATP, some anabolic (biosynthetic) processes also occur in these organelles. For instance, mitochondria contain DNA and the machinery needed for protein synthesis. Therefore they can make some of their own proteins.

Mitochondrial genes are located on a single DNA of 4–5 μm contour length. (In bacteria the DNA may be over 1000 μm.) This limited amount of DNA codes for mitochondrial rRNA and tRNA but probably less than a dozen different proteins. The proteins so far identified are subunits of the ATPase, portions of the reductase responsible for transfer of electrons from CoQ to the iron of cyt c, and three of the seven subunits in cytochrome oxidase. Altogether, no more than 5–10% of mitochondrial components can be attributed to mitochondrial genes. The majority of mitochondrial components are specified by nuclear genes and transported to the mitochondrion after synthesis.

In addition to carrying out some of the biosynthetic reactions necessary for their own maintenance and replication, mitochondria participate in a few biosynthetic pathways of primary benefit to the rest of the cell. For example, the synthesis of heme (needed for cytochromes and myoglobin as well as for hemoglobin) begins with a mitochondrial reaction catalyzed by the enzyme, δ-aminolevulinic acid synthetase.

Some of the early steps in the conversion of cholesterol to steroid hormones in the adrenal cortex are also catalyzed by mitochondrial enzymes. The inner membranes of mitochondria in these cells have a unique configuration consisting of tubules and vesicles rather than the platelike cristae we are accustomed to seeing. These "tubulovesicular" cristae resemble the smooth endoplasmic reticulum with which many of the later enzymes in the steroid pathways are associated (see Fig. 7-10).

Transport Properties. As noted earlier, the inner mitochondrial membrane is impermeable to all but selected metabolites. It must therefore contain transport systems to control passage in and out. ADP and ATP, for example, are exchanged one for the other by a specific carrier. There is also a phosphate-hydroxyl exchange and additional carriers for certain of the carboxylic acids.

Some substances for which one might expect easy mitochondrial entry do not get in at all. Thus, we stated that mitochondria can oxidize cytoplasmic NADH. That is not quite correct, for mitochondrial inner membranes are not permeable to NADH. They are, however, permeable to malate and aspartate, presumably through specific carriers. Because malate dehydrogenase is found both inside and outside mitochondria, "reducing equivalents" are made available to the mitochondria as follows:

$$H^+ + NADH + OAA \longrightarrow malate + NAD^+$$

$$\uparrow \qquad \qquad \downarrow$$

aspartate

↕ (inner mitochondrial membrane) outside / inside

aspartate

$$\uparrow \qquad \qquad \downarrow$$

$$H^+ + NADH + OAA \longleftarrow malate + NAD^+$$

Aspartate:
COO⁻ | HC—NH₂ | CH₂ | COO⁻

Malate:
COO⁻ | HC—OH | CH₂ | COO⁻

OAA (oxaloacetate):
COO⁻ | C=O | CH₂ | COO⁻

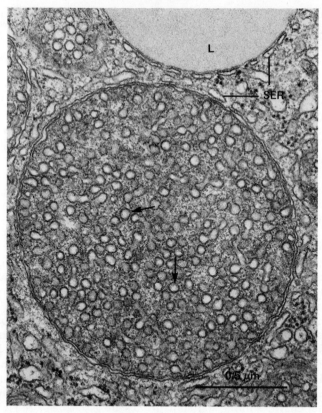

FIGURE 7-10 Mitochondrion from the Adrenal Cortex. These cells produce steroid hormones. Some of the steps are localized in mitochondria having "tubulovesicular" cristae (*arrows*) that look much like smooth endoplasmic reticulum (SER). Lipid droplet (L). [Courtesy of D. S. Friend and G. Brassil, *J. Cell Biol.*, **46**: 252 (1970).]

Note that OAA is converted to aspartate before membrane passage. Conversion is accomplished by *transamination* with another α-keto/α-amino pair: α-ketoglutarate/glutamate.

The *malate shuttle*, just described, is reversible and is widely distributed. Another mechanism for accomplishing the same task, the *glycerol phosphate shuttle*, is unidirectional and is found in fewer cell types. It has been studied mostly in insect flight muscles, where extramitochondrial NADH is used to reduce dihydroxyacetone phosphate (DHAP) to glycerol phosphate (Gly-P), which is then reoxidized within the mitochondrion by a flavoprotein.

$$\begin{array}{c} H_2C-OH \\ | \\ HCOH \\ | \\ H_2C-OPO_3^{2-} \\ \text{Glycerol-P} \end{array} \longrightarrow \begin{array}{c} H_2C-OH \\ | \\ C=O \\ | \\ H_2C-OPO_3^{2-} \\ \text{DHAP} \end{array} + 2H^+ + 2e^-$$

The complete reactions are

outside: $H^+ + NADH + DHAP \rightarrow Gly\text{-}P + NAD^+$
inside: $Gly\text{-}P + FAD \rightarrow DHAP + FADH_2$
sum: $H^+ + NADH + FAD \rightarrow NAD^+ + FADH$

The sequence has a large negative standard state free energy change ($\Delta G° = -38$ kJ/mole). Hence, it is not very efficient, but it is very effective at accumulating reducing power.

These transport systems, even the glycerol phosphate shuttle, technically are examples of facilitated diffusion. However, mitochondria also have several active transport systems, mostly driven by electrochemical gradients. The most important for our purposes is the one responsible for calcium uptake.

Ca^{2+} uptake by mitochondria can occur against significant gradients and at the expense of electron transport. In fact, active transport of Ca^{2+} suppresses ADP phosphorylation. This competition for a common source of energy (the electrochemical gradient) is further support for the chemiosmotic hypothesis.

Mitochondria of animal cells can accumulate very large amounts of Ca^{2+} and can store it in part as a calcium phosphate gel. Most of the Ca^{2+} in a typical cell is sequestered in this way and can be released in response to appropriate stimuli. The apparent role of cyclic AMP in this release was mentioned in Chap. 4 and will be part of our discussion several times again in later chapters.

7-4 THE CHLOROPLAST

The aerobic extension of glycolysis provided by mitochondria results in the complete oxidation of glucose to carbon dioxide and water, with a portion of the energy released captured in the form of ATP. The leaves of higher plants are able to carry out exactly the opposite sequence—i.e., the synthesis of glucose and other carbohydrates from carbon dioxide and water, evolving oxygen in the process. The energy required for this synthetic pathway is provided by light (hence the term *photosynthesis*), captured and transformed to a chemical equivalent by organelles called chloroplasts.

Plastids. Chloroplasts belong to a family of organelles known as *plastids*. Plastids are found only in plant cells. They are enclosed by double membranes, each membrane having the usual trilaminate structure, and they have characteristic pigments such as chlorophyll and carotenoids. The former is a heme-related lipid that is responsible for the green color of leaves. The carotenoids, which come in various shades of yellows and reds, as a family are related to vitamin A.

Proplastids (see Fig. 7-11) seem to be the undifferentiated precursors of other plastids, including the leukoplasts and chloroplasts. *Leukoplasts* are specialized organelles capable of accumulating and storing for future use various oils, fats, starch, and proteins. *Chloroplasts*, as already noted, are the major photosynthetic members of this family. They are bright green because of their high concentration of chlorophyll. If this pigment is lost, as when fruit ripens or leaves turn color in the fall, the other pigments are unmasked and the chloroplasts may become *chromoplasts*.

Plastids are self-replicating and contain DNA and a protein-synthesizing capacity comparable to that of mitochondria. The various mature plastids are also to some extent interconvertible.

Chloroplast Structure. Chloroplasts are usually somewhat larger than mitochondria—typically 2–3 μm thick by 5–10 μm long. They are also often banana-shaped or biconvex, like a lens. Some of the internal features of these organelles are large enough to be visible in the light microscope and hence were discovered during the last century. Their

CHAPTER 7
The Oxidative Organelles

details, of course, only became known with the advent of electron microscopy.

In 1833, A. Meyer reported that the internal structure of the chloroplast consists of dense cylinders which he named *grana*, embedded in a

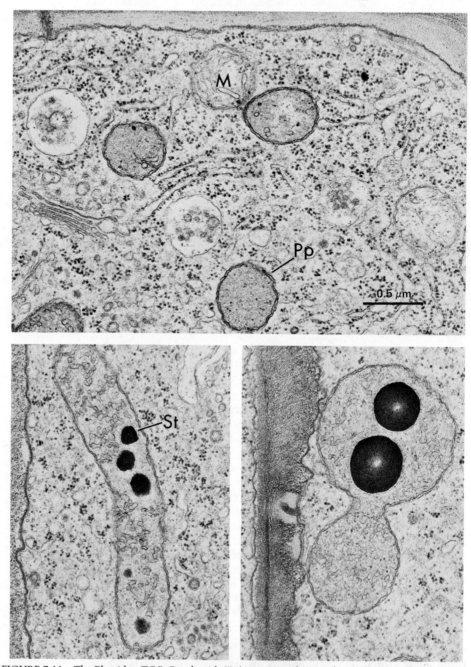

FIGURE 7-11 The Plastids. *TOP:* Proplastids (Pp) in a sieve element from a bean seedling. Note mitochondrion (M). Note also the numerous ribosomes and (at left) the dictyosome. *BOTTOM:* Sieve element plastids containing starch granules (St). The plastid on the right appears to be in the process of replication. [Courtesy of B. A. Palevitz and E. H. Newcomb, *J. Cell Biol.*, **45**: 383 (1970).]

SECTION 7-4
The Chloroplast

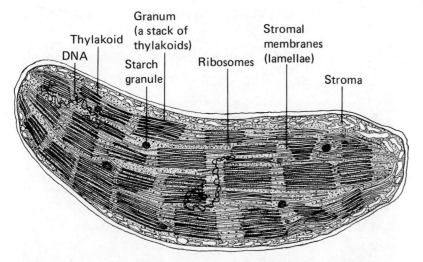

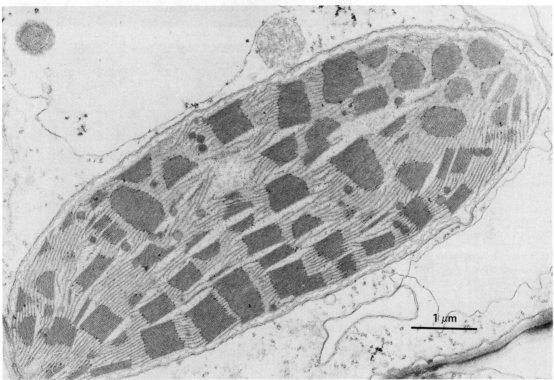

lighter material, the *stroma*. There are 40–80 grana in a typical chloroplast, each composed of layers of membranes that are paired off to form disks or sacks called *thylakoids* (Fig. 7-12). Each granum usually consists of five to thirty sacks, 0.25–0.8 μm in diameter with membranes about 0.01 μm (100 Å) thick. The membranes of the sacks are interconnected by tubes or lamellae that extend through the stroma to adjacent grana (see Fig. 7-13).

The above description is that of a typical chloroplast from most cells of higher plants. Other configurations are also seen. In particular, some

FIGURE 7-12 The Mature Chloroplast. *TOP:* Diagramatic representation. *BOTTOM:* A maize leaf chloroplast. [Micrograph from *Brookhaven Symp. Biol.*, **19**: 353 (1966), courtesy of L. K. Shumway, Program in Genetics and Dept. of Botany, Washington State Univ.]

311

CHAPTER 7
The Oxidative Organelles

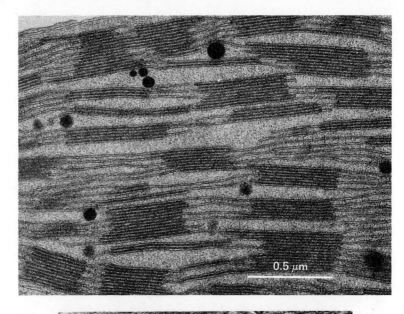

FIGURE 7-13 Thylakoids. The chloroplast granum is a stack of flattened sacks (disks) called thylakoids. *TOP:* Side view of several grana. Note the interconnections. *BOTTOM:* Face view of a single thylakoid after freeze-fracture. The particles seen here in an unusual paracrystalline array have been called quantasomes. They probably correspond to the ES particles to be described in Fig. 7-19. [Top photo courtesy of W. M. Laetsch; bottom courtesy of R. B. Park and the University of California Lawrence Berkeley Laboratory.]

chloroplasts (e.g., the red alga in Fig. 7-14) lack grana and utilize instead an extensive network of stromal lamellae.

Development and Distribution of Chloroplasts. As already noted, chloroplasts develop from proplastids. The inner membrane invaginates to form sacks, vesicles, and tubes that eventually rearrange to yield mature chloroplast grana. Development of this highly organized pattern is dependent upon adequate light. At low levels, a less organized pattern develops, often lacking grana and with aggregations of vesicles that form *prolamellar bodies*, one of which is seen in Fig. 7-15. An increase in illumination causes reorganization of these membranes and emergence of the usual mature structure.

SECTION 7-4
The Chloroplast

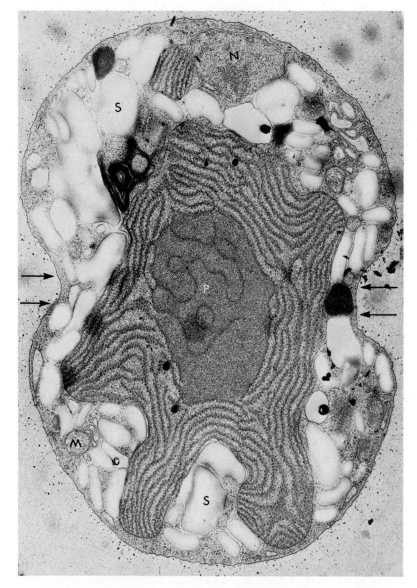

FIGURE 7-14 The Chloroplasts of Algae. The cytoplasm of this red alga *(Porphyridium cruentum)* is dominated by a single large red chloroplast. The spheres on the chloroplast lamellae have been named phycobilosomes because they contain accessory pigments called phycobilins. A prominent pyrenoid (P), which serves as a center for starch synthesis, is characteristic, as is the lack of grana. This particular cell was about to divide (arrows). One of the two daughter nuclei (N) is evident. The long axis of the cell is about 8 μm. Starch granule (S). Mitochondrion (M). [Courtesy of E. Gantt and S. F. Conti, *J. Cell Biol.*, **26**: 365 (1965).]

A fully developed cell from a leaf of higher plants will contain large numbers of chloroplasts (e.g., 40), so that hundreds of thousands are found beneath each square millimeter of surface. The position of chloroplasts within cells is not highly fixed, but may vary from time to time. They often move with cytoplasmic streams, for example, and in some cases seem to change orientation with respect to light.

The photosynthetic activities of chloroplasts will be discussed in two parts, the *light reaction* and the *dark reaction*. The important distinction between them was first recognized in the 1930s (see Table 7-2). In the former, light energy is used to oxidize water to oxygen, passing the electrons to $NADP^+$ to create NADPH and, at the same time, generating some ATP. The light reaction seems to be centered in the membranes. The dark reaction, which is catalyzed by soluble enzymes found in the

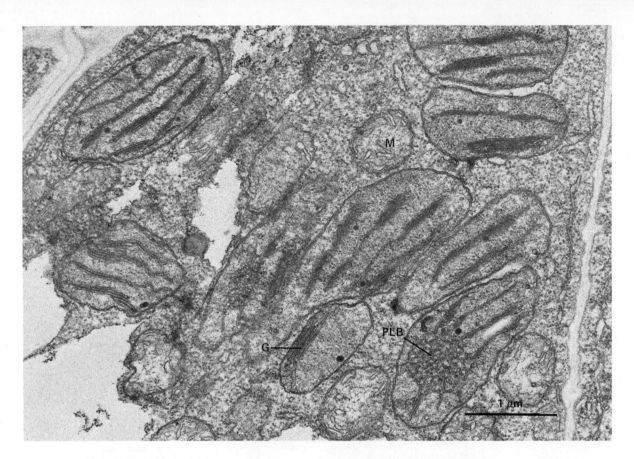

FIGURE 7-15 Immature Chloroplasts. In addition to their capacity for self-replication, chloroplasts may arise from proplastids. An early stage in that process is illustrated here (sugar cane mesophyll cell), showing some elementary grana (G) and a prolamellar body (PLB). Note mitochondrion (M). [Courtesy of W. M. Laetsch, *Sci. Prog. (Oxford),* **57**: 323 (1969).]

stroma, consists of those steps by which the "reducing power" of NADPH is used to turn CO_2 into carbohydrate.

7-5 THE PHOTOSYNTHETIC LIGHT REACTION

The oxidation of glucose to carbon dioxide and water is accompanied by a $\Delta G°$ of -2870 kJ/mole. Hence, that much energy is needed to synthesize glucose from carbon dioxide and water. In fact, the input of free energy must be even more if the equilibrium for the overall reaction is to lie significantly on the side of glucose formation. The energy comes from light, for the chloroplast is an energy transducer. It can convert light energy to chemical energy, much as a solar battery uses light to run a transistor radio.

Light is an electromagnetic radiation, different from radio, radar, infrared rays (heat rays), x rays, and gamma rays only in the frequency with which the source is oscillating. Light would fall between infrared and X rays in this list, which is ordered to reflect progressively faster frequencies or shorter wavelengths. The frequency of oscillation of the source, ν, is related to the wavelength, λ, of the emitted radiation by the following equation:

$$\lambda = \frac{c}{\nu} \qquad (7\text{-}10)$$

TABLE 7-2. The Early History of Photosynthesis

1727	Hales concludes that plants are nourished, in part, from the atmosphere.
1772	Priestly discovers the evolution of oxygen by plants. He also shows that oxygen is consumed by animals.
1779–96	Ingen-Housz shows that carbon dioxide is consumed by plants as oxygen is evolved, and that light is essential to the process. He associated these reactions with the green parts of plants.
1804	de Saussure notes and measures the fixed stoichiometry between CO_2 consumption and oxygen evolution in plants.
1837	Dutrochet recognizes that chlorophyll is essential to oxygen evolution by plants.
1862	Sachs proves that starch is synthesized by plants in a light-dependent reaction (photosynthesis).
1882	Engelmann shows that red light is the most efficient stimulator of photosynthesis.
1883	Meyer first describes details of chloroplast structure.
1913	Wilstätter and Stoll isolate chlorophyll and later determine its structure.
1919	Warburg finds that the efficiency of photosynthesis is higher with intermittent light.
1922	Warburg and Negelein first measure the efficiency of photosynthesis.
1923	Thunberg recognizes that carbon dioxide is reduced and water oxidized during photosynthesis.
1936	Wood and Werkman distinguish between the "light reaction" and the "dark reaction."
1938	Hill finds that isolated chloroplasts evolve oxygen when illuminated, provided that an appropriate electron acceptor is also made available.
1941	Ruben, Randall, Kamen, and Hyde show that the oxygen liberated during photosynthesis comes from water.
1948	Calvin and Benson show that phosphoglycerate is an early product of CO_2 fixation.

where c is the speed of light, 3×10^{10} cm/sec *in vacuo*. Frequency is measured by the rate at which electrical and magnetic fields rise and collapse as radiation passes a given point, while wavelength is the distance a wave train travels in one complete cycle, or the distance between conjugate points in the oscillating field.

The Photoelectric Effect. Einstein suggested in 1905 that light and other electromagnetic radiations travel in discrete packets called *photons*, and that when light interacts with matter it does so by annihilating complete photons, never a part of one. This hypothesis explained Philipp Lenard's 1899 observation that electrons ejected from a polished metal plate by light have velocities, and hence energies, that are independent of the intensity of the light. According to Einstein's *photoelectric theory*, it takes one photon to eject one electron. Thus, increasing the intensity of light, or flux of photons, only increases the number of electrons ejected, not their velocities. On the other hand, changing the wavelength of

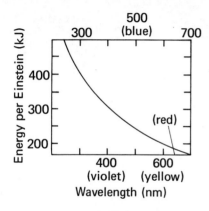

FIGURE 7-16 The Relationship between Wavelength and Energy.

light does change the velocities of ejected electrons, implying that the energy of a photon must be related to its wavelength. That relationship (see Fig. 7-16) was found by Max Planck in 1900 to be

$$E = h\nu = \frac{hc}{\lambda} \qquad (7\text{-}11)$$

where h is called *Planck's constant* and has a value of 6.6256×10^{-34} J-s. Avogadro's number (6.023×10^{23}, or a "mole") of photons is called an *einstein*. At $\lambda = 700$ nm (7000 Å or 7×10^{-5} cm, a dark red to the eye), one einstein of photons has an energy of about 170 kJ, calculated as follows:

$$N_{av}E = \frac{N_{av}hc}{\lambda} = \frac{(6.023 \times 10^{23})(6.626 \times 10^{-34})(3 \times 10^{10})}{7 \times 10^{-5}} = 171 \text{ kJ}$$

When a molecule absorbs a photon of light, it is absorbing a *quantum* of energy. Several things can happen to that energy: (1) It can be dissipated in molecular motion, manifest as heat. (2) It can be reemitted as a new photon of light at a longer wavelength, with the shift representing losses to other processes—if reemission occurs very quickly, it is called *fluorescence*; if there is a long lag (milliseconds to seconds) between absorption and reemission, the process is called *phosphorescence*. Or (3) the energy of light can cause a chemical change in the compound that absorbs it. It is this latter possibility that interests us, since it is one of the things that can happen when a molecule of chlorophyll absorbs a photon.

When chlorophyll absorbs a photon of light, an electron may be ejected. That electron may find its way to a molecule of NADP⁺ and, along with a second electron, cause its reduction to NADPH. Oxidized chlorophyll, of course, then needs an electron donor to resume its original state. In the blue-green algae and higher plants (but not in photosynthetic bacteria), the electron donor is water, which becomes oxidized to oxygen. Since this is a reaction of great importance to us, it will be worth our while to learn something about how it is accomplished.

Light Capture in Chloroplasts. According to the above statements, electrons ejected from chlorophyll cause the reduction of NADP⁺. Ejection is the result of light absorption and the photoelectric effect. We might expect, therefore, that the *action spectrum* of photosynthesis—the relative effectiveness of various wavelengths of light—should follow the absorption spectrum of pure chlorophyll. In fact, this is not the case, as you can see from the graphs in Fig. 7-17. There are two reasons for this discrepancy: (1) Chlorophyll *in vivo* does not have the same absorption spectrum as chlorophyll *in vitro*; and (2) other pigments are capable of transferring energy from absorbed light to chlorophyll, thus considerably broadening the action spectrum toward the middle wavelengths of visible light.

The transfer of energy from pigment to pigment is important to the efficiency of photosynthesis, because it means that a photon of light from almost any part of the visible spectrum (about 400 nm–700 nm) may be absorbed by a chloroplast and its energy utilized in photosynthesis—but not all of its energy, for the transfer is not 100% efficient. We said earlier that fluorescence is one of the options when a molecule absorbs a photon of light. Fluorescence refers to the emission of a new photon at a longer wavelength and therefore lower energy. (The shift to a longer

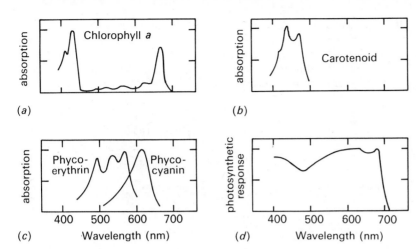

FIGURE 7-17 Light Absorption and the Photosynthetic Response. Chloroplasts contain a variety of pigments, each of which absorbs light maximally in a different region of the visible spectrum. (See (a)—(c) for examples.) Since the energy of captured photons can be transferred from pigment to pigment, photosynthesis can be sustained by light of any visible wavelength. Note the two absorption peaks of chlorophyll. The short-wavelength peak is known as the Soret band and is found also in heme compounds. Note also the sharp drop in photosynthetic response in the far red, at 700 nm.

wavelength, called a *Stokes shift,* is necessary because a certain amount of energy is lost to heat or entropy in the process.) As you can see from Fig. 7-17, the various pigments have quite different absorption maxima, meaning that a photon of given wavelength may be preferentially absorbed by a pigment other than chlorophyll. This "excited" molecule may transfer a portion of the absorbed energy to a pigment with a longer-wavelength absorption maximum, a process closely akin to fluorescence but without the production of light.

The transfer of an absorbed photon continues until the pigment with the longest-wavelength (lowest energy) absorption maximum is reached. If that pigment cannot do something useful with the energy, it will then be dissipated internally or lost by true fluorescence. In the chloroplast, this longest-wavelength absorber is a special fraction of the chlorophyll. This fraction is the only chlorophyll fraction that is capable of ejecting electrons.

Thus, energy from light absorbed at a wide range of wavelengths is funneled, in about 10^{-9} sec, through the pigments to a long-wavelength "trap" that utilizes it. According to this scheme, a low-energy photon absorbed by the trap molecule itself should be as effective as a higher-energy photon absorbed by another pigment and transferred to it. The efficiency of photosynthesis, therefore, depends on the wavelength of light used to support the process.

But what is the nature of this energy trap? According to the action spectrum, it must absorb light at or very near 700 nm, a wavelength that is much longer than the usual absorption maximum of chlorophyll. And yet the trap molecules are chlorophyll, distinguished from the bulk chlorophyll only by their environment.

The absorption spectrum of any molecule depends on both its structure and its environment. In the case of absorption by delocalized π electrons (as in chlorophyll) the size of the area over which they can move is proportional to the wavelength at which light will be absorbed. For example, the peptide bond absorbs at about 200 nm, the benzene rings of tyrosine or phenylalanine at about 280 nm, the carotenes of chloroplasts at 400–500 nm, the heme group at about 400 and again at about 550 nm, and chlorophyll (which has a more extensive side-chain structure) at 400 and again at about 650 nm. However, these figures can be altered by a change in the environment of the absorbing groups. Chlorophyll *a*, universal to eukaryotes, has an absorption maximum of 660 nm in solution, 670 in an amorphous monolayer, and 740 in a crystalline monolayer. The shift is due to increasing interactions among molecules as they become more closely packed.

$$R \begin{cases} -CH_3 \text{ in chlorophyll } a \\ -CHO \text{ in chlorophyll } b \end{cases}$$

Since the chlorophyll molecules in chloroplasts are not all in the same environment, we find that they absorb light over a rather wide range. Chlorophyll *a*, which has an *in vivo* absorption maximum at 675 nm, actually has significant absorption from about 660 to around 700 nm. This observation led Bessel Kok to suggest that the various fractions of chlorophyll constitute a cooperative unit, part of a *pigment system*. Only a few (about 0.3%) of the chlorophyll *a* molecules have an absorption maximum at 700 nm. It is these that constitute the long-wavelength energy trap, designated *P700*.

The Dual Pigment System. The organization of the chloroplast pigments, as we have developed it so far, fails to explain an interesting observation made at the University of Illinois by Robert Emerson and his colleagues in 1956. They found that while light of 700 nm is very inefficient in producing photosynthesis in chloroplasts by itself, the addition of a second beam of light at a shorter wavelength (they used 650 nm) greatly improves the overall efficiency. In other words two beams, one at 650 and one at 700 nm, cause more oxygen production per photon than the sum of the two used individually. But if all absorbed energy is funneled to P700, then one beam should have the same "quantum efficiency" as the other—i.e., it should produce as much oxygen for a given number of photons absorbed. The dual light, or "Emerson effect," led to the suggestion that at least two different traps exist, each with a pigment system feeding energy into it. Later experiments indicated that the second trap has an absorption maximum of 680–690 nm in some systems, leading to its designation as *P680* or *P690*.

The pigments in chloroplasts thus seem to be divided into two parts, called *pigment system I* (PS I) and *pigment system II* (PS II), each with a long-wavelength trap (*reaction center*) into which absorbed energy is funneled. In addition to accessory pigments absorbing at shorter wavelengths, each system in higher plants contains both chlorophyll *a* and chlorophyll *b*—the latter absorbs maximally at 650 nm. (Some lower plants do not contain chlorophyll *b*.) System I has chlorophyll *a* fractions exhibiting absorption maxima at several wavelengths between 650 and 700 nm, reflecting different environments of the protein-bound chlorophyll, in addition to some chlorophyll *a* that acts as P700. System II also contains several discrete chlorophyll absorption maxima and a trap, P690.

The two pigment systems are coupled chemically rather than through direct energy transfer. This was made clear by experiments in which the two beams of light used for the Emerson effect were flashed one at a time, with several seconds between them. That such a time lapse still allows the shorter-wavelength light to reinforce the longer-wavelength beam is interpreted to mean that a chemical intermediate is produced, for if coupling between the pigment systems were electrical, the transfer would occur a billion times faster than it does.

In addition to the pigments already discussed, chloroplast membranes contain cytochromes, a plastoquinone, and the metal-containing proteins plastocyanin and ferredoxin. By using light of the appropriate wavelength, L. N. M. Duysens and coworkers showed in 1961 that one of these components, cytochrome *f*, is preferentially oxidized when PS I is illuminated, while the illumination of PS II hastens its return to the reduced state. This experiment is possible because the oxidized form of the cytochrome absorbs less light at 420 nm than does its reduced form. (420 nm is a wavelength that is not readily absorbed by either pigment system in the red algae that they used.) The work was further extended to show that the chemical reaction by which the two pigment systems are coupled involves the reduction and oxidation of a whole set of components similar to the electron transport chain of mitochondria. The set includes the cytochromes, plastoquinones, and plastocyanin.

We can now fit some of the pieces together. There are two pigment systems and an electron transport chain between them (see Fig. 7-18). The direction of electron flow, as determined from the experiment described in the last paragraph, tells us that PS II must be responsible for catalyzing the removal of electrons from water to create oxygen and that PS I must be responsible for reducing $NADP^+$. Apparently electrons removed from water are transferred to the electron transport chain by PS II, and from the chain to $NADP^+$ by PS I. This latter step involves an intermediate, ferredoxin, which is reduced by illumination of pigment system I and which in turn reduces $NADP^+$ to NADPH.

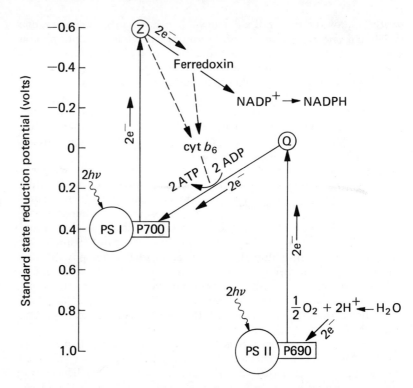

FIGURE 7-18 The Dual Pigment System. The shorter-wavelength trap, sometimes designated P690, captures light energy and uses it to oxidize water. The electrons are passed to an electron transport chain at some acceptor, Q, from which they cyclically oxidize and reduce plastoquinone, cytochrome f, and plastocyanin before being passed to PS I. An additional photon input, trapped by P700, "boosts" the electrons to a level from which NADP$^+$ can be reduced. Cyclic electron flow (dashed lines), which utilizes PS I only, provides a way of using light to make ATP. (Z designates the primary electron acceptor of photosystem I, which is an iron-sulfur protein. The symbol $h\nu$ represents a photon of light.)

The concept of a dual pigment system has received a good deal of support from the isolation of membrane fragments that exhibit only PS I or PS II activity. These structures have been associated with two classes of membrane particles revealed by freeze-fracture of the thylakoids (see Fig. 7-19): the smaller, more densely packed particles appear to have PS I activity while the larger, more widely spaced particles have PS II activity.

Photophosphorylation. The scheme outlined above explains the production of reduced nicotinamide nucleotides, using water as an electron donor and light as a source of energy. The nucleotides are used to reduce carbon dioxide to the carbohydrate needed to support synthesis of other cellular constituents, including the chloroplasts themselves. The process of producing carbohydrate requires ATP as well as NADPH. One might assume that all the ATP needed could be produced by mitochondria in the same cell, using reduced nicotinamide nucleotides as an energy source. However, measurements of the rate of carbon dioxide fixation show that mitochondria could not possibly provide ATP fast enough, implying that ATP must also be produced as a by-product of photosynthesis. This assumption was confirmed by measurements made on isolated chloroplasts.

ATP is produced by chloroplasts through electron transport. Although the process is comparable to that found in mitochondria, there are fewer intermediates involved in the transport chain and their order has been difficult to determine. Recent work indicates that no simple linear array of a few molecules is adequate to describe the chain; rather, electrons pass from PS II to a pool of plastoquinones and from there via cytochrome f to a pool of plastocyanin. The latter is able to pass electrons to the PS I complex.

SECTION 7-5
The Photosynthetic Light Reaction

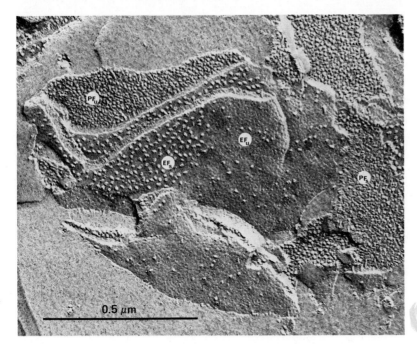

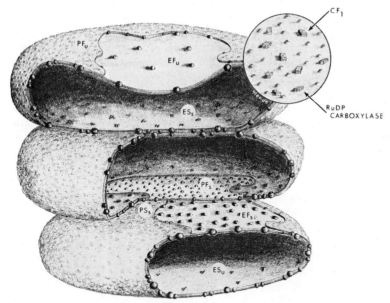

FIGURE 7-19 Freeze-fracture of Chloroplast Membranes. The nomenclature is as follows: F is a fracture face, S an unfractured surface; P is the "protoplasmic leaflet" or that half of the membrane bilayer (whether fractured or not) that is on the outside of the thylakoid, while E is the "exoplasmic leaflet" on the interior of the thylakoids; the subscript s is a thylakoid that is part of a granum stack while u refers to unstacked membranes. The smaller, closely spaced particles associated with protoplasmic (P) leaflets have PS I activity. The larger, more widely-spaced particles associated with exoplasmic (E) leaflets have PS II activity. The "quantasomes" seen in Fig. 7-13 appear to be an unusual array of PS II particles that protrude above the exoplasmic surface (ES). (CF_1 is discussed on p. 322 and RuDP carboxylase on p. 325.) [Courtesy of C. J. Arntzen and Paul Armond. From *Current Topics in Bioenergetics*, vol. VII, (R. Sanadi and L. Vernon, eds.). © 1977, Academic Press, N.Y.]

This electron transfer, which appears to involve also at least one b-type cytochrome, generates ATP, a coupling that is referred to as *photophosphorylation*. It is thought that the chemiosmotic hypothesis presented earlier as the probable mechanism for oxidative phosphorylation in mitochondria is equally applicable to photophosphorylation in chloroplasts.

The components responsible for photophosphorylation are comparable to those used by mitochondria for oxidative phosphorylation, but their arrangement is inside out. While mitochondria eject protons during oxidation, chloroplasts accumulate protons. The external medium thus becomes alkaline instead of acidic, with gradients up to 3.5 pH units having been measured. When artificially induced, these gradients have been shown to cause ATP formation in the dark as accumulated protons pass outward again.

Spherical particles some 90 Å in diameter have been identified as the chloroplast ATPase. The particles, called coupling factor 1 (CF 1), sit on the *outside* of the thylakoids rather than on the inside of the membrane as in mitochondria (see Fig. 7-19). This difference is, of course, consistent with the different direction of proton transfer during electron transport.

The synthesis of ATP by electron transport is an attractive prospect because it utilizes a portion of the 0.6-volt drop from PS II to PS I as the electrons make their way between the two pigment systems. This drop is much less than in the mitochondrial electron transport chain (1.14 volts) but it is equivalent to about 50 kJ/mole of electrons, an amount of energy that would be enough to produce one ATP per electron transferred. Whether the process is really that efficient, however, is not firmly established.

Cyclic Photophosphorylation. The scheme just outlined includes both NADPH and ATP production. ATP production can occur by itself, however. When the bulk of available NADP$^+$ is already in the reduced form, electrons can be shunted from the PS I receptor back to the main electron transport chain (see Fig. 7-18). The result is additional ATP production via photophosphorylation but without the release of oxygen and without reduction of NADP$^+$. This process, called *cyclic photophosphorylation*, is purely a light-driven ATP synthesizer.

The Stoichiometry and Efficiency of Photosynthesis. The implication so far has been that one photon can cause the ejection of one electron from either pigment system. This scheme is still challenged from time to time but we shall assume for the following discussion that it is entirely correct. In addition, we shall assume that every electron traversing the electron transport chain produces one ATP from ADP and P_i. The light reaction can then be summarized as follows, where $h\nu$ represents a photon:

(1) Light reaction II, catalyzed by PS II, is

$$H_2O \xrightarrow{2h\nu} \tfrac{1}{2}O_2 + 2H^+ + 2e^- \tag{7-12}$$

(2) Light reaction I, catalyzed by PS I, is

$$2e^- + H^+ + NADP^+ \xrightarrow{2h\nu} NADPH \tag{7-13}$$

(3) The total reaction, including ATP synthesis, is

$$H_2O + NADP^+ + H^+ + 2ADP^{3-} + 2P_i^{2-} \xrightarrow{4h\nu}$$
$$\tfrac{1}{2}O_2 + NADPH + 2ATP^{4-} \tag{7-14}$$

In other words, eight photons produce one molecule of oxygen (O_2) plus two molecules of NADPH with as many as four ATPs as a by-product of electron transport.

The air oxidation of NADPH has about the same change in energy as the air oxidation of NADH, so the reduction of one NADP$^+$ at the expense of water must have a $\Delta G°$ of about +218 kJ/mole. We determined earlier that light at 700 nm has an energy of about 170 kJ/einstein (Avogadro's number of photons). The four einsteins that it takes to produce one mole of NADPH thus represent about 680 kJ. If in addition to NADPH a full two moles of ATP are produced, requiring about 66 kJ, the overall process conserves 284 kJ (218 + 66) of the 680 available, or a little over 40%. The figure is less if ordinary daylight is assumed, since higher-energy photons are presumably no more effective than photons with wavelengths of 690 or 700 nm corresponding to the traps. In either case, however, the photosynthetic light reaction is a very efficient process.

Bioluminescence. Artificial pH gradients in vesicles made from chloroplasts can drive electron transport in the reverse direction, which in turn causes light emission. This reaction can be considered a form of *bioluminescence*, defined as the production of light by living things. Fireflies, certain fungi, beetles, bacteria, marine and terrestrial worms, and fish, including most deep-sea species living in total darkness, exhibit bioluminescence. The mechanism of light production is not the reverse of photosynthetic light capture, yet the similarities are there; instead of using light to drive reactions that are coupled to ATP synthesis and coenzyme reduction, ATP or reduced coenzymes are used to drive steps that result in light emission.

Raphael Dubois, experimenting in 1887 with a species of clam, discovered that light could be obtained from mixing two substances that he called *luciferin* and *luciferase* (after Lucifer, the Roman god whose torch brought each dawn). Although Dubois was not able to identify the chemical nature of these compounds, W. D. McElroy and his associates at Johns Hopkins University showed in 1940 that the corresponding luciferin of fireflies is a reduced phenolic compound whose oxidation in an ATP-driven, luciferase-catalyzed reaction results in light emission. The electron acceptor for the oxidation is O_2, and one photon is emitted for each luciferin molecule oxidized. The electronic configuration of luciferin is excited by the reaction; light emission occurs when the molecule decays back to the ground state. The overall reaction for firefly bioluminescence is thus

$$LH_2 + ATP + O_2 \xrightarrow[Mg^{2+}]{luciferase} H_2O + AMP + L^*$$

where LH_2 is reduced luciferin and L^* the oxidized, excited luciferin, which then decays:

$$L^* \rightarrow L + light$$

Bioluminescence is utilized by the firefly as a mating signal. Other organisms use bioluminescence to attract prey or for protection, and many different luciferin-luciferase systems have been identified (see Fig. 7-20). In some cases the luciferin and luciferase are separate, as they are in the firefly, but in other cases luciferin and luciferase are tightly bound together. An example of the latter is *aequorin*, a "photoprotein" from the jellyfish *Aequorea* that emits light in the presence of ATP and Ca^{2+}.

Bioluminescent reactions have been very useful to biochemists and cell physiologists. The firefly system, for example, has been used as a sensitive indicator for ATP, and aequorin as an indicator for either ATP or Ca^{2+}.

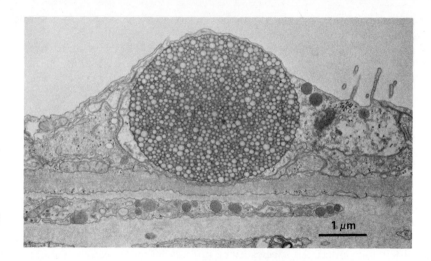

FIGURE 7-20 The Bioluminescent Organelle of a Sea Pansy. *Renilla mulleri* localizes its bioluminescent reactions to special cells (sometimes called photocytes) having a membrane-enclosed collection of small vesicles (0.1–0.2 μm diameter) or lumisomes. Light production is controlled by Ca^{2+} availability within the lumisomes. [Courtesy of B. Spurlock and M. Cormier, *J. Cell Biol.*, **64**: 15 (1975).]

7-6 THE PHOTOSYNTHETIC DARK REACTION

The light reaction just described accounts for the capture of radiant energy and explains how this energy is made available for biosynthesis in the form of NADPH and ATP, evolving oxygen in the process. The biosynthetic reaction most often associated with photosynthesis is, of course, the production of glucose from carbon dioxide and water. Since glucose synthesis can occur without light, as long as NADPH and ATP are available, it is termed the "dark reaction." Thus, when illuminated chloroplasts are transferred to the dark, carbon dioxide fixation continues to occur for some time, and can be prolonged even further by adding NADPH and ATP. In addition, we know that the light reaction is centered in the membranes while the dark reaction is enzymatically catalyzed within the stroma (see Fig. 7-21).

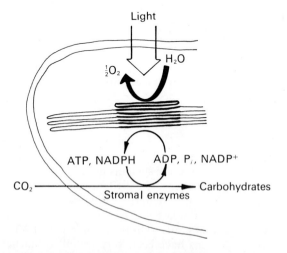

FIGURE 7-21 Localization of the Light and Dark Reactions. The light reaction is catalyzed by chloroplast lamellae, especially in the grana. The dark reaction, on the other hand, is catalyzed by enzymes of the stroma.

Carbon Dioxide Fixation. When radioactively labeled carbon dioxide ($^{14}CO_2$) is made available to illuminated chloroplasts, radioactive carbon appears very quickly in carbohydrate and later in other chemical fractions of the organelle. Melvin Calvin and his coworkers at the University of California at Berkeley traced the path of this fixed radioactive carbon in algae. They used a brief exposure to the isotope followed by rapid extraction of the chloroplast contents. The first stable product in which ^{14}C could be found was 3-phosphoglycerate (3-PGA), an intermediate in glycolysis from which any other chemical constituent of the cell can be made. Glucose, for example, is readily synthesized from 3-PGA by a reversal of the first six steps of glycolysis.

Closer examination of labeled PGA revealed that radioactivity appears first in its carboxyl carbon. Only after fairly long exposure to $^{14}CO_2$ does radioactive carbon appear in the other two positions in the molecule. This suggests that the fixation of carbon dioxide occurs by a carboxylation. One might assume that the carbon acceptor is a two-carbon compound, since PGA contains three carbons and CO_2 contains one. However, the primary carbon acceptor was soon identified as the five-carbon sugar derivative ribulose-1,5-diphosphate (RuDP). The result of carboxylation is a six-carbon addition compound that is immediately hydrolyzed to two molecules of PGA, thus partially offsetting an unfavorable reaction, carboxylation, with a favorable one, hydrolysis:

$$\begin{array}{c} H_2C-OPO_3^{2-} \\ | \\ C=O \\ | \\ HC-OH \\ | \\ HC-OH \\ | \\ H_2C-OPO_3^{2-} \\ RuDP \end{array} \xrightarrow[\text{carboxylation}]{^*CO_2} \begin{array}{c} H_2C-OPO_3^{2-} \\ | \\ {}^-OOC^*-C-OH \\ | \\ C=O \\ | \\ HC-OH \\ | \\ H_2C-OPO_3^{2-} \end{array} \xrightarrow[\text{hydrolysis}]{H_2O} \begin{array}{c} H_2C-OPO_3^{2-} \\ | \\ HC-OH \\ | \\ {}^*COO^- \end{array} + \begin{array}{c} COO^- \\ | \\ HC-OH \\ | \\ H_2C-OPO_3^{2-} \\ \text{3-PGA} \end{array}$$

The enzyme that catalyzes the carboxylation of RuDP is present in large quantities in green plants—up to about 15% of the soluble protein of spinach, for example. It is called *ribulose diphosphate carboxylase*, an enzyme that is found in aggregates loosely bound to membrane surfaces (see Fig. 7-19).

To prove their contention that RuDP is the carbon acceptor, Calvin and his colleagues followed the change in RuDP concentration when the supply of CO_2 was interrupted. The concentration of RuDP immediately increased while the concentration of PGA decreased, representing the expected buildup of substrate and removal of product from the blocked reaction.

The Calvin Cycle. It is obvious that some of the 3-PGA formed by the carboxylation and hydrolytic cleavage of RuDP must be used to regenerate RuDP, since otherwise carbon fixation would stop for lack of acceptor. Specifically, only the fraction corresponding to CO_2 addition, one carbon out of six, can be used for other purposes. The remaining five carbons are needed for RuDP regeneration.

The cyclic process whereby RuDP is regenerated is called the *Calvin cycle*, or the *Calvin-Benson-Bassham cycle*, and is represented in Fig.

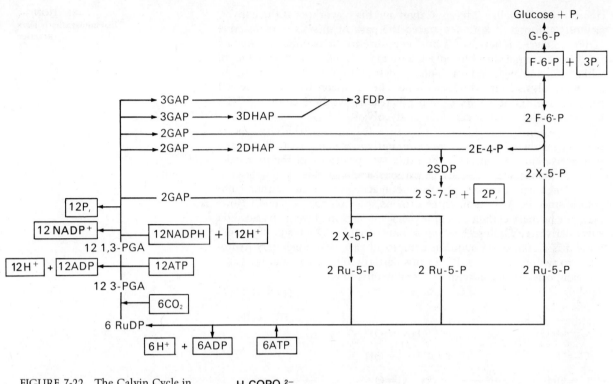

FIGURE 7-22 The Calvin Cycle in Spinach Chlorophasts. Input and output substances appear in boxes.

RuDP ribulose-1,5-diphosphate (Ru-5-P = ribulose-5-phosphate)

1,3-PGA 1,3-diphosphoglycerate (3-PGA = 3-phosphoglycerate)

GAP glyceraldehyde-3-phosphate

DHAP dihydroxyacetone-3-phosphate

F-6-P fructose-6-phosphate

X-5-P xyulose-5-phosphate

E-4-P erythrose-4-phosphate

SDP sedoheptulose-1-7-diphosphate (S-7-P = sedoheptulose-7-phosphate)

7-22. As you can see, 3-PGA is phosphorylated by ATP to 1,3-diphosphoglycerate, and then reduced to glyceraldehyde-3-phosphate (GAP). The reduction is accomplished by an NADPH-linked enzyme. Regeneration of RuDP starts with GAP and proceeds through four-, five-, six-, and seven-carbon intermediates.

The net carbon fixation reaction, neglecting RuDP, is

$$6CO_2 + 18ATP^{4-} + 12NADPH + 12H_2O \rightarrow$$
$$6H^+ + 18P_i^{2-} + 18ADP^{3-} + 12NADP^+ + C_6H_{12}O_6 \quad (7\text{-}15)$$

If we assign 33 kJ/mole as the energy of hydrolysis of ATP and 218 kJ/mole as the "reducing power" of NADPH, then glucose synthesis has available 3210 kJ/mole. Since the complete oxidation of glucose would yield 2870 kJ/mole, the process of glucose synthesis via the above reactions is very favorable, with an overall $\Delta G° = -340$ kJ/mole. Clearly, equilibrium lies strongly on the side of glucose formation.

According to the stoichiometry already discussed, Equation 7-15 requires at least 48 photons. Taking the most optimistic estimate of ATP production—one ATP per electron transported between pigment systems—we are left with 6 ATP after the 12 NADPH and 18 ATP are consumed. If this is in fact the *in vivo* situation, it represents an ATP surplus that would be available for other biosynthetic reactions. In addition, mitochondria continue to oxidize carbohydrates even in periods of illumination, thus generating still more ATP. (Mitochondria use much less oxygen than illuminated chloroplasts produce.) In times of darkness, of course, mitochondrial activity must provide all the energy necessary to run the cell, making the plant cell a little like a storage battery: It stores excess energy as carbohydrate during periods of illumination, and utilizes it when solar energy is not available.

Experiments with cell fractions indicate that carbohydrate synthesis can take place entirely within chloroplasts. In fact, protein synthesis and a certain amount of lipid synthesis—functions that we normally associate with other parts of the cytoplasm—can also take place in these organelles. As mentioned earlier, a portion of the protein synthesis in chloroplasts (e.g., some of the enzymes responsible for chlorophyll synthesis) is under the control of local (chloroplast) genes, though the cell nucleus remains by far the most important repository of genetic information.

Alternate Pathways of Carbon Fixation. The Calvin cycle just described is not the only carbon-fixing reaction sequence. In 1966 M. D. Hatch and C. R. Slack suggested an alternate pathway for carbon fixation in corn and certain other hot-weather plants having an unusual anatomy. These plants have thick-walled *bundle-sheath cells* interspersed between the vascular elements and *mesophyll cells* of their leaves.

Carbon dioxide, in this type of plant, diffuses into the leaf through openings called *stomata* and then is organically fixed in mesophyll cells by being added to phosphoenolpyruvate (PEP). The product is a four-carbon compound, normally an intermediate in the citric acid cycle, called oxaloacetate or OAA (see Fig. 7-23). Oxaloacetate or a closely related derivative then passes through channels (plasmodesmata) connecting adjacent cells to a bundle-sheath cell, the chloroplasts of which are often less mature looking and lack grana. (They are called *dimorphic chloroplasts*—see Fig. 7-24.) In the bundle-sheath chloroplasts, OAA re-

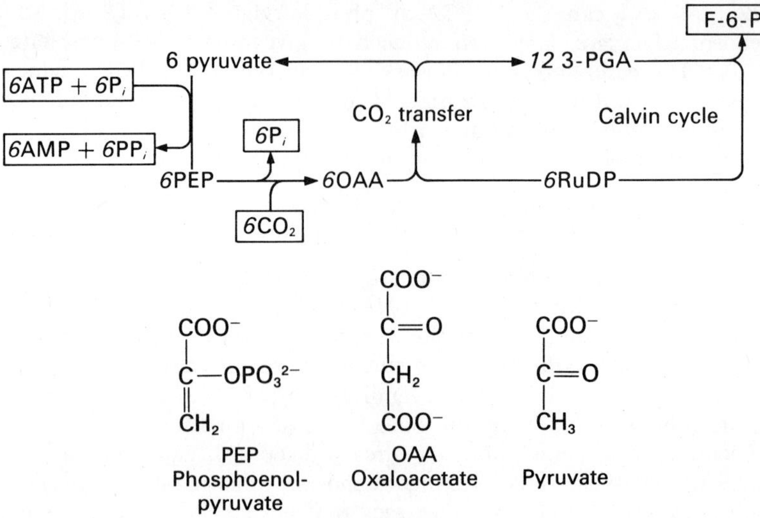

CHAPTER 7
The Oxidative Organelles

FIGURE 7-23 The Hatch and Slack (C4) Pathway for Carbon Fixation. The net reaction is the same as for the Calvin cycle except for the input of 6ATP → 6AMP, which is the equivalent of an additional 12 ATP consumed in the production of each molecule of glucose. Carbon fixation occurs in mesophyll cells near the leaf openings (stomata) while the Calvin cycle reactions are limited mostly to bundle-sheath cells of the leaf interior.

leases its new carbon as CO_2 again, leaving pyruvate. After transfer back to the original cell, pyruvate is phosphorylated to PEP at the expense of an ATP-to-AMP hydrolysis in a reaction that has no counterpart in animal cells. The regenerated PEP is then ready to accept another molecule of carbon dioxide.

The effect of the Hatch and Slack pathway, which is also called the C4 pathway since OAA is a four-carbon molecule, is to fix CO_2 by an enzyme having a very high affinity for this substrate and in a reaction that is energetically extremely favorable. As a result, stomata can be nearly closed and still admit enough CO_2 to support the full rate of photosynthesis. With their stomata partially closed, plants require less water. As an example, sorghum (a C4 plant) requires only about a third as much water per gram of growth as alfalfa, a *C3 plant* (i.e., one that utilizes only the Calvin cycle). Additional ATP is required (12 per molecule of glucose) for the C4 pathway, but if water is scarce and temperatures high, the C4 plants have a considerable advantage over their C3 counterparts.[3]

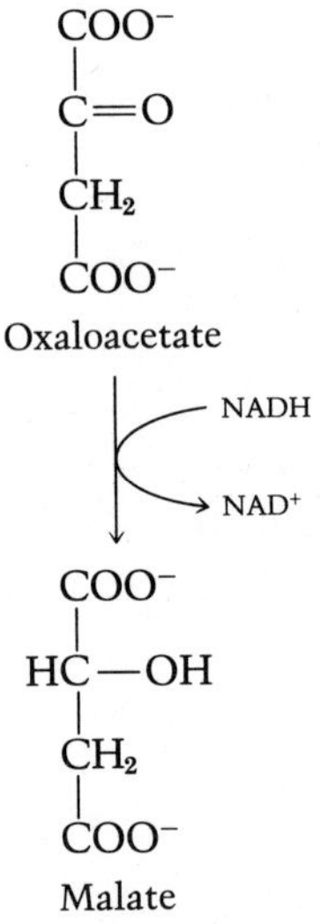

A third type of photosynthesis is found in succulents, such as cacti, which are even more efficient in their use of water than C4 plants. Succulents open their stomata only at night, when evaporative losses are minimal, and accumulate CO_2 by fixing it to PEP, the latter provided by the breakdown of starch. OAA produced in this way is accumulated in a reduced form as malate. The reduction, which requires NADH (presumably provided by mitochondria), is a very exergonic reaction with a $\Delta G°$ of -28 kJ/mole.

The next day, when the sun comes up, the stomata close and the process is reversed to make CO_2 available to the chloroplasts and their C3 pathway. This system, which is called *Crassulacean acid metabolism* (*CAM*—from the genus *Crassula*) thus separates carbon fixation from photosynthesis in time, whereas the C4 pathway separates the two spatially. The process is not efficient, but the problem for desert plants is not rapid growth but water conservation and survival, and this they do very well.

The Phosphogluconate Pathway of Animal Cells. The complicated Calvin cycle of chloroplasts is found almost intact in the cytosol of most

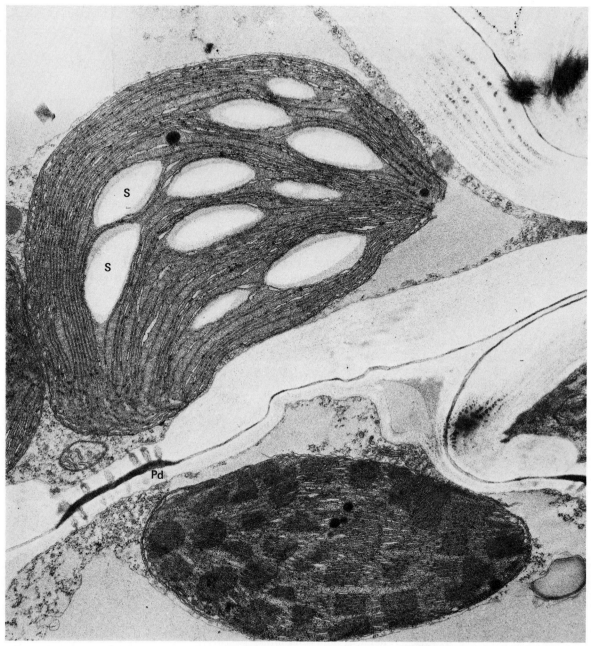

FIGURE 7-24 The Dimorphic Chloroplast. Plants that utilize the Hatch and Slack pathway also have dimorphic chloroplasts, notable for their lack of grana. One is shown here in a bundle-sheath cell of sugar cane (top). The cell is connected via a plasmodesma (Pd) to a mesophyll cell (bottom) containing the usual chloroplast configuration. Starch (S). [Courtesy of W. M. Laetsch, *Sci. Prog. (Oxford)*, **57**: 323 (1969).]

eukaryotic cells, but for a quite different purpose. It is referred to in animal cells as the *phosphogluconate pathway*, the *pentose phosphate pathway*, or the *hexose monophosphate shunt*.

We can think of the Calvin cycle (Fig. 7-22) as starting with six ribulose-5-phosphates (Ru-5-P), adding six CO_2, and ending up back at six Ru-5-P's again with the added carbons removed as glucose 6-phosphate (G-6-P). In the meantime 12 NADPH have been oxidized. The phosphogluconate pathway, using many of the same enzymes, starts with glucose 6-phosphate and ends up with CO_2. In the process, 12 $NADP^+$ are reduced:

$$G\text{-}6\text{-}P + 12NADP^+ + 7H_2O \rightarrow 6CO_2 + 12NADPH + 12H^+ + P_i \qquad (7\text{-}16)$$

Its major purpose is to generate NADPH, which is used for lipid (fatty acid and steroid) synthesis, to maintain membrane components in their reduced state, and presumably to provide additional reducing power for mitochondrial electron transport. Additional functions of the phosphogluconate pathway include converting hexoses to pentoses and vice versa.

The phosphogluconate pathway begins with the oxidation of G-6-P by G-6-P dehydrogenase (G6PDH), an $NADP^+$-linked enzyme. The hydrated product of this enzyme is 6-phosphogluconate, for which the pathway was named. Thereafter, 6-phosphogluconate is oxidized and decarboxylated to ribulose-5-phosphate, a Calvin cycle intermediate. The Calvin cycle enzymes (without RuDP carboxylase or Ru-5-P kinase) are then used to regenerate G-6-P. Hence the pathway is more accurately represented as beginning with six G-6-P and ending with five G-6-P and six CO_2.

7-7 THE EVOLUTIONARY ORIGIN OF MITOCHONDRIA AND CHLOROPLASTS

It is now considered probable that both mitochondria and the plastids are descended from prokaryotic cells that took up symbiotic residence in some ancient eukaryote. This hypothesis is based largely on striking similarities between mitochondria, chloroplasts, and aerobic and photosynthetic prokaryotes.

Photosynthesis in Prokaryotes. Photosynthesis occurs in only a few prokaryotes: blue-green algae and the green and purple bacteria. The light-driven ATP synthesis of the purple salt-lover, *Halobacterium*, should not be included as it is quite a different process (see Section 7-2). In the other cases, however, prokaryotic and eukaryotic photosynthesis have much in common.

The blue-green algae have a photosynthetic arrangement similar to that of higher plants, including dual pigment systems, each with an energy sink, and the capacity to use water as an electron donor, thus evolving oxygen. Blue-green algae contain cytoplasmic lamellae with which most of the photophosphorylation system is associated (see Fig. 7-25). As with chloroplasts, isolated particles of these lamellae are capable of photophosphorylation.

Photosynthetic bacteria, on the other hand, are quite different from either the blue-green algae or higher plants. Although they do contain cytoplasmic membranes (lamellae) with which photophosphorylation is associated, there is no oxygen evolution and no Emerson enhancement. It appears, therefore, that only one photosystem is used. Some of the pigments of this system are unique to bacteria, including a special chlorophyll called *bacterial chlorophyll*. The single photosystem appears to have an energy sink, P890, that is well into the red. It represents approximately 2% of the chlorophyll molecules.

SECTION 7-7
The Evolutionary Origin of
Mitochondria and Chloroplasts

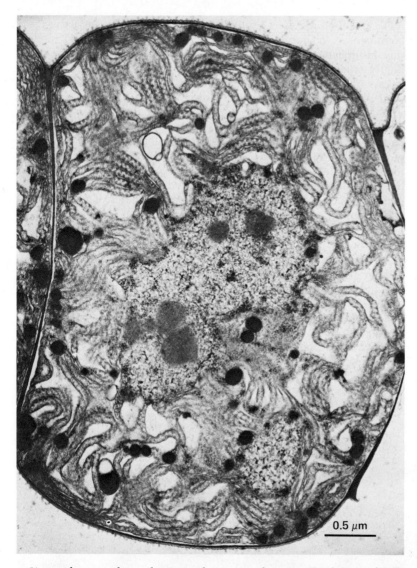

FIGURE 7-25 Blue-Green Alga. Note the photosynthetic lamellae inside this freshwater prokaryote, *Fremyella diplosiphon*. The spheres on the membranes appear similar to those on the photosynthetic membranes within the chloroplasts of some eukaryotic algae (see Fig. 7-14). The DNA is the lighter, granular material in the center. [Courtesy of E. Gantt and S. F. Conti, *J. Bacteriol.*, 97: 1486 (1969).]

Since photosynthetic bacteria do not evolve oxygen during photosynthesis they must have an electron donor other than water. Some use carbon monoxide, methane, or other organic compounds. Others use inorganic electron donors such as hydrogen gas or hydrogen sulfide (H_2S). The latter is utilized by the *purple sulfur bacteria*, which can also "fix" nitrogen from the atmosphere, reducing it to ammonia. They can therefore derive all their major atoms—C, H, O, and N—from air.[4]

Bacterial photosynthesis is represented by the general reaction

$$CO_2 + 2H_2A \xrightarrow{h\nu} (CH_2O) + 2A + H_2O \qquad (7\text{-}17)$$

where H_2A is oxidized to A and (CH_2O) represents the newly added carbohydrate unit. Photosynthesis in plants is the special case where H_2A is water and 2A is O_2.

Other points of similarity between prokaryotic and eukaryotic photosynthesis include a membrane-bound electron transport chain and an

331

CHAPTER 7
The Oxidative Organelles

ATPase that protrudes as a spherical complex above the surface of these membranes. In addition, coupling of electron transport to ATP production is probably by the chemiosmotic mechanism in all cases. And finally, studies on the amino acid sequence of cytochrome *f* from several eukaryotic and prokaryotic algae reveal striking similarities and no clear-cut distinction between the prokaryotic and eukaryotic groups.

Oxidative Metabolism in Prokaryotes. Electron transport and oxidative phosphorylation in bacteria are not substantially different from the same processes in mitochondria. Flavoproteins, iron-sulfur proteins, cytochromes, and ubiquinone or related species (e.g., vitamin K) are all present in bacteria, associated with the cell membrane. There are also particles comparable to the inner membrane spheres of mitochondria and having the same ATP-generating functions. ATP synthesis is apparently via chemiosmotic gradients created by oxidatively driven proton extrusion.

Protein Synthesis in Mitochondria and Chloroplasts. Both mitochondria and chloroplasts reproduce by binary fission (see Figs. 7-26 and 7-27) at rates that are not necessarily synchronized to that of their "host." Reproduction in both cases is supported in part by genes contained within the organelles.

For example, some of the subunits of mitochondrial cytochrome oxidase are specified by mitochondrial DNA and some by nuclear DNA. Similarly, the larger of the two subunits of ribulose diphosphate carboxylase (the catalytic subunit) is apparently coded by chloroplast DNA and synthesized within the organelle. The smaller subunit, probably regula-

FIGURE 7-26 The Replication of Chloroplasts. From a tobacco leaf. Note the prominent starch granule in each daughter plastid, and the light areas within the stroma that are probably DNA (arrows). [Courtesy of D. A. Stetler and W. M. Laetsch, *Amer. J. Botany,* **56**: 260 (1969).]

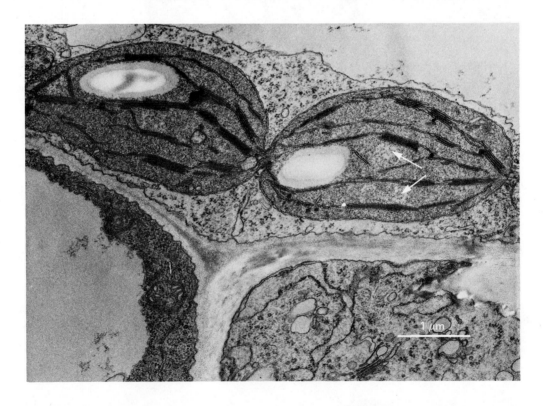

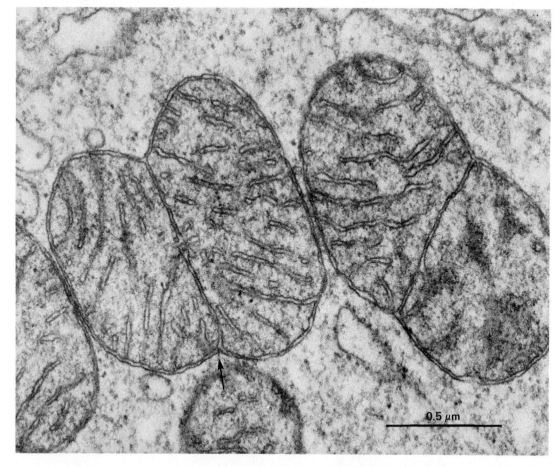

FIGURE 7-27 The Replication of Mitochondria. From the fat body of an insect, *Calpodes ethlius*. Note that the partitions (arrow) are constructed like cristae—that is, they are invaginations of the inner membrane. [Courtesy of W. J. Larsen, *J. Cell Biol.*, **47**: 373 (1970).]

tory in function, is specified by nuclear genes and synthesized outside the chloroplast.

The genes of mitochondria and chloroplasts are embodied in simple DNA strands like those of prokaryotes (see Fig. 7-28) rather than in the complex chromosomal structure of eukaryotes. Though there may be multiple copies (especially in the plastids), each closed circular molecule of DNA contains the entire genome. Significant genetic homologies in this DNA have been found between blue-green algae and some eukaryotic plastids.

The ribosomes of mitochondria and chloroplasts are unmistakably smaller than cytoplasmic ribosomes and resemble prokaryotic ribosomes in both size and antibiotic sensitivity. Chloramphenicol, for example, inhibits protein synthesis on these smaller ribosomes but not on cytoplasmic ribosomes, while cycloheximide has just the opposite spectrum of targets.[5]

Endosymbiotic Evolution of Mitochondria and Chloroplasts. One could construct from the observations on eukaryotic and prokaryotic photosynthesis an evolutionary progression: photosynthetic bacteria have the most primitive mechanism, with their single photosystem and variety of electron donors; the blue-green algae are also prokaryotes, but they have a dual photosystem and evolve oxygen; finally, the eukaryotic chloro-

CHAPTER 7
The Oxidative Organelles

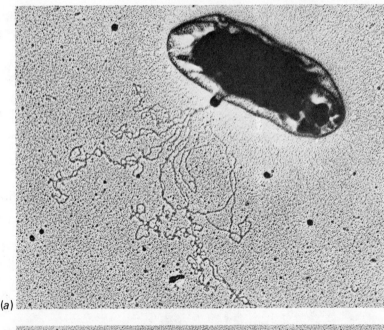

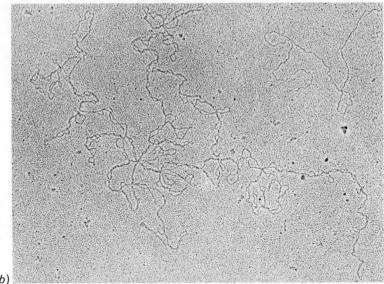

FIGURE 7-28 The DNA of Mitochondria, Chloroplasts, and Bacteria. (a) Bacterial DNA. Total contour length would be about 1100 μm. (b) DNA from a spinach chloroplast. Contour length = 42 μm. (c) Mitochondrial DNA from a mouse tumor cell. Contour length = 4.9 μm. [(a) Courtesy of M. M. K. Nass, from *Biological Ultrastructure*, P. J. Harris, ed., Corvallis: Oregon State Univ. Press, 1971; (b) and (c) courtesy of D. R. Wolstenholme, (b) from D. Wolstenholme, and O. Richards, *J. Cell Biol.* **53**: 594 (1972).]

plast has these features plus a more organized membrane system, providing internal compartmentation. The chloroplast in some respects thus resembles a metabolically simplified blue-green alga trapped in a membrane-enclosed compartment of its host cell.

There is a corresponding pattern in oxidative metabolism, which probably evolved later since it utilizes the oxygen produced by photosynthesis. Note in particular that the ATPase spheres found on mitochondrial inner membranes are found associated with the plasma membrane of aerobic prokaryotes. In some modern photosynthetic bacteria, this ATPase can be chemiosmotically driven to produce ATP by either oxidative or photosynthetic electron transport. It is likely, therefore, that aerobic electron transport evolved from photosynthetic electron transport, and it is reasonable to propose that mitochondria are the simplified remnants of an aerobic prokaryote that formerly resided within a vacuole of its host.

In support of the prokaryotic origin for mitochondria and chloroplasts, one notes that intracellular symbiosis (*endosymbiosis*) is a well-known phenomenon. For example, a certain species of paramecia (*Paramecium aurelia*) harbors several gram-negative bacteria. In addition, there are a very large number of symbiotic relationships involving simple aquatic animals and the prokaryotic blue-green algae. The advantages of an endosymbiosis involving photosynthetic prokaryotes are obvious, for the photosynthetic cell confers on its host an ability to capture light energy, secreting nutritive products of that capture into the host cytoplasm; in return, the photosynthetic cell secures a sheltered and rather constant environment in which to grow and reproduce.[6] However, in such an environment the prokaryote may have little use for a number of functions that are necessary for survival in the free state. Since the loss of these functions would confer no disadvantage (and perhaps a positive advantage in avoiding duplication), one can understand how a very old symbiotic relationship might have evolved to the present system of a eukaryotic cell with semiautonomous organelles.

Additional clues concerning the evolution of mitochondrial and chloroplast function have been derived from the study of a third type of oxidative organelle, the peroxisome. In the next section we shall advance the hypothesis that peroxisomes preceded the appearance of mitochondria and chloroplasts and gradually ceded to them functions that are more efficiently handled by these more recent and more complex organelles.

7-8 PEROXISOMES AND THE EVOLUTION OF OXIDATIVE METABOLISM

Organelles called *microbodies*, first described in mammalian kidney cells, are now known to be widely distributed in eukaryotic cells. Independent of their discovery by electron microscopy, organelles known as *peroxisomes* were isolated from a variety of cells by differential centrifugation and shown to be capable of producing and degrading hydrogen peroxide, H_2O_2. Cytochemical staining for this capacity has demonstrated that, with only a few exceptions, structures labeled microbodies are in fact also peroxisomes.

Peroxisomes represent a family of organelles with numerous morphological, physical, and biochemical similarities, including a seemingly

CHAPTER 7
The Oxidative Organelles

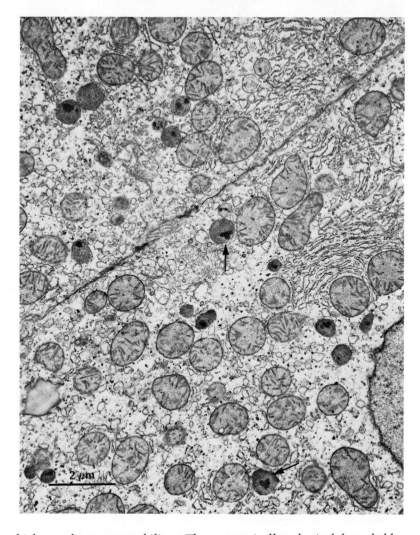

FIGURE 7-29 Rat Liver Peroxisomes. Peroxisomes are the heavily stained bodies (arrows) interspersed among the numerous mitochondria. See also Fig. 1-17. [Courtesy of F. Leighton, L. Loloma, and C. Koenig, *J. Cell Biol.*, **67**: 281 (1975).]

high membrane permeability. They are typically spherical, bounded by a single membrane, and have a dense, granular matrix (Fig. 7-29). When analyzed for specific enzymes, however, it is clear that these organelles do differ significantly from one cell type to another. The one thing that ties them together is the presence of catalase, an enzyme responsible for the breakdown of hydrogen peroxide

$$H_2O_2 \rightarrow H_2O + \tfrac{1}{2}O_2$$

Although the biological role of peroxisomes is incompletely understood, we shall present some of the functions they are known to catalyze and then speculate briefly on their evolutionary origin.

Peroxisomal Reactions. Molecular oxygen is used as an electron acceptor in certain extramitochondrial oxidations. Often, however, the reduction product of oxygen is not water, but hydrogen peroxide. This substance is highly toxic to cells because of its tendency to form chemically reactive free radicals capable of attacking and nonspecifically altering both nucleic acids and proteins. It is of some advantage, therefore, to confine peroxide-producing reactions to an organelle where the peroxide can be safely destroyed as it is made.

Examples of peroxide-producing reactions include flavoprotein-catalyzed amino acid oxidation. A typical case would be the deamination of alanine to form pyruvate.

$$\underset{\text{Alanine}}{\underset{|}{\overset{|}{\underset{CH_3}{H-C-NH_2}}}\overset{COO^-}{}} \xrightarrow{\tfrac{1}{2}O_2 \quad H_2O_2} \underset{\text{Pyruvate}}{\underset{|}{\overset{COO^-}{\underset{CH_3}{C=O}}}} + NH_3 \qquad (7\text{-}18)$$

Both L and D amino acid oxidases are found in some peroxisomes. Although other pathways are more important in the deamination of L amino acids, peroxisomal enzymes are the primary mechanism for utilizing D amino acids, obtained mostly from intestinal bacteria.

Another common peroxisomal enzyme is α-hydroxyacid oxidase, capable of converting α-hydroxy acids to α-keto acids. An example would be the oxidation of lactate to pyruvate:

$$\underset{\text{Lactate}}{\underset{|}{\overset{COO^-}{\underset{CH_3}{HC-OH}}}} \xrightarrow{\tfrac{1}{2}O_2 \quad H_2O_2} \underset{\text{Pyruvate}}{\underset{|}{\overset{COO^-}{\underset{CH_3}{C=O}}}} \qquad (7\text{-}19)$$

This reaction can salvage lactate or ethanol produced in glycolysis or fermentation and otherwise excreted. Its usefulness would presumably be greatest when mitochondrial function (which would inhibit glycolysis or fermentation) is absent for some reason.[7]

Many other peroxisomal reactions have been identified in one tissue or another. Their role in animal metabolism is certainly not fully understood. However, peroxisomes account for over 2% of rat liver protein and have rapid half-lives of less than two days, indicating an importance to mammals that is much greater than one would suspect from the reactions just discussed. Their importance to plants and some protozoa, on the other hand, is much better appreciated, as will be explained.

The Glyoxylate Cycle (Glyoxysomes). Higher animals lack an effective mechanism for converting fats to carbohydrate. Fats are degraded mostly to acetyl-SCoA. The latter cannot be converted to pyruvate and, when entered into the citric acid cycle, is soon lost as CO_2. Ordinarily, this limitation poses no problem, for conversion of fat to carbohydrate is seldom needed. It is, however, the basis for *ketoacidosis*, which is a life-threatening complication of severe, untreated diabetes mellitus.[8] Ketoacidosis would be no threat if we had the peroxisomal system common in plants and known as the glyoxylate cycle.

The early growth of a typical seedling is supported by fatty acid catabolism, some of which must be aimed at producing carbohydrates. Seedlings convert fats to carbohydrates by bypassing the CO_2-evolving steps of the citric acid cycle. The key enzymes are isocitrate lyase and malate synthase:

$$\text{Isocitrate} \quad \begin{array}{c} COO^- \\ HC-OH \\ {}^-OOC-CH \\ CH_2 \\ COO^- \end{array} \xrightarrow{\text{isocitrate lyase}} \begin{array}{c} COO^- \\ HC=O \\ {}^-OOC-CH \end{array} \text{Glyoxylate} \\ \begin{array}{c} CH_2 \\ COO^- \end{array} \text{Succinate}$$

$$\text{Acetyl-SCoA} \quad \begin{array}{c} CO-SCoA \\ CH_3 \end{array} \\ + \qquad \xrightarrow{\text{malate synthase}} \quad \begin{array}{c} COO^- \\ CH_2 \\ HC-OH \\ COO^- \end{array} \text{Malate (+ coenzyme A)} \\ \text{Glyoxylate} \quad \begin{array}{c} HCO \\ COO^- \end{array}$$

The position of these two enzymes in the glyoxylate cycle is shown in Fig. 7-30, with the heavy arrows indicating reactions that take place in the peroxisome itself. The rest of the reactions, except for the synthesis of carbohydrate from phosphoenolpyruvate (PEP) and oxaloacetate (OAA), normally take place in mitochondria. The synthesis of carbohydrate is largely the function of chloroplasts.

In prokaryotes that have the glyoxylate cycle some enzymes are shared directly with the citric acid cycle. In eukaryotes, however, there is compartmentation and partial duplication so that, for example, malate dehydrogenase (which converts malate to OAA) is found both in mitochondria and in those peroxisomes having the glyoxylate cycle. Because of the uniqueness and importance of the glyoxylate cycle, peroxisomes having it are often given the special name *glyoxysome*.

Photorespiration. Another important peroxisomal function in green plants has to do with a peculiar phenomenon known as photorespiration. Studies with radioactive tracers reveal that many plants take up O_2 and evolve CO_2 at the same time as photosynthesis is causing CO_2 to be fixed and oxygen to be evolved. The competing reaction is referred to as *photorespiration*.

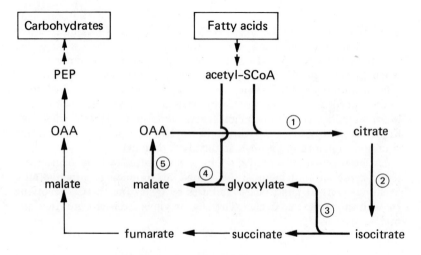

FIGURE 7-30 The Glyoxylate Cycle. The enzymes are (1) citrate synthase; (2) aconitase; (3) isocitrate lysase; (4) malate synthase; (5) malate dehydrogenase. All five are found within the special peroxisome known as a glyoxysome, and 1, 2, and 5 are found also in mitochondria.

A small fraction of the O_2 used and CO_2 evolved during periods of photosynthesis comes from mitochondrial reactions. By far the greater part, however, comes from the metabolism of glycolate, which in turn is produced by the oxidative degradation of ribulose diphosphate (RuDP). The cause of RuDP breakdown seems to be a sensitivity of RuDP carboxylase to oxygen, which competes with CO_2 for the catalytic site of the enzyme. In the presence of oxygen, RuDP carboxylase catalyzes the conversion of RuDP into 3-phosphoglycerate (3-PGA) and phosphoglycolate

$$\text{RuDP} \begin{array}{c} H_2C-OPO_3^{2-} \\ | \\ C=O \\ | \\ HCOH \\ | \\ HCOH \\ | \\ H_2C-OPO_3^{2-} \end{array} \xrightarrow{O_2} \begin{array}{c} H_2C-OPO_3^{2-} \\ | \\ COO^- \\ \\ COO^- \\ | \\ HCOH \\ | \\ H_2C-OPO_3^{2-} \end{array} \begin{array}{l} \text{Phosphoglycolate} \\ \\ \\ \text{3-PGA} \end{array} \quad (7\text{-}20)$$

One of the products of the RuDP oxygenase reaction, 3-PGA, is a usual intermediate in the Calvin cycle. The other product, phosphoglycolate, appears to undergo the following sequence of reactions, being passed in the process from one organelle to another:

(1) *In chloroplasts:* dephosphorlyation of phosphoglycolate to glycolate.

(2) *In peroxisomes:* the oxidation of glycolate to glyoxylate by a flavoprotein (glycolate oxidase) that couples the reaction to production of H_2O_2:

$$\begin{array}{c} COO^- \\ | \\ H_2COH \\ \text{Glycolate} \end{array} \xrightarrow[\text{oxidase}]{O_2 \quad H_2O_2 \atop \alpha\text{-hydroxyacid}} \begin{array}{c} COO^- \\ | \\ HC=O \\ \text{Glyoxylate} \end{array}$$

Some of the glyoxylate is aminated to glycine and some probably enters the glyoxylate pathway (Fig. 7-30), at least in certain plants. The rest may well be degraded by the action of H_2O_2 to yield CO_2 and formate ($HCOO^-$).

(3) *In mitochondria:* glycine salvaged from glyoxylate can be converted to serine, two glycines forming one serine plus CO_2.

(4) *In peroxisomes:* serine can be converted to glycerate (glyceric acid).

(5) *In chloroplasts:* glycerate can be converted to phosphoglycerate (3-PGA), which puts the salvaged carbons back into the Calvin cycle.

This close metabolic cooperation among the organelles is paralleled by close physical proximity as well (see Fig. 7-31). The result of this cooperation, for that fraction of the original glycolate that gets converted to serine, is to produce one molecule of glucose and two molecules of CO_2 from four molecules of glycolate. In the process, there is O_2 uptake by the peroxisomes.

Photorespiration, because of its loss of carbons as CO_2, is obviously a very inefficient process. In fact, up to half the photosynthetic carbon fixation and reducing power of chloroplasts may be wasted in this way. The most logical explanation currently available for the existence of these reactions is that photorespiration is an evolutionary accident. The primordial atmosphere is thought to have contained much CO_2 but little or no oxygen. As the atmospheric concentration of oxygen increased

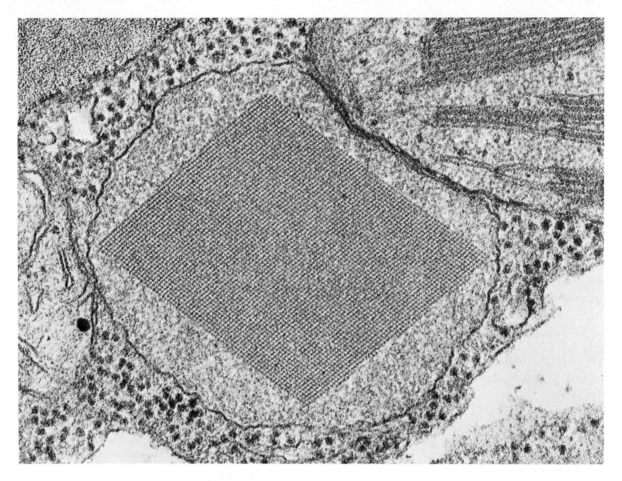

FIGURE 7-31 Plant Leaf Peroxisome. From a tobacco leaf. Note the crystalline inclusion (of catalase) and the close apposition to a chloroplast in the corner of the micrograph. [Courtesy of S. Frederick and E. H. Newcomb, Univ. of Wisconsin. See *Science*, **163**: 1353 (1969) and *J. Cell Biol.*, **43**: 343 (1969).]

after the evolution of photosynthesis, competition by O_2 for RuDP carboxylase became significant and glycolate formation the result. The original response of organisms was probably to excrete the glycolate, and some modern algal species still do this. The further reactions of glycolate outlined above make up a salvage pathway capable of preserving and using at least a portion of what would otherwise be wasted organic material.

In support of the evolutionary hypothesis for photorespiration, one notes that its significance varies directly with oxygen and CO_2 availability. If oxygen is low and CO_2 high, photorespiration is greatly inhibited and the overall efficiency of the plant improves accordingly.

Why hasn't a mechanism for avoiding photorespiration evolved in plants? We don't know, of course. However, maximizing efficiency seems to be less important for plants than for animals, which because of their mobility are much more active competitors with each other. (Modern plants are inefficient in many ways; the correction of these inefficiencies is an important area of modern agricultural research.) Aside from these excuses, it is important to note that in fact some plants *have* evolved a way of avoiding photorespiration.

Photorespiration is favored not only by high oxygen and low CO_2, but also by strong light and high temperatures. In describing photosynthesis, we noted that many of the plants adapted for sunny, arid regions have C4 metabolism rather than just the Calvin cycle. Such plants also have extremely low levels of photorespiration. You will recall that these C4 plants have smaller stomatal openings, and that they fix their carbon in mesophyll cells and then transfer it to bundle-sheath cells where RuDP carboxylase and the Calvin cycle operate. The oxygen content of bundle-sheath cells is quite low and the RuDP oxygenase reaction correspondingly low. Hence C4 plants, in spite of the extra ATP utilized in fixing their CO_2, are actually more efficient not only in their use of water but in overall photosynthesis. These plants also appear to be relatively recent evolutionary additions to the world's flora.

Evolution of Oxidative Metabolism. It has been proposed that modern peroxisomes are the evolutionary remnants of an ancient precursor that was the major site of oxidative metabolism in eukaryotes prior to the appearance of mitochondria (see reference by deDuve). According to this thesis, the ancient peroxisome carried out most of the functions now associated with mitochondria but, because of a lack of oxidative phosphorylation, with much less efficiency. We note, for example, that most peroxisomal reactions that use O_2 as an electron acceptor have mitochondrial counterparts that capture the electrons and use them to drive oxidative phosphorylation. In addition, several of the citric acid cycle enzymes of mitochondria are found also in peroxisomes, although in at least some cases the peroxisomal enzyme is an isozyme coded by its own gene.

According to this proposal for peroxisomal evolution, various functions were lost as more efficient alternatives became available. Different evolutionary lines thus retained different subsets of the original peroxisomal enzymes, leaving us with the current array of peroxisomes and their diverse but poorly understood functions.

SUMMARY

7-1 Mitochondria use molecular oxygen as an electron acceptor in the oxidation of fatty acids, carbohydrates, amino acids, and coenzymes (e.g., NADH). The process produces carbon dioxide and water, and releases a large amount of energy, a portion of which is captured in the form of ATP. Since ATP production is their major function, mitochondria tend to be concentrated in those aerobic cells and at those locations where large amounts of ATP are consumed. The actual production of ATP (phosphorylation of ADP) is catalyzed by a complex of proteins associated with mitochondrial "inner membrane spheres." The inner membrane also has associated with it most of the oxidative enzymes of mitochondria. To increase the surface available for those activities, the inner membrane is thrown into redundant folds called cristae that extend into the interior (matrix) of the organelle where many soluble enzymes are found.

7-2 Electron transport in mitochondria uses molecules which, through reversible oxidations and reductions, pass electrons from one component to another, terminating with the reduction of oxygen to water. Electron transport thus utilizes (via NADH or reduced flavins) some of the energy from oxidation to produce ATP. There are three prominent theories explaining how ATP synthesis may be coupled to electron transport: (1) direct chemical coupling, (2) chemiosmotic or electrochemical coupling, and (3) conformational coupling. At the present time, the chemiosmotic hypothesis seems to have the most experimental support. It can be demonstrated, for example, that electron transport produces both pH and electrical gradients, and that such gradients in isolated vesicles containing mitochondrial ATPase can be utilized to phosphorylate ADP. These observations are precisely what one would expect from the chemiosmotic theory.

7-3 The major function of mitochondria is to produce ATP via electron transport and oxidative phosphorylation. Most of the oxidations needed to supply the electron transport chain with reduced coenzymes are carried out by enzymes of the citric acid cycle. Entry to the cycle is normally through activated acetate (acetyl-SCoA) derived from pyruvate (hence from carbohydrates), from fatty acids, or from certain of the amino acids. The net yield from each acetyl-SCoA degraded is one GTP and enough reduced coenzymes to produce 11 additional ATP by oxidative phosphorylation. The effect is to complete the oxidation of acetate to carbon dioxide and water, with an energy yield (in the form of ATP) per glucose that is some 19 times greater than could be obtained anaerobically.

Mitochondria carry out many activities in addition to ATP production. These activities include heat production (especially in brown fat mitochondria), participation in heme and steroid biosynthesis, the synthesis of some mitochondrial proteins, and the active uptake of Ca^{2+}. The latter function allows mitochondria to serve as regulators of Ca^{2+} availability for the rest of the cell.

7-4 Plastids are plant organelles having a double membrane, various kinds of internal structure, and the capacity for semiautonomous replication using in part their own DNA and apparatus for protein synthesis. Proplastids are the undifferentiated precursors of the mature plastids, which consist of leukoplasts (a storage device for fats, oils, starch, or protein), chloroplasts, and the nongreen counterpart of the chloroplast called a chromoplast. The chloroplast is a photosynthetic organelle that is capable of using light energy to convert carbon dioxide to glucose. This conversion consists of two parts: The first is a light reaction carried out by proteins and pigments embedded in the membranous interior of the organelle (i.e., the *grana*—which are stacks of flattened sacks or *thylakoids*—and their interconnecting lamellae). The light reaction provides reducing power and ATP to fuel the dark reaction (carbohydrate biosynthesis) catalyzed by enzymes of the soluble part of the interior, or stroma.

7-5 Chlorophyll exhibits the photoelectric effect, which is actually a light-driven oxidation. The corresponding reduction is $NADP^+ \rightarrow NADPH$. In order to restore chlorophyll prior to its next oxidation, electrons are transferred to it from water, creating molecular oxygen. These light-dependent reactions seem to be divided into two parts via two pigment systems, or photosystems. Each system contains a few chlorophyll molecules that act as a long-wavelength (low-frequency or low-energy) "trap" capable of accepting energy captured by other molecules of chlorophyll or by accessory pigments. Photosystem II uses light energy to oxidize water and transfers the electrons to the first member of an electron transport chain. The chain terminates at photosystem I (PS I), which uses the electrons, energetically "boosted" by additional light capture, to reduce $NADP^+$. Electron transport itself results in the production of ATP (photophosphorylation), an activity that can be initiated by PS I alone through the transfer of electrons back to the chain instead of to $NADP^+$. This latter process is called cyclic photophosphorylation.

The photosynthetic light reaction, then, is a conversion of light energy to chemical energy. The opposite process, conversion of chemical energy to light energy, is called bioluminescence. It consists of a much simpler system: the oxidation of a reduced "luciferin" catalyzed by an enzyme, or "luciferase." Fireflies, for example, use a reduced phenolic compound as luciferin and ATP as a source of energy.

7-6 The dark reaction consists of the enzymatic synthesis of glucose from carbon dioxide, using NADPH for its "reducing power" and ATP to drive otherwise unfavorable steps. The fixation of carbon dioxide is catalyzed by ribulose diphosphate carboxylase, an enzyme that adds carbon dioxide (as a carboxyl group) to ribulose diphosphate (RuDP) and then cleaves the addition product to two molecules of 3-PGA (phosphoglyceric acid). The major fraction (five-sixths) of the PGA must be used to regenerate RuDP via the Calvin cycle so that more molecules of CO_2 can be captured. The other one-sixth of the PGA is available for conversion to carbohydrate, lipid, protein, or nucleic acid.

The Hatch and Slack, or C4, pathway is an alternate route for carbon fixation. It utilizes the carboxylation of phosphoenolpyruvate (PEP), creating oxaloacetate. The latter is transferred to another cell where it gives up its new carbon to the Calvin cycle. A major advantage is the highly favorable carboxylation step, which permits partial closure of stomata and hence water conservation. An even more water-efficient mechanism is found in succulents, which fix carbon at night and use it in the Calvin cycle during the day when stomata are closed.

Most Calvin cycle enzymes (with the exception of Ru-5-P kinase and RuDP carboxylase) are found also in animal cells where they support the phosphogluconate pathway. This pathway is an extramitochondrial mechanism for oxidizing carbohydrates to produce NADPH and provides a way of interconverting pentoses and hexoses.

7-7 Mitochondria and chloroplasts have many things in common with prokaryotes. This observation led to the suggestion that mitochondria and chloroplasts are evolutionary descendants of prokaryotic symbionts that lived in some ancient eukaryote. Mitochondria and chloroplasts (1) are self-replicating, dividing by binary fission as do prokaryotes; (2) contain genes responsible for synthesizing some of their own proteins; (3) have ribosomes similar to those of prokaryotes, but unlike other eukaryotic ribosomes; and (4) are similar in size to prokaryotic cells. In addition, the lamellae of chloroplasts bear some structural and functional resemblance to those of the blue-green algae, while mitochondrial cristae have surface features and enzymes similar to those reported for the plasma membrane of some aerobic bacteria.

7-8 Peroxisomes are small, membrane-enclosed organelles having a variety of enzymes, some of which are oxidases that use molecular oxygen as an electron acceptor and in the process produce H_2O_2. The latter substance is broken down by catalase, the enzyme by which peroxisomes are identified. The metabolic role of peroxisomes in multicellular animals is poorly understood, though their prevalence suggests a nontrivial function. In plants, however, at least two classes of peroxisomes have well-defined roles: the glyoxysome, which permits the conversion of fats to carbohydrates; and those peroxisomes that participate in the glycolate salvage pathway of photorespiration. Photorespiration may be a wasteful evolutionary accident caused by increasing atmospheric concentrations of oxygen and the sensitivity of ribulose diphosphate carboxylase to it. It is avoided in some of the more advanced plants that feature C4 metabolism because enzymes of the Calvin cycle are in interior cells of these plants, further away from outside air and its high concentrations of oxygen.

STUDY GUIDE

7-1 (a) What are the characteristic structural features of mitochondria that aid in their identification? (b) What is the major function of mitochondria? (c) What differences exist in structure and function between the inner and outer membranes? (d) Describe a respiratory complex and the function of the headpiece or inner membrane sphere.

7-2 (a) Define the following: electron transport; respiratory chain; oxidative phosphorylation; cytochrome oxidase. (b) Summarize the three schemes presented here for coupling electron transport to ADP phosphorylation. (c) In what way do experiments with the purple membrane of *Halobacterium* tend to support the chemiosmotic hypothesis? (d) A common membrane potential is about 100 mV. If that potential existed across the mitochondrial inner membrane (negative inside) what additional pH gradient, acid outside, would be needed to cause ATP synthesis? [Ans: about 1.8 pH units.]

7-3 (a) What is the significance of acetyl-coenzyme A and where does it come from? (b) What is an oxidative decarboxylation? Give an example. (c) Ethanol can be completely oxidized by the citric acid cycle after an activation sequence as follows:

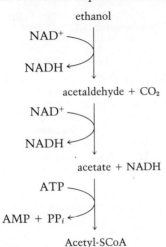

Is the yield of ATP per *gram* of ethanol (molecular weight 46) less or greater than the yield per gram of glucose (MW = 180)? By what fraction? [*Hint:* ATP → AMP should be counted as *minus* 2 ATP.] [Ans: 65% higher for ethanol. Now you know in part where a "beer belly" comes from.] (d) What is the special function of brown fat mitochondria and how is it carried out? (e) What biosynthetic activities of primarily extramitochondrial interest are catalyzed by mitochondrial enzymes? (f) What is the role of mitochondria in the regulation of free Ca^{2+}?

7-4 (a) What are some of the more obvious similarities and differences between chloroplasts and mitochondria? (b) Define the following: granum, thylakoid, leukoplast, proplastid. (c) The light reaction is associated with what part of the chloroplast? How about the dark reaction?

7-5 (a) Define the following: photon; a quantum of energy; an einstein; action spectrum; fluorescence. (b) What is the photoelectric effect? How would you increase the velocity of ejected electrons? How would you increase their number? (c) What are the possible fates (as discussed here) of an absorbed photon? Which is most applicable to the chloroplast? (d) What is "Emerson enhancement"? To what do we attribute it? (e) If absorbed light is utilized only by P700, what is the percent of energy lost in transferring it to P700 if the absorbed light is (1) yellow (λ = 600 nm), (2) blue (λ = 500 nm), or (3) violet (λ = 400 nm)? [Ans: 14%; 29%; 43%] (f) What is "bioluminescence?" Outline the bioluminescent mechanism of fireflies.

7-6 (a) What is the Calvin cycle, and what is its purpose? (b) Based on the stoichiometry suggested here, show why eight photons are needed per CO_2 molecule "fixed" to the oxidation level of glucose. (c) Write the reaction (according to the Calvin cycle) for the net production of one molecule of 3-PGA, regenerating any RuDP used in the process. (d)

What are the advantages of the C4 and CAM mechanisms for carbon accumulation? (e) What is the relationship of the Calvin cycle to the phosphogluconate pathway?

7-7 (a) The dual light effect (Emerson enhancement) is not observed with photosynthetic bacteria. Why? (b) What other differences are there between photosynthesis in bacteria and the mechanism used by blue-green algae or higher plants? (c) What features do mitochondria and chloroplasts have in common, and which of these are shared with prokaryotes but not found elsewhere in eukaryotes? (d) What advantage would be conferred by an endosymbiotic blue-green alga in an animal cell?

7-8 (a) Define the terms peroxisome and glyoxysome. (b) What is the function of the glyoxylate cycle? (c) What is photorespiration and how do peroxisomes contribute? (d) What would be the most likely role of peroxisomes in cells lacking mitochondria?

REFERENCES

ELECTRON TRANSPORT AND THE CONSERVATION OF ENERGY

AVRON, M., "Energy Transduction in Chloroplasts." *Ann. Rev. Biochem.* **46**: 143 (1977).

BOYER, P. D., "Conformational Coupling in Oxidative Phosphorylation and Photophosphorylation." *Trends Biochem. Sci.*, **2**(2): 38 (Feb. 1977).

BOYER, P., B. CHANCE, L. ERNSTER, P. MITCHELL, E. RACKER, and E. SLATER, "Oxidative Phosphorylation and Photophosphorylation." *Ann. Rev. Biochem.*, **46**: 955 (1977).

CIBA FOUNDATION, *Energy Transformation in Biological Systems*. New York: American Elsevier, 1975. Energy conversion in mitochondria, chloroplasts, and other systems.

DILLEY, R., and R. GIAQUINTA, "H+ Ion Transport and Energy Transduction in Chloroplasts." *Current Topics in Membranes and Transport*, **7**: 49 (1975).

EVANS, M., "The Mechanism of Energy Conversion in Photosynthesis." *Sci. Prog., Oxford*, **62**: 543 (1975).

FLATMARK, T., and J. PEDERSEN, "Brown Adipose Tissue Mitochondria." *Biochim. Biophys. Acta*, **416**: 53 (1975).

GREEN, D. E., "Mechanism of Action of Uncouplers on Mitochondrial Energy Coupling." *Trends Biochem. Sci.*, **2**(5): 113 (May 1977).

———, "The Electromechanical Model for Energy Coupling in Mitochondria." *Biochim. Biophys. Acta*, **346**: 27 (1974).

JONES, C. W., *Biological Energy Conservation*. New York: John Wiley, 1976. (Paperback.) On mitochondria and chloroplasts.

MALOFF, B., S. SCORDILIS, and H. TEDESCHI, "Membrane Potential of Mitochondria Measured with Microelectrodes." *Science*, **195**: 898 (1977). 16 mV, inside positive.

MARTONOSI, A., ed., *The Enzymes of Biological Membranes*, Vol. 4: *Electron Transport Systems and Receptors*. New York: John Wiley, 1976.

MITCHELL, PETER, "Coupling of Phosphorylation to Electron and Hydrogen Transfer by a Chemiosmotic Type of Mechanism." *Nature*, **191**: 144 (1966).

———, "Vectorial Chemistry and the Molecular Mechanics of Chemiosmotic Couplings." *Biochem. Soc. Trans.*, **4**: 399 (1976). Review by the originator of the chemiosmotic hypothesis.

NELSON, N., "Structure and Function of Chloroplast ATPase." *Biochim. Biophys. Acta*, **456**: 314 (1976).

NICHOLS, D. G., "The Bioenergetics of Brown Adipose Tissue Mitochondria." *FEBS Letters*, **61**: 103 (1976). Also see *Trends Biochem., Sci.*, June 1976, p. 128.

PACKER, L., "Membrane Structure in Relation to Function of Energy-Transducing Organelles." *Ann. N.Y. Acad. Sci.*, **227**: 166 (1974).

PACKER, L., and A. GÓMEZ-PUYOU, eds., *Mitochondria: Bioenergetics, Biogenesis, and Membrane Structure*. New York: Academic Press, 1976.

PALMER, J., "The Organization and Regulation of Electron Transport in Plant Mitochondria." *Ann. Rev. Plant Physiol.*, **27**: 133 (1977).

PANET, R., and D. R. SANADI, "Soluble and Membrane ATPases of Mitochondria, Chloroplasts, and Bacteria: Molecular Structure, Enzymatic Properties, and Functions." *Current Topics in Membranes and Transport*, **8**: 99 (1976).

PAPA, S., "Proton Translocation Reactions in the Respiratory Chains." *Biochim. Biophys. Acta*, **456**: 39 (1976). A review.

POSTMA, P., and K. VAN DAM, "The ATPase Complex from Energy Transducing Membranes." *Trends Biochem. Sci.*, **1**(1): 16 (Jan. 1976).

RAGAN, C. I., "The Reconstitution of Mitochondrial Energy Conserving Systems." *Int. J. Biochem.*, **7**: 1 (1976).

SCHULDINER, S., H. ROTTENBERG, and M. AVRON, "Membrane Potential as a Driving Force for ATP Synthesis in Chloroplasts." *FEBS Letters*, **28**: 173 (1972).

SLATER, E. C., "Mechanism of Phosphorylation in the Respiratory Chain," *Nature*, **172**: 975 (1953).

———, "The Mechanism of Energy Conservation in the Mitochondrial Respiratory Chain." *Harvey Lectures, Ser.* **66**(1970–1971): 19 (1972).

MITOCHONDRIA: STRUCTURE, TRANSPORT, AND BIOSYNTHESIS

AVADHANI, N. G., F. S. LEWIS, and R. J. RUTMAN, "Mitochondrial RNA and Protein Metabolism." *Sub-Cellular Biochem.*, **4**: 93 (1975).

BORST, P., "Structure and Function of Mitochondrial DNA." *Trends Biochem. Sci.*, **2**(2): 31 (Feb. 1977).

CHRISTENSEN, H. N., *Biological Transport*, 2d ed. Reading, Mass: W. A. Benjamin, 1975. See Chap. 10 on transport in mitochondria.

HARMON, H. J., J. D. HALL, and F. L. CRANE, "Structure of Mitochondrial Cristae Membranes." *Biochim. Biophys. Acta*, **344**: 119 (1974).

LLOYD, D., *The Mitochondria of Microorganisms.* New York: Academic Press, 1975.

MELA, L., Mechanism and Physiological Significance of Calcium Transport Across Mammalian Mitochondrial Membranes." *Curr. Topics Memb. Transport,* vol. 9, 1977.

MILNER, J., "The Functional Development of Mammalian Mitochondria." *Biol. Rev. (Cambridge),* **51**: 181 (1976).

MUNN, E. A., *The Structure of Mitochondria.* New York: Academic Press, 1975.

RACKER, E., "Inner Mitochondrial Membranes: Basic and Applied Aspects." In *Cell Membranes,* G. Weissmann and R. Claiborne, ed. New York: H. P. Publ., 1975, p. 135. Also see *Hospital Practice,* Sept. 1974, p. 87.

SACCONE, C., and E. QUAGLIARIELLO, "Biochemical Studies of Mitochondrial Transcription and Translation." *Int. Rev. Cytol.,* **43**: 125 (1975).

TANDLER, B., and C. L. HOPPEL, *Mitochondria.* New York: Academic Press, 1972. (Paperback.) Excellent micrographs.

TEDESCHI, H., *Mitochondria (Cell Biology Monograph #4).* New York: Springer-Verlag, 1975.

TZAGOLOFF, A., "Genetic and Translational Capabilities of the Mitochondrion." *BioScience,* **27**(1): 18 (1977). Semiautonomy of mitochondria.

———, "Genetic Origin of Cytochrome Oxidase." *Trends Biochem. Sci.,* **1**(6): 139 (June 1976).

VIGNAIS, P. V., "The Mitochondrial Adenine Nucleotide Translocator." *J. Bioenergetics,* **8**: 9 (1976). Also see *Biochim. Biophys. Acta,* **456**: 1 (1976). Reviews.

CHLOROPLASTS AND PHOTOSYNTHESIS

ANDERSON, J., "The Molecular Organization of Chloroplast Thylakoids." *Biochim. Biophys. Acta,* **416**: 191 (1975).

BARBER, J., ed., *The Intact Chloroplast.* New York: Elsevier/North-Holland, 1976. First of a series "Topics in Photosynthesis."

BASSHAM, J. A., "The Path of Carbon in Photosynthesis." *Scientific American,* June 1962. (Offprint 122.)

BERRY, J., "Adaptation of Photosynthetic Processes to Stress." *Science,* **188**: 644 (1975).

CHOLLET, R., "The Biochemistry of Photorespiration." *Trends Biochem. Sci.,* **2**(7): 155 (July 1977).

DEVLIN, R. M., A. V. BARKER, *Photosynthesis.* New York: Van Nostrand Reinhold, 1971.

DOWNTON, W., "The Occurrence of C4 Photosynthesis Among Plants." *Photosynthetica,* **9**: 96 (1975).

GANTT, E., "Phycobilisomes: Light-Harvesting Pigment Complexes." *BioScience,* **25**: 781 (1975).

GIVEN, C., and J. Harwood, "Biosynthesis of Small Molecules in Chloroplasts of Higher Plants." *Biol. Rev. (Cambridge),* **51**: 365 (1976).

GOVINDJEE, ed., *Bioenergetics of Photosynthesis.* New York: Academic Press, 1975.

GOVINDJEE, and R. GOVINDJEE, "The Absorption of Light in Photosynthesis." *Scientific American,* **231**(12): 68 (Dec. 1974).

GREGORY, R. P. F., *Biochemistry of Photosynthesis* (2d ed). New York: John Wiley, 1977.

HARTSOCK, T., and P. NOBEL, "Watering Converts a CAM Plant to Daytime CO_2 Uptake." *Nature,* **262**: 574 (1976).

HATCH, M., "C4 Pathway Photosynthesis: Mechanism and Physiological Function." *Trends Biochem. Sci.,* **2**(9): 199 (Sept. 1977).

HILLMAN, WM. S., ed., *Papers in Plant Physiology.* New York: Holt, Rinehart & Winston, 1970.

KELLY, G., E. LATZKO, and M. GIBBS, "Regulatory Aspects of Photosynthetic Carbon Metabolism." *Ann. Rev. Plant. Physiol.,* **27**: 181 (1977).

KUNG, S., "Tobacco Fraction 1 Protein: A Unique Genetic Marker." *Science,* **191**: 429 (1976). Origin of RuDP carboxylase subunits.

LEVINE, R. P., "The Mechanism of Photosynthesis." *Scientific American,* December 1969. (Offprint 1163.)

RABINOWITCH, E. I., and GOVINDJEE, "The Role of Chlorophyll in Photosynthesis." *Scientific American,* July 1965. (Offprint 1016.)

———, *Photosynthesis.* New York: John Wiley, 1969. (Paperback.)

SMITH, H., ed., *The Molecular Biology of Plant Cells (Botanical Monographs,* Vol. 14). London: Blackwell Scientific, 1977.

SMITH, B. N., "Evolution of C4 Photosynthesis in Response to Changes in Carbon and Oxygen Concentrations in the Atmosphere Through Time." *Biosystems,* **8**: 24 (1976).

TREBST, A., "Coupling Sites, Native and Artificial, in Photophosphorylation by Isolated Chloroplasts." *Trends Biochem. Sci.,* **1**(3): 60 (March 1976).

WERNER, D. D., ed., *The Molecular Biology of Plant Cells.* Oxford: Blackwell Scientific, 1977.

WHITTINGHAM, C. P., *The Mechanism of Photosynthesis.* New York: Elsevier, 1974. (Paperback.)

ZELITCH, I., "Pathways of Carbon Fixation in Green Plants." *Ann. Rev. Biochem.,* **44**: 123 (1975).

BIOLUMINESCENCE

CORMIER, M., J. LEE, and J. WAMPLER, "Bioluminescence: Recent Advances." *Ann. Rev. Biochem.,* **44**: 255 (1975).

DELUCA, M., "Firefly Luciferase." *Advan. Enzymol.,* **44**: 37 (1976).

JOHNSON, F. H., and O. SHIMOMURA, "Bioluminescence." *Trends Biochem. Sci.,* **1**(11): 250 (Nov. 1976).

JOHNSON, F. H., and O. SHIMOMURA, "Bacterial and Other 'Luciferins'." *BioScience,* **25**: 718 (1975).

OXIDATIVE AND PHOTOSYNTHETIC METABOLISM IN PROKARYOTES

DANON, A., and W. STOECKENIUS, "Photophosphorylation in *Halobacterium halobium.*" *Proc. Nat. Acad. Sci. U.S.,* **71**: 1234 (1974).

GEL'MAN, N. S., M. A. LUKOYANOVA, and D. N. OSTROVSKII, *Bacterial Membranes and the Respiratory Chain.* New York, Plenum Press, 1975. Translated from the Russian (*Biomembranes,* Vol. 6).

HADDOCK, B., and C. W. JONES, "Bacterial Respiration." *Bacteriol. Rev.,* **41**: 47 (1977).

MARRS, B., J. WALL, and H. GEST, "Emergence of the Biochemical Genetics and Molecular Biology of Photosynthetic Bacteria." *Trends Biochem. Sci.,* **2**(5): 105 (May 1977).

OELZE, J., and G. DREWS, "Membranes of Photosynthetic Bacteria." *Biochim. Biophys. Acta,* **265**: 209 (1972). For comparison with chloroplasts.

STOECKENIUS, W. et al., "Light Energy Transduction by the Purple Membrane of Halophilic Bacteria." *Fed. Proc.,* **36**: 1797 (1977). A symposium of eight papers. Also see *Scientific American,* June 1976, p. 38.

THAUER, R., K. JUNGERMANN, and K. DECKER, "Energy Conservation in Chemotropic Anaerobic Bacteria." *Bacteriol. Rev.,* **41**: 100 (1977). Includes a discussion of the evolution of oxidative metabolism.

MITOCHONDRIA AND CHLOROPLASTS: COMMON PROPERTIES AND EVOLUTION

AMBLER, R., and R. BARTSCH, "Amino Acid Sequence Similarity Between Cytochrome *f* from a Blue-green Bacterium and Algal Chloroplasts." *Nature,* **253**: 285 (1975). Amino acid sequence similarity.

BIRKY, C. W., Jr., "The Inheritance of Genes in Mitochondria and Chloroplasts." *BioScience,* **26**(1): 26 (1976).

BIRKY, C. W., Jr., P. S. PERLMAN, and T. BYERS, eds., *Genetics and Biogenesis of Mitochondria and Chloroplasts.* Columbus: Ohio State University Press, 1976.

BOGORAD, L., "Evolution of Organelles and Eukaryotic Genomes." *Science,* **188**: 891 (1975).

BRODA, E., *The Evolution of Bioenergetic Processes.* New York: Pergamon Press, 1975.

COHEN, SEYMOUR S., "Are/Were Mitochondria and Chloroplasts Microorganisms?" *Am. Scientist,* **58**: 281 (1970).

FLAVELL, RICHARD, "Mitochondria and Chloroplasts as Descendants of Prokaryotes." *Biochem. Genet.,* **6**: 275 (1972).

HALE, D., K. RAO, amd R. CAMMACK, "The Iron-Sulfur Proteins: Structure, Function and Evolution of A Ubiquitous Group of Proteins." *Sci. Prog. (Oxford),* **62**: 285 (1975).

HOLTZMAN, E., "The Biogenesis of Organelles" In *Cell Membranes,* ed. G. Weissmann and R. Claiborne, eds. New York: H. P. Publ., 1975, p. 153. Also see *Hospital Practice,* March 1974, p. 75.

LEE, S. G., and W. R. EVANS, "Hybrid Ribosome Formation from *Escherichia coli* and Chloroplast Ribosome Subunits." *Science,* **173**: 241 (1971). Functional hybrids indicate similarity of prokaryotic and chloroplast ribosomes.

MARGULIS, L. "The Origin of Plant and Animal Cells." *Am. Scientist,* **59**: 230 (1971). Model for the evolution of eukaryotes through a serial symbiosis.

OLSON, JOHN M., "The Evolution of Photosynthesis." *Science,* **168**: 438 (1970).

RAFF, R. A., and H. R. MAHLER, "The Non-Symbiotic Origin of Mitochondria." *Science,* **177**: 575 (1972). Not everyone agrees with the symbiosis theory.

STILLWELL, Wm., "On the Origin of Photophosphorylation." *J. Theoret. Biol.,* **65**: 479 (1977).

TZAGOLOFF, A., ed., *Membrane Biogenesis: Mitochondria, Chloroplasts, and Bacteria.* New York: Plenum Press, 1975.

PEROXISOMES

DE DUVE, C., "The Nature and Function of Peroxisomes (Introductory Remarks)." *Ann. N.Y. Acad. Sci.,* **168**: 211 (1969). Historical account of the discovery and elucidation of peroxisomes. Also see *Proc. Roy. Soc. (London), Ser. B,* **173**: 71 (1969)

———, "Evolution of the Peroxisome." *Ann. N.Y. Acad. Sci.,* **168**: 369 (1969).

HALLIWELL, B., "Photorespiration." *FEBS Letters,* **64**: 266 (1976). Summary of recent advances.

HRUBAN, Z., and RECHEIGL, Jr., "Microbodies and Related Particles." *Int. Rev. Cytol.,* **Suppl. 1**, (1969).

HUANG, A., and H. BEEVERS, "Localization of Enzymes Within Microbodies." *J. Cell Biol.,* **58**: 379 (1973).

MASTERS, C., and R. HOLMES, "The Metabolic Roles of Peroxisomes in Mammalian Tissues." *Int. J. Biochem.* **8**: 549 (1977).

MAZLIAK, P., *Lysosomes, Glyoxysomes, Peroxysomes.* Paris: Doin, 1975. (Paperback.)

MÜLLER, M., "Biochemistry of Protozoan Microbodies." *Ann. Rev. Microbiol.,* **29**: 467 (1975).

NOVIKOFF, A., and J. ALLEN, eds., "Symposium on Peroxisomes". *J. Histochem. Cytochem.,* **21**: 941 (1973). A collection of papers.

TOLBERT, N. E., "Microbodies—Peroxisomes and Glyoxysomes." *Ann. Rev. Plant Physiol.,* **22**: 45 (1971).

NOTES

1. When uncouplers were first discovered, there was a brief attempt to use them as reducing aids—food energy could be converted to heat without making ATP and therefore without making fat. The principle is sound enough, but the chemicals proved far too dangerous to be used in this way.

2. As a rule of thumb, each gram of protein or carbohydrate has a nutritional yield of roughly four Calories (kilocal) while each gram of fat yields about nine.

3. C4 plants apparently have an additional advantage in minimizing an inefficient process called photorespiration. The significance and probable evolution of this process is discussed in Section 7-8.

4. The fixation of nitrogen is extraordinarily difficult because of the very strong (940 kJ/mole) triple bond in N_2. The process is limited to only a few strains of microorganisms. All life depends on these, for other organisms are forever utilizing nitrogen compounds, releasing N_2 through oxidation.

5. The sensitivity of mitochondria to chloramphenicol may be the reason for the occasional severe side effects observed when this common antibiotic is used to control bacterial infections in humans.

6. Chloroplasts can be removed from plants and introduced into certain animal cells *in vitro,* where they continue to function for a considerable time, conferring on the

animal a capacity for photosynthesis. This was accomplished in 1969 by M. Nass of the University of Pennsylvania. She facetiously suggested that future astronauts should receive chloroplast implants, thus permitting them to recycle their respired carbon dioxide and reduce their food requirements.

7. The reaction is regularly used by trypanosomes, which are flagellated protozoa that cause widespread and serious human diseases, including African "sleeping sickness" and Chagas' disease of Central and South America. Mitochondria are visible in these cells only at certain stages of their life cycle, but are absent when the cells are in the bloodstream of their host. At this stage microbodies, which morphologically and biochemically resemble peroxisomes but lack catalase, are responsible for regenerating NAD^+ by a glycerol phosphate shuttle similar to that discussed in Section 7-3 but using an O_2-linked peroxisomal oxidation.

8. With an insulin deficit, glucose utilization is inhibited and fatty acid catabolism stimulated. The latter may overwhelm the citric acid cycle and result in release of acidic intermediates of fatty acid breakdown ("ketone bodies"), driving blood pH to dangerously low levels.

8

Excitable Cells

8-1 The Neuron and the Nerve Impulse 348
    Structure of the Neuron
    The Action Potential
    Initiation of the Action Potential
8-2 Propagation of Action Potentials 357
    Conduction by Local Currents
    Saltatory Conduction
    Intercellular Transmission by Acetylcholine
    Other Neurotransmitters
    Excitatory and Inhibitory Impulses
    Electrotonic Synapses

8-3 Other Excitable Cells 364
    Sensory Cells
    Photoreceptors
    Neurosecretion
Summary 369
Study Guide 370
References 370
Notes 372

An excitable cell is one that responds to external stimuli through a rapid and reversible alteration in membrane potential. The external stimulus can be mechanical (e.g., to provide touch, hearing, and information concerning the location of body parts), thermal, photic, or chemical. In higher animals, excitable cells convert the external stimulus into electrical signals that are transmitted by *peripheral nerves* to a *central nervous system* (CNS) for processing. The CNS, which in vertebrates consists of the brain and spinal cord, may then initiate responses that are transmitted outward again by other peripheral nerves to *effector cells*, which are responsible for activities such as secretion and movement.

Though many kinds of cells are electrically excitable, the prototype and best understood example is the *neuron*. In this chapter the structure and activity of this cell will be described, followed by a brief discussion of other types of excitable cells and of electrical communication between cells.

8-1 THE NEURON AND THE NERVE IMPULSE

Cells typically have a membrane potential of -60 to -90 mV, negative inside and maintained by a selectively permeable plasma membrane and the outward pumping of Na^+ (see Section 6-5). Excitable cells differ from their nonexcitable counterparts in their ability to utilize changes in this potential as a means of communication.

If the membrane potential of an excitable cell is reduced to a critical *threshold level*, a sequence of events is set in motion that results in a localized and temporary reversal of polarity lasting a millisecond or so. This response, which is called an *action potential*, will be propagated in all directions from the point of stimulation, like waves spreading outward from a pebble dropped into water. The neuron is a cell type whose special function is the propagation of these action potentials as a *nerve impulse*.

Structure of the Neuron. The main body of the neuron, containing its nucleus, is known as the *perikaryon*, *cyton*, or *soma*. Usually there are a number of short processes called *dendrites* and a single longer process called an *axon* radiating from the perikaryon (see Figs. 8-1 and 8-2). Either type of process may be highly branched, providing multiple connections to adjacent neurons.

The impulse in most neurons originates at the dendrites and travels past the cell body to the tip of the axon, from which it is then passed to the dendrites of another nerve cell or to a muscle cell membrane.[1] The junction between two neurons (or sometimes between any two excitable cells) is called a *synapse*. The connection between a neuron and a muscle cell is also referred to as a *neuromuscular junction* or *myoneural junction*. (The prefix "myo-" means muscle.)

Although processes, or *neurites*, of neurons are usually only a few μm in diameter, they may be several feet long in the larger animals—extending, for example, from a perikaryon near the base of the spine to the big toe. A process is usually surrounded and protected by a sheath of cells. In the case of peripheral nerves, these are called *Schwann cells* (Fig. 8-3); in the central nervous system they are called *glial cells*. The nerve cell process and its sheath together constitute a *nerve fiber*. Nerve fibers are commonly found in bundles, held together by connective tissue. It is these bundles that we identify as the anatomical unit called a *nerve* (see Fig. 8-4). It is not unusual to find a thousand fibers, each 10–20 μm in diameter, within a mammalian nerve.

The longer nerve cell processes of vertebrates are often myelinated. *Myelin* is the name given to the glistening, white, fatty covering about a nerve process. It is now known to be a tight spiral wrapping of the associated sheath cells. When the individual processes within a nerve are myelinated, the nerve itself is said to be myelinated or *medullated*. Myelinated nerves are found both in the central and peripheral nervous systems of vertebrates and comprise the white matter of the central nervous system. Duller, unmyelinated nerves and their associated cells form the gray matter.

Each individual Schwann cell of a myelinated peripheral neuron covers perhaps a millimeter of the process, with a space of about a μm between it and the next Schwann cell (see Fig. 8-5). These gaps where the process is relatively exposed are called *nodes of Ranvier*. In general, nerves that actuate skeletal muscles are heavily myelinated, those from sensory organs have lighter myelin, and those to visceral, involuntary muscles have very light myelin or none at all. As we shall see, the presence of myelin leads to a greatly increased velocity of nerve impulse conduction.

Invertebrates do not have myelin (compare Figs. 8-4 and 8-5*b*). They achieve rapid conduction by using processes with large diameters. An English zoologist, J. Z. Young, pointed out in 1933 that the swim reflex

(a)

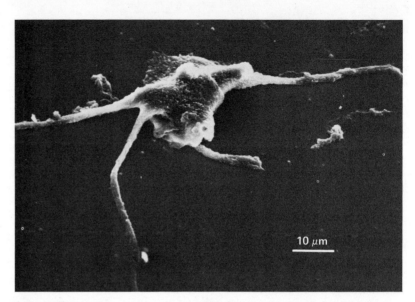

FIGURE 8-1 The Neuron. (a) Scanning EM of a rabbit neuron. Note the several long processes emanating from the cell body. (b) Neuroblastoma (tumor) cells, differentiated in culture. A unipolar (one-process) and bipolar neuron are shown. Note the numerous fine branches on all processes. [(a) Courtesy of A. Hamberger, H. Hansson, and J. Sjostrand, *J. Cell Biol.*, **47**: 319 (1970); (b) courtesy of J. Ross, J. Olmsted, and J. Rosenbaum, *Tissue and Cell*, **7**: 107 (1975).]

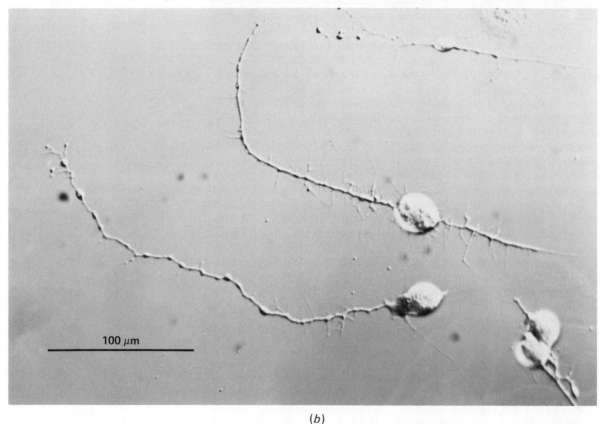

(b)

FIGURE 8-3 (Opposite page) The Schwann Cell. From the sciatic nerve of a newborn rat. The accompanying sheath cell (Schwann cell, in this case) wraps itself again and again about the axon. Successive stages are shown: axon A_1 is merely enclosed by the Schwann cell; there are $2\frac{1}{8}$ turns about axon A_2, and $3\frac{1}{3}$ turns about axon A_3. Note the prominent nuclei of the Schwann cells. [Courtesy of H. de F. Webster. Similar to *J. Cell Biol.*, **48**: 348 (1971).]

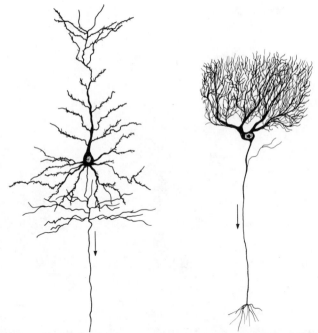

FIGURE 8-2 Multipolar Neurons. Neurons from the central nervous system have numerous branching processes, permitting contact with many other cells. Two examples are diagramed. The arrow indicates the axon and direction of impulse propagation.

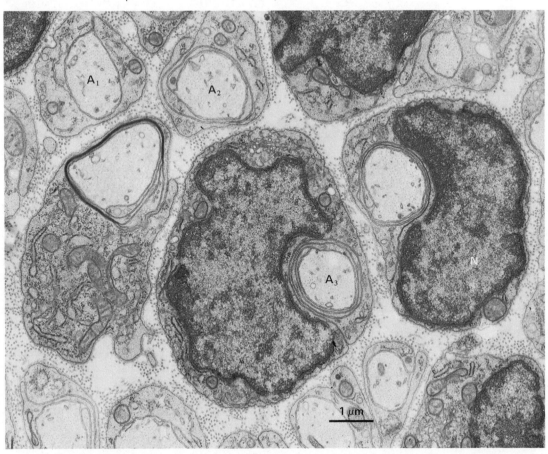

351

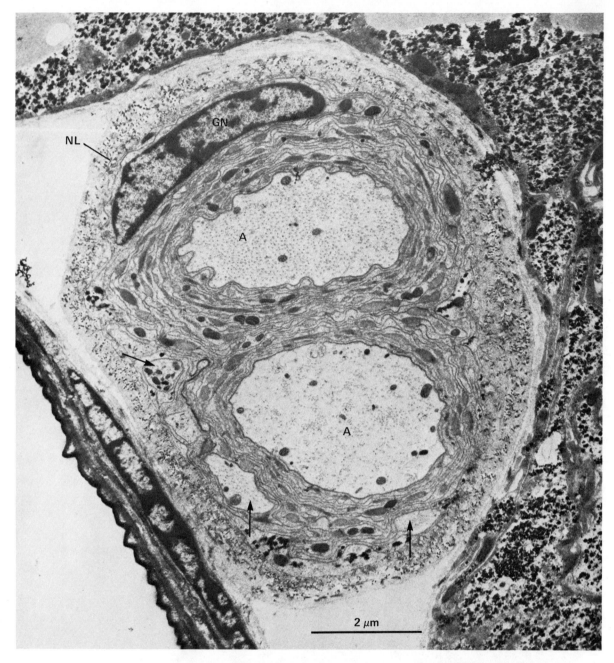

FIGURE 8-4 A Nerve. As with vertebrates, the axons in this cockroach nerve are invested with a sheath or neurolemma (NL) formed by associated cells. One can see two large axons (A), several smaller ones (arrows), and the prominent nucleus of one of the sheath cells (GN). [Courtesy of A. T. Whitehead, *J. Morph.*, **135**: 483 (1971).]

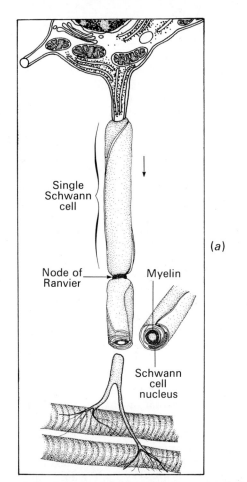
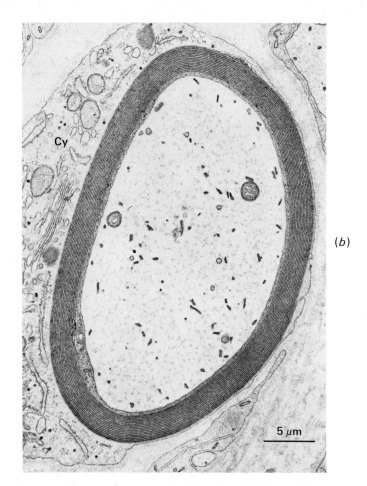

of the Atlantic squid, *Loligo*, is controlled by neurons with axon diameters that are often as much as 0.5 mm (500 μm). This was a fortunate observation, for the large size of these axons makes it possible to do experiments that would be technically very difficult or impossible with smaller cells. In addition, the squid's giant axon can remain excitable for hours after being dissected from the animal, providing only that it is tied at both ends to prevent loss of cytoplasm (called *axoplasm* in this case). In other words, an isolated giant axon can be stimulated to produce and conduct an impulse even though the axon is nothing more than a cell fragment. Most of what is known about the excitable properties of nerve cells has been learned through studies on this system.

The Action Potential. Recall that cells typically have high internal K^+ and low Na^+, maintained by ion pumps. In the case of the squid's giant axon, the Na^+, K^+, and Cl^- contents are, respectively, about 0.05 M, 0.40 M, and 0.05 M. In sea water, with concentrations of 0.46 M, 0.01 M, and 0.54 M for these three ions, the axon will have a membrane potential of −60 mV. This value is correctly predicted by the Goldman equation (6-16) with $b = 1/15$, and turns out to be slightly less negative than the K^+ equilibrium potential predicted by the Nernst equation (6-14). As usual, Na^+ is very far from equilibrium, entry being driven both by concentration and electrical gradients.

FIGURE 8-5 A Myelinated Neuron. Myelin is formed by sheath cells that tightly enwrap the nerve cell process. Adjacent sheath cells are separated by gaps called nodes of Ranvier. (a) This example is a motor neuron, innervating striated muscle. (b) A completed myelin sheath around a cat axon. Note the many distinct layers in the myelin wrapping. Note also the numerous filaments and microtubules within the axon, i.e., in the axoplasm. Schwann cell cytoplasm, Cy. Note that Fig. 8-3 (p. 351) actually shows early stages in the formation of myelin, which in mammals occurs largely after birth. [(b) Courtesy of T. H. Williams and J. Jew, *Tissue and Cell*, **7**: 407 (1975).]

Consider what would happen if a portion of the membrane surrounding the squid giant axon were suddenly to become permeable to sodium by the opening of some kind of channel or "gate." The inrush of ions would offset the negative potential in the region of the leak and, because of the greater concentration of Na^+ outside, would actually push the potential toward the Na^+ equilibrium point of $+56$ mV. If before that point is reached the Na^+ channel were closed again, diffusion and the Na^+ pump would eventually restore the original ionic concentration on both sides of the membrane, and with it the original -60 mV potential. However, if a channel for K^+ were to open at about the time the Na^+ channel closed, K^+ would leave the axoplasm, driving the potential rapidly toward the K^+ equilibrium value of about -90 mV and thus causing a quick return to the *resting potential*. At that point the original permeability might be reinstated.

The situation just described is a relatively accurate description of the way in which the excitable membranes of neurons and most muscles respond to the proper stimulus. That stimulus may be in the form of an electrically induced partial depolarization of perhaps 20 mV or more (e.g., a change from -60 to -40 mV), or it may be in the form of a chemical. In either case, the result is to open a channel for sodium and trigger the events as they were outlined. The resulting reversal of polarization and its restoration constitute the *action potential* (see Fig. 8-6). This wave of polarization reversal may be propagated as a nerve impulse.[2]

An action potential, then, is an electrical response of a membrane to certain kinds of stimulation. It consists of a swing in the voltage from some negative resting potential to zero and then through a positive overshoot, with a subsequent return to the original resting potential. The depolarization and positive overshoot are due to a sudden increase in an otherwise slow sodium influx; later return to the resting potential is greatly accelerated by a corresponding increase in potassium efflux beginning just before the sodium ion movement reaches its maximum.

K. S. Cole and H. J. Curtis, working in the mid-1930s at the marine biological laboratories in Woods Hole, Massachusetts, found that although the resistance of the squid axon membrane drops precipitously during the action potential there is no change in its capacitance. This observation rules out the possibility of a simple "short circuit" as an explanation for the permeability change. They concluded that the bulk of the membrane continues to act like a capacitor during an action potential, with the ionic flux occurring at isolated points (channels) that occupy altogether less than 1% of the surface area.

The capacitance of the squid's axon can be used to calculate the minimum transfer of sodium ions needed to account for the rising phase of

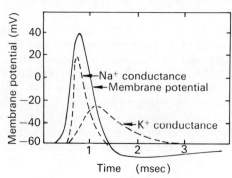

FIGURE 8-6 The Action Potential. The curves show relative changes in ion permeabilities and voltage as an action potential (also called a spike potential in this case) passes a pair of recording electrodes in a nerve fiber. Note the temporary hyperpolarization caused by K^+ efflux in the absence of Na^+ influx. The refractory period during which a second stimulation would produce no response is roughly 2 msec for this fiber. [Original data from A. L. Hodgkin, *Proc. Roy. Soc. (London), Ser. B*, **148**: 1 (1958).]

an action potential. The capacitance, C, of the squid giant axon is about one microfarad (μf) per cm² of surface, where

$$C = \frac{Q}{E} \qquad (8\text{-}2)$$

Here, Q is the charge in coulombs (a measure of the number of excess positive ions on one side or the number of excess negative ions on the other side) and E is the potential difference across the membrane, in volts. Thus, for a voltage change of about 0.1 (from 60 mV negative to 40 mV positive), there must be a corresponding change of about 10^{-7} coulomb of charge. Since there are 96,500 coulombs per mole of sodium ion (the Faraday constant), a transfer of 10^{-12} mole (or equivalent) of Na^+ is needed per cm² to cause the observed swing in potential. Actual measurements using radioactive Na^+ ($^{22}Na^+$ or $^{24}Na^+$) show that the value is really three or four times that amount, mostly because the above calculation did not take into consideration that K^+ efflux begins before Na^+ influx ends, and because it neglects the counteracting flow of Cl^-.

If, as suggested, the size of the action potential is due almost exclusively to changes in sodium ion permeability, then altering the concentration of sodium ion in the bathing fluid should affect the positive overshoot. This experiment was executed at the Plymouth, England, marine biological laboratories in 1947 by A. L. Hodgkin and Bernard Katz. They found that both the rate of voltage change and the magnitude of the positive overshoot vary with the external sodium ion concentration in accordance with predictions. The resting potential, on the other hand, is almost independent of Na^+ concentration in the bathing fluid. Rather, it depends strongly on the K^+ concentration, becoming more negative as the external K^+ is increased. This, too, is consistent with our model, which includes a greater permeability to potassium than to sodium.

An additional ion flux, the importance of which has become clear only in the last few years, is that of Ca^{2+}. The internal concentration of Ca^{2+} is very low compared to that of the surrounding fluid and the concentration of free Ca^{2+} is even lower, thanks in large part to mitochondrial uptake. The low axoplasmic Ca^{2+} is maintained by a constant efflux, much of it coupled to Na^+ influx by Na^+-Ca^{2+} exchange proteins. The stoichiometry of this exchange is thought to be about three Na^+ for one Ca^{2+}, utilizing the Na^+ gradient as a direct source of energy. During an action potential, however, there is marked Ca^{2+} influx, partly, it is thought, through the Na^+ channel and partly through separate, voltage-sensitive Ca^{2+} channels. This influx prolongs the action potential, with heart muscle cells being particularly affected. Calcium ions also alter the excitability of the membrane, with higher external levels stabilizing excitable membranes and low levels causing increased excitability.

The major features of the action potential, however, are clearly due to changes in permeability to Na^+ and K^+. Permeability to sodium increases first and normalizes first, but these changes are overlapped and followed by corresponding alterations in potassium ion permeability. The result is a return to the resting potential in a few milliseconds.

Initiation of the Action Potential. An action potential may be initiated by changes in membrane permeability caused by partial depolarization or by chemical stimulation. Each plays an important role in nervous function and integration.

Detailed study of permeability changes as a function of membrane potential was made possible by a device called the *voltage clamp*, developed by K. S. Cole in the 1940s. It provides a way of holding the membrane potential at a preset value so that conductances can be measured without triggering the explosive changes associated with an action potential. In the case of the squid axon, electrodes are inserted into the cut end of the axon and into the surrounding fluid, one set to supply an external voltage and a second set to monitor the membrane potential.

With the voltage clamp, it was found that a depolarization of only 10 mV increases the sodium permeability about eightfold. However, when the membrane potential is dropped to zero with the voltage clamp and held there, a transient inward flow of sodium ions is followed within a couple of milliseconds by an outward flow of potassium ions. The latter continues as long as the membrane is held in a depolarized state. In other words, whereas the induced increase in sodium permeability is self-correcting, potassium permeability follows the membrane potential (see Fig. 8-7). Since K^+ efflux normally drives the potential negative again following the Na^+ spike, it appears that potassium permeability is controlled by negative feedback and that the key to electrical initiation of the action potential is the voltage-sensitive opening of a Na^+ gate.

It is difficult to explain the molecular bases for these changes in ionic permeability because we do not know enough about the membrane's components or how they interact. However, measurements of several kinds indicate a considerable change in the conformation of membrane proteins during the action potential. Such ideas are not hard to reconcile with our experience, for there are many cases in which small conformational changes in proteins are known to be associated with large changes in biological activity. Since conformational shifts can be induced by a wide variety of stimuli, we can imagine that certain membrane proteins respond to changes in the membrane potential. Such changes could be responsible for controlling sodium and potassium permeability.

The initiation of an action potential through chemical, rather than electrical, stimulation puts us into more familiar territory, for the existence of conformational shifts in proteins due to the binding of small molecules was discussed earlier. The first direct demonstration that action potentials can be affected by chemicals was in 1921, when Otto Loewi used a pipet to pick up a small quantity of fluid from a frog's heart just after it had received a string of impulses from the vagus nerve. Loewi found that the effect of this fluid on a second heart was like that of the vagus nerve itself. The active substance in the fluid was later identified as acetylcholine.

$$CH_3-\overset{\overset{O}{\|}}{C}-O-CH_2-CH_2-\overset{+}{N}(CH_3)_3$$
Acetylcholine

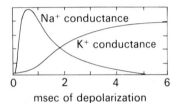

FIGURE 8-7 Ionic Conductances in a Voltage-clamped Axon. The axon was subjected to a constant depolarization via the voltage clamp. Note that Na^+ entry is transitory. Potassium conductance, on the other hand, varies with membrane potential. (Original data by A. L. Hodgkin and A. F. Huxley.)

Actually, acetylcholine does not initiate action potentials in the heart, for the rhythmic generation of these impulses originates within the heart itself. In fact, acetylcholine inhibits spontaneous contraction of the heart. It does, however, initiate action potentials in other excitable membranes, including those of many nerve cells. Acetylcholine increases the permeability of many nerve cell membranes to sodium ions, thus moving their potential toward the threshold level or beyond. Once the threshold is reached, an action potential is initiated and runs its normal course. The effect of acetylcholine on the membranes of heart muscle cells, however, is to increase their permeability to K⁺ rather than to Na⁺, thus making the initiation of action potentials more difficult.

Whether an action potential is initiated chemically or electrically, there is a period immediately after firing when neither type of stimulation can trigger a new impulse. This time interval, which is known as a *refractory period*, lasts until the resting potential is reattained. The maximum rate at which action potentials can be generated depends on the duration of the refractory period for a particular nerve. Furthermore, as the frequency of stimulation is increased, a maximum rate of firing is reached. Thereafter, the stimuli can no longer be distinguished as a string of separate events, for the effect will be the same as a continuous stimulation. This phenomenon, which is most noticeable to us in the perception of sights and sounds, is known as *flicker fusion*.

8-2 PROPAGATION OF ACTION POTENTIALS

We shall now consider ways in which action potentials can be *conducted* by an excitable membrane, and ways in which they may be *transmitted* from cell to cell. These two processes, conduction and transmission, are generally regarded as being fundamentally different.

Conduction in nearly all cases is an electrical process that is based either on *local currents* or on something called the *cable effect*. In the former mechanism, action potentials spread continuously outward from the point of initiation. The cable effect, on the other hand, can transmit the electrical disturbance to a relatively distant point in much the same way as electricity is carried by a wire. Transmission between cells, in contrast, is mostly via chemical intermediates, such as acetylcholine, that are released from one cell and diffuse across the intercellular gap to initiate a new action potential on the other side.

Conduction by Local Currents. Impulse conduction by simple electrical stimulation is attractive as a general mechanism, for it requires no metabolic involvement except through the maintenance of ion gradients. It suggests only that a partial depolarization may result from a larger depolarization at a nearby point (part *a* in Fig. 8-8). Such an effect is all that is necessary to ensure propagation, for once the threshold potential at that second point is reached a new action potential will be triggered, spreading the wave of depolarization across the surface.

The independence of impulse conduction from the metabolic machinery of the nerve cell was demonstrated in the early 1960s by two groups working independently at the marine biological stations in Plymouth, England, and Woods Hole, Massachusetts. They removed the axoplasm from an isolated squid axon and replaced it with solutions of various composition—a process called *perfusion*. As long as the perfusing fluid was chosen properly, including the requirement that it be high in K⁺ and low in Na⁺, the axon was capable of conducting hundreds of thousands of action potentials before failing. Only a thin layer of axoplasm adheres to the membrane of the perfused axon, clearly showing that the bulk of the axoplasm is not involved in conduction. Hence, conduction appears to be independent of the metabolic activities of the cell and dependent on the properties of the membrane itself.

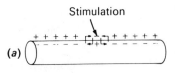
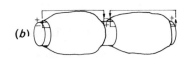

FIGURE 8-8 Conduction of the Impulse. (*a*) Conduction by local currents, in which an action potential depolarizes adjacent areas to push them past their threshold potential; (*b*) the cable effect, or saltatory propagation, thought to function in myelinated axons.

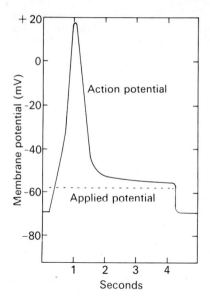

FIGURE 8-9 Action Potentials in a Synthetic Membrane. The independence of the action potential from metabolic function is emphasized by studies with synthetic membranes (lipid bilayers with added protein), for they can exhibit excitable properties remarkably like those of the living cell—even to reversible blockage by local anesthetics such as cocaine. A depolarization potential of about 11 mV was applied as indicated by the dotted line. The refractory period is about 10 s. [Data from P. Mueller and D. O. Rudin, *Nature*, **213**: 603 (1967).]

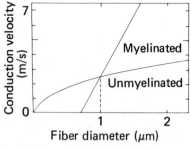

FIGURE 8-10 The Advantage of Myelination. An impulse can be propagated more rapidly in myelinated axons, presumably because it is conducted from node to node via the cable effect. In axons of very small diameter, with closely spaced nodes, that advantage is lost. [Data from W. Rushton, *J. Physiol.*, **115**: 101 (1951).]

Similar conclusions are reached in studies using synthetic membranes. For example, in 1967 P. Mueller and D. O. Rudin reported the generation of action potentials in a synthetic membrane consisting of a lipid bilayer, a crude protein extract from a bacterium, and protamine (a basic protein that takes the place of histones in sperm cells). Partial depolarization from an applied electrical source triggered voltage changes that resemble those seen in cellular systems (see Fig. 8-9). Experiments of this sort indicate that generation and conduction of action potentials is due to changes in the ionic conductivity of membranes, brought about no doubt by the direct effect of voltage changes on ionic channels.

The speed with which an action potential spreads by local currents can be calculated from the dimensions of the axon, its permeability during various stages of the cycle, and the properties of the ions involved. According to equations derived by A. L. Hodgkin and A. F. Huxley in 1952, the speed of conduction should be proportional to the square root of the diameter of the axon, a relationship that has been verified in a wide variety of unmyelinated nerves.

Saltatory Conduction. A second mechanism of impulse propagation, the cable effect, requires a medium of low resistance. This requirement is only marginally met by the nerve cell axon. Because of the electrical resistance of axoplasm, an action potential that is propagated solely by the cable effect should be quickly distorted and dissipated. The distance over which it survives can be demonstrated with the local anesthetic procaine, which eliminates action potentials wherever the drug is applied. To deaden an axon, at least a millimeter must be treated with procaine, implying that an impulse can jump such distances but no more. However, while the cable effect is thus eliminated as a general mode of conduction, it is still applicable to the special case of myelinated neurons.

A typical spacing between nodes of Ranvier in myelinated neurons is about a millimeter. Since action potentials cannot be detected between nodes, where the nerve process is insulated by myelin, it has been proposed that the potential jumps from node to node via a cable effect, a mechanism known as *saltatory conduction*. The impulse is renewed at each node as the propagated wave triggers a new potential in it, much as a submarine cable sends telephone messages from one repeater station to the next (part *b* in Fig. 8-8). One would expect the cable effect to conduct an impulse rapidly and, indeed, the fastest nerve fibers, which support velocities well in excess of 100 m/sec, are myelinated (see Fig. 8-10).

The length of time it takes an impulse to travel from node to node in a myelinated nerve is negligible compared to its regeneration time at a node. The conduction velocity of such a neuron will therefore depend on the distance between nodes, a value that for a given nerve is proportional to its diameter—hence, large humans with larger diameter axons may have reflex times that are just as fast as those of small humans, in spite of the greater distances that impulses must travel. This relationship between axon diameter and internodal distance exists because the number of Schwann cells does not increase with growth. Hence, as an axon increases in length, each Schwann cell covers a longer stretch. The fact that length, diameter, and internodal distance all vary together produces a linear relationship between conduction velocity and axon diameter. In the case of the human ulnar nerve, for example, the internodal distance of various sized axons is about 100 times their diameter.

Since each action potential must affect a finite length of the axon, myelination fails to confer any advantage in propagation to very small axons. Accordingly, one finds that in vertebrates, where myelination is used to increase the rate of conduction of larger axons, peripheral axons with diameters of less than about one μm are nevertheless unmyelinated (Fig. 8-10), although in the central nervous system axons as small as 0.2 μm may be myelinated. Where speed of conduction is vital, as with the neurons of a reflex arc, myelinated nerves with larger axons are found. Where a little delay in getting the message can be tolerated, as in the control of viscera, space seems to be conserved by using smaller fibers, often without myelination. Invertebrates, as noted earlier, rely on unmyelinated axons of extremely large diameters to provide rapid impulse conduction wherever that is vital. The squid's giant axon is in this group, since it is part of the motor nerve controlling the swim reflex.

Intercellular Transmission by Acetylcholine. Intercellular transmission of nerve impulses almost always requires a chemical intermediate. The best understood of these intermediates is acetylcholine.

The model for neurotransmission by acetylcholine consists of the following steps: (1) When an action potential reaches the end of an axon, it causes acetylcholine to be released from storage vesicles found there (see Fig. 8-11). (2) Acetylcholine diffuses across the synaptic gap. (3) At the postsynaptic membrane, the chemical encounters specific receptor sites to which it binds reversibly. (4) The appearance of acetylcholine at the postsynaptic receptors alters the ion permeability of the associated membrane, in some cases triggering a new action potential. (5) The reversible binding of acetylcholine means that each molecule is also unbound and exposed for a time to hydrolysis by an enzyme, acetylcholine esterase. (6) The hydrolysis products, acetate and choline, are inactive as neurotransmitters, but may diffuse back across the gap to be taken up again by the presynaptic neuron. (7) And finally, the amount of acetylcholine used in the process is replaced in the presynaptic neuron by transferring acetate from coenzyme A to choline and packaging the product into vesicles.

There is a good deal of support for this general mechanism, and a number of the details are now available. For example, from the work of Sir Bernard Katz and others, we have considerable information about the nature of acetylcholine release from its storage vesicles within the presynaptic neuron. Acetylcholine vesicles are constantly being opened at a slow rate (e.g., one per second), but the number of acetylcholine molecules contained in each vesicle (a few thousand) is only enough to cause a transient change of about 0.5 mV in the potential of the postsynaptic membrane, not enough to trigger a full action potential. However, when an impulse reaches the region of acetylcholine vesicles in the presynaptic neuron, several hundred packets (a number that varies with the magnitude of the potential) may be released within a millisecond or so from the axon. Since the degree of polarization in the target cell is proportional to the concentration of acetylcholine reaching it, a new action potential may be produced there.

Katz, working in 1965 with R. Miledi, also measured the amount of time that it takes acetylcholine to transmit a nerve impulse. They placed a microelectrode at a neuromuscular gap, and found two electrical disturbances for each nerve impulse. The first was the arriving nerve

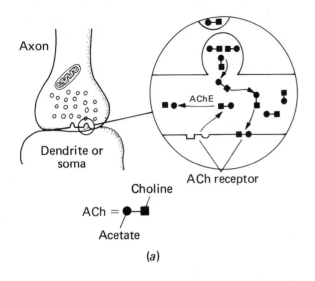

(a)

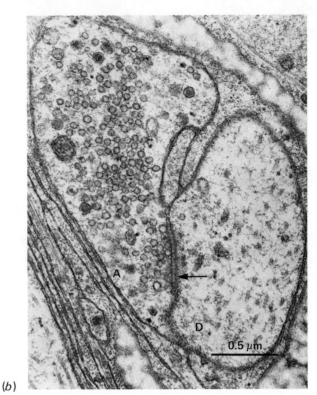

(b)

FIGURE 8-11 The Synapse. (a) Diagramatic representation showing synaptic vesicles emptying acetylcholine (ACh) into the synaptic cleft or space. The transmitter diffuses to the post-synaptic membrane where it reversibly binds to receptors until it is eventually destroyed by acetylcholine esterase (AChE). (b) From a cat, showing synaptic vesicles in the terminal portion of an axon (A) and the synaptic cleft (arrow) separating axon and dendrite (D). (c) Scanning EM of nerve terminals from a mollusk, *Aplysia californica*. The knobs, or *boutons*, are termini of nerve cell processes. Possible synapses are marked with arrows. [(b) Courtesy of J. J. Pysh and R. Wiley. From *Science* **176**: 191, copyright 1972 by the AAAS; (c) courtesy of E. R. Lewis, similar to *Science*, **165**: 1140 (1969).]

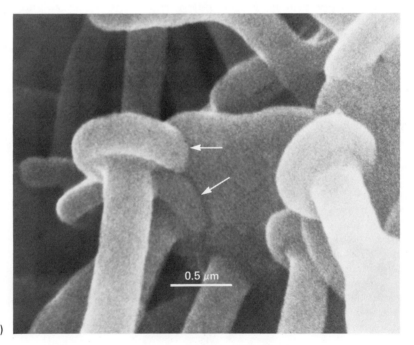

(c)

impulse itself. The second disturbance, 0.5 to 0.8 milliseconds later, was the initiation of the new action potential in the muscle cell membrane. This time lapse is consistent with the presence of a chemical transmitter, and is much too long to suggest any kind of direct, electrical transmission.

The actual events leading to emptying of the synaptic vesicles are not well understood. A widely held view at the present time attributes the release to Ca^{2+}-dependent changes in microfilaments and microtubules within the axon. The action potential is accompanied by a Ca^{2+} influx that somehow mobilizes these elements to cause movement of the synaptic vesicle toward the plasma membrane with which it fuses, releasing its contents of neurotransmitter. Cyclic AMP and possibly cGMP may also be involved in this process, perhaps indirectly through modulation of Ca^{2+} levels.

The fusion of synaptic vesicles with the plasma membrane of the axon adds lipid that must be removed again by formation of new vesicles if the axon size is to remain constant. Although the fate of these new vesicles is not clearly established, some workers have proposed that they are recycled within the nerve terminal itself to become new synaptic vesicles, filled with transmitter. However, while there is evidence for some local production and packaging of transmitter, most biosynthetic activities occur in the perikaryon (Fig. 8-12). Neurotransmitter molecules and the enzymes needed for additional neurotransmitter synthesis are packaged into vesicles derived from the Golgi apparatus of the perikaryon and transported outward along the axon to the nerve terminals by mechanisms that will be treated in Chap. 9.

The events at the postsynaptic membrane have been studied in several ways. For example, the nature of the acetylcholine binding sites has been examined indirectly through the use of inhibitors. These inhibitors include the curare family (e.g., *d*-tubocurarine) and atropine, both of which compete with acetylcholine for the membrane receptor sites, and eserine, which inhibits acetylcholine esterase. However, more direct studies have also been carried out, such as those by the Argentine scientist Eduardo De Robertis, who reported the isolation of an acetylcholine receptor from mammalian brain. The acetylcholine receptor molecules

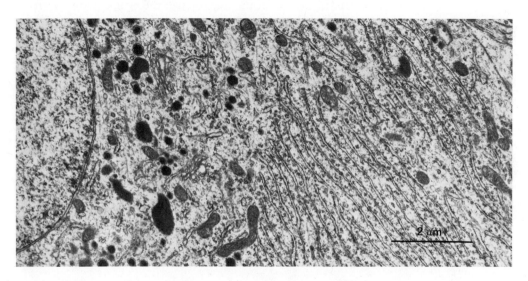

FIGURE 8-12 Perikaryon. Most biosynthesis is limited to the neuronal body or perikaryon. This section through the perikaryon of a mammalian neuron shows a portion of the nucleus on the left and stacked lamellae of rough endoplasmic reticulum (known to the light microscopist as Nissl substance) on the right. In between, one sees many dense vesicles being formed. Most will be transported out neuronal processes toward the terminals. [Courtesy of R. Clattenburg, Univ. of Western Ontario.]

Other Neurotransmitters. The discussion so far has been about *cholinergic neurons*—i.e., those that use acetylcholine as a transmitter. Many other neurotransmitters have also been identified, although a given neuron utilizes no more than one. Examples of other transmitters include amino acids such as glutamate, aspartate, and proline; amino acid derivatives such as γ-amino butyrate (GABA), derived from glutamate, serotonin (from tryptophan), and histamine (from histidine); and certain peptides, including "substance P" found in mammalian brains.

The mode of action of these neurotransmitters is much like that of acetylcholine. An important distinction, however, is that termination of action is not ordinarily by enzymatic destruction as it is with acetylcholine, but by reaccumulation in the presynaptic cell. The presynaptic cell can thus recycle the transmitter by packaging it into new vesicles.

What is the reason for so many neurotransmitters? One possibility is that different neurotransmitters in adjacent pathways help avoid interference. A more important advantage, however, is that several incoming neurons can affect the same target cell in different ways by using different neurotransmitters. Some transmitters, for example, increase excitability of the postsynaptic membrane, while others decrease it (see Fig. 8-13).

Excitatory and Inhibitory Impulses. The first example of dual excitatory and inhibitory neurotransmitters was provided by O. H. Loewi. After demonstrating that a chemical intermediate from the vagus nerve slows the heart rate of a frog, Loewi demonstrated that the chemical released from *sympathetic nerves* increases it. This second transmitter was later identified as *adrenaline* in the frog, though the sympathetic system in humans and most other animals uses mainly *noradrenaline*. These neurotransmitters, which are also known as *epinephrine* and *norepinephrine*, respectively, are by definition found in *adrenergic neurons*, a category that includes a third *catecholamine* transmitter named *dopamine* (see Fig. 8-14).[3]

In the discussion of synaptic transmission, it was explained that acetylcholine causes an increased permeability to sodium ions in the membrane of a neuron. Thus it causes a depolarization and possibly an action potential. In the heart, however, acetylcholine facilitates entry of K^+ rather than Na^+. This effect tends to move the resting potential closer to the K^+ equilibrium value, which is more negative than the normal resting potential. Permitting easier K^+ entry, then, causes a *hyperpolarization*, inhibits the generation of action potentials, and slows the heartbeat. The adrenergic transmitter, on the other hand, causes a loss of polarization by opening the sodium channel in heart muscle cell membranes, and thus increases the heartbeat.

The difference between an *excitatory impulse* and an *inhibitory im-*

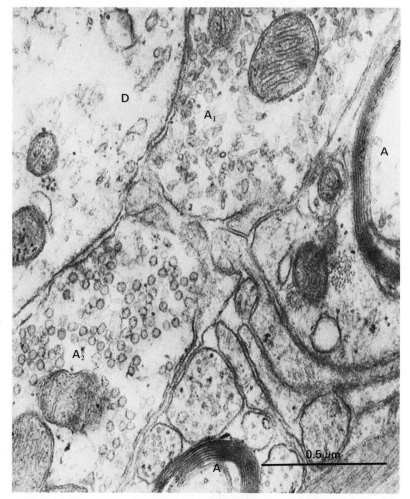

FIGURE 8-13 Excitatory and Inhibitory Synapses. In general, excitatory synapses seem to have round vesicles and relatively wide synaptic clefts as in axon A_2, whereas inhibitory synapses are thought to have ellipsoidal vesicles and narrower clefts as in axon A_1. This micrograph, from the spinal cord of a monkey, may thus represent the synapse of an excitatory and inhibitory axon on a common dendrite (D). Other axons (A) are seen at the edges of the micrograph. [Courtesy of D. Bodian. From *Science*, **151**: 1093, copyright 1966 by the AAAS.]

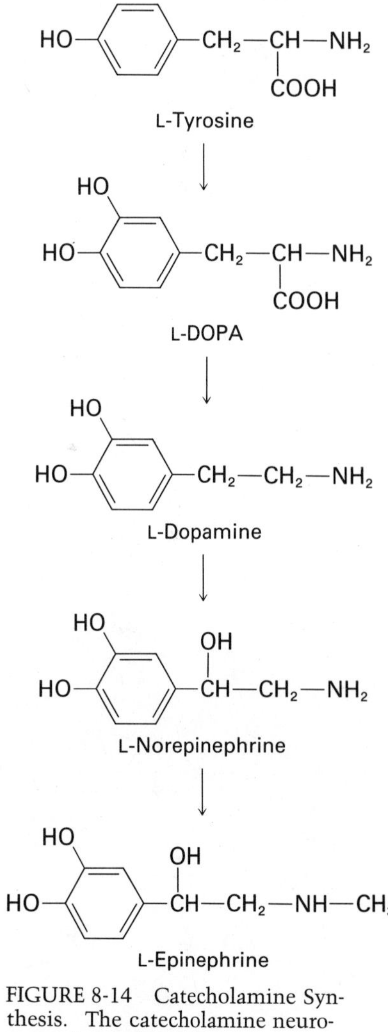

FIGURE 8-14 Catecholamine Synthesis. The catecholamine neurotransmitters dopamine, norepinephrine, and epinephrine are derived from tyrosine.

pulse in other systems can also be explained in many cases by an increased permeability to different ions. That is, the excitatory impulse makes the target cell's membrane more permeable to sodium, causing a depolarization, whereas the inhibitory impulse tends to stabilize the membrane at its resting potential or even to cause a slight hyperpolarization. The latter can be achieved by increasing permeability to K^+ or sometimes to Cl^-. Even where these ions are already at electrochemical equilibrium, increasing permeability to them allows a counterflow during sodium entry, partially canceling the depolarizing effect of Na^+ influx.

Excitatory and inhibitory synapses, along with other mechanisms, are employed by the brain to integrate and process nerve impulses and thus achieve the complex functions we associate with the central nervous system.

Electrotonic Synapses. Propagation of action potentials between cells can be electrical as well as chemical. Electrical coupling of excitable cells in mammals is limited to certain kinds of muscle, including the smooth muscle of the intestine. These cells are electrically connected via gap

CHAPTER 8
Excitable Cells

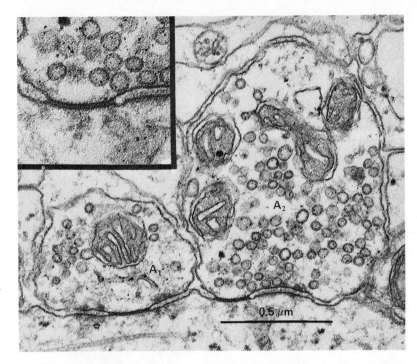

FIGURE 8-15 The Electrotonic Synapse. Ionic coupling, apparently via cytoplasmic bridges (see Section 6-2), provides a rapid feedback to synchronize the firing of some communicating neurons. The micrograph shows two axons (A_1 and A_2) of a spiny boxfish in apposition to a common nerve cell body. The presynaptic and postsynaptic membranes are joined by a gap junction, shown to better advantage in the inset. Mitochondria are seen in both axons. [Courtesy of M. Kriebel, M. Bennett, S. Waxman, and G. D. Pappas, *Science*, **166**: 520 (1969). Copyright by the AAAS.]

junctions (see Chap. 6) that permit cell-to-cell propagation of action potentials.

Electrical propagation between cells is fast but unselective—it is not subject to modulation and is bidirectional, unlike the unidirectional chemical synapse. It is therefore not used much by neurons, though there are some exceptions, one of which is seen in Fig. 8-15. This figure shows an *electrotonic synapse*, in which both synaptic vesicles and an electrically coupling gap junction can be seen. The advantage is speed of transmission, which in this particular case is presumably useful in helping the fish escape from predators.

8-3 OTHER EXCITABLE CELLS

The neuron is by no means the only excitable cell. Muscle cells, as already noted, fall into this category, and the sense organs provide many other examples.

Sensory Cells. Most of the information provided to the central nervous system comes from specialized, excitable cells that function as sense organs. Classically, the senses are divided into five groups, providing sight, taste, smell, touch, and hearing. To that one sometimes adds balance. However, from a molecular standpoint all the sense organs are transducers, converting energy from one form to another. It is more convenient, therefore, to consider them on the basis of the kinds of stimulation to which they respond. Thus, there are (1) thermal receptors; (2) mechanoreceptors; (3) chemoreceptors; and (4) photoreceptors.

In general, heat and cold are perceived with the help of axonal endings that generate action potentials at a temperature-sensitive rate. Touch, balance, hearing, and a knowledge of the position of body parts are provided by mechanical receptors—in the skin for touch, in the inner ear for balance and hearing, and as stretch receptors in muscles to provide a sense of position. In each case, mechanical deformation leads to changes in Na^+ permeability and either to the initiation of action potentials or to a change in their frequency. Chemoreceptors for taste and smell also function through generation of action potentials, but in response to binding of appropriate molecules.

We cannot go into these various cell types in detail. In most cases not enough is known about the molecular mechanisms involved to make inclusion worthwhile, even if space were available. Rather, we shall introduce here only one type of sensory transducer, namely the photosensitive cells of the vertebrate retina, and one effector cell, that responsible for neurosecretion. In the next chapter, however, another effector cell, muscle, will be treated in more detail.

Photoreceptors. As you might guess, the basis for sight is the absorption of light by chromophores (colored substances) within sensitive cells, and a corresponding change in ion permeability and resting potential. Vertebrates have two overlapping visual systems that respond in just this way. One system is provided by a group of photosensitive retinal cells called *cones,* and the other by cells called *rods* (Fig. 8-16). The names of the cells are derived from the shape of their light-sensitive tips (see Figs. 8-17 and 8-18).

Cones provide color vision, using three pigments that have absorption maxima spread through the visible region—one absorbs maximally in the blue-violet, a second in the green, and a third in the yellow part of the spectrum. Light at any wavelength from about 400–700 nm will be absorbed in varying degrees by one or more of these pigments. The pattern of stimulation is interpreted by the central nervous system as color. (In hereditary color blindness, one or more of these pigments is missing.) Rods, on the other hand, have a single pigment with an absorption maximum at about 500 nm, and are therefore quite insensitive to light near the red end (700 nm) of the visible spectrum (Fig. 8-16).

The rods and cones are adapted to different purposes. Rod cells are sensitive to very low light levels—too low to satisfactorily excite the cones. Therefore, rods supply night vision (*scotopic vision*), which, however, is nearly monochromatic because it is dominated by a single pigment. At light levels intense enough to stimulate cones easily, the pigment in the rods is bleached and hence does not absorb. This pigment, called *rhodopsin* or visual purple, is the dominant protein of the rod discs shown in Fig. 8-17. When light levels fall again, color returns to rhodopsin. We perceive this change as "dark adaptation." While the separation of function between rods and cones is not as clean as implied here, the model presented is a useful approximation to help us understand the system.

Although the rods and cones are different in physical appearance and in the pigments they contain, the conversion of light energy to nerve impulse appears to use the same mechanism in each case. The key to the process is the photoisomerization of 11-*cis*-retinal (a vitamin A derivative) to its *trans* configuration by rotation about the double bond at carbon 11 (see Fig. 8-19). A visual pigment such as rhodopsin is the combination of retinal and a protein called an *opsin*. There is a different

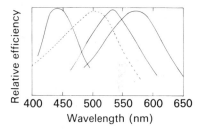

FIGURE 8-16. The Visual Spectrum. Solid lines show the responsiveness of the three human cone pigments. Dashed line is the corresponding curve for rhodopsin, the rod pigment.

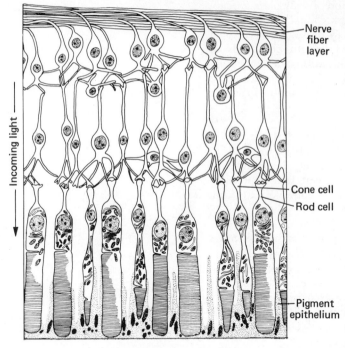
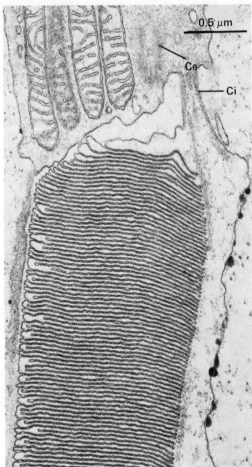

FIGURE 8-17 Photoreceptors. The diagram is a section through a vertebrate retina. Light passes through a nerve fiber layer to reach and excite rod and cone cells. These cells have stacks of light-sensitive discs in their outer segments—that is, in the portion of the cell that is embedded in the pigment epithelium at the back of the retina. The discs within rod cells, clearly seen in the micrograph, are constantly being formed by the cell body (or inner segment, which is connected to the outer segment by a modified cilium, Ci) and shed into the pigment epithelium at the outer extremity of the stack. Centriole (Ce). [Micrograph courtesy of R. W. Young, J. Cell Biol., **49**: 303 (1971).]

opsin for each pigment. The change in configuration of retinal during excitation by light causes it to dissociate from its opsin. A conformational change in the latter alters ion permeability of the disc membranes. In the vertebrate rod, for example, Ca^{2+} is released from the discs in the presence of light and H^+ is accumulated. The released Ca^{2+} is thought to mediate subsequent changes in the cell's plasma membrane, including a decreased Na^+ influx and hyperpolarization. (Invertebrate photoreceptors, on the other hand, are depolarized by light.) Translation of these changes into nerve impulses is a complex process and is different for different cells of the retina—some nerve fibers arising from the retina discharge constantly but are silenced by light, whereas others are normally quiet but discharge (either continuously or only briefly) when illuminated.

The actual perception of an optical image depends on the simultaneous response of millions of cells. A portion of the required processing takes place within the retina itself. The rest occurs within the brain after signals are carried to it by the optic nerve. Even with some prior processing, this nerve has a tremendous amount of information to transfer. Accordingly, the optic nerve consists of about 1200 bundles of myelinated fibers held together by associated glial cells. It is quite un-

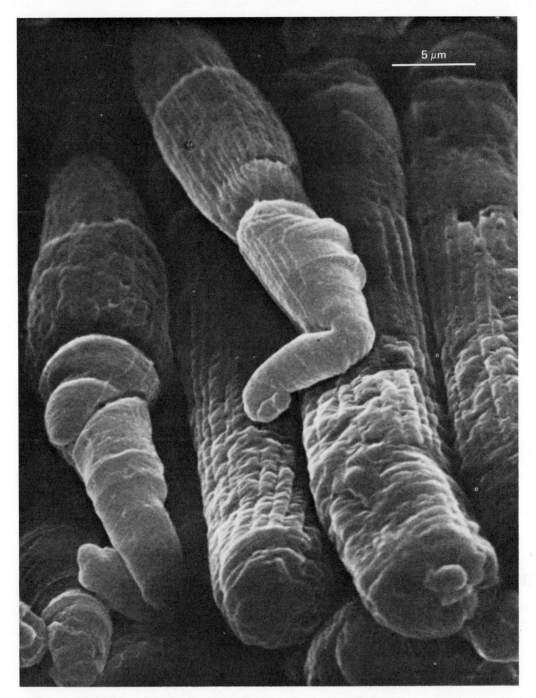

like peripheral nerves in both structure and function, and may be properly classed as an extension of the brain itself, from which it arises during embryonic development.

Neurosecretion. The central nervous system regulates activities in two ways, by direct nervous stimulation and by the release of hormones. For

FIGURE 8-18 Rods and Cones (Outer Segments) of the Mudpuppy (*Necturus*). Fixed and air dried. [Courtesy of E. R. Lewis; similar to E. Lewis, Y. Zeevi, and F. Werblin, *Brain Res.*, **15**: 559 (1969).]

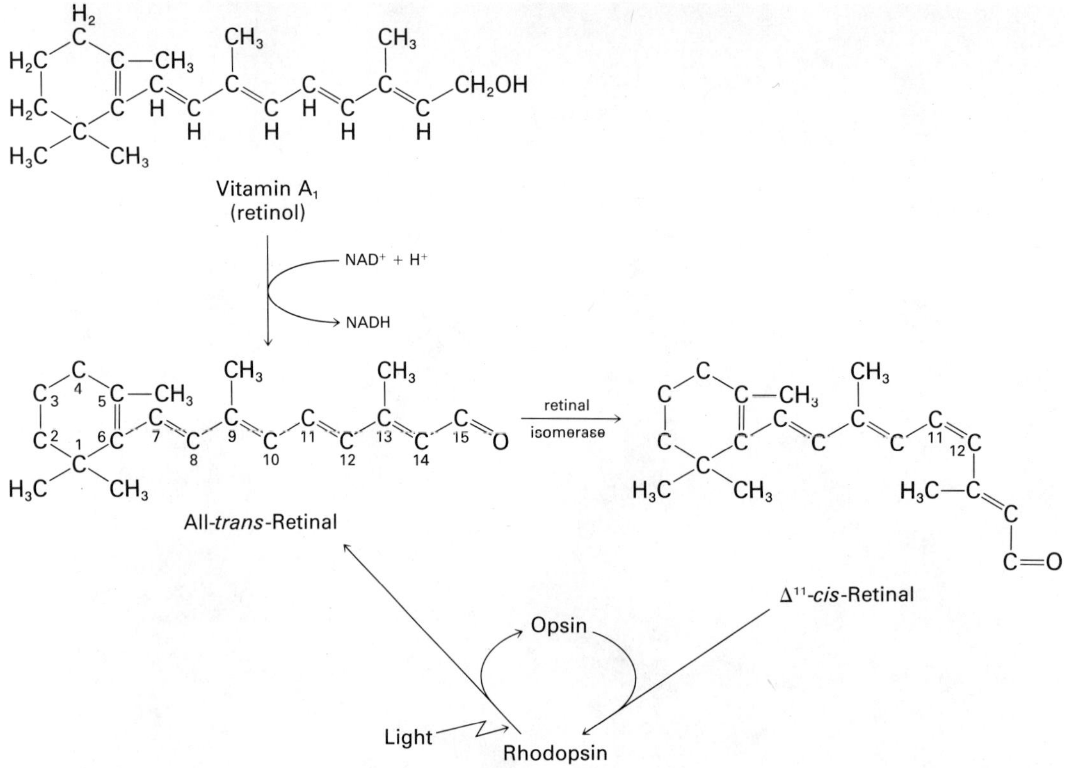

FIGURE 8-19 The Fate of Retinal During Illumination and Dark Recovery.

example, some neurons in the hypothalamus (at the base of the brain) release short peptides into the circulation, the function of which is to stimulate the anterior part of the pituitary gland to secrete various hormones. Other neurons in the hypothalamus have axons that extend into the posterior pituitary and themselves release the peptide hormones oxytocin and vasopressin. Thus, in both of these cases, neurons release hormones instead of neurotransmitters.

Several neurotransmitters can double as hormones. Notable among these are serotonin, epinephrine, and norepinephrine. The latter two are produced by specialized neurosecretory *chromaffin cells* (Fig. 8-20), so named because they develop a yellow-brown reaction product in the presence of chrome compounds such as potassium dichromate. Chromaffin cells are found in many isolated areas but the adrenal gland (specifically, the adrenal medulla) is the most important source in vertebrates. These excitable cells contain vesicles of hormone that are released directly into the bloodstream when the cells receive an action potential from incoming nerves.

Neurosecretion thus provides a connection between the nervous and endocrine systems. Other examples of secretion—in part stimulated by nerves—will be considered in Chap. 10. First, however, we shall consider in Chap. 9 another important type of excitable cell, the muscle fiber.

SECTION 8-3
Other Excitable Cells

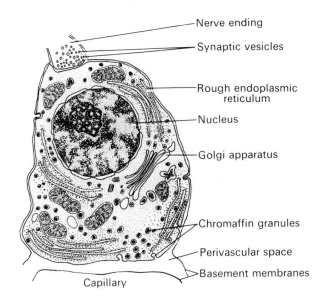

FIGURE 8-20 The Chromaffin Cell. This specialized neuron discharges its vesicles into capillaries instead of into synaptic clefts. Each cell contains vesicles (chromaffin granules) either of epinephrine or norepinephrine. The process of excitation and discharge is comparable to that occurring in true neurons.

SUMMARY

8-1 The basic unit of any nervous system is the individual neuron, which consists of a cell body (the perikaryon) containing a nucleus and generally one long and several short processes called axons and dendrites, respectively. Glial cells (also called Schwann cells in the peripheral system) are often found wrapped about nerve processes, presumably to protect and insulate them from other cells. When these cells are spirally wrapped in a tight sheath about an axon, the axon is said to be myelinated.

An action potential is initiated by an increase in permeability to Na^+, which enters in response to both electrical and concentration gradients. As the membrane potential becomes positive, K^+ efflux begins. The Na^+ channel is then closed, while K^+ efflux continues until the resting potential is reachieved. Since the positive overshoot is driven by Na^+ entry, its magnitude varies with the external Na^+ concentration. The resting potential, on the other hand, is much more sensitive to the external K^+ concentration, reflecting the much greater permeability of the membrane to K^+.

Except during the refractory period after the last firing, an action potential may be initiated by a slight (e.g., 20 mV) depolarization or by certain chemicals such as acetylcholine. In both cases, the stimulus acts by increasing Na^+ permeability—i.e., by opening a "Na^+ channel." The opening is temporary (self-correcting) even when the membrane potential is held constant by a voltage clamp. Permeability to K^+, on the other hand, varies with membrane potential.

8-2 There are three general mechanisms for the propagation of nerve impulses:

(1) Nerve impulses can be propagated by chemicals. Although the use of metabolic poisons and isolated, perfused axons indicates that chemicals are not commonly used for intracellular conduction, chemical propagation is the most common mechanism for intercellular transmission. In cholinergic nerves, for example, acetylcholine, which is released from storage vesicles within the presynaptic axon, diffuses across the synaptic gap to interact with receptor sites on the postsynaptic membrane. Other neurotransmitters are also known, among the most common of which are adrenaline and noradrenaline (epinephrine and norepinephrine), collectively referred to as adrenergic transmitters. Multiple transmitters permit different neurons to affect the same target cell in different ways—e.g., by causing increased Na^+ conduction and excitation or by increasing K^+ conduction to cause hyperpolarization and subsequent inhibition.

(2) Nerve impulses may be propagated by direct electrical stimulation in the form of local currents. These ionic currents induce enough depolarization at an adjacent site to trigger a new action potential there. This mode of propagation is commonly used for intracellular conduction in unmyelinated fibers.

(3) The cable effect refers to the ability of action potentials to be propagated directly to relatively distant points (saltatory conduction). This effect is apparently utilized by myelinated nerves to transmit action potentials from one node of Ranvier to the next. Since transmission is very rapid between nodes, myelinated nerves propagate impulses at great speeds. However, axons of a μm or less have nodes close enough together (because of their proportionately smaller Schwann cells) to eliminate the advantage gained by myelination. Hence, very small fibers are ordinarily unmyelinated.

8-3 Neurons are by no means the only excitable cells. There are, for example, thermal receptors for heat and cold sensation; mechanoreceptors for touch, balance, hearing, and a knowledge of the position of our body parts; chemoreceptors for taste and smell; and photoreceptors for vision.

The most thoroughly studied photoreceptor is the rod cell from the vertebrate retina. Stacks of membranous discs in these cells are light sensitive by virtue of the presence of rhodopsin, a chromophore that is bleached by light. Bleaching changes the ionic permeability of the membranes in which rhodopsin resides. Flow of ions (especially Ca^{2+}) in and out of the discs affects Na^+ permeability of the surrounding plasma membrane and thus alters the excitability of the cell.

Additional excitable cells are found among the effectors of central nervous system pathways. For example, chromaffin cells are modified neurons that secrete stored epinephrine or norepinephrine in response to action potentials. The most prevalent of the excitable effector cells, however, is the muscle fiber, which contracts in response to action potentials.

STUDY GUIDE

8-1 (a) What is the distinction between (1) An axon and a dendrite? (2) A Schwann cell and a glial cell? (3) A synapse and a neuromuscular junction? (b) Define the terms perikaryon, myelin, node of Ranvier. (c) There are an estimated 13 Na^+ channels per square μm of nerve membrane. How many sodium ions will enter through each channel during a typical action potential? [Ans: about 1400 to 1800] (d) How many action potentials can be initiated in an isolated squid axon with a diameter of 0.5 mm before the internal Na^+ concentration will be doubled? (Assume that the Na^+ pump has been poisoned.) What is the corresponding figure for a human axon, if the only difference between the two is an axon diameter of 10 μm in the latter? [Ans: about 150,000 and 3000, respectively] (e) Describe the sequence of events comprising an action potential. (f) What is a voltage clamp and for what purpose is it used? (g) The magnitude of the action potential depends mostly on Na^+ concentration, whereas the magnitude of the resting potential depends mostly on K^+ concentration. Why? (h) How was the effect of acetylcholine on action potentials discovered? (I) What is the refractory period, and how is it related to flicker fusion?

8-2 (a) What are the three basic mechanisms of action potential propagation? (Be careful to distinguish between the two varieties of electrical propagation.) (b) What is the significance of myelination? Why do very small fibers lack it? (c) What are the individual steps involved in cholinergic neurotransmission? (d) What are some of the noncholinergic neurotransmitters? (e) If a chemical were to double the permeability to K^+ in the squid axon, what change would be seen in the resting potential? (Use Equation 6-16 where $b = 1/15$.) If there is no change in the threshold potential at which the axon fires, would this chemical make the axon more or less excitable? [Ans: the resting potential would move from -60 to -70 mV] (f) What is the difference in the way an inhibitory and an excitatory impulse affect a cell?

8-3 (a) What are the four classes of stimuli to which our sense organs respond? Summarize the type of response expected in each case. (b) What is the difference in function between the rods and cones, and to what do we attribute the difference? (c) What is meant by neurosecretion? Where is it found in mammals? (d) What are chromaffin cells and what is their function?

REFERENCES

GENERAL

AIDLEY, D. J., *The Physiology of Excitable Cells.* Cambridge Univ. Press, 1971. An advanced, but excellent, treatment of nerve cells, muscle cells, and some sensory cells.

COLE, K. S., Membranes, Ions, and Impulses. Berkeley: Univ. of California Press, 1968.

DE ROBERTIS, E., *Cellular Dynamics of the Neuron.* New York: Academic Press, 1969.

HUNT, C. C., ed., *Neurophysiology* (MTP International Review of Science). Baltimore: Univ. Park Press, 1975.

KATZ, B., *Nerve, Muscle, and Synapse.* New York: McGraw-Hill, 1966. (Paperback.) An excellent introduction to neuronal function.

PAUL, D. H., *The Physiology of Nerve Cells.* Oxford: Blackwell Scientific Publ., 1975. (Paperback.)

SIEGEL, G., R. ALBERS, R. KATZMAN, and B. AGRANOFF, eds., *Basic Neurochemistry* (2d ed.). Boston: Little, Brown, 1976. (Paperback.)

WATSON, W. E., *Cell Biology of Brain.* New York: Halsted Press, 1976.

WORDEN, F., J. SWAZEY, and G. ADELMAN, eds., *The Neurosciences: Paths of Discovery.* Cambridge, Mass: MIT Press, 1975. (Paperback.) A history of modern concepts.

EXCITABILITY

BAKER, P. F., and A. L. HODGKIN, "Replacement of the Axoplasm of Giant Nerve Fibers with Artificial Solutions." *J. Physiol.*, **164**: 330 (1962).

BAKER, P. F., "The Nerve Axon." *Scientific American*, March 1966. (Offprint 1038.)

deJONG, R., "Neural Blockade by Local Anesthetics." *J. Am. Med. Assoc.*, **238**: 1383 (1977).

EISENBERG, M., and S. McLAUGHLIN, "Lipid Bilayers as Models of Biological Membranes." *BioScience*, **26**(7): 436 (July 1976).

HILL, B., E. SCHUBERT, M. NOKES, and R. MICHELSON, "Laser Interferometer Measurement of Changes in Crayfish Axon Diameter Concurrent with Action Potential." *Science*, **196**: 426 (1977). The change in diameter is about 18 Å.

HODGKIN, A. L., "The Ionic Basis of Nervous Conduction." *Science*, **145**: 1148 (1964). Nobel Lecture, 1963.

HODGKIN, A. L., and A. F. HUXLEY, "A Quantitative Description of Membrane Current and Its Application to Conduction and Excitation in Nerve." *J. Physiol.*, **117**: 500 (1952).

HUXLEY, A. F., "Excitation and Conduction in Nerve: Quantitative Analysis." *Science*, **145**: 1154 (1964). Nobel Lecture, 1963.

JUNGE, D., *Nerve and Muscle Excitation.* Sunderland, Mass: Sinauer, 1976. (Paperback.)

KEYNES, R. D., "The Nerve Impulse and the Squid." *Scientific American*, December 1958. (Offprint 58.)

LÜTTGAU, H., and H-G. GLITSCH, *Membrane Physiology of Nerve and Muscle Fibers.* Stuttgart: Gustav Fischer Verlag, 1976.

MARGOLIS, R. U., and R. K. MARGOLIS, "Metabolism and Function of Glycoproteins and Glycosaminoglycans in Nervous Tissue." *Int. J. Biochem.*, **8**: 85 (1977).

MONTAL, M., "Experimental Membranes and Mechanisms of Bioenergy Transductions." *Ann. Rev. Biophys. Bioeng.*, **5**: 119 (1976).

MONTAL, M., and P. MULLER, "Artificial Nerve Membrane." *Proc. Nat. Acad. Sci.*, **69**: 3561 (1973).

ROJAS, E., and C. BERGMAN, "Gating Currents: Molecular Transitions Associated with the Activation of Sodium Channels in Nerve." *Trends Biochem. Sci.*, **2**(1): 6 (Jan. 1977).

SEEMAN, P., "The Actions of Nervous System Drugs on Cell Membranes." In *Cell Membranes*, G. Weissmann and R. Claiborne, eds. New York: H. P. Publ., 1975, p. 239. Also see *Hospital Practice*, Sept. 1974, p. 93.

SHAMOO, A., ed., *Carriers and Channels in Biological Systems.* Ann. N.Y. Acad. Sci., Vol. **264** (1975). Symposium.

THE SYNAPSE

AXELROD, J. "Noradrenalin: Fate and Control of Its Biosynthesis." *Science*, **173**: 598 (1971). Nobel lecture, 1970.

———, "Neurotransmitters." *Scientific American*, **230**(6): 58 (June 1976). (Offprint 1297.)

BACQ, Z. M., *Chemical Transmission of Nerve Impulses: A Historical Sketch.* New York: Pergamon, 1975.

BARKER, J. L., "Peptides: Roles in Neuronal Excitability." *Physiol. Rev.*, **56**: 435 (1976).

BRINKLEY, F. J., Jr., ed., *Calcium and Magnesium Metabolism in Cephalopod Axons.* Fed. Proc., **35**: 2572 (1976). A symposium.

COLD SPRING HARBOR SYMPOSIA IN QUANTITATIVE BIOLOGY, Vol. **60**: *The Synapse.* Cold Spring Harbor (N.Y.) Laboratory, 1976.

DE ROBERTIS, E., *Synaptic Receptors. Isolation and Molecular Biology.* New York: Dekker, 1975.

———, "Synaptic Receptor Proteins, Isolation and Reconstitution in Artificial Membranes." *Rev. Physiol. Biochem. Pharmacol.*, **73**: 9 (1975).

ECCLES, JOHN C., "Ionic Mechanism of Post-Synaptic Inhibition." *Science*, **145**: 1140 (1964). Nobel Lecture, 1963.

———, "The Synapse." *Scientific American*, January 1965. (Offprint 1001.)

GREENGARD, P., "Possible Role for Cyclic Nucleotides and Phosphorylated Membrane Proteins in Postsynaptic Actions of Neurotransmitters." *Nature*, **260**: 101 (1976).

KATZ, B., "Quantal Mechanism of Neural Transmitter Release." *Science*, **173**: 123 (1971). Nobel Lecture, 1970.

KITA, H., "Mechanisms for Neurotransmitter Release." *BioScience*, **24**: 13 (1974).

KRNJEVIĆ, K., "Chemical Nature of Synaptic Transmission in Vertebrates." *Physiol. Rev.*, **54**: 418 (1974).

LESTER, H. A., "The Response to Acetylcholine." *Scientific American*, **236** (2): 107 (Feb. 1977).

LOEWI, OTTO, "On the Humoral Transmission of the Action of Heart Nerves." *Pflüger's Archiv.*, **189**: 239 (1921). Translated and reprinted in *Great Experiments in Biology*, ed. M. L. Gabriel and S. Fogel. Englewood Cliffs, N.J.: Prentice-Hall, 1955, p. 69.

———. "The Chemical Transmission of Nerve Action." In *Nobel Lectures, Physiology or Medicine, 1922–1941.* Amsterdam: Elsevier, 1965, p. 416. Nobel Lecture, 1936.

McLENNAN, H., *Synaptic Transmission.* Philadelphia: W. B. Saunders, 1969.

NATHANSON, J. A., "Cyclic Nucleotides and Nervous System Function." *Physiol. Rev.*, **57**: 158 (1977). See also *Scientific American*, August 1977, p. 108.

SCHORDERET, M., "Cyclic AMP in the Nervous System." *J. Physiol., (Paris)*, **68**: 471 (1975). A review.

VON EULER, U. S., "Andrenergic Neurotransmitter Functions." *Science*, **173**: 202 (1971). Nobel Lecture, 1970.

WHITTAKER, V. P., "Origin and Function of Synaptic Vesicles." *Ann. N.Y. Acad. Sci.* **183**: 21 (1971).

WHITTAKER, V. P., "Membranes in Synaptic Function." In *Cell Membranes*, G. Weissmann and R. Claiborne, eds. New York: H. P. Publ., 1975, p. 167. Also in *Hospital Practice*, April 1974, p. 111.

ZAIMIS, E., ed., *Neuromuscular Junction*. New York: Springer-Verlag, 1976. Expensive reference work.

SENSORY CELLS

CATTON, W. T., "Mechanoreceptor Function." *Physiol. Rev.*, **50**: 297 (1970).

DETHIER, V. G., "A Surfeit of Stimuli: A Paucity of Receptors." *Am. Scientist*, **59**: 706 (1971). Taste and olfactory systems of insects.

EBREY, T., and B. HONIG, "Molecular Aspects of Photoreceptor Function." *Quart. Rev. Biophys.*, **8**: 129 (1975).

GRANIT, RAGNAR, "The Development of Retinal Neurophysiology." *Science*, **160**: 1192 (1968). Nobel Lecture, 1967.

HARTLINE, H. K., "Visual Receptors and Retinal Interaction." *Science*, **164**: 270 (1969). Nobel Lecture, 1967.

HUBBARD, R., "100 Years of Rhodopsin." *Trends Biochem. Sci.*, **1**(7): 154 (July 1976). A review.

LOEWENSTEIN, W. R., "Biological Transducers." *Scientific American*, August 1960. (Offprint 70.) Touch.

MACNICHOL, E. F., Jr., "Three-Pigment Color Vision." *Scientific American*, December 1964. (Offprint 197.)

TOMITA, TSUNEO, "Electrical Activity of Vertebrate Photoreceptors." *Quart. Rev. Biophys.*, **3**: 179 (1970). An excellent review.

WALD, GEORGE, "Molecular Basis of Visual Excitation." *Science*, **162**: 230 (1968). Nobel Prize, 1967.

WOLKEN, JEROME J., *Invertebrate Photoreceptors: A Comparative Analysis*. New York: Academic Press, 1971.

YOUNG, RICHARD W., "Visual Cells." *Scientific American*, October 1970. (Offprint 1201.)

NOTES

1. Unfortunately, "axon" and "dendrite" are defined in two ways. In the functional definition, the impulse always passes from dendrites to the axon. But when dendrites and axon are defined on the basis of their structure, we may find the opposite direction of travel. Although the two definitions are compatible in most cases, some of the sensory neurons in the peripheral system are notable exceptions.

2. The explanation of action potentials in terms of changes in ionic permeabilities is by no means a new concept. As long ago as 1902, Julius Bernstein of the University of Halle, in Germany attributed the action potential to a transient change in sodium permeability. He erred mainly in failing to predict the positive overshoot as the Na^+ equilibrium potential is approached, and in failing to recognize the importance of K^+ efflux to the quick restoration of a resting potential.

3. Dopamine is an important neurotransmitter in the human brain. Underproduction of dopamine by certain pathways results in the movement disorder known as Parkinson's disease, treated with L-DOPA, a dopamine precursor. Overproduction of dopamine by other pathways may contribute to psychotic disorders, specifically schizophrenia, for which the dopamine-receptor blocking drug, chlorpromazine, is so useful. In fact, most mood-altering drugs seem to affect one or another of the brain's adrenergic pathways.

9

Contractility and Motility

9-1 Structure of the Muscle Cell   374
   Striated Muscle
   Organization of the Myofibril
   Smooth Muscle
   Invertebrate Muscle

9-2 The Mechanism of Contraction   378
   Actin Filaments
   Myosin Filaments
   Other Proteins
   The Sliding Filament Model
   Smooth Muscle

9-3 The Metabolism of Muscle   385
   ATP Maintenance
   Red vs. White Muscle

9-4 Contraction-Relaxation Control   388
   The Role of Calcium
   Control by Troponin
   Control by Myosin
   The Calcium Pump
   Excitation-Contraction Coupling
   Action Potentials and Innervation
   Hormonal Effects

9-5 Contractile Proteins of Non-muscle Cells   395
   Actin
   Myosin
   Other Proteins
   Control of Non-muscle Contraction

9-6 Cilia and Flagella   400
   Structure of the Axoneme
   Axonemal Movement
   Regulation of Movement
   Ciliogenesis
   Bacterial Flagella

9-7 Cell Shape and Cell Movement   408
   The Cytoskeleton
   Regulation of Cytoskeletal Changes
   The Erythrocyte
   Cytoplasmic Streaming
   Ameboid Locomotion
   Chemotaxis

Summary   424

Study Guide   425

References   426

Notes   429

If anything is characteristic of living things, it is movement: movement of cells or organisms from one place to another; the continual shuffling of organelles within the cytoplasm referred to as *cytoplasmic streaming*; the separation of chromosomes and pinching of a cell in two during division; changes in cell shape; the beating of cilia or flagella, and so on. These and other sorts of movements apparently arise through the activities of two basic systems, one of them involving microtubules and the other involving microfilaments. The latter system was the first to be studied in detail, because microfilaments are utilized by the most obvious motile system of all, that of animal muscle.

9-1 STRUCTURE OF THE MUSCLE CELL

Muscle cells of vertebrates may be classified according to their appearance in the light microscope as *striated* or *smooth*. Striated muscles, as seen with the phase contrast or polarizing microscope, have alternating light and dark bands or "striae" (see Fig. 9-1). The skeletal muscle of humans, which accounts for about 40% of body weight, is of this type. Cardiac (heart) muscle is also striated. Most of the other internal organs have unstriated, smooth muscle.

Striated Muscle. An individual muscle cell is called a *muscle fiber*. It may vary in diameter over a wide range, from 10 to 100 μm or more, with lengths that can run into centimeters. Each striated cell is composed of parallel bundles of filaments, called *myofilaments*.[1] Each bundle, or *myofibril*, is 1–2 μm in diameter. The myofilaments are responsible for contraction, which is achieved by sliding filaments past one another.

A network of membranes crisscrosses the muscle cell both longitudinally and transversely. The term sarcoplasmic reticulum (SR) has been used to describe the longitudinal component of this system, which consists of elongated sacks. The transverse system is comprised of tubules that span the cell from membrane to membrane. It is called the T system. These two components, the longitudinal and transverse systems, are not directly connected with each other, however, nor do they have the same origins or functions.

The longitudinal components of the sarcoplasmic membranes run parallel to myofibrils (see Fig. 9-2). These channels contain periodic enlargements called *terminal cisternae*. The sarcoplasmic reticulum and its terminal cisternae are apparently derived from the endoplasmic reticulum of the immature muscle cell, or *myoblast*.

The transverse component of the sarcoplasmic membranes, on the other hand, consists of regularly spaced tubules, often about 400 Å in diameter, that open onto the surface of the cell. This T system is, therefore, continuous with the cell membrane (sarcolemma) from which it is thought to be derived (see Figs. 9-2 and 9-3). At points where the longitudinal and transverse systems cross, one often sees a configuration called a *triad*, composed of two terminal cisternae lying on either side of a transverse tubule (see Fig. 9-4).

The nuclei of striated muscle cells are usually found just under the sarcolemma. Each skeletal muscle cell has many of them. This arrangement, called a *syncytium*, results from the fusion of myoblasts during development of the cell. Cardiac muscle, though striated, is clearly divided into individual cells, each apparently with a single nucleus.

Mitochondria (formerly called the sarcosomes of muscle) are found clustered near the sarcolemma and often in the interior of the cell, lying in rows parallel to the myofibrils. This is a useful arrangement since mitochondria supply much of the ATP needed to support contraction of the myofibrils.

Organization of the Myofibril. A nomenclature describing the appearance of the myofibril was developed from observations made originally with the polarizing light microscope (see Fig. 9-5). In this nomenclature, regularly spaced lines, called *Z lines* or *Z discs* (from the German word *zwischen*, meaning "between"), divide the myofibril into *sarcomeres*.

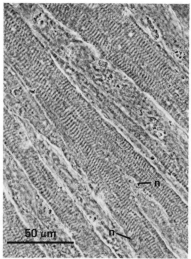

FIGURE 9-1 Striated Muscle. Fibers from a culture of chick embryo. Nuclei (n). Note the alternating light and dark bands in this phase contrast micrograph. [Courtesy of Y. Shimada, *J. Cell Biol.* **48**: 128 (1971).]

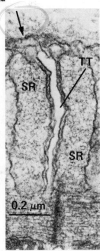

FIGURE 9-2 Myofibrils and the Sarcoplasmic Membranes. ABOVE: Thin section through frog skeletal muscle shows parallel rows of myofibrils (bundled filaments). One can also see a single row of mitochondria (M) and sections of the sarcoplasmic reticulum (SR) and transverse tubules (TT). The tubules run nearly perpendicular to the plane of the paper, except near the periphery, where some angle toward the cell surface and open onto it. (Because these peripheral SR and TT are not cut in cross section, they appear much larger than similar structures further into the cell.) RIGHT: The opening (arrow) of one tubule is seen at higher magnification. [Courtesy of C. Franzini-Armstrong, *J. Cell Biol.*, **47**: 488 (1970).]

There is a light region at either end of the sarcomere that is called an *I band* because it appears isotropic (the same in every direction) when viewed with crossed polarizers. Between the two I bands of each sarcomere is an *A band* (anisotropic), which is generally much darker, except for a region of intermediate density at its center called the *H zone* (from the German word *hell*, or "clear"). At the middle of the H zone, which is also the middle of the sarcomere, is a thin, darker *M line*.

Electron micrographs taken at M.I.T. in the early 1950s by H. E. Huxley and J. Hanson made it clear that these striations are due to a pat-

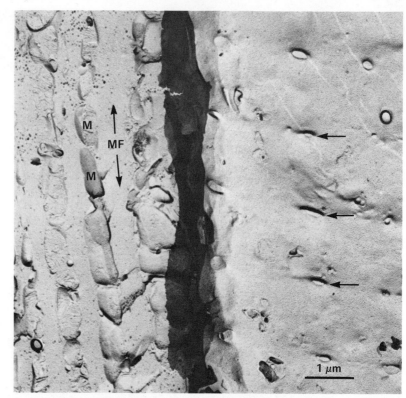

FIGURE 9-3 Freeze-fractured View of Guinea Pig Heart Muscle. The cell is fractured, showing rows of T-system apertures (arrows) on the surface and myofilaments (MF) beneath. Mitochondria (M).]Courtesy of G. Rayns, F. O. Simpson, and W. S. Bertrand, *Science*, **156**: 656 (1967). Copyright by the AAAS.]

tern of overlapping myofilaments of two types. The I bands contain only thin filaments, which are attached to the Z lines and extend toward the center of the sarcomere from either side. The A bands contain these thin filaments plus thicker filaments that interdigitate with thin filaments in the dark regions.

When viewed in cross section, the filaments of many muscles reveal a hexagonal arrangement, with each thick filament surrounded, in regions of overlap, by six thin ones (see Fig. 9-6). The ratio of thin to thick would therefore be about two to one for the myofibril as a whole. The

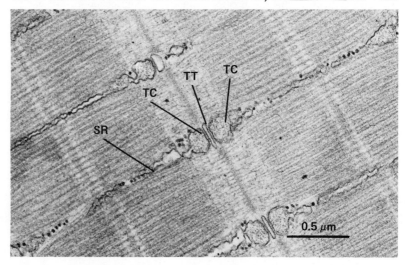

FIGURE 9-4 The Triad. The sarcoplasmic reticulum (SR) has a long axis that runs perpendicular to the transverse tubules (TT). Cross sections therefore show two terminal cisternae (TC) of the SR on opposite sides of a tubule, a composition that is known as a triad. (Some muscles have a different arrangement, leading to dyads, pentads, etc.) Numerous triads are also visible in Fig. 9-2, the enlargement of which shows a tangentially sectioned example. [Courtesy of C. Franzini-Armstrong, *J. Cell Biol.*, **47**: 488 (1970).]

interfilament spacing varies from one muscle to another, and increases as the muscle shortens—the spacing is 200 Å in extended frog muscle, but is more than 250 Å when the muscle is contracted.

Smooth Muscle. Most involuntary muscles, such as those controlling nearly all the internal organs except the heart, are smooth rather than striated. The cells have an elongated structure with a single nucleus and relatively little endoplasmic reticulum. As with striated muscle, myofilaments run parallel to the long axis, so contraction can be achieved only in this one direction. However, the myofilaments are not as highly organized as in striated cells and are usually more difficult to see, hence the lack of striations and term "smooth muscle" (see Figs. 9-7 and 9-8).

Contraction of smooth muscle is generally very much slower than contraction of striated muscle, and the same is true of the rate of relaxation. A typical value for striated muscle would fall in the range 40–100 ms for either contraction or relaxation, while smooth muscle might take anywhere from 200 to 30,000 ms or more.

Smooth muscle cells are arranged circumferentially around the lumen of blood vessels, the size of which can thus be regulated. Two layers are found in the intestines, one circumferential and one longitudinal. Together they provide the peristaltic (rhythmic) contractions by which food is propelled. And in the uterus and bladder, which must contract uniformly in all directions, smooth muscle cells are arranged tangentially but with their long axes almost at random with respect to each other, thus providing the generalized contraction required by those organs.

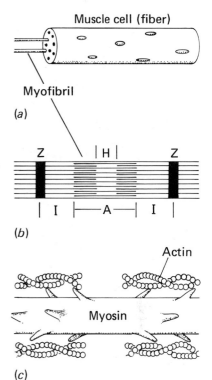

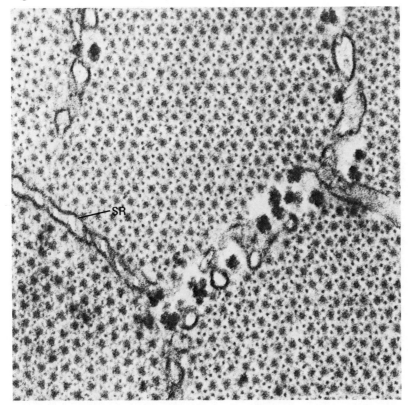

FIGURE 9-5 The Microanatomy of Skeletal Muscle. (a) Multinucleated muscle cell, called a muscle fiber, comprised of parallel myofibrils. (b) Myofibrils are divided by Z discs into sarcomeres. Each sarcomere consists of light and dark bands caused by overlapping myofilaments. The I region of the sarcomere contains thin filaments comprised mostly of a protein called actin. The H region contains thick filaments (mostly myosin) while the rest of the A region contains overlapping thin and thick filaments. (c) Thin and thick filaments near the center of the sarcomere. Note the spiral arrangement of cross-bridges.

FIGURE 9-6 The Myofibril in Cross Section. Longitudinal sections of the myofibril were seen in Figs. 9-2 and 9-4. This micrograph, part of a frog skeletal muscle cell, shows several myofibrils in cross section. Note the regular array of six thin filaments about each thick one. Note also sections of the sarcoplasmic reticulum (SR) between the myofibrils. The diameter of each complete myofibril is about 1 μm. (Courtesy of H. E. Huxley.)

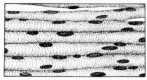
Skeletal muscle

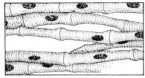
Cardiac muscle

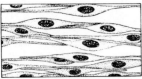
Smooth muscle

FIGURE 9-7 Striated and Smooth Muscle. Most muscle is striated, meaning that it has alternate light and dark bands when viewed in polarizing or phase contrast light microscopes. Although skeletal muscle and heart muscle are both striated, skeletal muscle cells are syncytial (many nuclei in a common cytoplasm), while heart muscle cells are mononucleated. Visceral involuntary muscle cells are also mononucleated, but they are smooth—i.e., without striations.

Invertebrate Muscle. The classification into smooth and striated, used for vertebrate muscle, is not sufficient to describe invertebrate muscle. In addition to smooth and striated, there are several variations known by different names. For instance, body muscle from the earthworm consists of *obliquely striated* muscle, also known as *helical smooth* muscle. In these cells, discrete myofibrils do not exist, but bundles of thick and thin filaments are arranged to provide striations that run obliquely through the muscle.

Obliquely striated muscle has a slightly different filament structure when compared with vertebrate muscle. The thick filaments of vertebrates, which are about 150 Å in diameter, are replaced in these cells by much thicker filaments containing large quantities of a protein called *paramyosin* (see Fig. 9-9).

Although there is much yet to be learned about the structure of invertebrate muscle, it appears that the basic features of its function are much the same as that of vertebrate muscle.

9-2 THE MECHANISM OF CONTRACTION

In the later decades of the nineteenth century, various laboratories demonstrated that part of the protein can be dissolved from muscle fibers by washing them in concentrated salt solutions. The resulting solution becomes turbid on the addition of more water, due to the precipitation of a substance that was given the name *myosin*. Working in Hungary during

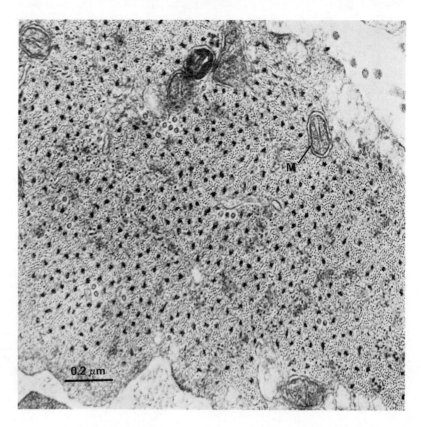

FIGURE 9-8 Smooth Muscle. This cross section through a smooth muscle cell (from a rabbit vein) reveals a regular array of thick filaments and a not-so-regular array of thin ones. The filaments are not organized into myofibrils, nor is there any significant amount of endoplasmic reticulum. Mitochondrion (M). [Courtesy of R. Rice, G. McManus, C. Devine, A. Somlyo, *Nat. New Biol.* **231**: 242 (1971).]

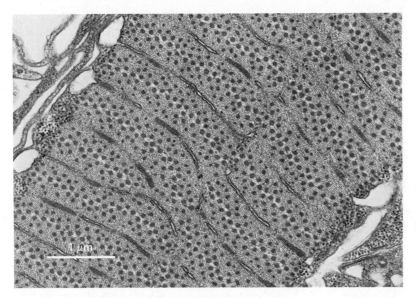

FIGURE 9-9 Obliquely Striated Muscle. Cross section, from the body of an earthworm. Note the large paramyosin filaments, the thinner actin-containing filaments, and the sections of a sarcoplasmic reticulum. [Photo by Y. Uehara. From N. Toida, H. Kuriyama, N. Tashiro, and Y. Ito, *Physiol. Rev.* **55**: 700 (1975).]

the early years of World War II, Albert Szent-Györgyi and his colleagues found that this crude preparation is actually composed of at least two kinds of protein. The name *myosin* was retained to identify one, while the other was called *actin*.

Szent-Györgyi found that when a dilute solution of myosin is mixed with a dilute solution of actin in the presence of the right kinds of salts, a precipitate called *actomyosin* is formed. A good molar ratio of actin to myosin was found to be 1:4, about the same ratio that the two proteins have in muscle. The gel formed by coprecipitation of the two proteins can be pulled into visible threads that are capable of contracting to about two-thirds of their original length. Contraction occurs when ATP is added along with Ca^{2+} and Mg^{2+}. It is accompanied by the hydrolysis of ATP to ADP and inorganic phosphate.

The demonstration that contraction can be achieved *in vitro* with relatively simple solutions puts the study of muscular contraction into the realm of molecular biology. However complicated the muscles themselves are found to be, and however involved the control devices employed to ensure useful and coordinated contraction, we are left with explaining the underlying phenomenon—the contraction itself—in terms of an interaction among a few kinds of protein, ATP, and inorganic ions. We shall proceed by describing some of the properties of actin, myosin, and certain other contractile proteins; then we shall go on to consider ways in which they cooperate to provide contractility.

Actin Filaments. Actin is a globular protein, about 55 Å in diameter, composed of one polypeptide chain with a mass of approximately 42,000 daltons. It polymerizes in the presence of salts into long filaments that are called *F-actin* to distinguish them from the depolymerized form, or *G-actin*. Each molecule of actin can tightly bind a molecule of ADP or ATP plus one divalent metal ion, usually Ca^{2+} or Mg^{2+}. In 0.1 M KCl plus Ca^{2+} and Mg^{2+}, G-actin polymerizes. This is accompanied by a stoichiometric hydrolysis of the bound ATP:

$$n(\text{G-actin}) + n\text{ATP} \rightarrow n(\text{G-actin-ATP}) \rightarrow (\text{F-actin-ADP}_n) + n\text{P}_i \quad \textbf{(9-1)}$$

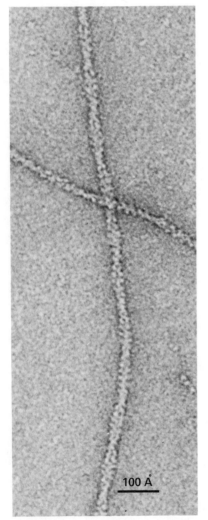

FIGURE 9-10 Actin. MARGIN: Polymerized actin is called F-actin. Note its double-helical structure. BELOW: Model for the actin filament, showing the association of troponin and tropomyosin (see text). [Margin photo, courtesy of H. E. Huxley, *J. Mol. Biol.*, **7**: 281 (1963).]

The equilibrium of this reaction lies far to the right in high salt. If the salt concentration is lowered, addition of more ATP reverses the polymerization by exchanging ADP for ATP:

$$(\text{F-actin-ADP}_n) + n\text{ATP} \rightarrow n(\text{G-actin-ATP}) + n\text{ADP} \qquad (9\text{-}2)$$

This ATP-mediated transition between G- and F-actin has sparked a great deal of interest, but its significance is unclear. ATP may facilitate or control polymerization *in vivo*, though it is reportedly not essential to polymerization *in vitro* under all conditions. Formation of actin filaments, except for the complicating factor of ATP hydrolysis, appears to be a simple entropy-driven self-assembly of the type discussed in Chap. 3.

Actin polymerizes to form a right-handed helix with a pitch of 700–800 Å. The completed filament has the appearance of two strings of beads wound about each other (see Fig. 9-10). At least two other proteins, called *tropomyosin* and *troponin*, are often associated with actin *in vivo*. Together, these proteins make up the thin filaments frequently just called actin filaments.

Tropomyosin is a protein of about 70,000 daltons, in two subunits of 35,000 each. It is a highly elongated molecule with a length of about 400 Å and a width of only 20–30 Å. Its two polypeptide chains are α helices that are wound about each other. According to a model of the actin filaments proposed by S. Ebashi and his colleagues of the University of Tokyo, tropomyosin molecules are stretched end to end along the groove of the actin helix, with molecules of troponin acting as dividers (Fig. 9-10). Although troponin is very similar in mass to tropomyosin, it is roughly spherical and therefore occupies much less space. The troponin-tropomyosin complex plays an important role in the contraction mechanism, as we shall see.

Myosin Filaments. Actin accounts for only about a quarter of the protein in myofibrils, and tropomyosin and troponin only a few percent each. More than half of the total protein in a myofibril is myosin, which is the major substance of the thick filaments.

Myosin is an unusual protein, with a bilobed head of about 200 Å and a rodlike tail perhaps 1300 Å (0.13 μm) long and 20 Å in diameter (Fig. 9-11). It has a total mass of about 470,000 daltons, including two *heavy chains* of about 200,000 daltons each plus several *light chains*. In rabbit skeletal muscle, there appear to be four light chains per molecule, two of which are probably essential and two of which can be removed *in vitro* without noticeable change in the properties of the molecule. The two heavy chains exist mostly as α helices wound about each other in a "coiled coil" such as one finds in tropomyosin. This coiled region constitutes the "tail" of the molecule.

The head region of myosin, together with a small portion (about 370 Å) of the tail, may be separated from the rest of the tail by the action of cer-

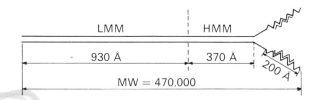

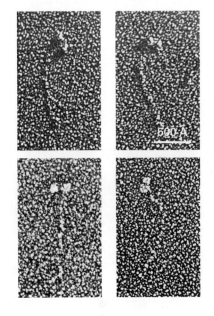

FIGURE 9-11 Myosin. ABOVE: Selective enzymatic cleavage of myosin by trypsin separates the molecule into heavy meromyosin (HMM) and light meromyosin (LMM). HMM has all the enzymatic activity and actin-binding capacity of myosin. RIGHT: Selected myosin molecules. FIRST ROW: Shadowed from various directions. SECOND ROW: Rotary shadowed. Note the bilobed head and the suggestion of flexibility in its connection to the shaft of the molecule. Each head is capable of independently binding actin. [Micrographs courtesy of H. S. Slayter and S. Lowey, *Proc. Nat. Acad. Sci. U.S.*, **58**: 1611 (1967).]

tain proteolytic enzymes. The resulting head fragment is called *heavy meromyosin*; the tail fragment is called *light meromyosin*. The enzymatic capacity of myosin, which consists of the ability to hydrolyze ATP to ADP and inorganic phosphate, is associated exclusively with heavy meromyosin. Thus, we know that the active site of the myosin ATPase is in the head region only.

Myosin is soluble in concentrated salt solutions (e.g., 0.6 M KCl) but polymerizes and precipitates when the salt concentration is lowered to about 0.16 M at physiological pH. Polymerization of myosin starts with a tail-to-tail association that produces a bare central shaft with the heads at either end. Additional molecules are added at the ends, with their tails toward the center of the growing filament. Several chains formed in this way may aggregate by partially overlapping one another. In rabbit muscle preparations, for example, one finds two heads on opposite sides of the filament at 143 Å intervals along the axis. There is a 120° rotation between adjacent pairs of heads in these preparations, so that a full revolution of the pattern is completed every 429 Å. The complete myosin filament is about 1.5 μm long, with a bare central shaft of 0.15–0.2 μm consisting of the tail-to-tail region (see Fig. 9-12), which is typically about 12 myosin molecules thick at its center.

Other Proteins. Several other proteins are also associated with myofilaments. One of these, called *C protein*, seems to bind specifically to myosin, especially in the bare central region of the filaments. A protein called α-*actinin*, in contrast, associates specifically with actin filaments. Alpha actinin is a rodlike protein that *in vitro* has been observed bridging adjacent actin filaments, especially near the Z region, of which it may be a structural part. Its function might be as a spacer to hold actin fila-

FIGURE 9-12 The Myosin Filament. Myosin polymerizes *in vitro* to form filaments that are symmetrical about a bare central shaft. Except in this bare region, projections, presumably due to myosin heads, are seen. [Courtesy of H. E. Huxley, *J. Mol. Biol.*, **7**: 281 (1963).]

ments in register where they enter the Z disc. A comparable spacer role has been proposed for C protein of myosin filaments, but at present the biological role of these proteins and several others found with them is not completely clear.

The Sliding Filament Model. The width of the A bands in striated muscle remains unchanged during contraction, while the I and H bands get narrower or disappear entirely. This observation led to the *sliding filament theory,* which attributes contraction to the movement of the thick and thin filaments past each other (see Fig. 9-13).[2]

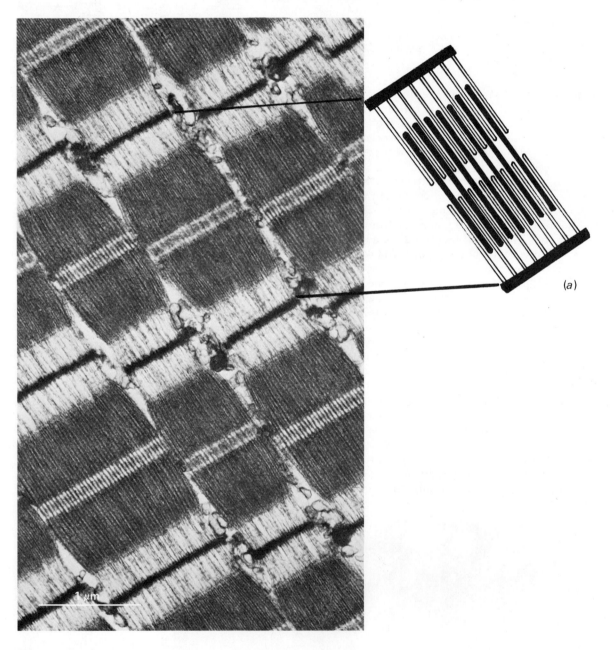

The generation of force in this system must be due to actin-myosin interactions. In fact, cross-bridges between the two kinds of filaments have been seen (Fig. 9-14). It is these cross-bridges, now known to be part of the myosin heads, that generate tension and thus pull the filaments past one another. If this model is correct, one would predict that the amount of overlap between the two types of filaments should be related in a simple way to the amount of tension developed. This prediction is verified by experiments such as the one shown in Fig. 9-15, where tension is plotted as a function of sarcomere length in frog muscle. The results are interpreted as follows:

SECTION 9-2
The Mechanism of Contraction

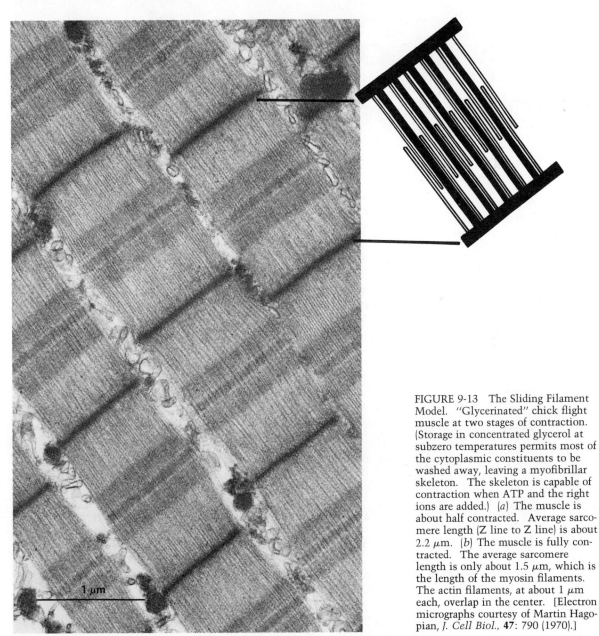

FIGURE 9-13 The Sliding Filament Model. "Glycerinated" chick flight muscle at two stages of contraction. (Storage in concentrated glycerol at subzero temperatures permits most of the cytoplasmic constituents to be washed away, leaving a myofibrillar skeleton. The skeleton is capable of contraction when ATP and the right ions are added.) (a) The muscle is about half contracted. Average sarcomere length (Z line to Z line) is about 2.2 μm. (b) The muscle is fully contracted. The average sarcomere length is only about 1.5 μm, which is the length of the myosin filaments. The actin filaments, at about 1 μm each, overlap in the center. [Electron micrographs courtesy of Martin Hagopian, *J. Cell Biol.*, **47**: 790 (1970).]

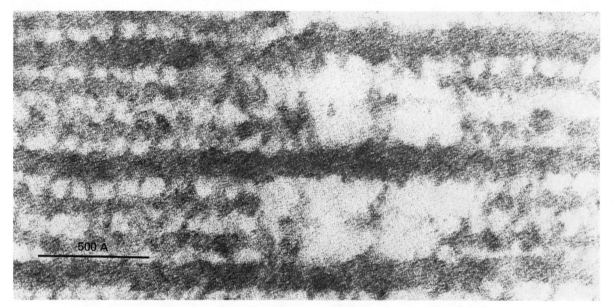

FIGURE 9-14 Myosin-Actin Interaction. Longitudinal section through a myofibril, showing cross-bridges between the actin and myosin filaments. The thickening in the middle of the myosin shafts is due to the M line protein, now thought by some to be creatine kinase. [Courtesy of H. E. Huxley, from *J. Biochem. Biophys. Cytol.* (now *J. Cell Biol.*), **3**: 631 (1957).]

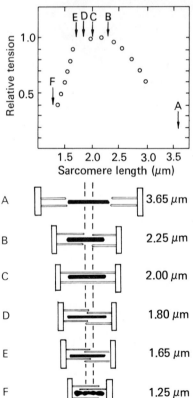

FIGURE 9-15 The Relationship Between Tension and Sarcomere Length. See text for discussion. [Described by A. M. Gordon, A. F. Huxley, and F. J. Julian, *J. Physiol.*, **184**: 170 (1966).]

(A) Stretched to a sarcomere length of 3.65 μm, the actin and myosin filaments of a frog's muscle no longer overlap. Hence no cross-bridges can form and no tension can develop.

(B) Between 3.65 and 2.25 μm, there is an increasing overlap between actin and myosin filaments, with an increasing number of cross-bridges and hence an increasing tension.

(C) Between 2.25 and 2.00 μm, tension remains constant since the myosin heads are all occupied. The tips of the actin filaments are passing the headless region in the center of the myosin filaments.

(D and E) At sarcomere lengths shorter than 2 μm, actin filaments begin to overlap at the center, interfering with actin-myosin bridges and so reducing the tension.

(F) At 1.25 μm, contact with Z lines (or Z discs) has compressed the myosin filaments enough to make further shortening impossible.

The sliding filament model also explains why the relaxation of a fully contracted muscle usually results in a spontaneous lengthening to its *slack length*, for this is the distance where the actin filaments just meet without crowding each other. That value is 2.00 μm in frog muscle.

The significance of the bilateral symmetry of the myosin filaments should now be clear, for the two halves must pull actin in opposite directions. That is, the actin filaments must be pulled toward each other along their longitudinal axes to meet at the center of the myosin filaments. Further examination reveals that actin filaments, too, have a bipolar symmetry reflected about the Z line. When H. E. Huxley investigated the binding of trypsin-produced heavy meromyosin to actin, he found that all myosin heads align themselves in the same direction on a given actin filament, and that the direction is the same for all the filaments on one side of a Z line (see Fig. 9-16).

These and other experiments lead to the following model of the contraction process: Actin and myosin filaments are parallel and connected by cross-bridges consisting of the enlarged ends of the myosin molecules. Each actin molecule is capable of binding one of these projections. Contraction occurs when myosin releases its contact with the adjacent actin, forms a new bridge with the next actin molecule in the chain, then reshortens to bring the newly contacted actin into the position occupied by the old one. Thus, the filaments move past each other in a ratchet-like manner. At any one time there must be enough stable bridges to prevent backsliding, but the force required to move the filaments would come from the alternate extension-contraction cycle of each of the myosin projections, powered in some way by the hydrolysis of ATP. Relaxation, in this scheme, requires that all the bridges be broken at once so that filaments are left free to slide past each other without tension as they are pulled apart by the contraction of opposing muscles.

Smooth Muscle. It is clear that smooth muscle also contracts by the sliding filament mechanism. The unit of contraction, however, is not the sarcomere but the entire cell. At least some of the actin filaments are anchored directly to the cell membrane, which thus replaces Z lines. Contraction in these long, spindle-shaped cells causes them to shorten and get rounder. Since adjacent cells are commonly stuck together by various junctions or interdigitating cell membranes, a coordinated change in shape of a layer of smooth muscle becomes possible.

The organization of myofilaments in smooth muscle cells has been difficult to determine. The thick myosin filaments are relatively labile and special care must be taken during preparation for microscopy if they are to be seen at all. When all precautions are taken, it still appears that there are a dozen or more thin filaments for each thick one (the ratio is two to one in striated muscle), which if correct means that all actin filaments cannot be pulled upon simultaneously.

The actin of smooth muscle, and the filaments formed from it, are very much like their striated counterpart even to the inclusion of tropomyosin in the actin filament grooves. Troponin is also present in some smooth muscles (e.g., uterine) but reportedly absent from others. Smooth muscle myosin, however, is a different protein from the one found in skeletal muscle, though the structure and subunit composition is similar.

Complicating our attempts to interpret smooth muscle is the presence of considerable quantity of 100-Å (10 nm) filaments. These filaments, which are neither myosin nor actin, are found in many cell types other than smooth muscle though not, apparently, in adult striated muscle. The 100-Å filaments are unambiguously identified only when cut in cross section, as they then appear as hollow, tubular structures comprised of four 20–30 Å subfilaments arranged around a hollow core. Their function is uncertain. However, they may form a supporting framework for the actin and myosin filaments in smooth muscle and, as we shall see, possibly in nonmuscle cells as well.

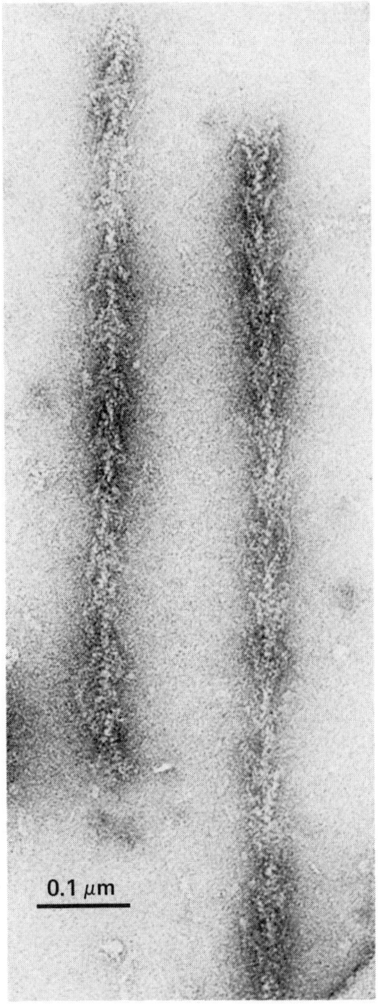

9-3 THE METABOLISM OF MUSCLE

Muscle is a remarkably efficient transducer of chemical energy to mechanical energy. Estimates of the efficiency under ideal conditions

FIGURE 9-16 Polarity of Actin. When heavy meromysin (HMM) is allowed to bind to isolated actin filaments, the HMM aligns in only one direction on a given filament giving the appearance of a row of arrowheads all pointing the same way and thus demonstrating that actin filaments have a definite polarity, caused by "head-to-tail" polymerization. The two "HMM-decorated" actin filaments shown here have opposite polarities. (Courtesy of H. E. Huxley, from *J. Mol. Biol.*, 7: 281 (1963).)

range up to 80%. The immediate source of this energy is the hydrolysis of ATP.

ATP Maintenance. The problem of correlating ATP utilization with work long baffled physiologists and biophysicists, for the amount of ATP used during a contraction may be more than the total amount found in the cell before contraction. In fact, E. Lundsgaard observed in 1930 that the ATP content of working muscles remains almost constant until they are near exhaustion. If ATP is the source of energy, then it must be replaced as fast as it is used. Metabolism does not produce ATP fast enough to be the immediate source of supply during sudden, strenuous activity. In addition, measurements of ATP levels during contraction in the presence of various metabolic poisons show that a relatively constant level can be maintained for a significant period of time.

As it turns out, there are two immediately available ways in which ATP can be resynthesized during contraction quite apart from the catabolism of foodstuffs. The first is an equilibrium involving creatine phosphate (also called phosphocreatine or phosphorylcreatine) and ADP:

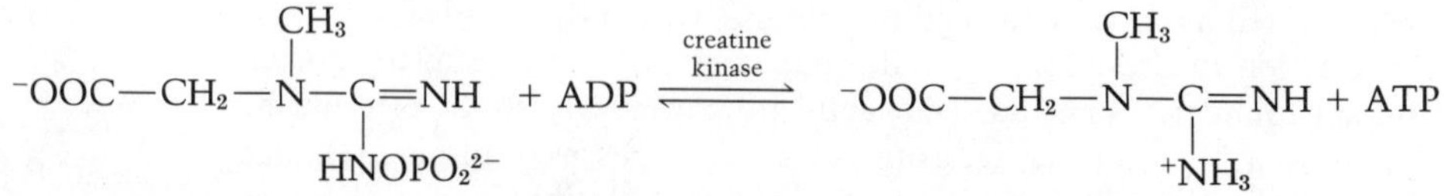

$$^-OOC-CH_2-\underset{|}{N}(CH_3)-\underset{|}{C}(HNOPO_2^{2-})=NH + ADP \underset{}{\overset{\text{creatine kinase}}{\rightleftharpoons}} {}^-OOC-CH_2-\underset{|}{N}(CH_3)-\underset{|}{C}(^+NH_3)=NH + ATP$$

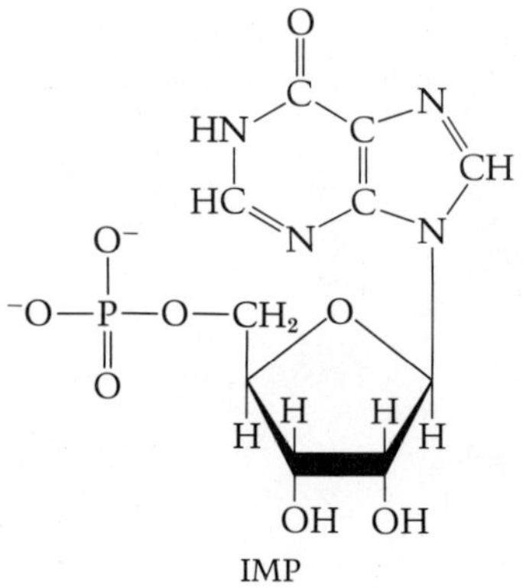

Creatine phosphate Creatine

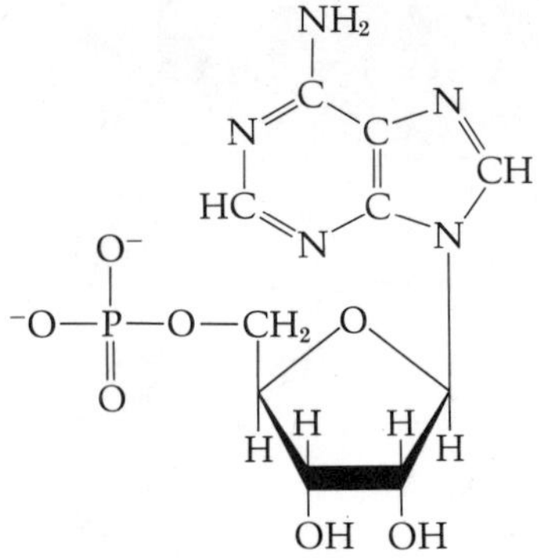

The hydrolysis of creatine phosphate has a free energy change that is even more favorable than that of ATP itself.

Invertebrates typically use arginine phosphate instead of creatine phosphate, but the type of bonding and hence the energetics of the reaction are much the same as with creatine. (There must, however, be some undiscovered advantage in using creatine, which is synthesized from amino acids and is clearly a recent evolutionary substitution.) The equilibrium with ATP of either arginine phosphate or creatine phosphate lies far to the right. Thus, most of the adenosine will be maintained as ATP until creatine or arginine phosphate is nearly gone.

The second method for maintaining ATP levels is an equilibrium catalyzed by *adenylate kinase*, which is widely distributed among various cell types. In muscle, it is also called *myokinase*. This enzyme permits the phosphate anhydride bond of ADP to be used by transferring it to a second ADP in the following reaction:

$$2ADP \overset{\text{adenylate kinase}}{\rightleftharpoons} ATP + AMP$$

Although this reaction has an equilibrium constant near unity, muscle also has large quantities of *adenylate deaminase*, which removes AMP from the equilibrium through an essentially irreversible deamination to inosine monophosphate (IMP), thus driving the adenylate kinase reaction toward the production of more ATP. During rest periods, metabolic energy is used to resynthesize AMP from IMP; the adenylate kinase equilibrium is tipped the other way to produce ADP from AMP; ADP is phosphorylated to ATP in the mitochondria or by glycolysis; and finally ATP is used to restore the pool of creatine phosphate (see Fig. 9-17).

Red vs. White Muscle. The major metabolic pathway by which ATP is replaced varies somewhat from one muscle cell to another. For example, skeletal muscle of vertebrates is broadly classified into two categories with quite different metabolism. These categories are called *red* and *white*, depending on appearance (like the red or dark meat and white meat of chicken).

Red fibers are known also as "slow twitch" fibers. Muscles comprised of them, which have been described by W. Mommaerts as "pay-as-you-go muscles," typically have an especially abundant blood supply. The muscle cells themselves are relatively small, with large surface-to-volume ratios, a great deal of myoglobin for oxygen storage and transfer, and numerous mitochondria (see Fig. 9-18). Their color is mainly due to myoglobin. These muscles, which are designed for relatively continuous use, oxidize fatty acids as a primary source of energy. White (or "fast twitch") fibers on the other hand are larger, with less access to the circulatory system, relatively little myoglobin, and few mitochondria. Their main source of energy is glycolysis, which because it is relatively inefficient can sustain activity for only brief periods before rest is required to remove accumulated lactic acid and to obtain more carbohydrate. Mommaerts refers to these as "twitch-now-pay-later muscles."

White meat in domestic fowl is the well-developed *pectoralis*, or flight muscle. In spite of the size of this muscle, these birds can only flutter along for a few yards before stopping to rest, a pattern of usage that is characteristic of white muscles. On the other hand, anyone who eats domestic fowl knows that their leg muscles, which get almost continuous use, are predominantly red. In migratory birds, however, the situation is reversed: very red flight muscles provide the endurance for sustained flight, and little-used leg muscles are relatively white. The same distinction is found in mammals, where intermittently used skeletal muscles are relatively white compared with the heart or other muscles designed for more continuous use.

The basis for differentiation into red and white fibers became clear around 1960. Experiments by J. C. Eccles, A. J. Buller, W. Mommaerts, and others indicated that a given cell is not inflexibly determined to become red or white in the way that we might at first expect. Rather, the

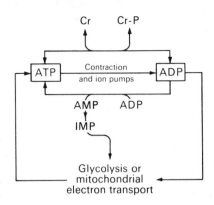

FIGURE 9-17 ATP Regeneration and Utilization in Muscle.

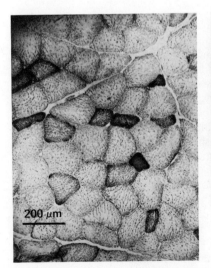

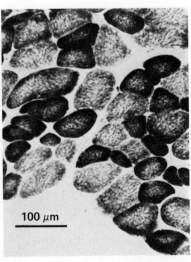

FIGURE 9-18 Red and White Muscle. LEFT: A white muscle (brachioradialis, from the arm of a rhesus monkey), stained for succinate dehydrogenase, a mitochondrial enzyme and therefore a marker for red fibers. Note the predominance of large white fibers, with a few smaller, red (stained) ones. RIGHT: Sartorius muscle (from the thigh) of a rhesus monkey, stained for succinate dehydrogenase. This muscle is decidedly mixed in its fiber type, and would be classed as pink or red by visual inspection alone. [Left photo, courtesy of C. Beatty, G. Basinger, C. Dully, and R. Bocek, *J. Histochem. Cytochem.*, **14**: 590 (1966); right photo, courtesy of R. Bocek and C. Beatty, *J. Histochem. Cytochem.*, **14**: 549 (1966). Copyright by the Williams & Wilkins Co., Baltimore.]

very considerable differences between these two types of fibers are due to the influence of nerves, a conclusion that was reached from experiments in which nerves were switched between red and white muscles. Cross-innervation turned white muscles into red ones and influenced red muscles to take on many of the properties of white muscles.

More recent work has shown that the change from red to white or vice versa is a result of the frequency of stimulation, which differs from one nerve to another. For example, a change from white (fast) to red (slow) fibers can be achieved through long-term, low-frequency stimulation by an external electrical device. The changes in gross muscle fiber characteristics are accompanied by changes in the molecular structure of myosin; the light chains of red and white muscle myosin differ from one another, and a histidine of the heavy chain is methylated in white fibers but not in red ones.

It appears, then, that the metabolic characteristics of red and white muscle develop as an adaptation—i.e., as a secondary consequence of the use to which they are put—and are not derived from any intrinsic differences in gene expression that arise during differentiation.

9-4 CONTRACTION-RELAXATION CONTROL

Contraction in most muscles is initiated by nerves, through the action of neurotransmitters such as acetylcholine, epinephrine, and norepinephrine. Contraction in smooth muscle can instead be due to certain hormones—the ability of oxytocin from the pituitary to initiate contractions of the uterus is an example. In some smooth muscle, mechanical stretching is enough to initiate contraction. And in heart muscle and many smooth muscles, contractions occur spontaneously with a frequency that is intrinsic to the cells, though capable of being modified by nerves or hormones.

Whatever the source of stimulation, the course of a contraction is determined by the release and removal of calcium ions within the cell. This is a function of specialized membranes and the enzymes they contain.

The Role of Calcium. Muscle contraction follows depolarization of the sarcolemma, or plasma membrane, but there is a certain lack of synchrony between the two that indicates the existence of a chemical intermediate. That is, contraction not only lags behind excitation by a significant interval of time (milliseconds), but may last much longer than depolarization. These observations suggest that a chemical, released by excitation, must diffuse over a short distance before meeting the contractile filaments and stimulating them to interact.

Calcium ions have long been suspected of being the transitory agent directly responsible for contraction. In 1883 S. Ringer demonstrated that Ca^{2+} is essential for the activity of isolated frog heart muscle. Other substances are also necessary, of course, but in the late 1940s L. V. Heilbrunn and F. J. Wiercinski at the University of Pennsylvania found that Ca^{2+} is the only common element capable of causing contraction when injected into isolated muscles with micropipets.

The observation that injected Ca^{2+} causes a muscle fiber to contract is consistent with the observation that Ca^{2+} is essential for the contrac-

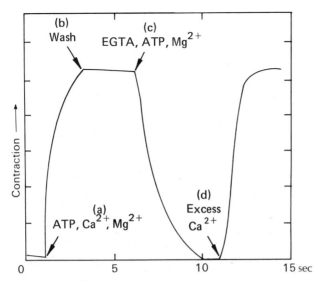

FIGURE 9-19 The Role of Ca^{2+} in Contraction. (a) A solution of ATP, Ca^{2+}, and Mg^{2+} was added to washed, glycerinated fibers. Contraction followed. (b) At the peak of the contraction the fibers were washed, removing Mg^{2+} and ATP but not all the Ca^{2+}, some of which was bound to the fibrils. With ATP gone, the fibers enter a state of rigor. (c) New ATP and Mg^{2+} were added along the EGTA, a chelating agent for Ca^{2+}. With ATP but no Ca^{2+}, the fibers relax. (d) Addition of excess Ca^{2+}, swamping the EGTA, initiated a new contraction. (Described by J. B. Bendall, in *Muscles, Molecules, and Movement*. New York: American Elsevier Publ. Co., Inc., 1969.)

tion of isolated myofibrils. Mg^{2+} is also required, because ATP is utilized as an ATP-Mg chelate. Under normal conditions, both Mg^{2+} and ATP are present in relatively large quantities (millimolar concentrations) in muscles, so their availability does not control contraction. However, the availability of Ca^{2+}, the concentration of which varies within wide limits, can be correlated with the contractile activity of a fiber: when both Ca^{2+} and ATP are present, the muscle contracts; when ATP is present without Ca^{2+}, the muscle relaxes. ATP itself cannot be the controlling element because when ATP is absent, with or without Ca^{2+}, the muscle enters a state of *rigor* (see Fig. 9-19). Rigor is why animal muscles become stiff several hours after cessation of heartbeat (rigor mortis, or the rigor of death); it lasts for up to 24 hours, when proteins and cells begin to break down from release of lysosomal enzymes.

The first study in which the internal availability of Ca^{2+} was correlated with the development of tension following a normal excitation was described in a 1966 paper by F. F. Jöbsis and M. O'Connor of Duke University. They found that by using dimethyl sulfoxide (DMSO) to increase membrane permeability, they could cause some toad muscles to take up significant quantities of Tyrian purple (murexide), a dye that changes color in the presence of calcium ions. The incorporated dye changed color immediately following an excitation, and then recovered slightly ahead of the development of peak tension.

A year later, E. B. Ridgway and C. C. Ashley made a similar observation with a different muscle and a different indicator. They used the protein aequorin, obtained from a jellyfish, to detect calcium in muscle fibers from giant barnacles. These fibers are typically 40 mm long by about 2 mm in diameter. (Vertebrate fibers rarely exceed 0.1 mm in diameter.) Aequorin produces a bluish luminescence, the intensity of which is directly related to the concentration of Ca^{2+} available to it. It was found that a barnacle muscle fiber that has been injected with aequorin glows briefly following excitation, but that the light fades again before maximum tension is reached (see Fig. 9-20).

The preceding studies reveal that a transient availability of Ca^{2+} is associated with excitation and the resulting contraction. Coupled with other observations, some of which have been mentioned, it is clear that

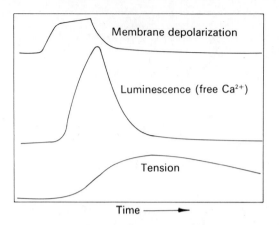

FIGURE 9-20 Excitation-Contraction Coupling. Electrical stimulation of a barnacle muscle fiber produces an action potential (top). (The sharp fall in depolarization marks the time when stimulation ceased.) The temporary appearance of free Ca^{2+} in the sarcoplasm is detected by luminescence from injected aequorin (middle). The fall in free Ca^{2+} is due to its binding by the myofibrils, resulting in tension in the fiber (bottom). Tension will last until Ca^{2+} has been reaccumulated by the sarcoplasmic reticulum. [Described by C. Ashley and E. Ridgway, in *J. Physiol.*, **209**: 105 (1970).]

contraction is a direct result of the appearance of free Ca^{2+} in the vicinity of the myofilaments, and that relaxation results from the removal of these ions. The slight delay between the appearance of Ca^{2+} at the myofilaments and development of maximum tension is caused by the damping effect of elastic elements both within and outside the muscle. Stretching these elements distorts the time course of tension development and decay, prolonging both, and reduces the measured value of the tension to 20% or less (in barnacle fibers) of the amount that would be felt if there were no give in the fiber. When this effect is considered, it is possible to show that activity of the contractile machinery is directly and immediately related to the availability of free Ca^{2+}.

Control by Troponin. It is one thing to correlate tension with Ca^{2+} availability, and quite another to explain the basis for the correlation. What evidence we have comes from studies with synthetic actomyosin. Pure actomyosin fibers contract in the presence of Mg^{2+} and ATP, but impure actomyosin requires Ca^{2+} also. S. Ebashi and his colleagues found in the early 1960s that impure actomyosin contains troponins and tropomyosin. These proteins, which are associated with the actin filaments, confer Ca^{2+} sensitivity on actomyosin, with troponin containing the actual Ca^{2+} receptor sites. The cooperative interaction between actin and myosin needed to hydrolyze ATP and power contraction is prevented in some way by troponins and tropomyosin in the absence of Ca^{2+} but is allowed to occur when Ca^{2+} is bound to a troponin.

X-Ray diffraction studies have led to a specific model for the Ca^{2+} regulation of actin-myosin interaction. It has been shown that in the presence of Ca^{2+}, tropomyosin changes its position in the groove of the actin filaments, moving some 15 Å deeper into the groove—a significant distance in a filament only 40–60 Å in diameter. This movement is presumed to expose myosin binding sites on the actin subunits, permitting myosin heads to attach. Actin is then pulled a short distance along its axis by the myosin head. If ATP is present, breakage of the myosin-actin interaction follows and the myosin head extends, accompanied by ATP hydrolysis, preparatory to another round. (Review Fig. 9-10.)

The key to Ca^{2+} regulation, then, is exposure of actin binding sites when tropomyosin changes position. That positional change, in vertebrate striated muscle at least, is caused by conformational changes in troponins that occur when Ca^{2+} is bound.

Control by Myosin. Although the preceding scheme for calcium mediation of excitation-contraction coupling is widespread, a 1973 paper by Andrew Szent-Györgyi and colleagues indicates that the scheme is not universal. These workers were investigating the regulatory properties of scallop muscles which, like those of other molluscs, contain tropomyosin but not troponin. The lack of troponin means that their actin filaments have no calcium receptors. Instead, Ca^{2+} sensitivity is conferred by myosin itself, which in this case consists of two heavy and three light chains. One of the light chains is the specific calcium receptor. When this regulatory subunit is removed, calcium no longer stimulates contraction; when the light chain is added back, sensitivity to Ca^{2+} is restored.

A similar myosin-mediated calcium sensitivity has now been demonstrated in many other invertebrate muscles. In fact, except for molluscs, most invertebrates have both regulatory systems, myosin and troponin, within the same muscle. Calcium thus controls both types of filaments in these muscles. Vertebrate striated muscle is apparently peculiar in relying solely on regulation by troponins. However, there is evidence that at least some vertebrate smooth muscle is at the other extreme, like molluscan muscle, and has only myosin regulation.

The Calcium Pump. Where does the free Ca^{2+} come from, and where does it go? If one divides the total Ca^{2+} content of muscle by its cytoplasmic volume, it is apparent that the average concentration of Ca^{2+} is roughly $10^{-3} M$. However, experiments in which solutions of the ion are injected into fibers, and experiments in which internal indicators are used to detect its level, indicate that the concentration of free Ca^{2+} near the myofibrils of a resting muscle is less than $10^{-8} M$. Furthermore, a full contraction can be achieved with a concentration of $5 \times 10^{-6} M$ or less. Most of the ions must normally be bound to some internal structure, and available neither to myofilaments nor to indicators.

Since some Ca^{2+} enters the fiber from outside the cell when it is excited, the rise in internal concentration could come from that source. However, the size of vertebrate striated fibers and the rate of diffusion of Ca^{2+} require other explanations—the distance between the surface and the innermost parts of the fiber is simply too great to be covered by a diffusing ion in the time alloted, at least in striated muscle. In addition, that myofibrils all seem to contract at about the same time argues against a diffusion gradient and in a favor of a coordinated Ca^{2+} release from numerous internal stores. And finally, the use of $^{45}Ca^{2+}$ (a radioactive isotope) shows that ions entering the cell during excitation are quickly sequestered and added to the stores of unavailable calcium.

It is now understood that the bulk of Ca^{2+} in striated muscle is bound by the longitudinal component of the internal membranes, and especially in the terminal cisternae of that system. These membranes are endowed with an effective *Ca^{2+} pump* (or *Ca^{2+}-activated ATPase*), which, like the Na^+ pump discussed earlier, is capable of translocating selected ions against an electrochemical gradient using the hydrolysis of ATP to power the process. (A similar calcium pump has also been identified in many other cell types.) In addition, a second mechanism for the rapid removal of free calcium may be provided by mitochondria lying adjacent to the myofibrils. Mitochondria can actively accumulate Ca^{2+} (in a one-to-one ratio with phosphate) during electron transport, at the rate of

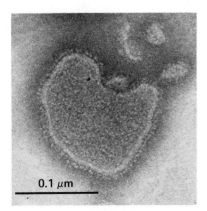

FIGURE 9-21 Ca^{2+}-activated ATPase. This vesicle was formed with purified Ca^{2+}-ATPase (calcium pump) from the sarcoplasmic reticulum. Note the projections, resembling the ATPase from mitochondrial inner membranes. [Courtesy of P. Stewart and D. H. MacLennan, *J. Biol. Chem.*, **249**: 985 (1974).]

about five calcium ions per electron pair transferred from NADH to O_2. Calcium accumulation and oxidative phosphorylation seem to be mutually exclusive events, for electron transport can furnish the energy for only one of these processes at a time.

Available evidence derived from studies with fragments of sarcoplasmic reticulum *in vitro* indicate that the calcium pump is capable of transporting about two calcium ions per ATP hydrolyzed. This coupling between Ca^{2+} transport and ATP hydrolysis is normally quite tight. In fact, reversal of the Ca^{2+} flow in isolated vesicles (see Fig. 9-21) reportedly can cause ATP synthesis. As with other ion pumps, however, chemical uncouplers have been discovered. One of these is caffeine, which explains the ability of caffeine to cause muscle contraction *in vitro*.

Efficiency of the calcium pump is enhanced by the sequestering of accumulated Ca^{2+} within the sarcoplasmic reticulum, probably as a sparingly soluble calcium phosphate gel. This gel reduces the concentration of free calcium within the sarcoplasmic reticulum and permits Ca^{2+} uptake to effective concentrations many thousandfold greater than the concentration within the rest of the cytoplasm.

It has been estimated that the release of 2×10^{-6} mole of Ca^{2+} per kilogram of muscle results in a strong twitch. The reaccumulation of that much calcium requires only about 10^{-6} mole of ATP. At -33 kJ/mole for the standard state free energy change of ATP hydrolysis, we conclude that only about 10^{-5} kilocalorie (Calorie), or 4×10^{-5} kJ, of metabolically supplied energy per kilogram of tissue is needed by the Ca^{2+} pump during each twitch. This seems like a very modest requirement.

Excitation-Contraction Coupling. A contraction-relaxation cycle, then, is controlled by the amount of free calcium available for binding to myofilaments at receptor sites located on troponin or myosin molecules. The variation in free Ca^{2+} near these sites in striated muscle is a result of release and reaccumulation of the ion by sarcoplasmic reticulum. Accumulation appears to be a continuous process, opposed by a rapid, transient release of Ca^{2+} in response to an excitation of the sarcolemma.

A depolarization of the sarcolemma is conducted inward by the transverse tubules. The rate of conduction in frog muscles is about 8 cm/s, which should allow it to reach the center of a 0.1-mm diameter cell in much less than a millisecond. As a wave of depolarization passes near terminal cisternae of the longitudinal system (at the triads, which lie near the junctions of the A and I bands in mammals), the longitudinal system becomes much more permeable to calcium, releasing ions into the vicinity of the myofibrils.

Muscle cells of all types appear to behave in much this same way. There are, however, certain variations. For example, white striated fibers have more extensively developed sarcoplasmic membranes than most red fibers. The additional conducting and Ca^{2+} storage capacity may account for their "fast twitch" response. At the other extreme, smooth muscle cells have sparse membrane systems. Contraction in these muscles, as noted earlier, is very slow. Calcium storage and release by mitochondria is thought to be more important in these cells, and diffusion of Ca^{2+} inward from the plasma membrane may contribute significantly to contraction of both smooth and cardiac muscle. In fact, when different smooth muscles are examined for their dependence on extracellular calcium, those with the most well-developed sarcoplasmic reticulum have least need for external Ca^{2+}. Control, however, does not rely on transverse tubules. Although tubules are absent in smooth muscle cells, close apposition of plasma membrane and sarcoplasmic reticulum has been noted in electron micrographs. This association supplies the necessary structural feature for an excitation-contraction coupling like that of skeletal muscle.

Heart muscle, as noted above, also depends in part on external Ca^{2+}. Therefore if the level of calcium in blood gets too low, the heart will stop. Similarly, too high a level may lead to inappropriate contractile activity that can be equally deadly.

Action Potentials and Innervation. The general features of action potentials were covered in Chap. 8. They are much the same in muscle and nerve, though certain characteristics are peculiar to muscle. Heart muscle cells, for instance, exhibit a gradual depolarization that, if allowed to continue, eventually reaches the threshhold potential and creates an action potential. Thus, if the pacemaker cells of the heart (those that control the rate of contraction) are damaged, or if conduction from them to the rest of the heart is impaired by scar tissue following a heart attack, the most rapidly depolarizing cells that are still intact may take over the pacing function and keep the patient alive. The risk in this situation is in getting multiple action potentials coming too close together from different sites. The resulting chaotic activity results in an uncoordinated twitching called *fibrillation* that will lead to certain death unless stopped—sometimes a sharp blow to the chest will do it; more often complete depolarization of all cells at once with an electrical shock is required.

Not all muscle cells can propagate an action potential. For instance, many invertebrates have nerve endings at multiple points along a muscle fiber, so that a synchronous activation is achieved without the necessity for action potential propagation. This arrangement allows a graded response with an intensity that is proportional to the frequency of incoming impulses, moderated of course by the activity of inhibitory neurons. Many vertebrate smooth muscles share these characteristics, though propagation of action potentials in vertebrate smooth muscle is also common. Intestinal smooth muscle, for instance, propagates action potentials from cell to cell apparently with the help of intercellular gap junctions and the same is true to some extent in cardiac muscle.

To obtain a full contraction in striated muscles (also called *twitch muscles*), a string of impulses must arrive in rapid succession. The ability to add the effects of several impulses allows a graded response, so that many degrees of contraction can be achieved. Maximum tension is generated in a striated muscle only during *tetanus*, when the impulses are arriving too fast to permit any relaxation between twitches (see Fig. 9-22). As an example, twitches in frog sartorius muscle at 0 °C just fuse to form a tetanus when their rate is 15 per second.

The ability of striated muscles to give a graded response to varying frequencies of incoming impulses is a consequence of the Ca^{2+} mediation between excitation and contraction. Calcium release from the reticulum is directly related to the degree and duration of depolarization. The amount of available Ca^{2+}, in turn, affects the degree of tension. In a muscle that generates action potentials, depolarization of the SR can be controlled by the frequency of incoming action potentials. But such muscles are also capable of giving a graded response even to

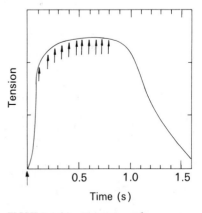

FIGURE 9-22 Tetanus. When excitations (arrows) fall too close together, twitches fuse into a tetanus, defined as a state of continued, complete contraction. (Continued partial contraction is a *tonus*.) The tension is greater than that produced in a single twitch, because there is time for elastic elements in the fiber to become maximally stretched. (Data from A. V. Hill.)

subthreshold depolarizations, a fact exploited by A. F. Huxley and R. E. Taylor in 1958 to define the points at which a depolarization can be conducted into the interior. They found that a very small electric stimulus, which gives highly localized depolarizations insufficient to trigger an action potential, can elicit a small twitch when applied at some points on the sarcolemma but not at others. The sensitive points were found to be arranged in a regular pattern across the surface, and were later identified as the points at which the sarcolemma invaginates to form transverse tubules.

Smooth muscle cells, unlike striated muscles, do not need a continuing input of nerve impulses to maintain a state of partial contraction. This ability to sustain a partially contracted state with very little expenditure of energy is called *muscle tone*. The extreme example of a tonic contraction is found in the paramyosin *catch muscles* of molluscs and annelids, such as those that hold a clamshell shut.

It appears that the ability to hold a partial contraction in smooth muscle and certain invertebrate muscle is due to a peculiarity of the Ca^{2+}-regulatory system. In vertebrate skeletal muscle, removal of Ca^{2+} following a twitch prevents actin-myosin interaction. In those muscles exhibiting significant tone, the actin-myosin cross-bridges are left intact when Ca^{2+} is removed, much as they would be in skeletal muscle during a state of rigor. This feature has obvious advantages, as it maintains tension without expenditure of energy.

Hormonal Effects. Smooth muscle differs from skeletal muscle in being responsive to a variety of hormones and peptides.

Angiotensin, for example, is produced when kidneys experience a decreased blood flow, as with rapidly falling blood pressure. One of its effects is to cause contraction of smooth muscle in the walls of blood vessels, thus reducing the vascular volume and increasing blood pressure. Oxytocin is another example, one that causes contraction of uterine smooth muscle, and thus can be used clinically to induce labor in a pregnant woman.

In explaining how these and other agents actually cause contraction in smooth muscle, one must take into account that smooth muscle from different sources may behave quite differently. Thus, epinephrine causes contraction of blood vessels in the skin (an "alpha adrenergic" effect) but relaxation of vessels in most other locations, as well as relaxation of smooth muscle in the intestine and around airways in the lung (a "beta adrenergic" effect).

The response of various smooth muscles to different agents is due to specific receptors. Cellular receptors, when occupied, either directly or indirectly affect Ca^{2+} availability. In many cases, cyclic nucleotides (cyclic AMP and cyclic GMP) may be the link. Thus oxytocin, acetylcholine, prostaglandin F, and serotonin all cause contraction of uterine smooth muscle and all are said to raise cGMP and to lower cAMP in these cells. Beta adrenergic agents such as epinephrine, on the other hand, have just the opposite effects.

Cyclic nucleotides may affect calcium availability in several different ways. Among the possibilities are changes in permeability of the plasma membrane, mitochondria, or sarcoplasmic reticulum. Another way in which hormones, and hence cyclic nucleotides, might alter contraction is by phosphorylation of key proteins. It is known for example, that myosin light chains are subject to phosphorylation by a specific kinase. The role that this step plays in regulation of contraction has not, however, been established.

These hormonal controls seem to be largely absent from the striated cells of skeletal muscle. Heart muscle, on the other hand, is subject to many of the same controls as smooth muscle. Thus, epinephrine increases not only the rate of contraction (explainable by depolarization as noted in the previous chapter) but the force of contraction. The latter appears to be mediated by cAMP, perhaps through Ca^{2+} release from mitochondria, thus raising the working level of free Ca^{2+}. Acetylcholine has opposing effects, including hyperpolarization of the membrane, rise in cGMP, fall in cAMP, and decrease in contractile activity.

9-5 CONTRACTILE PROTEINS IN NON-MUSCLE CELLS

Cytoplasmic streaming, ameboid locomotion, and other movements of non-muscle cells have fascinated scientists for two centuries. As long ago as 1835 F. Dujardin proposed that all cells are composed of a substance called "sarcode" that has both structural and contractile properties. Only recently, however, have we begun to appreciate how nearly right Dujardin was.

Once the individual proteins of the muscle sarcomere were identified and purified, it became possible to ask whether the same or closely related proteins are present in non-muscle cells and, if so, the extent to which they are responsible for cell shape and cell movements. It appears now that nature has once again proven to be conservative, for the proteins known to be responsible for contraction of muscle are indeed present in cells of a great many different types. Furthermore, though their function in non-muscle cells is not as well understood, there is little doubt that these proteins have comparable roles wherever they are found.

Actin. The first muscle protein to be found in non-muscle cells was actin. Microfilaments, long observed by microscopists and long associated with cell movements, are comprised of actin. To be sure, there are some differences between muscle and non-muscle actin, both in chemical composition and in the properties of the filaments formed from them. The two proteins are, in fact, products of different genes. However, they are closely related genes, no doubt sharing a common ancestor, and they produce closely related proteins.

Actin has now been found in all the major branches of the phylogenetic tree, including protozoa, lower and higher plants, and many cell types from higher animals. Nor is it present in just trace amounts; it often constitutes 2–4% of the total protein and up to 15% in some non-muscle cells.

Filaments formed from this non-muscle actin (microfilaments) closely resemble the actin filaments of muscle. They are more labile—that is, they polymerize and depolymerize more readily—but in the filamentous form they look like actin and, when myosin is added, they act like actin. These filaments can be "decorated" with heavy meromyosin (HMM) just as we described earlier for actin filaments from muscle, including the

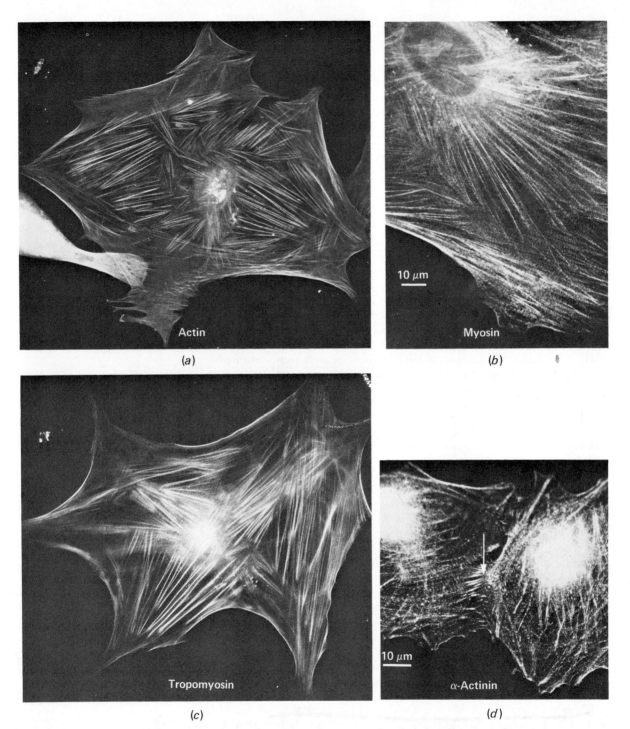

FIGURE 9-23 Contractile Proteins in Non-muscle Cells. Light micrographs of whole cells using indirect immunofluorescence and antibodies specific for the designated protein (see text). (a) Actin. (b) Myosin. (c) Tropomyosin. (d) α-actinin. Note the presence of fluorescent fibers at the region of cell-cell contact in (d) arrow). (a)–(c) are human fibroblasts. (d) is from a mouse embryo culture. [(a) and (c) courtesy of E. Lazarides, *J. Cell Biol.* **65**: 549 (1975); (b) courtesy of K. Weber and U. Groeschel-Stewart, *Proc. Nat. Acad. Sci. U.S.*, **71**: 4561 (1974); (d) courtesy of E. Lazarides and K. Burridge, *Cell,* **6**: 289 (1975).]

same evidence for polarity. And like muscle actin, the interaction with myosin stimulates myosin ATPase activity.

But where is actin located? That is difficult to say for unpolymerized actin, but actin filaments can be identified in several ways. The classic method is to find 60-Å filaments that can be decorated with HMM. Another approach is to use fluorescent-labeled antibodies that are specific for actin. The location of the antibodies, and hence of actin, can then be revealed by illumination with ultraviolet light in a microscope. The procedure is called *direct immunofluorescence.* The word "direct" is added to distinguish it from a variation that is actually more common, *indirect immunofluorescence.* In the latter procedure, actin antibodies are not labeled. Their location in the cell is revealed by adding fluorescent-labeled antibodies that bind not to actin but to the first antibodies—goat antibodies to rabbit gamma globulin, for instance, a product that is commercially available. Results with this technique are shown in Fig. 9-23.

Indirect immunofluorescence of actin shows that the protein is located in long fibers, often stretching the length of the cell. Most of these fibers are apparently near the surface of the cell, for electron microscopy of sectioned cells places the bulk of the visible actin microfilaments in the *cortex*—that is, just under the cell membrane. This region, which is also known as the *ectoplasm,* is generally free of organelles, probably because of this web of filaments.

Myosin. Actin, of course, does not have an intrinsic capacity to contract. It is required, however, to give myosin something on which to pull. If the analogy between muscle and non-muscle contraction is complete, then where actin is found, myosin should also be found. And, indeed, myosinlike molecules have now been identified chemically and immunologically in a variety of cell types (see Fig. 9-24). Discrete myosin filaments are not usually seen by routine electron microscopy, however. This should not bother us too much, for it was not until the late 1960s that investigators found a way to stabilize and thus visualize myosin filaments even in smooth muscle. If myosin filaments are more unstable in smooth than striated fibers, perhaps they are that much more unstable in non-muscle cells. In addition, there is not much myosin present—typically less than 1% of the total protein in non-muscle cells compared with 35% of the protein in rabbit skeletal muscle.

Even where discrete myosin filaments are not seen, fluorescent antibodies reveal large fibers that stain for myosin (Fig. 9-23b). It does not follow automatically that these fibers contain myosin filaments, of course. Individual myosin molecules could be scattered along actin strands and give the same fluorescent staining pattern. In fact, while myosin isolated from most vertebrate non-muscle cells can form thick filaments that look like those of muscle, and behave the same with actin, such is not the case with myosins from all sources. The soil ameba *Acanthamoeba castellani,* for instance, has a myosinlike molecule only 40% as large as vertebrate myosin that does not form filaments *in vitro.* However, it does form a gel with actin that has contractile properties similar to skeletal muscle actomyosin.

In striated muscle, a filament of myosin can pull on two actin filaments with opposite polarities and in this way cause contraction. This arrangement may have evolved to provide the large number of actin-

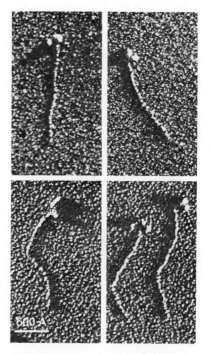

FIGURE 9-24 Non-muscle Myosin. TOP: from chicken brain. BOTTOM: from chicken blood platelets (a cell fragment that contracts during blood clotting). Note the bilobed head, precisely like the skeletal muscle myosin in Fig. 9-11. [Courtesy of A. Elliott, G. Offer and K. Burridge, *Proc. Roy. Soc.* (London), Ser. B, **193**: 45 (1976).]

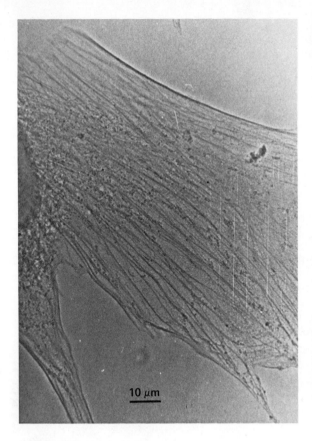 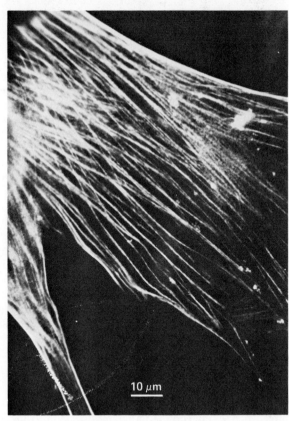

FIGURE 9-25 Stress Fibers. These bundles, often seen in living cells and known to contain microfilaments, are seen here in a cultured human fibroblast. Nucleus is at the left edge of the photos. LEFT: Phase contrast optics. RIGHT: Same cell, with indirect immunofluorescence staining for myosin. Note that the same fibers are visible in both photos. [Courtesy of Klaus Weber and U. Groeschel-Stewart, *Proc. Nat. Acad. Sci. U.S.*, 71: 4561 (1974).]

myosin cross-bridges needed to generate considerable tension. In nonmuscle cells, however, the amount of tension generated is not a problem, for the cells are not pulling on tendons and bones but merely using a force-generating system to translocate organelles or to change shape and move along a surface. A few myosin molecules linked tail-to-tail and pulling on long strands of actin may be all that is required. Indeed, examination of myosin immunofluorescence at high magnification shows the pattern to be interrupted at regular intervals, unlike the continuous immunofluorescence found in fibers stained for actin.

When a cell prepared for myosin immunofluorescence is also photographed with phase contrast optics, the myosin is seen to be associated with long *stress fibers* of the kind seen in living cells and known already to contain actin microfilaments (see Fig. 9-25). Hence, these stress fibers appear to be comprised, in part at least, of actomyosin complexes.

Other Proteins. Tropomyosin (see Fig. 9-23c) is clearly identified with the stress fibers and probably lies in the grooves of actin filaments just as it does in muscle cells. α-Actinin, on the other hand, is spaced irregularly along the fibers and is also found between fibers (Fig. 9-24d). This long, slender protein in muscle is part of the Z discs, but *in vitro* can cross-link parallel actin filaments. The protein may therefore be a structural component of the non-muscle fibers, acting as spacers for the actin filaments. The presence of α-actinin at cell-cell contacts is particularly interesting, for the structural similarities between Z discs of muscle and the desmosomes that link many cells has long been noted. In fact, the special desmosomes ("intercalated discs") that help glue heart muscle cells together have been shown to react with antibodies specific for α-actinin.

Several other proteins have also been associated with the contractile system of non-muscle cells although their functions are not well understood as yet. In addition, another type of filament, the 100-Å filament, is also present in a great many cells. In smooth muscle, 100-Å filaments probably serve as structural components to help organize actin and myosin. Since the contractile machinery of smooth muscle cells is functionally a compromise between that of striated muscle and non-muscle, 100-Å filaments may serve a similar role in many cell types. Other possibilities for this filament have been suggested, however, as we shall see.

Control of Non-muscle Contraction. Although the site of Ca^{2+} action is not well defined as yet, it is clear that calcium can control at least some non-muscle contractile systems in much the same fashion as muscle is controlled.

The effects of Ca^{2+} on contractility have been demonstrated with several *in vitro* non-muscle systems. For example, if cells of the small ameba *Acanthamoeba castellani* are extracted in the cold, the extracted solution is found to gel when warmed. Warming causes some of the solubilized proteins, especially actin, to polymerize. (Recall from Chap. 3 that this sort of temperature effect is expected when proteins polymerize via hydrophobic bonding.) As demonstrated by T. D. Pollard and colleagues at Harvard Medical School, this gel is a contractile system having the properties of actomyosin. That is, it is able to contract in the presence of Mg-ATP and Ca^{2+}. The rate of contraction was found to be exquisitely sensitive to minute changes in Ca^{2+} concentration—from very slow contraction at $10^{-8}\ M$ to a violent shrinkage and fragmentation at $2 \times 10^{-7}\ M$.

Non-muscle contractile systems have also been demonstrated in higher animals. For example, M. Mooseker and L. G. Tilney of the University of Pennsylvania isolated the microvillus "brush border" from intestinal absorptive cells of chicken and demonstrated calcium-controlled contractility *in vitro*. Microvilli are the fingerlike projections of the cell membrane that extend into the intestinal lumen, greatly increasing the absorptive surface of the cell. These projections were shown to contain a core of actin filaments stretching between the microvillus tip and a dense complex of filaments and fibers (the *terminal web*) that lies in the cytoplasm below the microvilli (Fig. 9-26). In the presence of Mg-ATP, Ca^{2+} causes these isolated microvilli to contract. In life, the motion of microvilli might help to stir intestinal contents and hence hasten absorption.

If Ca^{2+} availability is going to control complex movements in a living cell, it must be possible to have very localized areas of higher Ca^{2+} concentration. Calcium is normally kept at a low level by Ca^{2+} pumps in mitochondrial and plasma membranes. The opening of Ca^{2+} channels in either place permits a rapid flow of the ion down its electrochemical gradient and into the cytosol (i.e., into the soluble part of the cytoplasm). It could then initiate local contractile activity.

An increase in Ca^{2+} level at one point in the cytoplasm need not spread to other areas. This has been directly demonstrated by B. Rose

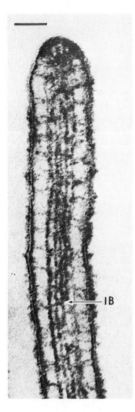
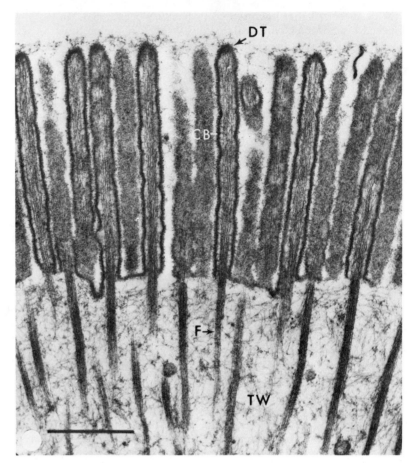

FIGURE 9-26 The Brush Border. Microvilli extend into the intestinal lumen. Their core is comprised of actin filaments (F, microfilaments) all having the same polarity—that which permits a pull on the microvillus tip from a region of the cytoplasm called the terminal web (TW). Cross-bridges (CB) between membrane and filaments, and a dense matrix (DT) where the filaments are anchored at the tip by α-actinin (Z line protein) are best seen in the enlargement, which also reveals interfilament bridges (IB) within the actin bundle. Bar on main photo is 0.5 μm, on enlargement, 0.05 μm. [Courtesy of M. S. Mooseker and L. G. Tilney, *J. Cell Biol.*, **67**: 725 (1975).]

and W. Loewenstein of the University of Miami, who used aequorin luminescence to show that when Ca^{2+} is injected carefully into a cell at any point, it is rapidly sequestered, apparently by mitochondria, before it can diffuse to other parts of the cell. Thus, localized differences in calcium ion availability could turn contractile systems on at one point in the cell while leaving other parts unaffected.

We shall apply these concepts to such things as ameboid locomotion later in this chapter. First, however, quite a different Ca^{2+}-controlled contractile system must be introduced, that due to microtubules.

9-6 CILIA AND FLAGELLA

If you've ever used a microscope to watch protozoa swimming, you must surely have been impressed by the activity of these little animals. They are able to swim as they do because of cilia and flagella, which are whiplike appendages protruding from their surfaces. If there is only one such appendage it is called a flagellum; if there are many, they are called cilia. Although cilia tend to be shorter and to have a somewhat different pattern of movement, the cilia and flagella of eukaryotes are really different versions of the same organelle. What is more, they are identical in all major respects whether they are propelling protozoa or mammalian sperm, sweeping water past the gills of sea animals or sweeping debris out of human lungs.

Structure of the Axoneme. The cilium or flagellum is comprised of a 0.2-μm diameter core of microtubular structures collectively called the *axoneme.* The axoneme extends outward from a cytoplasmic *basal body* and is surrounded by an extension of the cell membrane. Most cilia and flagella have a "9 + 2" axoneme, comprised of nine outer *doublets* that are arranged about a central core containing two individual microtubules (see Fig. 9-27b). Each doublet consists of one complete microtubule called *subfiber A* to which is attached an incomplete C-shaped *subfiber B.* The A subfibers and core microtubules, like most other microtubules, are each comprised of 13 parallel *protofilaments.* The B subfibers have either 10 or 11 protofilaments (see Fig. 9-28).

The axonemal structures so far described are polymers of α and β tubulin, the spontaneous polymerization of which was discussed in Chap. 3. Tubulin is actually a family of closely related proteins. The α and β tubulins of the A subfibers, for example, are demonstrably different from their counterparts in B subfibers.

FIGURE 9-27 Structure of a Cilium and Basal Body. (a) Longitudinal section of a cilium and associated basal body from the protozoan *Tokophrya infusionum.* (b) Cross section of the cilium. (c) Cross section at the axosome (A), or region where the basal body and cilium join. (d) Cross section taken below the basal body, showing microtubules. (e), (f), (g) Three levels of the basal body in cross section. In (e) only the basic structure is present. In (f) a striated fiber (SF) and microtubular fiber (MF) (probably anchors) arc from the basal body toward the cell membrane. Note the cartwheel structure at level (g). [Courtesy of Lyndell Millecchia. In part from L. Millecchia and M. A. Rudzinska, *J. Cell Biol.*, **46**: 553 (1970).]

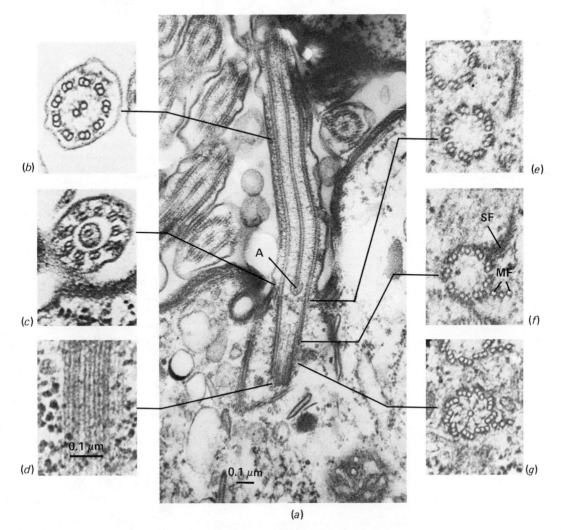

401

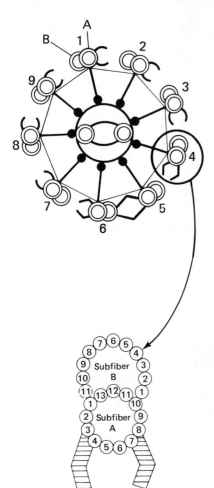
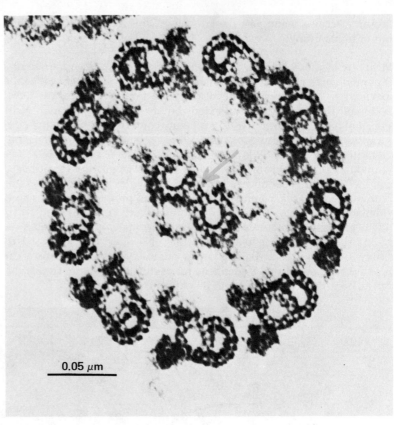

FIGURE 9-28 Structure of the Axoneme. The drawing and micrograph from the sea urchin *Lytechinus* show the nine outer doublets, each with two clawlike *dynein arms* on the A subfiber and a *radial spoke* connecting the doublet to the *central sheath*. Note also the fine *interdoublet links*, and the bridge between doublets 5 and 6 and between the two *central microtubules*. The detail shows the protofilament substructure. As drawn, one is looking from the base toward the tip of the axoneme. The direction of beating is in a plane perpendicular to the central microtubular bridge. [Micrograph courtesy of K. Fujiwara and L. G. Tilney, *Ann. N.Y. Acad. Sci.*, **253**: 27 (1975).]

Several other proteins in addition to tubulin are present in the axoneme, giving it a complex architecture that first started to become clear in the 1950s and early 1960s, largely due to the electron microscopy of B. Afzelius in Sweden, I. Manton in England, and D. Fawcett and K. Porter in the United States. It is now clear, for instance, that each A subfiber has rows of clawlike arms that reach toward the back of the adjacent B subfiber. These arms, which consist of a protein are called *dynein*, are essential to movement. In many cells, one set of arms seems to be permanently attached to the next doublet. This bridge is by definition between doublets number 5 and 6, making it possible to unambiguously identify each of the nine doublets (see Fig. 9-28) and thus define how their relationship to one another changes during movement.

Other features of the axoneme include radial spokes connecting doublets to the central pair, thin interdoublet links that seem to connect the outer ends of the spokes, and a *central sheath* of fibers connecting and surrounding the central pair of microtubules, all readily visible in the diagram in Fig. 9-28.

Axonemal Movement. Cilia and flagella move by sliding adjacent doublets past each other. This model appears to have been first suggested by Bjorn Afzelius of the University of Stockholm in 1959. However, direct evidence was not available until the mid-1960s when Peter Satir and his associates, then at the University of Chicago and later at the University of California, Berkeley, found a way to demonstrate by serial sections on fixed material that indeed there is displacement of doublets during bending, with the amount of displacement proportional to the degree of bend (see Fig. 9-29). In fact, this sliding has now been observed while actually in progress. For example, brief exposure of sperm tails to detergent and trypsin causes enough breakdown in the internal structure that, when ATP is added, the axonemes fall apart by sliding their doublets past each other. Under normal conditions, of course, internal constraints would convert this sliding into bends.

The force needed to cause displacement between adjacent doublets and hence the bending of cilia and flagella is provided by hydrolysis of ATP (actually Mg^{2+}-ATP). This hydrolysis and conversion of chemical to mechanical energy is made possible by the dynein arms on subfiber A. The role of dynein was demonstrated by Ian Gibbons and his colleagues at the University of Hawaii who found that dynein could be extracted selectively from axonemes and in some cases put back again. When it was gone, so were the arms on the A subfibers and so was the ability of the axonemes to hydrolyze ATP and to move.

More recently, a small group of male patients being examined because of sterility were found by Afzelius to have immotile sperm that lack dynein arms in their flagella. Most of these patients also suffer from chronic cough and recurrent lung infections due to poor ciliary action in the airways and consequent failure to clear efficiently mucous and inhaled particles from their lungs.[3] The importance of dynein to the axoneme is thus quite clear.

Dynein can be thought of as a myosinlike ATPase. Instead of pulling on actin, it pulls instead on the adjacent doublet. Thus microtubules are forced to slide past each other. Each cycle of cross-bridge attachment and release performed by dynein appears to be accompanied by the hy-

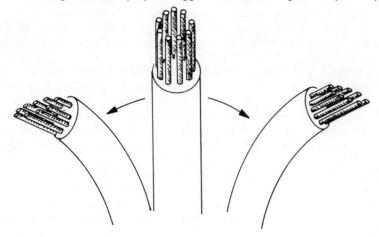

FIGURE 9-29 Sliding Microtubules. The cilium is cut off and then stripped back, leaving only the nine outer doublets exposed, here represented by single tubules for clarity. Note the sliding of the doublets relative to each other as the cilium bends. It is this sliding that causes the bending.

CHAPTER 9
Contractility and Motility

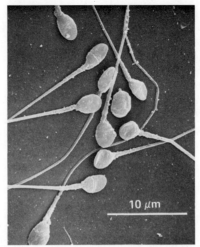

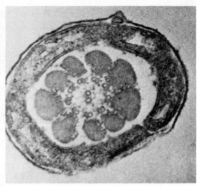

(a)

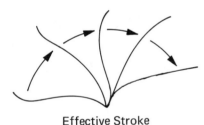

Effective Stroke

(b)

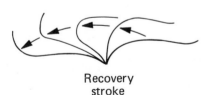

Recovery stroke

(c)

FIGURE 9-31 Movement of Cilia and Flagella. (a) Typical wave-form propagation of a flagellum, such as that found on sperm. As shown, it would drive the cell headfirst through water. (b) Effective stroke common with cilia of all types. (c) Recovery of cilium prior to the next effective stroke.

FIGURE 9-30 Spermatozoa. LEFT: Scanning electron micrograph of normal human spermatozoa. The thickened part of the flagellum near the head is called the midpiece. It is here that mitochondria surround the axoneme. RIGHT: Transmission EM through the midpiece of bull sperm. Note the heavy *dense fiber* associated with each doublet in the axonemal cross section. These fibers, common to most vertebrates and replaced by extra tubules in insect sperm, are thought to stiffen the flagellum and thus shape the form of the beat. [SEM courtesy of Dr. Nancy Alexander, Oregon Regional Primate Research Center; TEM courtesy of C. B. Lindemann and I. R. Gibbons, *J. Cell Biol.*, **65**: 147 (1975).]

drolysis of one ATP. If ATP is removed, the cross-bridges form but do not release, yielding a rigid axoneme comparable to the rigor of ATP-depleted muscle.

The ATP needed for ciliary movement is generated by mitochondria or by glycolysis in the cytoplasm and diffuses outward within the axoneme. This geometry is a little different for the spermatozoa of most species, however, for in those cells ATP generation is via mitochondria that surround a portion of the axoneme itself. This segment of the sperm flagellum is called the midpiece (see Fig. 9-30).

The long diffusion pathway for ATP is partly compensated for by a mechanism that uses not just the terminal, but both high-energy phosphate bonds of ATP. This is achieved by the presence within the axoneme of adenylate kinase, an enzyme described earlier as catalyzing the reaction 2ADP → AMP + ATP.

Two common types of movement achieved by sliding microtubules are diagramed in Fig. 9-31. Part (a) shows propagation of a sinusoidal wave, more or less all in one plane moving from the base (or head) toward the tip. The organism using this pattern of movement—commonly a uniflagellate sperm—is thus propelled head first through the water. Some microorganisms (e.g., *Crithidia*), propagate the wave in the opposite direction and thus swim tail first.

Parts (b) and (c) of Fig. 9-31 show the typical beating pattern of a cilium, with its straight-arm effective stroke (power stroke) and subsequent recovery. This movement either forces an organism through the water or, in the case of ciliated tissues, sweeps fluid and particles past the cell.

Regulation of Movement. Flagella and cilia commonly beat continuously at 40–60 times per second. Coordination of cilia is necessary, however, and both the rate and direction can be altered in some cases.

Reversal of the ciliary beat in *Paramecia*, for example, occurs when the animal bumps into an object. Reversal has been shown to be caused by Ca^{2+} influx, associated with the depolarization of a Ca^{2+}-dependent action potential. In other experiments, striking the animal from behind produced K^+ influx, followed by hyperpolarization of the plasma membrane and an increased forward-propelling beat frequency.

Cilia typically beat in a coordinated fashion that causes visible waves (called *metachronal waves*) to move across the surface like stalks of wheat bending in the wind (see Figs. 1-21 and 9-32). The basis for this coordination is not clear but is assumed by many to be mechanical coupling. For example, when cilia that have been paralyzed by low pH and K^+ are moved slightly with a microneedle, a single new beat is initiated. It is thus possible that the movement of a metachronal wave is due to displacement and consequent excitation of those cilia just ahead of the wave. In other words, there may be a brief rest period at the end of the recovery stroke that is cut short when fluid forced ahead of an advancing wave front reaches and excites the resting cilia.

In other cases, however, there is some kind of direct coupling that ensures coordinated beating. The two flagella of the alga *Chlamydomonas*, for example, continue to beat synchronously even when the apparatus containing the flagella and its basal bodies is isolated from the cell.

Ciliogenesis. Cilia and flagella arise from basal bodies (see Fig. 9-27). These structures are said to have "9 + 0" construction because they have nine outer groups of tubules (commonly triplets instead of doublets) and lack the central pair found in cilia. In addition to the basal body there is frequently some sort of anchoring structure (striated fibers, basal micro-

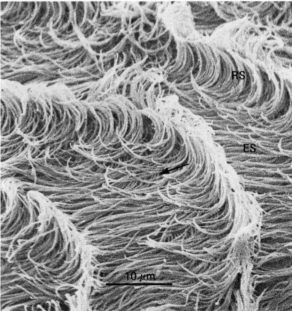

FIGURE 9-32 Coordination of Cilia. Details from a protozoan, *Opalina ranarum* (see Fig. 1-21), showing clear metachronal waves. *Arrows* indicate direction of wave propagation. (The metachronal wave is perpendicular to the direction of the effective stroke.) Cilia caught toward the end of their effective stroke are labeled ES; those in their recovery stroke, RS. [Courtesy of G. Horridge and S. L. Tamm, *Science*, **163**: 817 (1969). Copyright by the AAAS.]

FIGURE 9-33 Ciliary Anchors. Many cilia have some sort of structure associated with their basal bodies that appears to lend stability. (a) Striated and dense fibers on the right and left, respectively, of a protozoan basal body (the ciliate *Tokophyra*). (b) Rootlet from a coelenterate, the "sea pansy," *Renilla*. See also Fig. 1-10c. [(a) courtesy of L. Millecchia and M. Rudzinska, *J. Cell Biol.*, **46**: 553 (1970); (b) courtesy of B. Spurlock and M. Cormier, *J. Cell Biol.*, **64**: 15 (1975).]

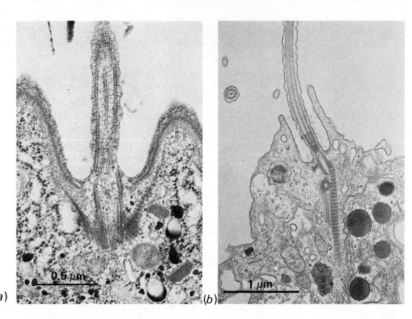

FIGURE 9-34 Basal Body Replication and Ciliogenesis in a Protozoan. *Tokophyra*, a suctorian (class Ciliatea) is motile in its immature form, with hundreds of cilia. However, at maturity it has no cilia and few basal bodies. Replication of the cell therefore includes massive basal body synthesis. Each basal body is formed as a short probasal body (PBB) at right angles to a mature basal body (BB). The new basal body then tilts upright before generating the cilium (Fig. 9-27). Basal microtubules (BMT) may help to anchor the BBs together. [Courtesy of L. Millecchia and M. A. Rudzinska, *J. Cell Biol.*, **46**: 553 (1970).]

tubules, basal foot, rootlets, etc.) that presumably give the cilium stability (see Figs. 9-27 and 9-33).

Basal bodies and centrioles appear to be one and the same structure put to different uses. That is, basal bodies serve as microtubular organizing centers (see Chap. 3) for cilia and flagella while centrioles organize the microtubules of nuclear spindles during cell division. In fact, in the formation of the single flagellum of sperm, one of the two centrioles of the cell serves a dual purpose and becomes the basal body that organizes the flagellar axoneme.

Because of the essential role of the basal body, cilia formation (ciliogenesis) can be said to begin with formation of the basal body. Examples of protozoan and mammalian basal body replication are summarized in Figs. 9-34 and 9-35. Note in Fig. 9-34 that new basal bodies form at right angles to old ones, then tip upright to initiate axonemal for-

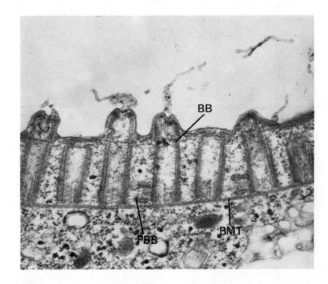

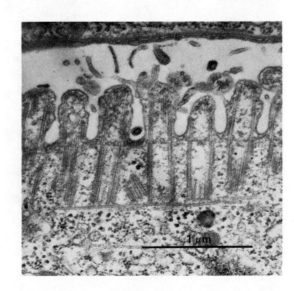

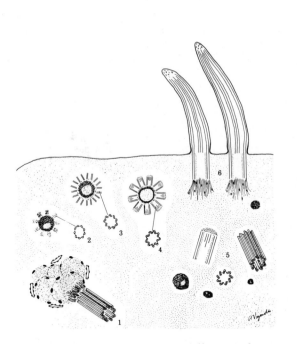
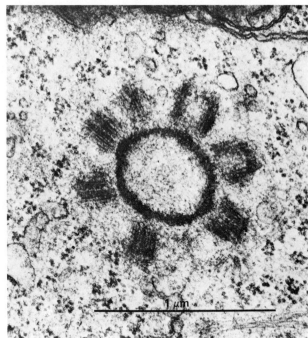

mation. In Fig. 9-35, quite a different scheme is shown. Here basal body replication is taking place in cells lining the oviduct of a mouse soon after birth. (In humans it occurs at about the middle of gestation.) The basal bodies are formed in clusters initiated by a mature centriole.

Axonemal doublets appear from microscopic studies to be direct extensions of the outer tubules of the basal body. However, the central pair of microtubules in the axoneme appears to spring from a less well-defined structure, the axosome (see Fig. 9-27). In the case of the doublets, at least, growth occurs at the tip of the elongating axoneme via proteins that either diffuse or are transported outward through it.

The controls on this process of ciliogenesis are not well understood, but eventually a rather characteristic length is reached and polymerization stops. Once formed, the axonemal structure is quite stable but not permanent. For example, if one of the two flagella are broken off the alga *Chlamydomonas*, the intact flagellum shrinks temporarily during regeneration of the damaged one. Since regeneration occurs to a considerable degree even when new protein synthesis is blocked, one is left with the impression that both a tubulin pool in the cytoplasm and tubulin from the intact flagellum can be used for construction of the growing appendage.

Bacterial Flagella. Prokaryotic and eukaryotic flagella have little in common. Bacterial flagella are helical, ropelike strands comprised of individual filaments of *flagellin*. According to Howard Berg and his associates at the University of Colorado, they propel the bacterial cell like a propeller moves a ship—i.e., by a rotating movement powered by some sort of membrane-bound "motor" that seems to use chemiosmotic gradients for energy.

Bacteria with multiple flagella, like the *Salmonella* seen in Fig. 9-36, execute a "random walk" swimming pattern. The flagella, rotating together in a bundle, drive the cell through the water for a second or so.

FIGURE 9-35 Basal Body Replication in the Mouse Oviduct. Diagram of the process: (1) Large amounts of microtubular precursor protein are synthesized next to a mature centriole. (2) A portion of the precursor protein condenses and probasal bodies form from it, initially with nine singlet microtubules in their walls. (3) A second microtubule is added to each singlet at the expense of the centrally located percursor protein. (4) The probasal bodies have added a third microtubule to each doublet. (5) The basal bodies are released to take up positions at the cell membrane. (6) They generate cilia. The micrograph is approximately a stage three. Note the ribosomal clusters. [Courtesy of E. R. Dirksen, *J. Cell Biol.*, **51**: 286 (1971).]

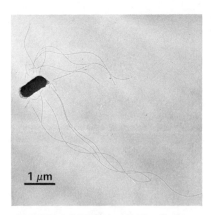

FIGURE 9-36 Bacterial Flagella. *Salmonella typhimurium*, a bacterium capable of associating with and invading the intestinal lining to cause a diarrheal illness. [Courtesy of G. Tannock, R. Blumershine, and D. Savage, *Infection and Immunity*, **11**: 365 (1975).]

Then the bundle comes apart and for perhaps a tenth of a second the cell tumbles wildly ("twiddles"), a period that ends when the flagella are able to resynchronize to drive the cell off randomly in a new direction. If the cell is moving up a gradient of desirable substances (a chemical attractant), the runs between twiddles are longer than they would be otherwise. Hence, the random walk becomes biased and the cells tend to accumulate in more desirable environments. This directed movement is called *chemotaxis*. It is prominent also in eukaryotic systems, as we shall see in the next section.

9-7 CELL SHAPE AND CELL MOVEMENT

The two motile systems just discussed, the actin-myosin system and the microtubule-dynein system, are apparently responsible for a variety of movements, both of material from one part of the cell to another and of cells themselves. A closely associated function, again apparently involving both systems, is the maintenance of cell shape.

The Cytoskeleton. Most cells have an intricate cytoplasmic network of microfilaments, 100-Å filaments, and microtubules, much of it lying just under the cell membrane. This network has been referred to as a *cytoskeleton* (see Figs. 9-37 and 9-38). It is the cytoskeleton that is pre-

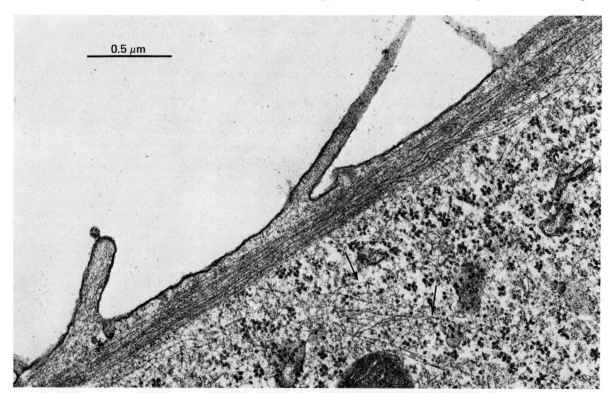

FIGURE 9-37 Cortical Filaments. Conventional thin section of a cultured fibroblast showing a bundle of parallel microfilaments in the cortex (just under the plasma membrane). These bundles are sometimes called stress fibers. A few 100-Å filaments seem to be associated with the stress fiber and extend deeper into the cytoplasm (*arrows*). The fingerlike surface extensions (small microvilli) contain only microfilaments. [Courtesy of R. B. Evans, V. Morhenn, A. Jones, and G. Tomkins, *J. Cell Biol.*, **61**: 95 (1974).]

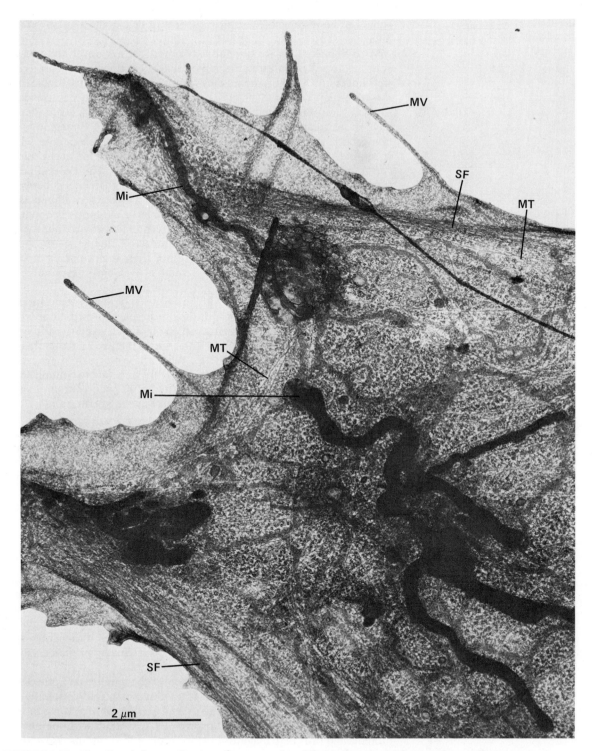

FIGURE 9-38 The Cytoskeleton. Portion of an unsectioned, critical point dried rat embryo cell, demonstrating the intricate network of filaments and microtubules that extends throughout the cytoplasm. Microvilli (MV). Stress fiber of microfilaments (SF). Microtubules (MT). Snakelike mitochondria (Mi). [Courtesy of Ian Buckley, *Tissue and Cell*, **7**: 51 (1975).]

sumed to be responsible for most aspects of cell shape and cell movement.

Microfilaments, as we have already noted, are a form of non-muscle actin. The individual subunits polymerize like actin but, for reasons not well understood, the filaments they form are a great deal less stable than the actin filaments of muscle. That is, they more readily come and go. In fact, L. G. Tilney and associates at the University of Pennsylvania have shown that in certain invertebrate spermatozoa, polymerization of non-muscle actin can occur with explosive rapidity. For example, when sperm of the sea cucumber, *Thyone*, comes in contact with an egg, a long *acrosomal process* shoots out from the sperm head. This process penetrates the protective material around the egg so that fusion between membranes of the process and egg can occur. The acrosomal process in this system is 15 times as long as the sperm head and forms in about 10 seconds. When examined by appropriate techniques, it was found to be packed with parallel actin filaments. Unreacted *Thyone* sperm, however, has only nonfilamentous actin. Since there is little or no myosin present in these cells, actin polymerization is apparently responsible by itself for the acrosomal reaction.

Other situations involving the polymerization or depolymerization of actin are not as dramatic as the acrosomal reaction. Nevertheless, when a living cell is viewed with the microscope it is seen to constantly form surface processes of various shapes and sizes, then retract them again only to form new ones (see Fig. 9-39). When these protrusions are examined by appropriate techniques they invariably are found to contain actin filaments (see Fig. 9-39*d* on p. 412). Whether polymerization of actin actually causes the surface to bulge, as in *Thyone* sperm, is not clear. Probably several different mechanisms are at work.

It has been suggested, for example, that blebs such as those seen in Fig. 9-39*a* result from herniation of cytoplasm out through a break in the microfilamentous web underlying the cell membrane. In support of this hypothesis, it has been noted that certain cytochalasins (cyto-ka-lā'-sins) greatly increase the rate of blebbing. Cytochalasins are drugs, some of which rather specifically depolymerize microfilaments, leaving microtubules and 100-Å filaments unaffected.

In other situations, changes in the membrane itself may be responsible for the change in shape. Thus, adding lipid specifically to the outer half of a portion of the membrane bilayer would cause the cell membrane to bulge outward at that point to form a microvillus or some other structure. Conversely, adding lipid to the inner half of the bilayer would cause the membrane to dimple. This phenomenon of differential membrane growth and associated change in shape has actually been demonstrated in erythrocytes by S. J. Singer and his associates at the University of California, San Diego. Its importance to systems other than erythrocytes, however, is yet to be determined.

Whatever the mechanism of formation of the smaller cell processes, microtubules are probably not involved directly. They generally lie deeper in the cytoplasm and are relatively large (Fig. 9-40). Some microfilaments and microtubules do, however, make contact with membrane proteins and thus help to control the location of the proteins, as we shall see in Chap. 10.

Larger cytoplasmic extensions, on the other hand, may depend directly on microtubules for support. Thus, nerve cell processes (axons and

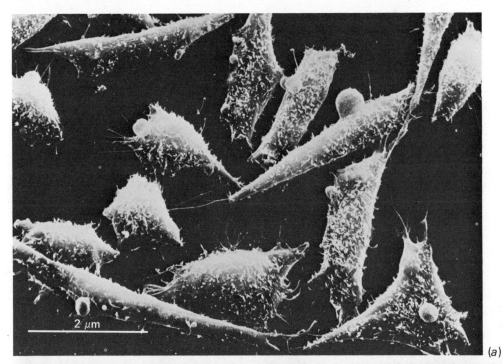

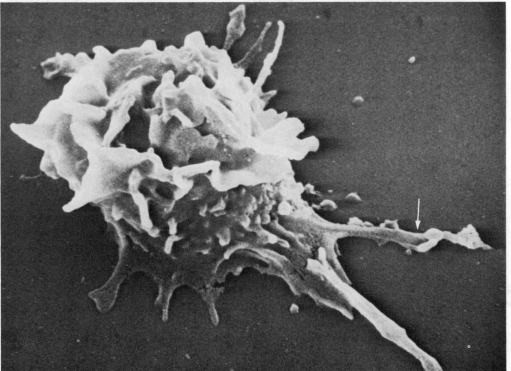

FIGURE 9-39 Surface Specialization. (a) Scanning electron micrograph of Chinese hamster ovary cells. Note the fingerlike projections (microvilli) and rounded knobs or *blebs*. (b) Scanning electron micrograph of a rabbit neutrophil (one of the white blood cells) showing multiple lamellar (platelike) processes and retraction fibers or filopodia (*arrow*) left behind as the cell moves away from a point of attachment. Particles trapped in the folds between lamellae are readily engulfed (see "phagocytosis," Chap. 10). (Parts *c* and *d* of Fig. 9-39 appear on the next page.)

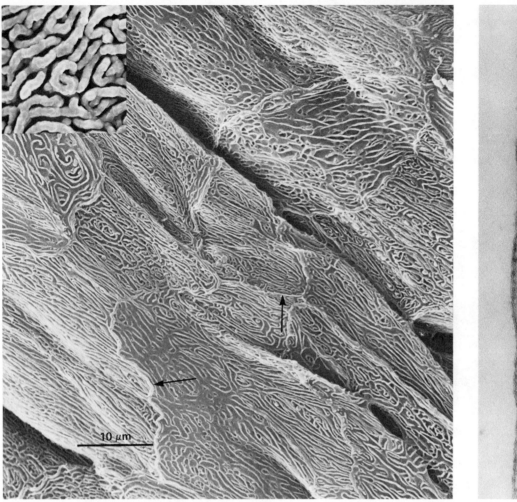
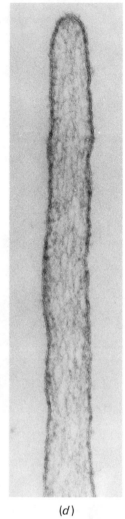

(c)   (d)

FIGURE 9-39 (c) Ridgelike folds called *microplicae* from the inner cheek of a rhesus monkey. Cell boundaries marked by *arrows*. INSET shows the microplicae to better advantage. (d) Microvillus (diameter about 0.12 μm) extending from the cell body of a neuron (actually cultured neuroblastoma) showing a core of microfilaments. Microfilaments are found generally at the core of all sorts of small surface extensions. [Micrographs courtesy of (a) D. Billen and A. Olson, *J. Cell Biol.*, **69**: 732 (1976); (b) P. Armstrong and J. Lackie, *J. Cell Biol.*, **65**: 439 (1975); (c) Peter Andrews, *J. Cell Biol.*, **68**: 420 (1976); (d) J. Ross, J. Olmsted, and J. Rosenbaum, *Tissue and Cell*, **7**: 107 (1975).]

dendrites) have a large number of microtubules. What is more, agents that are known specifically to depolymerize microtubules—low temperature, high pressure, or the drug colchicine—cause retraction of nerve cell processes, which then grow out again when conditions are favorable for microtubule repolymerization. A comparable rounding of other cells in the presence of colchicine or hydrostatic pressure has also been seen, implying a wide role for microtubules in determining the grosser aspects of cell shape.

Microtubules also form the core of certain processes peculiar mostly to protozoa. For example, in the class of protozoa called suctorians, the mature animal has no cilia but does have long tentacles for capturing prey. These tentacles have a highly ordered core (also called an axoneme) of microtubules that changes configuration in a specific way during the process of contraction and expansion of the tentacles. This role is, of course, reminiscent of the role microtubules play in cilia and flagella (see Fig. 9-41).

SECTION 9-7
Cell Shape and
Cell Movement

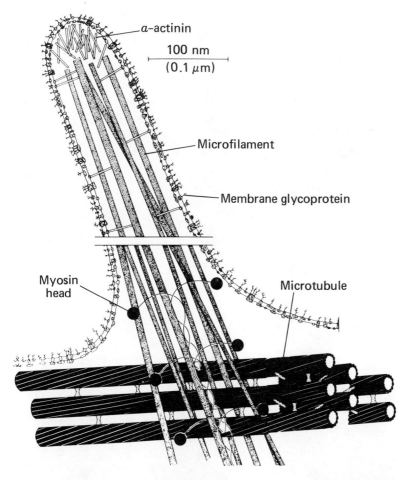

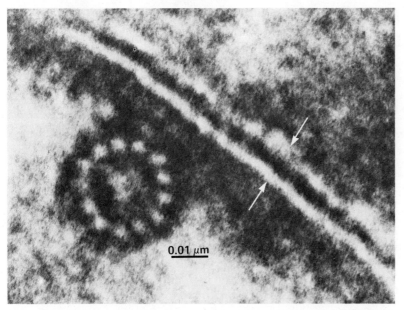

FIGURE 9-40 The Relative Size of Cytoskeletal Components. ABOVE: The diagram shows the ultrastructure of the tip and base of a microvillus (i.e., with the middle cut out as shown). Major components are drawn to scale. Note glycoproteins (with abbreviated carbohydrate chains) embedded in the plasma membrane and connections between microfilaments and membrane proteins. Alpha actinin forms a non-muscle "Z line" to anchor the microfilaments. Myosin molecules are shown distributed along the base of the microfilaments. Microtubules run parallel to the cell surface. LEFT: The micrograph is a cross section through the plasma membrane (*arrows*) and an adjacent microtubule from a plant (juniper). Note the relative sizes. [(a) Courtesy of Francis Loor. Adapted from *Nature,* **264**: 272 (1976). (b) Courtesy of M. C. Ledbetter, *J. Agr. Food Chem.,* **13**: 405 (1965), copyright by the American Chemical Society.]

CHAPTER 9
Contractility and Motility

FIGURE 9-41 Tentacles of a Protozoan. (a) The suctorian *Heliophyra erhardi.* Phase contrast micrograph showing tentacles, mostly expanded. (b) Cross section through the midregion of an expanded tentacle shaft. (c) Cross section at the base of a contracted tentacle. Note the way in which the array of microtubules collapses. Note also the bridges between microtubules, reminiscent of the dynein arms seen in cilia. [Courtesy of M. Hauser and H. Van Eys. (b) and (c) from *J. Cell Sci.,* **20**: 589 (1976).]

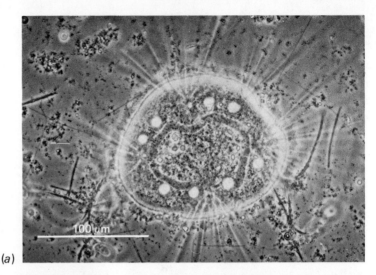

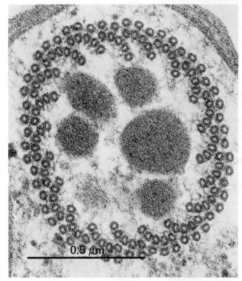

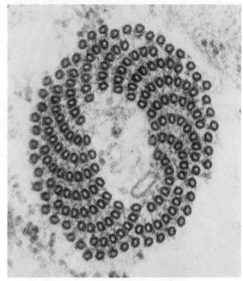

Regulation of Cytoskeletal Changes. If cytoskeletal components are responsible for intracellular movement and changes in shape, it should be possible to identify factors that alter these components and thereby regulate the changes for which they are responsible. Unfortunately we only have clues, no details, to the regulatory processes involved.

The probable regulation of non-muscle actin-myosin interactions via changes in Ca^{2+} was mentioned in Section 9-5. The same ion also has effects on microtubules—a little is required for polymerization, too much causes depolymerization. Thus, calcium-transporting ionophores have been used to show that microtubules of certain protozoan processes can be reversibly disassembled and reassembled by changing Ca^{2+} levels; these changes are accompanied by collapse and regrowth of the processes.

Other examples of Ca^{2+}-mediated changes in cell shape and activity will be noted in the next two chapters. In many of these examples, the level of Ca^{2+} is controlled by the cell membrane, with an increased permeability following stimulation permitting Ca^{2+} influx. In other cases the changes seem to be mediated by cyclic nucleotides, which may have direct effects on cytoskeletal components (e.g., by phosphorylation of their proteins) or which may affect Ca^{2+} levels through changes in permeability of mitochondrial or cell membranes.

It is known, for example, that cyclic AMP affects the shape and motility of a variety of cultured cells. Some cells elongate in the presence of added cAMP to take on the spindle-shape of a fibroblast, others are caused to flatten. A common factor in these and other changes is an altered intracellular distribution of microfilaments and microtubules. Specifically, cAMP causes an increase in both microfilaments and microtubules near the plasma membrane. It also causes microtubule accumulation in parallel arrays near the center of the long processes of those cells that assume a spindle shape (see Fig. 9-42). Whether these changes

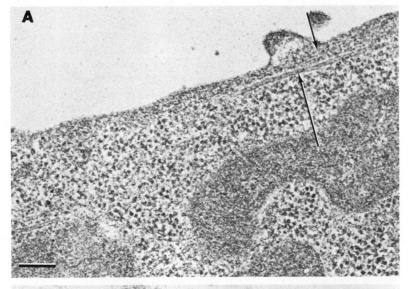

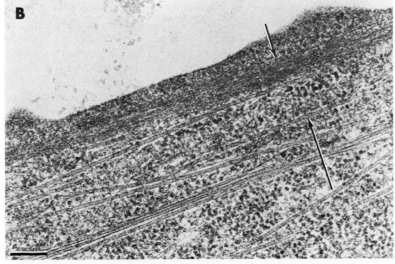

FIGURE 9-42 The Effect of Cyclic AMP on Microtubules and Microfilaments. (A) Control cell (cultured line L929) without added cAMP. (B) Same type of cell after 24-hour exposure to dibutyryl cyclic AMP, a lipid-soluble cAMP derivative that penetrates the cell better. Note the many microtubules (long arrows) and bundles of microfilaments (short arrows). Bar = 0.2 μm. [Courtesy of M. C. Willingham and Ira Pastan, J. Cell Biol., **67**: 146 (1975).]

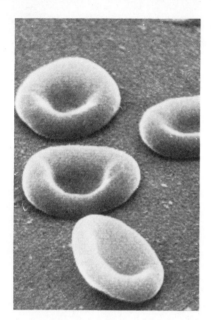

FIGURE 9-43 Human Erythrocytes. [Scanning electron micrograph courtesy of Dr. R. F. Baker.]

occur as a direct result of the presence of cAMP or through its effects on Ca^{2+} levels, however, is uncertain.

The Erythrocyte. The mature mammalian erythrocyte is a unique cell in many respects. It contains no nucleus, no internal organelles, and has a highly characteristic biconcave shape (see Fig. 9-43). There has long been speculation on the mechanism whereby this shape is maintained. The nucleated erythrocytes of non-mammalian vertebrates have a band of microtubules that appear to be responsible for their shape, which is oblong rather than biconcave (see Fig. 9-44a). The shape of the mammalian erythrocyte, however, seems to be a property of its membrane proteins, a situation that is not unique to these cells (see Fig. 9-44b, c).

When human erythrocytes are hemolyzed and the "ghosts" examined, it is found that a long, slender, phosphate-containing protein named *spectrin* (from "spector," meaning ghost) comprises 20–25% of the total membrane protein. Spectrin seems to be distantly related to muscle myosin—its polypeptide chains are about the same size as the large chains of myosin; the proteins appear to have similar shapes; spectrin has weak Ca^{2+}-stimulated ATPase activity; and it is reported to cross-react weakly with antibodies to smooth muscle myosin. Spectrin also interacts with actin. However, it seems to impede actin polymerization and may account for why long microfilaments are not seen in erythrocytes in spite of the demonstrated presence of actin in those cells.

Spectrin forms a fuzzy coat that lines the inside (cytoplasmic) surface of the erythrocyte membrane. Although it is not a true membrane protein, it seems to control the location of membrane proteins, presumably by linking together their cytoplasmic ends. Thus when the spectrin network on the inside of the cell is disturbed, protein-attached carbohydrates on the outer surface also rearrange. This cross-linking of membrane proteins by spectrin—probably also involving in some way the cell's actin—is thought to be responsible for the biconcave shape of the mammalian erythrocyte. The musclelike properties of spectrin may also be responsible for the great deformability of the cell. Although the diameter of the erythrocyte is about 7 μm, it easily squeezes through capillaries only 2 μm in diameter and, until it gets old and stiff, even through 1-μm openings in the spleen.

A number of experiments and observations support the above role for spectrin. For instance, when erythrocytes are heated gradually, they suddenly become spherical at 48–50 °C. This is the temperature at which spectrin (and apparently only spectrin) denatures. And in a condition called *hereditary spherocytosis* the red cells are rigid and round instead of flexible and biconcave. They therefore get trapped in the spleen, causing a severe anemia for which the only treatment at present is removal of the spleen. These cells contain a spectrin that fails to aggregate as readily as does normal spectrin. Other hereditary conditions involving abnormal spectrin behavior include certain types of *muscular dystrophy* (which may mean that a spectrinlike system exists in muscle cells) and a condition called *stomatocytosis*. Stomatocytes are cup-shaped, leaky erythrocytes in which there is an apparent defect in spectrin cross-linking. Remarkably, chemicals that artificially cross-link proteins *in vitro* seem to partially replace the spectrin and return some of these stomatocytes to near-normal configuration and permeability.

As with so many of the other previously described systems, intracellular Ca^{2+} seems to be the key to the normal functioning of erythrocyte

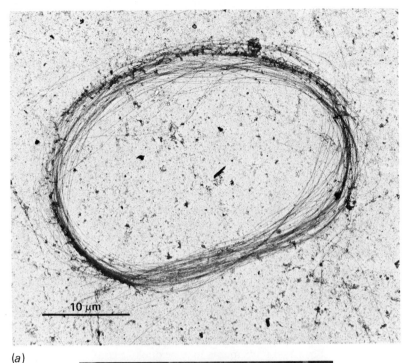

FIGURE 9-44 Determinants of Shape. The shape of a cell can be determined by underlying structural elements or in some cases by the membrane itself or closely associated proteins. (a) Isolated microtubular band from a newt erythrocyte. This band, which is not found in mammalian erythrocytes, is presumably responsible for the oval shape of the cells. (b) *Euglena*, a unicellular organism having both plant and animal characteristics. (c) The detail shows a cross section of its cell membrane (pellicle). This unique shape is maintained in isolated fragments even when the underlying microtubules (arrows) are removed, suggesting a role for membrane-associated proteins in establishing shape. [Courtesy of (a) B. Bertolini and G. Monaco, *J. Ultrastruct. Res.*, **54**: 59 (1976). © Academic Press, N.Y.; (b and c) C. Hofmann and G. B. Bouck, *J. Cell Biol.*, **69**: 693 (1976).]

spectrin. ATP-requiring Ca^{2+} pumps normally keep the cytoplasmic level very low. Starvation or calcium-transporting ionophores raise this level and reduce deformability even before any change in cell volume. If caught early the changes are reversible, but high levels of Ca^{2+} seem eventually to cause irreversible polymerization of spectrin. This phenomenon may explain why some erythrocytes from patients with sickle cell anemia or sickle cell trait (see p. 187) do not regain their original shape after an episode of sickling; stretching of the membrane into the sickled configuration may cause it to become leaky and admit too much Ca^{2+}.

The extent to which spectrinlike systems operate in other cells is at present unknown, but there have been many reports of similar proteins isolated from a variety of sources. It seems likely, therefore, that some of the observations made on mammalian erythrocytes will be applicable to other cell types as well.

Cytoplasmic Streaming. This term refers to the spontaneous and continuous movement of cell organelles and other inclusions. It is easiest to observe in plant cells having a large central vacuole (see Fig. 1-23), for in these cells the streaming part of the cytoplasm (the *endoplasm*) is limited to a thin layer trapped between the vacuole and the organelle-free *ectoplasm* just under the cell membrane. Organelles in the endoplasm seem to move endlessly around the cell, a circular pattern that is also described as *cyclosis*. When measured, streaming velocities in some of these cells have been recorded at rates up to 80 μm/s, which is greater than the maximum speed of movement of actin filaments past myosin in striated muscle.

Many proposals have been put forth to explain cytoplasmic streaming. For example, the fastest movements may be due to a continuous wave of Ca^{2+}-controlled contraction that passes through the ectoplasm, pushing endoplasm ahead of it much as food is forced through intestines by a peristaltic wave. In other cases, organelles may be pulled by an actin-myosin interaction. (Other components of the cytoplasm would be dragged along passively.) Pull would be against anchors attached to the ectoplasmic web. In fact, the required organelle-attached filaments have actually been observed (see Fig. 9-45). There are, of course, many other

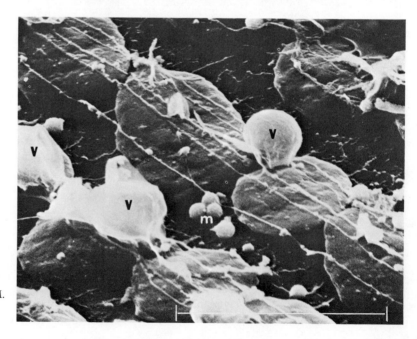

FIGURE 9-45 Microfilaments and Cytoplasmic Streaming. Streaming in green algae and higher plants seems to occur along microfilament bundles, seen here attached to a row of chloroplasts from the alga *Chara australis*. Vesicles (v); mitochondria (m). Bar = 10 μm. [Courtesy of Y. M. Kersey and N. K. Wessells, *J. Cell Biol.*, **68**: 264 (1976).]

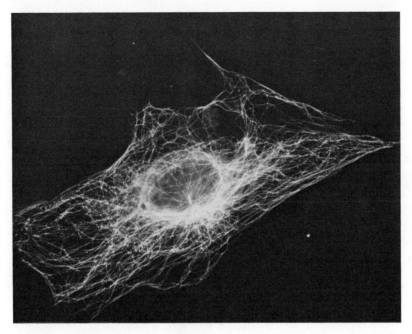

FIGURE 9-46 Cytoplasmic Microtubules. Indirect immunofluorescence was used to locate tubulin in a cultured animal cell. Note the complex cytoplasmic network. Note also the greater density near the nucleus. The major microtubule organizing center in these cells seems to be the centrioles or a nearby structure. Microtubules grow from these centers toward the plasma membrane. [Courtesy of M. Osborn and K. Weber, *Proc. Nat. Acad. Sci. U.S.*, **73**: 867 (1976).]

feasible mechanisms, some of which involve translocations by microtubules and 100-Å filaments instead of microfilaments.

Microtubules are, of course, present in most if not all cells, and they are widely distributed in the cytoplasm (see Fig. 9-46). The best evidence for their capacity to participate in cytoplasmic streaming comes from the rather specialized version of that phenomenon known as *axoplasmic transport*. This term refers to the translocation of organelles (including at least some of the synaptic vesicles) and proteins from the cell body of a neuron where they are produced out the axon toward its tip. There is also a corresponding return flow of materials to be recycled.

Axoplasmic transport has a slow and a fast component, and probably uses separate mechanisms for each. The slow transport system moves at only 1–3 mm/day. There is some indication that much of this slowly moved material consists of structural proteins (tubulin, for example), a proposal that agrees with the observation that a cut axon regrows from the cell body at a rate comparable to that of slow axoplasmic transport. (Try figuring out how long it would take to get feeling back in your big toe after severing the appropriate sensory nerve near your spine.) The mechanism of slow axoplasmic transport is not understood, but may involve only diffusion.

Fast axoplasmic transport, on the other hand, takes place at 100–400 mm/day, utilizes ATP generated locally along the axon, and is readily inhibited by colchicine. The latter observation suggests an involvement by microtubules, a proposition for which there is now rather direct evidence in the form of micrographs showing close association and sometimes actual contacts between axonal microtubules and organelles that are presumably being transported (see Fig. 9-47).

The arms that link axonal organelles to microtubules remind us of the dynein arms on the microtubular structures of cilia and flagella. Since dynein catalyzes an ATP-powered sliding of one tubule past another, comparable proteins on cytoplasmic microtubules might pass an

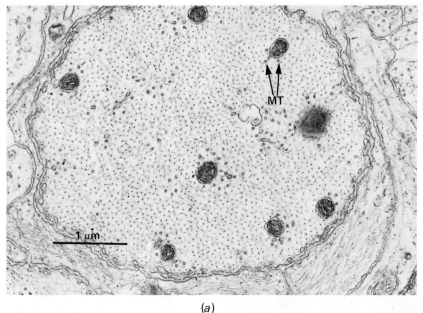

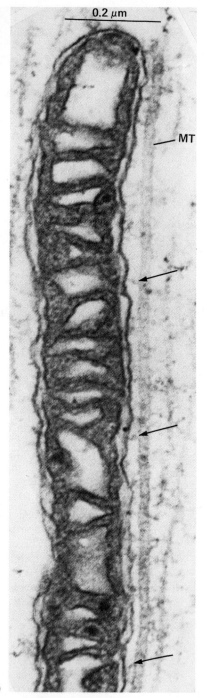

FIGURE 9-47 Axoplasmic Transport. (a) Cross section of an axon from a lamprey showing microtubules (MT) in close proximity to mitochondria. The latter may well be in transit down the axon, assisted in the process by microtubules. Other microtubules are scattered throughout the axoplasm, which is also studded with numerous 100-Å (10 nm) filaments, called in this context *neurofilaments*. (b) Detail from a longitudinal section, apparently showing bridges (*arrows*) between a mitochondrion and adjacent microtubule. [Courtesy of D. S. Smith, U. Järlfors, and B. Cameron, *Ann. N.Y. Acad. Sci.*, **253**: 472 (1975).]

organelle from one of these arms to the next and hence be responsible for axoplasmic transport. Projections of this sort are seen also on the microtubules of protozoan tentacles (Fig. 9-41), where again they are implicated in movement, and on reconstituted microtubules *in vitro* (Fig. 9-48). In the latter case, they have been identified as the "high molecular weight" components known to facilitate tubulin polymerization.

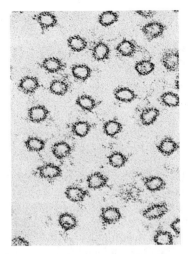 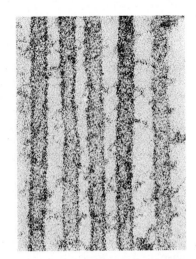

FIGURE 9-48 Microtubules *in vitro*. Cross section and longitudinal section through microtubules repolymerized *in vitro*. The arms project at 32.5-nm intervals from the microtubular walls and have been identified as a "high molecular weight" protein complex that copurifies with tubulin. [Courtesy of D. Murphy and G. Borisy, *Proc. Nat. Acad. Sci. U.S.*, **72:** 2696 (1975).]

An alternate hypothesis for axoplasmic transport involves not microtubules but the 100-Å filaments that are so numerous in axons (Fig. 9-47). This proposal lacks the same degree of experimental support, however, and for the present it seems more likely that 100-Å filaments in axons play some sort of structural role. On the other hand, there are situations in which 100-Å filaments do seem to be likely candidates for organelle translocation. One of these situations is melanosome transport in human skin.

Melanosomes are membrane-enclosed pigment granules that are produced by cells called *melanocytes*. Melanocytes are found in the same concentration in human skin of all colors although the amount of pigment, *melanin*, that they produce is highly variable from the very dark skin of certain African tribes to the colorless skin of *albinos*, which produces little or no melanin.[4] Comparable cells, more often called *chromatophores*, are also responsible for the rapid and reversible color changes of many lower animals; to cause a color change in the surrounding tissue, chromatophores send their pigment granules away from the cell body out into long cellular processes called dendrites. The granules can also be recalled again in order to change back to the original color.

Microtubules have been repeatedly implicated in pigment granule movement by chromatophores. Human melanocytes, however, behave a little differently and may use a different transport system. For one thing, human melanocytes do not ordinarily recall their melanosomes. After stimulation (by sunlight, for example), melanosomes travel out the dendrites and are then passed to adjacent epidermal cells (*keratinocytes*) where they will remain until the keratinocytes mature, die, and are sloughed a month or so later, taking your tan with them. This one-way dispersion of melanosomes by human melanocytes is accompanied by a parallel dispersion of bundles of 100-Å filaments that otherwise occupy a position near the nucleus, as they do also in many other cells. These 100-Å filaments are thought to be somehow responsible for the movement of melanosomes in humans. Unfortunately, we don't know enough about these filaments to propose a mechanism, but it is clear that in discussing contractile and motile systems, such as cytoplasmic streaming, we must keep in mind the possibility that 100-Å filaments may in some instances take an active part.

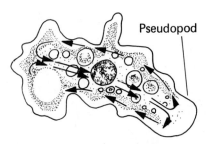

FIGURE 9-49 Ameboid Locomotion. A forward-streaming, contracting endoplasm is converted to a gellike state in the pseudopod. A given particle works its way back through the ectoplasmic layer to be resolubilized in the "tail" (via relaxation of contractile elements) and then brought forward again.

Ameboid Locomotion. This term includes a variety of different movements. Although the classical example is the movement of *Amoeba proteus,* other protozoa, many mammalian cells, and some plant cells can exhibit much the same pattern of locomotion.

The anterior, or forward tip of a moving *Amoeba proteus* consists of a large pseudopod having a clear, microfilament-containing hyaline cap (*hyalos* = glass). This cap is continuous with a narrow *hyaline ectoplasm* that coats the cytoplasmic side of the cell membrane. Just inside the hyaline ectoplasm is a layer of *granular ectoplasm,* and inside the latter is a core of endoplasm that can be seen streaming forward toward the anterior pseudopod. Just before this stream of endoplasm reaches the hyaline cap, however, it sprays out like a fountain to merge with the granular ectoplasm (Fig. 9-49).

A particle carried forward in this endoplasmic stream appears to become incorporated into the granular ectoplasm when it reaches the forward tip. Then as more material is brought forward, the particle gets displaced and works its way back toward the "tail" or *uroid* region of the cell (see Fig. 9-49). If the ameba has a substrate to which it can attach, the streaming just described will result in forward movement. The pattern is like that of the treads on a tank or tractor: a given point moves forward at twice the overall speed of the machine, then becomes attached to the ground while the machine moves past it, and finally is picked up and moved forward again. Some sort of contractile system must, of course, be responsible for this movement. In fact, isolated cytoplasm continues to stream in this way as long as Ca^{2+} and Mg-ATP are present, and may even move across a surface.

R. D. Allen and D. L. Taylor, working in part at the Marine Biological Laboratory in Woods Hole, Massachusetts, have formulated a model that appears to explain the major features of ameboid locomotion. They view the cytoplasm as undergoing a cycle between fluid and gel, corresponding to a relaxed and contracted state, respectively, of cytoplasmic actomyosin. Contracted, gellike ectoplasm in the tail is relaxed and solubilized to become part of the stream of endoplasm. It is then pulled forward as it contracts again, reaching the fully contracted, gellike state when it is reincorporated into the ectoplasm at the front of the cell. The control over this cycle must be the availability of Ca^{2+}—low at the tail to facilitate relaxation and increasing toward the front, thus causing contraction.

Many of the cells from higher organisms are, of course, confined to tissues and thus do not have the opportunity to exhibit the kinds of movements just described. During embryonic development, however, or when cells are isolated and grown in tissue culture, such movements are common. Even in adult mammals, we can find a number of examples of ameboid locomotion—fibroblasts as they move over a wound and lay down collagen to form a scar, the neutrophils and macrophages (white blood cells) that leave the bloodstream to fight infections, and so on. Although the appearance of these cells during movement differs from that of amebae (they typically use broad, flat *lamellipodia* such as those seen in Fig. 9-39b instead of the rounded pseudopods of amebae), the streaming pattern is much the same.

Chemotaxis. Ameboid locomotion is not necessarily random. Most motile cells appear able to respond to various external chemical agents

SECTION 9-7
Cell Shape and
Cell Movement

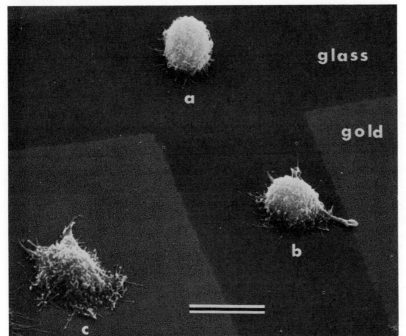

(a)

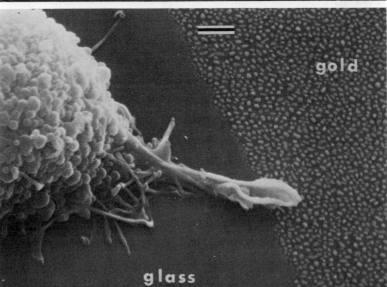

(b)

FIGURE 9-50 Surface Preference of Mouse Fibroblasts. Cultured cells (3T3 cells) were allowed to settle on a glass surface partly coated with gold. (a) 30 min after plating, cell **a** is still rounded. A filopodium of cell **b** has contacted the gold and a lamellipodium is now forming. Cell **c** has spread onto the gold surface and appears to be no longer motile. Bar = 20 μm. (b) Higher magnification, showing ruffling of the extending lamellipodium. Bar = 2 μm. [Courtesy of G. Albrecht-Buehler, *J. Cell. Biol.*, **69**: 275 (1976).]

by extending pseudopodia (or lamellipodia) preferentially on the side of the cell sensing the highest concentration of the agent, thereby assuring that the cell will move toward the stimulus. This directed movement is called *chemotaxis*. Chemotaxis helps amebae find food, causes neutrophils and macrophages to accumulate at the site of infection (peptides released from the serum proteins called complement may be the major attractants), and probably helps to shape a growing embryo.

Motile cells also seem able to explore their environment physically and to move preferentially toward a more desirable surface. The cultured mouse fibroblasts shown in Fig. 9-50, for example, send out long, thin filopodia, seemingly as feelers. For some reason, these cells prefer gold to glass. Hence, when a filopodium contacts a gold-studded surface, a lamellipodium is extended in the same direction and the cells congregate on the gold, a behavior that should be readily appreciated by humans.

SUMMARY

9-1 All muscle may be classified as either striated or smooth, depending on its appearance in the polarizing light microscope. In general, voluntary muscles of vertebrates are striated while involuntary muscles (except for the heart) are smooth. Striations result from the organization of filaments (myofilaments) into discrete bundles, or myofibrils. Each myofibril is further divided into segments (called sarcomeres) by thin, dark Z lines (or Z discs). Within each sarcomere one finds two kinds of filaments, thick and thin. Thick filaments occupy the A zone in the middle of the sarcomere. Thin filaments extend into the A zone from each Z disc, interdigitating with thick filaments.

The cytoplasmic membranes of striated muscle are divided into two components, longitudinal and transverse. The longitudinal component runs parallel to the myofilaments and has periodic enlargements or cisternae. The transverse system is a periodic set of tubules opening onto the surface of the cell. Where the systems cross, one frequently sees two cisternae flanking a tubule, a configuration that is called a triad.

Smooth muscle has very little reticulum, and also differs from striated muscle (except cardiac muscle) in being mononucleate rather than a syncytium.

Invertebrate muscle differs from vertebrate muscle in appearance (e.g., it has oblique striations) and often in filament structure, with large paramyosin filaments replacing the thick (myosin) filaments of vertebrate muscle.

9-2 The thick filaments of myofibrils are composed of myosin, which is a large, highly elongated molecule with a thickened, bilobed "head" at one end. The molecules are oriented within a filament with their heads away from the center leaving a bare, "headless" central shaft. The thin filaments are basically two strands of a smaller, globular protein called actin. The strands are wound about each other, leaving a groove on either side of the resulting helix. These grooves may be occupied by rodlike tropomyosin molecules oriented end to end and separated by globular (spherical) troponin molecules.

According to the sliding filament theory, contraction is achieved by moving filaments past one another. Movement must be accomplished by a cyclical extension and contraction of the interfilament contacts, or bridges, which are known to be part of the myosin heads, accompanied by ATP hydrolysis.

Careful measurements of the energetics of contraction confirm the essential aspects of the sliding filament theory. For example, the sliding filament model predicts that tension should be directly proportional to the number of myosin-actin bridges, and hence to the degree of overlap between the two kinds of filaments, exclusive of the bare central shaft of myosin. Observations of tension as a function of sarcomere length follow these predictions quite closely.

The sliding filament model of contraction applies also to smooth muscle, though the less ordered arrangement of filaments makes activity harder to interpret. In addition, smooth muscle has an abundance of intermediate-sized, 100-Å (10-nm) filaments. Their function is not known for certain, though they may play a structural role in supporting and organizing the other filaments.

9-3 Intact muscle maintains a nearly constant ATP level during exercise, at least until a state of near-exhaustion is reached. This is accomplished through a replacement of the ATP used, mainly from the creatine kinase (or equivalent invertebrate) reaction but additionally from the adenylate kinase (myokinase) equilibrium. The latter is an effective way of tapping most of the available pyrophosphate-type linkages, since the equilibrium can be pulled toward ATP by removing AMP through deamination to inosine monophosphate.

Replacement of energy stores is achieved during periods of rest. The metabolic mechanism, however, depends on the type of cell. For example, a fully differentiated striated fiber can be classed as red or white. Red and white fibers are distinctly different from each other in color and size, and in their metabolism and contractile characteristics. This aspect of the differentiation of a muscle cell depends on which nerves innervate it, as one can demonstrate by switching nerves between a red and white muscle.

9-4 Contraction is initiated by a depolarization of the sarcolemma. Depolarization in striated cells is conducted into the interior by transverse tubules, and is transmitted to the longitudinal system at the point where the two systems cross (often at a triad), causing Ca^{2+} to be released from the longitudinal system. The free ions are then bound by troponin, which is associated with tropomyosin on the actin filaments. In the presence of Ca^{2+}, these proteins lose their capacity to prevent the cooperative actin-myosin interaction that causes ATP hydrolysis and contraction. This control by troponin is replaced (or augmented) in certain invertebrate muscles, and apparently in some vertebrate smooth muscle, with a control system that uses one of the myosin light chains as a Ca^{2+} sensor. In any case, relaxation occurs when Ca^{2+} is removed by calcium pumps in the sarcoplasmic reticulum and (especially in some smooth muscle) by mitochondria.

A single action potential in striated fibers initiates the above sequence and produces a twitch. Repeated action potentials are required in these cells to sustain a contraction. This is not so in many smooth and invertebrate muscles, which can maintain tension via fixed actin-myosin cross-bridges even between contractions.

Smooth muscle also differs from skeletal muscle (not so much from cardiac muscle) in being responsive to a variety of hormones and peptides. These substances regulate Ca^{2+} availability to the myofibrils, in some cases via cyclic nucleotide intermediates.

9-5 Proteins that are closely related to the major proteins of muscle sarcomeres (actin, myosin, tropomyosin, and α-actinin) have been found in a variety of non-muscle cells. They have been localized to long "stress fibers" that come and go within the cytoplasm and which may therefore constitute the contractile system of non-muscle cells. The exact organization is not clear—actin filaments (i.e., microfilaments) are much the same, but continuous myosin filaments are rarely seen. The properties of the overall system, however, including sensitivity to Ca^{2+}, are much the same as one would expect based on experiments with muscle.

9-6 The cilia and flagella of eukaryotes are essentially identical structures. They are composed of an axoneme that consists in nearly all cases of nine outer microtubular doublets and a central pair of single microtubules ("9 + 2"). Motion in these structures, which is a wavelike beating, is conferred by microtubules that slide past each other, powered by ATP hydrolysis and catalyzed by dynein (seen as arms on subfiber A of each doublet). Dynein, like myosin, converts chemical to mechanical energy. The beating continues at a more-or-less constant rate, though in some systems changes in Ca^{2+} availability can alter the pattern of the beat.

Cilia and flagella are organized by basal bodies, the structure of which is nine outer triplets without the central pair ("9 + 0"). Basal bodies and centrioles are identical structures put to different uses.

The eukaryotic axoneme has no prokaryotic counterpart. Bacterial flagella, for example, appear to be relatively simple, helical structures that propel the cell via rotations instead of by beating.

9-7 The cytoplasmic elements responsible for cell shape are the microfilaments and associated proteins (myosin, tropomyosin, etc.), probably the 100-Å (10 nm) filaments, and certainly microtubules. These components are collectively referred to as the cytoskeleton. Small changes in cell shape seem most often to involve microfilaments alone, whereas major changes (formation of nerve cell axons, for example) clearly involve microtubules as well. A possible exception, though it most nearly resembles the actin-myosin system, is maintenance of the biconcave shape of mammalian erythrocytes by a network of membrane-associated components consisting mostly of a protein called spectrin.

Cell movements seem also to rely on cytoskeletal filaments and tubules. Cytoplasmic streaming and ameboid locomotion, for example, are most often attributed to an actin-myosin system, although specific cases of organelle translocation have also been attributed to microtubules and to 100-Å filaments.

The control over most movements and changes in shape seems to be availability of Ca^{2+}, at least when the actin-myosin or microtubular systems are involved.

STUDY GUIDE

9-1 (**a**) Define the terms myofilament, myofibril, sarcoplasmic reticulum, terminal cisternae, T-system, and sarcomere. (**b**) The sarcoplasmic membranes of striated cells have two unconnected components. What are they, and how are they related? (**c**) How do striated and smooth muscle cells compare (1) in the arrangement of myofilaments? (2) in the extent of their reticulum? (3) in function? (**d**) What will happen to the size of the I, A, and H zones as a sarcomere shortens when the two kinds of filaments slide past each other? (**e**) How does the structure of invertebrate muscle differ from that of vertebrates?

9-2 (**a**) What is actomyosin, and what relevance

does it have to the study of muscle contraction? (b) What is the probable physical relationship of troponin, tropomyosin, and actin in the thin filaments? (c) What is the structure of the myosin filaments? (d) The myosin filaments have a bilateral symmetry about their midpoints. The actin filaments are symmetrical about the Z lines. What is the importance of these configurations to the contractile mechanism? (e) What features are unique to smooth muscle cells? (f) Suppose that the myosin filaments have a length of 1.5 µm and a "headless" central shaft of 0.2 µm, and that the actin filaments are each 1.1 µm in length. What is the longest sarcomere length at which *maximum* tension can be generated? [Ans: 2.4 µm] (g) Using the dimensions from part (f), determine what will be the fraction of maximum tension generated at sarcomere lengths of (1) 3.7 µm; (2) 3.5 µm; (3) 3.0 µm. [Ans: 0, 0.15, 0.54]

9-3 (a) Consider only the reaction

$$\text{Cr-P} + \text{ADP} \rightleftharpoons \text{Cr} + \text{ATP} \qquad K_{eq} = \frac{[\text{Cr}][\text{ATP}]}{[\text{CrP}][\text{ADP}]}$$

where Cr represents creatine and the brackets indicate molar concentrations. A $\Delta G°$ of -12.6 kJ/mole for this reaction is equivalent to $K_{eq} = 132$ at 37 °C. Neglecting AMP, calculate the fraction of adenosine nucleotides in the form of ATP (1) when 90% of the creatine is phosphorylated, (2) when 50% is phosphorylated, and (3) when 10% is phosphorylated. [Ans: 99.9%; 99.2%; 93.6%] (b) How does the deamination of AMP make it possible to prolong a period of muscular exertion? (c) Frogs, fish, and rabbits are examples of animals in which the muscles tend to be very white. Dogs and ducks, on the other hand, tend to have many red fibers. Can you identify any behavioral patterns of these two groups of animals that might have led you to predict these differences?

9-4 (a) What are the various stimuli by which a contraction may be initiated? (b) Early attempts to find a relaxing factor—a substance that would terminate a contraction—resulted in the isolation of a material that did have such an effect when injected into a contracted muscle cell. The substance (called the Marsh factor, or Marsh-Bendall factor) looks like membrane fragments in the electron microscope. What do you think it is, and how does it work? (c) How do the contractile properties of smooth muscle differ from those of striated cells, and what are some possible reasons for the differences? (d) Frog muscle undergoes a prolonged contracture when bathed in concentrated KCl solutions provided Ca^{2+} is also present, but not if Ca^{2+} is absent. Explain both the K^+-dependent and Ca^{2+}-dependent aspects of this observation. (e) It has been found that a portion of the Ca^{2+} in striated muscles can be readily exchanged for other ions during rest, and that still another fraction is available for exchange only when the muscle contracts. Explain. (f) How does a tonic contraction differ from a tetanus? (g) What is the probable mechanism of hormonal control of smooth muscle? How do epinephrine and acetylcholine affect cardiac cells?

9-5 (a) Which muscle proteins have been demonstrated to exist in non-muscle cells and what are their presumed functions? (b) Describe the technique of direct and indirect immunofluorescence. (c) Where are the cortex and ectoplasm of a cell? (d) Describe two calcium-sensitive non-muscle contractile systems that have been studied *in vitro*. (e) Describe the system for controlling Ca^{2+} levels in the cytoplasm.

9-6 (a) Describe the structure of the axoneme. (b) What features distinguish subfiber A from subfiber B? (c) Describe the sliding microtubule model of axonemal movement. (d) Flagella and cilia, though identical in structure, commonly exhibit a quite different pattern of movement. Describe the two patterns and conditions under which one or the other would be more appropriate. (e) Describe the basal body and its role in ciliogenesis. (f) How do bacterial and eukaryotic flagella differ from each other in structure and movement?

9-7 (a) What are the major elements of the so-called cytoskeleton and in what kinds of movements are each thought to be involved? (b) What systems appear to determine shape in mammalian and non-mammalian erythrocytes, respectively? (c) What is meant by cytoplasmic streaming? (d) Describe ameboid locomotion as it occurs in *Amoeba proteus*, and the model proposed to explain it. (e) How may Ca^{2+} control such things as cytoplasmic streaming and ameboid locomotion? What is the possible involvement of cyclic AMP? (f) What is chemotaxis?

REFERENCES

MUSCLE CELLS: STRUCTURE, METABOLISM, AND STIMULATION

(Also see volume **37** (1972) of the *Cold Spring Harbor Symp. Quant. Biol.*)

ALTURA, et al., "Comparative Cellular and Pharmacologic Actions of Neurohypophyseal Hormones on Smooth Muscle." *Fed. Proc.*, **36**: 1840 (1977). Symposium on the evolution and action of oxytocin and associated pituitary hormones.

BENDALL, J. R., *Muscles, Molecules and Movement. An Essay in the Contraction of Muscles.* New York: American Elsvier, 1968. A fine introductory text, concentrating on the energetics of contraction.

CARLSON, F. D., and D. R. WILKIE, *Muscle Physiology.* Englewood Cliffs, N.J.: Prentice-Hall, 1973. A small book, covering initiation, control, and energy conversion in muscle.

CLOSE, R. I., "Dynamic Properties of Mammalian Skeletal Muscles." *Physiol. Rev.,* **52**: 129 (1972). Properties of red and white muscle.

COOKE, PETER, "A Filamentous Cytoskeleton in Vertebrate Smooth Muscle Fibers." *J. Cell Biol.,* **68**: 539 (1976). 100 Å filaments.

FAY, F., and C. DELISE, "Contraction of Isolated Smooth-Muscle Cells—Structural Changes." *Proc. Nat. Acad. Sci. U.S.,* **70**: 641 (1975).

GOLDBERG, N. D., "Cyclic Nucleotides and Cell Function." In *Cell Membranes,* G. Weissmann and R. Caliborne, eds. New York: H. P. Publ., 1975, p. 185. Also see *Hospital Practice,* May 1974, p. 127.

HILL, A. V., *First and Last Experiments in Muscle Mechanics.* Cambridge: Cambridge Univ. Press, 1970.

HUDDART, H., and S. HUNT, *Visceral Muscle. Its Structure and Function.* New York: John Wiley, 1975.

HUXLEY, A. F., and R. M. SIMMONS, "Proposed Mechanism of Force Generation in Striated Muscle." *Nature,* **233**: 533 (1971).

HUXLEY, H. E., "The Mechanism of Muscular Contraction." *Science,* **164**: 1356 (1969).

———, "The Mechanism of Muscular Contraction." *Scientific American,* December 1965. (Offprint 1026.)

———, "The Structural Basis of Muscular Contraction." *Proc. Roy. Soc. (London) Ser. B,* **178**: 131 (1971). Croonian Lecture, 1970.

LIPMANN, FRITZ, "Discovery of Creatine Phosphate in Muscle." *Trends Biochem. Sci.,* **2**(1): 21 (Jan. 1977).

MALCOLM, R. Muscle. New York: Halsted Press, 1977. Paperback.)

MARGARIA, RODOLFO, "The Sources of Muscular Energy." *Scientific American,* March 1972. (Offprint 1244).

MOREL, J., and I. PINSET-HARSTROM, "Ultrastructure of the Myofibril and Source of Energy." *Biomedicine,* **22**: 88 (1975).

SALMONS, S., and F. SRÉTER, "Significance of Impulse Activity in the Transformation of Skeletal Muscle Type." *Nature,* **263**: 30 (1976). Red vs. white muscle.

SHOENBERG, C. F., and D. M. NEEDHAM, "A Study of the Mechanism of Contraction in Vertebrate Smooth Muscle." *Biol. Rev. (Cambridge),* **51**: 53 (1976).

SMITH, DAVID S., *Muscle.* New York: Academic Press, 1972. (Paperback.) Primarily concerned with ultrastructure.

STEPHENS, N. L., ed., *The Biochemistry of Smooth Muscle.* Baltimore: Univ. Park Press, 1977.

SZENT-GYÖRGYI, A. G., "Nature of Actin-Myosin Complex and Contraction." In *Physiology and Biochemistry of Muscle as a Food,* Vol. I, E. J. Briskey, R. G. Cassens, and J. C. Trautman, eds. Madison: Univ. of Wisconsin Press, 1966. p. 287.

TAKEUCHI, A., and N. TAKEUCHI, "Actions of Transmitter Substances on the Neuromuscular Junctions of Vertebrates and Invertebrates." *Advan. Biophys.* **3**: 45 (1972). Nice review.

TOIDA, N., H. KURIYAMA, N. TASHIRO, and Y. ITO, "Obliquely Striated Muscle." *Physiol. Rev.* **55**: 700 (1975). Invertebrate muscle.

TOMITA, T., "Electrophysiology of Mammalian Smooth Muscle." *Progr. Biophys. Mol. Biol.,* **30**: 185 (1975).

VERMA, S., and J. MCNEILL, "Cardiac Histamine Receptors and Cyclic AMP." *Life Sci.,* **19**: 1797 (1976).

EXCITATION-CONTRACTION COUPLING

AKERA, T., and T. BRODY, "Inotropic Action of Digitalis and Ion Transport." *Life Sci.,* **18**: 135 (1976). How to get increased contractility from heart muscle by inhibiting the Na^+-K^+ pump.

COHEN, C., "The Protein Switch of Muscle Contraction." *Scientific American,* **233**(5): 36 (Nov. 1975).

COSTANTIN, L. L., "Contractile Activation in Skeletal Muscle." *Prog. Biophys. Mol. Biol.,* **29**: 197 (1975). A review.

EBASHI, S., "Excitation-Contraction Coupling." *Ann. Rev. Physiol.,* **38**: 293 (1976).

ENDO, M., "Calcium Release from the Sarcoplasmic Reticulum." *Physiol. Rev.,* **57**: 71 (1977).

FOZZARD, H., "Heart: Excitation-Contraction Coupling." *Ann. Rev. Physiol.,* **39**: 201 (1977).

HADDY, F. J., et al., "Excitation-Contraction Coupling in Cardiac and Vascular Smooth Muscle." *Fed. Proc.,* **35**: 1272 (1976). Proceedings of a symposium.

HOYLE, GRAHAM, "How Is Muscle Turned On and Off?" *Scientific American,* April 1970. (Offprint 1175.)

KENDRICK-JONES, J., and R. JAKES, "The Regulatory Function of the Myosin Light Chains." *Trends Biochem. Sci.,* **1**(12): 281 (Dec. 1976).

KOTANI, M., ed., *Advances in Biophysics,* Vol. 9. Baltimore: Univ. Park Press, 1977. See the chapter on actinins by Maruyama.

KRETSINGER, R. H., "Evolution and Function of Calcium-binding Proteins." *Int. Rev. Cytol.,* **46**: 323 (1976).

KURIYAMA, H., T. OSA, Y. ITO, and H. SUZUKI, "Excitation-Contraction Coupling Mechanism in Visceral Smooth Muscle." *Advan. Biophys.,* **8**: 115 (1976).

LANGER, G., "Ionic Basis of Myocardial Contractility." *Ann. Rev. Med.,* **28**: 13 (1977).

LEHMAN, W., "Phylogenetic Diversity of the Proteins Regulating Muscular Contraction." *Int. Rev. Cytol.,* **44**: 55 (1975).

LEHMAN, W., and A. SZENT-GYÖRGYI, "Regulation of Muscular Contraction. Distribution of Actin Control and Myosin Control in the Animal Kingdom." *J. Gen. Physiol.,* **66**: 1 (1975).

LEHNINGER, A. L., "Mitochondria and Calcium Ion Transport." *Biochem. J.,* **119**: 129 (1970). Includes their possible contribution to muscle relaxation.

MACLENNAN, D. H., and P. C. HOLLAND, "Calcium Transport in Sarcoplasmic Reticulum." *Ann. Rev. Biophys. Bioeng.,* **4**: 377 (1975).

MURRAY, J. M., and A. WEBER, "The Cooperative Action of Muscle Proteins." *Scientific American,* **230**(2): 58 (Feb. 1974).

PORTER, K. R., and C. FRANZINI-ARMSTRONG, "The Sarcoplasmic Reticulum." *Scientific American,* March, 1965. (Offprint 1007.)

ZACHAR, J., *Electrogenesis and Contractility in Skeletal Muscle Cells.* Baltimore: Univ. Park Press, 1972. Exhaustive but clearly written study on skeletal muscle action.

CILIA AND FLAGELLA

AFZELIUS, B. A., "A Human Syndrome Caused by Immotile Cilia." *Science,* **193**: 317 (1976). Also see *J. Cell Biol.,* **66**: 225 (1975) and *New Engl. J. Med.* **297**: 1 (1977).

BLAKE, J. R., and M. A. SLEIGH, "Mechanics of Ciliary Motion." *Biol. Rev. (Cambridge),* **49**: 85 (1974).

HYAMS, J. S., and G. G. BORISY, "Flagellar Coordination in *Chlamydomonas reinhardtii:* Isolation and Reactivation of the Flagellar Apparatus." *Science,* **189**: 891 (1975). Synchronous beating of the isolated apparatus.

INOUÉ, S., and R. E. STEPHENS, eds., *Molecules and Cell Movement.* New York: Raven Press, 1975. Includes reviews on cilia and flagella.

MOHRI, H., "The Function of Tubulin in Motile Systems." *Biochim. Biophys. Acta,* **456**: 85 (1976).

SALISBURY, G., R. HART, and J. LODGE, "The Spermatozoon." *Perspectives Biol. Med.,* **20**: 372 (1977). Structure and motility.

SATIR, PETER, "How Cilia Move." *Scientific American,* **231**(10): 45 (Oct. 1974).

SLEIGH, M., ed., *Cilia and Flagella.* New York: Academic Press, 1974.

SUMMERS, K., "The Role of Flagellar Structures in Motility." *Biochim. Biophys. Acta,* **416**: 153 (1975).

WIEDERHOLD, M. L., "Mechanosensory Transduction in 'Sensory' and 'Motile' Cilia." *Ann. Rev. Biophys. Bioeng.,* **5**: 39 (1976). A good review of structure and control.

THE CYTOSKELETAL AND CONTRACTILE SYSTEMS OF NON-MUSCLE CELLS

ALLEN, R. D., "Evidence for Firm Linkages between Microtubules and Membrane-Bounded Vesicles." *J. Cell Biol.,* **64**: 497 (1975). Transport by microtubules.

ALLEN, R. D., and D. L. TAYLOR, "The Molecular Basis of Ameboid Movement." In *Molecules and Cell Movement,* ed. S. Inoué and R. E. Stephens. New York: Raven Press, 1975, p. 239.

ARMSTRONG, P., and J. LACKIE, "Studies on Intercellular Invasion *in vitro* Using Rabbit Peritoneal Granulocytes (PMNs)." *J. Cell Biol.,* **65**: 439 (1975). A good description of ameboid locomotion in a mammalian cell.

BHISEY, A., and J. FREED, "Possible Role of Microtubules and Microfilaments in Cell Locomotion." In *Regulation of Growth and Differentiated Function in Eukaryote Cells,* ed. G. P. Talwar. New York: Raven Press, 1975, p. 155.

CLARKE, M., and J. SPUDICH, "Nonmuscle Contractile Proteins: The Role of Actin and Myosin in Cell Motility and Shape Determination." *Ann. Rev. Biochem.,* **46**: 797 (1977).

DANIELS, M., "The Role of Microtubules in the Growth and Stabilization of Nerve Fibers." *Ann. N.Y. Acad. Sci.,* **253**: 535 (1975).

GASKIN, F., and M. SHELANSKI, "Microtubules and Intermediate Filaments." *Essays Biochem.,* **12**: 115 (1976).

GOLDMAN, R., T. POLLARD, and J. ROSENBAUM, eds., *Cell Motility.* Cold Spring Harbor (N.Y.) Laboratory, 1976. In 3 volumes.

HEILMEYER, L., J. RUEGG, and Th. WIELAND, eds., *Molecular Basis of Motility.* New York: Springer-Verlag, 1976.

HESLOP, J. P., "Axonal Flow and Fast Transport in Nerves." *Advan. Comp. Physiol. Biochem.,* **6**: 75 (1975).

INOUÉ, S., and R. STEPHENS, eds., *Molecules and Cell Movement (Soc. Gen. Physiol. Series,* Vol. **30**). New York: Raven Press, 1975. Nonmuscle movement.

JIMBOW, K., and T. B. FITZPATRICK, "Changes in Distribution Pattern of Cytoplasmic Filaments in Human Melanocytes During Ultraviolet-Mediated Melanin Pigmentation." *J. Cell Biol.,* **65**: 481 (1975). Role of the 100-Å filaments.

KORN, E. D., "Cytoplasmic Actin and Myosin." *Trends Biochem. Sci.,* **1**(3): 55 (March 1976).

MANNHERZ, H., and R. GOODY, "Proteins of Contractile Systems." *Ann. Rev. Biochem.,* **45**: 427 (1976).

MARX, JEAN, "Actin and Myosin: Role in Nonmuscle Cells." *Science,* **189**: 34 (1975). A brief review.

MOOSEKER, M. S., "Brush Border Motility." *J. Cell Biol.,* **71**: 417 (1976). Ca^{2+}-activated contraction.

PERRY, S., A. MARGRETH, and R. ADELSTEIN, eds., *Contractile Systems in Non-Muscle Tissues.* New York: Elsevier/North Holland, 1977.

ROSE, B., and W. R. LOEWENSTEIN, "Calcium Ion Distribution in Cytoplasm Visualized by Aequorin: Diffusion in Cytosol Restricted by Energized Sequestering." *Science,* **190**. 1204 (1975).

STORTI, R., D. COEN, and A. RICH, "Tissue-Specific Forms of Actin in the Developing Chick." *Cell,* **8**: 521 (1976). Embryonic muscle has non-muscle type actin. A gradual switch to its own version occurs during development.

TILNEY, L. G., "Actin Filaments in the Acrosomal Reaction of *Limulus* sperm." *J. Cell. Biol.,* **64**: 289 (1975). See also his article on the book edited by Inoué and Stephens.

THE ERYTHROCYTE

JACOB, H. S., "Pathologic States of the Erythrocyte Membrane." In *Cell Membranes,* G. Weissmann and R. Claiborne, eds. New York: H. P. Publ., 1975, p. 249. Also see *Hospital Practice,* Dec. 1974, p. 47.

KIRKPATRICK, F. H., "Spectrin." *Life Sci.,* **19**: 1 (1976). Review.

MARCHESI, V. T., "The Structure and Orientation of a Membrane Protein." In *Cell Membranes: Biochemistry, Cell Biology, and Pathology,* G. Weissmann and R. Claiborne, eds. New York: H. P. Publ., 1975, p. 45. See also *Hospital Practice,* June 1973, p. 76.

MENTZER, W., B. LUBIN, and S. EMMONS, "Dimethyl Adipimidate for Permeability Defect in Hereditary Stomatocytosis." *New Engl. J. Med.,* **294**: 1200 (1976). Correcting a spectrin-related hereditary defect.

SHEETZ, M. P., R. G. PAINTER, and S. J. SINGER, "Biological Membranes as Bilayer Couples. III. Compensatory Shape Changes Induced in Membranes." *J. Cell Biol.,* **70**: 193 (1976).

SHEETZ, M. P., R. G. PAINTER, and S. J. SINGER, "Relationships of the Spectrin Complex of Human Erythrocyte Membranes to the Actomyosin of Muscle Cells." *Biochemistry,* **15**: 4486 (1976).

MOVEMENT IN PROKARYOTES

ALDER, J., "The Sensing of Chemicals by Bacteria." *Scientific American*, **234**(4): 40 (April 1976).

BERG, H. C., "How Bacteria Swim." *Scientific American*, **233**(2): 36 (Aug. 1975).

CALLADINE, C. R., "Construction of Bacterial Flagella." *Nature*, **255**: 121 (1975).

KOSHLAND, D. E., Jr., "Bacterial Chemotaxis as a Simple Model for a Sensory System." *Trends Biochem. Sci.*, **1**(1): 1 (Jan. 1976). See also *Science*, **193**: 405 (1976).

NOTES

1. The prefixes "myo-" and "sarco-" are often used to define features of muscle cells that we now recognize to be the same for all cell types. Thus, the *sarcolemma* is the plasmalemma, or cytoplasmic membrane. A *sarcosome* is a mitochondrion, the *sarcoplasm* is the cytoplasm, and so on.

2. The sliding filament hypothesis first arose from observations made with the light microscope by A. F. Huxley and R. Niedergerke, and was published at about the same time (1954) as the early electron microscopy of Hanson and H. E. Huxley. These two Huxleys, it should be noted, are not related, though A. F. Huxley is from the prominent English family of authors and scientists that includes Aldous, Sir Julian, and Thomas Henry Huxley. The 1963 Nobel Prize for physiology and medicine was shared by A. F. Huxley, A. L. Hodgkin, and J. C. Eccles for their studies on the excitable membranes of nerve and muscle.

3. Tobacco smoke and a factor present in the serum of patients with cystic fibrosis also inhibit airway cilia, leading eventually to chronic lung problems.

4. Melanin is black, like the top of a blackhead (comedone), which isn't dirt at all, but melanin. However, when superimposed on the yellow-pink of skin, melanin can also appear as any shade of tan or brown, depending on concentration of the granules. It can also look blue when it is very deep in the skin, as in a blue nevus (a type of birthmark). The common brown or black nevus ("mole") is an abnormal but harmless collection of melanocytes. A malignant melanocyte on the other hand, although it only rarely arises in a common nevus, may spread to become a devastating cancer known as malignant melanoma.

10

Bulk Transport and Cellular Recognition

10-1 The Plasma Membrane 431
  The Cell Surface
  Blood Types
  HLA Types
  Lymphocyte Surface Receptors
  Lectins

10-2 Hormone Receptors 438
  Receptor Localization
  Hormone-Receptor Interaction
  The Mobile Receptor Hypothesis
  Cholera Toxin as a Hormone Model
  Receptor Regulation

10-3 Endocytosis and Lysosomal Digestion 443
  Mechanism of Endocytosis
  Selectivity of Endocytosis
  Lysosomes
  Killing and Digestion
  Defects in Lysosomal Function
  Autophagy
  Residual Bodies

10-4 The Nuclear Envelope 452
  Structure of the Nuclear Envelope
  Nuclear Pores
  Nucleocytoplasmic Exchange
  Relationship to the Cytoskeleton
  Nuclear Blebbing

10-5 Endoplasmic Reticulum 458
  Structure of the Endoplasmic Reticulum
  Functions of the Smooth ER
  Functions of the Rough ER
  Synthesis of Secretory Proteins
  Isolation of Secretory Proteins by the ER
  Glycosylation of Proteins

10-6 The Golgi Apparatus 466
  Morphology
  Function of the Golgi Apparatus

10-7 Secretion 469
  Types of Secretion
  The Pancreatic Acinar Cell
  Vesicle Transport
  Insulin Secretion
  Stimulus-Secretion Coupling
  Mast Cells
  Role of Cyclic Nucleotides in Secretion
  Fate of the Secretory Vesicle

10-8 Synthesis and Turnover of Membrane Components 482
  Turnover of Membranes
  Synthesis of Membranes
  Lipid Exchange

Summary 485

Study Guide 487

References 488

Notes 490

Membranes and membrane-associated phenomena have occupied most of the preceding four chapters. In Chap. 6, the fluid mosaic membrane was introduced, along with the mechanism of molecular transport across it. The next chapter dwelt in part on the membranes of mitochondria and chloroplasts and how they may function as energy transducers. Excitable membranes were treated in Chap. 8, which included an introduction to the neurotransmitters and how they alter permeability of membranes. Then, in Chap. 9, the concept of a cellular cytoskeleton was described. The cytoskeleton clearly influences the position of membrane proteins. Locomotion and other membrane-dependent functions appear also to depend on an intact cytoskeleton, which is an obvious way to explain how these functions consume metabolically supplied energy.

In this chapter, we shall deal at more length with the structure and function of membranes. Although other activities of the involved membranes will also be considered, we shall be concerned mostly with the process of *bulk transport*, which is transport of materials in membrane-enclosed vesicles. It is the major process by which substances are moved from one part of the cytoplasm to another. It is also the major process by which substances move into and out of cells. Bulk transport is often highly selective, which raises the question of how membranes can be recognized and differentiated from other membranes. It is there that we shall start, using as a basis for our discussion the membrane about which most is known, the plasma membrane.

10-1 THE PLASMA MEMBRANE

It can truly be said of living cells, that by their membranes ye shall know them.

E. N. Harvey, 1943

The plasma membrane, mostly due to its accessibility, is the most thoroughly studied of all the cellular membranes. Although much of what has been learned about the plasma membrane is equally true of other membranes as well, the plasma membrane, or cell membrane, has a great many functions peculiar to itself. Among those functions is cellular recognition, meaning the recognition by cells of each other and the recognition of target cells by hormones and other outside mediators of cell function.

The Cell Surface. Isolated plasma membranes typically contain from 2–10% carbohydrate, probably all of it on the outside surface and most of it contributed by membrane glycoproteins. The glycoproteins vary from one cell type and species to another, and hence so does the carbohydrate, which serves to give the cell a unique identity—a "sugar coating" sometimes referred to as its *glycocalyx* (see Fig. 10-1). For example, human blood types (type A, type O, etc.) are determined by terminal sugars present on otherwise identical glycolipid and glycoprotein carriers.

The glycocalyx participates in most of the functions of the plasma membrane. Receptors for hormones and other modulators of cellular function are commonly glycoproteins, the carbohydrate portion of which is often essential for recognition. Some (but not all, as we shall see) of these receptors span the plasma membrane, putting them in a position to translate external events into cytoplasmic changes.

Another function of the cell surface is intercellular adhesion. The first clues to the participation of carbohydrate in this process came from experiments with marine sponges carried out by T. Humphreys and A. A. Moscona in the early 1960s. It had long been known that when these simple multicellular animals are disaggregated, the individual cells are able to move about, find each other, and reassociate. When the cells

FIGURE 10-1 The Erythrocyte Glycocalyx. On this cell we find an unusually lush, 1400-Å thick coat of oligosaccharide filaments, 12–25 Å in diameter. [Courtesy of H. Latta, W. Johnson, and T. Stanley, *J. Ultrastruct. Res.* **51**: 354 (1975) © Academic Press, N.Y.]

from two unrelated strains of sponge are mixed, they even sort themselves out during reassociation, each to its own kind. The critical principle in this recognition and adhesion are high-molecular-weight glycoprotein *aggregation factors* made and secreted by the cells. These factors are able, in the presence of Ca^{2+}, to glue cells together by binding to receptors on the cell surface.

Aggregation factors have also been identified in several tissues of higher animals, including mammals. It is probable that they play a role in embryonic development, when cells differentiate and segregate into specific tissues, and that they participate in cellular adhesion in some adult tissues. However, there is also evidence that cellular recognition and adhesion can be a property of the plasma membrane alone and of the carbohydrates attached to it. In this latter instance, each membrane would have to have both receptors and donors, whereas with a soluble aggregation factor only receptors are required (Fig. 10-2).

The presence of carbohydrate-specific surface receptors on mammalian cells was demonstrated with agarose beads containing covalently attached sugars. Mouse fibroblasts were found to adhere to beads containing galactose but not to beads containing glucose or N-acetylglucosamine, thus revealing the presence of galactose receptors on the cell surface. These receptors normally interact with galactose-containing oligosaccharides protruding from the surface of other cells, thus gluing the cells together. When the carbohydrate surface of the cells is damaged, adhesion is prevented until new sugars can be synthesized and used to restore the normal glycocalyx.

A model for this sort of donor-receptor interaction has been offered by Saul Roseman. He has proposed that the sugar receptors are membrane-bound enzymes called *glycosyl transferases*, which have a high degree of specificity for oligosaccharides. Since these same enzymes are involved in the synthesis of surface carbohydrates, they would be able to recognize and bind to oligosaccharides of the type they participate in making, even though the oligosaccharide is found on a different cell. In addition to this specificity, an attractive feature of the proposal is that binding would be reversible. Release of a bound oligosaccharide would come when the energy-requiring enzymatic activity of the transferase is initiated, the results of which elongate the bound oligosaccharide by one more sugar unit and reduce its affinity for the enzyme's active site.

There is no direct proof of the glycosyl transferase hypothesis as yet, though a good deal of indirect evidence has accumulated. Even if it is ultimately proven wrong in detail, the essential features will probably be found valid—that is, there will be specific oligosaccharides and oligosaccharide receptors, and a reversible interaction that depends on the cell's metabolism for continuation, all of which is characteristic of the glycosyl transferase system.

In any case, it is evident that cells do recognize each other and that recognition involves the carbohydrate coat. Unfortunately, we know little as yet about the specific molecules that confer this identity to cells. Some glycoproteins have been purified that appear to be specific for a given tissue, but most of the work has been on surface structures that identify a species or individual rather than a specific type of cell or tissue. Examples are the blood group and human lymphocyte antigens.

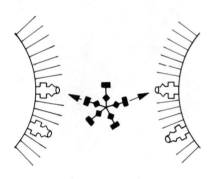
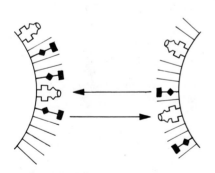

FIGURE 10-2 Cellular Adhesion. TOP: Extracellular adhesion (aggregation) factors. BOTTOM: The receptor and donor units are both attached to each cell.

Blood Types. The first recognition that specific antigens exist on the surface of cells, and that they may vary from one individual to another, was due to the work of Karl Landsteiner in about 1900. Landsteiner found that proteins from the blood of some people cause clumping or agglutination of washed red cells from other people. In subsequent work he was able to establish the presence of two antigens, which he called A and B, either one or both of which could be present on red cells. People could be classified into four groups: A, B, AB, or O, the latter having neither antigen. The observed agglutination reactions occur because humans have antibodies against whichever antigen is missing on their red cells. Thus, for example, type AB individuals have neither antibody while type O individuals have both, though the source of the antigenic exposure causing these antibodies is not clear.

It is now known that blood group antigens are glycolipids or glycoproteins present in the red cell membrane, and that there are antigens similar or identical to some of the blood group antigens on many other cell types as well. The system is much more complex than originally defined, however, with additional red cell antigen families bearing names like Rh, MN (which seems to reside on the erythrocyte protein called glycophorin), Lewis, Kell, etc. Even the ABO system has a twist, for there are two major A types called A_1 and A_2.

Although the blood group antigens are glycolipids or glycoproteins, it is the terminal sugars in the carbohydrate chains that determine antigenic specificity. The ABO system, for example, is built on an erythrocyte glycolipid called *substance H*. Type O erythrocytes carry unmodified substance H. People with the B gene have an enzyme capable of adding galactose, while an A gene causes, instead, N-acetylgalactosamine to be added. Type AB erythrocytes have some H substance modified in one way, some in the other. The four responsible genes, A_1, A_2, B, and O, are alleles (see Chap. 5). A person inherits one of the four genes from his or her mother, and on the same location in the homologous chromosome is the paternally inherited allele. A child will have type A blood if either or both A alleles are present without B, type B blood if B but no A alleles are inherited, and so on (Table 10-1).

A comparable situation exists for the other blood group antigens, but not all are the result of variation at a single locus. The *Rh system*, first described in the Rhesus monkey, is particularly complex. It appears to consist of three closely linked (i.e., closely spaced) genes, each with two

TABLE 10-1. Inheritance of Blood Types

Inherited alleles	Apparent blood type[a]	Serum antibodies
A O A A	A	anti-B
B O B B	B	anti-A
A B	AB	neither
O O	O	both

[a] For white North Americans, the frequency of the four blood types is approximately 42%, 9%, 3%, and 46%, respectively.

alleles, one of which expresses a surface antigen and one of which does not. To be Rh negative the inactive allele has to be present at each of the major ("D") loci, as is true for some 15% of Caucasians.[1]

HLA Types. The antigenic pattern known as H-2 in the mouse and HLA in humans is even more complex than the Rh system. The term HLA stands for *human lymphocyte antigen*, but in fact these antigens are found on apparently all nucleated human cells. They are surface glycoproteins, constantly being synthesized and either shed or destroyed. Mature red blood cells, because they have no nucleus and thus no protein synthesis, are unable to replace HLA antigens as they are lost, and so erythrocytes are probably the only cells without HLA antigens.

The HLA antigens are products of a group of closely linked genes on chromosome number 6. This cluster of genes is referred to as the *major histocompatibility complex* or *MHC*. The prefix "histo" means tissue, so these are the major genes that establish whether tissues are compatible with each other during transplants, which is why they have received so much attention. They have also been enormously useful for other purposes. Because of their variety they are the fingerprints of the cell, with few people outside the same family likely to have exactly identical HLA patterns. (Fortunately for kidney transplant patients, some of the antigens are more important than others in rejection of foreign tissue.) And because they are genetically determined, they have also been used by anthropologists in studies on the relationships among various groups of people, and by mothers in paternity suits.

HLA antigens are the products of at least four genes in the major histocompatibility complex. These genes are referred to as A, B, C, and D (see Fig. 10-3). Over 20 alleles are known for each of the first two genetic loci. While fewer alleles have been discovered at the HLA-C and D loci, new ones are being described at a regular rate.

The four HLA loci are closely linked. There is less than a 1% chance of crossover (rearrangement) between adjacent genes during egg or sperm formation. Hence in almost every case, the HLA pattern (called a *haplotype*) on one chromosome number 6 of a child can be clearly identified as maternal and the haplotype of the homologous chromosome as paternal. For example, if the father has HLA haplotype arbitrarily designated "W" on one chromosome and "X" on the homologue, while the mother is "Y" and "Z," then with few exceptions each child has an equal chance of getting one of the four combinations WY, WZ, XY, and XZ. It follows that there is a 25% chance that two siblings will be HLA-identical but none will be identical to a parent, a factor that influences to some extent the survival of grafted kidneys and other organs.

The function of the HLA antigens is unknown. We get a clue, however, from the observation that certain diseases are more common in the presence of one or another HLA antigen. The most striking correlation is between the presence of the HLA-B27 antigen (product of the 27th numbered allele at the B locus) and a condition called ankylosing spondylitis. This condition usually affects young men, leaving them with a rigid spine and hunched over posture. People with HLA-B27 have about 120 times the usual incidence of this disease. Other diseases that are markedly more frequent with a particular HLA antigen include multiple sclerosis, childhood rheumatoid arthritis, and certain types of psoriasis and diabetes. The thing that these and other HLA-related diseases seem

FIGURE 10-3 Relative Spacing of the HLA Genes. Note the very close linking between B and C. The alleles inherited at these four loci on one chromosome constitute a haplotype.

to have in common is an autoimmune component. That is, they seem to involve a defect in the immune system and in many cases include the production of antibodies to one's own tissues. Hence, it is possible that an important function of the HLA antigens is to provide recognition of self to the immune system—that is, to provide a surface pattern on cells that is recognized as normal, in order to be able to distinguish foreign cells or, quite probably, cancerous cells, and hence eliminate them from the body. That sort of *immune surveillance* involves the activities of lymphocytes, the study of which has revealed some important and unexpected properties of the plasma membrane and the receptors it contains.

Lymphocyte Surface Receptors. The lymphocyte surface differs from other cell surfaces in containing receptors capable of recognizing and binding a wide variety of foreign antigens. Bone marrow or (in birds) bursa-derived lymphocytes called B cells respond to a foreign antigen by transformation and proliferation into antibody-secreting plasma cells (see Chap. 3). Antibodies, which circulate in the blood, attach to and thus help to neutralize the foreign antigen. A given lymphocyte will respond to only one specific type of antigen, since it has only one type of receptor on its surface. In the case of B cells, the receptor is an antibody of the same specificity that the progeny of that cell will produce. The receptor on thymus-derived lymphocytes, or T cells, has not been identified as yet, but is presumed to be a comparable molecule.

When a specific antigen is encountered, a series of events takes place in the lymphocyte that leads eventually to activation. These events begin with a change in ionic permeability—specifically, an influx of Ca^{2+}, to which the membrane is normally relatively impermeable, and an increase in activity of the Na^+-K^+–stimulated sodium pump. Subsequent events leading to cell division are considered in the next chapter. What we would like to know here, is how binding of antigen to membrane-bound receptors (antibodies) is able to bring about the initial changes in permeability of the cell membrane.

The study of lymphocyte activation was greatly facilitated by an observation of Stewart Sell and Philip Gell of the University of Birmingham (England), who demonstrated in 1965 that rabbit lymphocytes can be stimulated to divide by exposure to serum ("antiserum") from another species of animal that had been injected with rabbit immunoglobulins used as an antigen. The resulting antibodies are able to recognize and bind to the constant region of rabbit immunoglobulins (Chap. 3), including those found on the surface of rabbit lymphocytes. The lymphocytes are stimulated to divide by this interaction (see Fig. 10-4).

In addition to permitting wholesale activation of a population of lymphocytes, the use of anti-immunoglobulin antiserum permitted the distribution of receptors on the lymphocyte surface to be directly visualized. Two different techniques were used; either the antiserum proteins were coupled covalently to ferritin, an iron-storage protein having a dense 55-Å iron-containing core that is easily visible in the electron microscope, or the antiserum proteins were labeled with a fluorescent molecule that permitted direct visualization in the light microscope using ultraviolet stimulation. The results were startling. The receptors are initially diffusely scattered over the surface in a relatively random distribution, a pattern that can be visualized without disturbing it by using ferritin or fluorescence-labeled *monovalent* antibodies (Fab frag-

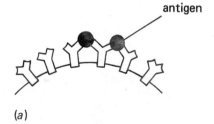

(a)

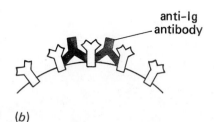

(b)

FIGURE 10-4 Receptor Cross-linking. The immunoglobulin that serves as a surface receptor for B-type lymphocytes can be cross-linked and the cells activated (a) by antigen or (b) by antibody specific for the surface immunoglobulin. Fab fragments (which are monovalent) or very small antigens (haptens) fail to cross-link receptors and fail to activate lymphocytes.

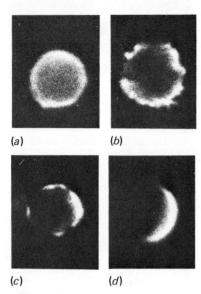

ments containing a single combining site). When normal, divalent antibodies or the specific antigen itself is used, receptors are cross-linked and are then seen to form distinct patches (Fig. 10-5). Patch formation reflects receptor mobility in the fluid mosaic membrane and does not require energy. (It takes place even in poisoned cells.) In normal cells, patching is followed in a few minutes by *capping*, in which the labeled receptors move to form a single cluster at one side of the cell.

The process of capping is energy dependent and requires an intact cytoskeleton. When both colchicine and cytochalasin B are added to dismantle microtubules and microfilaments, respectively, capping is strongly hindered. Colchicine by itself either has little effect or may enhance capping, implying a stabilizing role for microtubules in holding the receptors apart, a point to which we shall return shortly.

Capping is followed usually by internalization (endocytosis) of the capped receptors (Fig. 10-6), or sometimes by shedding. In either case, the cell is left with only a small number of receptors on the surface until, in 6–20 hours, new receptors can be synthesized to take their place. Capping and endocytosis were thought at first to be the critical events leading to activation of lymphocytes, but they probably represent instead a mechanism for cleaning and renewing the surface. (Different surface antigens are independently subject to capping by antibody cross-linking in a way that leaves most others undisturbed.) The key to lymphocyte

FIGURE 10-5 Patching and Capping of Lymphocyte Receptors. These cells were isolated by using the affinity of their surface receptors for a specific antigen and then capped by the same antigen. (*a*) Uniform distribution of fluorescent antigen; (*b*) patching; (*c*) early capping; (*d*) complete capping. [Courtesy of G. J. V. Nossal and J. E. Layton, *J. Exp. Med.*, **143**: 511 (1976).]

FIGURE 10-6 Capping of Radioactively Labeled Surface Proteins. Lymphocyte surface proteins (including but probably not limited to immunoglobulin) were labeled with radioactive iodine (^{125}I), then the cells were exposed to anti-immunoglobulin antibody. (*a*) Uniform distribution of a label on a T cell, which does not cap with this procedure; (*b*) capping of a B cell (the peculiar shape is characteristic of a capped cell); (*c*) internalization of the cap (*arrow*). Bar = 1 μm. [Courtesy of N. Gonatas, J. Gonatas, A. Stieber, J. Antoine, and S. Avrameas, *J. Cell Biol.*, **70**: 477 (1976).]

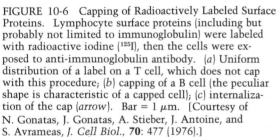

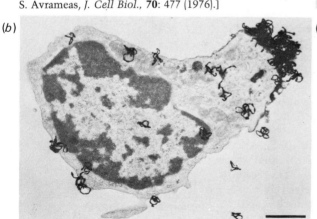

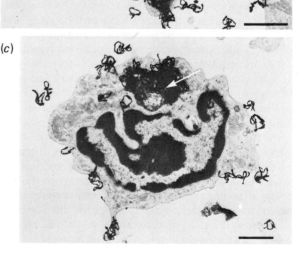

activation is probably more subtle, involving only the effects of local cross-linking. Evidence for that position comes from studies with cross-linking agents called lectins that are less specific and are not antibodies at all.

Lectins. Lectin is a word derived from Latin meaning "to pick out" or "to choose." The name recognizes that lectins are proteins that behave very much like antibodies in their ability to bind to specific chemical configurations. They are, however, derived mostly from plants, not animals. Although their natural function is unknown, their specificity for plasma membrane sugars leads one to speculate that they may have a role in cellular adhesion in higher plants, much like the aggregation factors described earlier for animals. In any event, because they have more than one combining site, they are able to cross-link surface antigens and thus cause patching, capping, cell agglutination (some are used commercially to type blood) and, in the proper concentration, lymphocyte activation.

Several hundred lectins have been described. The two that have been most widely used in cell biology are *phytohemaglutinin* and *concanavalin A*. Phytohemaglutinin, or *PHA*, is a protein obtained from the red kidney bean. It is a tetramer, with each subunit capable of binding N-acetylgalactosamine. Concanavalin A, or just *Con A*, is obtained from jack beans. It, too, is a tetramer under most conditions, but its binding site seems to be specific for α-D-mannose or α-D-glucose.

When small amounts (roughly 1 μg/ml) of Con A are added to lymphocytes, stimulation occurs and the cells subsequently synthesize DNA and undergo cell division in much the same way as antigen-stimulated cells. However, when larger amounts of Con A are used (above 10 μg/ml), patching, capping, and activation are prevented even when specific antibodies are added. These concentrations of Con A, not enough to saturate the surface, cause a general though reversible immobilization of surface receptors. Even when Con A is added to only one part of the cell (for example, by exposing the cells to solid supports having Con A attached) receptors are immobilized on other parts of the surface.

It appears that receptors responding to Con A are anchored to the underlying cytoskeleton, and that cross-linking of a few receptors by Con A causes activation. Immobilization of a larger part of the receptor network by cross-linking with Con A is sufficient to immobilize all of it and prevent activation. This hypothesis is supported by the observation that Con A inhibition of receptor patching and capping (i.e., receptor immobilization) did not occur in the presence of colchicine or at low temperatures, both of which cause depolymerization of microtubules. Anchoring receptors to an underlying cytoskeleton provides an obvious mechanism for transmitting to the interior information about surface changes. These cytoskeletal anchors may also mediate stimulatory signals from low doses of Con A and from antigen specific for surface receptors. The increase in Ca^{2+} influx and sodium pump activity that precedes lymphocyte activation are probably direct consequences of this signal transmission.

Obviously much has yet to be learned about the way signals are transmitted from the type of surface receptor just discussed. Another type of surface receptor, which is better understood, mediates the response to certain hormones. This type of receptor will be described next.

10-2 HORMONE RECEPTORS

Lipid-soluble hormones such as steroids penetrate cells to interact with cytoplasmic receptors while water-soluble hormones typically interact with surface receptors. This latter group of hormones is a diverse family including simple molecules such as epinephrine and acetylcholine, oligopeptides such as oxytocin and vasopressin, as well as proteins such as insulin and growth hormone. These and many other hormones interact with cells chiefly by binding to specific receptors on the cell membrane. The receptors, thus activated, regulate the levels of *second messengers* such as cyclic AMP and cyclic GMP.

Some of the effects of the cyclic nucleotides on cellular activities were discussed in Chaps. 4 and 5. We are now in a position to consider the properties and regulation of the receptors themselves.

Receptor Localization. Insulin has received more than its share of attention from workers interested in hormone receptors, a natural consequence of its economic and medical importance in the treatment of the insulin deficiency disease, diabetes mellitus ("sugar diabetes"). Many diabetics require daily injections of insulin in order to survive, for its presence alters a great many activities in many kinds of cells. These activities include an increase in the uptake of amino acids and glucose, the synthesis of macromolecules of all kinds and, in fat cells, a decrease in the breakdown of triglycerides and decreased release of fatty acids. Some, but probably not all, of these processes are mediated through an increase in cyclic GMP and fall in cyclic AMP.[2]

It has been amply demonstrated that insulin causes its many changes by binding to receptors on the surface of cells. For example, insulin attached to insoluble beads retains its ability to stimulate cells, even though there is no possibility for uptake. And when autoradiography is used to localize radioactive insulin, or electron microscopy is used to visualize insulin-ferritin complexes, the hormone is found only on the surface (see Fig. 10-7).

These experiments demonstrate that insulin does not need to enter cells to be active. But if it did get in, would it find more receptors there? When cells are fragmented and the various internal membranes examined, an ability to bind insulin cannot be demonstrated. Furthermore, when vesicles made from isolated plasma membrane of fat cells are examined, insulin binding occurs only if the outer surface of the vesicle is the former surface of the plasma membrane. Inside-out vesicles do not bind insulin. It appears that insulin receptors are limited to the plasma membrane and are oriented only to the outside, with little or no tendency to rotate or flip, whether or not insulin is present.

When similar experiments are carried out with other nonlipid hormones, the results are much the same. There are, however, reports of receptors for catecholamines (e.g., epinephrine) and some of the peptide hormones on internal organelles. Whether this finding implies a biological role for these internal receptors is not clear, for they could be newly synthesized receptors in transit to the surface or old receptors removed from the surface. Overall, however, the generalization that water-soluble hormones carry out their biological role by attachment to receptors on the cell surface seems adequately supported.

The specific location of hormone receptors on the cell surface can also be determined. Not surprisingly, the most asymmetric distribution of

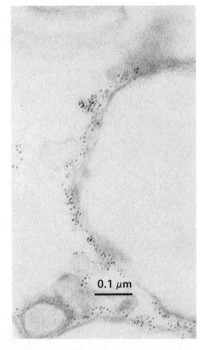

FIGURE 10-7 Insulin Receptors. Insulin, conjugated to ferritin to make it visible, is seen (as small black dots) only on the external surface of an isolated fat cell plasma membrane. [Courtesy of L. Jarett and R. M. Smith, *J. Biol. Chem.*, **249**: 7024 (1974).]

receptors is at neuronal synapses and neuromuscular junctions. For example, receptors for acetylcholine reach a density as high as 100,000 per μm^2 at neuromuscular junctions in frogs. This density corresponds to nearly complete occupation of the membrane by receptor, almost to the exclusion of lipid. However, when the incoming neuron is severed, there is rapid dispersal of receptors, lessening sensitivity in the synaptic region and increasing it elsewhere.

Hormone-Receptor Interaction. The binding of insulin to its receptor is rapid, specific, and very tight. The free energy of binding is about -60 kJ/mole and the half-life is several minutes. Such behavior makes one think of an enzyme-substrate interaction and, in fact, many hormone receptors are known to be glycoproteins that, when complexed with their hormone, cause an increased enzymatic production of some substance such as cAMP. It would be natural to assume that the receptors themselves are also enzymes and in some cases they probably are. In many other cases, however, the link between hormone binding and enzymatic activity is somehow indirect.

If the receptor for a hormone were also its effector—that is, if the receptor had enzymatic activity, say that of adenylate cyclase—then one would expect the following to be true: (1) The enzymatic activity should last only as long as the receptor is occupied. (2) The total amount of enzymatic activity should be proportional to the number of receptors occupied. (3) Chemical alteration of the hormone should reduce its biological activity in direct proportion to the reduction in its capacity to bind to receptors. There are some important deviations to this expected pattern of behavior.

For example, the ability of a hormone to bind to its receptor is not necessarily related to its ability to turn on the associated biological activity. With epinephrine and the other catecholamines activity depends on the catechol ring (i.e., on the dihydroxybenzene) whereas binding is a function of the "tail" of the molecule (see structure). Hence, it is possible to synthesize compounds that bind very tightly but have no biological activity, thus blocking the receptor, a finding that has strong medical application for states of excess catecholamine output such as hyperthyroidism.

The proposed direct relationship between receptor binding and biological activity is not always valid even for the natural hormone. For example, there is a strong negative cooperativity (see Chap. 4) to catecholamine binding. This means that after a very small proportion of sites are occupied, binding to additional sites becomes distinctly more difficult. Negative cooperativity is a potentially useful feature, as it provides sensitivity to low concentrations of the hormone while buffering a cell against rapid changes in level above this point. It is not, however, what one would expect from isolated and independent receptors distributed over the surface of a cell, for cooperativity implies a considerable interaction among receptor sites. Since negative cooperativity is seen also with insulin binding and with some other hormones, it is not an isolated oddity of catecholamine receptors but requires a more general explanation.

An additional deviation from the simple model of hormone receptor-effector interaction is the apparent excess of binding sites in some systems. In isolated fat cells, insulin needs to occupy only 2–3% of the available sites to achieve nearly full stimulation of glucose utilization. The "spare" receptors do not differ physically from occupied ones, hence it appears that any 2 or 3% will do. Comparable observations have been made with many other hormones.

None of these observations is impossible to rationalize with the simple "receptor equals enzyme" theory. Nevertheless, their overall effect is to make one look for important modifications of that theory. A model for hormone receptor activity has been offered by P. Cuatrecasas and others that seems to provide a better explanation for the observed behavior of many systems. That model has been termed the *mobile receptor hypothesis*.

The Mobile Receptor Hypothesis. This proposal takes advantage of the properties of the fluid mosaic membrane. In brief, it suggests that receptors by themselves do not always have effector ability. Rather, they are free to diffuse about the membrane, floating partially submerged with their binding sites exposed to the surface. Adenylate cyclase or other effectors partially penetrate the same membrane but from the other side. When diffusion brings an occupied receptor into contact with an effector, activation may take place (see Fig. 10-8).

The mobile receptor hypothesis offers an especially satisfying explanation for the apparent excess of hormone receptors. Thus, maximum cyclic AMP-mediated response would occur when all the adenylate cyclase effectors were in contact with an occupied receptor, at which point there might be a large number of receptors left over. Note, too, that sensitivity to a hormone could be altered by controlling the number of receptors only, without the added problem of changing the number of effectors.

This latter point, the ability to change the number of receptors and effectors independently, is an important part of the mobile receptor model, for we already know that a number of hormones may work through the same effector. In fat cells, for example, at least eight hormones are known to affect adenylate cyclase, including glucagon, the catechola-

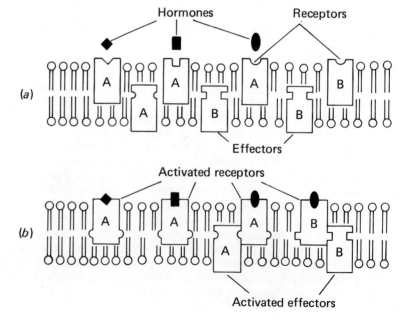

FIGURE 10-8 The Mobile Receptor Model. Schematic representation showing receptors for a variety of hormones capable of activating two different types of effectors, labeled A and B. Effector activation occurs only upon contact with an occupied receptor. Note that one hormone may have more than one type of response (part b). Conversely, many hormones may have the same response.

mines, and certain prostaglandins. Since these hormones do not have identical responses in the cell, it makes sense to propose both a family of receptors and a family of effectors (some of them probably undiscovered), with different combinations resulting in one or another type of response. The advantages in terms of conserving genetic material and simplifying control are great, for only the receptor concentration needs to be regulated in order to change sensitivity to a given hormone. The several effectors responsive to that hormone need not be altered, but may remain available for other hormone-receptor complexes. The existence of this sort of overlap is supported by the observation that a cultured cell line deficient in adenylate cyclase has been shown to be unresponsive to not just one, but at least four hormones that normally use adenylate cyclase as an effector.

Strong support for the mobile receptor hypothesis has come also from observations made on a protein that is not a hormone at all, but a toxin. Specifically, it is the toxin that produces the symptoms of cholera, an epidemic intestinal disease.

Cholera Toxin as a Hormone Model. Cholera toxin is a protein of about 84,000 daltons made by the bacterium *Vibrio cholerae* (see Fig. 10-9). The toxin binds to cells lining the intestine where it increases their adenylate cyclase activity and causes unregulated water and salt secretion. The resulting massive diarrhea is often fatal if medical attention is not available. While cholera toxin is not a hormone, the fact that it binds to specific receptors (a particular ganglioside) and activates adenylate cyclase after doing so, makes it a useful model with which to study hormone action. It differs from hormones in that binding is essentially irreversible. This poses a medical problem as the patient must be kept alive for the three days or so that it takes for these cells to die and be replaced (a continual process, anyway) but it also makes many experiments easier.

Cholera toxin contains two kinds of subunits, light and heavy (L and H). Isolated L subunits by themselves bind to the receptor but cannot activate adenylate cyclase, a function that is relegated to the H subunit. (The latter is found also in the toxins of several other intestinal bacteria including pathogenic strains of *E. coli*.) One can imagine that L subunits

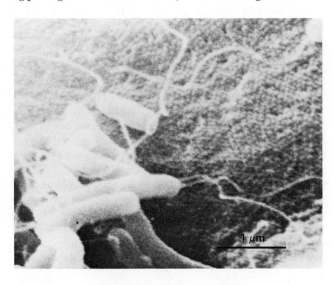

FIGURE 10-9 *Vibrio cholerae* Adherent to Rabbit Intestine. The toxin produced by these bacteria causes severe diarrhea and subsequent dehydration by activating adenylate cyclase. [Courtesy of E. Nelson, J. Clements, and R. Finkelstein, *Infection and Immunity*, **14**: 527 (1976).]

recognize the receptor, while H subunits interact with adenylate cyclase. In fact, the H subunit alone is active with lysed cells, where it has direct access to adenylate cyclase located on the under side of the membrane.

Another feature of cholera toxin that supports the mobile receptor hypothesis is the presence of a significant lag of some minutes between binding and measurable increases in cAMP. Though exaggerated when compared to hormones, the lag would be consistent with the need to diffuse about the surface until a suitable effector is found. As one would expect, lowering the temperature to below the phase transition of the membrane greatly extends the lag time.

Thus, the mobile receptor hypothesis offers a satisfying explanation for the behavior of hormones and this hormone analog. At the same time, it lends further support to important features of the fluid mosaic model of membranes.

Receptor Regulation. The turnover of membrane components will be considered as a general phenomenon later in this chapter. Receptor turnover is singled out for discussion here because it represents an important regulatory feature, the simplicity of which, in current thinking, is due to the mobile receptor hypothesis.

The point has already been made that a cell could change its sensitivity to a hormone by altering the number of specific receptors without necessarily changing the effectors, and therefore without changing the level of maximum response. A higher concentration of receptors would make activation more likely from a given amount of hormone. Since all cells are exposed to the same concentration of hormone, the ability to alter the sensitivity of different cell types independently might be of considerable importance, adding another layer of regulation to that inherent in controlling the level of hormone itself, the affinity of the receptors (negative and positive cooperativity), the number or sensitivity of effectors, or modulation of events that take place in the cell after activation.

Turnover (that is, loss and replacement) of receptors has been demonstrated in many systems. For example, insulin receptors in cultured lymphocytes have a half-life of about 35 hours. This value is not fixed, but can be changed by altering either the rate of synthesis or the rate of degradation. Thus, when quiescent, normal lymphocytes are stimulated to divide, there is a marked increase in the rate of receptor synthesis. And when cells are chronically exposed to high levels of insulin, the rate of receptor destruction is increased and the number of receptors per cell reduced thereby. (Since insulin levels increase with overeating, obese individuals frequently have poor regulation of blood glucose levels secondary to a relative lack of receptors on certain cells. Such individuals may be mildly diabetic until their weight becomes more normal.)

Decrease in receptor concentration as a consequence of increased hormone is an example of negative feedback and seems to be common to many hormone-receptor systems. In the case of insulin, the mechanism appears to be an increased susceptibility to destruction of receptor when it is occupied. In such systems, drugs that block protein synthesis prevent recovery of the original receptor concentration. A second mechanism, which applies at least to some catecholamine receptors, is independent of protein synthesis and may therefore involve reversible inactivation by the hormone.

Receptor modulation by negative feedback leads to *desensitization*, so that increasing amounts of the hormone are required to have the same cellular effect. Sudden removal of the extra hormone may lead temporarily to a deficiency condition. This mechanism may explain some cases of *tolerance* to drugs and *withdrawal* symptoms on their sudden elimination, though enzyme induction could also lead to the same changes. For example, receptor modulation was thought to be the mechanism of addiction to the pain-relieving narcotics, morphine, heroin, and codeine. Although more recent work indicates that the concentration of receptors (which after all are designed for some natural substance, not products of the opium poppy) is not affected. However, the activity of adenylate cyclase is increased in addicted cells to compensate for suppression of cAMP by narcotics. The relatively slow rate of change in adenylate cyclase activity and the adjustments required to accommodate to that change could therefore explain the tolerance, addiction or dependency, and withdrawal symptoms associated with narcotics and other drugs (including possibly caffeine) that act through adenylate cyclase.

10-3 ENDOCYTOSIS AND LYSOSOMAL DIGESTION

There is at bottom only one genuinely scientific treatment for all diseases, and that is to stimulate the phagocytes.

George Bernard Shaw, 1906
from *The Doctor's Dilemma*

Phagocytosis refers to the enclosure of extracellular particles by the cell membrane and the subsequent internalization of those particles without breaking the continuity of the membrane (see Fig. 10-10). The particles are left in a vesicle or vacuole called a *phagosome,* surrounded by a piece of the former plasma membrane. *Pinocytosis* is the same process, except that the new vesicles are smaller and contain no obvious particles. The prefixes phago- and pino- mean, respectively, to eat and to drink. Either process can be referred to as *endocytosis.*

The prevalence of endocytosis was long underestimated because the smaller vesicles are often only a few hundred angstroms in diameter and are therefore not visible with the light microscope. It is now clear, however, that the process is an extremely important one and that pinocytosis is more or less continuous in a wide variety of cells. Phagocytosis, as such, is common in unicellular animals and some plant cells, but in only a few cell types of higher animals *in vivo*. However, it is readily demonstrated in a great many others when they are grown *in vitro*. Much of what we know about phagocytosis has been learned by studying amebas and the major phagocytes of higher animals, the macrophages and neutrophils.

Mechanism of Endocytosis. Phagocytosis and pinocytosis are processes that clearly involve the cytoskeleton described in the preceding chapter. To ingest large particles, membrane-covered extensions of the cytoplasm reach out to surround the target particle and then fuse on the far side to enclose the particle in a vacuole lined by the former surface of the plasma membrane (see Fig. 10-10). Alternately, and especially with pinocytosis,

CHAPTER 10
Bulk Transport and Cellular Recognition

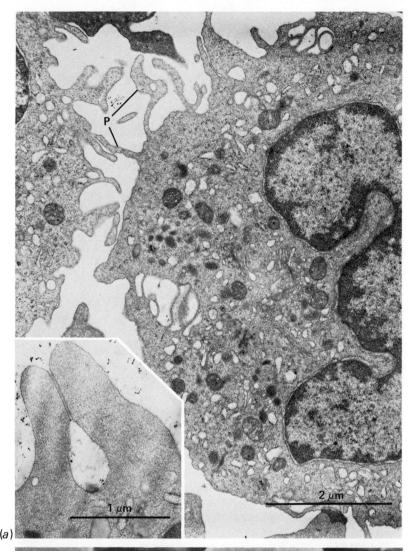

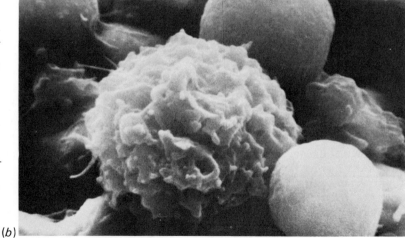

FIGURE 10-10 Phagocytosis. (a) The undulating surface membrane of these malignant monocytes (called *macrophages* after they leave the vascular system) demonstrates the mechanism of endocytosis. Note pseudopodia (P) and many vacuoles. INSET: Another cell, showing formation of a phagocytic vacuole. (The septate junction is unusual.) (b) Scanning electron micrograph of a normal human monocyte, showing the undulating surface to better advantage. The smooth cells in this field are lymphocytes, which are not phagocytically active—something you might have guessed just by looking at them. [(a) Courtesy of F. T. Sanel. Large photo from *Science*, **168**: 1458 (1970). Copyright by the AAAS. (b) Courtesy of Aaron Polliack.]

the vacuole or vesicle is created by a depression in the cell membrane, which then closes over the top (Fig. 10-11). Both mechanisms depend on local changes in microfilaments and their associated proteins.

To support the involvement of cytoplasmic contractile systems in endocytosis, one notes that (a) the process is energy dependent, requiring active metabolism and a source of ATP; (b) it requires the extracellular presence of Ca^{2+}, entry of which may act as a trigger; and (c) cytochalasin B, which inhibits microfilament function, in many cases inhibits endocytosis as well.[3]

Microtubules are also involved in some way during endocytosis, though just how is not clear. It is known that surface membrane is rather effectively cleared of transport proteins and some hormone receptors before being internalized. This clearance is inhibited by the addition of colchicine, which indicates that one role of microtubules is to control the location of membrane proteins. It seems likely, however, that microtubules are also involved in other ways, not yet understood.

Selectivity of Endocytosis. Endocytosis, especially phagocytosis, is not necessarily a random event but can be a specific cellular response to external stimuli. For example, phagocytosis in amebas is stimulated more by positively charged proteins and dyes than by neutral or negatively charged ones, and mammalian cells in tissue culture discriminate in favor of poly-D-lysine over poly-L-lysine, or arginine-rich histones over lysine-rich histones. Amebas, of course, use phagocytosis as a way of obtaining food and so may tolerate a relatively low degree of discrimination. Mammalian phagocytes, on the other hand, are charged with the task of removing debris, bacteria, and foreign substances, tasks that may require a high degree of selectivity.

The major mammalian phagocytes are the *macrophage* ("big eater"—Fig. 10-10) and the *neutrophil* (Fig. 10-12). Both arise by differentiation of precursor cells in the bone marrow, both then circulate in the blood for a few days (at this stage, the macrophage is called a monocyte—see Fig. 10-10), and then both pass from the blood into the tissues where they carry out their assigned functions. Neutrophils are the more numerous, accounting by themselves for 60% of circulating white blood cells in spite of a half-life in the blood of only 6–7 hours.[4] However, the macrophage, also called a histiocyte, is larger, longer-lived, and more complex. In addition to its role as a phagocyte, the macrophage assists T- and B-type lymphocytes in establishing an immune response (in ways that are only partially understood) and in return may be stimulated to greater phagocytic activity by lymphocyte secretions.

Both the macrophage and neutrophil are known to have plasma membrane receptors for the Fc portion of IgG antibodies and for the activated form of C3, which is one of the serum proteins collectively referred to as "complement." Once an antibody response has been established, these receptors are important in clearing the circulation and tissues of antigen-antibody complex. For example, bacteria coated with specific IgG antibody are much more rapidly phagocytosed than uncoated bacteria. (The antibody is said to be an *opsonin*, which, according to George Bernard Shaw, is "what you butter the disease germs with to make your white corpuscles eat them.")

Additional clues to how phagocytic cells may recognize their potential targets came from work on the clearance of proteins from serum. In

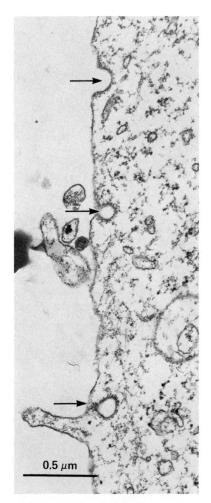

FIGURE 10-11 Pinocytosis. Portion of a neuroblastoma (malignant neural cell) in culture showing pinocytosis. Note how the vesicle forms by invagination of the cell membrane (consecutive stages top to bottom). Note also the dense coating of the pinocytic vesicles, the composition and function of which is not clear. [Courtesy of J. Ross, J. Olmsted, and J. Rosenbaum, *Tissue and Cell*, 7: 107 (1975).]

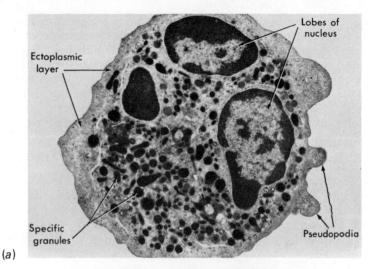
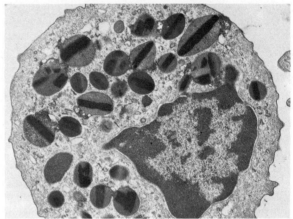

FIGURE 10-12 Granulocytes. The major phagocytic cells of higher animals are the monocyte-macrophage (Fig. 10-10) and the neutrophil or heterophil. The neutrophil is part of the polymorphonuclear *granulocyte* series having lobulated nuclei and "specific granules" containing digestive enzymes. (a) Guinea pig neutrophil. (b) Guinea pig eosinophil (i.e., an eosinophilic leukocyte). The eosinophil is much less common than the neutrophil, but actively phagocytoses antigen-antibody complexes. [Courtesy of W. Bloom and D. W. Fawcett, *A Textbook of Histology*, 10th ed. Philadelphia: W. B. Saunders, 1975.]

1968, A. G. Morell and his colleagues were studying the clearance from blood of a glycoprotein called ceruloplasm, a transporter of copper ions. They found that treating ceruloplasm with neuraminidase, which removed terminal sialic acid (neuraminic acid) caused liver cells (hepatocytes) to destroy the protein by phagocytosis in about 30 minutes instead of the usual time of several days. However, it turns out that removal of sialic acid was not the major factor. Rather, it was exposure of the underlying oligosaccharide chain. This conclusion was demonstrated by showing that when the underlying oligosaccharide was altered chemically, there was resumption of a near-normal circulating lifetime of ceruloplasm even in the absence of sialic acid.

The principle of using carbohydate residues as recognition markers for phagocytosis is now well established. They serve as an accurate way of directing substances to specific phagocytic cells. The protective effect of sialic acid or comparable surface feature is also quite general. Thus, some bacteria (e.g., *Streptococcus pneumoniae*) are notoriously resistant to phagocytosis as long as they retain their polysaccharide capsule. The mammalian erythrocyte is an additional example, for it is rapidly phagocytosed by cells of the spleen if sialic acid is first removed from the sur-

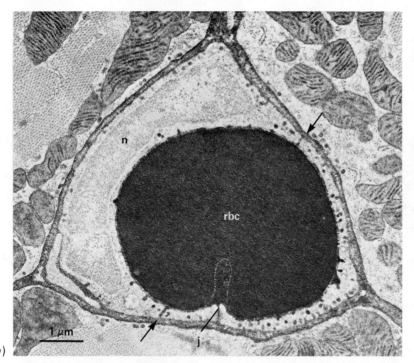
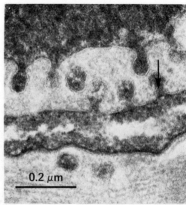

FIGURE 10-13 Endocytosis in a Blood Capillary. (a) Capillary from rat diaphragm 2 min after injection of a heme-containing peptide to provide visualization. Note the presence of two complete channels (arrows), temporarily created by endocytic vesicles, across the endothelial cell that forms the capillary. Nucleus (n); junction (j) where the cell meets itself after forming the lumen; trapped erythrocyte (rbc). (b) Detail of the endothelial cell, lumen at the top, showing vesicles forming, in transit, and emptying into the space around the capillary (arrow). [(a) Courtesy of N. Simionescu, M. Simionescu, and G. Palade *J. Cell Biol.*, **64:** 586 (1975); (b) courtesy of N. Simionescu and M. Simionescu, *J. Cell Biol.* **70:** 608 (1976).]

face of the erythrocyte. And since malignant cells have very little sialic acid on their surface to begin with, we may also have an explanation for their destruction by the immune system, in most cases before a cancer can be established.

Specificity of phagocytosis, then, may be explained by a combination of positive and negative signals. Positive signals cause binding to specific target cells and subsequent uptake of the particle by those cells. Negative signals, on the other hand, prevent activation of phagocytosis. Without these signals, both phagocytosis and pinocytosis are relatively nondiscriminatory. Thus, macrophages ingest a wide variety of particulate matter, including latex spheres, carbon, and iron filings. And pinocytosis probably occurs more or less continuously in most cells as a general mechanism for bulk transport across the cell membrane.

An example of a relatively continuous pinocytic process is seen in Fig. 10-13, which shows an endothelial cell lining a blood capillary. These cells use pinocytosis to transport serum from within the capillary to the surrounding space. The process in endothelial cells, however, is unusual in that vesicles are delivered to the other side of the cell unchanged. In most cell types, vesicles produced by phagocytosis or pinocytosis fuse with lysosomes to allow digestion of the vesicle's contents. That subject is considered next.

Lysosomes. Endocytosis is normally followed by fusion of the new vesicle with a lysosome. Lysosomes are small (typically 0.5 μm), membrane-enclosed bodies that contain destructive enzymes called hydrolases capable of breaking down a wide variety of particulate materials, including all the classes of macromolecules and many small molecules as well. Products of this breakdown are able to pass out of the lysosome to become available to other areas of the cytoplasm. Altogether, more

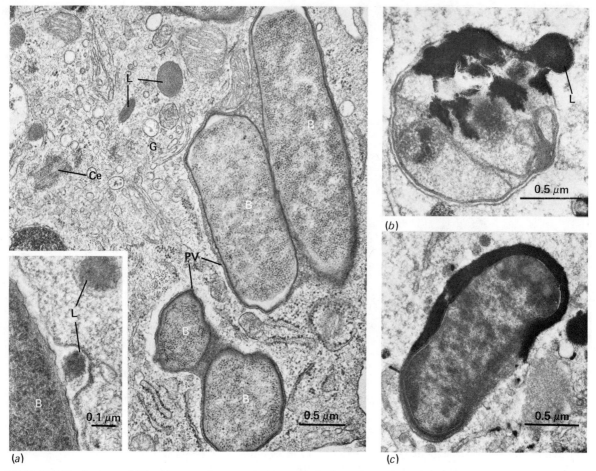

FIGURE 10-14 Lysosomal Digestion. (a) Portion of a guinea pig macrophage, containing ingested bacteria (B) in phagocytic vacuoles (PV, also called phagosomes). Lysosomes (L) are seen in the Golgi region (G). Centriole (Ce). INSET: Fusion of a lysosome and bacteria-containing phagosome to form a digestive vacuole. (b) Portion of a rabbit macrophage, again showing fusion of lysosome and bacteria-filled phagosome. Staining for the lysosomal enzyme aryl sulfatase reveals its spread. (c) The lysosomal enzyme has spread nearly around the trapped bacterium. [Courtesy of B. Nichols, D. Bainton, and M. Farquhar, *J. Cell Biol.*, **50**: 498 (1971).]

than 40 different hydrolases have been identified. The one most often used as a "marker," because of the ease with which its activity can be localized in electron microscopy, is acid phosphatase. Another is called aryl sulfatase. The need to protect the cytoplasm is presumed to be the reason for confining these enzymes to their own organelle.

Lysosomes are probably present in all animal cells except the mature mammalian red blood cell, and they are present in many plant cells as well. They are formed principally as vesicles pinched off from Golgi membranes but in some cells a specialized portion of the endoplasmic reticulum called GERL (*G*olgi-associated *ER* from which *L*ysosomes arise) is probably also involved.

Newly formed lysosomes are called *primary lysosomes*. Fusion with an endocytic vacuole produces a *secondary lysosome* or *digestive vacuole* (see Fig. 10-14). In neutrophils (and eosinophils) this step is preceded by fusion of specific granules with the phagocytic vacuole, so that a digestive vacuole in these cells contains enzymes from two sources.

The objective of secondary lysosome formation is to mix lysosomal hydrolases with the ingested material so that breakdown of the latter can occur. The digestion products (amino acids, nucleotides, etc.) leave the lysosome to become sources of energy or raw materials for biosynthetic pathways. As products leave, the lysosome often diminishes in size.

This change requires a loss of surface area, which can be achieved by an inward budding of the limiting membrane in much the same way as pinocytic vesicles are formed. The internalized membrane is then digested. This process, which is particularly prominent in the digestion of pinocytic vesicles, leads to an electron micrographic picture of a vacuole (the secondary lysosome) containing numerous small vesicles and fragments, a configuration known as a *multivesicular body* (see margin).

Fusion of a lysosome with an endocytic vacuole is a selective process. The vacuoles do not fuse with endoplasmic reticulum, mitochondria, etc. Nor does the lysosome fuse with plasma membrane under ordinary circumstances, even though plasma membrane and phagocytic vacuole are made of the same stuff. However, in the presence of cytochalasin B, extracellular release of lysosomal enzymes does occur readily, indicating that perhaps the microfilamentous web beneath the plasma membrane normally protects it from fusion with lysosomes.

There is some evidence also for involvement of microtubules in guiding endocytic vacuoles, lysosomes, and specific granules, and there is evidence for regulation by cyclic nucleotides, probably via their effect on microtubules. In certain cells, at least, colchicine and cAMP seem to inhibit and cGMP to promote digestive vacuole formation.

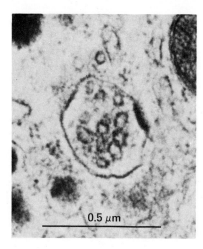

FIGURE 10-14(d) The multivesicular body. The digestive vacuole or secondary lysosome shrinks by inward budding of its limiting membrane to yield the picture of tiny vesicles within a vacuole as shown here. [Courtesy of J. Braaten, U. Jarlfors, David Smith, and D. Mintz, *Tissue and Cell*, 7: 747 (1975).]

Killing and Digestion. Simple enzymatic digestion by hydrolases is not the only process that occurs in secondary lysosomes. Many microorganisms, while they are viable, resist attack by hydrolases. Hence, mechanisms have evolved specifically to kill ingested organisms prior to their breakdown.

Lysosomes typically contain *lysozyme*, an enzyme that degrades bacterial cell walls, especially the simpler walls of gram-positive bacteria. The lysosomes of neutrophils also contain lactoferrin, which binds iron and other metal ions tightly, thus inhibiting growth of the ingested organisms. The contents of the digestive vacuoles are usually very acidic (less than pH 4 in neutrophils), which by itself inhibits growth of many bacteria. And finally, oxidases, thought to be present at the surface of the cell membrane and thus also on the inside surface of digestive vacuoles, produce hydrogen peroxide from molecular oxygen. Hydrogen peroxide is microbicidal in its own right, but in neutrophils and most macrophages it is also utilized by an enzyme called myeloperoxidase to covalently fix Cl^- or I^- onto organic targets such as bacteria. This halogenation greatly facilitates killing of the organism.

Defects in Lysosomal Function. The importance of hydrogen peroxide in phagocytic killing of microorganisms is illustrated in humans by an X-linked recessive condition called chronic granulomatous disease. A granuloma is a collection of dead and dying macrophages. The disease is a deficiency of the oxidase that produces hydrogen peroxide in the digestive vacuole. The macrophages and neutrophils of these individuals are unable to kill effectively even common bacteria, which may live and multiply inside the cell, eventually killing it instead. Such people are prone to all sorts of fungal and bacterial infections, in spite of an immune system that is otherwise normal.

Other types of defects in lysosomal function include failure to fuse with the endocytic vacuole and failure to digest certain classes of chemical substances due to a lack of one or more critical enzymes.

For example, fusion of lysosome and phagocytic vacuole is sometimes inhibited, for reasons that are not understood, when the vacuole contains live bacteria responsible for tuberculosis or leprosy, or when it contains live toxoplasma, a common protozoan parasite contracted from raw meat or cat feces.[5] Lack of fusion permits these organisms to live within the phagocytic cells and often to destroy them and to establish active disease. Algae that live symbiotically in protozoa seem to be similarly protected. The only clue to how fusion might be avoided in these cases comes from the observation that concanavalin A-induced endocytosis also results in vacuoles that frequently fail to fuse with lysosomes. Hence, the distribution of membrane glycoproteins and the effect this has on underlying filaments and tubules may in some way be involved.

Fusion of phagocytic vacuoles with lysosomes and specific granules may also come too soon. If fusion occurs before the plasma membrane has closed over the forming vacuole, contents of the lysosomes and specific granules may *regurgitate* through the vacuole and into the extracellular space. Because of the destructive character of these enzymes, a local inflammation (redness, swelling, tenderness, etc.) may result. This process, especially common with neutrophils, is important in localizing an infection and hence not always undesirable. However, it is probably also responsible for some of the degenerative joint changes in rheumatoid and gouty types of arthritis. The effectiveness of colchicine in the treatment of gout is probably due to its inhibition of lysosome and granule movement and fusion, which are processes that involve microtubules.

Another type of defect in lysosomal function occurs if an essential hydrolase is missing. Because there are many hydrolases, there are many (over two dozen so far defined) of these *lysosomal storage diseases*. They are almost always inherited as autosomal recessive traits because one good gene out of the two alleles inherited from the two parents is usually enough to prevent symptoms. The diseases have names like Tay-Sachs, Hunter's, Hurler's, Niemann-Pick, and so on, because they were described long before lysosomal function was understood. Most result in accumulation of undigested lipids or polysaccharides, which eventually fill the cell and thus compromise its function. The type of cell most affected will be determined by which enzyme is faulty, but the central nervous system is commonly affected, resulting in progressive lack of body control and loss of mental function leading to a slow death. It may take months or years to accumulate enough debris to cause significant impairment. Hence, the affected infant may at first seem perfectly normal and the parents quite unaware of the tragedy that lies ahead.

The lysosomal storage diseases have attracted much attention in spite of the fact that they are relatively uncommon. They are of particular interest because understanding them biochemically leads to a better understanding of lysosomes; because carriers can often be detected by virtue of reduced but not absent enzyme activity; because the child can be diagnosed by amniocentesis (puncture of the womb with a needle) early enough in pregnancy to consider an abortion; and most recently because for the first time there is at least some hope for eventual treatment. That hope stems from the discovery that liposomes (described in Chap. 3) can be used as carriers of the missing lysosomal enzyme. When liposomes are phagocytosed, fusion with the defective lysosome dissolves the liposome and releases the needed enzyme. So far it only works well in tissue culture, but someday, perhaps, these unfortunate children may be spared.

Autophagy. Although lysosomes are present in probably all nucleated cells of both animals and plants, we have described their activities only in the major phagocytic cells. Their function in other cells is much the same. That is, they fuse with endocytic (including pinocytic) vesicles to destroy the vesicle and its contents. Lysosomes are involved not only in this process of digesting extracellular materials, known as *heterophagy*, but also in the digestion of intracellular substances, an activity known as *autophagy*.

In plants, for example, the cytoplasm frequently has large vacuoles containing reserve food supplies in a stored form. These storage polymers can be broken down to useful size by fusion of the vacuole with lysosomes.

Lysosomes are similarly responsible for turnover of other organelles in both plants and animals, a process that is accelerated during involutional states—in a muscle that is not being used, during starvation, after injury, or during the remodeling that occurs with cellular differentiation. Even in unstressed normal cells, the turnover of organelles by autophagy is relatively rapid. The half life of liver cell mitochondria, for example, is five to six days, for peroxisomes it is one to two days, and for ribosomes about five days.

The autophagy of cytoplasmic organelles starts with an *isolation envelope*, which is a sack of smooth-surfaced membrane that wraps around the material to be removed and separates it into a sort of phagocytic vacuole known as an *isolation body* (Fig. 10-15). The source of the isolation membrane, at least in damaged liver and kidney, seems to be the endoplasmic reticulum. However, recent work by Michael Locke and his associates at the University of Western Ontario indicates that the isolation membrane in insect cells during metamorphosis, and in some other animal cells, is derived instead from the Golgi apparatus. In any case, lysosomes can fuse with the isolation body to create an autophagic digestive vacuole and release breakdown products to the cytoplasm to be used as an emergency source of food or as raw material for cellular remodeling.

Residual Bodies. Although killing and breakdown in the lysosomes is efficient, some materials are simply not digestible. These substances may be released from the cell by fusion of the digestive vacuole with the cell membrane (exocytosis). This process is very efficient in protozoa, but not so efficient in higher animals. Hence, inactive lysosomes containing debris may accumulate. (See Fig. 10-16.)

These debris-filled organelles, called *residual bodies*, are of no consequence to cells like macrophages or neutrophils, because the cells don't live long enough for significant amounts to accumulate. In cells that do not divide, and are not replaced, substantial quantities of this material may accumulate. In neurons and muscle cells, especially, there is a steady increase in the number of these bodies, which are called in this context *lipofuscin granules*. The age of an animal can often be estimated from their concentration, but whether enough ever accumulates in a normal lifespan to compromise the function of the cell is unclear.

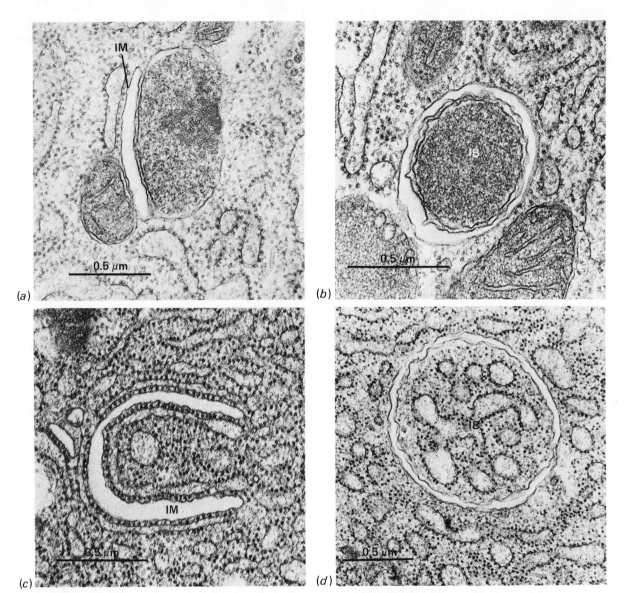

FIGURE 10-15 Autophagy. Remodeling of fat body cells of the butterfly *Calpodes* during preparation for pupation occurs by autophagy. Organelles are surrounded by isolation membranes (IM) in the form of an envelope. This creates an isolation body (IB) where lysosomal digestion can occur. (a), (b) Two stages in the isolation of microbodies (peroxisomes). (c), (d) Isolation of a portion of the rough endoplasmic reticulum. [Courtesy of M. Locke. (a) from M. Locke and J. McMahon, *J. Cell Biol.*, **48**: 61 (1971); (b)–(d) from M. Locke and A. Sykes, *Tissue and Cell* **7**: 143 (1975).]

10-4 THE NUCLEAR ENVELOPE

So far in this chapter, we have dealt mostly with the plasma membrane, its specialized properties and functions. The discussion of endocytosis, however, brought us into the cell and led to consideration of an important internal organelle, the lysosome.

In this and the next several sections, other internal organelles of eukaryotes are examined in a sequence that follows the synthesis and secretion of a protein. By the time we get back to the plasma membrane again in the bulk transport process of secretion itself, all of the major internal organelles that are common to animal and plant cells will have

SECTION 10-4
The Nuclear Envelope

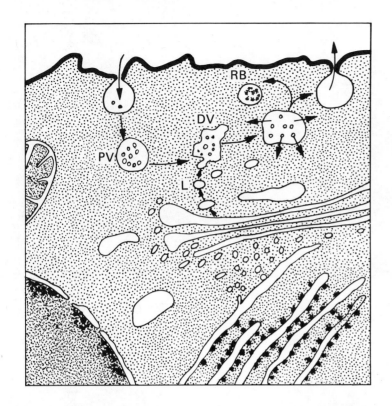

FIGURE 10-16 Fate of the Digestive Vacuole. The entire cycle is diagramed, including phagocytosis and fusion of phagocytic vacuole (PV) with lysosome (L) to create a digestive vacuole (DV). The digestive vacuole may be eliminated by reinsertion into the plasma membrane. Often, especially in cells of higher organisms, it becomes instead an inert residual body (RB), containing undigestible debris. Depending on appearance, the residual body may also be called a lipofuscin granule. Note flow of vesicles from a smooth-surface area of ER toward the Golgi saccules.

been discussed. Missing from this chapter, of course, are the mitochondrion, chloroplast, and peroxisome, which were treated together in Chap. 7. Also missing, due to a limitation of space, are the various storage organelles mentioned in Chap. 1, and several specialized organelles found in only a few cell types.

We begin with the nuclear envelope, through which messenger RNA, synthesized according to the scheme outlined in Chap. 5, must pass.

Structure of the Nuclear Envelope. The nucleus is the heart of the cell. It is here that almost all of the cell's DNA is confined, replicated, and transcribed. The nucleus is bounded by an envelope of two membranes, fused at frequent intervals to create pores, 500–800 Å in diameter, that permit communication between the nucleoplasm and cytoplasm (see Fig. 10-17).

The outer membrane of the nuclear envelope in most cell types is directly connected to the endoplasmic reticulum, and like the endoplasmic reticulum often has ribosomes attached to it. Because of this connection, the space between the inner and outer nuclear membranes, called the *perinuclear space* or *perinuclear cisterna*, is often continuous with the cisternae of the endoplasmic reticulum (Fig. 10-18).

The inner nuclear membrane has no ribosomes attached (though it is capable of binding them *in vitro*) but instead is connected to chromatin fibers. The attachments to chromatin seem to be primarily limited to regions near but not at nuclear pores. The possible significance of that attachment will be considered in the next chapter, when DNA replication is discussed.

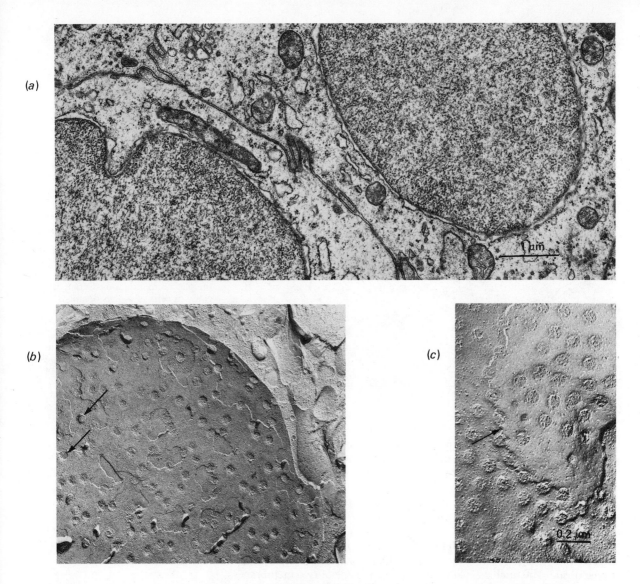

FIGURE 10-17 The Nuclear Envelope. (a) Rat parathyroid gland showing two adjacent cells. Note the wavy outer nuclear membrane and pores (areas of fusion of inner and outer membranes). (b) Freeze-fracture view of a rat kidney nucleus, revealing numerous pores (arrows). The outer membrane is broken away in patches. (c) Nuclear membrane of *Tetrahymena pyriformis*, a protozoan. Note the greater pore density in these rapidly growing cells. The fracture line (arrow) is such that both the particle-studded inner membrane and much smoother outer membrane are seen. [(a) Courtesy of N. Nakagami, W. Warshawsky, and C. P. Leblond, *J. Cell Biol.* **51**: 596 (1971); (b) courtesy of G. Maul, J. Price, and M. Lieberman, *J. Cell Biol.* **51**: 405 (1971); (c) courtesy of V. Speth and F. Wunderlich, *J. Cell Biol.*, **47**: 772 (1970).]

Because of the close association between membranes of the endoplasmic reticulum (ER) and nuclear envelope—an association that is emphasized by the realization that new nuclear envelopes form from ER after cells divide—it is not surprising that there are few chemical or physical differences between them. Both types of membrane are somewhat thinner (6–8 nm) than other cellular membranes and the two have nearly identical proteins and lipids, though not necessarily in the same

SECTION 10-4
The Nuclear Envelope

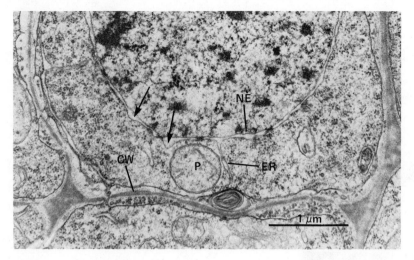

FIGURE 10-18 Continuity of Outer Nuclear Membrane and Endoplasmic Reticulum. Androgonial cell from a liverwort (*Basia pusilla*). The nuclear envelope (NE) also appears to be forming surface projections, or blebs (*arrows*). Cell wall (CW). Plasmid (P). These cells occasionally show continuity between the ER and plasma membrane, a very rare occurrence. [Courtesy of Z. Carothers, *J. Cell Biol.* **52:** 273 (1972).]

quantity. An exception to the similarity between nuclear envelope and ER are those enzymes associated with DNA replication, which are found attached to the inner nuclear membrane only.

Nuclear Pores. Since RNA synthesis occurs within the nucleus while protein synthesis takes place in the cytoplasm, it is evident that substances must be able to move outward. In fact, there is a flow of material in both directions through the nuclear pores, which, in spite of their size, are not wide open channels. Rather, they seem to exercise at least some degree of selectivity over what gets through them. This observation has prompted much study of their structure.

In thin sections such as Fig. 10-19a, the pores are seen to be formed by fusion of the outer and inner membranes. The pore opening contains fibers and granules. Some of the latter can be identified as ribosomal precursors in transit to the cytoplasm. Much of the material, however, is a structural part of the *pore complex*. Careful examination with special techniques reveals eight granules, each 10–20 nm in diameter, around the pore margin on the inside face, another eight around the outside face, and eight less well-defined structures in the opening itself. There is often one more particle in the exact center. This eightfold symmetry and complex architecture seems to be nearly universal in eukaryotic cell nuclei of both plants and animals (see Fig. 10-19b and c).

Pore size and the number of pores per nucleus can vary even within the same cell at different times in its life cycle. Often, the number increases with the activity of the cell. In lymphocytes stimulated to divide, for instance, there is a rapid increase in pore number, though the mechanism of formation remains unclear (Fig. 10-19a). In some cells there are several thousand pores per nucleus, occupying altogether as much as 30% of the total surface.

Nucleocytoplasmic Exchange. Small molecules up to about 1000 daltons in mass are commonly found distributed in equal concentrations in the cytoplasm and nucleoplasm and appear *in vivo* to be relatively free to pass back and forth. In spite of these observations, isolated nuclei reportedly have membrane potentials and considerable electrical resis-

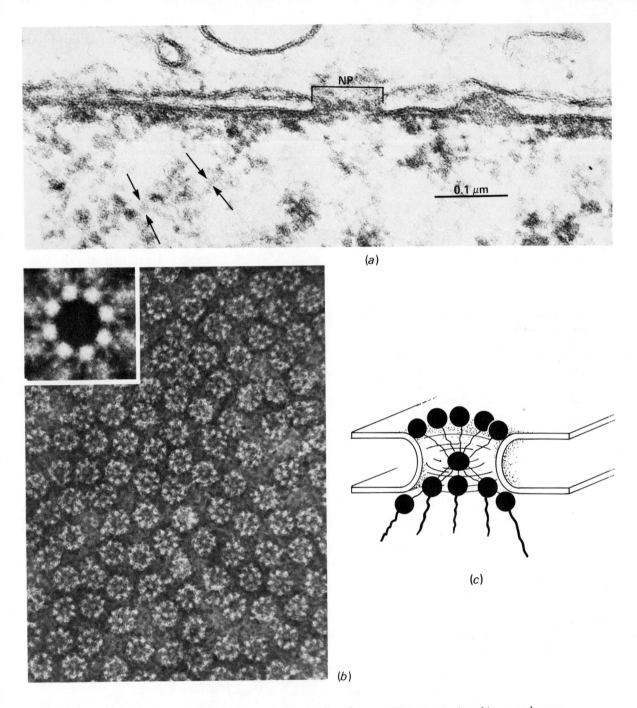

FIGURE 10-19 Nuclear Pores. (a) Thin section of a human lymphocyte. Note continuity of inner and outer membrane at the pore (NP) and the adjacent indentation that may be a forming pore. *Arrows* outline chromatin. (b) Face view, negatively stained, showing pores in an oocyte (forming egg cell) nucleus of the newt *Taricha*. Pore diameter is about 100 nm. The ring of eight particles is called an annulus. The eightfold symmetry of the pore complex is clearly seen in the INSET, which shows a pore from an onion root tip nucleus using rotational averaging, a photographic technique referred to as a Markham projection. (c) Diagram of the nuclear pore complex. [(a) Courtesy of G. Maul, J. Price and M. Lieberman, *J. Cell Biol.* **51**: 405 (1971); (b) courtesy of A. Faberge, *Z. Zellforsch.* **136**: 183 (1973); INSET courtesy of W. Franke, *Phil. Trans. Roy. Soc. London, Ser. B*, **268**: 67 (1974).]

tance, a situation that can only be explained by some restriction to the free flow of ions through the pores.

Thus, the situation regarding nuclear permeability is not at all clear. There is general agreement that active transport across nuclear pores has not been demonstrated. And it is obvious that, regardless of the selectivity conferred by the pore complex, there is considerable traffic through it. Nucleotides, in particular, must be able to get in readily and messenger RNA and immature ribosomes must get out. In addition, the use of radioactive tracers shows that histones and other nuclear proteins, which are synthesized outside the nucleus, rapidly and efficiently collect in the nucleoplasm.

Relationship to the Cytoskeleton. Microfilaments and microtubules are frequently seen in the vicinity of, and are sometimes attached to, the nuclear membranes. In simpler eukaryotic cells, special areas of the nuclear membrane clearly serve as microtubule organizing centers to assemble the spindle apparatus needed for chromosome separation during cell division (see Chap. 11). The nuclear membrane thus serves in place of centrioles in these cells. In addition, microtubules can be assembled on the outer membrane of many nuclei and extend from there into the cytoplasm. Although the function of these cytoskeletal attachments to the nucleus is not known, they may well direct cytoplasmic streaming of substances toward and away from the nucleus, and could serve also to anchor the nucleus in place.

Nuclear Blebbing. An irregular ruffling of the outer nuclear membrane is often observed and, in some instances (especially in plants), blebbing has been seen (Fig. 10-20). Blebs are fingerlike extensions of the outer

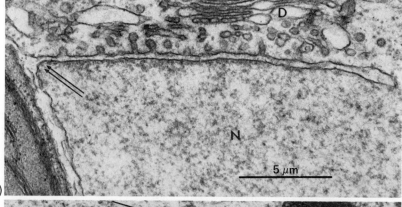

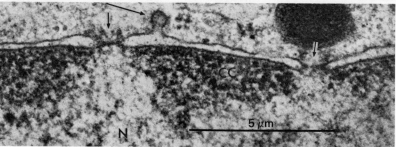

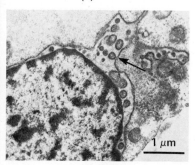

FIGURE 10-20 Nuclear Blebbing. Pouches, or blebs, in the nuclear membrane are sometimes pinched off as vesicles. (a) An alga, *Botrydium granulatum*. Note adjacent dictyosome (D) which may be recipient of the vesicles. Nucleus (N). *Double arrow* points to a continuity between the perinuclear cisterna and the ER. (b) Rat adrenal cortex. Note the single vesicle, nearly completed (*long arrow*), and adjacent pores (*short arrows*). Chromatin (CC). (c) Cell from the lung of a human fetus. The outer membrane balloons out into the cytoplasm as dilated rough ER that appears to fold back into itself (*arrow*) to form vesicles. [(a) and (b) courtesy of W. Franke, *Int. Rev. Cytol.*, **Suppl. 4**: 71 (1974), © Academic Press, N.Y.; (c) courtesy of J. McDougall, *J. Cell Sci.*, **22**: 67 (1976).]

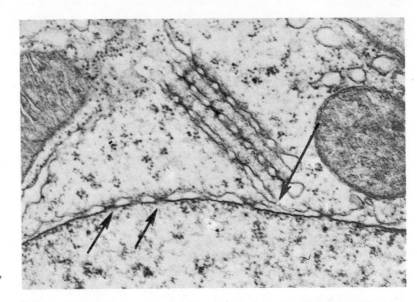

FIGURE 10-21 Annulate Lamellae. Chick embryo liver cell after two days of organ culture. One can see continuity with the rough ER at one end of the annulate lamellae and continuity with the nuclear envelope at the other end (*long arrow*). Note nuclear pores (*short arrows*). [Courtesy of Camillo Benzo, State Univ. of N.Y., Upstate Medical Center.]

membrane that may be pinched off to form vesicles. Many workers believe that those regions of the outer nuclear membrane exhibiting this activity (regions that are always free of ribosomes) represent sites of new membrane formation, a function otherwise associated with the ER itself, and that the vesicles are destined to coalesce to form new lamellae of the ER or Golgi apparatus.

In some cells *annulate lamellae* (Fig. 10-21) are seen near the nuclear envelope. These flattened sacks have fenestrae ("windows") similar in structure to nuclear pores. There is some evidence that these lamellae are derived from the nuclear envelope by a process akin to blebbing and that they subsequently mature to form endoplasmic reticulum.

10-5 ENDOPLASMIC RETICULUM

The endoplasmic reticulum (ER) is a network of interconnecting tubes and channels found in all nucleated eukaryotic cells. We have already discussed its role in regulating Ca^{2+} availability in muscles, mentioned briefly its contribution to the nuclear envelope during cell division, and noted its probable role in giving rise to peroxisomes and possibly autophagic isolation membranes. In addition to these activities, numerous enzymes, either associated with the membranes or enclosed by them, support lipid synthesis, drug detoxification, and certain steps in the synthesis of carbohydrate and its addition to proteins. And finally, the endoplasmic reticulum is involved in transport and packaging of products for secretion, a function in which the Golgi apparatus (Section 10-6) also participates.

Structure of the Endoplasmic Reticulum. The membranes of the endoplasmic reticulum are found in two basic configurations, lamellae (flattened sacs) and tubes. Either form may have attached ribosomes, though ribosomes are only rarely found on tubular ER. The presence or absence of ribosomes defines the *rough* and *smooth ER*, respectively. Much of what we know about the structure and biochemistry of these

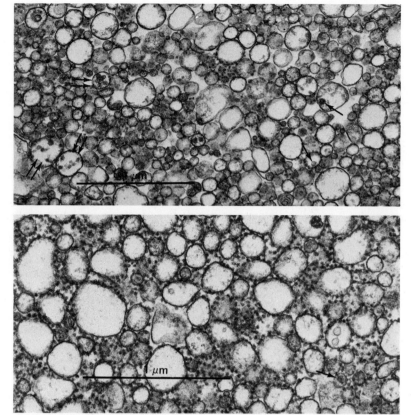

FIGURE 10-22 Rat Liver Microsomes. These vesicles are produced from the ER when the cell is disrupted. The contents of the microsomal vesicles (in this case lipoprotein—*arrows*) reflect that of the original ER cisternae. (a) Smooth microsomes. (b) Rough microsomes. [Courtesy of H. Glaumann, A. Bergstrand, and J. Ericsson. Similar to *J. Cell Biol.*, **64**: 356 (1975).]

membranes has been established by studies on isolated fragments known as *microsomes*.

When a cell is disrupted by homogenization, even rather gently, much of the ER breaks up and pinches off as vesicles, which can then be collected as microsomes and analyzed for their content and composition (Fig. 10-22). Analysis of microsomal contents is meaningful because little exchange occurs between the lumen of the ER and cytoplasm during microsome formation, which means that lumenal contents are trapped in microsomal vesicles. The cell type most studied in this way is the mammalian liver parenchymal cell (hepatocyte)—a large, complex cell that comprises the bulk of the liver. The supply is plentiful because the liver is a large organ and because the hepatocyte ER is extensive, with a surface area 30 to 40 times that of the plasma membrane itself.

Chemical and structural analysis of liver microsomes shows them to be composed of membrane that is not strikingly different from other membranes. The membranes are about two-thirds protein and one-third lipid. Most of the lipid, as usual, is phospholipid and much of the protein is glycoprotein, though the fraction of carbohydrate is small. Freeze-fracture and etching shows that some of the proteins extend well into the membranes and probably through to the other side. Other proteins are confined largely to one surface or the other, an asymmetry that is also true of the lipids.

The quantity of ER in a cell is not fixed. There is constant synthesis and turnover, making it possible to increase or decrease the amount in response to need. For example, regular feeding of phenobarbital, a common sedative and antiepileptic drug, doubles the amount of smooth ER in the livers of experimental animals and no doubt of humans, too. The reason for this increase is that phenobarbital and many other drugs and foreign substances are metabolized and thus detoxified by the liver. Drug metabolism, however, is only one of the diverse functions of the smooth ER.

Functions of the Smooth ER. It is a little artificial to try to separate the activities of the rough and smooth ER because of their numerous interconnections. One obvious difference, of course, is participation of the rough ER in protein synthesis by virtue of its associated ribosomes. Smooth ER, on the other hand, is more involved with lipid synthesis.

Cells concerned largely with synthesis of lipids have well-developed smooth ER just as cells that synthesize and secrete proteins have well-developed rough ER. Thus, cells of the adrenal cortex (which synthesize steroid hormones) have abundant smooth ER (see Fig. 10-23). So do intestinal absorptive cells, for the smooth ER supports triglyceride synthesis from the mixture of fatty acids, monoglycerides, and diglycerides taken up by these cells from the intestine. (Dietary fat is broken down by pancreatic lipase to provide these raw materials.) Fat droplets (chylomicrons), comprised almost exclusively of triglyceride, can be readily identified within the smooth ER near the intestinal lumen shortly after a meal.

The smooth ER of liver cells has functions that, while not unique to the liver, are certainly more prominent there. These functions include regulation of glycogen breakdown (via glucose 6-phosphatase, an ER-associated enzyme), and the conjugation and oxidation of a variety of naturally occurring and foreign substances, the purpose of which is usually

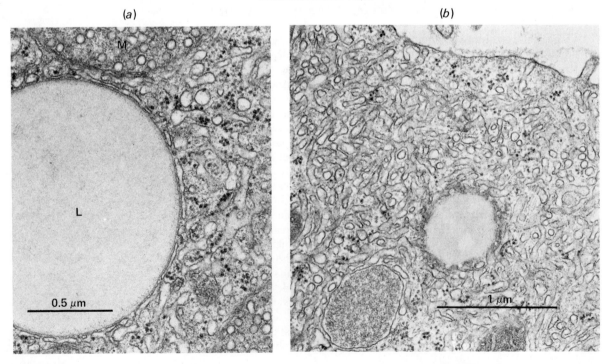

FIGURE 10-23 Smooth Endoplasmic Reticulum. (a) From the adrenal cortex of a rat. Cisternae of smooth ER are aligned along the lipid droplet (L) and are seen as tubular structures between the mitochondrion (M) and lipid droplet. Note the unusual, smooth ER-like cristae in the mitochondrion. Mitochondria participate in some of the intermediate steps in steroid synthesis in these cells. (b) From a pig testis. Note tubular smooth ER in this steroid-hormone–producing cell. [Courtesy of Daniel Friend. (a) from D. Friend and G. Brassil, *J. Cell Biol.*, **46**: 252 (1970).]

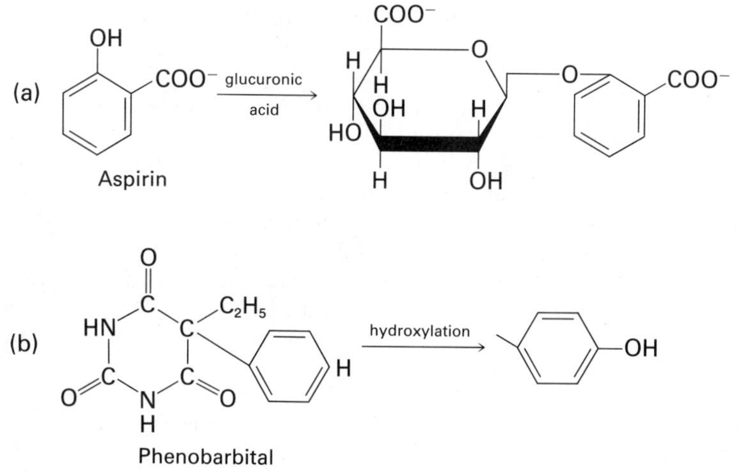

FIGURE 10-24 Biotransformation by the Endoplasmic Reticulum. Two of the more common reactions are shown. (a) Glucuronic acid conjugation of aspirin (acetylsalicylic acid); (b) hydroxylation of phenobarbital.

to terminate their biological activity. When applied to drugs, this process is referred to as *detoxification*.

The reactions that detoxify substances consist of oxidations, reductions, hydrolyses, or a covalent linking (conjugation) to soluble small molecules, particularly conjugation to glucuronic acid, a sugar derivative (Fig. 10-24). These changes either inactivate the substance or make it more soluble and hence more readily eliminated by the kidneys. A wide variety of natural and artificial chemicals are affected in this way, including environmental pollutants and many drugs. Birth control pills, for example, were made possible by the discovery of drugs having some of the activities of estrogens and progesterone but which are not rapidly inactivated by the liver as are the natural steroid hormones.

In general, these reactions of the smooth ER serve a valuable protective function and also play a role in the normal handling of fatty acids, bile salts, steroids, and heme recovered from hemoglobin breakdown. The responsible enzymes are associated with endoplasmic reticulum in several different cell types, and related systems have been identified even in prokaryotes. They are especially prominent, however, in liver.

Most of the biotransformations in this category are oxidations, primarily because of the multitude of ways in which organic molecules can be oxidized. The first key to understanding the mechanism was a proposal by Howard S. Mason of the University of Oregon Medical School for the existence of "mixed function oxidases" utilizing both NADPH and molecular oxygen. The principal enzyme is cytochrome P450, a heme protein so named because it absorbs light maximally at 450 nm when in the reduced form. This protein is capable of binding substrate and O_2, passing one of the oxygen atoms of O_2 to the substrate in the form of an OH group (an oxidation of the substrate) while the other oxygen atom combines with H^+ to form water:

$$S + P450(Fe^{3+}) \xrightarrow{O_2 + 2e^-} S\text{—}P450(Fe^{2+})\text{—}O_2^- \xrightarrow{2H^+} P450(Fe^{3+}) + S\text{—}OH + H_2O \quad (10\text{-}1)$$

Here, S is the substrate and Fe^{3+} the heme-associated iron of P450. The electrons are delivered to P450 mostly from NADPH by an endoplasmic-reticulum–associated electron transport chain involving flavoproteins and in some cases another cytochrome known as cytochrome b_5.

Cytochrome P450 and other elements of the *microsomal electron transport chain* seem to be membrane-bound proteins present mostly on the cytoplasmic surface of the smooth endoplasmic reticulum. Their concentration, however, varies with need. As mentioned earlier, phenobarbital can cause the amount of smooth ER to more than double; but the amount of P450 per cell may at the same time increase fivefold. Although there are apparently several versions of P450 with somewhat different specificities, there is broad overlap so that induction with one substrate increases the ability of the ER to detoxify other substrates handled by the same system even though the other substrates may not themselves be effective inducers of P450. Thus, for example, treatment of a human with more than one ER-metabolized drug at a time often requires careful adjustment of dosage.

Functions of the Rough ER. It has long been assumed that proteins destined for secretion from the cell are synthesized on ER-bound ribosomes while cytoplasmic proteins are translated for the most part on free ribosomes (see Fig. 10-25). This separation is not hard and fast, for even cells that secrete little or no protein typically have significant amounts of rough ER. However, when radioactive amino acids are injected into a cell that makes secreted proteins, the radioactivity very quickly becomes recoverable in the rough microsomal fraction, thus supporting our generalization. What is more, further investigation has revealed that this rough ER-associated protein is immune to proteases. Not only does translation occur on membrane-bound ribosomes, but the secretory protein, instead of passing into the cytoplasm, appears to pass instead into the cisterna of the rough ER, and hence into microsomes where it is protected from experimentally added proteases.

From the rough ER, secreted proteins pass sequentially to smooth surfaced ER, Golgi apparatus, secretion vesicles, and finally to the exterior of the cell. The questions to be answered here are (1) how does mRNA

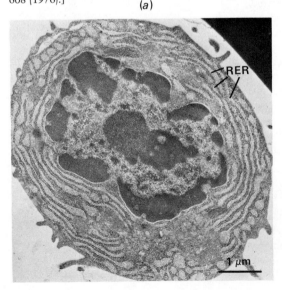

FIGURE 10-25 Rough Endoplasmic Reticulum. (a) Rabbit plasma cell. The swollen cisternae of the rough endoplasmic reticulum (RER) contain antibodies, which these cells secrete. (b) Rough ER from an acinar cell of rat pancreas. Intracisternal space (IS). Ribosomes line the surface of the membranes, which have a clear trilaminate structure. [(a) Courtesy of F. Gudat, T. Harris, S. Harris, and K. Hummeler, *J. Exp. Med.*, **132**: 448 (1970); (b) courtesy of N. Simionescu and M. Simionescu, *J. Cell Biol.* **70**: 608 (1976).]

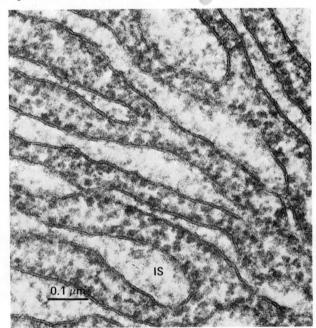

of the secretory proteins bind specifically to membrane-bound ribosomes, and (2) how does the protein get from the ribosome to the interior of the ER? The answers to those critical questions appear to be at hand, thanks in large part to pioneering experiments performed by Gunter Blobel at Rockefeller University and David Sabatini at New York University School of Medicine.

Synthesis of Secretory Proteins. The obvious first question is whether there is anything different about membrane-bound ribosomes. If they are the same as free ribosomes, then the specific association between mRNA and rough ER must be due to a feature peculiar to the mRNA or to the protein made from it. In fact, that is where we must look, because it appears that ribosomes are selected nonspecifically for membrane association.

Electron microscopy reveals that membrane-bound ribosomes are attached by their large, 60S subunit, with the small or 40S subunit sitting on top like a snowman. There is now considerable evidence for the existence of specific binding sites within the membrane capable of attaching only to the 60S subunit.

Evidence for specific ribosomal binding sites was obtained by experiments in which rough microsomes were stripped of their ribosomes and then exposed to free ribosomes obtained either in this way or from the cytoplasmic pool. The result was reattachment of ribosomes to roughly the original concentration. Equivalent treatment of smooth microsomes results in much lower rates of ribosomal attachment, though the affinity of the attached ribosomes is comparable, leading one to conclude that the same type of binding site is involved but present in much lower concentration. Control experiments with plasma membranes and Golgi membranes showed weak interaction with ribosomes and low capacity for attachment. Still further evidence for specific binding sites was obtained in experiments where the stripped microsomes were exposed to very mild proteolytic activity, destroying their capacity to accept ribosomes by damaging the ribosomal binding sites.

The presumed ribosomal binding sites (or receptors) appear as membrane proteins on freeze-fracture, extending well into and possibly through the lipid bilayer. These proteins, like most other membrane proteins, enjoy considerable freedom to diffuse laterally, so that "patching" and even a sort of "capping" of bound ribosomes can occur *in vitro* at temperatures above the phase transition of the membrane. This mobility, of course, would facilitate formation of the polysome and probably translation, which requires that the mRNA and ribosome move with respect to each other. Absolute freedom of movement is apparently not allowed *in vivo*, however, otherwise the distinction between rough and smooth ER would disappear.

It is probable, based on other experiments, that the association between mRNA and ribosomes—i.e., polysome formation—begins while ribosomes are free. They later attach to the ER, depending on availability of ribosome binding sites. This association, according to some data, is fostered by binding of the mRNA to membrane; other data point to association between the new polypeptide chain and membrane. Both factors may be important.

Work from Sabatini's laboratory demonstrated that the 3' end of mRNA from membrane-bound polysomes can remain attached to the membrane even after the ribosomes have been dissociated and stripped away. Other experiments suggest that certain mRNAs can associate with microsomes formed from stripped rough ER even in the absence of ribosomes. The 3' end of mRNA is, of course, the end that gets translated first. It is also the end carrying, in eukaryotic cells, a 150–200 nucleotide segment of poly A. However, the poly A is found on mRNA of both types, free as well as membrane bound. Hence, it is probable that mRNA coding for secreted proteins contains near this segment a sequence that binds specifically to ER—or, more likely, to protein receptors within the ER.

Another line of investigation, carried out in large part by Blobel and his colleagues, points to an affinity between membrane and an N-terminal segment of those polypeptide chains destined for secretion. Translation of the immunoglobulin light chain, for instance, produces first a 20–amino-acid segment that is later cleaved from the chain. This segment is largely hydrophobic in character, giving it an affinity for membrane. Comparable segments, also later sacrificed, have been identified at the N-terminus of other incomplete secretory proteins. Removal of these segments, at least in the case of the immunoglobulin light chain, is catalyzed by enzymes of the rough ER and takes place even before completion of the polypeptide chain.

These two proposals for directing secretory protein mRNA to rough ER—namely, attachment of the mRNA itself, and affinity between the nascent (incomplete) polypeptide and membrane—are not mutually exclusive. Both may function to help ensure translation of these mRNAs specifically at the rough ER.

Isolation of Secretory Proteins by the ER. The next problem is how the polypeptide chain that is destined for export gets into the cisternae of the ER. It does so very quickly, as demonstrated by Sabatini and Blobel in 1970, when they showed that mild proteolysis of rough microsomes did not destroy the nascent polypeptide chains but divided them into two parts. The incomplete C-terminal segment was protected by the ribosome on which it was being synthesized. The N-terminal segment was protected by the membrane of the ER. Hence, passage into the ER cisterna takes place during translation, leaving only a small segment exposed to the cytoplasm at any one time.

How the polypeptide chain gets through the lipid bilayer is not so clear, but it is quite reasonable to propose that the membrane protein serving as a ribosomal receptor also has a channel through its core that opens into the cisterna of the ER. The channel would have to be no larger than some of the water and cation pores associated with certain proteins of the plasma membrane discussed earlier, for there is free rotation about most of the nonpeptide bonds in the peptide chain. Hence the chain has great flexibility, permitting the amino acids to snake their way single file through the proposed pore (see Fig. 10-26).

Folding into secondary and tertiary structures also begins while the polypeptide is still in the process of synthesis. Formation of such structures would prevent return of the peptide through the narrow orifice from whence it came, thus trapping it in the cisterna of the rough ER.

Glycosylation of Proteins. Nearly all proteins destined for secretion are glycoproteins. (A notable exception is albumin, actually the most

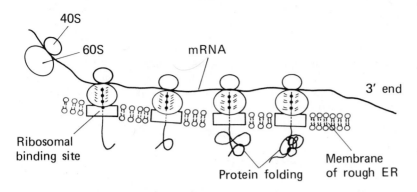

FIGURE 10-26 Synthesis of Secretory Proteins. Polysomes carrying the message for secretory proteins are attracted to the rough ER, where ribosomes attach to specific receptors. (Side view shown, with the cisternae of the ER below.) The nascent (incomplete) chain seems to pass into the cisterna as it is being synthesized. Folding and glycosylation typically start before the chain is completed.

common protein in serum.) Glycosylation—i.e., the addition of carbohydrate—begins even before the polypeptide chain is completed. The sugar molecules appear to be added one at a time, transferred usually from the nucleotide UDP, which serves as a carrier. Glycosylation is ordinarily still in progress when the polypeptide chain is completed and released into the cisterna of the rough ER. However, if the glycoprotein is to contain a terminal galactose, fucose, or sialic acid, those sugars are added in the Golgi apparatus where the appropriate sugar transferases are localized.

The stepwise addition of sugars to proteins from a UDP-sugar intermediate is well documented. However, in recent years it has become clear that animal cells, like cells of bacteria and plants, have a second and quite different way of glycosylating proteins. In this second pathway, the sugar chains are assembled in a stepwise manner from nucleotide-sugars, but the assembly is not on the protein itself. Rather, it is on a lipid, either *retinol* (vitamin A) or on one of a family of lipids called *dolichols*. The latter system has been particularly well studied.

In the lipid-linked assembly of certain oligosaccharides, transfer is initially to dolichol phosphate, creating a high-energy pyrophosphate (—P—O—P—) linkage. The first sugar in the chain is typically N-acetylglucosamine (GlcNAc). Thus we have (where **P** is phosphate):

$$UDP-GlcNAc + dol-P \longrightarrow dol-P-P-GlcNAc + UMP$$

Thereafter the chain is lengthened one sugar at a time. When the chain is nearly complete, it is transferred intact to the protein, using cleavage of the pyrophosphate to drive the reaction.

Why should two so very different mechanisms for protein glycosylation exist? All the evidence is not in, but it appears probable that the lipid-linked system functions in the *glycosylation of membrane proteins.* The assembly of carbohydrate chains onto dolichol seems to take place, at least in part, on the cytoplasmic side of the membrane, rather than in the cisternae of the ER. Although membrane proteins seem to be synthesized largely on membrane-bound ribosomes (one reason for the presence of rough ER in nonsecretory cells), one can imagine that their hydrophobic nature causes them to become incorporated into the membranes of the endoplasmic reticulum rather than confined to the cisternae within. Transfer from dolichol might then take place at the surface of the ER, which would also explain how membrane proteins come to be glycosylated in such an asymmetric way—i.e., on only one side.

One of the functions of carbohydrate on secretory proteins, then, might be to provide a hydrophilic coating to help keep the proteins in the cisternae of the ER.

In any case, membrane proteins, like those destined for export, pass from the rough ER to smooth surfaced ER adjacent to the Golgi apparatus and then ordinarily to the Golgi apparatus itself where terminal sugars may be added in a stepwise fashion. That organelle, the Golgi apparatus, is our next topic.

10-6 THE GOLGI APPARATUS

"Apparatus" seems like a peculiar designation for a cellular organelle, bringing to mind as it does a complex piece of machinery. However, the word is also applied in physiology to a group of organs that work together for a common purpose, as with the "digestive apparatus." Camillo Golgi first applied the word apparatus to the organelle now named in his honor, describing in 1898 an "internal reticular apparatus." He couldn't have known at the time how apt the designation really was; it is indeed a group of apparently separate entities working together for a common purpose. It is also a complex piece of machinery.

Morphology. The typical Golgi apparatus is a stack of three to seven flattened sacs (or *saccules*) that look a little like smooth endoplasmic reticulum, although sometimes fenestrae are seen having a structure similar to nuclear pores. The sacs are ordinarily cup-shaped and often they are located near the nucleus with the open end of the cup facing toward the cell surface (Fig. 10-27). However, occasional cell types (e.g., neurons) have a Golgi apparatus that enfolds the nucleus like a catcher's mitt.

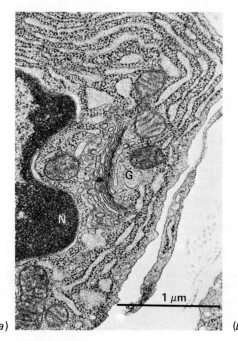
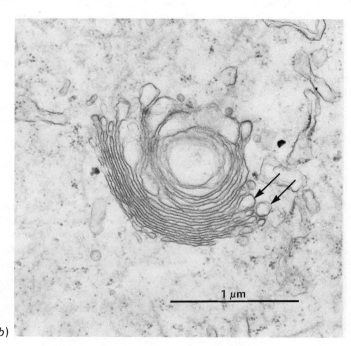

FIGURE 10-27 The Golgi Apparatus. (a) Mouse plasma cell. Note the swollen cisternae of rough ER (as in Fig. 10-25a) and numerous small vesicles near the proximal (left) face of the Golgi apparatus (G). These vesicles may contribute to formation of Golgi saccules. Nucleus (N). (b) Golgi apparatus from an ameba, with vesicles being pinched off from saccules (arrows). Golgi saccules in these cells have been shown to disappear when the nucleus is removed, thus stopping protein synthesis, and to return when the nucleus is replaced. [(a) Courtesy of A. P. Somlyo, R. Garfield, S. Chacko and A. V. Somlyo, *J. Cell Biol.* **66**: 425 (1975); (b) courtesy of C. J. Flickinger, *J. Cell Biol.* **49**: 221 (1971).]

The most complex Golgi configurations are seen in higher plants, where there may be many (even hundreds) of well-developed stacks, unconnected to each other and distributed throughout the cytoplasm. The term *dictyosome* has been applied to the individual stacks in these cases. At the other extreme, insects have many small stacks with few saccules and in some lower plants the apparatus may contain no more than one saccule. In all cases, careful observation frequently reveals tubules or vesicles at the outer margins of the apparatus, often connected by thin stalks to the saccules.

The different saccules of the Golgi apparatus are not identical to each other. In order to discuss this point, it is necessary to have a name for the two faces of the apparatus. The convex surface, typically nearest the nucleus, will be referred to as the *proximal* face. As we shall see, it is developmentally the youngest. Other designations for this face include outer, forming, and cis. The generally concave surface, developmentally older, is also called the inner, maturing, trans, or *distal* face.

There is evidence for a gradient of maturation in Golgi stacks as one proceeds from proximal to distal. In the fungus *Pythium ultimum*, for example, the proximal saccule has thin membranes similar to the nucleus and ER, whereas the distal saccule has thicker membranes more like the plasmalemma. There is a gradient through the intermediate levels that gives one the impression of a maturation process as a saccule moves outward through the stack.

Progressive displacement of saccules across the stack has been postulated for plants by H. Mollenhauer and D. Morre and for animal cells by M. Neutra and C. P. Leblond. Experiments by the latter workers on goblet cells of intestine (Fig. 10-30a, p. 470.) after injection of ^{3}H-glucose led them to estimate that a Golgi saccule could be converted to secretion vesicles each 2–4 minutes, implying a constant and rapid renewal.

Such observations lead to a picture wherein proximal Golgi saccules are formed by fusion of ER-derived vesicles while distal saccules "give their all" to vesicle formation and disappear. This model is not universally accepted and is conceivably applicable to some cell types and not to others. However, many electron microscopic studies reveal a proximal face that is close to a smooth-surface segment of ER (actually considered in most cases to be a part of the rough ER, in spite of the lack of ribosomes) with many small vesicles (*transition vesicles*) in between (Fig. 10-28). Because of their intimate structural and functional relationship, this segment of the ER, together with the transition vesicles and the Golgi saccules themselves, are perhaps more accurately thought of as a single unit, the *Golgi complex*.

Function of the Golgi Apparatus. The function of the Golgi apparatus is to accept vesicles from the ER, to modify the vesicular membranes and their contents, and then to distribute the products to other parts of the cell. The modified and distributed vesicles may be identified as lysosomes, secretory vesicles (see Fig. 10-28), or (at least in insects) as isolation envelopes needed to support autophagy.

Most Golgi-derived vesicles are from the distal saccules of the apparatus. Lysosomes are typical, though in a few cell types they seem to be derived instead from nearby smooth-surface ER. Another exception is blood neutrophils, where two kinds of Golgi-derived vesicles are involved in phagocytosis: the large, "specific granules" that usually fuse

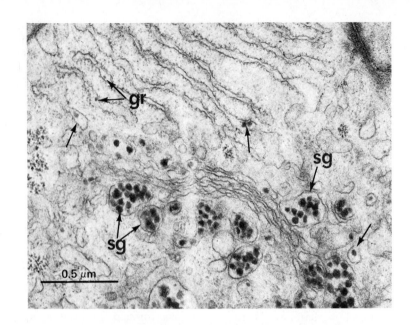

FIGURE 10-28 Secretion Vesicles and the Golgi Apparatus. Rat liver cell showing small grains (gr) of secretory product (lipoprotein) in the rough ER, large particles in smooth-surfaced terminal ends of ER cisternae (arrows) near the proximal face of the Golgi apparatus, and finished particles in secretory granules (sg) arising from the Golgi saccules. Note the small vesicles, presumably transition vesicles, bridging the space between the smooth-surfaced ER and Golgi saccules. [Courtesy of H. Glaumann, A. Bergstrand, and J. Ericsson, *J. Cell Biol.*, **64**: 356 (1975).]

first with the phagocytic vacuole are derived from proximal saccules, while lysosomes usually come from the distal saccules. The proximal saccules seem also to be actively producing vesicles in cells of the corn rootcap, which secrete a polysaccharide mixture called *root cap slime*. In these cells, all the cisternae of the Golgi apparatus appear to be enlarged and active in polysaccharide synthesis and vesicle formation (see Fig. 10-29).

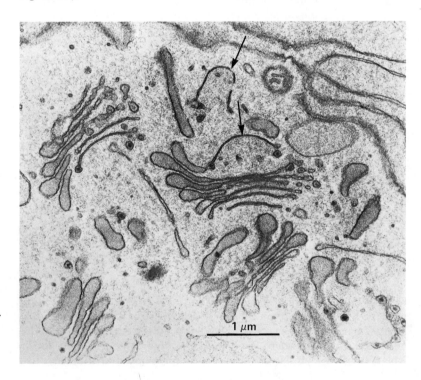

FIGURE 10-29 Dictyosomes. From a maize root cap cell. Note the swollen cisternae in this actively secreting cell. Residual parts of the saccules seem to get sloughed (arrows) after secretion vesicles have formed. [Courtesy of H. H. Mollenhauer, *J. Cell Biol.*, **49**: 212 (1971).]

The production of secretory vesicles was the first known function of the Golgi apparatus. Its major role in that process was once thought to be carbohydrate addition to secretory proteins. We now know, however, that secreted glycoproteins receive most of their carbohydrate in the endoplasmic reticulum, with only the finishing touches (specifically sialic acid, fucose, and galactose) added in Golgi cisternae. It is also now clear that albumin, which is synthesized and secreted by liver cells, has no carbohydrate yet is processed through the Golgi apparatus in the same sequence as exportable glycoprotein.

When the secreted material is primarily polysaccharide, rather than glycoprotein, synthesis often takes place entirely in the Golgi apparatus. Thus, when radioactive sugars are added to a cell to follow the synthesis of mucopolysaccharides such as hyaluronic acid or chondroitin sulfate (Chap. 3), radioactivity seems to appear first in the Golgi apparatus. The polysaccharide scales found on some algae are also assembled in the Golgi apparatus, as is a portion of the cell wall of higher plants. Cellulose, however, is an exception, as its synthesis seems to take place *extracellularly* at the plasma membrane. The enzymes responsible for cellulose synthesis, however, are membrane-bound and transported to the surface in Golgi-derived vesicles.

The role of the Golgi apparatus in membrane synthesis is less well understood. Although many membrane glycoproteins receive their terminal sugars in Golgi cisternae, lipid synthesis itself seems to be associated primarily with endoplasmic reticulum, especially the smooth ER. The ER can thus grow by incorporation of components made there. However, there is also some evidence to suggest transfer of vesicles to the Golgi apparatus and then back to the ER, presumably to allow for glycoprotein processing.

Because of this constant flow of membrane in all directions, the Golgi complex and its products, together with the endoplasmic reticulum and associated components (e.g., peroxisomes and the nuclear envelope), constitute a single *vacuolar system* through which all cellular membranes (with the possible exception of the mitochondria and chloroplasts) are interrelated. By incorporation into the plasma membrane of Golgi-derived vesicles on the one hand (discussed in the next section) and fusion of endocytic vesicles with Golgi-derived lysosomes on the other, the vacuolar system includes even the plasma membrane itself.

10-7 SECRETION

Secretion is an activity carried on by a great many different cells. For unicellular organisms, especially, it is a way of getting rid of waste products. For cells of higher plants and animals, it provides an extracellular matrix: the basement membrane on which many cells rest; the protein fibers on which the calcium salts of bone are deposited; rootcap slime; the noncellulose portion of plant cell walls and the enzymatic machinery for assembling the cellulose portion; and the protein component of arthropod cuticles, to name just a few. In addition, some cells are specialized for the purpose of secretion. Exocrine and endocrine glands secrete hormones, neurons secrete neurotransmitters, mucous cells lining the airways and gastrointestinal tract secrete protective polysaccharides and glycoproteins, etc. (Examples of secretory cells are shown in Fig.

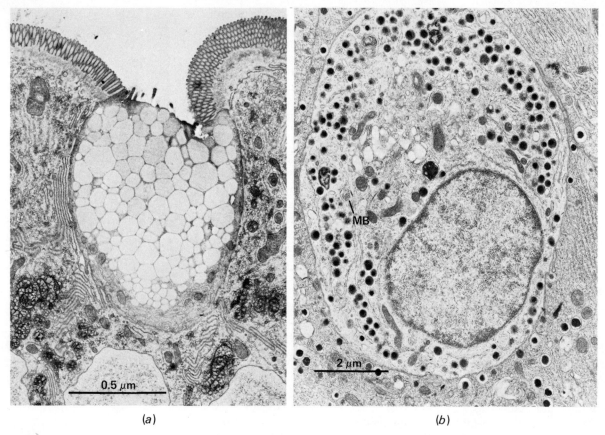

(a) (b)

FIGURE 10-30 Secretory Cells. (a) Goblet cell from rat intestine, flanked on either side by microvillus-coated absorptive cells. Goblet cells secrete a protective mucopolysaccharide that coats the intestinal walls. Note that the cytoplasm of this small cell is packed with secretion vesicles, crowding other structures into lower parts of the cell. (b) A cell from rat pancreas. The secretion product of this endocrine cell is the hormone glucagon. Note the more even distribution of darkly staining secretion granules. Multivesicular body (MB). [(a) Courtesy of N. and M. Simionescu, *J. Cell Biol.* **70**: 608 (1976); (b) courtesy of J. Braaten, U. Jarlfors, David Smith, and D. Mintz, *Tissue and Cell*, **7**: 747 (1975).]

10-30.) This list could be much longer; fortunately, however, the mechanisms used for secretion by these cells are satisfyingly few. It is our purpose here to consider them.

Types of Secretion. Molecular transport and diffusion across membranes account for the transfer of some materials from cytoplasm to extracellular space. Those topics were discussed in Chap. 6 and will not be treated further here. Nor are we concerned, except to mention it, with *holocrine secretion*, where the entire cell fills with secretory product and then is broken up and sacrificed. An example is the sebaceous gland, which secretes a fatty discharge onto adjacent hair and skin. It is this gland that, when overstimulated, gets plugged to form blackheads and ultimately the pimples of adolescent acne. The types of secretion of most interest here are the bulk transport processes of apocrine and merocrine secretion.

In *apocrine secretion*, the secretion vesicle buds through the plasma membrane carrying with it a portion of that membrane and hence a thin layer of cytoplasm. The secretion vesicle is thus covered by two layers of membrane, its own plus the plasma membrane. A good example of the process in a nonsecretory cell is shown in Fig. 10-31a and b, which reveals the nucleus being extruded from a human erythrocyte during the process of its maturation in bone marrow. Another example is the viral budding seen in Fig. 1-34. In true secretory cells, however, apocrine secretion is uncommon. One of the few examples of it is the *apocrine*

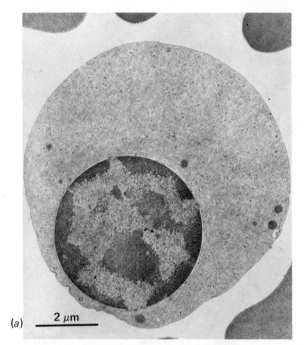
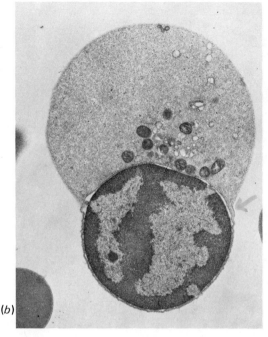
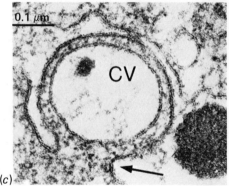

FIGURE 10-31 Apocrine Secretion. (a), (b) Extrusion of nucleus by a human erythrocyte precursor, utilizing apocrine type secretion. (c) Secretion of a casein vesicle (CV, containing milk protein) in a lactating rat mammary gland. This is not the usual mode of secretion for these vesicles, but a good demonstration of the mechanism of apocrine secretion. Note that the vesicle buds through the surface (arrow), surrounded by plasma membrane and some cytoplasm. [(a), (b) Courtesy W. Bloom and D. W. Fawcett. *A Textbook of Histology*, 10th ed., Philadelphia, W. B. Saunders, 1975; (c) courtesy of W. Franke, M. Lüder, J. Kartenbeck, H. Zerban, and T. Keenan, *J. Cell Biol.*, **69**: 173 (1976).]

sweat gland, which becomes active in the armpits and around the anus at puberty. And even in this case there has been much controversy over whether or not the secretion is purely apocrine.

The more common mode of secretion is called *eccrine* or *merocrine*. The process itself is also called *exocytosis* because it is the functional opposite of endocytosis. That is, the secretion vesicle fuses with the cell membrane and opens to the exterior (see Fig. 10-32). It is the usual mode of secretion of most glands and of most other cells, both animal and plant. It is, for example, the mode of secretion of the true sweat glands, which are distributed over most of the human skin.

The Pancreatic Aciner Cell. The synthesis and secretion of digestive enzymes by the exocrine portion of the pancreas was the first secretory pathway to be investigated by modern methods. These cells secrete trypsin, chymotrypsin, and other digestive enzymes in an inactive, or *zymogen*, form. The stimulus for secretion can come either from the

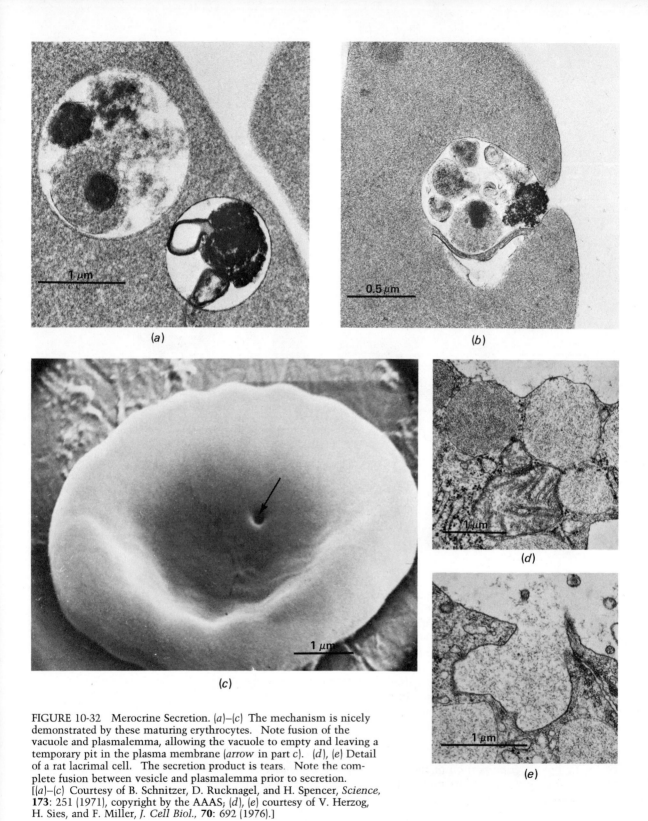

FIGURE 10-32 Merocrine Secretion. (a)–(c) The mechanism is nicely demonstrated by these maturing erythrocytes. Note fusion of the vacuole and plasmalemma, allowing the vacuole to empty and leaving a temporary pit in the plasma membrane (arrow in part c). (d), (e) Detail of a rat lacrimal cell. The secretion product is tears. Note the complete fusion between vesicle and plasmalemma prior to secretion. [(a)–(c) Courtesy of B. Schnitzer, D. Rucknagel, and H. Spencer, *Science*, **173**: 251 (1971), copyright by the AAAS; (d), (e) courtesy of V. Herzog, H. Sies, and F. Miller, *J. Cell Biol.*, **70**: 692 (1976).]

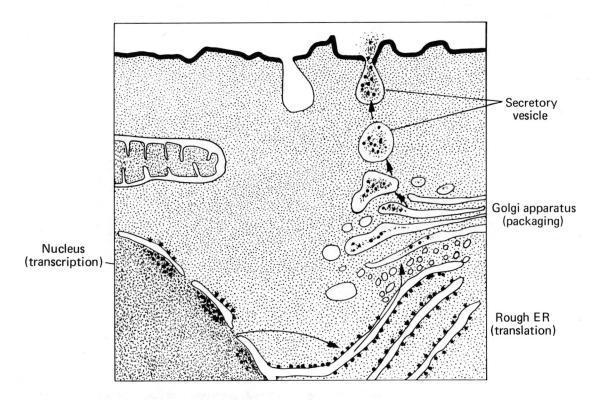

FIGURE 10-33 Synthesis and Secretion. A summary of the process, including the path of mRNA from the nucleus to the rough endoplasmic reticulum, and progress of the polypeptide chains through the ER, transition vesicles, Golgi apparatus, and secretory vesicles.

nervous system or from hormones produced in the gastrointestinal system after a meal. George Palade and Lucien Caro took advantage of this fact and of the recent availability of radioactively labeled amino acids in a pioneering set of experiments carried out at Rockefeller University in the mid-1950s.

Caro and Palade injected radioactive amino acids into fasted and re-fed guinea pigs. Injection was followed in three minutes by a much larger dose of unlabeled amino acid in order to stop uptake of the radioactivity (i.e., this was a "pulse-chase" experiment). Animals were then sacrificed at various times. Within a few minutes from the time of injection, radioactivity was found in the rough endoplasmic reticulum of the pancreatic cells; by 20 minutes it was found also in the Golgi apparatus and nearby vesicles; and by two hours it was localized to the secretory vesicles or *zymogen granules*. It was by this experiment and others like it that the sequence whereby secretory proteins move from rough ER to Golgi to secretory vesicle was deduced. That process can be summarized as follows (see Fig. 10-33):

Secretory proteins are synthesized on membrane-bound ribosomes and pass immediately into the cisternae of the rough ER for proteolytic processing and carbohydrate addition. They then transfer to the contiguous smooth-surface ER; from there they travel, by ER-derived vesicles, to the Golgi saccules, where they may get their terminal carbohydrate. The probable passage of a saccule through the Golgi stack was discussed in the previous section; new saccules form from ER vesicles on one side and old ones yield their membrane to secretory vesicles on the other. The secretory vesicles fuse with the plasma membrane to empty themselves and complete the process.

The questions that remain to be answered are (1) how is flow from the ER to the Golgi apparatus and from the Golgi apparatus to the appropriate part of the cell membrane directed, (2) what are the controls on the process to ensure secretion at the proper time, and (3) what happens to the secretory vesicle once it is emptied?

Vesicle transport. The first requirement for a secreted material, once it has left the protection of the endoplasmic reticulum in vesicles derived therefrom, is to find the Golgi apparatus. That task would be easier if the vesicles formed only from a portion of the interconnected ER that happened to lie close to a Golgi saccule—i.e., from smooth-surface ER associated with the Golgi complex. In fact, Michael Locke and P. Huie of the University of Western Ontario have recently found evidence in several types of animal cells for rings of beadlike structures in this part of the ER near the Golgi apparatus (Fig. 10-34). These structures appear to mark a specialized region of the ER where vesicle formation and flow to the adjacent Golgi apparatus is permitted. This flow is dependent on a source of energy, though it is uncertain whether the requirement is for vesicle formation alone or for transport as well.

The trip from Golgi apparatus to cell membrane is longer and demands a more definite travel plan. Exit from the cell is commonly permitted only in one specified region of the plasmalemma—at the lumen of an excretory duct in the case of exocrine glands or at the adjacent capillary in the case of endocrine glands, for example. The distance between vesicle formation and site of secretion can be very long (see Fig. 10-35). In neurosecretory cells of the human brain, the hormones oxytocin or vasopressin are packaged into secretory vesicles in the cell body, which resides in the hypothalamus at the lower part of the brain, but are secreted into the blood at the pituitary gland several centimeters away.

In all cases, secretory granules collect rather specifically at the membranes adjacent to where they will be emptied and in many cases microtubules can be seen along the path of vesicle movement. It has been natural to assume a role for microtubules in vesicle translocation, and there is now much evidence in support of that assumption.

Insulin Secretion. The process of secretion by β cells of the pancreas has received much attention. β cells are the endocrine cells that produce insulin and release it into the bloodstream. Release occurs at a low *basal rate* that is capable of being greatly accelerated by the proper stimulus. An increase in blood glucose to high levels is an obvious choice of stimulus and one that is easy to use experimentally.

FIGURE 10-34 ER-Golgi Transition Region. Rings of beadlike structures, revealed by special stains, are found on the smooth-surfaced ER near the proximal face of the Golgi apparatus and may mark the point of transfer between ER and Golgi saccules. From the silk gland of the butterfly, *Calpodes*. (a) Side view, showing a line of beads on the surface of the ER facing the Golgi apparatus. Transition vesicles (TV) bridge the space between ER and the outer (proximal) saccule of the Golgi apparatus (OS). Inner (distal) saccule (IS). Rough endoplasmic reticulum (RER). (b) Detail, showing formation of a transition vesicle from a ribosome-free portion of the ER. This section was stained to show ribosomes (r) rather than Golgi-complex beads (b). (c) Face view, demonstrating that the Golgi complex (GC) beads are found in rings, each of which may define an outlet from the ER. [Courtesy of M. Locke and P. Huie, *J. Cell Biol.*, **70**: 384 (1976).]

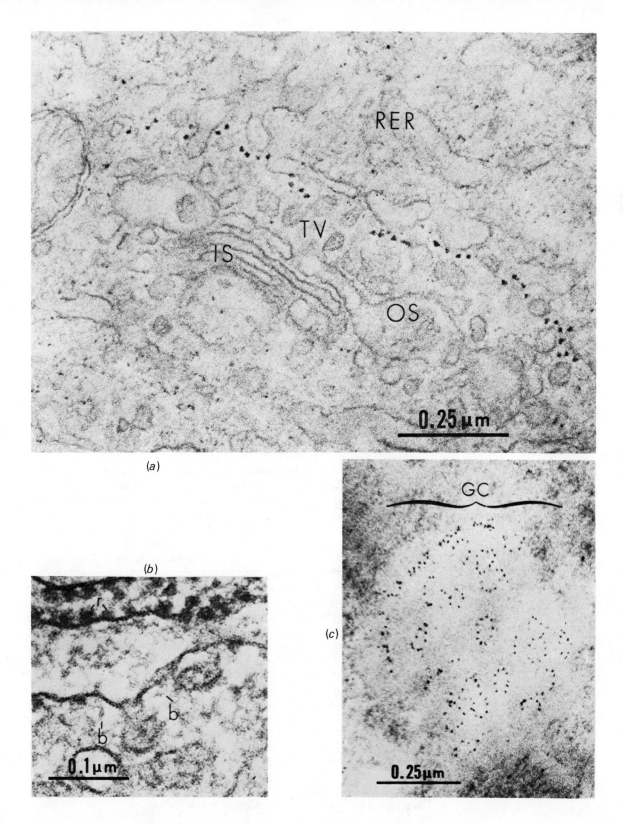

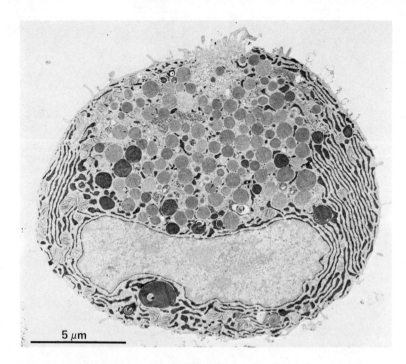

FIGURE 10-35 Isolated Secretory Cell from a Rat Lacrimal Gland. The product, tears, contains peroxidase, here revealed by cytochemical stain in the ER (including the perinuclear cisterna), Golgi apparatus (not clearly seen), and all secretory vesicles. Note that vesicles must traverse a major part of the cytoplasm to reach the single site (at top) where secretion is permitted. [Courtesy of V. Herzog, H. Sies, and F. Miller, *J. Cell Biol.*, **70**: 692 (1976).]

Under certain conditions, the secretory vesicles in rat pancreatic β cells appear to be in relatively well-ordered arrays and often in intimate association with microtubules (Fig. 10-36). The association between vesicle and microtubule is related to the mechanism of insulin release, as was shown by pretreatment with high levels of colchicine or vinblastine, agents that disrupt microtubules. Under these conditions, the response to glucose was greatly diminished, giving the impression that microtubules are guiding the vesicles toward the secretion site. From what we have already learned about microtubules in Chap. 9, it is natural to assume that they furnish the motive force as well.

The response of β cells to glucose stimulation is a two-phase event. There is an initial rapid release occurring within the first minute and lasting 5–6 minutes. This phase is followed by a decrease in output that gradually rebuilds over the next 10 minutes or so to reapproach the peak levels (Fig. 10-37b). This second phase lasts as long as the glucose stimulus. The apparent explanation for the two phases is that secretory vesicles are in two groups, one immediately available for exocytosis and the other a larger group of reserves.

A more careful examination of the effects of colchicine on insulin release revealed that at low levels of the drug (1 mM), only the second phase of the release was inhibited. Only this second phase, then, seems to be intimately dependent on microtubules.

When cytochalasin B was applied to β cells, the drug actually enhanced secretion of insulin. Cytochalasin B disrupts microfilaments, which in these as in so many other cells form a web just under the plasma membrane. It appears to be generally true that one of the functions of this web is to prevent contact between the cell membrane and internal organelles, and the possible fusion between organelle and cell membrane that could follow. This pattern appears to be true of β cells, for disrupting the web with cytochalasin B enhanced insulin secretion.

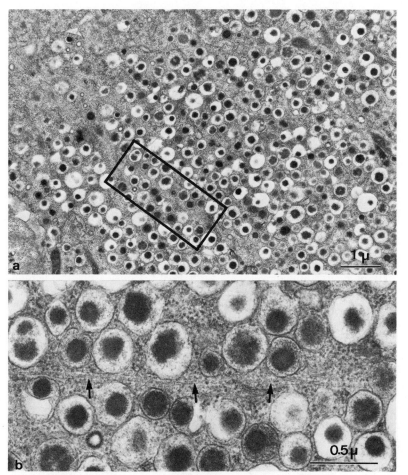
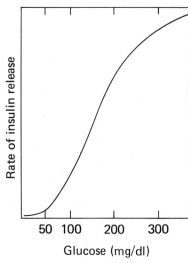
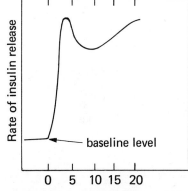

FIGURE 10-36 Vesicle Transport. Rat pancreatic β cell showing rows of secretory vesicles which, in the detail, seem to be associated with microtubules (*arrows*). [Courtesy of W. Malaisse, F. Malaisse-Lague, E. Van Obberghen, G. Somers, G. Devis, M. Ravazzola, and L. Orci, *Ann. N.Y. Acad. Sci.* **253**: 630 (1975).]

The model for insulin secretion that emerges from these studies is as follows: Secretory vesicles are moved from the Golgi apparatus toward a designated portion of the plasmalemma by microtubules. Movement stops when the microfilamentous web is reached. Changes in the web structure following glucose stimulation result in exocytosis of these vesicles, furnishing the early phase of the secretory response. Continued secretion requires the mobilization of more distant vesicles by microtubules and their energy-dependent transport to the secretory site.

There is good reason to believe that this model for secretion applies not only to pancreatic β cells, but also to most secretory cells. The involvement of microtubules and microfilaments also gives us a logical place to look for control mechanisms that ensure release at the proper time.

Stimulus-Secretion Coupling. In 1968, W. W. Douglas suggested that the coupling event between stimulus and secretion in a wide variety of

FIGURE 10-37 Dynamics of Insulin Secretion. (*a*) Rate of insulin release increases with glucose concentration if the latter is greater than the threshold of about 50 mg/dl (i.e., 50 mg per deciliter, or per 100 cc). Above about 300 mg/dl there is little additional effect. (*b*) Sudden application of stimulus (in this case 300 mg glucose/dl) results in a two-phase release, the first presumably due to vesicles already at the cell membrane and the second due to transport of newer vesicles from deeper in the cytoplasm.

cells is an increase in available Ca^{2+}, just as Ca^{2+} is also responsible for excitation-contraction coupling in skeletal muscle cells. The calcium pool may be extracellular, as in nerve cells, where depolarization leads to Ca^{2+} influx and exocytosis of neurotransmitter vesicles. The calcium pool might also be intracellular (blood platelets are an example) or there could be a combination of intra- and extracellular pools.

Secretion by β cells of the pancreas in response to glucose is dependent on the presence of Ca^{2+} in the surrounding fluid. By using radioactive calcium ($^{45}Ca^{2+}$) a two-way flux has been demonstrated, with an inward "leakage" countered by an ATP-driven calcium pump. Secretion of insulin follows a net increase in cytoplasmic calcium, whether that increase is caused by a greater influx or decreased efflux. The effect of glucose seems to be the latter—namely, inhibition of efflux. High levels of glucose (but not its analog 2-deoxyglucose) cause a drop in the efflux rate of Ca^{2+} from β cells and subsequent Ca^{2+} accumulation. Accumulation can also be caused by using ionophores (see Chap. 6) that facilitate Ca^{2+} entry. Either approach stimulates insulin secretion.

Controlling secretion by varying Ca^{2+} levels is the functional counterpart of controlling movements with Ca^{2+}. Considering the demonstrated effect of Ca^{2+} on both microfilament-mediated and microtubule-mediated movement discussed in Chap. 9, the parallel should not be surprising.

Mast Cells. Involvement of Ca^{2+} has been demonstrated in numerous secretory processes. For example, *mast cells* are common in many vertebrate tissues. These cells, which are derived from blood basophils (i.e.,

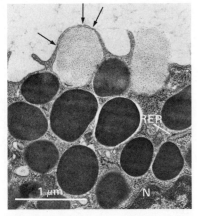
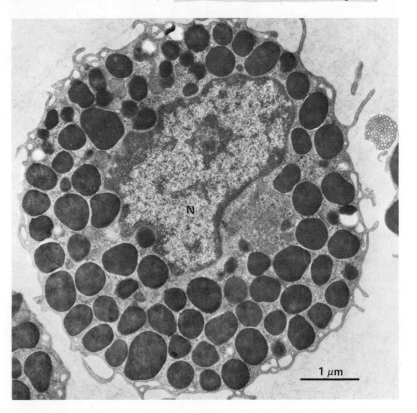

FIGURE 10-38 The Degranulation of Mast Cells. Mast cells from the peritoneal cavity of rats. Degranulation (loss of granules via discharge of their contents) was induced *in vitro*, a process that takes only a few seconds. RIGHT: Normal mast cell. ABOVE: Surface granules are being released. Some deeper granules empty by fusing with those already opened to the outside. Nucleus (N). Rough endoplasmic reticulum (RER). *Arrows* mark points of fusion between granular and plasma membranes. [Courtesy of P. Röhlich, P. Anderson, and B. Uvnäs, *J. Cell Biol.*, **51**: 465 (1971).]

basophilic granulocytes, one of the three classes of polymorphonuclear leukocytes) have enormous secretory granules containing mostly histamine (Fig. 10-38). This chemical, which is the amino acid histidine without its carboxyl group, causes, among other things, many of the unpleasant effects of hay fever. (Antihistamines are competitive inhibitors of histamine receptors.)

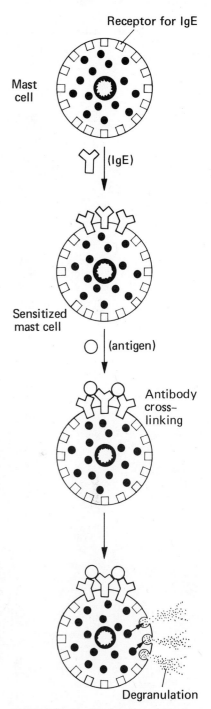

$$\begin{array}{c} H_2N \\ | \\ HC \leftarrow COOH \\ | \quad \text{(histidine)} \\ CH_2 \\ \diagup \diagdown NH \\ N = \end{array}$$

Histamine

Histamine is released from mast cells via an antibody-antigen reaction (see Fig. 10-39). The cells are sensitized specifically by having IgE antibody from the blood attach to receptors on the mast cell surface. Then, if an antigen appears that is capable of cross-linking that antibody, degranulation by exocytosis follows. The coupling event between IgE cross-linking and exocytosis is an increase in cellular Ca^{2+}. In this case, however, accumulation of Ca^{2+} is caused by a temporary increase in Ca^{2+} permeability following antigen cross-linking, so that extracellular Ca^{2+} can enter in response to its electrochemical gradient. (Contrast this with Ca^{2+} accumulation in pancreatic β cells caused by glucose-inhibited efflux.) Support for the link between calcium accumulation and exocytosis in mast cells comes from experiments showing a dependence on extracellular calcium and from experiments demonstrating that Ca^{2+}-transporting ionophores cause secretion in the absence of antigen.

Many other examples of Ca^{2+}-mediated stimulation-secretion coupling could be offered. In addition, you will recall that earlier in this chapter the response of lymphocytes to antibody or lectin was attributed to Ca^{2+}. In Chap. 11, the role of Ca^{2+} in cell division will be discussed. It appears that nature, like the great composer Bach, capitalizes on a good idea by using it over and over.

Role of Cyclic Nucleotides in Secretion. Calcium mediation of stimulus-secretion coupling is modulated in many cells by the cyclic nucleotides cAMP or cGMP. In β cells of the pancreas, for example, cyclic AMP or agents that increase cyclic AMP also increase the magnitude of insulin secretion in response to glucose. β cells respond to a great many substances other than glucose, some of which cause and some of which inhibit insulin release in the presence of a given quantity of glucose. The cyclic nucleotides offer a way of modulating secretion in response to these additional signals, which include among them several different hormones.

There are, of course, numerous possibilities for cyclic nucleotide modulation, for example by phosphorylating or dephosphorylating this or that protein and by regulating Ca^{2+} levels. The latter can be accomplished by changes in calcium permeability of mitochondrial membranes (see Chaps. 4 and 7). In the case of mast cells, at least, there is also evidence for a direct effect of cAMP on Ca^{2+} permeability of the cell membrane.

FIGURE 10-39 Mechanism of Mast Cell Degranulation. Mast cells become sensitized by IgE attachment at receptor sites specific for the Fc portion of IgE. Degranulation follows if an antigen capable of cross-linking the IgE appears.

Antigen stimulation of IgE-sensitized mast cells results not only in an influx of Ca^{2+} but also in a fall in the level of cytoplasmic cAMP. The fall is temporary, returning to normal in about five minutes, at which time degranulation also stops. (The antigen, of course, is still tightly bound to surface IgE.) The cell is then at rest, ready to respond to a second provocation.

There are two kinds of experiments that suggest a Ca^{2+}-channel closure in mast cells that is fostered by cyclic AMP. In the first place, elevation of cyclic AMP by drugs such as theophylline, which inhibits cAMP breakdown, prevent antigen-induced Ca^{2+} influx and subsequent degranulation. (The rationale for the use of theophylline in allergic asthma is thus obvious.) Conversely, β-adrenergic agents (e.g., epinephrine), prostaglandin E_2, or histamine itself cause a fall in the level of cAMP and inhibit degranulation. Acetylcholine has the same effect, achieved by elevating the level of cGMP.

The second line of evidence linking cAMP to Ca^{2+} influx is the observation that calcium ionophores, which cause direct calcium influx and bypass the usual Ca^{2+} channels, result in a degranulation that is independent of cAMP level. It thus appears that cAMP tends to close the calcium channels opened in response to antigen stimulation.

Fate of the Secretory Vesicle. The last step in the process of secretion is disposal of the secretory vesicle. With apocrine secretion there is no problem, of course, because the vesicle is gone. With merocrine secretion, there are two possibilities: the empty vesicle can detach from the plasma membrane and return to the cytoplasm for destruction or recycling; or the vesicle can merge with, and become part of, the plasma membrane. The first option, returning the empty vesicle to the cytoplasm, is apparently used by some cell types—reportedly in the adrenal medulla where catecholamines are released, and in some insect cells. It is an uncommon choice, however; the usual fate of the vesicle is incorporation into the cell membrane.

Incorporation of secretory vesicles into the plasma membrane, though common, creates a problem. An active secretory cell may in this way add each day to the cell membrane a surface many times the original area of the membrane. For a short time, the membrane can accommodate new surface by formation of additional microvilli or other redundant structures. However, in the long run, if explosive growth of the cell surface is to be avoided, there must be some compensatory removal of membrane.

At least one cell type solves the problem very neatly; the secretory cells of lactating mammary glands have two products, casein (the most abundant protein in milk) and fat droplets. Casein is ordinarily secreted by the merocrine mechanism, which adds membrane to the surface, whereas fat droplets are secreted in the apocrine fashion, which removes membrane from the surface (Fig. 10-40). The two processes can balance each other under some conditions to keep the cell size constant.[6]

Most secretory cells, however, must balance the addition of new membrane caused by secretion with removal of membrane via pinocytosis. This process has been amply demonstrated in many cell types by adding a marker, peroxidase or ferritin for example, to the bathing solution and then observing its appearance in pinocytic vesicles following a secretory event. Thus when a nerve cell is stimulated, a compensatory pinocytosis quickly follows. In fact, externally added markers are frequently

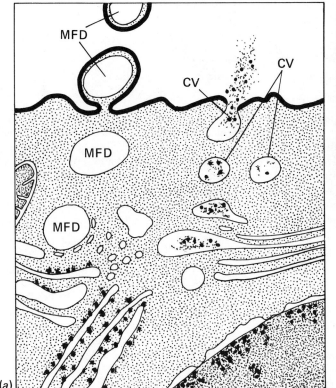

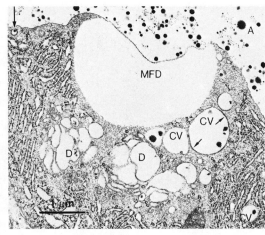

FIGURE 10-40 Lactating Mammary Gland. Milk fat droplets (MFD) are released by the apocrine mechanism, whereas casein vesicles (CV) normally release their protein by merocrine secretion. The former removes plasma membrane, the latter adds to it. (a) Diagramatic representation. (b) Actual cell from a rat mammary gland. Golgi region dictyosomes with swollen cisternae (D). Note "bristle coat" patches (arrows), that seem to aid exocytosis of the casein vesicles. [(b) Courtesy of W. Franke, M. Lüder, J. Kartenbeck, H. Zerban, and T. Kennan, *J. Cell Biol.*, **69**: 173 (1976).]

not necessary, for the pinocytic vesicles often have a "coating" of unknown composition that makes them identifiable in the electron microscope (see Fig. 10-11, p. 445). Many experiments have been designed to establish whether the pinocytic vesicles are preferentially formed from newly added membrane, or whether there is general mixing of membrane after exocytosis, followed by random removal during endocytosis. Some data have been accumulated indicating that a selective removal is possible, but it is too soon to say with certainty how general that phenomenon is.

While secretory cells that utilize exocytosis face the problem of removing excess cell membrane, apocrine secretors or phagocytic cells have the opposite problem—that of adding new membrane. For instance, the phagocytically active soil ameba, *Acanthamoeba castellani*, is thought to ingest 5 to 50 times its surface area per hour, much of it by pinocytosis. It has been proposed that some of these small vesicles might be emptied and returned directly to the cell membrane. However, phagocytic vacuoles probably are not. Thus, a human macrophage after ingesting a particularly large particle loses its surface microvilli, rounds up, and is thereafter phagocytically inactive for 5–6 hours. Macrophages, like virtually all other cells (possibly excluding its sister phagocyte, the mature neutrophil) can make and incorporate new plasma membrane. Hence, it gradually returns to its former shape and activity. The process that makes this recovery possible—membrane addition—is part of an ongoing turnover of membrane components, to be discussed as our next topic.

10-8 SYNTHESIS AND TURNOVER OF MEMBRANE COMPONENTS

All things flow and change . . . even in the stillest matter there is unseen flux and movement.

Attributed to Heraclitus
ca. 500 B.C.

It should be abundantly clear by this time that nothing in the cell is fixed or stable. There is constant movement, and there is constant turnover. The overall structure of the cell may not change visibly, but with a molecule removed and replaced here, another removed and replaced there, it is only a matter of days before a typical cell—even a stable, long-lived cell such as an hepatocyte—is completely renewed.

This constant synthesis and turnover is the topic to be discussed. Specifically, where and how are cellular components made and assembled, and where do the materials come from?

Turnover of Membranes. Turnover of the plasma membrane was treated indirectly in the preceding section as a necessary consequence of endocytosis or secretion. The rate of turnover obviously depends on the rate of those other processes, and may be as short as a few minutes in certain particularly active amebas. It doesn't necessarily follow, however, that all components of the plasma membrane turn over at the same rate. We have seen earlier how a section of membrane might be cleared of various proteins prior to endocytosis. The cell is thus saved the expense of replacing cell membrane proteins at the same rate as its lipids. The lipids, of course, might be rather easily reclaimed for insertion into new membrane, whereas replacing degraded proteins requires synthesis from messenger RNA.

Note that if insertion of new lipid occurs at one side of a cell while removal occurs preferentially on the opposite side, the result would be a constant flow of lipid across the cell surface. Membrane proteins, normally anchored to the underlying cytoskeleton, need not be included. This flow could explain certain surface phenomena such as the capping of receptors on lymphocytes, if the receptors are freed of their cytoskeletal connections by cross-linking (see reference by Bretscher).

There is marked heterogeneity in the lifetimes of plasma membrane components and the same is true for constituents of internal membranes. Thus, one estimate places the half-life of the lipid component of liver endoplasmic reticulum at 1–2 days but the protein component at 2–5 days. These are overall averages, produced by nonspecific incorporation of a radioactive label for a brief period of time (a "pulse") followed by observation of the rate at which label disappears from the structure in question. When one begins to look at specific proteins of the endoplasmic reticulum, however, half-lives are found to vary from a few hours to more than two weeks.

Unfortunately, we do not know the reason for the rapid turnover of most cellular components. We do know that it is not just a process of molecular aging, for destruction seems to be a random event. That is, newly labeled and hence newly made proteins do not last a given period of time and then begin disappearing from a membrane. Rather, both an old and a new protein have the same chance of surviving each time

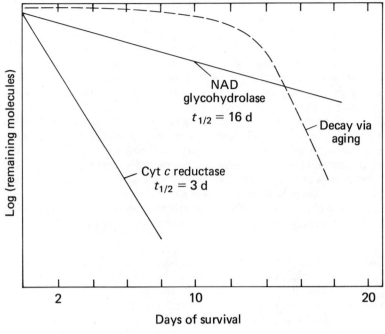

FIGURE 10-41 Turnover of Membrane Proteins. Decay curves and half-lives ($t_{1/2}$) of two proteins of the endoplasmic reticulum. This turnover by random destruction contrasts markedly with the dynamics of destruction due to "old age" (dashed line).

period examined, yielding a logarithmic decay of pulse-labeled proteins similar to that seen in Fig. 10-41.

The mechanism of turnover is not always clear, although lysosomal digestion is the probable route, of course. Components released from the digestive vacuoles could thus be made available for new synthesis. Whatever the mechanism, it is clear that one benefit of turnover is the ability to alter composition in response to changing needs, either by controlling the rate of synthesis or the rate of degradation, or both.

Synthesis of Membranes. A good deal of what we know about the synthesis of cellular membranes has already been discussed. Thus we know that probably most membrane-bound proteins are synthesized on membrane-bound ribosomes (i.e., rough endoplasmic reticulum) and appear to be immediately incorporated into the membrane of the ER. Carbohydrate addition, if indicated, may occur on the cytoplasmic surface of the ER via transfer of the oligosaccharide chain from a lipid intermediate, a mechanism that appears to be different from the stepwise addition of sugars to secretory proteins that occurs within the cisternae of the ER. From the site of synthesis there appears to be lateral movement to a segment of smooth-surface ER, taking advantage of the fluid nature of membranes and the multiple interconnections of endoplasmic reticulum. From this point, however, there is by no means universal agreement among workers as to which path is followed.

If a membrane glycoprotein requires the addition of galactose, fucose, or sialic acid it presumably must be transferred to the Golgi apparatus, for the enzymes capable of adding those sugars seem to be confined to Golgi cisternae. From there, the finished protein could be transferred to the plasma membrane or back to the endoplasmic reticulum in a Golgi-derived vesicle. But if the membrane protein doesn't need that final processing, must it still go to the Golgi apparatus? Some experiments indicate not. A significant fraction of plasma membrane protein appears in some labeling studies to be rapidly transferred to the plasma membrane, possibly by direct transfer of vesicles derived from the ER. It is also probable that some proteins destined for a final home in the endoplasmic reticulum simply stay there once synthesized.

Lipid components of membranes, like other lipids, are synthesized largely by enzymes associated with the smooth ER, though the Golgi apparatus is also implicated in some cases. It is logical to assume that with both protein and lipid synthesis occurring in the ER, those membranes may be largely self-perpetuating. The Golgi and nuclear membranes, on the other hand, are put together from ER-derived vesicles, as are the membranes of the large storage vacuoles of plant cells. Plasma membrane is derived either directly from ER or (more commonly) from ER via the Golgi apparatus. And finally, vesicles derived from the plasma membrane or from other membranes are subject to degradation by lysosomes with the products released to start the cycle over.

Lipid Exchange. Lipid transfer through the cytoplasm certainly occurs by way of cytoplasmic vesicles. It could conceivably also occur by micelle formation (Chap. 3). Although the lipid composition of cellular membranes is such that micelles probably do not form, transfer of lipid via vesicles is apparently not the only alternative. A more rapid movement of lipid, and one that requires no metabolic energy (which vesicle formation presumably does), is thought to be catalyzed in many cells by specific cytoplasmic proteins. These *lipid-exchange proteins* are capable of reversibly binding and releasing specific lipids. They diffuse about the cytoplasm, carrying lipid molecules from one membrane to another. Net transfer of lipid by lateral diffusion within the ER or by vesicle formation is superimposed on this protein-catalyzed exchange of lipids.

The concept of lipid-exchange proteins has been particularly important in explaining the origin and composition of mitochondrial membranes, since nearly all of their lipids are cytoplasmic in origin. The outer membranes of the mitochondria are often continuous at some point with the ER and hence could receive lipid by lateral diffusion. There are, however, distinct differences in protein content between the ER and outer mitochondrial membrane, so it would be inaccurate to think of one as an extension of the other. And the inner mitochondrial membrane, except at a few points of junction (the significance of which is unknown), is separate from either the ER or outer membrane. The existence of lipid-exchange proteins helps explain how mitochondrial inner membranes are able to participate in the general turnover of cellular lipids.

The situation with chloroplasts is probably similar, though they are able to synthesize a significant fraction of their own lipids.

The concept of lipid exchange applies not just to intracellular membranes but to cells as well. At least in mammals, serum lipoproteins (Chap. 3) act as lipid-exchange proteins capable of transferring lipids from cell to cell. This activity is particularly noticeable in the composition of erythrocyte membranes, which varies widely with the serum cholesterol level. (Nearly all serum cholesterol is carried by the lipoproteins.)

Serum lipoproteins contribute to membrane composition in other ways as well. It has been suggested that mammalian erythrocytes are

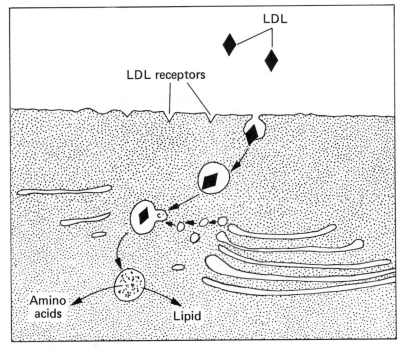

FIGURE 10-42 Serum Lipoprotein Uptake. Cells of higher animals get most or all of their cholesterol and much of their phospholid from lipoproteins made by the liver. Specific receptors for low density lipoprotein (LDL) aid uptake.

able to incorporate serum lipoproteins directly into their membranes, thus converting the lipoproteins into new membrane proteins. More widespread, however, is the presence of surface receptors that are specific for serum lipoproteins, especially for the low density lipoproteins (LDL). Binding of LDL to receptors is followed by endocytosis and lysosomal digestion that breaks down the protein, freeing cholesterol and phospholipid to be used in membrane synthesis (see Fig. 10-42). This mechanism, established by Michael Brown and Joseph Goldstein at the University of Texas, is the way in which most mammalian cells outside the liver obtain their cholesterol.[7]

The importance of intercellular lipid exchange and extracellular lipoproteins to the lower animals is not yet known. However, there is no reason to suspect that the mechanisms are very different. Cholesterol, for example, is known to be a dietary requirement in insects, implying an inability of their cells to synthesize the substance and hence the existence of an extracellular transport system. Most of the work to date has been on mammalian systems because of the need to understand better the way cholesterol is handled by humans and how it is related to human heart and blood vessel disease.

SUMMARY

10-1 The plasma membrane contains numerous glycoproteins and glycolipids that have carbohydrate chains extending into the extracellular space. These carbohydrates allow the cell to be recognized by other cells, either directly or through soluble aggregation factors. Examples of recognized surface carbohydrate include the blood type factors, especially the ABO system where inherited genes may cause either an A or a B type modification of a surface glycolipid called substance H. Another system that has received much attention involves the human lymphocyte antigens (HLA). The HLA glycoproteins, through many alleles at each of four known loci, provide all nucleated cells with a readily established identity that may be important in immune surveillance and subsequent elimination of foreign or malignant cells.

Although HLA antigens are not limited to lymphocytes, the latter do have unique surface antigens. Specifically, the B cells have a surface immunoglobulin that serves as an antigen receptor. When cross-linked, molecules of surface immunoglobin aggregate into patches and later into caps at one end of the cell; the cap is either internalized or shed and then replaced. Capping has been noted with many different receptors (which often cap independently of each other). Although capping is not necessary for lymphocyte activation, both are due to cross-linking of receptors, either by antigen, by lymphocyte-specific antibodies, or by carbohydrate-specific lectins.

10-2 Receptors for nonlipid hormones are typically glycoproteins located within or on the plasma membrane. They are usually mobile, and able to move laterally within the lipid bilayer. In some cases, the receptor is intimately associated with effector (ion channels are an example) but in many cases the receptor and effector appear to be two different proteins, capable of interacting only when hormone is bound to receptor. This "mobile receptor hypothesis" offers attractive possibilities, through a variable receptor number, for regulating hormone sensitivity and for integrating the actions of many different hormone receptors and effectors in the same cell.

10-3 Endocytosis of particles is called phagocytosis ("eating"). Endocytosis of liquids is called pinocytosis ("drinking"). Endocytosis can be very specific, occurring in response to surface contact with a particle, sometimes at a specific receptor. It also occurs nonspecifically and can be prevented apparently by the presence of certain surface carbohydrates on what would otherwise be a potential target. In any case, energy is required, presumably to power the responsible microfilaments and microtubules.

Endocytosis or autophagy (engulfment of cellular components by an ER or Golgi-derived isolation membrane) is usually followed by lysosomal digestion of the material in the vacuole. In order to accomplish this, fusion of vacuole and lysosome occurs, creating the digestive vacuole. In addition to the effect of lysosomal hydrolases thus introduced, killing of microorganisms is greatly facilitated by an acidic pH, production of hydrogen peroxide, and in some cells by a myeloperoxidase-induced halogenation of the microorganism. The products of digestion diffuse out of the digestive vacuole. Undigestible debris may be expelled outside the cell (exocytosis) or accumulate to convert the digestive vacuole into an inert residual body.

10-4 The nucleus is bounded by a double membrane called its envelope. There are frequent openings or pores in this envelope. These pores have a highly organized structure, with octagonal symmetry conferred by two rings of eight particles each—one ring on the inner membrane and one on the outer membrane. Pores provide a way to get materials in and out, apparently exercising some selectivity in the process. The membranes of the nuclear envelope are otherwise very similar to membranes of the endoplasmic reticulum. This similarity is especially true for the outer nuclear membrane, which may be in continuity with the ER, may have ribosomes attached, and can often be seen in the process of vesicle formation via blebbing.

10-5 Smooth endoplasmic reticulum is most often seen as tubules or vesicles. Smooth ER is concerned primarily with lipid synthesis and certain aspects of carbohydrate metabolism. The biotransformations that result in drug detoxification—conjugations, oxidation by P450, etc.—are also associated with the smooth ER, though not necessarily limited to it. Rough ER, on the other hand, is usually seen as ribosome-studded flattened sacs. Its major function is synthesis of secretory (and probably membrane) proteins. These proteins are translated specifically by ribosomes bound to receptors within the membrane of the ER. The association between ribosome and membrane appears to occur after polysome formation has begun, fostered either by an affinity between membrane and the 3' end of mRNA, or by affinity between membrane and a temporary N-terminal segment of the forming peptide. Both factors may be at work. In any case, the polypeptide appears to pass directly through a pore into the cisterna of the ER where folding and glycosylation take place immediately.

Glycosylation, for many proteins, is a stepwise addition of sugars. For others, the oligosaccharide is transferred to protein intact from a lipid carrier (e.g., dolichol) on which it was assembled. This latter process may be more important for glycoproteins destined to become part of membranes, rather than for secretory proteins. In either process, certain terminal sugars are added in a stepwise fashion only later, after transfer to the Golgi apparatus.

10-6 The Golgi apparatus is a stack of usually smooth-surfaced saccules. The membranes of these saccules can give rise to vesicles of various sorts, primarily lysosomes and secretion vesicles. Products destined for secretion are commonly processed by Golgi-associated enzymes, many of which are concerned with carbohydrate synthesis or terminal carbohydrate addition to proteins.

10-7 There are three generally recognized modes of secretion: (1) holocrine, in which the whole cell is sacrificed; (2) apocrine, in which the secretory vesicle buds through the cell surface, carrying with it a portion of the plasma membrane; and (3) merocrine or eccrine secretion, in which the vesicle fuses with the plasma membrane. Merocrine secretion (exocytosis) is most common. It typically involves microtubule-guided transport of Golgi-derived vesicles to the microfilament web at the cell surface and release when the web is altered by some stimulus. The coupling between stimulus and release in many cases appears to be related to the level of Ca^{2+}, which may be changed directly by the stimulus (e.g., by increased influx) or indirectly through cAMP or cGMP. The secretory vesicle, once empty, may be returned to the cytoplasm or added to the cell membrane. In the latter case, the resulting increase in surface area must be compensated for by endocytic removal of plasma membrane in order to maintain a constant cell size.

10-8 There is constant synthesis and turnover of all cell components, including the lipids and proteins of membranes. The lifetimes of various components differ markedly, however, including vast differences from one membrane protein to another, the destruction of which seems random (rather than age-related) for proteins of a given type. Replacement proteins may insert directly into the endoplasmic reticulum after synthesis or be transferred by vesicles derived from the ER or from the Golgi apparatus. Lipids may be transferred by vesicles, though transfer catalyzed by specific cytoplasmic lipid-exchange proteins is also important, especially to mitochondria and chloroplasts. Lipid exchange between plasma membrane and the serum lipoprotein of mammals has also been documented. However, in growing cells, at least, the traffic is relatively one-way, from serum lipoprotein to cell, fostered and controlled by specific membrane receptors for serum lipoproteins, which thus furnish much of the lipid (especially the cholesterol) needed by the cells.

STUDY GUIDE

10-1 (a) Define and describe the glycocalyx. (b) Cellular adhesion must be preceded by a specific aggregation. Describe two models for that process. (c) Describe the ABO blood group system and the surface modifications responsible for it. (d) Describe the inheritance of HLA antigens. Why are HLA haplotypes normally transmitted to progeny intact? (e) Describe the phenomena of patching and capping. Which part requires metabolic energy? (f) What is a lectin?

10-2 (a) What evidence points to a surface localization of the insulin receptor? (b) Receptor occupancy is not always proportional to hormone effect. What sorts of deviations are observed? (c) Explain the "mobile receptor hypothesis." How is it supported by observations made on cholera toxin? (d) Densensitization to constant levels of a hormone is a common event. What mechanisms might be responsible?

10-3 (a) Distinguish phagocytosis from pinocytosis. (b) What mechanisms are known to account for the selectivity often associated with phagocytosis? (Include the role of antibodies.) (c) Describe the digestive events that usually follow phagocytosis. (d) What mechanisms assist in the killing of microorganisms after phagocytosis? (e) Describe the process of autophagy. (f) What is the ultimate fate of the digestive vacuole?

10-4 (a) Describe the nuclear envelope and the structure of its pores. (b) What sorts of observations point to the pores as open channels, and what observations seem to contradict this view? (c) What is meant by nuclear blebbing? (d) What similarities between nuclear envelope and endoplasmic reticulum were discussed?

10-5 (a) What functions seem relegated mostly to smooth ER? (b) What is the function of microsomal electron transport? (c) Proteins destined for secretion are translated primarily by the rough ER instead of by free ribosomes. What factors probably account for this selectivity? (d) If a polypeptide chain passes readily into the cisterna of the rough ER during synthesis, why doesn't it later leak out through the same passageway? (e) There are two types of glycosylation systems for proteins. Describe them. What are their differences and what is the probable reason for having two systems?

10-6 (a) Describe the Golgi apparatus. Which is the proximal and which is the distal face? (b) What types of vesicles arise from Golgi membranes? (c) There appears to be a regular turnover of Golgi membranes. According to available evidence, where do the membranes come from, and what happens to them?

10-7 (a) Give an example of each of the three major modes of secretion. (b) At which stages in secretion is energy specifically required, and for what purposes? (c) Describe and give the presumed basis for the biphasic release of insulin from pancreatic β cells. (d) How is calcium involved in the secretory process? (e) Secretion in mast cells is affected by IgE antibody, antigen, histamine, prostaglandins, epinephrine, and acetylcholine. In general terms, what is the apparent mode of action of each? (f) What are the three possible fates of the secretory vesicle? (*Hint:* Include apocrine secretion.) What compensatory changes in cell membrane are required by each?

10-8 (a) Describe the removal and replacement of plasma membrane. What is the likely explanation for the longer survival of plasmalemma proteins compared to lipids? (b) Membrane proteins do not just "grow old and die." How do we know that? (c) What is the significance of lipid-exchange proteins? (d) There are at least two mechanisms whereby a growing animal cell can get needed cholesterol without synthesizing it. What are they?

REFERENCES

THE CELL SURFACE

Ashwell, G., and A. Morell, "Membrane Glycoproteins and Recognition Phenomena." *Trends Biochem. Sci.*, **2**(4): 76 (April 1977).

Bach, F. H., "Genetics of Transplantation: The Major Histocompatibility Complex." *Ann. Rev. Genet.*, **10**: 319 (1976).

Bach, F. H., and J. van Rood, "The Major Histocompatibility Complex—Genetics and Biology." *New Engl. J. Med.*, **295**: 806, 872, and 927 (1976).

Boyse, E., and H. Cantor, "Surface Characteristics of T-Lymphocyte Subpopulations." *Hospital Practice*, April 1977, p. 81.

Crumpton, M., D. Allen, J. Auger, N. Green, and V. Maino, "Recognition at Cell Surfaces: Phytohaemagglutinin-lymphocyte Interaction." *Phil. Trans. Roy. Soc. London, Ser. B.*, **272**: 173 (1975). Ca^{2+}-activation of lymphocytes.

Cunningham, B., "The Structure and Function of Histocompatibility Antigens." *Scientific American*, **237**(4): 96 (Oct. 1977).

Fox, C. F., ed., *Biochemistry of Cell Walls and Membranes*. Baltimore: Univ. Park Press, 1975. An excellent series of reviews.

Gold, E. R., and P. Balding, *Receptor-Specific Proteins. Plant and Animal Lectins*. New York: Elsevier, 1975.

Greaves, M. F., *Cellular Recognition*. New York: Halsted Press, 1975. (Paperback.)

Hughes, R. C., "The Complex Carbohydrates of Mammalian Cell Surfaces and Their Biological Roles." *Essays Biochem.*, **11**: 1 (1975).

———, *Membrane Glycoproteins: A Review of Structure and Function*. Boston: Butterworth, 1976.

Katz, D., "Genetic Controls and Cellular Interactions in Antibody Formation." *Hospital Practice*, Feb. 1977, p. 85.

Lloyd, C. W., "Sialic Acid and the Social Behaviour of Cells." *Biol. Rev. (Cambridge)*, **50**: 325 (1975).

Luft, J. H., "The Structure and Properties of the Cell Surface Coat." *Int. Rev. Cytol.*, **45**: 291 (1976).

Marchalonis, J., "Lymphocyte Surface Immunoglobulins." *Science*, **190**: 20 (1975).

Miller, J., "Blood Groups: Why do they Exist?" *BioScience*, **26**: 557 (1976). Some ideas, but no proof as yet.

Nicolson, G. L., "The Interactions of Lectins with Animal Cell Surfaces." *Int. Rev. Cytol.*, **39**: 89 (1974).

———, "Transmembrane Control of the Receptors on Normal and Tumor Cells. I. Cytoplasmic Influence over Cell Surface Components." *Biochim. Biophys. Acta*, **457**: 57 (1976). See also **458**: 1 (1976) for part II.

Poste, G., and G. L. Nicolson, eds., *Dynamic Aspects of Cell Surface Organization* (*Cell Surface Reviews*, Vol. 3). New York: Elsevier/North Holland, 1977.

Raff, M. C., "Cell Surface Immunology." *Scientific American*, **234**(5): 30 (May 1976).

Rowlands, D., Jr., and R. Daniele, "Surface Receptors in the Immune Response." *New Engl. J. Med.*, **293**: 26 (1975).

Sasazuki, T., F. Grumet and H. McDevitt, "The Association of Genes in the Major Histocompatibility Complex and Disease Susceptibility." *Ann. Rev. Med.*, **28**: 425 (1977).

Schmell, E., B. Earles, C. Breaux, and W. Lennarz, "Identification of a Sperm Receptor on the Surface of the Eggs of the Sea Urchin *Arbacia punctulata*." *J. Cell Biol.*, **72**: 35 (1977).

Sharon, N., "Lectins." *Scientific American*, **236**(6): 108 (June 1977).

Shur, B. D., and S. Roth, "Cell Surface Glycosyltransferases." *Biochim. Biophys. Acta*, **415**: 473 (1975). Possible role in cellular recognition.

Sundqvist, K-G., and A. Ehrnst, "Cytoskeletal Control of Surface Membrane Mobility." *Nature*, **264**: 226 (1976). The control is there, but the details just are not yet clear.

HORMONE RECEPTORS

Bar, R., and J. Roth, "Insulin Receptor Status in Disease States of Man." *Arch. Internal Med.*, **137**: 474 (1977).

Bengelsdorf, I., "Cell Membrane Receptors." *BioScience*, **26**: 400 (1976). Brief review.

Bennett, V., E. O'Keefe, and P. Cuatrecasas, "The Mechanism of Action of Cholera Toxin and the Mobile Receptor Theory of Hormone Receptor-Adenylate Cyclase Interactions." *Proc. Nat. Acad. Sci. U.S.*, **72**: 33 (1975).

Catt, K., and M. Dufau, "Peptide Hormone Receptors." *Ann. Rev. Physiol.*, **39**: 529 (1977).

Cuatrecasas, P., and M. Greaves, eds., *Receptors and Recognition*, Series A. New York: Halsted Press, Vol. 1, 1976; Vol. 2, 1977. Reviews on surface receptors.

Cuatrecasas, P., and M. D. Hollenberg, "Membrane Receptors and Hormone Action." *Advan. Protein Chem.*, **30**: 252 (1976). Includes a good summary of Cuatrecasas's mobile receptor hypothesis.

Cuatrecasas, P., M. Hollenberg, K-J. Chang, and V. Bennett, "Hormone Receptor Complexes and Their Modulation of Membrane Function." *Rec. Progr. Hormone Res.*, **31**: 37 (1975).

Dumont, J., and J. Nunez, *Hormones and Cell Regulation*. New York: Elsevier/North Holland, 1977. Symposium.

Helmreich, E., H. Zenner, and T. Pfeuffer, "Signal Transfer From Hormone Receptor to Adenylate Cyclase." *Current Topics Cell. Reg.*, **10**: 41 (1976).

Kahn, C. R., "Membrane Receptors for Hormones and Neurotransmitters." *J. Cell Biol.*, **70**: 261 (1976).

Kahn, C. R., et al., "Receptors for Peptide Hormones." *Ann. Internal Med.*, **86**: 205 (1977).

Lefkowitz, R. J., "The β-Adrenergic Receptor." *Life Sci.*, **18**: 461 (1976). A brief, very readable review.

Lefkowitz, R. J., L. Limbird, C. Mukherjee and M. Caron, "The β-Adrenergic Receptor and Adenylate Cyclase." *Biochim. Biophys. Acta*, **457**: 1 (1976).

Levey, G. S., ed., *Hormone-Receptor Interaction. Molecular Aspects.* New York: Dekker, 1976.

Lockwood, D., "Insulin Resistance in Obesity." *Hospital Practice*, Dec. 1976, p. 79.

Papaconstantinou, J., ed., *The Molecular Biology of Hormone Action.* New York: Academic Press, 1976.

Raff, Martin, "Self Regulation of Membrane Receptors." *Nature*, **259**: 265 (1976). A short review.

Snyder, S. H., "Opiate Receptors and Internal Opiates." *Scientific American*, **236**(3): 44 (March 1977).

van Heyningen, S., "The Structure of Bacterial Toxins." *Trends Biochem. Sci.*, **1**(5): 114 (May 1976). Cholera toxin.

ENDOCYTOSIS AND LYSOSOMAL DIGESTION

Ashwell, G., and A. G. Morrell, "The Role of Surface Carbohydrates in the Hepatic Recognition of Circulating Glycoproteins." *Advan. Enzymol.*, **41**: 99 (1974).

Dean, R. T., and A. J. Barrett, "Lysosomes." *Essays Biochem.*, **12**: 1 (1976).

Holtzman, E., *Lysosomes: A Survey.* New York: Springer, 1976. (Cell Biology Monographs, Vol. 3).

Kolodny, E., "Lysosomal Storage Diseases." *New Engl. J. Med.*, **294**: 1217 (1976).

Morrison, M., and G. Schonbaum, "Peroxidase-Catalyzed Halogenation." *Ann. Rev. Biochem.*, **45**: 861 (1976).

Neufeld, E., T. Lim, and L. Shapiro, "Inherited Disorders of Lysosomal Metabolism." *Ann. Rev. Biochem.*, **44**: 357 (1975).

Pitt, D., *Lysosomes and Cell Function.* New York: Longman, 1975. (Paperback.)

Roos, D., "Oxidative Killing of Microorganisms by Phagocytic Cells." *Trends Biochem. Sci.*, **2**(3): 61 (March 1977).

Silverstein, S., R. Steinman, and Z. Cohn, "Endocytosis." *Ann. Rev. Biochem.*, **46**: 669 (1977).

Stossel, T. P., "Phagocytosis." *New Engl. J. Med.*, **290**: 717, 774, & 833 (1974). A very nice three-part series on the normal and abnormal.

Weissmann, G., "Experimental Enzyme Replacement in Genetic and Other Disorders." *Hospital Practice*, Sept. 1976, p. 49.

THE NUCLEAR ENVELOPE AND ENDOPLASMIC RETICULUM

Blobel, G., and B. Dobberstein, "Transfer of Proteins Across Membranes." *J. Cell Biol.*, **67**: 835 (1975).

Bourne, G. H., J. F. Danielli, and K. W. Jeon, eds., *International Review of Cytology.* Supplement 4. Aspects of Nuclear Structure and Function. New York: Academic Press, 1974.

Cardell, R., Jr., "Smooth Endoplasmic Reticulum in Rat Hepatocytes during Glycogen Deposition and Depletion." *Int. Rev. Cytol.*, **48**: 221 (1977).

Cooper, D. Y., O. Rosenthal, and R. Snyder, eds., *Cytochromes P450 and b_5.* New York: Plenum Press, 1975.

Depierre, J., and G. Dallner, "Structural Aspects of the Membrane of the Endoplasmic Reticulum." *Biochim. Biophys. Acta*, **415**: 411 (1975). A lengthy review covering, in addition to the title topics, protein glycosylation.

Franke, W. W., "Nuclear Envelopes. Structure and Biochemistry of the Nuclear Envelope." *Phil. Trans. Roy. Soc. London, Ser. B*, **268**: 67 (1974). Architecture of the pore in various cell types.

Kappas, A., and A. Alvares, "How the Liver Metabolizes Foreign Substances." *Scientific American*, **232**(6): 22 (June, 1975).

Lande, M., M. Adesnik, M. Sumida, Y. Tashiro, and D. Sabatini, "Direct Association of Messenger RNA with Microsomal Membranes in Human Diploid Fibroblasts." *J. Cell Biol.*, **65**: 513 (1975).

Lennarz, W. J., "Lipid Linked Sugars in Glycoprotein Synthesis." *Science*, **188**: 986 (1975). Sugar addition to membrane glycoproteins.

McIntosh, P., and K. O'Toole, "The Interaction of Ribosomes and Membranes in Animal Cells." *Biochim. Biophys. Acta*, **457**: 171 (1976). A review.

Molnar, J., "A Proposed Pathway of Plasma Glycoprotein Synthesis." *Mol. Cell. Biochem.*, **6**: 3 (1975).

Northcote, D. H., "Complex Envelope System. Membrane System of Plant Cells." *Phil. Trans. Roy. Soc. London, Ser. B*, **268**: 119 (1974). A very nice review.

Parodi, A., and L. Leloir, "Lipid Intermediates in Protein Glycosylation." *Trends Biochem. Sci.*, **1**(3): 58 (March 1976).

Sabatini, D., G. Ojakian, M. Lande, J. Lewis, W. Mok, M. Adesnik, and G. Kreibich, "Structural and Functional Aspects of the Protein Synthesizing Apparatus in the Rough Endoplasmic Reticulum." In *Control Mechanisms in Development*, ed. R. Meints and E. Davies. New York: Plenum Press, 1975, p. 151. A very readable review.

Shore, G. and J. Tata, "Functions for Polyribosome-Membrane Interactions in Protein Synthesis." *Biochim. Biophys. Acta*, **472**: 197 (1977).

Wunderlich, F., R. Berezney, and H. Kleinig, "The Nuclear Envelope." In *Biological Membranes*, Vol. 3, C. Chapman and D. Wallach, eds. New York: Academic Press, 1976.

SECRETION

Baker, P. F., "Calcium and the Control of Neuro-Secretion." *Sci. Progr., Oxford*, **64**: 95 (1977).

Beaven, M. A., "Histamine." *New Engl. J. Med.*, **294**: 30 and 320 (1976). Synthesis, release from mast cells, and actions.

Cerasi, E., "Insulin Secretion: Mechanism of the Stimulation by Glucose." *Quart. Rev. Biophys.* **8**: 1 (1975).

Chrispeels, M. J., "Biosynthesis, Intracellular Transport, and Secretion of Extracellular Macromolecules." *Ann. Rev. Plant Physiol.*, **27**: 19 (1976).

Cook, G. M. W., "The Golgi Apparatus: Form and Function." In *Lysosomes in Biology and Pathology*, Vol. 3, J. Dingle, ed. New York: American Elsevier, 1973, p. 237.

Douglas, W. W., "Stimulus-Secretion Coupling." *Brit. J. Pharmacol.*, **34**: 451 (1968). Early proposal for the role of Ca^{2+}.

Foreman, J., L. Garland, and J. Mongar, "The Role of Calcium in Secretory Processes: Model Studies in Mast Cells." *Soc. Exp. Biol.*, Symposium No. **30**, p. 193 (1976).

Gillespie, E., "Microtubules, Cyclic AMP, Calcium, and Secretion." *Ann. N.Y. Acad. Sci.*, **253**: 771 (1975).

Hamburger, R. N., "Allergy and the Immune System." *Am. Scientist*, **64**: 157 (1976). Mostly on IgE.

Ishizaka, K., "Structure and Biologic Activity of Immunoglobin E." *Hospital Practice*, Jan. 1977, p. 57.

Judah, J., and P. Quinn, "On the Biosynthesis of Serum Albumin." *Trends Biochem. Sci.*, **1**(5): 107 (May 1976).

Lambert, A. E., "The Regulation of Insulin Secretion." *Rev. Physiol. Biochem. Pharmacol.*, **75**: 97 (1976).

Leppard, G., L. Sowden, and J. R. Colvin, "Nascent Stage of Cellulose Biosynthesis." *Science*, **189**: 1094 (1975).

Locke, M., and P. Huie, "The Beads in the Golgi Complex-Endoplasmic Reticulum Region." *J. Cell Biol.*, **70**: 384 (1976). Possible site of transfer from ER to Golgi.

Norman, P. S., "The Clinical Significance of IgE." *Hospital Practice*, Aug. 1975, p. 41. The mast cell and its antibody-antigen trigger.

Palade, G., "Intracellular Aspects of the Process of Protein Synthesis." *Science*, **189**: 347 (1975). Nobel Lecture, 1974.

Patton, S., and T. W. Keenan, "The Milk Fat Globule Membrane." *Biochim. Biophys. Acta*, **415**: 273 (1975).

Satir, Birgit, "The Final Steps in Secretion." *Scientific American*, **233**(4): 28 (Oct. 1975).

Sharoni, Y., S. Eimerl, and M. Schramm, "Secretion of Old Versus New Exportable Protein in Rat Parotid Slices." *J. Cell Biol.*, **71**: 107 (1976). First made is first out.

Whaley, W. G., *The Golgi Apparatus* (Cell Biology Monograph No. 2). New York: Springer-Verlag, 1975.

Whaley, W. G., M. Dauwalder, and T. Leffingwell, "Differentiation of the Golgi Apparatus in the Genetic Control of Development." *Current Topics Develop. Biol.*, **10**: 161 (1975). A general review of the Golgi apparatus.

MEMBRANE TURNOVER

Bretscher, M., "Directed Lipid Flow in Cell Membranes." *Nature*, **260**: 21 (1976). An explanation for capping.

Brown, M. S., and J. L. Goldstein, "Receptor-Mediated Control of Cholesterol Metabolism." *Science*, **191**: 150 (1976). See also *New Engl. J. Med.*, **294**: 1386 (1976) and *Trends Biochem. Sci.*, **1**(9): 193 (Sept. 1976).

De Camilli, P., D. Peluchetti, and J. Meldolesi, "Dynamic Changes of the Luminal Plasmalemma in Stimulated Parotid Acinar Cells." *J. Cell Biol.*, **70**: 59 (1976). Secretory vesicle membrane may be selectively removed by pinocytotic vesicles.

Cook, J. S., ed., *Biogenesis and Turnover of Membrane Macromolecules* (Soc. Gen. Physiol. Series, Vol. 31). New York: Raven Press, 1976.

Langdon, R. G., "Serum Lipoprotein Apoproteins as Major Protein Constituents of the Human Erythrocyte Membrane." *Biochim. Biophys. Acta*, **342**: 213 (1974).

Martonosi, A. M., *The Enzymes of Biological Membranes*. Vol. 2: *Biosynthesis of Cell Components*. New York: John Wiley, 1976.

Meldolesi, J., "Membranes and Membrane Surfaces. Dynamics of Cytoplasmic Membranes in Pancreatic Acinar Cells." *Phil. Trans. Roy. Soc. London, Ser. B.*, **268**: 39 (1974). Recycling of secretory granule membrane.

Poste, G., and G. L. Nicolson, eds., *The Synthesis, Assembly, and Turnover of Cell Surface Components* (Cell Surface Reviews, Vol. 4). New York: Elsevier/North Holland, 1977.

Schimke, R. T., "Turnover of Membrane Proteins in Animal Tissues." In *Biochemistry of Cell Walls and Membranes*, C. F. Fox, ed. Baltimore: Univ. Park Press, 1975, p. 229.

Steinman, R., S. Brodie, and Z. Cohn, "Membrane Flow During Pinocytosis." *J. Cell Biol.*, **68**: 665 (1976).

Teichberg, S., E. Holtzman, S. Crain, and E. Peterson, "Circulation and Turnover of Synaptic Vesicle Membrane in Cultured Fetal Mammalian Spinal Cord Neurons." *J. Cell Biol.*, **67**: 215 (1975). Endocytotic vesicles are transported to the soma.

Wirtz, K., "Transfer of Phospholipids Between Membranes." *Biochim. Biophys. Acta*, **344**: 95 (1974).

Wirtz, K., and L. van Deenen, "Phospholipid-Exchange Proteins." *Trends Biochem. Sci.*, **2**(3): 49 (1977).

NOTES

1. Unlike the ABO system, antibodies to missing Rh antigens are not usually present. They can be produced, however, following transfusion of Rh$^+$ blood into an Rh$^-$ donor or during birth when an Rh$^-$ mother has an Rh$^+$ baby. In the latter case, subsequent Rh$^+$ fetuses are threatened with hemolysis, a situation that is preventible by giving the mother anti-Rh antibodies soon after the first birth to remove fetal Rh antigens from the mother's circulation.

2. Although the level of glucose in the blood is only one of the parameters affected by insulin, it is a sensitive one and is easily measured; hence it is commonly used to adjust insulin dosage in diabetes. Even more convenient is the urine test. When the blood level of glucose is greater than about twice normal, it exceeds the ability of the active transport mechanisms of the kidney to reclaim it and spills over into the urine.

3. Neutrophils and macrophages (except macrophages in the lung) depend largely on anaerobic glycolysis for energy. This is significant because of poor oxygen availability in many sites outside the circulatory system, especially in the presence of a local infection and pus formation. It is thought that this dependence on glucose is responsible for the increased risk of infection experienced by untreated diabetics and, in part, by those who are receiving long-term cortisone therapy. Diabetics may lack the insulin needed

to promote glucose uptake by neutrophils and macrophages, while cortisone and its related compounds tend to antagonize insulin-stimulated glucose transport.

4. White blood cells, or leukocytes, are divided into two classes; "mononuclear" cells (40% of the total in humans) consisting of lymphocytes and a few *monocytes*; and *polymorphonuclear cells* or *PMNs* having lobulated nuclei and large cytoplasmic granules. The neutrophil, also called a heterophil, is the most common cell in this latter group, the others being eosinophils and basophils, so named because of their staining properties (eosin is an acidic, red dye).

5. Toxoplasma normally causes a very mild disease in humans that often goes unnoticed. By adulthood, a large fraction of the population has already been exposed. However, a woman who becomes newly infected early in pregnancy runs a serious risk of miscarriage or bearing a deformed child.

6. Recall that there are two types of sweat glands, apocrine and merocrine (eccrine). That the mammary gland should combine these two modes of secretion is interesting in light of the developmental history of the mammary gland, which is a modified sweat gland.

7. These findings have an important clinical application in understanding familial hypercholesterolemia, an inherited disease marked by extraordinarily high levels of serum cholesterol, early atherosclerosis, heart attacks, and strokes. The defective gene appears to result in nonfunctional or missing LDL receptors. Without receptors for obtaining an outside source of cholesterol, fibroblasts and smooth muscle cells of the arteries produce their own cholesterol, but have no effective feedback system to keep from overproducing it.

11

Cell Division

11-1 Replication of DNA 492
    Semiconservative Replication
    Replicases
    Replication Models
    Errors in Replication
    Mutagens
    Repair
    The Replication of Chromatin
11-2 Mitosis 510
    The Cell Cycle
    Prophase
    Metaphase
    Anaphase
    Telophase
    The Evolution of Mitosis
11-3 Meiosis 517
    The First Meiotic Division
    The Second Meiotic Division
    Gametogenesis
11-4 Spindle Dynamics 523
    The Dynamic Equilibrium of Spindle Fibers
    The Mitotic Mechanism
    Microtubule Disassembly and Chromosome Movement
    Sliding Microtubules
    Conclusions
11-5 Cytokinesis 528
    Synchrony between Nuclear and Cytoplasmic Division
    The Contractile Ring
    Membrane Changes During Cytokinesis
    Cytokinesis in Plants
11-6 Cell Cycle Controls 536
    Cell Growth and Cell Division
    Checkpoints
    G_2 Arrest
    G_0 Cells
    Growth Factors
    Cyclic AMP and Other "Second Messengers"
    Lymphocyte Activation
    Steroid Hormones
    Mitotic Inhibitors
Summary 545
Study Guide 546
References 547
Notes 549

For life to continue, cells must reproduce. Most do so by *binary fission*, yielding two daughter cells each of which is identical to the parent (see Fig. 11-1*a* and *c*). Other cells *bud* to produce a daughter that is initially much smaller than the parent cell (Fig. 11-1*b*). And in still other cases *spores* are formed that, until they germinate, are quite unlike the cells from which they were derived. Since we cannot describe in detail all modes of cellular replication, we shall concentrate instead on the replication of genetic material and on binary fission as it occurs in most prokaryotes and higher eukaryotes.

11-1 REPLICATION OF DNA

Watson and Crick pointed out that genetic information could be conserved during cellular replication if each strand of the parent molecule

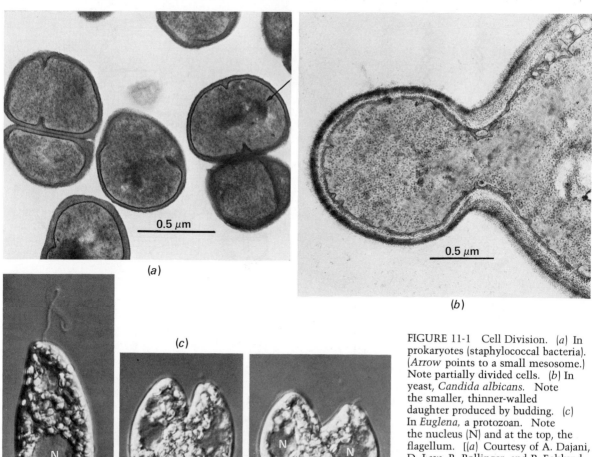

FIGURE 11-1 Cell Division. (a) In prokaryotes (staphylococcal bacteria). (*Arrow* points to a small mesosome.) Note partially divided cells. (b) In yeast, *Candida albicans*. Note the smaller, thinner-walled daughter produced by budding. (c) In *Euglena*, a protozoan. Note the nucleus (N) and at the top, the flagellum. [(a) Courtesy of A. Dajani, D. Law, R. Bollinger, and P. Ecklund, *Infection and Immunity*, **14**: 776 (1976): (b) Courtesy of W. Djaczenko and A. Cassone, *J. Cell Biol.*, **52**: 186 (1972); (c) Courtesy of C. Hofmann and B. Bouck, *J. Cell Biol.*, **69**: 693 (1976).]

were to serve as a template for the polymerization of a complementary strand in a daughter molecule. Thus, the sequence of bases would be reproduced.

If the two parent strands separate completely so that each daughter molecule consists of one new and one old strand, the mode of replication is called *semiconservative*. On the other hand, one could imagine polymerization occurring with a temporary and localized separation of parent strands at the point where new DNA is being put together. This mechanism would lead to a *conservative* replication, producing a daughter molecule that is entirely new. An experiment capable of distinguishing between the two modes of replication was designed by M. Meselson and F. W. Stahl at the California Institute of Technology in 1958.

Semiconservative Replication. Meselson and Stahl utilized the fact that any solute not isodense with its solvent experiences a net force in a gravi-

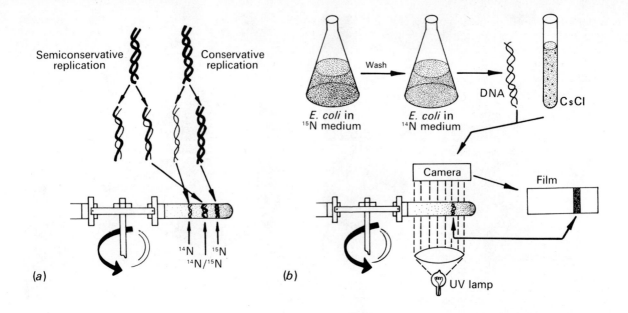
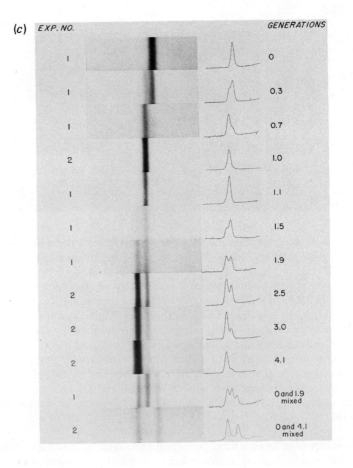

FIGURE 11-2 The Semiconservative Replication of DNA. (a) DNA labeled with heavy nitrogen (^{15}N, dark strands) is allowed to replicate once in the presence of the common isotope of nitrogen, ^{14}N. The diagram shows the results expected from conservative and semiconservative replication. (b) The experimental protocol used by Meselson and Stahl. (c) The results of different trials, showing the banding pattern revealed by the camera after various periods of growth on ^{14}N, plus plots of film densities. (Dense CsCl is to the right.) [(c) Courtesy of M. Meselson and F. W. Stahl, *Proc. Nat. Acad. Sci. U.S.*, **44**: 671 (1958).]

tational or centrifugal field. In a strong centrifugal field, there may be a noticeable tendency for even small solutes to float or sink, depending on whether they are less or more dense than the surrounding solvent. Meselson and Stahl designed an experiment in which newly synthesized DNA would have a buoyant density different from its parent (or template) and thus be readily identified (see Fig. 11-2).

Cultures of *E. coli* were grown in a medium containing heavy nitrogen (^{15}N), washed, and then allowed to continue growing in a normal medium (^{14}N). At the time of transfer, isolated DNA was added to a dense (1.7 g/ml) solution of CsCl. The tube was spun in a centrifuge to redistribute the salt in a density gradient ranging from about 1.75 g/ml at the bottom to 1.65 g/ml at the top. In this gradient, added DNA "bands" at the level at which it is isodense with the salt solution. After one generation of growth in the normal medium, isolated DNA formed a band that fell halfway between the isodense point for fully labeled ^{15}N-DNA and the position expected for normal, light DNA. After one generation of growth, then, each DNA molecule was half original and half new. Heating the DNA to separate its strands made it clear that these molecules contained one parental (fully ^{15}N) and one newly synthesized (completely ^{14}N) "subunit," which was later determined to be a single polynucleotide strand. In other words, the DNA was replicated semiconservatively.

The semiconservative replication of DNA raises some perplexing problems (see Fig. 11-3). For example, since the length of the *E. coli*

SECTION 11-1
Replication of DNA

FIGURE 11-3 The Replication Problem. This bacterium, an *Hemophilus influenzae* spheroplast, is sitting in the middle of DNA extruded from it—some 832 μm in a single molecule. Replicating a molecule so enormous, and then segregating the daughters into separate cells at division, would seem to present a staggering mechanical problem. [Courtesy of L. A. MacHattie. For details, see *J. Mol. Biol.*, **11**: 648 (1965).]

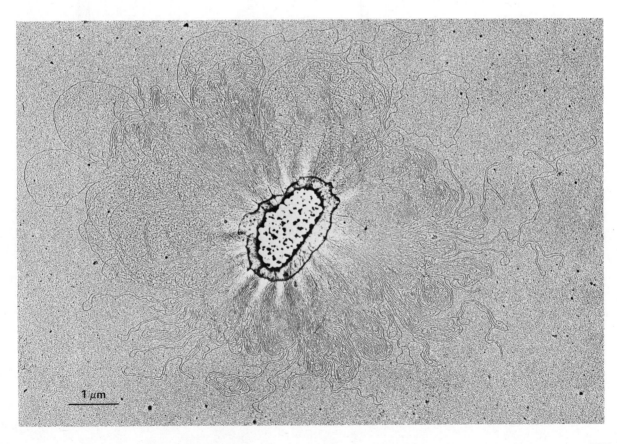

chromosome is about a millimeter, a pitch of 34 Å per turn of the DNA helix means that a third of a million twists are required to separate the two strands during replication. Since replication can occur in about 40 minutes, a twist rate of roughly 7500 revolutions per minute is implied—truly a staggering number. Additional problems with the semiconservative model arise when one attempts to reconcile the properties of those enzymes responsible for replication with observations made on replicating chromosomes.

An Australian scientist, John Cairns, used *autoradiography* to capture *E. coli* chromosomes in the process of replication. First, he fed growing cells tritiated (^{3}H) thymidine. Tritium is a radioactive form of hydrogen that emits weak β rays (electrons) and thymidine of course is a DNA nucleoside. Next, the cells were placed in a cellulose casing and disrupted with detergent. When the solution was removed from the casing, some DNA molecules were left behind, adsorbed to the surface of the container. The casing was then transferred to a microscope slide and coated with a fine-grain photographic emulsion. As the tritium atoms decayed, the released electrons caused the conversion of silver halide in the emulsion to silver grains, in much the same way as photons of light would do. After several months, the emulsion was developed to reveal a trail of silver grains visible in a microscope and corresponding to the positions of the decaying molecules (Fig. 11-4). Thus, the shape of the la-

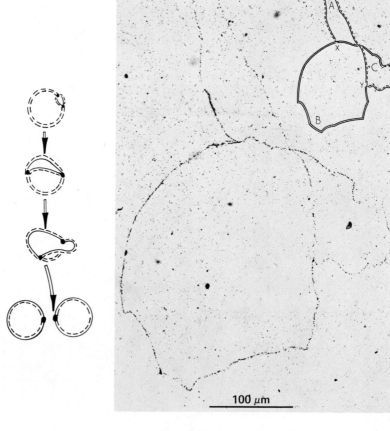

FIGURE 11-4 Autoradiography of Replicating DNA. RIGHT: *E. coli* DNA in the process of replication. The inset is a schematic representation of the autoradiogram. The dashed line in regions A and C represents one parent strand, one-fourth tritiated (solid portion) from the previous replication. The other parent strand and the new daughter strands (solid lines) are fully tritiated. X and Y are the "replication forks."
LEFT: The replication scheme suggested by autoradiography. Here, the addition of tritium is synchronized with the beginning of replication. [Courtesy of John Cairns, from *Cold Spring Harbor Symp. Quant. Biol.*, **28**: 43 (1963).]

beled DNA was visualized, although with rather poor resolution: the silver grains, after all, are much larger than the diameter of the DNA helix.

This process, autoradiography, allowed Cairns to see the shape of the complete chromosome. It turned out to be a closed circle having a circumference of 1100–1400 μm even during replication. It appears as if new molecules are separated from the original in the same way that a closed zipper with its two ends sewed together would yield two circles when unzipped. There is a point on the molecule, called the *origin,* from which replication starts. Typically, a *replication fork* takes off from this point in each direction. When the two replication forks meet on the other side of the circle, replication is complete and two daughter chromosomes, each containing one new and one old strand, are formed.

Replicases. An enzyme capable of catalyzing the replication of DNA (a *replicase*) was discovered in *E. coli* by Arthur Kornberg in 1955. It polymerizes a DNA strand complementary to an existing strand (the *template*) creating the double-stranded molecule required by semiconservative replication. Kornberg's "DNA-dependent DNA polymerase," now called *polymerase I,* is capable of adding a nucleotide to the 3' hydroxyl of the last deoxyribose in an existing chain. Thus one strand of DNA serves as an essential *primer* for the enzyme while its longer complementary strand serves as template.

The new nucleotide, initially a triphosphate, loses its terminal pyrophosphate (PP_i) upon addition to the chain (see Fig. 11-5). Pyrophosphate is then cleaved by a pyrophosphatase, making the overall reaction thermodynamically favorable. To add a guanine nucleotide to the chain, for example, involves two steps:

$$(\text{polynucleotide})_n + \text{GTP} \rightarrow (\text{polynucleotide})_{n+1} + PP_i$$
$$PP_i \rightarrow 2P_i$$

The first step has a standard state free energy change of about 2 kJ/mole, whereas that of the second step is about −33 kJ/mole. Thus, when considered as a whole the combined reaction is very favorable.

But note that polymerase I adds only to the 3' end of an existing chain, while Cairns's data lead us to expect the simultaneous replication of two antiparallel chains at a single replication fork. The autoradiographs make it appear that one of the new strands is polymerized in the 5' → 3' direction, consistent with the mode of action of Kornberg's enzyme, while the other strand is being synthesized from its 3' to its 5' end. It was first assumed that another polymerase, undiscovered, catalyzes the 3' → 5' mode, but all attempts to identify such an enzyme have failed.

The discovery in 1969 of a mutant *E. coli* that has only about 1% of the normal polymerase I activity but grows nonetheless, rekindled the search for additional replicases. In fact, two other major replicases, called DNA polymerase II and III, have been isolated. Both, however, have the same limitation as polymerase I—they can elongate an existing strand from its 3' end but they can neither polymerize DNA in the opposite direction nor initiate a new strand without a primer.

Replication Models. Thus we have three problems to consider in explaining DNA replication: (1) how a high-speed twisting of a very long molecule might be initiated and sustained; (2) what mechanism might be

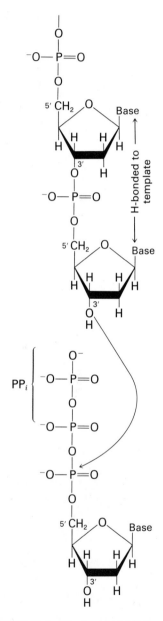

FIGURE 11-5 Nucleotide Addition to DNA. The new nucleotide is chosen to be complementary to the corresponding nucleotide of the template strand.

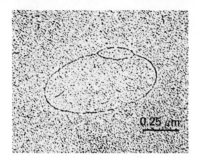
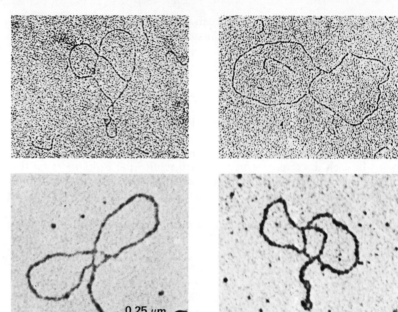

FIGURE 11-6 Supercoiling and the Replication of Circular DNA. From SV40, a cancer-causing animal virus. TOP ROW: Metal-shadowed specimens showing a replication sequence similar to that of Fig. 11-4. Note the beginning of a supercoil in one frame. SECOND ROW: In most specimens the supercoil is very tight, as in these two preparations stained with uranyl acetate. [Shadowed specimens courtesy of Bernard Hirt, *J. Mol. Biol.*, **40**: 141 (1969); stained specimens courtesy of N. P. Salzman et al., *J. Virol.*, **8**: 478 (1971).]

responsible for lengthening the strand that grows in the 3' → 5' direction; and (3) how initiation of DNA synthesis begins if replicases can only elongate existing strands. There are now apparently satisfactory answers to these problems. DNA replication, it turns out, requires not just a replicase but a whole family of proteins.

First, the twisting problem is neatly solved by an enzyme that creates a transient break (a "nick") in one strand safely ahead of the replication fork. The nick permits swiveling about the P—O—P bonds in the intact chain without the necessity for twisting the whole molecule. We know that the break is very temporary because in small circular DNAs, intermediates are regularly seen that have supercoiled configurations like those in Fig. 11-6. These coils are produced in the absence of a swivel when untwisting the parent molecule during replication causes a compensatory coiling ahead of the replication forks. (If you have trouble following that argument, twist two strings into a helix and then tie the helix into a circle. As you try to separate the strings, you will produce configurations like those shown in Fig. 11-6.) This supercoiling is relieved by the nick, which is quickly repaired before the replication fork can move into it.

The second problem, that of chain growth in the 3' → 5' direction (opposite to the direction catalyzed by the known replicases), is solved by discontinuous synthesis of this strand. That is, while one new strand grows more or less continuously from its 3' end at a replication fork, the other new strand is being put together from small pieces, each synthesized in the 5' → 3' direction by a replicase and then laced together to provide an overall direction of growth that is 3' → 5' (see Fig. 11-7). Autoradiography and other techniques insensitive to very small changes in the size of DNA would overlook these pieces, called *Okazaki fragments* (in honor of their discoverer), and see only a smooth elongation. The pieces are joined together by an enzyme capable of forming

SECTION 11-1
Replication of DNA

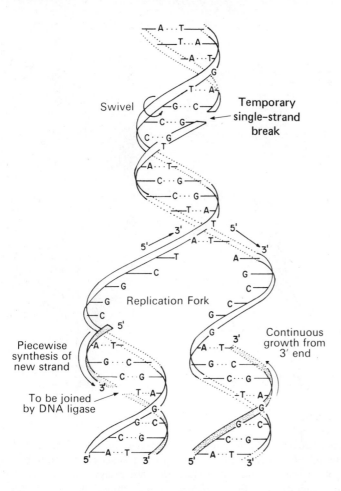

FIGURE 11-7 The Discontinuous Synthesis of DNA. Discontinuous synthesis, or backfilling, could account for a net chain elongation in the 3' → 5' direction using enzymes that can add only at a free 3' hydroxyl.

phosphodiester bonds between two nucleotide monophosphates. This enzyme is called *DNA ligase* (it ligates, or ties together, the strand).

The proposal for discontinuous synthesis of DNA is supported by examination of systems having delayed backfilling, such as one sees in Fig. 11-8. It is also supported by experiments with *E. coli* mutants having a DNA ligase that is unusually temperature sensitive. At slightly elevated temperatures, newly synthesized DNA in these mutants is found only in small fragments. At lower temperatures, the fragments rapidly become incorporated into DNA of normal size.

The replication schemes considered so far show one closed circle peeling off another. An alternate scheme, called the *rolling circle model*, is also dependent on backfilling (see Fig. 11-9). In it, the 5' end at a nick is peeled off the circle and a complementary strand laid down on it by piecewise backfilling. Meanwhile, the circle retains its double-stranded character by growth from the free 3' end. This mechanism is probably used during transfer of DNA from one bacterium to another during mating, and it is thought to account for gene amplification in nucleoli of eukaryotic cells (Chap. 5). Note that the circle need not stop after one turn, but can continue to peel off the new strand for many turns, thus creating multiple copies of itself end-to-end on the linear tail. This may be a mechanism for temporarily increasing the number of ribo-

499

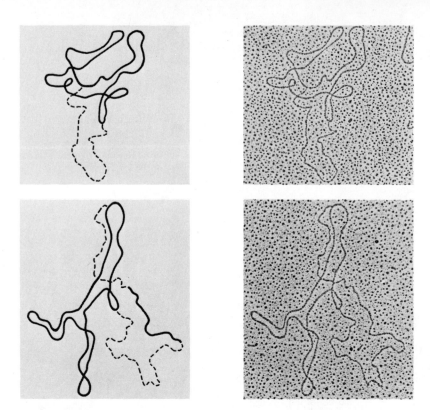

FIGURE 11-8 Delayed Backfilling. In mitochondrial DNA from mouse L cells, a cultured tumor strain. Unit length is 5 μm. Note the delayed backfilling, leaving long stretches of single-stranded DNA (light areas in micrograph, dashed line in drawings). TOP: No backfilling of the displaced strand. BOTTOM: 15% backfilled. [Courtesy of D. L. Robberson, H. Kasamatsu, and J. Vinograd, *Proc. Nat. Acad. Sci. U.S.*, **69**: 737 (1972).]

somal genes and hence the rate of ribosomal synthesis in nucleoli of eukaryotic cells.

The third problem mentioned at the outset of this section is how a new segment of DNA can be initiated using replicases that add only to the 3' end of an existing strand. That problem is apparently solved by initiating each piece of DNA with RNA and using a DNA-dependent RNA polymerase to do so. The replicase then uses this short piece of RNA as a primer.

Invoking RNA primers requires that still another protein be added to the replication arsenal. In fact, even more are needed. The unwinding of parental DNA ahead of the replication fork, for example, appears to be the work of specific proteins that bind only to single strands of DNA. They catch the two strands as their hydrogen bonds "breathe," and by coating the single strands prevent them from coming together again.

The reasons for the existence of three DNA polymerases is also becoming clear. Polymerase III is thought to be the main replicase. Without it, cells cannot reproduce. The other two are thought to be associated with "proofreading" and repair processes, to be explained shortly. In addition, however, DNA polymerase I (the original Kornberg polymerase) plays an essential role in linking the DNA fragments.

DNA polymerase I turns out to be a remarkable enzyme, having at least two different catalytic sites and three known enzymatic activities. These are (1) the elongation of an existing strand, as already noted; (2) an *exonuclease* activity that degrades a strand from its exposed 3' end (i.e., 3' → 5' exonuclease activity); and (3) a 5' → 3' exonuclease activity. The first two activities are dispensable—that is, a cell can survive without them, as demonstrated by the original polymerase I mutants

SECTION 11-1
Replication of DNA

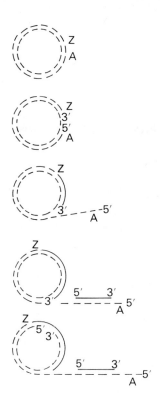

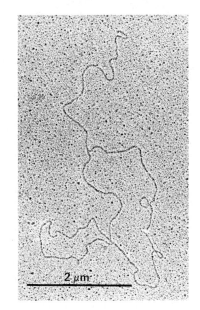

FIGURE 11-9 The Rolling Circle Model. The circle remains complete by elongation of the 3' end left when one strand is broken and peeled off. Backfilling is needed to produce the new strand on the "tail." The micrograph is from replicating DNA of bacteriophage lambda, showing the predicted configuration. [Micrograph courtesy of J. A. Kiger, Jr. and R. L. Sinsheimer, *Proc. Nat. Acad. Sci. U.S.*, **68**: 112 (1971).]

mentioned earlier. These mutants turn out to have normal 5' → 3' exonuclease activity, the catalytic site for which is on a distinctly different part of the molecule from the polymerase function. Other mutants of this enzyme have been found that lack 5' → 3' exonuclease activity, a situation that turns out to be lethal to the cell.

A major function of polymerase I appears to be replacement of the RNA primers at the 5' end of each newly synthesized fragment. As it removes the ribose nucleotides, it adds the appropriate deoxynucleotides as shown in Fig. 11-10. This function would permit a ligase to join deoxynucleotide to deoxynucleotide and hence complete the new strand.

Errors in Replication. Accurate transmission of genetic information requires accurate replication of DNA. Mistakes in the process may, if not corrected, yield mutant daughter cells. A low level of mutation is a desirable feature in the overall scheme of things, because without it a species could not evolve. Too high a level, however, would make life impossible since most mutations are harmful. DNA replication must therefore be very accurate, but not perfect.

We can usually detect a new spontaneous mutation about once in a million gene replications. Considering the average size of a gene, the probability of making a mistake during replication is probably about 10^{-9} per nucleotide per replication, a rather impressive record.

There are several ways in which mistakes can occur during replication of DNA. The wrong base might be inserted in the daughter strand because of the tendency of the bases to spend a very small amount of time in alternate tautomeric forms. In the case of thymine, for example, tau-

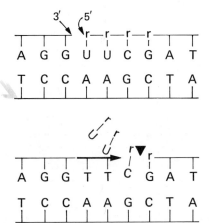

FIGURE 11-10 Ribonucleotide Replacement. The ribonucleotide primer (r) must be replaced by deoxyribonucleotides. Both the excision (a 5' → 3' exonuclease activity) and 3' elongation can be catalyzed in *E. coli* by polymerase I. A DNA ligase is necessary to close the break.

Adenine

Thymine
(normal keto form)

Guanine

Thymine
(rare enol form)

tomerism could result in a satisfactory hydrogen bonding to guanine instead of to adenine, as shown above.

However, when the frequency with which this type of error should occur was calculated from the amount of time that bases spend in their "wrong" isomeric form, it was found to predict an error rate very much higher than that actually observed. The reason, it now appears, is that DNA synthesis is a two-step process: (1) the new DNA strand is elongated using its complementary strand as a template; (2) then the added base or bases are checked for accuracy. It is possible that by taking a "second look," time is allowed for the isomers to return to their usual configuration. If a mistake is found, the bad nucleotide is excised and replaced. All the DNA polymerases are probably capable to some extent of fulfilling this proofreading function, but there is reason to suspect that polymerase I is the most important.

When a mistake of the type just described gets by this error-checking procedure, a *point mutation* results. This term means that a single base pair has been substituted for another, yielding either a single amino acid replacement or premature termination of the polypeptide chain. Such mutations have a low but definite frequency of *back mutation* to the wild type. (Hence, this type of mutant would not be satisfactory candidate for a live-virus vaccine, for example.) More substantial and virtually irreversible mutations can also occur. These are the deletions and insertions, schematically shown in Fig. 11-11. Though they can occur during replication, this type of mutation is probably more common during *crossover*, which is a physical exchange of DNA between molecules (Fig. 11-12). Crossover occurs frequently between the chromosomes of replicating bacteriophage, between bacterial chromosomes, and in eukaryotic cells during meiosis (see Section 11-3). It also accounts for the insertion into host DNA of certain viral genomes (specifically the lysogenic bacteriophages and at least some animal cancer viruses) as shown schematically in Fig. 11-13.

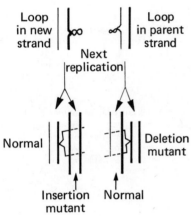

FIGURE 11-11 Insertions and Deletions. The dark lines represent newly synthesized DNA.

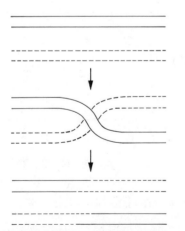

FIGURE 11-12 Crossover. It consists of an actual breakage and reunion, thus exchanging parts of two DNA molecules.

Mutagens. Errors in replication occur spontaneously, as already noted. However, there are many agents that increase the natural rate of mutation. Such substances are called *mutagens*. Many of them increase the risk of cancer, hence are also *carcinogens*.

Certain chemicals alter the organic bases directly, thus causing faulty replication. The alkylating agents, developed from the chemical warfare gas, nitrogen mustard, act in this way. They are now widely used as anticancer agents because of their ability to interfere with cell division. Some other chemical mutagens are flat molecules that are able to slip ("intercalate") between adjacent base pairs, and distort the helix so that errors in replication occur. Acriflavins and, at least in bacterial systems, caffeine may fall into this category. The list could go on to include other types of mutagenic chemicals, some of them quite common—even vitamin C and oxygen are each mutagenic to tissue culture cells when present in high concentration. In most cases, however, chemical mutagens have so many different effects that it is difficult to pinpoint exactly their mutagenic mechanism. Not so for another common mutagen, ultraviolet (UV) light. The mechanism of its action is reasonably well understood.

Ordinary sunlight has much of its ultraviolet portion screened out by the upper atmosphere, largely by the ozone therein. Nevertheless, some UV light gets through. At a wavelength of 300 nm (well past the short end of the visible spectrum, which is at 400 nm), light has an energy of about 400 kJ per Avogadro's number of photons. Thus, at this and shorter wavelengths, each photon has an energy similar to the energy of a covalent bond. Although the energy from absorbed light is usually lost again through increased mechanical motion, by "dilution" over a number of bonded groups, by transfer to another molecule through collisions, or by radiative loss in the form of an emitted photon of somewhat longer wavelength—i.e., through fluorescence or phosphorescence—ultraviolet light can and does cause bond rearrangements in biological molecules. The nucleic acids are the most vulnerable because they absorb light strongly at short wavelengths and because damage to them alters genetic information.

The pyrimidine bases are particularly susceptible to UV damage. At a wavelength of 254 nm, irradiation can cause upwards of one base change per thousand photons absorbed under favorable conditions. The two changes most apt to be observed are a hydration of the 5,6 double bond and the linking of pyrimidines (especially thymine) that lie adjacent to each other in the same chain (see structures in margin).

Pyrimidine hydration is readily reversed *in vivo*, however, and so it is probably not a major cause of mutations. On the other hand, although thymine dimers may be reversed in a light-catalyzed reaction—a process called *photoreactivation*—they cause enough distortion in DNA and are stable enough to be the source of replication errors.

In addition to damage from chemicals and ultraviolet light, errors in replication may result from the effects of *ionizing radiation*. Ionizing radiation may consist of very short wavelength electromagnetic radiation (X rays or γ rays, depending on the source) or particulate radiations, namely β rays (electrons) and α rays (accelerated helium nuclei). These radiations have in common the ability to rip electrons from atoms to create ions and free radicals. Free radicals are by far the most dangerous to the cell because they are extremely reactive—two of them may get together to pair their odd electrons in a covalent cross-link; a free radical may add to another molecule, especially at a double bond; free radicals may cause the hydrolysis of macromolecules or the deamination of proteins; they may cause the oxidation of sulfhydryl groups, and so on.

The cell is equipped to protect itself from moderate amounts of free radicals. Protection is necessary because cells are constantly exposed to low levels of background radiation and because free radicals arise as a direct or indirect consequence of certain oxidative reactions. They arise directly from the one-electron transfers to or from flavins and quinones

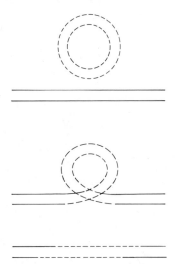

FIGURE 11-13 Insertion of Plasmid DNA into a Chromosome. The plasmid, which is a small, extrachromosomal circle of DNA (e.g., of viral origin) is inserted by double crossover.

Thymine dimer

Hydroxycytosine

in the mitochondria and chloroplasts, for example. And they arise indirectly from peroxides produced by amino acid oxidases in peroxisomes (see Chap. 7).

To cope with free radicals produced by these mechanisms, the cell is equipped with antioxidants in the form of vitamins C and E, the quinones, and sulfhydryl compounds like glutathione (a tripeptide, glu-cys-gly). These substances mediate or even repair some free radical damage. They do so by forming stable free radicals and, perhaps, by obtaining electrons from NADPH or NADH to match the unpaired electron of a free radical. Nevertheless, the large quantities of free radicals that appear following exposure to significant levels of ionizing radiation are apt to result in serious cellular damage, especially to genes.

Repair. Certain kinds of genetic damage can be repaired. There are several kinds of observations that support such a conclusion. For instance, one finds that many biological effects of radiation, ultraviolet or otherwise, are strongly dependent on dose rate and timing. At high dose rates, the fraction of surviving cells can be predicted from the probability of hitting one or more "targets" within the cell, any one of which will be lethal. At low dose rates, on the other hand, the fraction of survivors will be greater than that predicted (see Fig. 11-14). In addition, a given total quantity of radiation often produces less damage as it is applied in smaller doses spaced further apart.

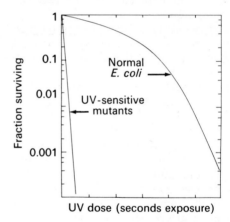

FIGURE 11-14 Ultraviolet-sensitive Mutants. The survival of normal cells after ultraviolet radiation is better at low doses than one would predict from UV sensitivity of the same cells at high doses. The existence of mutants in which killing is a logarithmic function of dose indicates the loss of an enzymatic repair mechanism.

We know, too, that repair mechanisms are inherited. In the case of *E. coli*, several genes have been identified which, when damaged, increase radiation sensitivity. The comparable situation in humans is the hereditary disorder, xeroderma pigmentosum, a condition that renders cells especially sensitive to UV light and leads to a high incidence of skin cancer in afflicted individuals.

Another inherited condition associated with an increased rate of malignancy is called ataxia telangiectasia (ataxia is a loss of coordination; telangiectasias are dilated blood vessels usually seen as red streaks on the skin). Cells from these patients show a greatly increased rate of mutation from ionizing radiation but *not* from UV. Hence, different repair mechanisms must be at work for these two types of damage. In fact, there are several different repair mechanisms in both prokaryotic and eukaryotic cells, some of which are more efficient than others.

The best known case of repair synthesis is *dark repair* in *E. coli* by the excision of thymine dimers. Given enough time between damage and replication (i.e., before the damaged area produces mutant progeny), the cell can often employ nucleases to excise the portion of the DNA strand containing a dimer, polymerize a new strand by elongation of the 3' end exposed by the cut, and then join the new string of nucleotides at the other side of the repaired region through the action of a DNA ligase (Fig. 11-15). Mutant *E. coli* that lack DNA polymerase I are compromised in their ability to carry out this process.

There is evidence for a similar "cut and patch" repair of eukaryotic DNA as well. However, as these repair mechanisms take a certain amount of time to function, there is always the danger that a cell will embark on an ill-fated replication cycle before repair is complete, locking the mistake into newly synthesized DNA. Because of this possibility, rapidly proliferating cells are especially prone to radiation damage. That fact is exploited in the use of ionizing radiation (X rays or γ rays from cobalt 60) and chemical mutagens to treat human malignancies.

The existence of repair mechanisms is partly responsible for the very considerable controversy that surrounds the question of what is a "safe" dose of radiation to a human. If, following some accidental or wartime exposure, one person in a hundred later (usually many years later) develops a cancer, can we assume that 1/1000 as much radiation would yield 1 cancer per 100,000 people? Some experiments seem on the surface to support this kind of linear extrapolation and thus imply that no dose is really safe. If that is true, then people who live at high altitudes or in new concrete buildings (which leak radon gas) are incurring an increased risk of getting cancer. However, it is also reasonable to maintain that life evolved in a sea of natural radiation, and that as long as one doesn't exceed the background levels by too much, our repair mechanisms will protect us. The animal experiments needed to settle the question can't be done properly because of the numbers involved, and attempts to correlate cancer rates in millions of humans with small changes in exposure are fraught with problems because there are so many

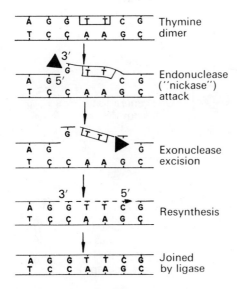

FIGURE 11-15 DNA Repair. Thymine dimers are repaired by a "cut and patch" mechanism, involving an endonuclease ("nickase") attack to create a single strand break, the hydrolysis of a portion of the damaged strand by an exonuclease, its replacement by newly synthesized DNA, and the joining of the new nucleotides to the existing strand by a ligase.

other factors that might be responsible for the observed changes. Hence, no one really has an answer as to how much radiation is "safe."

The Replication of Chromatin. The experiments of Meselson and Stahl demonstrated that DNA replication is semiconservative. It does not automatically follow that eukaryotic DNA, which is complexed with protein to form chromatin, is similarly replicated. However, soon after Meselson and Stahl's experiments were published, workers were able to use similar techniques to show that DNA replication is almost certainly semiconservative in every cell type, eukaryotic as well as prokaryotic.

The replication of eukaryotic DNA occurs when chromatin is dispersed within the nucleus between cell divisions. The chromatin will later condense into compact, visible chromosomes, in order to permit accurate distribution to daughter cells. Each newly condensed chromosome will ordinarily contain two identical strands, or *sister chromatids*, which will end up in separate daughter nuclei. Experiments conducted at Columbia University in 1958 by J. Herbert Taylor demonstrated

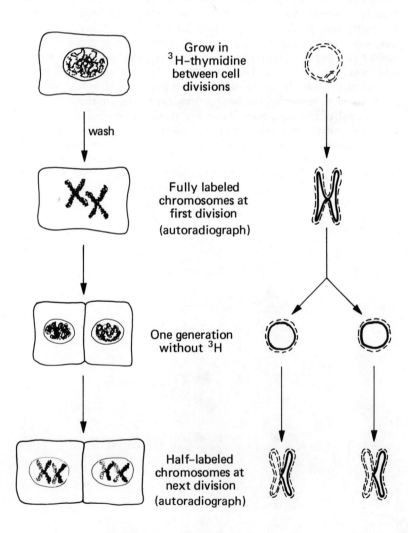

FIGURE 11-16 Replication of the Eukaryotic Chromosome. Cells are grown in ^{3}H-thymidine only between divisions. Both chromatids are found to be labeled at the first division (mitosis). At the second mitosis after removing tritium, one chromatid is labeled and one is not. The diagrams on the right interpret these findings by assuming that each chromatid consists of one long double-stranded DNA molecule, although it would naturally have a much more complex folding pattern than shown here.

that the two chromatids result from semiconservative replication of the single chromatid left after the previous cell division.

Taylor incubated the root cells of a plant (*Bellavalia romana*) in radioactive thymidine (^{3}H) between cell divisions, and found that when the chromatin formed visible chromosomes, the two sister chromatids were equally labeled (Fig. 11-16). When cells treated in this way were allowed to proceed through a second division cycle, this time without radioactivity, the chromosomes appeared with one labeled chromatid and one chromatid free of label (see Fig. 11-17). These experiments suggest that each chromatid might contain only a single long molecule of DNA.

Autoradiography of nuclei exposed to isotopes for strictly limited periods of time (*pulse labeling*) provides measurements of the rate of chromatin replication and reveals the pattern of replicated material after the latter forms condensed chromosomes. Measurements by Cairns in 1966 yielded values of about 0.5 μm/min for the rate of DNA replication in cultured human cells. (These were *HeLa cells*, originally derived from the cervical cancer of one *Henrietta La*cks.) This rate is very much slower than the 20–30 μm/min that he found for *E. coli* chromosomes a couple of years earlier, but apparently the value is not unusual, for rates of 0.2–2 μm are commonly found in eukaryotes.

The relatively slow rate of DNA replication in eukaryotes tells us that replication must be proceeding at a number of points at once. For example, the human diploid nucleus has a total DNA content of about 175 cm. Thus, a complete replication in 10 hours (a typical value *in vitro*), even at the relatively rapid rate of 2 μm/min (about 100 nucleotide pairs per second), would require about 1500 simultaneously replicating points. The actual number turns out to be much higher.

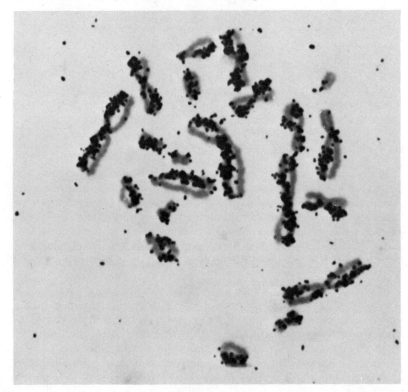

FIGURE 11-17 Autoradiography of Replicated Chromosomes. Superimposed light micrograph shows distribution of label at the second division. In a few cases, label appears on the sister strand as well, a result of an interchange (crossover) between the strands called a *sister chromatid exchange*. [Courtesy of G. Marin and D. M. Prescott, *J. Cell Biol.*, **21**: 159 (1964).]

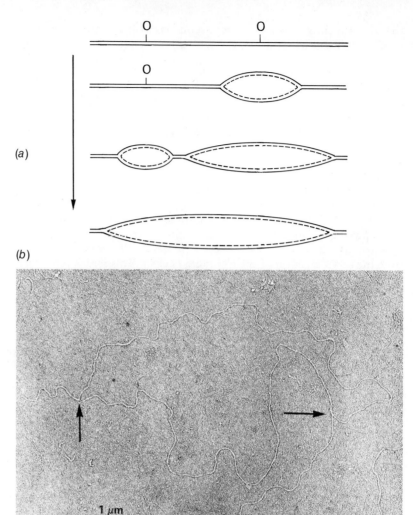

FIGURE 11-18 Replicons. (a) Eukaryotic DNA is replicated in multiple discrete units, or replicons, having origins (O). The sequence shows the replication and joining of two replication units. Dashed line represents newly synthesized DNA. (b) Replication of yeast chromosomal DNA, revealing a configuration similar to one of the intermediate steps in part a. The arrows point to the replication forks. [(a) original description by J. Huberman and A. Riggs, *J. Mol. Biol.*, **32**: 327 (1968); (b) courtesy of C. Newlon, T. Petes, L. Hereford, and W. Fangman, *Nature*, **247**: 32 (1974).]

It appears, then, that eukaryotic DNA is replicated in discrete units, called *replicons*. Autoradiographs support this concept and place the size of the mammalian replicon at 15–100 μm, with similar lengths for other animals. Replication forks are thought to proceed from either end of each replicon towards its middle. When they meet, the replicon will have been duplicated. When all replicons in a DNA molecule have been duplicated in this fashion, two complete daughter molecules are formed (see Fig. 11-18).

Replicons are not all activated at once in most chromosomes, though they may be in polytene chromosomes. Rather, the pattern is highly specific. In the ciliate *Euplotes*, for instance, replicons in the macronucleus are activated sequentially, starting at both ends of the macronucleus and proceeding (at 7 μm/hr) toward the center. (See Fig. 11-19.)

Although the pattern of replicon activation is more complicated in human cells, it is still highly reproducible. While some human chromo-

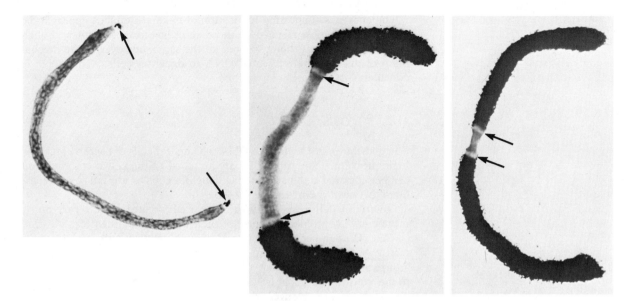

somes, notably the heterochromatic X chromosome of females, are replicated hours later than the others, autoradiography reveals no simple pattern for the nucleus as a whole nor for individual chromosomes. However, the chromatin near the centromere often seems to be replicated later than other chromatin within the same chromosome. This observation, along with other data, led E. J. DuPraw to propose in 1965 that replication stops before it gets to the actual region where the two chromatids are joined. This unreplicated region may hold the two chromatids together, so that separation of sister chromatids might be accomplished either by hydrolyzing or by replicating this last little bit of DNA. More recently, however, interarm fibers have been identified at numerous points, including the ends of the chromosomes (Fig. 11-20), suggesting the need for a somewhat more complex mechanism.

FIGURE 11-19 The Organized Replication of DNA. Macronucleus of *Euplotes*, a ciliated protozoan. Size about 140 µm. The autoradiographs (made using ³H-thymidine) show nuclei isolated during successive stages of its 10-hour replication phase. Note sequential activation of replicons (*arrows*) proceeding toward the center. [Courtesy of D. M. Prescott, *J. Cell Biol.*, **31**: 1 (1966).]

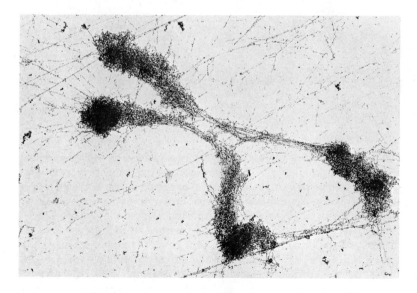

FIGURE 11-20 Interarm Fibers. Chromosome from a human lymphocyte, treated with trypsin to dissolve much of its protein and reveal the DNA strands. Note the existence of numerous interarm fibers, especially at the centromere and at the ends of the arms. [Courtesy of D. E. Moore and J. G. Abuelo, *Nature*, **234**: 467 (1971).]

The enzymes supporting replication of eukaryotic DNA appear to be comparable in most respects to those described for prokaryotes, though less is known about them. Not at all the same, however, is the mechanism of distributing replicated DNA to daughter cells. That topic is considered in the following section.

11-2 MITOSIS

The distribution of replicated DNA to daughter cells of bacteria presents little problem. The molecules are attached to the cell membrane and can be separated and distributed to daughter cells by growth of the membrane between them (see Fig. 11-21).

Eukaryotic cells, especially the higher eukaryotes, have a very much more complex problem because of their multiple nonidentical chromosomes. Human cells, for example, have the considerable task of replicating 46 chromosomes and then separating them into two identical groups prior to cell division. The process is remarkably accurate thanks to the evolution of a sophisticated mechanism called *mitosis*. Mitosis is part of the overall *cell cycle* that includes replication of DNA, division of the nucleus (*karyokinesis*), and division of the cell itself, or *cytokinesis*.

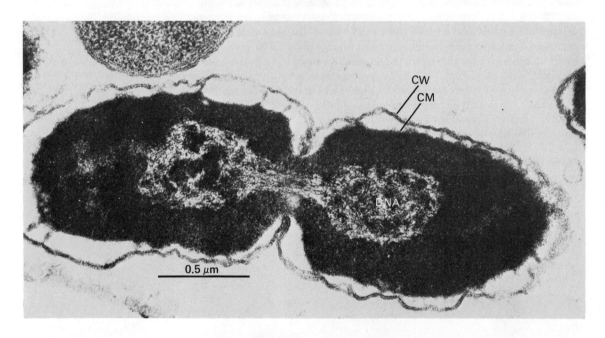

FIGURE 11-21 Division of a Bacterial Cell. *E. coli*, plasmolysed with sucrose to show invagination of the cell wall (CW) and cell membrane (CM). The large light areas in the cytoplasm are the DNA molecules being distributed to daughter cells. [Courtesy of M. E. Bayer and Cambridge Univ. Press. From *J. Gen. Microbiol.*, **53**: 395 (1968).]

The Cell Cycle. A complete cell cycle usually consists of five stages, labeled G_1, S, G_2, M, and D. The first three of these together were originally believed to be a single *interphase*, or period of rest between cell division. This concept arose because visible activity is confined to mitosis (M), during which daughter nuclei form, and to cytokinesis itself, labeled D for division. (Mitosis is sometimes defined to include cytokinesis, eliminating the separate D stage.) However, it is now recognized that most DNA, RNA, and protein is made during interphase, which therefore is the period when metabolic activity is greatest.

Although protein and RNA synthesis continue throughout interphase, DNA and histone synthesis is limited to the S period (S for synthesis). Once a cell enters its S stage, it usually proceeds inexorably through the others until it is back at G_1 again—i.e., back at the "first gap," or the one preceding DNA synthesis. An animal cell in tissue culture might repeat this sequence every 24 hours, spending approximately 10, 9, and 4 hours in the G_1, S, and G_2 stages, respectively (see Fig. 11-22). Mitosis and cytoplasmic division together may be completed in the remaining hour. Cells not destined for an early repeat of the division cycle are commonly arrested at the G_1 phase or, according to some systems of nomenclature, at a "G_0 stage" between D and G_1.

Mitosis itself was described by cytologists in the latter part of the nineteenth century. The most detailed reports came from Walter Flemming of the German University of Prague in the 1880s. Since then, the electron microscope and other tools of molecular biology have allowed us a much more intimate look at the process, although the original nomenclature is still retained for most of the structures and events involved. According to this nomenclature, mitosis is divided rather arbitrarily into four phases called *prophase, metaphase, anaphase,* and *telophase.* (The prefixes mean "before," "between," "back," and "end," respectively.) Cytokinesis usually takes place in synchrony with the events of telophase. (See Fig. 11-23.)

Prophase. Prophase consists of five events: (1) chromatin condensation; (2) nucleolar dispersal; (3) centriole migration; (4) fragmentation of the nuclear envelope; and (5) initiation of spindle apparatus assembly.

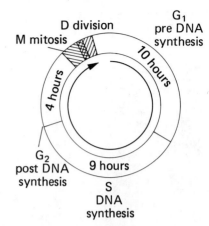

FIGURE 11-22 The Cell Cycle. The time shown is consistent with many lines of cultured mammalian cells.

FIGURE 11-23 Mitosis in an Animal Cell. The mechanism of cell division and the presence of centrioles and asters (starlike configuration of microtubules around the centrioles) identify this as an animal cell. (Note the haploid number, $n = 2$.) **1.** Interphase. **2–4.** Prophase. **5.** Metaphase. **6.** Anaphase. **7–8.** Telophase.

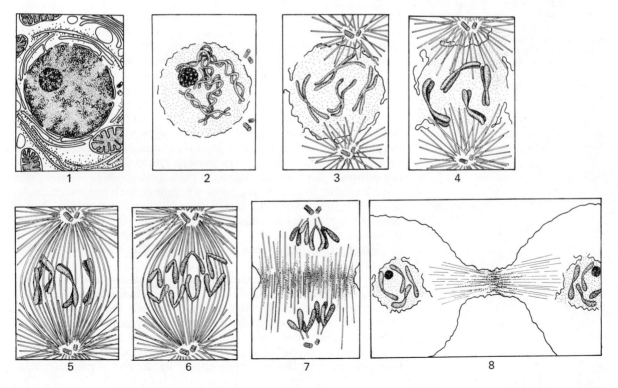

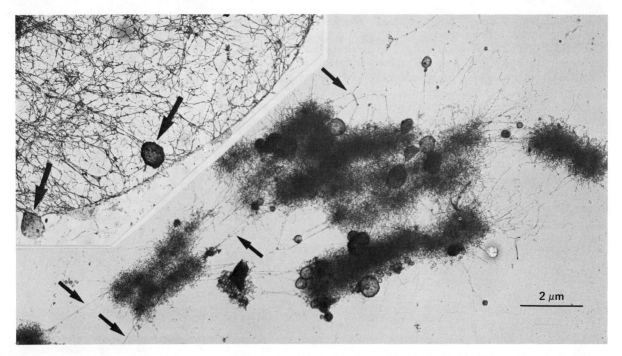

FIGURE 11-24 Chromatin Condensation. Human lymphocyte. The dispersed chromatin characteristic of interphase (inset) is gathered into discrete chromosomes during prophase, accompanied by fragmentation of the nuclear envelope. The large arrows in the inset point to remnants of the nuclear membrane, broken by surface tension forces in the whole-mount preparation technique. The smaller arrows point to single chromatin fibers extending between chromosomes. All chromosomes seem to be interconnected and to be still connected at many points to fragments of the nuclear envelope. [Courtesy of F. Lampert, *Humangenetik*, **13**: 285 (1971).]

When a dividing cell is viewed with the light microscope one first sees thin threads within the nucleus, then discrete chromosomes, each composed of two identical or "sister" chromatids. Sister chromatids are held together by their centromeres. Because chromatin is attached to the inner membrane of the nuclear envelope (Fig. 11-24), the progressive coiling and folding that produces chromosomes also causes the forming chromosomes to move toward the envelope, leaving the center of the nucleus relatively empty. During this time, the fibers of the nucleoli usually disperse, so that discrete nucleoli disappear from view.

Centrioles, in those cells that have them (namely, almost all animal cells plus those of the lower plants) migrate during prophase to take up positions on opposite sides of the nucleus. Centrioles, as noted in Chaps. 1 and 9, are identical in structure to basal bodies and, like basal bodies, can serve as microtubular organizing centers. Each daughter cell from a division gets one pair of centrioles, arranged perpendicularly to each other. During the G_1 stage, prior to mitosis, the two members of the centriole pair separate slightly. Then, near the beginning of the S stage a smaller *procentriole* appears adjacent to each mature centriole, again with the two at right angles to each other (Fig. 11-25a). During prophase, procentrioles elongate (Fig. 11-25b) and the two pairs of centrioles separate and migrate. By the end of prophase there is a pair of full-length centrioles on each side of the nucleus, defining the *poles* of the cell.

The nuclear envelope begins to break up during early prophase, starting at the poles (see Fig. 11-26). Remnants of the nuclear envelope remain more or less in place throughout prophase. They are still readily identified by their pores, though they otherwise resemble the rest of the endoplasmic reticulum, often including the presence of ribosomes on both surfaces. (Recall that in the intact nucleus, ribosomes may be found on the cytoplasmic surface only.) By the end of prophase, frag-

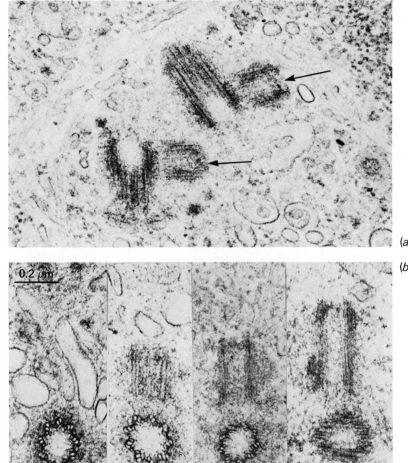

FIGURE 11-25 Centriole Replication. (a) Four hours after mitosis in a cultured mammalian cell (L cell). Two procentrioles (arrows) about 0.25 μm long have formed, one adjacent and at right angles to each mature centriole. (b) Centriole elongation. From left: early G_1 stage; S stage; prophase; metaphase. [Courtesy of J. Rattner and S. G. Phillips, J. Cell Biol., **57**: 359 (1973).]

mentation of the nuclear envelope is complete and, in most cells, the fragments are incorporated into the ER. Since pores are no longer visible, nuclear envelope remnants at this point are indistinguishable from the rest of the ER.

Spindle formation also begins during prophase, although the spindle is not fully functional until metaphase.

Metaphase. Metaphase is a period of chromosome alignment and completion of the *spindle apparatus*. The spindle apparatus is a microtubular structure that extends from pole to pole.[1] It is organized, apparently, by the centrioles or by the ill-defined aggregate of granules and vesicles that replace centrioles in higher plants.

Those cells that have centrioles typically also have asters ("stars"), which are fibers radiating out in all directions from the poles (Fig. 11-27). Each of the astral and spindle fibers are actually bundles of microtubules.

The microtubular proteins that comprise the spindle and asters were synthesized in part during the immediately preceding G_2 stage. However, there is ample reason to believe that a major fraction of the proteins involved were present in the cell for a longer period of time, and were probably a part of the spindle during the previous mitosis. When one

CHAPTER 11
Cell Division

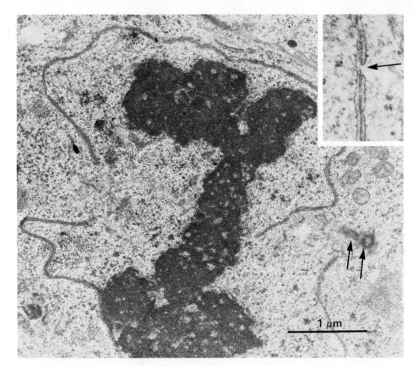

FIGURE 11-26 Fragmentation of the Nuclear Envelope. Lung epithelial cell at prophase. (From a midtrimester human fetus.) Nuclear fragmentation has begun at the poles. Note that one set of centrioles has been caught in the plane of section (*arrows*). These rapidly dividing cells seem to produce extra nuclear envelope prior to division, presumably to facilitate reassembly of daughter nuclei. In this case the extra envelope is folded back on itself, as is seen clearly in the detail. The detail also shows pores (*arrow*.) [Courtesy of Jennifer McDougall; small photo from *J Cell Sci.*, **22**: 67 (1976).]

considers the amount of material involved, this a highly practical mechanism; the assembled spindle often accounts for 10% or more of the total cellular protein, and in some cases for more than a quarter of the cell's dry weight. The characteristic rounding of cells prior to mitosis may be a consequence of depolymerization of the microtubule skeleton used to maintain shape, and to the utilization of its subunits in the construction of spindle fibers.

During metaphase, chromosomes become attached to spindle fibers.[2] More specifically, each chromatid has near its centromere a dense

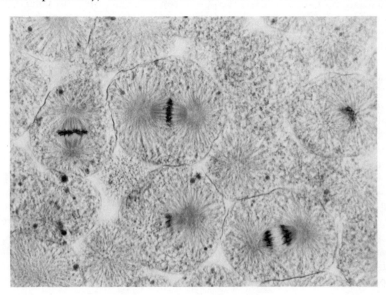

FIGURE 11-27 Asters. From a whitefish embryo. Complete mitotic figures can be seen in two metaphase cells and in one anaphase cell (lower right). [From W. Etkin, R. Devlin, and T. Bouffard, *A Biology of Human Concern*. Philadelphia: J. B. Lippincott, 1972. By permission.]

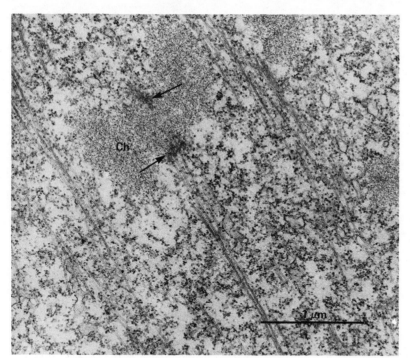

FIGURE 11-28 The Kinetochore. Metaphase chromosome of a fertilized sea urchin egg. Note the dense kinetochores (*arrows*), one on either side of the chromosome (Ch). Spindle microtubules are attached to both kinetochores. [Courtesy of Patricia Harris, *J. Cell Biol.*, **25**: 73 (1965).]

granule called a *kinetochore* to which spindle fibers are attached (Fig. 11-28). The two sister chromatids, by virtue of their kinetochores and attached fibers, are pulled toward opposite poles. Because they are held together at the centromere—i.e., because the two chromatids still comprise a single chromosome—there is pull in both directions at once, resulting in oscillatory movements that eventually align the chromosomes all in one plane. This plane is called the *equatorial plane* or *metaphasic plate* (see Fig. 11-27).

Anaphase. Anaphase starts with centromere division. The two chromatids, now free of each other, move toward their respective poles at a velocity typically between 0.2 and 4 μm/min. Because they are being pulled by their kinetochores, the chromatids assume a V shape (Fig. 11-29). At the same time as the chromatids move toward their poles, the poles themselves move further apart. As will be explained later, the poles are pushed apart by spindle fibers that do not end at chromosomes but instead stretch more or less from pole to pole.

Telophase. Telophase finds the two diploid sets of daughter chromosomes (or chromatids) gathered at opposite poles. During this period chromosomes and spindle fibers disperse and disappear from view, nucleoli condense and reappear, and new nuclear envelopes are assembled. Cytokinesis (discussed in Section 11-5) also accompanies telophase in most cells.

The nuclear envelope is assembled from elements of the endoplasmic reticulum—in part, no doubt, from fragments added to the ER during prophase. As the sacs of ER align about the still-condensed chromosomes, ribosomes are lost from the inner (nuclear) surface and, at the same time, pores begin to appear. The stimulus for these two changes in

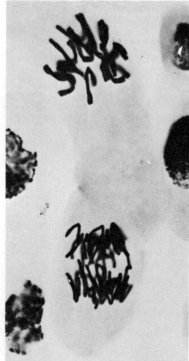

FIGURE 11-29 Anaphase. From a plant *(Alium)* root tip. One cell is clearly in anaphase, with the arms of the chromosomes dragging behind the kinetochores. The spindle is not stained. [From the Turtox Collection, courtesy of General Biologicals, Inc.]

the membranes is unknown but may be associated with reattachment of chromatin to the inner surface. When new nuclei are completed by fusion of adjacent ER membranes, mitosis is complete.

The pattern of mitosis just given is that of higher eukaryotes. Examination of some of the lower eukaryotes is instructive because it allows us to see how this pattern has evolved.

The Evolution of Mitosis. As noted earlier, bacterial chromosomes are attached to membranes and are separated by growth of the membrane between them. Eukaryotic DNA is also attached to membrane (the nuclear envelope) and when one looks at some of the lower eukaryotes a mitotic mechanism very much like that of prokaryotes can be found.

For example, in the dinoflagellates (an order of biflagellate protozoa) chromosomes are anchored permanently to the nuclear envelope, which remains intact through mitosis. As in bacteria, chromosomes become separated as the envelope elongates and is pinched in two. This pinching-off process seems to be due to growth of microtubules outside the envelope, a feature that has been retained through evolution to shape the nucleus of forming spermatozoa. These external microtubules in the dinoflagellates may also help separate daughter nuclei.

The marcronucleus of ciliated protozoa seems to share with dinoflagellates the mitotic mechanism just described. In the *micro*nucleus of these same cells, however, both chromosomal and pole-to-pole spindle fibers are seen. This latter pattern is found also in the slime mold *Polysphondylium*, as seen in Fig. 11-30. Pole-to-pole fibers seem to push the

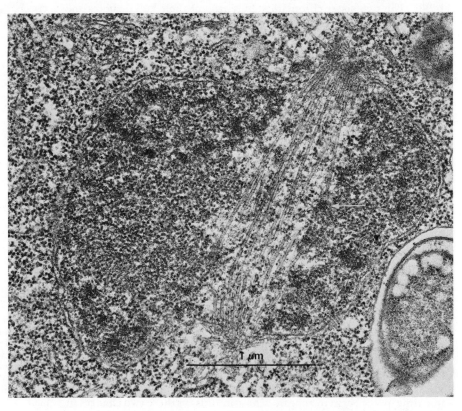

FIGURE 11-30 The Mitotic Apparatus of a Slime Mold (*Polysphondylium*). Note the intact nuclear envelope and the presence of both pole-to-pole and chromosome-attached (*arrow*) spindle microtubules. These plants do not have centrioles. Instead, the poles are an ill-defined cluster of vesicles and granules just outside the nuclear envelope. The microtubules penetrate into the nucleus through windows in the envelope. [Courtesy of U.-P. Roos, *J. Cell Biol.*, **64**: 480 (1975).]

poles apart as the fibers grow, causing elongation of the nucleus in preparation for its division into two daughter nuclei.

These and other observations allow us to put together an evolutionary progression. In the most primitive eukaryotic mitoses, the nuclear envelope remains intact and chromosomes are separated by growth of the nuclear envelope. As one proceeds up the evolutionary ladder, microtubules appear within the nucleus. At first they seem to be relatively unoriented, but in other cells clear attachments to chromosomes are found. The pattern seems to be one of decreasing importance of membrane attachment and increasing importance of the spindle as a mechanism of chromosome separation.

As the need for a nuclear envelope (as a means of separating chromosomes) during mitosis diminishes with evolutionary progression, so does its integrity. Cells at intermediate stages in the evolutionary progression, for example, have an envelope that develops tears but remains largely intact. Almost all higher eukaryotes, however, have a completely open mitosis, where the envelope breaks down into small fragments at prophase, allowing mixing of nucleoplasm and cytoplasm.

The advantage of using a spindle instead of membranes to separate chromosomes is clear, at least where the diploid number of chromosomes is large. The advantage of nuclear fragmentation, with its requirement for reassembly and migration of nucleoplasmic components back into daughter nuclei, is not so clear. One possibility is that assembly of the spindle from cytoplasmic proteins is facilitated by absence of a barrier between cytoplasm and nucleoplasm. It may be, however, that the major importance of nuclear fragmentation is found not in mitosis itself but in the closely related mechanism of meiosis or, even more likely, during fertilization (sexual reproduction) where there is complete mixing of two nuclei.

11-3 MEIOSIS

Meiosis (which comes from the Greek work *meioun*, to diminish) produces a total of four haploid cells from each original diploid cell. These haploid cells either become or give rise to *gametes*, which through union (fertilization) support sexual reproduction and a new generation of diploid organisms.

Meiosis superficially resembles two mitotic divisions without an intervening period of DNA replication. (See Table 11-1 and Fig. 11-31.) In the first division, two homologous but not identical haploid cells are produced. Each of these cells then goes through a mitosislike division to produce two daughter cells. Most of the unique features of meiosis take place during the first of these two divisions.

The First Meiotic Division. This first division is called the *reduction division* because it produces haploid daughter cells (or just nuclei) from the diploid parent. Reduction is the result of chromosome behavior during a complex *prophase* that is traditionally described in five stages:

(1) *Leptonema* (from the Greek *leptas*, thin; *nema*, thread) is characterized by the condensation of chromosomes to the point where they become visible as long filaments (Fig. 11-31, part 1), often with periodic, beadlike thickenings.

TABLE 11-1. A Summary of Mitosis and Meiosis

	Mitosis	Meiosis — First division	Meiosis — Second division
Interphase	(Subdivided into G_1, S, and G_2 stages). Chromosome replication occurs during S phase.	Chromosome replication occurs.	Interkinesis: no replication.
Prophase	Each chromosome is visible as two identical chromatids moving toward center of cell. Spindle apparatus forms. Nuclear envelope fragments.	Each chromosome is visible as two identical chromatids. Homologous pairs synapse to form four-stranded tetrads. A spindle apparatus forms and the nuclear envelope fragments.	The chromosomes reappear as double-stranded structures, but their number is haploid. A new spindle apparatus forms and the nuclear envelope fragments.
Metaphase	The double-stranded chromosomes align at the center of the nucleus and attach to spindle fibers. Chromatids separate by centromere division, with sister chromatids under the influence of opposite poles.	Tetrads align at center, but without centromere division. Homologous chromosomes—each still consisting of two chromatids—are attached to fibers from opposite poles.	Double-stranded chromosomes align at the center and attach to spindle fibers. This time centromeres divide, leaving two identical haploid sets in each of the dividing cells.
Anaphase	Chromosomes, which are now single-stranded, move toward opposite poles. Cleavage furrow (animals) or cell plate (plants) begins to form.	The chromosomes, which are double-stranded, move toward opposite poles. Cleavage begins.	Each haploid set moves toward its own pole, and the second cleavage begins.
Telophase	Spindle dissolves and new nuclear envelopes form. Division of nucleus (karyokinesis) is usually accompanied by a division of the cytoplasm (cytokinesis) to yield two identical, diploid daughter cells.	Spindle dissolves and new nuclear envelopes form. Two nonidentical cells with haploid chromosome number are formed, but each chromosome still consists of two identical (sister) chromatids.	Spindle dissolves and new nuclear envelopes form. Meiosis is complete, leaving a total of four haploid cells of two allelic types.

FIGURE 11-31 Meiosis. The lack of centrioles and asters, plus the mode of cell division, identify this as a plant cell. **1–5.** The first meiotic division. Note synapsis at step 3. A protein framework called the *synaptonemal complex* is found between paired chromosomes at this stage. Note also that two haploid cells are produced by the first meiotic division. The cells contain homologous but not identical chromosomal sets. The chromosomes remain two-stranded—i.e., with two identical chromatids. **6.** Interphase. **7–9.** The second meiotic division doubles the number of haploid cells to four. (Note that the haploid chromosome number is two.)

(2) *Zygonema* (from the Greek *zygon*, adjoining) is distinguished by *synapsis*, which is a specific pairing of homologous chromosomes, the alignment of which is usually gene-for-gene exact. The paired chromosomes are joined by a roughly 0.2-μm thick, protein-containing framework called a *synaptonemal complex* (see Fig. 11-32). This complex extends along the whole length of the paired chromosomes and is usually anchored at either end to the nuclear envelope. (The X and Y chromosomes pair off and become joined by their short arms but a synaptonemal complex does not form between them.) The complex is penetrated at frequent intervals by threads of DNA, apparently from only one chromatid in each case. It is this DNA that makes possible a specific alignment of the chromosomes. Synaptonemal pairing produces an opportunity for *crossover*, which is a physical exchange of DNA between participating chromatids and hence exchange of allelic genes (recombination) between homologues. The actual event of crossover, however, appears to take place not during zygonema but during the next stage, pachynema.

(3) *Pachynema* (*pachus*, thick) generally lasts much longer than either of the two previous stages and is distinguished by greater condensation of

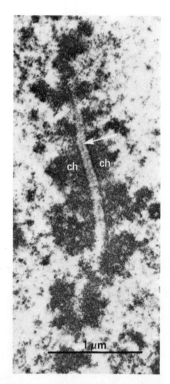

FIGURE 11-32 The Synaptonemal Complex. From the first meiotic prophase in a plant cell, *Endymion nonscriptus* (bluebell). Homologous chromosomes (ch) are held together by a synaptonemal complex having a prominent *medial element* (arrow) running parallel to the paired chromosomes. [Courtesy of E. Jordan and B. Luck, *J. Cell Sci.*, 22: 75 (1976).]

the chromosomes so that individual chromatids start to become visible. Each synaptonemal pair at this point is commonly referred to as a *bivalent* because it consists of two visible chromosomes, or as a *tetrad* because of the four visible chromatids (see Fig. 11-31, part 3). As noted, crossover probably occurs at this stage. It is a random event that can occur anywhere along the paired chromosomes, and may occur at several locations simultaneously in the same pair.

(4) In *diplonema* the chromatids of each tetrad are usually clearly visible, but the synaptonemal complex appears to have dissolved, leaving participating chromatids of the paired chromosomes physically joined at one or more discrete points called *chiasmata*. These points are where crossover took place. Often there is also some unfolding of the chromatids at this stage, allowing for RNA synthesis and cellular growth.

(5) During *diakinesis*, the chromosomes resume a tighter folding pattern. The chiasmata often seem to work their way toward the ends of the chromatids during this period, apparently as the two chromatids that are joined by crossover begin to untangle from each other. Eventually each bivalent may be held together only by the tips of two chromatids (Fig. 11-31, part 4). During this stage, the nucleoli disperse and the nuclear envelope breaks down. Metaphase follows.

Metaphase I consists of spindle fiber attachment to chromosomes and chromosomal alignment at the equator. The two chromosomes in each pair are under the influence of opposite poles but they are still held together by their chiasmata.

At *anaphase I* homologues are freed of each other and move toward their respective poles. Their arrival at the poles defines the onset of *telophase I*, during which nuclei are reassembled. As with mitosis, cell division usually begins at telophase. Unlike mitosis, however, the first meiotic division produces daughter cells in which the chromosomal sets are haploid—they are homologous but they are not identical.

The Second Meiotic Division. The period between the first and second meiotic divisions is called *interkinesis* or *interphase*. No DNA replication occurs, so that chromosomes at the second prophase are the same double-stranded structures that disappeared at the first telophase. The second meiotic division proceeds just as normal mitosis would, but the cell is working with only half as many chromosomes. Since centromere division takes place this time, the second meiotic division produces a total of four haploid cells from the single premeiotic diploid parent. Two of these haploid cells carry copies of one set of allelic genes while the other two haploid cells have copies of the homologous allelic set.

Gametogenesis. The products of meiosis in animals are gametes—i.e., sperm and eggs (see Fig. 11-33). For example, in *spermatogenesis* the four haploid products of meiosis differentiate into *spermatozoa*. The parent cell—the one that undergoes the first meiotic division—is called the primary *spermatocyte*. (Its own precursor, from which it arose via mitosis, is called a *spermatogonium*.) The result of the first meiotic division is two secondary spermatocytes, one carrying each allelic set. The second meiotic division, then, yields four cells called *spermatids*, and it is these that grow a tail and differentiate into spermatozoa (see Fig. 11-34). Differentiation of the spermatids is remarkable for its synchrony, probably because of the existence of persistent cytoplasmic

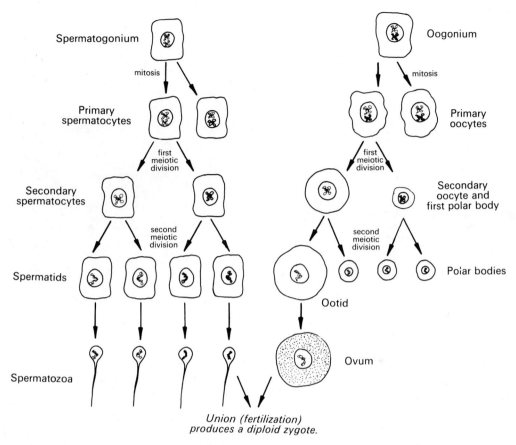

connections between the cells. (This was perhaps the first known example of direct communication between animal cells via cytoplasmic channels.)

The formation of eggs (*oogenesis*) uses the same type of nomenclature. That is, an *oogonium* undergoes mitosis to yield the primary *oocytes*, which produce secondary oocytes as a consequence of the first meiotic division. They, in turn, produce *ootids* at the second division, which

FIGURE 11-33 Gametogenesis. Note the parallel between spermatogenesis and oogenesis. The diagram is drawn with a haploid chromosome number equal to one in both cases.

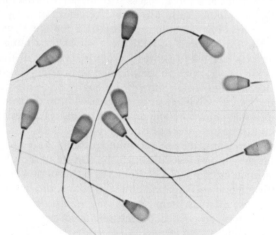

FIGURE 11-34 Bull Sperm. (From the Turtox Collection, courtesy of General Biologicals, Inc.)

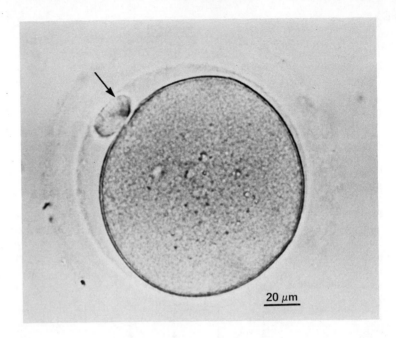

FIGURE 11-35 A Human Oocyte and First Polar Body. The oocyte and its small polar body (*arrow*) are surrounded by a transparent covering, called the *zona pellucida*. Fertilization, if it were to occur, would take place at this stage. [Courtesy of J. F. Kennedy and R. P. Donahue, *Science*, **164**: 1292 (1969). Copyright by the AAAS.]

differentiate into the mature ova, or eggs. The most pronounced difference between this process and spermatogenesis is that cytokinesis divides the cytoplasm unequally between two daughter cells. Only the larger of the two cells produced by the first meiotic division is called the secondary oocyte; the other cell is called a *polar body* (see Fig. 11-35). Both cells complete the meiotic cycle, but the secondary oocyte yields a single ootid and another polar body. In other words, one primary oocyte gives rise to a total of three polar bodies and a single ovum. This unequal division has the advantage of providing the ovum with more than its share of the stored food and other materials necessary to sustain a new life during its early stages.

The timing of oogenesis is very different from the timing of spermatogenesis. The primary oocytes in female mammals are all produced before the animal is born, whereas a male's production of spermatocytes and sperm continues at a more or less regular rate throughout his adult life. In a typical female, only a few oocytes are maturing at any one time. In humans, for example, oocytes in a female fetus reach diplonema by the fifth month of fetal life. There they remain arrested (this is sometimes referred to as the *dictyotene stage*) until after puberty. Each month thereafter, from puberty to menopause, several oocytes are stimulated to begin growing. However, usually only one oocyte completes the first meiotic division and gets released from the ovary at ovulation. (Figure 11-35 shows an oocyte at that point.) The second meiotic division does not occur unless fertilization takes place.

Note that oocyte nuclei are tetraploid during their growth phase, meaning that they have passed through the S stage of their cell cycle to double the normal diploid DNA content. It is this period of growth that features multiple nucleoli and, in some organisms, the lampbrush chromosomes. These changes help supply the ribosomes and "maternal messages" that will be needed by the fertilized egg during its initial period of development.

SECTION 11-4
Spindle Dynamics

Gamete formation in plants also involves meiosis, of course, but the process is complicated in each of the higher plants by a life cycle that consists of both a multicellular diploid and a multicellular haploid stage (Fig. 11-36), called its *sporophyte* and its *gametophyte* generation, respectively. In the common garden plants, for example, it is the diploid stage that forms the structure with which we are familiar, whereas the haploid organism is tiny and must be nurtured within specialized organs of the mature diploid plant. In some plants, however (green algae, mosses, and so on), the multicellular haploid stage is the dominant one. In both cases, the usual pattern is for male and female haploid cells, called *spores*, to be produced by meiosis in the diploid (sporophyte) organism. Spores grow into multicellular male and female haploid structures, which through mitosis produce haploid cells corresponding to the actual gametes. (A pollen grain is a male gamete.)

In both animals and plants, male and female gametes unite during fertilization to produce a *zygote* in which the diploid chromosome number is restored. In animals and simpler plants, the zygote matures to a new diploid organism. In the seed-producing plants, development is arrested at an early multicellular stage as a *seed*, which may remain stable for long periods of time before germination permits a continuation of growth.

While gametogenesis is thus seen to differ somewhat between plants and animals, the meiotic process itself is clearly identical in all major respects. Also the same (or virtually the same) in both plants and animals is the spindle apparatus used to separate chromosomes during mitosis and meiosis. We pause now for a closer look at that complex device, the mechanism of which we are finally beginning to understand.

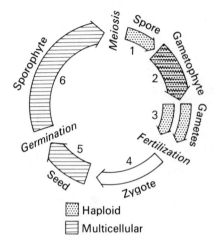

FIGURE 11-36 Alternation of Generations. The higher plants have a life cycle consisting of a mitotically proliferating haploid stage as well as a diploid stage.

11-4 SPINDLE DYNAMICS

The spindle is responsible both for chromosome separation during mitosis and meiosis, and for increasing pole-to-pole distance, the latter so that two daughter nuclei have room to form. As noted earlier, spindle fibers are actually bundles of microtubules (Fig. 11-37). Some of these bundles, containing anywhere from one (in certain fungi) to over a

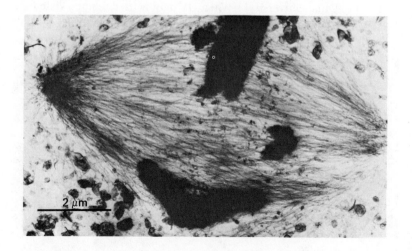

FIGURE 11-37 An Isolated Spindle. From a cultured mammalian cell. In spite of the obvious density of the spindle microtubules, microtubules themselves appear to constitute only a minority of the total protein of a spindle *in vivo*. Little is known about the rest. [Courtesy of J. R. McIntosh, Z. Cande, J. Snyder, and K. Vanderslice, *Ann. N.Y. Acad. Sci.*, **253**: 407 (1975).]

hundred microtubules, extend from pole to kinetochore. Others extend from a pole past the chromosomes and into the opposite half-spindle. The latter type of fiber is generally referred to as a pole-to-pole fiber, though it now appears that few, if any, microtubules actually make it the entire distance from one pole to the other.

The behavior of these microtubular bundles, and in particular the way they generate force to separate poles or move chromosomes, has fascinated cell biologists for nearly a century. Only in very recent years, however, have we begun to understand these processes at the molecular level.

The Dynamic Equilibrium of Spindle Fibers. Individual microtubules can be seen with the electron microscope after carefully fixing and sectioning the cell. They are not visible in the light microscope. However, the microtubular density in most cells is great enough to provide considerable birefringence (see Appendix, Section A-1). Hence, when a dividing cell is placed in the proper orientation on a microscope stage between crossed polarizers, it appears bright on a black background. The degree of birefringence (and hence brightness) correlates well with the density of the spindle fibers and thus provides a way of observing changes in microtubular density in the spindles of living cells (see Fig. 11-38).

With this technique, called *polarization microscopy*, it can be shown that spindle microtubules are in dynamic equilibrium between the polymerized and depolymerized states. (See Chaps. 3 and 9 for additional discussion of microtubule dynamics.) Conditions that reversibly depolymerize microtubules (e.g., the drug colchicine, low temperature, or high pressure) cause reversible shrinkage of the spindle, which then returns to normal when the depolymerizing agent is removed (see Fig. 11-39). Additional measurements using light scattering indicate that these changes in the spindle are due primarily to the amount of tubulin in the fibers, not to the number of microtubules. In other words, the polymerization-depolymerization involves addition and removal of tubulin from existing microtubules, changing their length, not their number.

Another way of demonstrating the dynamic nature of spindle fibers is with microbeams of ultraviolet light (see, for example, the reference by A. Forer of Duke University). When UV light is used to put a visible defect in a spindle fiber, the defect can be observed to flow toward the pole

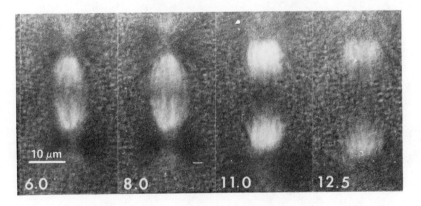

FIGURE 11-38 Birefringence of the Spindle. Mitosis in the fertilized egg of a sea urchin (*Lytechinus variegatus*) as seen by polarization microscopy. Numbers are minutes after nuclear envelope breakdown. [Courtesy of G. Sluder, *J. Cell Biol.*, **70**: 75 (1976).]

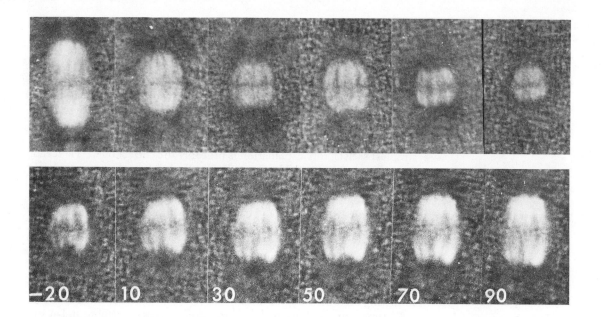

to which that fiber is attached, and to do so at a rate equivalent to the rate of chromosome movement in the same type of cell. Before the chromosomes themselves start moving (i.e., prior to anaphase) the defect alone travels toward the pole. Once anaphase begins, defect and chromosome travel at the same speed, thus maintaining their separation.

The Mitotic Mechanism. From these and other observations, we can piece together the following picture of the mitotic mechanism:

(1) The spindle forms from a pool of preexisting subunits. Assembly is catalyzed by microtubular organizing centers (see Chap. 3) consisting of spindle poles (centrioles when present) and kinetochores.

(2) Kinetochores, possibly because they participate in microtubule organization, fall under the influence of whichever pole they face and experience a pull toward that pole transmitted by the spindle fibers.

(3) Each chromosome has two oppositely facing kinetochores, hence is pulled into a midplane position to form the metaphasic plate. (In the first meiotic division, each of the two chromosomes falls completely under the influence of one pole.) During this period of alignment, tubulin is constantly being added to the microtubules at the middle of the spindle and removed at the poles.

(4) At centromere division (separation of bivalents in meiosis), tubulin addition at the kinetochores stops. Kinetochores, their chromatids trailing behind them, are pulled toward the poles as their spindle fibers dissolve by continued polar disassembly.

(5) In most cells, chromosome movement is accompanied by lengthening of the pole-to-pole distance—as much as a fivefold or sixfold increase in the total span of the mitotic apparatus. It is as if tubulin that is removed from chromosomal spindle fibers is added instead to the pole-to-pole fibers. Thus, one set of fibers shrinks as the other grows, permitting the chromosomes to gather at two widely separated points in the cell.

FIGURE 11-39 Reversible Disassembly of a Spindle. Same type of cell as in Fig. 11-38 but treated with 10^{-6} M colcemid, a derivative of colchicine. TOP ROW: Metaphase spindles treated for 0, 1, 2, 3, 4, and 5 minutes, respectively left to right, beginning 20 minutes before nuclear envelope breakdown. The colcemid reduces the size of the tubulin pool available for spindle assembly. BOTTOM ROW: Inactivation of the colcemid by irradiation with 366-nm UV light. First frame: metaphase spindle before irradiation. Subsequent frames at 10, 30, 50, 70, and 90 s after irradiation as indicated. Note rapid growth of the spindle from the newly available tubulin. [Courtesy of G. Sluder, *J. Cell Biol.*, **70**: 75 (1976).]

The whole mitotic process therefore depends on the dynamic nature of spindle microtubules—that is to say, it depends on a shifting equilibrium between assembled microtubules and their pool of subunits. This view is now rather widely accepted. Still controversial, however, is the origin of the force responsible for movement.

The amount of force needed to move chromosomes is not great (probably less than 10^{-8} dynes each) but its origin is obscure. There is no dearth of theories, however, ranging from satisfyingly simple to incredibly complex. The principle known as Occam's Razor states that the simplest satisfactory explanation is probably the correct one.[3] Hence, of the many choices available, we shall describe briefly only two—both elegant, both simple, and both probably correct in principle if not in detail.

Microtubule Disassembly and Chromosome Movement. The concept of a dynamic equilibrium of spindle microtubules evolved over a number of years. One of the more important formulators of that concept is S. Inoué of the University of Pennsylvania, who also suggested that the rate of disassembly may determine the rate of chromosome movement. In fact, disassembly may even be responsible for chromosome movement, as the following argument makes clear.

Picture a microtubule anchored at one end to a kinetochore and at the other to a spindle pole. Then remove a tubulin subunit from a region near the pole. If the microtubule is a dynamic structure, as we have supposed, there will be a rearrangement of those subunits left behind, thus filling in the hole. In the process of rearrangement, however, the tubule is shortened by a distance equivalent to the lost subunit. Conversely, of course, adding a subunit to the pole-to-pole fibers would drive the poles apart.

The amount of force on a chromosome that could be generated by removing a microtubular subunit depends on how much attraction the other subunits have for each other, since it is this attraction that will cause remodeling and hence shortening. According to Inoué's estimates, the attraction is sufficient to cause chromosomal movement.

Note, however, that if disassembly causes movement and movement takes energy, then disassembly must take energy. Where that energy comes from is not clear. However, it might well be supplied by an ATP-driven phosphorylation of the subunits, causing a conformational change favoring disassembly.

Support for the idea that spindle disassembly causes chromosomal movement comes in part from observations made on developing oocytes of the marine worm, *Chaetopterus*. The spindle poles in this organism are anchored through asters to the cell cortex, or perhaps to the cell membrane itself. If astral and spindle microtubules are slowly depolymerized with any of the agents mentioned previously, the poles and attached chromosomes are transported away from each other and toward the cell membrane. Removing the depolymerizing agent allows regrowth of the spindle and asters and subsequent return of the chromosomes to their original position.

At least under these artificial conditions, then, spindle disassembly can cause chromosomal movements comparable to the movements seen in mitosis. It is possible, however, that under ordinary conditions disassembly merely accompanies, but does not cause, chromosomal movement. The origin of the force in that case must come from another source.

Sliding Microtubules. In Chap. 9 we discussed how filaments slide past each other to generate force in muscle contraction, how microtubules use the same technique to cause movement of cilia and flagella, and how microtubules may serve to transport organelles. These ideas can be applied also to spindle fiber dynamics, a position first advocated by J. R. McIntosh of the University of Colorado.

Movement of chromosomes could be explained by having kinetochore microtubules slide along pole-to-pole fibers the way microtubules in a cilium slide past each other. A similar sliding of pole-to-pole against pole-to-pole microtubules could widen the spindle (see Fig. 11-40). Intermicrotubular bridges that might be dyneinlike (or myosinlike) arms have been noted in the spindle. These arms may be responsible for the proposed sliding of microtubules. As an alternate possibility, these arms may work not against other microtubules but against actin filaments, also known to be present in the spindle. In fact, microtubules may even be structural, rather than motile, components of the spindle, with force generation arising from myosin-actin interactions rather than tubulin-tubulin or tubulin-actin interactions. Unfortunately, we cannot at the present time distinguish with certainty among these possibilities.

Conclusions. The complexity of the spindle apparatus and an inability to make it function *in vitro* have frustrated our desire to know in detail how it works. However, it is clear that spindle microtubules are in dynamic equilibrium with their unpolymerized subunits, that tubulin can "flow" down an apparently stationary fiber, and that chromosomal movement is accompanied by fiber disassembly. We do not know whether disassembly itself accounts for the movement or, as seems more likely, force is generated by sliding two elements past each other. That choice will be based on the results of future experiments.

FIGURE 11-40 Microtubule Overlap and Spindle Elongation. The increasing distance between spindle poles at anaphase may be due to microtubules from the two poles pushing against each other. That proposition is consistent with these micrographs showing overlap of microtubules originating from the two sides. (a) From the diatom *Pinnularia* at telophase. Note that chromatin (ch) surrounds the spindle near the poles. There is clear overlap of pole-to-pole fibers at the center of the spindle. The complex polar structures that serve in lieu of centrioles are clearly seen (arrows). Pole-to-pole distance is about 9 μm. (b) Thin section of an isolated midbody (overlap region) from a human cell spindle. [(a) Courtesy of J. Pickett-Heaps; see *BioScience*, **26**: 445 (1976); (b) courtesy of J. McIntosh, Z. Cande, J. Snyder, and K. Vanderslice, *Ann. N.Y. Acad. Sci.*, **253**: 407 (1975).]

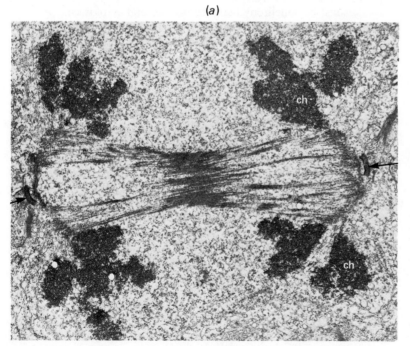

(a)

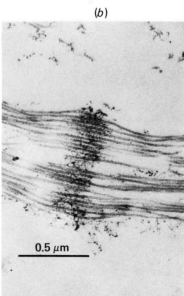

(b)

0.5 μm

11-5 CYTOKINESIS

Mitosis or meiosis divides the nuclear material and encloses the two groups of chromosomes in the envelopes of daughter nuclei. Thereafter, cytokinesis—if it occurs—takes place at a point that distributes the daughter nuclei to different cells.

Synchrony between Nuclear and Cytoplasmic Division. Cytokinesis is generally, but not always, synchronized with anaphase and telophase in both meiosis and mitosis. A notable exception occurs after fertilization of *Drosophila* eggs, for some 4000 nuclei are produced by mitoses without any cell division at all. Then, in a remarkable demonstration of synchrony, each nucleus is partitioned into a separate cell by multiple cytokineses. On the other hand, many fungi and some green algae regularly undergo nuclear division without cytokinesis to form a multinucleated cell called a *coenocyte*. Animal cells with more than one nucleus are also seen, especially in the liver. (Skeletal muscle cells have multiple nuclei, but they got that way by cell fusion rather than by nuclear proliferation.) These multinucleated cells are the exceptions rather than the rule. Most eukaryotic cells couple cytokinesis to karyokinesis.

In those animal cells in which karyokinesis and cytokinesis are normally synchronized, the point where the cells actually divide is clearly influenced by the position of the center of the spindle. For example, in 1919 E. G. Conklin found that when the spindle apparatus of an oocyte was displaced by centrifugation, so was the site of division, altering the relative size of the polar body and ovum. In fact, the mitotic apparatus can be experimentally shifted to and fro, causing new *division furrows* (indentations) to form in the cell membrane and old ones to regress. The mitotic apparatus is clearly responsible in some way for inducing the formation of these furrows; it determines both their location and it establishes the synchrony between mitosis and cytokinesis.

The mitotic apparatus itself, however, is not responsible for cytokinesis, only for initiating it. Once the chain of events leading to cytokinesis has been established, the spindle can be surgically removed or dissolved without substantially altering the subsequent division (see Fig. 11-41).

It now seems probable that asters are responsible for establishing the division furrow. (Higher plants, of course, do not have asters but neither do they have division furrows. They divide by cell plate formation, as will be explained.) Cytokinesis seems to start at a point on the cell membrane midway between the two asters, and therefore at a point affected equally by them. Experiments have been carried out in which the spindle was surgically removed while leaving asters intact, and in which cells were distorted to bring two asters into close proximity without an intervening spindle. Division furrows in these cases formed midway between the asters.

Fibers radiate out in all directions from asters, often reaching the cytoplasmic membrane. It may be this connection between aster and membrane that is responsible for the aster's influence. In cells with very small asters and a large spindle, however, spindle fibers themselves may be responsible for initiating division. This difference is a minor one

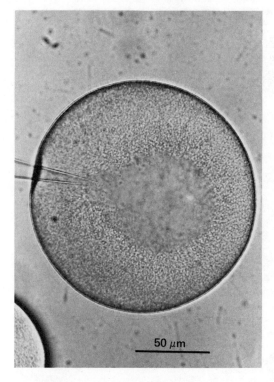 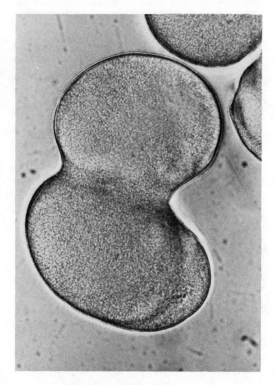

since astral fibers and spindle fibers, as far as we can tell, are composed of the same stuff.

The Contractile Ring. The cytoplasm directly beneath a division furrow contains a dense ring of microfilaments arranged circumferentially around the cell like a purse string (see Fig. 11-42). This collection of filaments is referred to as the *contractile ring.*

The contractile ring is assembled when a division furrow forms, and it disappears again when the cells are divided. Direct evidence that this ring actually has contractile properties has been obtained from observations on certain abnormal grasshopper spermatocytes where the ring fails to stick to the cell membrane. Instead, it constricts during meiosis, compressing the cytoplasm and mitotic spindle but without creating a division furrow in the cell membrane.

Under normal circumstances, we envision the contractile ring as growing ever smaller, dragging the cell membrane with it until division is complete. If this is what actually happens, we would expect the cell membrane to be puckered where it is drawn into the division furrow. Scanning electron micrographs frequently support this expectation (see Fig. 11-43).

Assembly of the contractile ring, as already noted, is usually initiated by the presence of a nearby mitotic apparatus. Once assembly has begun, however, it seems to be self-propagating. In those cells having a mitotic apparatus that is located much closer to one part of the cell membrane than another, the ring forms at the near point (midway between the asters) and then grows in both directions to encircle the cell.

FIGURE 11-41 Cytokinesis without a Spindle. LEFT: The spindle apparatus (clear area in center) of the sea urchin *(Clypeaster)* egg was dissolved by injecting less than a microliter of sucrose solution. RIGHT: Cytokinesis occurred on schedule, and in the position that it would have taken had the spindle been left intact. [Courtesy of Y. Hiramoto, *J. Cell. Biol.*, **25**: 161 (1965).]

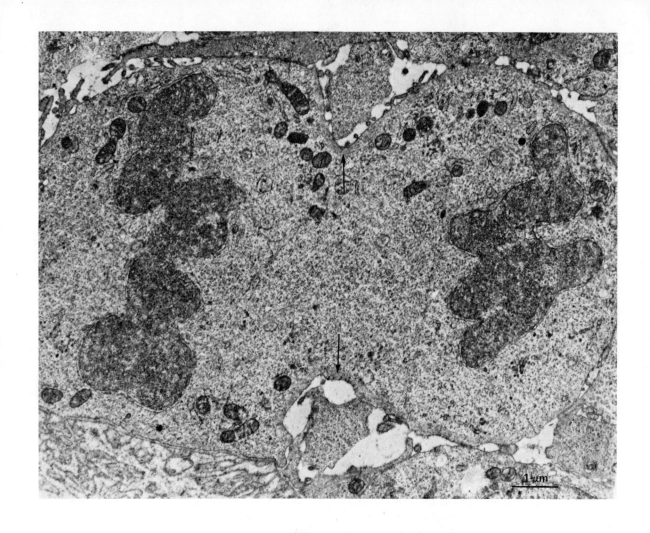

FIGURE 11-42 The Division Furrow and Contractile Ring. TOP: Early telophase in a mouse mammary epithelial cell. Arrows mark the furrow. (Note the irregularly shaped nuclei, newly formed around the still partially condensed chromatin.) BOTTOM: Detail of one furrow. The microfilaments are cut in cross section, but they are packed densely enough to be easily visible. [Courtesy of D. G. Scott and C. W. Daniel, *J. Cell Biol.*, **45**: 461 (1970).]

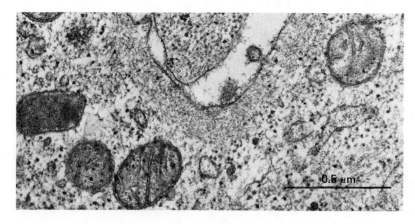

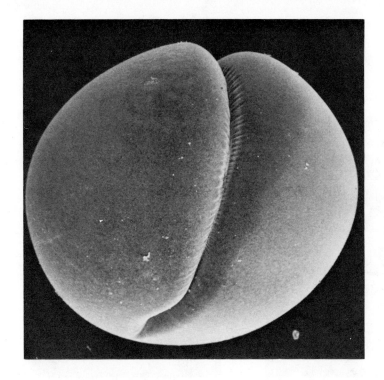

FIGURE 11-43 Cleavage. A fertilized frog egg (diameter about 150 μm) is undergoing its first cytokinesis ("cleavage"). Note the advancing division furrow near the bottom of the cell and the puckered "stress lines" in the furrow itself. The stress lines are better seen in the detail, which also reveals numerous small microvilli, which make the surface look fuzzy. [Courtesy of H. W. Beams and R. G. Kessel, *Am. Scientist*, **64**: 279 (1976).]

The composition of the contractile ring is not completely known. Microfilaments (i.e., actin filaments) were the first components to be identified because they are visible by electron microscopy. However, where we find actin and movement we have come to expect the presence also of myosin and, in fact, immunofluorescence has been used to confirm its presence in the contractile ring (see Fig. 11-44).

Probably, then, the contractile ring is another example of actin-myosin interaction (Chap. 9) and it is no doubt controlled in much the same way as other contractile systems. Thus, when glycerol is used to extract the bulk of the cytoplasm from cells, division furrows can still be induced to form by the addition of Ca^{2+} and Mg^{2+}-ATP. Furthermore, cortical contractions resembling early furrows can be induced in some cells by electrical stimulation, providing that Ca^{2+} is available in the bathing medium. And finally, the closely related process of wound closure in the cell membrane also involves the appearance of a ring of filaments around the hole, which is then sealed as the ring contracts, an activity that again requires Ca^{2+}.

Thus calcium is probably the key to contractile ring assembly and movement, just as it controls other contractile systems. However, we should not presume from this similarity that contractile rings behave exactly like muscle sarcomeres. They do not. Muscle cells maintain a constant volume as they contract by sliding their filaments past each other. The contractile ring, on the other hand, typically retains much the same width and thickness as it contracts. Hence, the total amount of protein in the ring must decrease. It appears, therefore, that the filaments in the ring disassemble as the ring contracts.

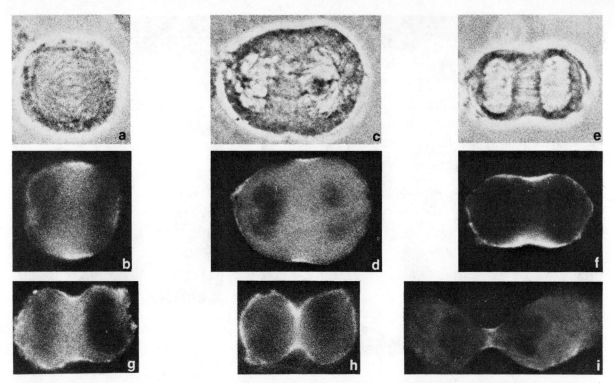

FIGURE 11-44 Myosin and the Contracile Ring. The contractile ring of microfilaments also contains myosin, here demonstrated by indirect immunofluorescence. TOP ROW: Dividing human cells by phase contrast microscopy. SECOND ROW: Same cells using fluorescence microscopy and myosin staining. BOTTOM ROW: Further stages in cytokinesis. Note localization of myosin fluorescence mostly to the site of the contractile ring. [Courtesy of K. Fujiwara and T. D. Pollard, *J. Cell Biol.*, **71**: 848 (1976).]

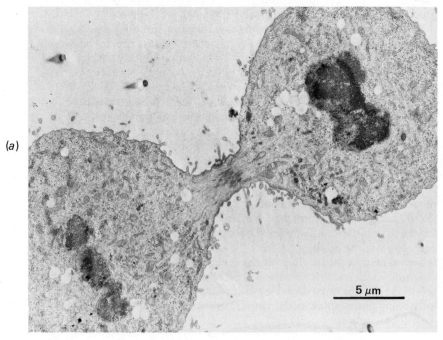

FIGURE 11-45 The Midbody and Telophase Neck. The contractile ring may compress pole-to-pole spindle fibers into a compact midbody that sometimes lasts for hours. (a) Human fibroblast (WI-38 cell) at telophase. (b) and (c) appear on the opposite page.

Disappearance of the ring when its job is completed is decidedly less dramatic than its sudden appearance. In many cells, remnants of the mitotic apparatus (the *midbody*) persist in the region below the ring and get compressed into a long-lasting "telophase neck" (see Fig. 11-45). During this time the contractile ring becomes less and less prominent. The purpose of this persistent midbody is uncertain. However, as it contains an overlap of microtubules from the two poles of the former spindle (see Fig. 11-40b), the midbody may elongate and serve to help separate the new cells and thus prevent their accidental fusion.

Membrane Changes During Cytokinesis. When a spherical cell divides its cytoplasm among two spherical daughter cells, the total surface area of the daughters will be 26% larger than the original area of the parent cell. Either new membrane must rapidly be made and inserted during division, or the existing membrane must stretch, or our assumption of spherical cells is not quite accurate. The latter appears to be the answer.

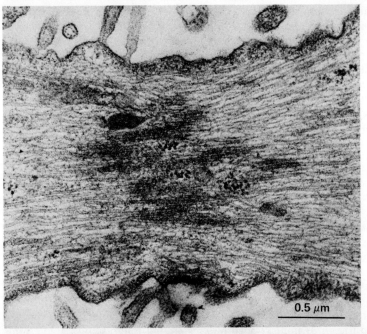

(b)

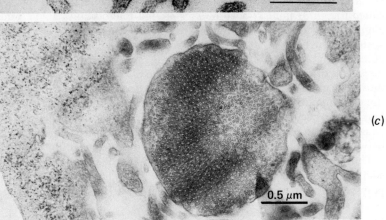

(c)

FIGURE 11-45 (*b* and *c*) Longitudinal and cross-sectional views of the telophase neck, showing the array of microtubules. [Courtesy of J. R. McIntosh and S. C. Landis, *J. Cell Biol.*, **49**: 468 (1971).]

Some cells remain roughly spherical throughout their cycle (at least as long as they are not allowed to settle onto a surface) while others round up only prior to division. Examination of cell surfaces with the scanning electron microscope, however, reveals that even very round cells have numerous surface projections consisting of microvilli, filipodia, lamellipodia, and so on (see Chap. 9). These projections significantly increase the total surface area of the cell. As a consequence, there is typically more than enough surface membrane in the parent cell to accommodate the two daughters.

From this model, one would expect newly formed daughter cells to be relatively smooth and the amount of surface projections (especially microvilli) to be at a maximum just before division. There has been considerable debate about the validity of these predictions, but there are studies that certainly seem to support them (see Fig. 11-46).

New cell membrane seems to be made and inserted throughout interphase. Calculations of the exact rate, and how that rate changes with the cell cycle, are complicated by a turnover that constantly removes membrane from the surface (see Chap. 10). However, in cells that are growing and dividing, the rate of turnover is very much less than in nongrowing cells. Hence, it appears that membrane synthesis is a continuous process. In nongrowing cells, destruction balances synthesis; growing cells, on the other hand, accumulate most of their new membrane as microvilli that are unfolded to support cell division.

Cytokinesis is associated with some other changes in the cell membrane, the significance of which is not so clear. For example, the ability of animal cells to be agglutinated by plant lectins is greatest during mitosis and the early G_1 phase (an ability that is present in malignant cells all the time). Since agglutination by lectins requires surface receptor mobility, this aspect of membrane structure seems to vary with the cell cycle.

Cytokinesis in Plants. Most cells from higher plants are surrounded by a dense cell wall that establishes shape. The cell membrane is attached to this wall at numerous points. Hence, the division furrow used by animal cells would not be an appropriate mechanism of cytokinesis in plants.

Cytokinesis in plants occurs by formation of a *cell plate*, usually at the equatorial plane of the spindle (see Fig. 11-47). A collection of vesicles called the *phragmoplast* accumulates around microtubules at the midplane of the spindle after chromosomes have been pulled away. These vesicles, which are thought to be derived at least in part from Golgi saccules and to be brought into position by the microtubules, fuse to form a disc. The disc then grows outward by fusion of more vesicles at its edges until the cell membrane is reached. Finally, the growing disc fuses with the cell membrane. When this process is completed, the vesicle membranes have become inward extensions of the plasma membrane and two daughter cells have been created. The vesicle contents are trapped between the two daughters as the beginning of the new cell wall.

The wall of plant cells grows and matures with the cells. It does so by addition of components carried to it in Golgi-derived vesicles, and by local polysaccharide synthesis using enzymes and raw materials transported

SECTION 11-5
Cytokinesis

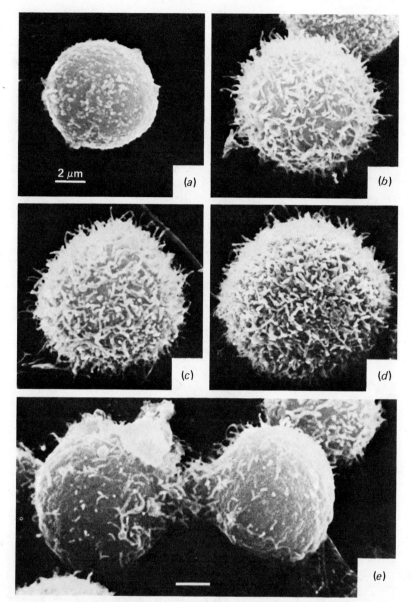

FIGURE 11-46 The Plasma Membrane During Growth and Division. Cultured mouse mastocytoma cells (tumor cells). Micrographs (a)–(d) were taken at identical magnification to show growth in size of cells at early G_1, late G_1, S, and G_2 phases, respectively. Note also the increasing density of microvilli on the surface as cells proceed through the cycle. (e) shows a cell at the later stages of cytokinesis. Note the abrupt loss of most microvilli, which apparently have unfolded to support the increased surface of two daughter cells. Bar = 2 μm. [Courtesy of S. Knutton, M. Sumner, and C. Pasternak, *J. Cell Biol.*, **66**: 568 (1975).]

to it from the Golgi apparatus (see Chap. 10). The new wall is rarely completely solid. Rather, as the vesicles coalesce during cell plate formation, they leave channels of cytoplasm. These channels continue to extend between, and thus connect, daughter cells after cell wall formation. They are the *plasmodesmata* (Fig. 11-47) that transfer substances and signals from cell to cell presumably to coordinate growth and function. Plasmodesmata thus replace in part the gap junctions and circulatory system by which intercellular communication is achieved in animals.

535

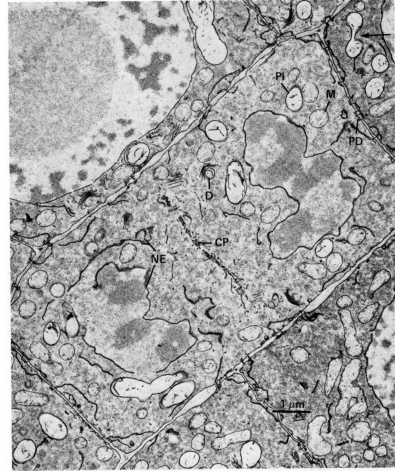
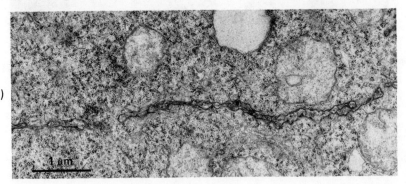

FIGURE 11-47 Cell Plate Formation. (a) A dividing cell of a maize root tip is shown in the process of assembling new nuclear envelopes (NE) around its chromosomes, which are still condensed. A cell plate (CP) is also forming. Both seem to be put together from endoplasmic reticulum. The specimen was fixed with potassium permanganate ($KMnO_4$), which preserves membranes but little else. Mitochondria (M). Plastids (Pl). Note the dividing plastid (arrow), the plasmodesma (PD, a cytoplasmic channel to the next cell), and a huge nucleus in the upper left, with a prominent nucleolus. Dictyosome (D). (b) Cell plate formation in *Equisetum arvense*. The walls will form on either side of the membrane-enclosed cell plate. [(a) courtesy of Dr. Hilton H. Mollenhauer; (b) Courtesy of Dr. Patricia Harris.]

11-6 CELL CYCLE CONTROLS

The cell cycle was introduced in Section 11-2 as consisting of five stages: G_1, S, G_2, M, and D, the latter being division or cytokinesis. The rate at which cells traverse this cycle is highly variable. Yeast cells can make it in three hours. Cells lining the small intestines of mammals take 17 or 18 hours to traverse the cycle, but some mammalian cells never divide. Skeletal muscle cells and neurons are in this latter category. Even the most stable cells, however, are capable of some degree of progress through the cycle when properly stimulated, indicating the presence of some sort of regulatory process that normally holds the cells in check.

There has been much interest in these regulatory processes in recent years, in part because we would so desperately like to know why cancer cells don't respond to them. In this section, we shall try to summarize some of what is presently known about mitotic regulation, concentrating on mammalian cells since they are the subject of most of the current work.

Cell Growth and Cell Division. The arithmetic of binary fission leads to exponential growth. That is, if we go from one cell to two, then four, eight, sixteen, and so on, we get

$$N = N_0 2^n = N_0 2^{t/g}$$

where N is the number of cells after n generations of growth requiring time t. The original inoculum is given as N_0, and the time per generation is represented as g. If N is plotted on a logarithmic scale against time, during ideal conditions a straight line should be observed:

$$\log N = \log N_0 + \frac{\log 2}{g} t$$

This exponential (or logarithmic) increase with time holds only as long as there is no change in g, which, however, will vary in response to changing conditions.

The rate at which a cell grows, hence the rate at which a critical size needed for division is reached, obviously depends on the rate of metabolism. That, in turn, will be influenced by such things as the surface-to-volume ratio, which limits the capacity to bring in food and get rid of wastes. (The ratio becomes less favorable as the cell gets larger.) Metabolic rate is also dependent on temperature and other environmental factors, including the supply of oxygen and nutrients. The actual rate of growth itself will be a function both of metabolic activity and of the kinds of organic molecules the cell must make, as opposed to those that are supplied by the growth medium. All these things, then, influence the generation time.

Even under constant conditions, however, the generation time for a given cell may vary widely in response to external controls. Most of the variation occurs in the length of time the cell spends in the G_1 stage. Very rapidly growing cells may have no noticeable G_1 stage at all. At the other extreme, some cells seem to be more or less permanently arrested in this phase. These observations have led to the concept of a G_1 "checkpoint" at which a cell may be temporarily or permanently halted, thus regulating the cycle time.

Checkpoints. Our overall view of the cell cycle, either prokaryotic or eukaryotic, is that of an ordered sequence of events, where the occurrence of one leads naturally to the next. A checkpoint is a critical event in this sequence that represents a barrier to further progress through the cycle.

The checkpoint in the bacterial cell cycle is apparently the initiation of DNA synthesis, because once a replication fork has been initiated DNA synthesis will continue until the chromosome has been duplicated, and that event in normal cells invariably leads to cytokinesis. In *E. coli*, replication of DNA takes a rather constant 40 minutes and cytokinesis follows about 20 minutes after its completion. Hence, the rate of cell division depends on the rate of initiation, which in *E. coli* can be as often as every 20 minutes (requiring obviously more than one replicating chromosome per cell).

In most eukaryotic cells, the major checkpoint is in the G_1 stage. Once it has been passed, the S, G_2, M, and D stages usually take place without hesitation and in a length of time that is, with few exceptions, relatively fixed for cells of a given species. In one study, for example, mouse epithelial cells from various locations in the animal had generation times ranging from 41 to 8000 hours, with the variation almost exclusively attributable to variation in the length of the G_1 stage.

Traversal of the G_1 checkpoint appears to be a random event, the probability of which can be increased or decreased by external controls. By random event, we mean that the chance of getting through this point in the next moment of time (i.e., the *transition probability*) is not affected by how long a cell has already been there.

The situation is like that of a chemical reaction having an energy of activation. Reactants all have an equal chance of achieving the necessary activation energy, but the number that do so per unit time depends on the size of the energy barriers. If the reaction is catalyzed by an enzyme, the effective size of the barrier depends upon the catalytic effectiveness of the enzyme. And that, in turn, is subject to external controls.

If this model for the G_1 checkpoint is correct, then the length of time a cell stays at the checkpoint, and hence remains in the G_1 stage, depends on the energy of activation needed to get through it. Controlling the rate of replication, then, becomes a matter of varying this energy of activation.

Though we do not know with certainty what the G_1 checkpoint actually is, passing it results in accumulation in the cytoplasm of a substance that makes possible entry into the S stage. The presence of this substance can be demonstrated by cell fusion and nuclear transplants: (1) When a nucleus is transplanted from a cell in the G_2 stage to a cell in the S stage, the transplanted nucleus may begin to replicate its DNA again; (2) when a G_1 and S stage cell are fused, the G_1 nucleus enters the S stage at once; and (3) when mouse and hamster cells are fused, both nuclei enter S stage at the time expected for the hamster cells, which have a much shorter G_1 stage than mouse cells. In fact, it is generally true that in multinucleate cells all nuclei enter the S stage simultaneously, apparently in response to the appearance of some cytoplasmic factor.

G_2 Arrest. A second checkpoint exists in the G_2 stage. Cells that are arrested at this G_2 transition will be tetraploid—that is, having passed the S stage they will have twice the usual amount of DNA.

Tetraploid nuclei are occasionally found in a number of different tissues; they are fairly common in the liver, and they are the rule in heart muscle cells of adult humans. In heart muscle, the conversion from diploid to tetraploid occurs during childhood, presumably to support the metabolism of the enlarging cells, which do not divide. This reasoning is supported by the observation that when adults experience congestive heart failure from abnormal enlargement of heart cells (for example, in response to long-standing high blood pressure), many of the nuclei pass through still another S stage to become octoploid and a few become 16- or 32-ploid.

Whereas the critical event at the G_1 checkpoint probably has something to do with initiating DNA replication, the critical event at the G_2 checkpoint may be concerned with the condensation of chromatin, which necessarily precedes mitosis. At least we know that traversal of the G_2 checkpoint leads to accumulation of a cytoplasmic substance capable of causing chromatin condensation.

The existence of a chromatin condensing factor was demonstrated by R. T. Johnson and P. N. Rao at the University of Colorado Medical Center by using Sendai virus to fuse HeLa cells at different stages in their cell cycle. When a mitotic cell is fused with a cell in the G_2 stage, chromatin condensation takes place at once in the G_2 nucleus, producing normal-looking chromosomes. When a mitotic cell is fused instead with a cell at its G_1 stage, chromatin condenses in that nucleus, too, but since DNA replication has not occurred only single-stranded chromatids are seen (Fig. 11-48). Even S-stage nuclei can undergo chromatin condensation in this way, although multiple fragments of condensed chromatin instead of complete chromatids are seen.

Chromatin condensation is thought to be associated with, and perhaps due to, phosphorylation of histone H1. Histone H1, like other histones, is synthesized at the same time as the DNA to which it binds (in contrast to nonhistone chromosomal proteins, which are synthesized at a regular rate throughout the cell cycle). The enzymes capable of phosphorylating H1, however, have a peak of activity late in G_2. The phosphate content of both new and old H1 rises sharply just prior to chromatin condensation and decreases again when the chromatin disperses after mitosis. If the relationship between H1 phosphorylation and chromatin condensation is one of cause and effect, as seems highly probable, then some aspect of this event becomes a good candidate for the G_2 checkpoint.

G_0 Cells. The term G_0 refers to cells that have completed mitosis but are not proceeding toward another one. These are quiescent cells—small lymphocytes, liver cells, skeletal muscle cells, neurons, and so on. Whether one refers to them as being in G_1 arrest or in G_0 is largely a matter of semantics. However, it is maintained by some workers that these cells are different from cells that merely have very long G_1 stages.

The transition from G_0 to G_1 by definition requires not just the passage of time but a specific stimulus—i.e., a *mitogenic* stimulus, or one that induces mitosis. Mitogenic stimulation is accompanied by biochemical changes in the cell that include increased transcription of DNA and increased levels of nonhistone chromosomal proteins.

To take an example of a G_0 to G_1 transition, consider the liver parenchymal cell (hepatocyte) of the rat. This highly differentiated cell rarely divides. Nevertheless, surgical removal of part of the liver ("partial hepatectomy") is followed by an abrupt change in the biochemical activities of the remaining cells. By 24 hours, mitotic figures are already seen. Their number peaks rapidly and then drops to almost nothing by 72

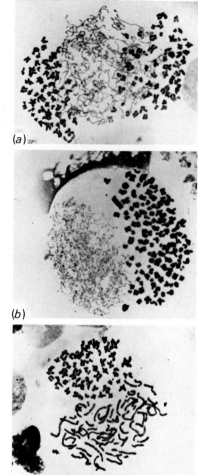

FIGURE 11-48 Premature Chromosome Condensation. HeLa cells fused at different stages of the cell cycle. (a) Fusion of metaphase and G_1 cells. Note the normal metaphase chromosomes plus the condensed single–stranded chromatids derived from the G_1 nucleus (center of figure). (b) Metaphase plus S-phase hybrid. The S-phase chromatin, caught in the process of replication, condenses irregularly into fragments of heterochromatin. (c) Metaphase plus G_2 hybrid. The G_2 chromatin condenses into chromosomes having the usual two chromatids, although the chromosomes are not as compact as in the metaphase cell. [Courtesy of Dr. P. N. Rao. From R. T. Johnson and P. N. Rao, *Nature*, **226**: 717 (1970).]

TABLE 11-2. Stimulation of Quiescent Cells

Cell	Stimulus
Liver parenchymal	partial hepatectomy; albumin depletion
Kidney tubule	removal of other kidney
Uterine smooth muscle and endometrium	estrogens
Mammary gland	estrogens
Prostate gland (castrated rat)	5-α-dihydrotestosterone
Adrenal gland	ACTH
Salivary gland	catecholamines
Growth cartilage (to lengthen long bones)	growth hormone/somatomedins
Fibroblasts	epidermal growth factor; fibroblast growth factor
Lymphocytes	lectins; receptor antibodies; specific antigens

hours. The newly divided cells then enlarge and differentiate so that even with removal of two-thirds of the liver, the original mass is regained in about two weeks.[4]

Other examples of quiescent cells and the stimuli that can cause them to replicate are given in Table 11-2, which is by no means complete. Unfortunately, we do not know in detail how any one of those stimuli actually results in restarting the cell cycle. In spite of the gaps in our knowledge, however, there has been much progress in recent years.

Growth Factors. Nearly all cultured animal cells require at least a small amount of blood serum in their culture medium. In an attempt to find out why, a number of substances have been identified in serum that are not nutrients in the usual sense (most never even gain entry to the cytoplasm) but which nevertheless are needed by the cells. These substances are called *growth factors*. The typical growth factor can act as a *mitogen* for specific types of quiescent cultured cells. That is, it can cause G_0 cells to reenter the cell cycle and divide.

When a culture of nonmalignant cells is permitted to grow under favorable conditions, replication usually stops when the cells have approximately covered the dish to a depth of one cell. This phenomenon was originally called *contact inhibition* but is now more often referred to as *density-dependent inhibition of growth*, because it seems to have more to do with the ability to transport nutrients into the cell than with cell-to-cell contact as such. The cells in this condition appear to have reverted to the quiescent G_0 phase, a situation from which they can often be rescued by the addition of an appropriate growth factor, as well as by certain less specific manipulations, including damage of membrane proteins with trypsin. This ability to rescue G_0 cells is commonly used to identify growth factors.

Most growth factors are peptides or small proteins. *Insulin* is the prototype, though it is a rather weak mitogen in mammalian systems. Many of the others seem to be evolutionarily related to insulin, however.

In fact, one is even called *NSILA*, which stands for "nonsuppressible insulin-like activity."

Growth hormone was one of the first growth factors other than insulin to be investigated. It was long assumed to be directly responsible for cellular proliferation in growth cartilage, and hence for elongation of bones during childhood. It is certainly true that a deficiency of growth hormone during childhood results in dwarfism while an excess results in a pituitary giant. However, it now seems that growth hormone is only indirectly responsible—it causes release from other tissues of a family of proteins called *somatomedins*. It is these secondary proteins that appear to be the true mitogens.

Erythropoietin is often listed as a growth factor though its target cell, the proerythroblast (precursor of the erythrocyte), is not absolutely dependent on erythropoietin. Nevertheless this glycoprotein, which appears to be made by the kidneys in response to oxygen deprivation, can cause red cell production to be increased as much as sevenfold or more.

Other growth factors having specific tissues as their target include *nerve growth factor*, which is an essential mitogen for embryonic sympathetic nerves and necessary for growth and maintenance of these same cells later in life; *epidermal growth factor*, which causes epithelial cell proliferation in a variety of locations; and *fibroblast growth factor*, which stimulates not only fibroblasts, but cells lining small blood vessels. The list is long, and continues to enlarge at a rapid rate.

Nonprotein growth factors have also been identified. Some are steroid hormones, consideration of which will come later. A particularly interesting nonprotein growth factor is the polyamine *putrescine*, which has been found in the growth medium of some cultured human cells. This putrid-smelling substance is produced by the enzyme, *ornithine decarboxylase* (ODC), mentioned in Chap. 5 as having the extraordinarily short half-life of five minutes under some conditions. The short half-life of ODC permits rapid fluctuations in its concentration, and in fact the levels of this enzyme may vary rapidly over a 20-fold range.

Interest in the polyamines (see Fig. 11-49) as possible regulators of the cell cycle *in vivo* comes from the apparent ability of putrescine to act as a growth factor and from a number of other provocative observations. For example, a rise in ornithine decarboxylase activity is one of the first events to be measurable in a mitogenically stimulated cell. In addition, it appears that spermidine (but not so much the other polyamines) fluctuates in concentration with the cell cycle and is essential for the G_1 to S transition. Beyond these facts, and the observation that polyamines

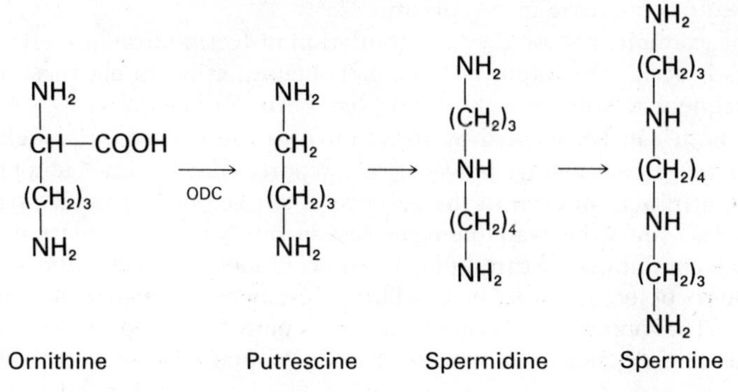

FIGURE 11-49 Polyamines. The control enzyme in the pathway is ODC, ornithine decarboxylase.

bind to DNA, we are as yet ignorant of their function, a situation that, in view of current efforts, promises to be temporary.

Putrescine and the steroids penetrate cell membranes and so are able to interact with cytoplasmic receptors. The other growth factors mentioned, however, are polypeptides or proteins and hence should interact with target cells by binding to specific receptors on the cell surface (see Chap. 10). Their mitogenic effects must therefore be mediated by some kind of cytoplasmic "second messenger." Prime candidates for that role include the cyclic nucleotides and Ca^{2+}.

Cyclic AMP and Other "Second Messengers." cAMP levels in many cells vary inversely with growth rate; that is, rapidly dividing cells have low levels while stationary cultures have high levels. The direct addition of cAMP or agents that raise cAMP levels (e.g., theophylline or caffeine) often slow growth rates, even in some cancer cells. Conversely, agents associated with a fall in cAMP (e.g., insulin) are often promoters of growth and cellular replication.

Certainly the known capacity of cAMP to stimulate protein kinases lends itself readily to a regulatory role, for not only enzymes but gene regulators may be controlled in part by phosphorylation and dephosphorylation. It is also possible that gene repressors having direct need for cAMP will be found in eukaryotes as they have been in prokaryotes.

Another candidate for second messenger is cGMP, the levels of which also fluctuate during the cell cycle and can be increased by external stimuli. Still another is Ca^{2+}. It is often hard to separate the role of Ca^{2+} from that of the cyclic nucleotides, since there is a strong interaction. Nevertheless, there are instances, ranging from the activation of frog eggs to the activation of human lymphocytes, where the mitogenic role of Ca^{2+} seems to be relatively well defined.

Lymphocyte Activation. When B lymphocytes come in contact with the right antigen, they may be stimulated to leave their G_0 state. Initially a cell with little cytoplasm and a dense, shrunken nucleus, the lymphocytes very quickly change into a more active-appearing cell with increased amounts of cytoplasm and less condensation of their chromatin. This transformation is called *blastogenesis* (see Fig. 11-50). It leads to a cell capable of dividing and of differentiating into a mature antibody-producing plasma cell. This change can be brought about by specific antigens, by antibodies to surface receptors, and by certain lectins, especially phytohemagglutinin (see Chap. 10). The key to blastogenesis in each case, however, appears to be cross-linking of surface receptors and a subsequent increase in cytoplasmic Ca^{2+}.

For example, a slow Ca^{2+} accumulation of lectin-stimulated lymphocytes is seen. The importance of that accumulation to blastogenic and subsequent mitogenic stimulation has been demonstrated by showing that both can be blocked by injection into the cell of Ca^{2+} chelators. Even more convincing, however, are reports that (1) Ca^{2+} ionophores cause stimulation even in the absence of antigen or lectin; (2) extracellular Ca^{2+} must be available regardless of the type of stimulation used; and (3) combining subthreshold amounts of ionophore with subthreshold amounts of lectin results in an additive response and mitogenic stimulation. The ionophore, of course, increases permeability of the cell membrane to Ca^{2+}; hence, it appears that Ca^{2+} uptake is the event causing

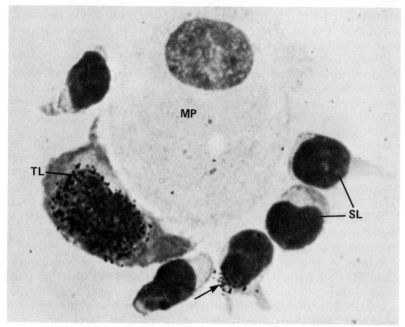

FIGURE 11-50 Lymphocyte Blastogenesis. Shown is a rosette containing a macrophage (MP) and several small lymphocytes (SL). Normally, small lymphocytes can exist for many years without replicating their DNA or dividing. Here ^{3}H-thymidine uptake and autoradiography reveal blast transformation preparatory to mitosis and differentiation. Note the many silver grains over the nucleus of the already transformed lymphocyte (TL) indicating ^{3}H uptake and DNA replication. The process is just beginning in one of the small lymphocytes (arrow). [Courtesy of J. M. Hanifin and M. J. Cline, J. Cell Biol., **46**: 97 (1970).]

stimulation. How it does so, on the other hand, is not clear, though activation of protein kinases is an often-mentioned possibility.

Steroid Hormones. The steroid hormones, as explained in Chap. 5, are able to enter cells directly, complex with cytoplasmic receptors, and then move into the nucleus where the expression of target genes is altered. Steroids apparently do not need cyclic nucleotides or Ca^{2+} to mediate their effects.

Of the various steroid hormones, the estrogens in particular are known for their mitogenic effects on target tissues. Smooth muscle of the rodent uterus, for example, can be induced to proliferate in the presence of estrogen. And the normal rise and fall of estrogen levels in menstruating women causes, in the first half of the menstrual cycle, rapid proliferation of cells (endometrium) lining the uterus. These cells will be caused to differentiate by progesterone in the latter half of the cycle and then sloughed at menstruation. Similarly, estrogens cause proliferation of ductal cells in the mammary glands, which then differentiate in response to progesterone during pregnancy.

The opposite effect, mitotic inhibition, is more often associated with the adrenal glucocorticoids such as cortisol or cortisone.[5] Lymphocytes are especially sensitive, which is the reason for the widespread use of glucocorticoids to treat severe allergies and inflammations, prevent rejection of transplanted organs, and treat lymphocytic leukemia.

The mode of action of the glucocorticoids seems, in part at least, to be due to suppression of the glucose transport system (an "antiinsulin" effect). This response is shared by a great many cell types other than lymphocytes. In fact, children with chronically high glucocorticoid levels, whether produced in response to long illness or administered as medicine, experience growth retardation. The effect is usually reversible, however. If glucocorticoid levels are reduced while growth cartilage is still present, these children commonly experience a "catch-up period" of rapid growth that leaves them almost as tall as they would have been without growth suppression.

In at least one situation, cortisone has a more direct antimitotic effect. That situation is the proliferation of liver cells in a young animal, where the inhibition seems specifically to affect nuclear DNA replication without affecting RNA or protein synthesis. The cells continue to grow, but they cannot divide.

Mitotic Inhibitors. Glucocorticoids are thus mitotic inhibitors of certain target cells. High levels of these hormones, however, represent an abnormal condition. We are more interested in what causes cells to stop replicating under normal conditions. The answer, in many cases, seems to be a negative feedback loop.

In general, cells that secrete a substance into the extracellular medium are often found to be mitotically inhibited by high extracellular levels of the substance. Thus when the level of that substance falls, new cells can be made to share the work of reestablishing the proper concentration. The feedback pathway is sometimes indirect, as when pituitary factors cause cellular proliferation in the adrenal glands after a fall in adrenal-produced cortisol. In other cases, the product of the secreting cells is thought to act directly on the cells that made it. In fact, it appears that some cells make and secrete a substance expressly for the purpose of regulating cell division.

Attempts to determine why damage of cells in the skin stimulates their division, thus permitting wound healing, led W. S. Bullough and E. B. Laurence of London's Birbeck College to the discovery in 1960 of a substance that was termed a *chalone*. According to Bullough, the name comes from a Greek word meaning "to slack off the main sheet of a sloop to slow the vessel down." Chalones, most of which are peptides or glycoproteins, share the following characteristics:

1. They inhibit mitosis both *in vivo* and *in vitro*.
2. They are tissue-specific, but they are remarkably the same from one species to another.
3. The tissue affected by a particular chalone is also a tissue that makes the same chalone.
4. Mitotic suppression by chalones is reversible, apparently without harm to the cell.

Substances fitting this description have now been isolated from a number of animal tissues.

How chalones are able to exert their mitotic inhibition is not clear, but cells often seem to be stimulated to divide when chalone production by nearby cells is compromised, as when tissue is damaged. Inhibition is reestablished when an adequate concentration of healthy, chalone-producing cells is again present.

The subject of mitotic inhibitors is receiving a good deal of attention from medically orientated scientists because of its obvious application to the treatment of tumors. Malignant cells, as you might expect, tend to be much less sensitive to mitotic inhibition by chalones and other natural inhibitors, but more sensitive to drugs that tend to have greater effects on rapidly dividing cells. Malignant cells, however, also differ from normal in many other respects as well, as we shall see in the next chapter.

SUMMARY

11-1 DNA is replicated semiconservatively in both prokaryotes and eukaryotes. In other words, parental strands are separated by a replication fork, with each strand serving as a template for a complementary daughter strand. Progeny cells each get a molecule composed of one old and one new strand. The enzymes responsible for DNA replication include unwinding proteins, enzymes capable of creating and repairing single-strand breaks that serve as a swivel ahead of the replication fork, and a family of DNA polymerases. The polymerases have in common a need for templates to determine the base sequence of the new strand, and for a primer that can be elongated at the 3' end. The polymerases thus support only $5' \rightarrow 3'$ growth. However, net growth of one of the new strands must be in the $3' \rightarrow 5'$ direction, achieved apparently by producing short $5' \rightarrow 3'$ segments, each started from an RNA primer. The short RNA primers are then replaced with DNA and the segments of DNA laced together by a DNA ligase. The system just described appears to be the same in eukaryotes, though the great length of eukaryotic DNA dictates a need for multiple independent units of replication called replicons.

Replication is a complicated process with many opportunities for errors. Most of these errors, however, are caught and corrected by a proofreading system during replication. Others can be repaired by a "cut and patch" mechanism. Still others, especially the long insertion and deletion mutants (which are more apt to occur during crossover) cannot be repaired. Uncorrected errors create a mutation. The rate of mutation can be greatly enhanced by chemical mutagens, ultraviolet light (which produces covalent pyrimidine dimers), or by ionizing radiation.

11-2 The process of cellular replication (the cell cycle) is divided into five stages: G_1; S (for DNA synthesis); G_2; mitosis (i.e., nuclear division or karyokinesis); and cell division or cytokinesis. Mitosis in higher eukaryotes consists of fragmentation of the nucleus and chromatin condensation into visible chromosomes during prophase; alignment of chromosomes at the equatorial plane of the spindle during metaphase; separation of chromosomes into individual chromatids by centromere division at anaphase; and assembly of daughter nuclei during telophase. Cytokinesis usually, but not invariably, starts during telophase.

11-3 Meiosis resembles two mitotic divisions. However, the first is a reduction division in which homologous chromosomes (instead of chromatids) are separated and incorporated into haploid daughter nuclei. Each of these haploid daughter cells undergoes a second meiotic division (with no intervening S stage) to produce a total of four haploid cells. Most of the unique features of meiosis take place during the first prophase, where synapsis and crossover (recombination) occur. This first prophase can be greatly prolonged—in human females it lasts from the fetal period to adulthood.

The haploid cells produced by meiosis either directly or after more divisions (depending on the organism) give rise to sex cells or gametes. Restoration of the diploid chromosome number comes with fertilization, or union of a male and female gamete to produce a zygote.

11-4 Spindle microtubules seem to be in dynamic equilibrium between a polymerized and depolymerized state. Their formation is catalyzed by microtubule organizing centers at kinetochores and poles, placing each kinetochore under the influence of the pole that it faces. The pull that a kinetochore experiences toward that pole will be responsible for chromosome movement after centromere division (mitosis) or separation of bivalents (first meiotic division). The force-generating mechanism is, however, controversial. It may come from microtubule disassembly ("melting") at the poles as chromosomes move) or by the sliding of two elements past each other (chromosomal fibers past either pole-to-pole fibers or past actin filaments, for example). Similarly, the increase in pole-to-pole distance at anaphase could come from microtubular assembly and elongation or by sliding microtubules.

11-5 Cytokinesis and karyokinesis are normally synchronized. Both the timing and site of cytoplasmic division seem to depend on the mitotic spindle. In animal cells, asters are probably the critical portions. At a point midway between the asters, a ring containing microfilaments forms just beneath the cell membrane. This ring then contracts, dragging cell membrane with it to create a division furrow. The construction of the contractile ring and its dependence on Ca^{2+} and ATP seem to be comparable to other cytoplasmic contractile systems. The additional cell membrane needed to accommodate two daughter cells as the ring contracts appears to come from an unfolding of surface redundancies, especially microvilli.

Cytokinesis in plants, at least in those having walls and lacking asters, takes place by cell plate formation. Coalescence of vesicles at the equatorial plane of the spindle creates a disc that grows as more vesicles fuse with it. When the cell membrane is reached by the growing disc, the two daughter cells are complete and, except for plasmodesmata, isolated from each other.

11-6 The time it takes a eukaryotic cell to make one complete trip around the cell cycle is highly variable, mostly due to changes in the amount of time spent in the G_1 stage. It appears that some kind of "checkpoint" exists in the G_1 stage, the passage of which is a random event. The probability of transition, however, can be altered, and with it the average length of time for a cell cycle.

The G_1 checkpoint seems to be an event associated with DNA replication and entry into the S stage. A comparable, though less widely used, checkpoint exists in the G_2 stage. It appears to be associated with chromatin condensation (possibly due to phosphorylation of histone H1) and hence with entry into mitosis.

While the amount of time spent at the G_1 checkpoint is highly variable, some cells seem instead to be arrested at a G_0 stage from which they make no progress through the cycle. Reactivation requires some specific event—addition of a growth factor, for instance. Some of these factors are steroid hormones or other small molecules capable of penetrating or being transported across the cell membrane. Many, however, are proteins that require a membrane receptor and an intracellular "second messenger" to mediate their mitogenic effect. Candidates for this role of second messenger include cAMP, cGMP, and Ca^{2+}. The importance of the latter is especially prominent in the blastogenic response of lymphocytes.

Growth factors are mitotic stimulators. Naturally occurring mitotic inhibitors are also known. For many cell types, mitotic inhibition is achieved by a negative feedback loop that senses the extracellular concentration of some secretory product. Often the secretory product (a hormone, for example) serves a function of general importance to the organism. In other cases, negative feedback seems to be established by a substance, called a chalone, that has no other obvious function aside from mitotic inhibition and hence regulation of cellular replication.

STUDY GUIDE

11-1 (a) What is the difference between conservative and semiconservative replication of DNA? Describe an experiment that supports the semiconservative mechanism. (b) A satisfactory model for DNA replication must explain the rapid unwinding of the parent molecule, the obligate need for a template, and the unidirectional 5' → 3' chain growth catalyzed by the known replicases. Explain how those problems are thought to be solved by current replication models. (c) What is meant by "proofreading" of replication and why is it important? (d) What is meant by "crossover"? (e) What is a mutagen? What is the mechanism of mutagenesis by UV light? by ionizing radiation? (f) What is "cut and patch" repair? (g) What is a replicon and how is it relevant to the replication of eukaryotic DNA?

11-2 (a) Name the stages of the cell cycle. Which is usually the longest stage? (b) What are the major features of each mitotic phase? (c) Describe the breakdown and reassembly of the nuclear envelope during mitosis. (d) Describe the behavior and presumed role of centrioles during mitosis. (e) Compare the mechanism of chromosome separation during karyokinesis in higher eukaryotes with the mechanism used by prokaryotes. What intermediate evolutionary stages are seen?

11-3 (a) Which phases of meiosis are the same as the corresponding mitotic phases and which are different? In what ways do they differ? (b) Summarize the events of the first meiotic prophase. (c) Define or describe the following: synapsis; synaptonemal complex; and chiasmata. (d) Summarize the steps in spermatogenesis and oogenesis. (e) Fertilization of the human ovum, if it occurs, takes place at what phase of meiosis?

11-4 (a) What do we mean by "dynamic equilibrium of the spindle fibers"? What experiments establish that concept? (b) What experiments point to microtubule disassembly as the origin of force needed for chromosome movement? (c) What observations make sliding microtubules a possible explanation for chromosome movement? Compare and contrast this role for microtubules with their presumed function in ciliary beating and axoplasmic transport. (d) Chromosome movements during anaphase are accompanied by increased separation of the poles. What is the presumed purpose and mechanism of this movement?

11-5 (a) What is a division furrow and contractile ring? (b) What influences appear to be instrumental in establishing a contractile ring? (c) If one creates two spheres out of one, additional surface would be needed. In cell division, where does this additional surface come from? (d) Describe the process of cell plate formation. (e) What are plasmodesmata and how are they formed?

11-6 (a) Define logarithmic growth. (b) What is a "cell cycle checkpoint?" (c) Describe the transition probability model for cell cycle control. (d) What events seem to be most directly dependent on passing the G_1 and G_2 checkpoints? (e) What is meant by a "G_0 cell?" Growth factor? Mitogen? (f) Describe the events leading to lymphocyte blastogenesis. (g) What are some cellular effects of estrogens and glucocorticoids? (Give specific examples.) (h) Describe mitotic regulation by negative feedback. How do chalones fit into this scheme?

REFERENCES

REPLICATION OF DNA

BOLLUM, F. J., "Mammalian DNA Polymerases." *Progr. Nucleic Acid Res. Mol. Biol.*, **15**: 109 (1975).

CAIRNS, J., "The Bacterial Chromosome." *Scientific American*, January 1966. (Offprint 1030.)

CHARGAFF, E., "Initiation of Enzymic Synthesis of Deoxyribonucleic Acid by Ribonucleic Acid Primers." *Progr. Nucleic Acid Res. Mol. Biol.*, **16**: 1 (1976).

CLARKE, C. H., "Mutagenesis and Repair in Micro-Organisms." *Sci. Progr. (Oxford)*, **62**: 559 (1975).

DRESSLER, D., "The Recent Excitement in the DNA Growing Point Problem." *Ann. Rev. Microbiol.*, **29**: 525 (1975).

EDENBERG, H., and J. HUBERMAN, "Eukaryotic Chromosome Replication." *Ann. Rev. Genet.*, **9**: 245 (1975).

GEFTER, M., "DNA Replication." *Ann. Rev. Biochem.*, **44**: 45 (1975).

GEIDER, K., "Molecular Aspects of DNA Replication in *E. coli* Systems." *Current Topics Microbiol. Immunol.*, **74**: 55 (1976).

GRAHM, DOUGLAS, "Genetic Effects of Low Level Irradiation." *BioScience*, **22**: 535 (1972).

HANAWALT, P. C., and R. H. HAYNES, "The Repair of DNA." *Scientific American*, February 1967. (Offprint 1061.)

HANAWALT, P. C., and R. B. SETLOW, eds., *Molecular Mechanisms for Repair of DNA*. New York: Plenum Press, 1975. In two volumes.

HELINSKI, D. R., "Plasmid DNA Replication." *Fed. Proc.*, **35**: 2026 (1976).

HUBERMAN, J. A., and A. D. RIGGS, "On the Mechanism of DNA Replication in Mammalian Chromosomes." *J. Mol. Biol.*, **32**: 327 (1968). Bidirectional replication in each replicon.

JACOB, F., A. RYTER, and F. CUZIN, "On the Association Between DNA and Membrane in Bacteria." *Proc. Roy. Soc. (London), Ser. B*, **164**: 267 (1966).

KOLBER, A. R., and M. KOHIYAMA, eds., *Mechanism and Regulation of DNA Replication*. New York: Plenum Press, 1974.

KONDO, S., "DNA Repair and Evolutionary Considerations." *Advan. Biophysics*, **7**: 91 (1975).

KORNBERG, A., "The Biological Synthesis of Deoxyribonucleic Acid." *Science*, **131**: 1503 (1960) and in *Nobel Lectures, Physiology or Medicine, 1942–1962*. Amsterdam: Elsevier, 1964, p. 665. Nobel Lecture, 1959.

KORNBERG, A., *DNA Synthesis*. San Francisco: Freeman, 1974.

LEHMAN, I., and D. UYEMURA, "DNA Polymerase I: Essential Replication Enzyme." *Science*, **193**: 963 (1976). A nice review.

LEHMAN, I., "DNA Ligase." *Science*, **186**: 790 (1974).

LIEBERMAN, M. W., "Approaches to the Analysis of Fidelity of DNA Repair in Mammalian Cells." *Int. Rev. Cytol.*, **45**: 1 (1976).

LINDAHL, T., "New Class of Enzymes Acting on Damaged DNA." *Nature*, **259**: 64 (1976).

MATSUSHITA, T., and H. KUBITSCHEK, "DNA Replication in Bacteria." *Advan. Microbiol Physiol.*, **12**: 247 (1975).

POLLARD, E. C., "The Biological Action of Ionizing Radiation," *Am. Scientist*, **57**: 206 (1969).

PRYOR, WM. A., "Free Radicals in Biological Systems." *Scientific American*, August 1970. (Offprint 335.)

RICHARDS, E. G., "Complementary Mispairs." *Nature*, **263**: 369 (1976). A brief review.

SAUERBIER, W., "UV Damage at the Transcriptional Level." *Advan. Radiation Biol.*, **6**: 50 (1976).

SÖDERHÄLL, S., and T. LINDAHL, "DNA Ligases of Eukaryotes." *FEBS Letters*, **67**: 1 (1976).

TAYLOR, J. H., "The Duplication of Chromosomes." *Scientific American*, June 1958. (Offprint 60.)

TAYLOR, J. H., "Units of DNA Replication in Chromosomes of Eukaryotes." *Int. Rev. Cytol.*, **37**: 1 (1974).

TOPAL, M. D., and J. R. FRESCO, "Complementary Base Pairing and the Origin of Substitution Mutations." *Nature*, **263**: 285 (1976). Also see p. 289, same issue.

WARD, J. F., "Molecular Mechanisms of Radiation-Induced Damage to Nucleic Acids." *Advan. Radiation Biol.*, **5**: 182 (1975).

WEISSBACH, A., "Vertebrate DNA Polymerases." *Cell*, **5**: 101 (1975). See also *Ann. Rev. Biochem.*, **46**: 25 (1977).

WHITSON, G. L., ed., *Concepts in Radiation Cell Biology*. New York: Academic Press, 1972.

WINTERSBERGER, E., "DNA-Dependent DNA Polymerases from Eukaryotes." *Trends Biochem. Sci.*, **2**(3): 58 (1977).

MITOSIS, MEIOSIS, AND CYTOKINESIS

ALBERSHEIM, P., "The Wall of Growing Plant Cells." *Scientific American*, **232**(4): 80 (April 1975). (Offprint 1320.)

BEAMS, H. W., and R. G. KESSEL, "Cytokinesis: A Comparative Study of Cytoplasmic Division in Animal Cells." *Am. Scientist*, **64**: 279 (1976). Nice review.

BRINKLEY, R. S., and E. STUBBLEFIELD, "Ultrastructure and Interaction of the Kinetochore and Centriole in Mitosis and Meiosis." *Advan. Cell Biol.*, **1**: 119 (1970).

FLEMMING, WALTER, "Contributions to the Knowledge of the Cell and Its Processes." Translated from an 1880 article and reprinted in *J. Cell Biol.*, **25** (1, pt 2): 3 (1965). An early description of mitosis.

FORD, E. H. R., *Human Chromosomes*. New York: Academic Press, 1973. Structure, morphology, behavior at mitosis and meiosis, abnormalities, and other topics.

FORER, A., "Local Reduction of Spindle Fiber Birefringence in Living *Nephrotoma suturalis* (Loew) Spermatocytes Induced by Ultraviolet Microbeam Irradiation." *J. Cell Biol.*, **25**(1, pt 2): 95 (1965). Spindle fiber movements.

FULLER, M. S., "Mitosis in Fungi." *Int. Rev. Cytol.*, **45**: 113 (1976).

GRAHAM, J., M. SUMNER, D. CURTIS, and C. A. PASTERNAK, "Sequence of Events in Plasma Membrane Assembly During the Cell Cycle." *Nature*, **246**: 291 (1973).

HEPLER, P., J. MCINTOSH, and S. CLELAND, "Intermicrotubule Bridges in Mitotic Spindle Apparatus." *J. Cell Biol.*, **45**: 438 (1970).

Heywood, P., and P. Magee, "Meiosis in Protists." *Bacteriol. Rev.*, **40**: 190 (1976).

Inoué, S., and R. Stephens, eds., *Molecules and Cell Movement* (Soc. Gen. Physiol. Series, Vol. 30). New York: Raven Press, 1975. Includes reviews on the spindle apparatus and mechanism of cytokinesis. Highly recommended.

John, B., "Myths and Mechanisms of Meiosis." *Chromosoma*, **54**: 295 (1976).

Kubai, D. F., "The Evolution of the Mitotic Spindle." *Int. Rev. Cytol.*, **43**: 167 (1975).

Leibowitz, P. J., and M Schaechter, "The Attachment of the Bacterial Chromosome to the Cell Membrane." *Int. Rev. Cytol.*, **41**: 1 (1975).

Mohri, H., "The Function of Tubulin in Motile Systems." *Biochem. Biophys. Acta*, **456**: 85 (1976).

Phillips, D. M., *Spermiogenesis*. New York: Academic Press, 1974. (Paperback.) Excellent micrographs.

Raikov, I., "Evolution of Macronuclear Organization." *Ann. Rev. Genet.*, **10**: 413 (1976).

Rappaport, R., "Cytokinesis in Animal Cells." *Int. Rev. Cytol.*, **31**: 169 (1971).

Rattner, J., and S. G. Phillips, "Independence of Centriole Formation and DNA Synthesis." *J. Cell Biol.*, **57**: 359 (1973). A nice summary of centriole behavior in mammalian cells. Also see *J. Cell Biol.*, **70**: 9 (1976).

Ris, H., et al., "Symposium on the Evolution of Mitosis in Eukaryotic Microorganisms." *Biosystems*, **7**(3 and 4) (1975).

Robards, H., "Plasmodesmata." *Ann. Rev. Plant Physiol.*, **26**: 13 (1975).

Salmon, E. D., "Pressure-Induced Depolymerization of Spindle Microtubules." *J. Cell Biol.*, **65**: 603 (1975).

Sanger, J. W. "Presence of Actin During Chromosomal Movement." *Proc. Nat. Acad. Sci. U.S.*, **72**: 2451 (1975). Actin is present in the spindle, localized to kinetochore-attached fibers.

Solari, A. J., "The Behavior of the XY Pair in Mammals." *Int. Rev. Cytol.*, **38**: 273 (1974).

Wettstein, R., and J. Sotelo, "The Molecular Architecture of Synaptinemal Complexes." *Advan. Cell Mol. Biol.*, **1**: 109 (1971).

Yasuzumi, G., "Electron Microscope Studies on Spermiogenesis in Various Animal Species." *Int. Rev. Cytol.*, **37**: 53 (1974).

Yeoman, M. M., ed., *Cell Division in Higher Plants*. New York: Academic Press, 1976.

CELL CYCLE CONTROLS

Baserga, R., *Multiplication and Division in Mammalian Cells*. New York: Dekker, 1976.

Berridge, M. J., "Control of Cell Division: A Unifying Hypothesis." *J. Cyclic Nucleotide Res.*, **1**: 305 (1975).

———, J. "Calcium, Cyclic Nucleotides, and Cell Division." *Soc. Exp. Biol.*, **30**: 219 (1976).

Blenkinsopp, W. K., "Cell Proliferation in the Epithelium of the Esophagus, Trachea, and Ureter in Mice." *J. Cell Sci.*, **5**: 393 (1969). Generation times from 4 to 8000 hours because of variability in G_1.

Bourne, H., P. Coffino, and G. Tomkins, "Somatic Genetic Analysis of cAMP Action: Characterization of Unresponsive Mutants." *J. Cell Physiol.*, **85**: 611 (1975). cAMP is not essential to the cell cycle.

Bradbury, E. M., "Histones, Chromatin Structure, and Control of Cell Division." *Current Topics Devel. Biol.*, **9**: 1 (1975).

Bullough, W. S., "Mitotic Control in Adult Mammalian Tissues." *Biol. Rev. (Cambridge)*, **50**: 99 (1975). Chalones.

Clarkson, B., and R. Baserga, eds., *Control of Proliferation in Animal Cells*. Cold Spring Harbor (N.Y.) Lab. Quant. Biol., 1974.

Crumpton, M., D. Allan, J. Auger, N. Green, and V. Maino, "Phytohemagglutinin-Lymphocyte Interaction." *Phil. Trans. Roy. Soc. London, Ser. B*, **272**: 173 (1975).

Cuatrecasas, P., and M. Greaves, eds., *Receptors and Recognition* (Series A, Vol. 2). New York: Halsted Press, 1977. See the article "Calcium and Cell Activation" by Comperts.

Donachie, W., K. Begg, and M. Vicente, "Cell Length, Cell Growth and Cell Division." *Nature*, **264**: 328 (1976). On bacterial systems.

Gospodarowicz, D., and J. Moran, "Growth Factors in Mammalian Cell Culture." *Ann. Rev. Biochem.*, **45**: 531 (1976).

Gross, D., "Growth Regulating Substances of Plant Origin." *Phytochemistry*, **14**: 2105 (1975).

Holley, R. W., "Control of Growth of Mammalian Cells in Cell Culture." *Nature*, **258**: 487 (1975).

Houck, J., ed., *Chalones*. New York: Elsevier/North Holland, 1976.

Johnson, J., D. Epel, and M. Paul, "Intracellular pH and Activation of Sea Urchin Eggs after Fertilization." *Nature*, **262**: 661 (1976). pH and Ca^{2+} levels increase prior to division.

Konyshev, V., "Chemical Nature and Systematization of Substances Regulating Animal Tissue Growth." *Int. Rev. Cytol.*, **47**: 195 (1976).

Krishnamoorthy, H. N., *Gibberellins and Plant Growth*. New York: Halsted Press, 1975.

Lauf, P. K., "Antigen-Antibody Reactions and Cation Transport in Biomembranes." *Biochem. Biophys. Acta*, **415**: 173 (1975). See p. 205ff for Ca^{2+} activation of lymphocytes.

Loeb, J., "Corticosteroids and Growth." *New Engl. J. Med.*, **295**: 547 (1976).

Lozzio, B., C. Lozzio, E. Bamberger, and S. Lair, "Regulators of Cell Division: Endogenous Mitotic Inhibitors of Mammalian Cells." *Int. Rev. Cytol.*, **42**: 1 (1975).

Luft, R., and K. Hall, eds., *Advances in Metabolic Disorders*, Vol. 8. New York: Academic Press, 1975. Articles on somatomedins and the various growth factors.

Marks, D., W. Paik, and T. Borun, "The Relationship of Histone Phosphorylation to Deoxyribonucleic Acid Replication and Mitosis." *J. Biol. Chem.*, **248**: 5660 (1973).

Martz, E., and M. S. Steinberg, "The Role of Cell-Cell Contact in 'Contact' Inhibition of Cell Division: A Review and New Evidence." *J. Cell. Physiol.*, **79**: 189 (1972).

Mazia, D., "The Cell Cycle." *Scientific American*, **230**(1): 54 (Jan. 1974).

NICOLSON, G. L., "The Interaction of Lectins with Animal Cell Surfaces." *Int. Rev. Cytol.*, **39**: 89 (1974). Lymphocyte activation.

ORD, M., L. STOCKEN, and S. THROWER, "Histone Phosphorylations and Their Disparate Roles in Interphase." *Sub-Cellular Biochem.*, **4**: 147 (1975). H1 and the mitotic transition.

OSBORNE, S., "Ethylene and Target Cells in the Growth of Plants." *Sci. Progr. (Oxford)*, **64**: (1977). Ethylene is a plant hormone.

PARKER, C., "Control of Lymphocyte Functions." *New Engl. J. Med.*, **295**: 1180 (1976).

PASTAN, IRA, "Regulation of Cellular Growth." *Advan. Metab. Disorders*, **8**: 377 (1975).

PRESCOTT, D. M., "The Cell Cycle and the Control of Cellular Reproduction." *Advan. Genet.*, **18**: 99 (1976). A lengthy review.

PRESCOTT, D. M., *Reproduction of Eukaryotic Cells.* New York: Academic Press, 1976.

RAO, P. N., and R. T. JOHNSON, "Induction of Chromosome Condensation in Interphase Cells." *Advan. Cell. Mol. Biol.*, **3**: 136 (1974).

REBHUN, L., "Cyclic Nucleotides, Calcium, and Cell Division." *Int. Rev. Cytol.*, **49**: 1 (1977).

ROBINSON, J., J. SMITH, N. TOTTY, and P. RIDDLE, "Transition Probability and the Hormonal Density-Dependent Regulation of Cell Proliferation." *Nature*, **262**: 298 (1976).

SHALL, S., *The Cell Cycle.* New York: Halsted Press, 1977. (Paperback.)

SLATER, M., and M. SCHAECHTER, "Control of Cell Division in Bacteria." *Bacteriol. Rev.*, **38**: 199 (1974).

TALWAR, G. P., *Regulation of Growth and Differentiated Function in Eukaryotic Cells.* New York: Raven Press, 1975.

THORNLEY, A., and E. LAURENCE, "The Present State of Biochemical Research on Chalones." *Int. J. Biochem.*, **6**: 313 (1975).

WILLINGHAM, M., "Cyclic AMP and Cell Behavior in Cultured Cells." *Int. Rev. Cytol.*, **44**: 319 (1975).

NOTES

1. Cells may be blocked at metaphase by applying microtubule-inhibiting drugs—e.g., colchicine, vinblastine, or podophyllum derivatives. Chromosomes condense to their metaphase configuration under these conditions but a spindle does not form.

2. It is also common to describe a *prometaphase* between prophase and metaphase consisting of nuclear breakdown and attachment of spindle fibers to centromeres.

3. William of Occam (or Ockham), c. 1280–1349. What he is actually supposed to have said is "Pluralities non est ponenda sine necessitate."

4. Human livers have a similar regenerative capacity. Consider, for example, Prometheus who, for having stolen fire from heaven and given it to man, was chained by Zeus to a mountain where his liver was daily devoured by a vulture only to regenerate at night. Actually, mortal humans can't regenerate their livers as fast as Prometheus, but the rate is impressive nevertheless. Chronic injury, on the other hand, as seen with alcohol abuse and certain diseases, can lead to fibrous scarring that frustrates regeneration, producing liver failure.

5. Because of reduced glucose utilization, glucocorticoids cause an increase in blood glucose levels. Thus their name—they are *glucose*-increasing *steroids* from the adrenal *cortex*.

12

Cellular Differentiation

12-1 Gene Expression in Differentiated Cells 550
Cloning of Plants and Animals
Cytoplasmic Influence on Nuclear Function
Nuclear Origin of Cytoplasmic Factors
Maternal Messages

12-2 Stability of Differentiation: Transdetermination, Regeneration, and Malignancy 555
Transdetermination
Regeneration
Tumors

Characteristics of the Malignant Cell
Tumorigenesis
Differentiation of Tumor Cells
Is Cancer Contagious?

12-3 Cellular Senescence 567
Theories of Aging
Programed Death

Summary 571
Study Guide 571
References 572
Notes 574

Differentiation in its broadest sense refers to the appearance of any new cellular property, whether it involves a change in morphology (form and structure) or in chemical composition. The induction and repression of enzymes thus represent simple kinds of differentiation. We are much more interested, however, in the complicated sequence of events by which a cell becomes something quite different from its parent.

The most dramatic examples of cellular differentiation occur during *embryogenesis*, which is the formation and development of the embryo from a single cell to a complete multicellular organism. These and less dramatic changes are due to the induction and repression of genes—that is, *epigenetic changes*. This chapter is not concerned with embryogenesis as such, but with the way genetic programing is changed during differentiation; with the stability of those changes, using as examples limb regeneration in amphibia and the transformation to malignancy; and with the final steps in cellular development, namely old age and death.

12-1 GENE EXPRESSION IN DIFFERENTIATED CELLS

One of the earliest theories of cellular differentiation was proposed in 1892 by August Weismann, who suggested that differentiation is due to a progressive loss of genes. That is, the chromosomes become less complete as the cell becomes more highly specialized, with the particular developmental pathway being determined by which genes are lost. This suggestion seemed at first to be quite reasonable, for we know that different genes are expressed in different cell types, resulting in varying enzyme patterns and physical properties. It is now perfectly clear, however, that an identical set of genes is found in almost every somatic cell of most organisms—that is, in all but germ cells. Differentiation must therefore be the result of changes in the pattern of gene expression—i.e., epigenetic changes—and this, of course, implies the existence of specific repressors.

Cloning of Plants and Animals. That even highly differentiated cells contain a full set of genes has been dramatically demonstrated both for animals and for plants. In 1964 F. C. Steward and his colleagues grew entire carrot plants from cultures containing individual cells and small cell aggregates taken from the phloem tissue of a carrot root. The next year V. Vasil and A. C. Hildebrandt isolated single cells from a tissue culture of tobacco cells and grew complete plants from them. In both cases normal, viable seed was produced by the plants. It is clear, therefore, that each of the highly differentiated cells from which these plants were started contained all the genes needed for every cell type found in the mature plant. In fact, each cell may have contained more than enough genes, for in 1969 Nitsch grew a complete haploid tobacco plant from a haploid spore, showing that only one gene from each homologous pair is needed to cause normal differentiation.

Though the cloning of plants (that is, the asexual formation of a line of descendants genetically identical to the original) was a spectacular achievement, the parallel experiment carried out with frogs was even more impressive because of the greater complexity of that organism. J. B. Gurdon and his colleagues at Oxford succeeded in cloning frogs in the mid-1960s. They used intestinal cells taken from tadpoles of the South African clawed frog, *Xenopus laevis*, and implanted the cells in unfertilized eggs from which the nucleus had been removed mechanically or destroyed by ultraviolet light. Some of these hybrid cells developed into fertile adult frogs (Fig. 12-1). In order to be sure that development was controlled by the transplanted nucleus rather than by the original one, two strains of frog were used. One contained the normal complement of two nucleoli per nucleus, while the other strain was a mutant with only one nucleolus per nucleus. The resulting adult always reflected the nucleolar arrangement of the strain that donated the nucleus to the hybrid, not the strain that donated the egg.

The ability to produce a complete frog from the nucleus of an already differentiated cell created a great deal of interest, both theoretical and practical. On the practical side, embryologists could now foresee producing clones of genetically identical domestic animals, thus preserving particularly valuable traits instead of taking chances with the genetic mixing inherent in sexual reproduction. (Extended to humans, this raises the possibility of a woman giving birth to her own—or to her husband's—identical twin.) On the theoretical side, nuclear transplant experiments have satisfied most scientists that differentiation is, indeed, the result of a particular pattern of gene expression in a cell that carries the full complement characteristic of the species. The question, then, is why one set of genes is expressed in one type of differentiated cell and quite a different set in another cell.

CHAPTER 12
Cellular Differentiation

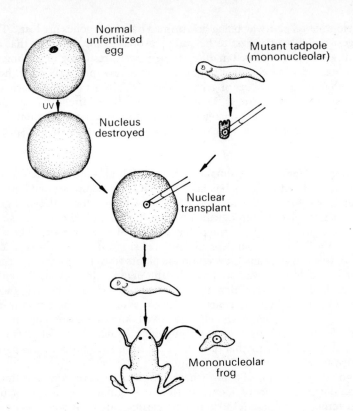

FIGURE 12-1 The Multipotential Nucleus. Differentiation does not ordinarily involve the loss of genes. That is demonstrated by the experiment described here, in which the nucleus of a normal, unfertilized egg is destroyed and replaced by the nucleus of a differentiated intestinal epithelial cell. The latter was taken from a mutant strain of tadpoles having but one nucleolus per diploid nucleus. Some of the hybrids developed into complete frogs, but always with the mutant nucleolar count. [Described by J. B. Gurdon and V. Uehlinger, *Nature*, **210**: 1240 (1966).]

Cytoplasmic Influence on Nuclear Function. The full genetic potential of the transplanted tadpole nucleus was realized only when the latter was introduced into the cytoplasm of an unfertilized egg. Apparently, cytoplasmic factors can affect the function of a nucleus. Experiments in which specialized and unspecialized cells from the same animal are fused tend to support that thesis, for the hybrid cells are generally without specialized function. It is possible, therefore, that differentiation is achieved by loss of cytoplasmic factors supplied by the unspecialized cell.

This conclusion might also be reached from experiments reported by Henry Harris of Oxford. Bird erythrocytes retain shrunken, silent nuclei in the mature state. During differentiation of these cells, genes are sequentially turned off. But when bird erythrocytes are fused with human HeLa cells, a poorly differentiated tumor strain, reactivation of formerly repressed genes is seen in the bird nuclei.

Other examples of cytoplasmic control over nuclear activity come from work with a single-celled alga of the genus *Acetabularia*, starting with experiments by J. Hämmerling in the 1930s. This graceful tropical seaweed grows to a height of several centimeters, with a funnellike cap on one end of a long slender stalk and an anchoring organ, the rhizoid, on the other end (Fig. 12-2). Each plant is a single cell, with a single nucleus located in the rhizoid, large enough to be visible to the naked eye. At the time of cap formation the nucleus becomes greatly enlarged, but when the cap is cut off a cell at this point, the nucleus immediately shrinks back to its former compact size (Fig. 12-3g). If the mature cap is

FIGURE 12-2 *Acetabularia mediterranea*. Each plant is a single cell, up to several centimeters in length. Occasionally one can see a rhizoid (attachment organ, arrow) on the end of the stem. This is where the nucleus is found. (From the Turtox Collection, courtesy of General Biologicals, Inc.)

transplanted to a young cell, the nucleus of the young cell undergoes the meiotic divisions necessary to create gametes (Fig. 12-3h). In both cases, the cytoplasm clearly affects the activity of the nucleus. Unfortunately, however, the nature of the gene regulators that mediate this control remains unknown.

Nuclear Origin of Cytoplasmic Factors. Although it is clear that differentiation involves a reciprocal interaction between nucleus and cytoplasm, it is the nucleus that ultimately makes possible differentiated function in *Acetabularia* as in other organisms.

When the stalk of a mature plant that is about to undergo spore formation is removed, the cap grows only a new stalk while the rhizoid regenerates a complete new plant. (The isolated stalk itself shows no growth.) The information necessary to produce a complete new cell is apparently confined to the nuclear region (Fig. 12-4, left). In addition, when the nucleus of one species is grafted to the severed stalk of a second species, a cap grows that is always characteristic of the species to which the nucleus belonged.

Apparently the nucleus controls differentiation through the manufacture and release of some diffusible substance that we might expect to be a messenger RNA. There are several kinds of experiments that are consistent with that supposition: (1) If the severed cap of a mature *Acetabularia* cell is treated with UV light or ribonuclease to destroy RNA, the expected regeneration of stalk does not occur. (2) If the stalk and rhizoid are left intact for several days after severing the cap, the stalk develops the ability to grow a new cap, independent of the rhizoid (Fig. 12-4, right). However, if actinomycin D (which prevents RNA synthesis) is injected into the rhizoid at the time of cap removal, regeneration of the cap is prevented. (3) And finally, if the nucleus is removed from a young plant before cap formation, the enucleated plant may still form its cap—though, of course, no spores can be produced. Since cap formation is accompanied by the synthesis of several new proteins, the nucleus must produce some long-lived, stable, but inactive messenger RNAs that are

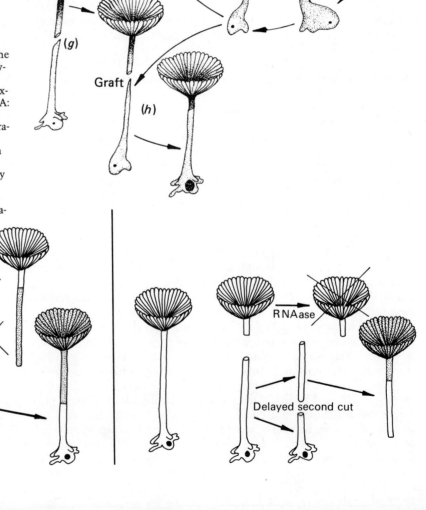

FIGURE 12-3 Cytoplasmic Control of Nuclear Function in *Acetabularia*. The life cycle of *Acetabularia* begins with (a) the fusion of two gametes to form (b) a zygote, which (c) grows a stalk and (d) a cap. (This process, (a) to (d) takes about two months.) (e) Maturation is complete when the nucleus undergoes multiple divisions to produce new haploid gametes. The latter migrate into the cap and are packaged and released in cysts (f) to start a new cycle. See text for an explanation of the grafting experiment, also diagramed. [Described by J. Haemmerling, *Ann. Rev. Plant Physiol.*, **14**: 65 (1963).]

FIGURE 12-4 Nuclear Control of Cytoplasmic Function in *Acetabularia*. LEFT: A severed stalk has no regenerative capacity, a severed cap only a limited amount. However, the nuclear-containing rhizoid, when severed, can regenerate a complete new plant. RIGHT: Nuclear control is exercised through the synthesis of RNA: if a severed cap is treated with ribonuclease or UV light, no stalk regeneration occurs; and a stalk that is allowed to remain on the rhizoid for a time after cap removal develops the ability to grow a new cap, apparently through the appearance of nuclear-derived RNA. (See J. Brachet, *Biochemical Cytology*. New York: Academic Press, 1957.)

554

translated long after they are made. This phenomenon is quite general in differentiating systems, both animal and plant.

Maternal Messages. Newly made RNA in a fertilized frog egg remains untranslated up to approximately the midblastula stage (about 10 cell divisions). Hence, fertilized frog eggs that have been treated with actinomycin D to block RNA synthesis generally undergo division, but stop developing at the blastula stage. The same developmental stage is reached in frog eggs that contain a nucleus transplanted from another species, or that have had their nucleus removed before being artificially stimulated to develop. Apparently the embryo soon reaches a point at which new gene products become necessary.

Even though translation of newly made mRNA during early growth is limited, many new proteins appear during this time. Those that are not the result of postfertilization mRNA synthesis must have been formed from mRNA that was present at fertilization. These latter RNAs are the *maternal messages* that are made during oogenesis and deposited in the cytoplasm as inactive mRNA-ribosome complexes. The activation of maternal messages, then, controls the earliest events of development, at least to about the blastula stage when proteins produced by newly synthesized mRNA become essential.

The maternal messages are probably not apportioned equally when the egg divides. Hence, as the messages are activated they cause different gene programs to be established in different cells. Some of these cells become the key to further development by virtue of their ability to affect the nuclear programing of other cells. They do so by the production of embryonic *inducers*. The change that takes place in recipient cells that receive those inducers is a process called *determination*. The location of an undetermined cell in the embryo when inducers are made establishes which inducers it receives and hence its future. The determined cell can then be transplanted to another location in the embryo without change in the form or function of its descendants. Determination could well be the activation of a single gene that initiates a cascade of later gene activations. The changes, spread over many generations, gradually lead to a differentiated terminal cell type.

12-2 STABILITY OF DIFFERENTIATION: TRANSDETERMINATION, REGENERATION, AND MALIGNANCY

The picture developed so far is one in which cells become determined, establishing at that time the nature of their differentiated progeny. We have seen how nuclei from differentiated cells can give rise to a variety of other differentiated cell types under the proper conditions. The genetic programing leading to differentiation is therefore not irreversible. In this section, the stability of determination and differentiation will be further examined from three aspects: (1) the stability of determination itself, as deduced from experiments with imaginal discs of the fruit fly, *Drosophila melanogaster*; (2) the stability of differentiated tissue during limb regeneration in amphibians; and (3) the loss of differentiated functions (and the gain of new properties) seen in the growth of plant and animal tumors.

Transdetermination. The fruit fly, *Drosophila melanogaster*, like other insects goes through four distinct phases during its life cycle: embryo, larva, pupa, and adult (also called an *imago*). During pupation, most of the differentiated cells of the larva dissolve and their components are utilized to form adult structures (Fig. 12-5). However, the adult cells themselves are descendants of embryonic tissue carried by the larva, not of the mature cells found therein. This embryonic tissue is present in discrete bundles called *imaginal discs*. Although proliferation of these cells takes place during the larval stage, their differentiation does not begin until it is triggered by hormones produced only during pupation.

One can demonstrate that cells of the imaginal discs are already determined in the larva by removing them from their host and placing them in the abdominal cavity of a second larva. When the new host undergoes pupation, both its own and the transplanted imaginal discs differentiate. The result is a fly with accessory organs. A transplanted leg disc, for example, produces leg structures in the abdomen of the host; an eye disc produces eye tissue in the host, and so on. In fact, it can be shown that the cells of a particular disc may be individually determined—that is, certain cells in the leg disc will form hairs, others muscle, etc.

When imaginal discs are transplanted from a larva to the abdomen of an adult, the cells proliferate but, due to the lack of proper hormones, they do not differentiate. By transplantation of a portion of this tissue from one fly to another, the cells can be kept viable for many generations. Their capacity to differentiate into mature structures is retained during these transfers, as may be shown by transferring a portion back to a larva about to undergo pupation.

Although determination is thus seen to be a condition that can be passed from a cell to its progeny, structures formed from the transplanted cells do occasionally change. In investigating this phenomenon, Ernst Hadorn found that *transdetermination* of the various cells in a genital disc results in a new pattern of development according to the following outline:

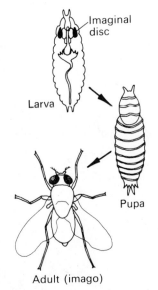

FIGURE 12-5 *Drosophila*. Most of the cells of the larva break down during pupation. Adult structures emerge from determined but undifferentiated larval cells found in clusters called imaginal discs.

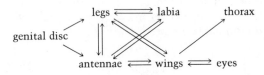

The arrows indicate the direction in which transdetermination can occur. For example, after a change in determination from presumptive genitalia to antennae, a further change can produce labia, legs, or wings, but not thorax or eye tissue. The latter, however, can arise from cells that have already undergone the transition from presumptive antennae to wings. Hadorn also found the probability that one of these transitions will take place increases with the number of cell divisions that occur before the cells are allowed to differentiate.

Transdetermination produces a new genetic programing, changing the capacity of the cells to differentiate. Determination is normally a relatively stable condition—that is, the state of determination is normally passed on to progeny as the cells proliferate. When transdetermination does occur, it follows only preset pathways, suggesting a certain rigidity, or pattern, to the way in which genes can be turned on and off. This restriction is consistent with our view of the developmental process as the sequential activation of genes, with only a restricted number of possible variations in order.

Regeneration. While transdetermination gives us some idea of both the stability and capacity for change present in determined cells, it tells us nothing about the corresponding capacity for change in already differentiated tissue. We normally think of differentiation as a one-way process. And it is true that the most highly specialized cells (e.g., nerve and muscle) are also the most stable. Since DNA replication and cell division seem to be prerequisites to change in the determined or differentiated condition, the stability of nerve and muscle probably arises from the fact that these cells rarely, if ever, divide. This is an important consideration, for a mitotic accident in highly specialized cells is apt to be fatal for the animal. One way to prevent such accidents is to design cells so that replication is not necessary, and that is apparently what has evolved.[1]

However, many fully differentiated cells *can* undergo mitosis, and the loss and regain of differentiated function can and do occur in certain instances. For example, when the limb of a newt is amputated, epithelium closes over the wound in a matter of hours. But underneath this seal a mound of embryonic cells called a *blastema* forms. The cells of the blastema proliferate and differentiate into a new limb just like the one that was lost (see Fig. 12-6). The trigger and control for this process apparently involve the presence of nerve axons that regrow from the still viable nerve cell bodies in or near the central nervous system, for if innervation is prevented, regeneration does not take place and scar tissue forms instead. (Innervation is not essential to the original, embryonic formation of a limb, however.)

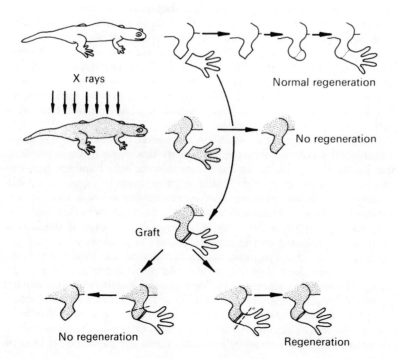

FIGURE 12-6 Regeneration. When a newt limb is amputated, embryonic cells (a regeneration blastema) appear at the site of injury and, in a few weeks, differentiate to produce a new limb. The cells of the blastema are locally derived, as was demonstrated by grafting a limb to an irradiated newt. Whether or not one sees regeneration after a second amputation depends on whether or not unirradiated tissue is left after the second cut. [Described by E. G. Butler, *Anat. Record*, **62**: 295 (1935).]

If a newt is heavily irradiated with X rays before amputation, regeneration is prevented. Yet if a new limb is grafted to the irradiated animal and then subsequently amputated, a regeneration blastema may form depending on where the cut is made. If all of the grafted tissue is removed during amputation, no new growth occurs; however, if even a small amount of grafted tissue is left behind at the second amputation, blastema formation and regeneration take place normally. This and other experiments lead to the conclusion that the embryonic cells of which the blastema is composed are locally derived from dedifferentiation of mature structures, including muscle, skin, etc.

Thus, like determination, differentiation itself is a process that can sometimes be reversed. In addition, it is clear that the pattern of inductions and determinations that lead to development is not a unique feature of embryogenesis or metamorphosis, but can be initiated in mature tissue.[2]

Tumors. The ability of cells to lose their differentiated properties during limb regeneration can be a useful feature in those animals that have it. In other cases, however, a change in the state of differentiation may have drastic consequences, as with the formation of malignant tumors.

A tumor is a *neoplasm*, or abnormal growth of new tissue. Such growths can arise anywhere in the body and may be quite *benign*, or harmless (e.g., the common wart). On the other hand, a tumor may become malignant, in which case it invades and destroys the supporting tissue on which it grows, eventually causing the death of the host. In addition, a malignant tumor, or *cancer*, may slough cells that are carried to other parts of the body where they establish new colonies of malignant cells. This process, called *metastasis*, spreads the tumor and hastens the time when a vital function will be interrupted by one of the growths. It also makes surgical removal and other therapies less effective.

Cancers are named according to the tissues from which they arise, usually with the suffix "-oma," added. Most common are the *carcinomas*, derived from epithelial cells, and the *sarcomas*, which originate in connective tissue. In addition, one may see reference to *lymphomas* (from lymphatic tissue), *lipomas* (from adipose, or fatty, tissue), and so on. However, not all malignancies result in solid tumors. *Leukemia*, for example, is a malignant disease involving the proliferation of leukocytes, or white blood cells.

Cancer has received a great deal of attention from biological and medical scientists not only because the disease is a major medical problem, but because it represents an aberration in some very basic mechanisms involving the regulation of cellular activity, including growth and differentiation. If we knew exactly how the replication of normal cells is controlled, we would have some important clues as to why this control is not exercised in malignant cells. Conversely, if we knew in detail how cells were converted to the malignant state (a process called *transformation*), it would tell us important facts concerning normal cell control.

Our objective here is to use some of the observations made on cancer cells to further illuminate the problem at hand, namely the mechanisms involved in achieving and maintaining a differentiated state. Cancer cells have failed in one or the other of these tasks. The failure could be due to the loss or gain of one or more genes, or it could be the result of changes in the way normal cellular genes are expressed—i.e., epigenetic changes.

Characteristics of the Malignant Cell. Let us start by describing some of the more important characteristics of the malignant cell (see Fig. 12-7). Although these characteristics vary somewhat from one cancer to another, and between various stages in the same disease, one of the earliest and most outstanding features is the capacity of the malignant cell for autonomous growth. That is, the cells exist and reproduce independently of normal regulatory mechanisms. (Malignant cells grow no faster than normal cells; they just don't know when to stop.) Other changes include a loss of contact inhibition and a decline in the selective adhesiveness of the cells. Whereas mixing cells from two different tissues usually results in reassociation each to its own kind, cancer cells adhere indiscriminately to all other cell types present.

The loss of contact inhibition and lack of discrimination in adhesiveness imply a change in the surface properties of the malignant cell and an accompanying breakdown in intercellular communication. Some of these same changes, including stimulation to divide, can be temporarily induced in normal cells by treatment with proteolytic enzymes that damage surface proteins. This observation is particularly interesting in view of the fact that many malignant cells, but not their normal counterparts, secrete proteolytic enzymes. Surface changes of this kind are thought by some investigators to be the key to the behavior of malignant

SECTION 12-2
Stability of Differentiation

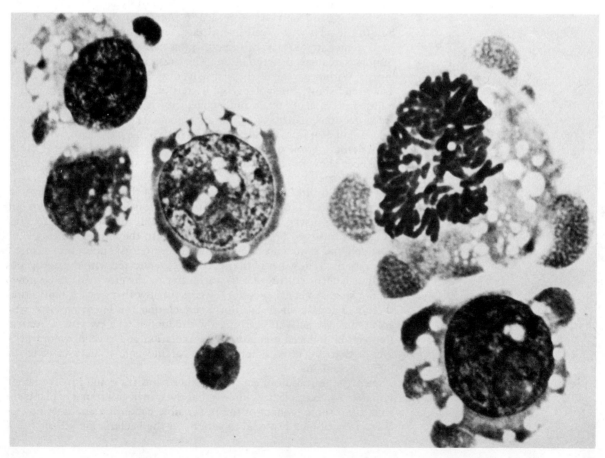

FIGURE 12-7 Cancer Cells. From a mouse mammary tumor grown *in vitro*. Note the large nucleus, the many chromosomes in the mitotic cell, and the outgrowths, or blebs, on the cell surface. (Photo by A. Rivenson. From W. Etkin, R. Devlin, and T. Bouffard, *A Biology of Human Concern*. Philadelphia: Lippincott, 1972.)

cells. That may be the case, but surface changes are only secondary effects of changes in gene expression, for it is the genes that determine the composition and properties of the cell surface.

Some of the characteristics of cancer cells, namely their ability to invade the supporting tissues on which they grow and their tendency to metastasize, are present to some extent in all cancers but normally appear later in the disease and may vary in degree. Even the capacity for autonomous growth, which is the most basic feature of any tumor, is sometimes conditional. For example, a young prostate or breast tumor may grow only as long as the proper hormonal stimulus is present, although it often becomes independent of hormonal control in its more advanced stages. In addition to maintaining vestiges of their former sensitivity to normal growth controls, many tumors retain some of the differentiated properties of the normal tissue from which they were derived (i.e., they undergo only a "partial dedifferentiation"). Thus, tumors growing out of an endocrine gland may continue to secrete hormones, leading to gross aberrations in the normal functioning of the body. On the other hand, tumors may also develop differentiated features of other cell types, such as the secretion of hormones not ordinarily made by the tissue from which the tumor is derived.

Tumorigenesis. The search for the causes of all these changes in malignant cells has been complicated by what appears to be a multiplicity of unrelated mechanisms. In the first decade of this century, V. Ellermann and O. Bang of Denmark succeeded in transferring leukemia from one bird to the next with a cell-free filtrate. At about the same time, P. F. Rous demonstrated that cell-free filtrates can also transmit some solid tumors (sarcomas) between birds. The causative agent in both cases is a virus. Within a year of these discoveries, Erwin F. Smith and C. O. Townsend showed that a specific bacterium, not a virus, is responsible for the initiation of a plant tumor called crown gall disease. And we now recognize that malignancies can arise not only from certain infectious agents, but also from a wide variety of radiations, chemicals, and physical irritants. Some plastics, and even silicon and stainless steel films, have been shown to elicit tumors in laboratory animals when implanted in their tissues.

It is hard to find a common thread in all of these multiple causes of tumors. If there is one—and many investigators believe there is not—it is probably viruses. We know that it is possible for the genetic material of some viruses to become incorporated into the chromosomes of the host cell and to remain there in a latent (provirus) state through many generations. The genes of the provirus get replicated with the host genes so that an entire culture of cells, each carrying the provirus, can be grown from a single infected ancestor. We also know that at each replication there is a certain small but finite chance that the latent provirus will spontaneously activate, causing the production of new virus. Activation can be induced prematurely by radiation and certain chemicals— agents that are often demonstrably carcinogenic (cancer causing) in higher organisms.

Even the initiation of cancer by other parasites could involve latent viruses. For example, it is known that the crown gall tumor of plants results from some product of the responsible bacterium and does not require a continuing bacterial infection. If the bacteria are killed after a few days of infection, before any demonstrable malignant transformation takes place, the tumor may still develop. This, of course, sounds very much like a chemical carcinogenesis.

Although it is probably incorrect to think that viruses are involved in all cancers, the discovery of cancer viruses has inspired more effort and more hope (because of the possibility of immunization) than any other aspect of tumorigenesis. But while one can demonstrate that leukemias and a wide range of solid tumors can be initiated in animals by infection with the proper virus, it is very difficult to prove the participation of a virus in a tumor that develops in response to some other stimulus. If the mode of action is primarily through a provirus state, it would not be necessary to suppose that fully developed viruses are present at any stage after infection first takes place.

In fact, Robert Huebner and George Todaro of the National Cancer Institute proposed in 1969 that an oncogenic (cancer causing) provirus exists in all normal cells, a result of vertical transmission from parent to child over countless generations since the initial infection. The viral genes, according to this theory, are normally repressed by genetic regulators. One of the viral genes, the *oncogene*, produces a substance that can transform a cell to the malignant state. Cancer is the result of the accidental derepression of this gene. Viral replication, on the other hand, requires that the entire provirus be derepressed. According to some reports, viruses have in fact been isolated from cell lines with no known prior exposure to the virus.

The oncogene hypothesis was proposed to explain by a single model most of the observations made concerning tumorigenesis. Other models have been proposed and modifications of the oncogene theory have been suggested, but the assumption of a latent virus remains central to much of the current research on cancer. Work is proceeding by studying in detail the mode of action of cancer viruses already identified, in the hope of gaining clues that will point to common mechanisms and to possible points where therapy might interrupt the process.

The most widely studied tumor viruses are the avian and murine (bird and mouse, respectively) leukemia viruses; the Rous sarcoma virus (RSV) mentioned above; and a pair of very similar viruses that cause solid tumors in animals, SV40 (for Simian Virus 40) and polyoma. These latter two viruses are unusual in their ability to infect a number of different animals, though not all hosts are equally susceptible to a neoplastic transformation by them. Polyoma and SV40 contain only enough DNA to code for 5–10 proteins, about half of which are essential in the establishment of malignancy.

Polyoma and SV40 multiply in the nucleus of an infected cell. However, even in a susceptible host there is only about a 1% chance that an infected cell will respond with immediate replication of the virus and subsequent death of the cell to release new viral particles. The chance of producing a malignant cell by infection is even lower, or about one in a hundred thousand. In most cases, the infecting virus just seems to "disappear." Although it is an unusual response, transformation is of course the process that interests us most.

Cells may also be transformed by certain RNA viruses such as RSV or the leukemia viruses (see Fig. 12-8). This observation immediately raises the question of how an RNA virus might function from a provirus state since there is no reason to believe that RNA can incorporate

CHAPTER 12
Cellular Differentiation

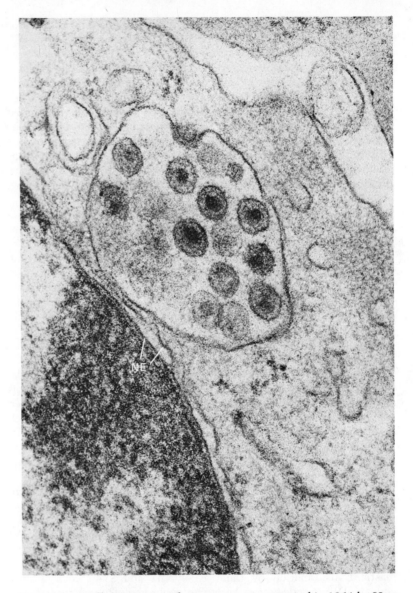

FIGURE 12-8 Formation of Avian Leukemia Virus. The viral configuration you see here is called a *type C particle*. The micrograph illustrates type C particles budding into a vacuole (*arrow*) just as they do from the cell membrane (see also Fig. 1-34). Note the nearby nucleus with its double-layered envelope (NE) and darkly stained chromatin. [Courtesy of Z. Mladenov, U. Heine, D. Beard, and J. W. Beard, *J. Nat. Cancer Inst.*, **38**: 251 (1967).]

directly into cellular DNA. The answer was suggested in 1964 by Howard Temin, who shared the 1975 Nobel Prize with fellow cancer virologists Renato Dulbecco and David Baltimore. Temin proposed, and with others later proved, that certain virus-specific enzymes exist that can polymerize DNA complementary to an RNA template. With these RNA-dependent DNA polymerases, or *reverse transcriptases*, one can postulate that all oncogenic viruses insert new genetic information into the chromosome of the host. The added genes may result in altered cellular properties (such as a loss of contact inhibition), either directly or by interfering with the normal scheme of gene expression in the host.

In the case of RSV, a single gene (the *src*—for sarcoma—gene) seems to be the key to malignant transformation. The product of this gene is thought to be a protein essential for the malignant change (see Fig. 12-9). Without the *src* gene, RSV can multiply but cannot transform its host.

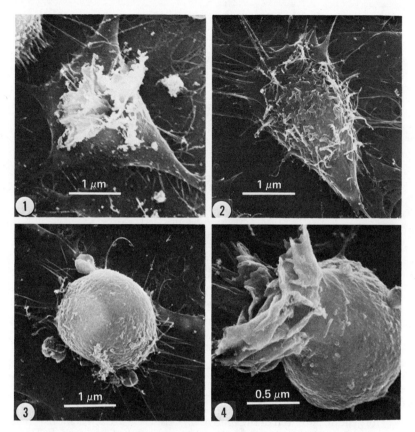

FIGURE 12-9 Transformation of a Normal Cell into a Cancer Cell. A temperature-sensitive *src* mutant of Rous sarcoma virus (RSV) causes transformation of chick cells at 37 °C but not at 41 °C. The micrographs show changes at various times in the course of malignant transformation following downshift from 41 °C to 37 °C, presumably the result of activation of a critical virus-specified protein. (1) At one hour, showing flowerlike ruffles in the cell membrane. (2) At two hours, long retraction fibers appear along with numerous microvilli. (3) By 24 hours the original spindle-shaped cells are converted to spherical cells. Note the surface blebs. Viruses bud from the cell surface at this stage and occasionally new ruffles appear as in photo (4). [Courtesy of E. Wang and A. Goldberg, *Proc. Nat. Acad. Sci. U.S.*, **73**: 4065 (1976).]

Malignancy, then, may result from loss of the associated repressor. *Src* thus fits the general concept of an oncogene.

This model of a repressed oncogene can be tested by fusing normal and malignant cells, for malignancy would thus be a recessive condition capable of being prevented by cytoplasmic components supplied by the normal cell. In fact, when fusions between normal and malignant cells from a variety of sources are carried out, the hybrids, as predicted, behave as normal cells. As the hybrids grow, however, chromosomes are lost. When a specific chromosome is lost, malignancy reappears, presumably due to loss of capacity to make the oncogene repressor.

Differentiation of Tumor Cells. The nuclei of tumor cells may have a full set of normal genes in addition to whatever viral genes they harbor, if any. This condition has been demonstrated by transplanting nuclei (see Fig. 12-10) and by causing the tumor cells themselves to differentiate. For example, Armin C. Braun of Rockefeller University succeeded in causing complete new plants to develop from single cells taken from a portion of a crown gall tumor. He proceeded by forming a clone of the malignant cell and then grafting a small amount to the cut stem of a tobacco plant tip. Under the influence of the host plant, the grafted tissue regained a portion of its ability to become mature structures. When some of this still-abnormal growth was grafted to a second tobacco plant

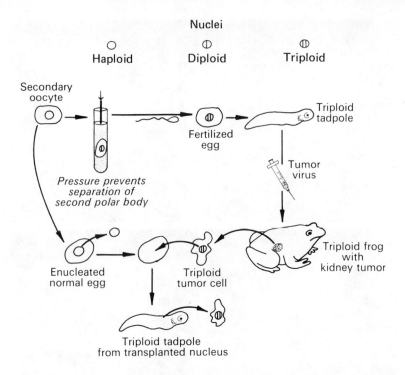

FIGURE 12-10 The Developmental Potential of a Tumor Nucleus. Hydrostatic pressure was used to produce a diploid egg which, after fertilization, became a triploid frog. Nuclear transplantion from a cancer cell of this frog to an enucleated but otherwise normal egg produced a triploid tadpole, thus demonstrating the developmental potential of the tumor nucleus. [Described by R. McKinnel, B. Deggins, and D. Labat, *Science*, 165: 394 (1969).]

even more function was regained, culminating in the production of seeds from which perfectly normal, healthy plants were grown.

The genetic potential of the malignant cell has also been demonstrated with amphibian regeneration. When a fragment of the Lucké adenocarcinoma (an adenocarcinoma is a cancer of gland tissue) is transplanted from the leopard frog, *Rana pipiens*, to the regeneration blastema of an adult newt amputee, the tumor cells differentiate into muscle, cartilage, and connective tissue as the new limb forms. To establish that tumor cells themselves are involved in the differentiation, M. Mizell in 1965 amputated a newt tail by cutting through a transplanted tumor already growing in the region. The tumor cells had been heavily labeled with tritium before transplantation, so autoradiographs of the regenerating tissue could be used to determine the fate of the tumor nuclei. In this way he left no doubt that tumor cells had been influenced to differentiate into a variety of normal tissues.

It is more difficult to demonstrate the genetic potential of malignant cells in mammals because of their relative lack of regenerative power. However, differentiated (and therefore nonmalignant) cell types are often associated with, and derived from, certain kinds of malignancies. For example, teratocarcinoma, which is a cancer derived from embryonic tissue, commonly has associated with it differentiated cells of many types (muscle, teeth, hair, fat, etc.). Clones derived from single teratocarcinoma cells show this same pluripotential capacity when transplanted to susceptible hosts.

Spontaneous reversion of malignant cells to mature cell types is seen with many other tumors. A cancer that has received much attention because of this capacity is neuroblastoma, which is one of the more common cancers of small children. (Neuroblasts are precursors of ma-

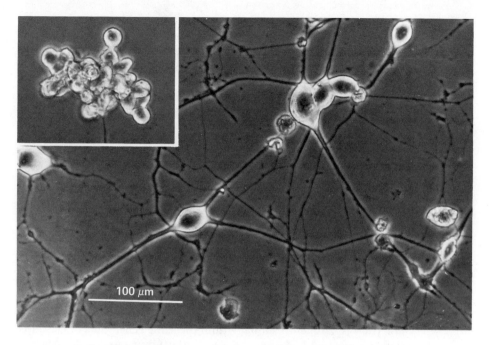

ture neurons.) When grown in culture, neuroblastoma cells can be caused to revert to cells having many characteristics of mature neurons, including the production of neurotransmitters and formation of cell-cell contacts (see Fig. 12-11). This system is particularly interesting since the changes are completely reversible—that is, cells *in vitro* can be cycled back and forth between differentiated and undifferentiated states by manipulation of the culture medium—and because neurite elongation does not seem to require new RNA. It appears that the genes conferring differentiated properties are always on in neuroblastoma, but that some step subsequent to RNA synthesis is blocked when the differentiated state is not expressed.

Although it seems certain that malignancies can be due to mutation of genes, it is thus probable that at least some cancers represent failures in the genetic regulatory mechanisms that govern normal differentiation, and are not the result of gene loss or mutation. The frequent reference to a tumor as "dedifferentiated" tissue, or tissue that has reverted to the embryonic state, may be an oversimplified interpretation. However, to the extent that malignancy can be attributed to epigenetic changes rather than to mutation, the description is fairly apt. In support of this interpretation, there have been reports of an immunological similarity between fetal antigens of unknown origin and new antigens (carcinoembryonic antigens) appearing with certain cancers.

These views have led to research on the possibility of curing cancers not by killing the malignant cells, but by causing them to differentiate to mature, benign cell types. Unfortunately, progress is impeded by an incomplete understanding of the mechanisms by which differentiation is achieved.

Is Cancer Contagious? The question is not directly related to cell biology, but the subject deserves consideration. Since cancer is the second most common cause of death in the United States, accounting by itself

FIGURE 12-11 Neuroblastoma. Phase contrast micrographs of neuroblastoma (clone N18), caused to differentiate *in vitro* by lowering the serum concentration in the culture medium. *INSET* shows undifferentiated neuroblastoma cells. Note the round perikaryon without neurites. (See also Fig. 8-1b.) [Courtesy of J. Morgan and N. W. Seeds, *J. Cell Biol.*, **67**: 136 (1975).]

for about 20% of the total,[3] it has become a much-feared disease. Its victims typically feel frightened, depressed, and alone. Just at this time in their lives when contact with other humans is so important, they are too often shunned by friends who worry about "catching" the disease. This tragedy might be averted if people understood that there is *absolutely no evidence* for person-to-person transmission of any human malignancy in spite of much careful research into the question.

If cancer were a contagious disease physicians, for example, would be expected to have an increased risk of contracting it. In fact, except for those who have had unusual exposure to X rays (e.g., radiologists before they became aware of the problem), physicians die from cancer at about the same rate as everyone else, and from some types of cancers (e.g., lung cancer) even less often. Spouses of cancer victims would also be expected to run a greatly increased risk, but the very small increase they actually experience is much easier to explain on the basis of common exposure to environmental carcinogens.

We know that viruses are involved in many animal cancers and there is no reason to believe humans are unique in this respect. Although some types of human cancers probably are caused by viruses, making them an infectious disease, an infectious disease is not necessarily a contagious disease. Even with known tumor viruses, horizontal transmission (that is, animal to animal within the same generation) of cancer is very rare among laboratory animals. Vertical transmission, however, is common (see Fig. 12-12).

In experimental studies with tumor viruses, animals typically have to receive very large doses of virus by direct injection in order to contract the disease, and even then only newborn animals (with immature immune systems) are usually susceptible. Among the exceptions is the feline leukemia virus, which can be spread from cat to cat. Even in this case, a healthy animal living in the same household as an infected one runs no more than one chance in five of contracting the disease in a three-month period according to one study. In fact, if human cancers were readily passed from person to person, the disease would be less of a problem than it is, for identification of viruses and the production of vaccines would become possible. (Vaccines are already being used for Marek's disease of chickens, a kind of leukemia, and fibropapillomatosis of cattle.)

The situation for most cancers is complicated in that responsible viruses are much more widespread than the cancers themselves. The virus most consistently implicated in human malignancy is the Epstein-Barr virus (EBV), which belongs to the herpes group (as does Marek's virus). EBV is associated with Burkitt's lymphoma, a malignancy of B-type lymphocytes in children that is common only in certain parts of Africa. It is also associated with nasopharyngeal carcinoma, a rare tumor of the back of the nose. EBV definitely transforms B-type lymphocytes in culture, is directly oncogenic for monkeys, and EBV genes are consistently found in cancer cells of the above two types. However, EBV is an exceedingly common virus—in areas where Burkitt's lymphoma is endemic, virtually all children are infected with EBV early in life, as determined by the presence of antibodies in their blood. In Northern Europe and the United States, most adults also have antibodies to the virus, but the first appearance comes typically as a teenager

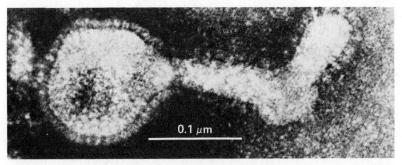

FIGURE 12-12 Mouse Mammary Tumor Virus. This odd-shaped particle is passed from mother to offspring by breast milk. The very high risk of breast cancer among female offspring in these strains is essentially eliminated by using a foster mother from a noninfected strain to nurse them. Vertical transmission (from parent to offspring) is common among cancers in laboratory animals, and may be among human cancers too, but it is usually via latent viruses carried with the genes, not via breast milk. Similar particles have been found in some human milk, though their relationship, if any, to human breast cancer is unclear and certainly complex. That is, breast cancer in humans is not simply determined by suckling history, as it is in these mice. [Micrograph by N. H. Sarkar. Provided courtesy of D. H. Moore and J. Charney, *Am. Scientist*, **63**: 160 (1975).]

or young adult and is associated in a fraction of the cases with the development of a quite benign disease, infectious mononucleosis.[4]

Thus, even where viruses are involved in oncogenesis, other factors must also be involved. These factors include age, health, genetics (cancers are demonstrably more common in some families than others), and environment. What is more, many cancers are probably quite independent of viruses. Lung cancer, for example, is the single most common cause of death from cancer in the United States, both because of a high incidence and because treatment is so unsatisfactory (10% survival at five years). Since lung cancer is rare among nonsmokers, cigarettes can be directly blamed for nearly 20% of cancer deaths. Other identified carcinogens include asbestos, the long-term abuse of alcohol, a number of industrial chemicals, and repeated exposure to certain kinds of radiation.

The surprising fact, in view of the above discussion, is that cancer is not more common among humans than it is. A widely accepted view is that we probably do produce malignant cells with great regularity but because these cells are antigenically different from their normal counterparts, they are soon destroyed by the immune system. Most human tumors appear to be clones, descended from a single cell. It is probably only the occasional lapse in immune surveillance, either by chance or because of an immune defect, that permits one of these aberrant cells to grow into a clinically detectable tumor. At that point, both specific antibodies and T cells sensitized against the tumor can usually be detected in large numbers, but the immune system is no longer equal to the task and the tumor continues to grow and spread. Current therapy in such cases is usually aimed at reducing the tumor bulk by surgery and/or radiation, then hitting the small inapparent metastases, if there is any reason at all to believe they might exist, with antimitotic chemicals. The objective is to reduce the total number of tumor cells to a point where the natural immune system can once again get the upper hand. Unless it does, there is apt to be no cure.

12-3 CELLULAR SENESCENCE

It should be the function of medicine to have people die young as late as possible.

       Ernst L. Wynder

The last stage in the development of many cells is senescence, or old age, culminating in death. It is a phenomenon that seems to be limited to the more complicated organisms, however, for there is no obvious counterpart in prokaryotes or simple eukaryotes; a bacterial cell, for instance, continues to divide as long as the environment is favorable. It is only cells with a high degree of specialization that have a life cycle that terminates in death.

Theories of Aging. The most general assumption concerning the process of aging is that it is due to accumulated insults, such as random "hits" from background radiation or free radicals produced in other ways. Even if most such damage is repaired (see Chap. 11), one may expect a gradual loss in ability to synthesize proteins accurately. This theory, which is essentially the "error catastrophe" theory of aging proposed by L. E. Orgel in 1963, predicts a buildup of debris within the cell, some of which may form a part of the lysosomal age pigment, or *lipofuscin*, that one sees in older cells (Fig. 12-13).

Accumulated genetic damage cannot be the whole story of aging, however, for different nucleated cell types have vastly different average survival times (Table 12-1). Yet each cell in a given animal has the same set of genes, and therefore potential access to the same repair mechanisms. (The mammalian erythrocyte and gametes are obvious exceptions.) We must thus consider the possibility that cells have a built-in obsolescence —i.e., that something "wears out" after a predetermined life span.

Programed Death. The concept of programed death certainly has ample precedent in embryology, for it has long been known that the death of certain cells is a necessary part of embryonic development. For example, separation of the fingers and toes of animals is due to the death of the tissue between them. Similarly, the chick wing is shaped in part by the death of selected cells. In an elegant set of experiments with this

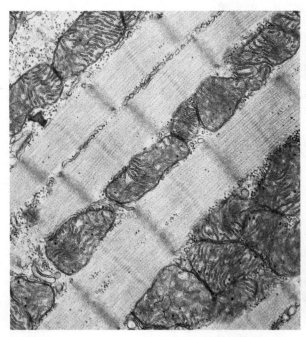
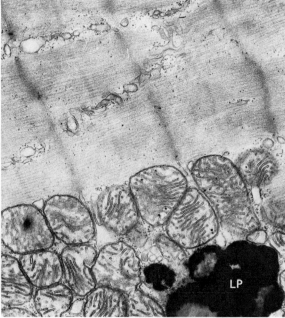

FIGURE 12-13 The Accumulation of Age Pigment in Rat Cardiac Muscle. LEFT: From a 90-day-old rat. RIGHT: From a 1004-day-old rat. The age pigment (lipofuscin, LP) is the dark area. (The contractile filaments seem discontinuous only because they are not parallel to the plane of section.) [Courtesy of N. M. and D. F. Sulkin, *J. Gerontol.*, **22**: 485 (1967).]

TABLE 12-1. Cellular Life Spans in the Adult Mouse[a]

Near or equal to that of the animal itself
    Neurons and associated cells
    Muscle cells of all types
    Brown fat cells
    Osteocytes (bone cells)
    Kidney medullary tubule cells
    Cells of the adrenal medulla
    Stomach zymogen cells

Slow renewal (more than 30 days, but less than the mean life span of the animal)
    Respiratory tract epithelium
    Kidney cortical cells
    Adrenal cortical cells
    Liver hepatocytes
    Pancreas acinar and islet cells
    Salivary gland cells
    Skin connective tissue cells
    Stomach parietal cells

Fast renewal (less than 30 days)
    Skin (epidermis)
    Cornea
    Epithelium of the mouth and gastrointestinal tract
    Precursors of red and white blood cells

[a] Most data from I. L. Cameron, *Texas Rep. Biol. Med.*, **28:** 203 (1970).

latter system, J. W. Saunders, Jr., demonstrated that the time of death of these cells is established at a given point during development, in much the same way as other characteristics are determined at given times. In other words, if cells that are destined to die during the shaping of a wing are transplanted before a certain stage of development, they may either die on schedule or survive as well as other cells, depending on the site of transplantation. After a certain point in the developmental scheme, however, death occurs on schedule no matter where they are moved, or even if they are transferred to tissue culture.

The metamorphosis of insects is another example of programed cell death, as is the metamorphosis of amphibians (e.g., the absorption of a tadpole's tail). It is possible, then, that the average life span of an animal or plant is not just the result of random damage, but due to something more subtle.

One could form a convincing teleological argument for programed death, for a species that is immortal is a species that cannot evolve once the upper limit of its population is reached. It is only through successive generations that natural selection can improve a species or make it better suited to a changing environment. One would expect, therefore, that the most favorable life span would be a period long enough to ensure procreation, but little more (see Fig. 12-14). On this basis, we would predict that animals with long gestation periods, long periods of offspring dependency, or low offspring survival rates, should either live longer or pro-

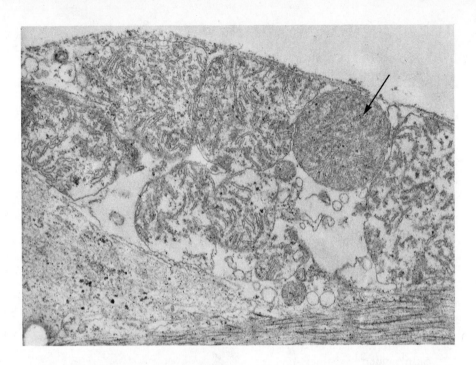

FIGURE 12-14 Mitochondrial Breakdown in Postspawning Salmon. Salmon seem to die a programed death. After three or four years in the open sea, and still apparently young and vigorous, they enter fresh water to spawn. Thereafter, the process of aging occurs at an enormous rate. Their bones turn to cartilage (loss of $CaPO_4$), their skin peels, and their liver turns green from the decomposition of hemoglobin. Even mitochondria degenerate, as seen here. (A normal mitochondrion, for comparison, is marked with an arrow.) [Courtesy of Andrew A. Benson, Scripps Inst. of Oceanography, and *Oceans Magazine*.]

duce huge numbers of descendants. To a reasonable degree, this does seem to be the case.

Further support for the concept of programed death comes from the work of Leonard Hayflick at Stanford University. He found that cultured human fibroblasts (the cells that produce collagen) obtained from embryonic tissue undergo only about 50 doublings before the culture dies out. When fibroblasts are obtained from persons of various ages, the decline in the number of doublings was about 0.2 per year of the donor's life. Furthermore, when cultured cells are frozen, even for a period of years, they "remember" their generation number. That is, when they are thawed and allowed to continue growing, they complete the originally assigned number of generations, then die on schedule.

It does not appear that accumulated radiation damage is the reason for the death of these cultures, for that should continue to some degree even in the frozen state. Hayflick also eliminated the possibility that a buildup of waste products kills the culture, for when a male population at the 40th generation was mixed with a female population at the 10th generation, both ran their normal course, oblivious to the other cells. This, plus careful control of the culture medium and population density, indicates that death is not a result of the gradual poisoning of the culture by a waste product, nor does it come from depletion of an essential nutrient.

Although the length of survival of cells in tissue culture is quite variable, depending on the source and type of cell being cultured, it is true that most normal human cells cannot be grown indefinitely. When a permanent cell line is established, it is found to contain variants that are somehow immune to normal aging. Tumor cells are particularly easy to culture because the transformation to an "immortal" condition seems already to have taken place in them, and represents another consequence of their release from normal controls on growth. HeLa cells, for example, which were originally derived from a carcinoma of the cervix, are decidedly abnormal in many respects. For one thing, the cells are *mixoploid* or *aneuploid*. That is, they have an abnormal chromosome number; instead of the normal human complement of 46 chromosomes, HeLa cells may have anywhere from 50 to about 350 chromosomes per cell.[5]

If normal human cells cannot be grown indefinitely in tissue culture and malignant human cells can be, we are tempted to conclude either that examination *in vitro* is not a fair test of life span or that normal human cells are programed to die. If the latter is true, then we must also conclude that the programing is subject to reversal. The slow decline in the efficiency of bodily functions with age might therefore be traced to a built-in rate of cell death. If we could but identify the process that appears to result in programed death, it might be possible to interfere with that process and, for better or worse, dramatically extend the human life span.

SUMMARY

12-1 Differentiation normally results from epigenetic changes rather than from an irreversible loss of genes. This can be shown by cloning experiments in which complete animals and plants are grown from single cells, where either the cell itself (in plants) or the nucleus (in animals) is derived from highly differentiated tissue. However, though nuclear genes are clearly responsible for the emergence of adult features, they are influenced by cytoplasmic factors, the earliest of which, in the case of animals, must be present in the egg before fertilization. Differentiation is thus a reciprocal gene/cytoplasm interaction in which regulators that control one stage of cellular development are produced by the activity of genes during the previous stage.

12-2 Though we normally think of them as stable conditions, neither determination nor differentiation is irreversible. Transdetermination in *Drosophila*, for instance, is due to an occasional accidental change in the determination of cells as they proliferate. The pattern of change, however, seems to be highly restricted, an observation that suggests an inherent limitation in the sequence of developmental determinations.

The conditional stability of fully differentiated tissue is revealed by the regeneration of limbs in amphibia and by transformation to malignancy. Limb regeneration involves the dedifferentiation of mature tissue to an embryonic condition and its subsequent proliferation and redifferentiation. This process also emphasizes the importance of tissue interactions during development, as the capacity to form the proper structures is obviously inherent in the cells and requires nothing that is unique to the embryonic environment.

Tumors also result from changes in differentiated properties, for besides losing their growth restraints, transformation to a malignant state may be accompanied by the loss (or sometimes the gain) of specialized functions.

12-3 The last stage in the development of specialized cells is old age, or senescence. It is accompanied by profound changes in the properties of the cell, including alterations in both structure and enzymatic activities. It is not clear, however, why these changes should occur, as they are not present in the simplest forms of life nor in mammalian cells that have been adapted for growth *in vitro*.

Cell death is not always associated with old age, but is a normal part of the embryonic shaping of hands, feet, and wings. The cells involved seem to be determined for this fate at a particular stage in development. Thus, it is possible that eventual senescence is programed into a cell in much the same way as any other function.

STUDY GUIDE

12-1 (a) What experiments suggest that the pattern of gene expression in a highly differentiated cell is reversible, and that most such cells still contain a complete gene set characteristic of the species? (b) What experiments, in animals and plants, indicate that nuclear activity is under the control of cytoplasmic factors? (c) What experiments indicate that the cytoplasmic factors themselves result from an earlier RNA synthesis in the nucleus?

12-2 (a) What is transdetermination, and what does

it tell us about the stability of determination? (b) What are the origin and ultimate fate of the cells in a regeneration blastema? (c) Why is it believed that at least some malignant cells (plant and animal) still carry a full complement of normal genes?

12-3 (a) Eventual senescence is less common in cells that reproduce themselves as opposed to those that result from the differentiation of a simpler stem cell. Give examples that support this statement. (b) What role might programed death play in normal development?

REFERENCES

CELLULAR DIFFERENTIATION

APPELS, R., and N. R. RINGERTZ, "Chemical and Structural Changes Within Chick Erythrocyte Nuclei Introduced into Mammalian Cells by Cell Fusion." *Current Topics Devel. Biol.*, **9**: 137 (1975).

BERNHARD, H., "The Control of Gene Expression in Somatic Cell Hybrids." *Int. Rev. Cytol.*, **47**: 289 (1976).

BRACHET, J., and S. BONOTTO, *Biology of Acetabularia.* New York: Academic Press, 1970.

BRYANT, S., and V. FRENCH, "Biological Regeneration and Pattern Formation." *Scientific American,* **273** (1): 66 (July 1977).

CARLSON, P., and J. POLACCO, "Plant Cell Cultures: Genetic Aspects of Crop Improvement." *Science,* **188**: 622 (1975). Application of cloning.

CIBA FOUNDATION, *Cell Patterning* (Symposium No. 29). Amsterdam: Elsevier, 1975. Imaginal discs, chimeras, regeneration, etc.

CRICK, F. H. C., and P. A. LAWRENCE, "Compartments and Polyclones in Insect Development." *Science,* **189**: 340 (1975). Imaginal discs.

DAVIDSON, E. H., *Gene Activity in Early Development,* 2d ed. New York: Academic Press, 1977.

EARLEY, K., "Harvesting the Cell." *The Sciences* **15**(4): 6 (May/June 1975). Practical results of cloning plant tissues.

GIBOR, AHARON, "Acetabularia: A Useful Giant Cell." *Scientific American,* November 1966. (Offprint 1057.)

GOODWIN, B. C., *Analytical Physiology of Cells and Developing Organisms.* New York: Academic Press, 1977. A quantitative approach.

GORDON, SAIMON, "Cell Fusion and Some Subcellular Properties of Heterokaryons and Hybrids." *J. Cell Biol.,* **67**: 257 (1975). A review.

GRAHAM, C. F., and P. F. WAERING, eds., *Developmental Biology of Plants and Animals.* Philadelphia: W. B. Saunders, 1976. Molecular orientation.

GURDON, J. B., *The Control of Gene Expression in Animal Development.* Cambridge, Mass.: Harvard Univ. Press, 1974. Animal cloning.

———, "Transplanted Nuclei and Cell Differentiation." *Scientific American,* December 1968. (Offprint 1128.)

———, "Nuclear Transplantation and the Control of Gene Activity in Animal Development." *Proc. Roy. Soc. (London) Ser. B,* **176**: 303 (1970). Also see **198**: 211 (1977).

GURDON, J. B., and V. UEHLINGER, "Fertile Intestine Nuclei." *Nature,* **210**: 1240 (1966).

GUSTAFSON, TRYGGVE, and MARK I. TONEBY, "How Genes Control Morphogenesis." *Am. Scientist,* **59**: 452 (1971).

HADORN, ERNST, "Transdetermination in Cells." *Scientific American,* November 1968. (Offprint 1127.)

HILDEBRANDT, A. C., "Growth and Differentiation of Plant Cell Cultures." In *Control Mechanisms in the Expression of Cellular Phenotypes* (Symp. Int. Soc. Cell Biol., vol. 9). H. A. Padykula, ed. New York: Academic Press, 1970, p. 147. Whole plants from single cells.

LASH, J., and M. BURGER, eds., *Cell and Tissue Interactions* (Soc. Gen. Physiol. Series, Vol. 32). New York: Raven Press, 1977. Cell interactions in normal development and in malignancy.

LAST, J., ed., *Eukaryotes at the Subcellular Level. Development and Differentiation.* New York: Dekker, 1976.

LOOMIS, W. F., ed., *Papers on Regulation of Gene Activity During Development.* New York: Harper & Row, 1970.

MACLEAN, N., *The Differentiation of Cells.* London: Edward Arnold, 1976. (Paperback.)

MATTSON, P., *Regeneration.* Indianapolis: Bobbs-Merrill, 1976. (Paperback.)

MEINTS, R., and E. DAVIES, eds., *Control Mechanisms in Development.* New York: Plenum Press, 1975.

MOSCONA, A. A., ed., *The Cell Surface in Development.* New York: John Wiley, 1974.

NITSCH, J. P., and C. NITSCH, "Haploid Plants from Pollen Grains." *Science,* **163**: 85 (1969). Tobacco plants can be grown from haploid cells and made to mature and flower, but not produce seed.

PASTERNAK, C. A., *Biochemistry of Differentiation.* New York: John Wiley, 1970. (Paperback.)

POSTE, G., and G. L. NICOLSON, eds., *The Cell Surface in Animal Embryogenesis and Development* (Cell Surface Reviews, Vol. 1). New York: Elsevier/North Holland, 1977.

PAUL, J., ed., *Biochemistry of Cell Differentiation.* Baltimore: Univ. Park Press, 1974. MTP International Review of Science.

PUISEUX-DAO, S., *Acetabularia and Cell Biology.* London: Logos Press, 1970.

RAGHAVAN, V., *Experimental Embryogenesis in Vascular Plants.* New York: Academic Press, 1977. Cloning.

REINERT, J., and H. HOLTZER, eds., *Cell Cycle and Cell Differentiation.* New York: Springer-Verlag, 1975.

RINGERTZ, N., and R. E. SAVAGE. *Cell Hybrids.* New York: Academic Press, 1976.

ROCKSTEIN, M., and G. T. BAKER, eds., *Molecular Genetic*

Mechanisms in Development and Aging. New York: Academic Press, 1972. A symposium.

STEWARD, F. C., "From Cultured Cells to Whole Plants: The Induction and Control of their Growth and Morphogenesis." Proc. Roy. Soc. (London), Ser. B, 175: 1 (1970). Croonian Lecture, 1969.

STEWARD, F. C., H. ISRAEL, R. MOTT, H. WILSON, and A. KRIKORIAN, "Observations on Growth and Morphogenesis in Cultured Cells of Carrot." Phil. Trans. Roy. Soc. London, Ser. B., 273: 35 (1975).

SUBTELNY, S., "Nucleocytoplasmic Interactions in Development of Amphibian Hybrids. Int. Rev. Cytol., 39: 35 (1974).

SUSSMAN, MAURICE, Developmental Biology: Its Cellular and Molecular Foundations. Englewood Cliffs, N.J.: Prentice-Hall, 1973. (Paperback.)

TALWAR, G. P., ed., Regulation of Growth and Differentiated Function in Eukaryote Cells. New York: Raven Press, 1975.

THOMAS, E., and M. R. DAVEY, From Single Cells to Plants. New York: Springer-Verlag, 1975. (Paperback.)

VASIL, V., and A. C. HILDEBRANDT, "Differentiation of Tobacco Plants from Single Isolated Cells in Culture." Science, 150: 889 (1965). Reprinted, in part, in JRW.

WEBER, R., ed., The Biochemistry of Animal Development. Vol. 3. Molecular Aspects of Animal Development. New York: Academic Press, 1975.

WIGLEY, C. B., "Differentiated Cells in vitro." Differentiation, 4: 25 (1975).

WILDERMUTH, H., "Determination and Transdetermination in Cells of the Fruitfly." Sci. Progr. (Oxford), 58: 329 (1970).

THE MALIGNANT CELL

AMERICAN CANCER SOCIETY, "Cancer Statistics, 1977." Ca–A Cancer Journal for Clinicians, 27: 26 (1977). Annually publishes mortality statistics, emphasizing cancer.

AARONSON, S., and J. STEPHENSON, "Endogenous Type-C RNA Viruses of Mammalian Cells." Biochim. Biophys. Acta, 458: 328 (1976).

ABLASHI, D., J. EASTNO, and J. GUEGAN, "Herpes Viruses and Cancer in Man and Subhuman Primates." Biomedicine, 24: 286 (1976).

AMBROSE, E. J., and F. J. C. ROE, eds., Biology of Cancer 2d ed. New York: Halsted Press, 1975.

AUGUSTI-TOCCO, G., "Neuroblastoma Culture: An Experimental System for the Study of Cellular Differentiation." Trends Biochem. Sci., 1(7): 151 (July 1976).

BALTIMORE, DAVID, "Viruses, Polymerases, and Cancer." Science, 192: 632 (1976). Nobel lecture, 1975.

BENDITT, E. P., "The Origin of Atherosclerosis." Scientific American, 236(2): 74 (Feb. 1977). Tumorlike growth.

BOUCK, N., and G. DI MAYORCA, "Somatic Mutation as the Basis for Malignant Transformation of BHK Cells by Chemical Carcinogens." Nature, 264: 722 (1976).

BURCH, P., The Biology of Cancer. Baltimore: Univ. Park Press, 1976.

BURNET, F. M., Immunology, Aging, and Cancer. San Francisco: Freeman, 1976. (Paperback.)

CAIRNS, JOHN, "The Cancer Problem." Scientific American, 233(5): 64 (Nov. 1975). Environmental influences.

CAMPBELL, A. M., "How Viruses Insert Their DNA into the DNA of the Host Cell." Scientific American, 235(6): 102 (Dec. 1976).

DULBECCO, R., "From the Molecular Biology of Oncogenic DNA Viruses of Cancer." Science, 192: 437 (1976). Nobel lecture, 1975.

———, "The Control of Cell Growth Regulation by Tumor-Inducing Viruses." Proc. Roy. Soc. (London), Ser. B., 189: 1 (1975).

FAREED, G., "Molecular Biology of Papovaviruses." Ann. Rev. Biochem., 46: 471 (1977). Polyoma and SV40.

FIALKOW, P. J., "Clonal Origin of Human Tumors." Biochim. Biophys. Acta, 458: 283 (1976). Most tumors are clones, descended from a single malignant cell.

FOULDS, L., Neoplastic Development. New York: Academic Press, 1976. Volume 1 provides a general theory. Volume 2 considers specific examples.

GALLO, R. C., ed., Cancer Research: Cell Biology, Molecular Biology and Tumor Virology. Cleveland, Ohio: CRC Press, 1976.

GREEN, MAURICE, "Viral Cell Transformation in Human Oncogenesis." Hospital Practice, 10(9): 91 (Sept. 1975).

GROSS, L., "The Role of C-Type and Other Oncogenic Virus Particles in Cancer and Leukemia." New Engl. J. Med., 294: 724 (1976).

HARRIS, H., J. MILLER, G. KLEIN, P. WORST, and T. TACHIBANA, "Suppression of Malignancy by Cell Fusion." Nature, 223: 363 (1969).

HARRIS, HENRY, "The Expression of Genetic Information: A Study with Hybrid Animal Cells." In The Harvey Lectures, 1969–1970 (Ser. 65). New York: Academic Press, 1971, p. 1.

HEHLMANN, R., "RNA Tumor Viruses and Human Cancer." Current Topics Microbiol. Immunol., 73: 141 (1976).

HIATT, H., J. D. WATSON, and J. WINSTEN, eds., Origins of Human Cancer. Cold Spring Harbor (N.Y.) Lab. of Quant. Biol., 1977.

HYNES, R. O., "Cell Surface Proteins and Malignant Transformation." Biochim. Biophys. Acta, 458: 73 (1976).

KLEIN, G., "Analysis of Malignancy and Antigen Expression by Cell Fusion." Fed. Proc., 35: 2202 (1976).

KLEIN, G., "The Epstein-Barr Virus and Neoplasia." New Engl. J. Med., 293: 1353 (1976).

KLEIN, P., and R. T. SMITH, "The Role of Oncogenic Viruses in Neoplasia." Ann. Rev. Med., 28: 311 (1977).

KOLATA, G. B., "Cell Surface Protein: No Simple Cancer Mechanisms." Science, 190: 39 (1975). A brief review.

LEVY, J., "Endogenous C-Type Viruses: Double Agents in Natural Life Processes." Biomedicine, 24: 84 (1976). Our relationship with some cancer viruses may be more symbiotic than parasitic.

MARTIN, G., "Teratocarcinomas as a Model System for the Study of Embryogenesis and Neoplasia." Cell, 5: 229 (1975). A review.

MAUGH, T. H., II, and J. L. MARX, Seeds of Destruction. New York: Plenum Publ., 1976. A report on cancer research by two experienced science writers.

NICOLSON, G. L., "Surface Changes Associated with Transformation and Malignancy." *Biochim. Biophys. Acta*, **458**: 1 (1976). A review.

NICOLSON, G. L., and G. POSTE, "The Cancer Cell: Dynamic Aspects and Modifications in Cell-Surface Organization." *New Engl. J. Med.*, **295**: 197 and 253 (1976).

NOWELL, P., "The Clonal Evolution of Tumor Cell Populations." *Science*, **194**: 23 (1976).

OLD, L. J., "Cancer Immunology." *Scientific American*, **236**(5): 62 (May 1977).

PASTAN, I., "Cyclic AMP and the Malignant Transformation of Cells." *Advan. Metab. Disorders*, **8**: 359 and 377 (1975). A review.

PIERCE, G. BARRY, "Differentiation of Normal and Malignant Cells." *Fed. Proc.*, **29**: 1248 (1970).

POSTE, G., and G. L. NICOLSON, eds., *Virus Infection and the Cell Surface* (*Cell Surface Reviews*, Vol. 2). New York: Elsevier/North Holland, 1977.

PRASAD, K. N., "Differentiation of Neuroblastoma Cells in Culture." *Biol. Rev. (Cambridge)*, **50**: 129 (1975).

SARASIN, A., and M. MEUNIER-ROTIVAL, "How Chemicals May Induce Cancer." *Biomedicine*, **24**: 306 (1976).

STANBRIDGE, E., "Suppression of Malignancy in Human Cells." *Nature*, **260**: 17 (1976). By cell fusion.

TEMIN, H., "RNA-Directed DNA Synthesis." *Scientific American*, January 1972. (Offprint 1239.)

———, "The DNA Provirus Hypothesis." *Science*, **192**: 1075 (1976). Nobel lecture, 1975.

———, "The Relationship of Tumor Virology to an Understanding of Nonviral Cancers." *BioScience*, **27**: 170 (1977).

TOOZE, J., ed., *Selected Papers in Tumour Virology*. Cold Spring Harbor (N.Y.) Lab. Quant. Biol., 1974. (Paperback.) Original papers.

VERMA, I. M., "The Reverse Transcriptase." *Biochim. Biophys. Acta*, **473**: 1 (1977).

YAMADA, K., and I. PASTAN, "Cell Surface Protein and Neoplastic Transformation." *Trends Biochem. Sci.*, **1**(10): 222 (Oct. 1976).

ZIEGLER, J., I. MAGRATH, P. GERBER, and P. LEVINE, "Epstein-Barr Virus and Human Malignancy." *Ann. Internal Med.*, **86**: 323 (1977).

CELLULAR SENESCENCE

ALDER, WM., "Aging and Immune Function." *BioScience*, **25**: 652 (1975). Also see comment by D. E. Harrison in **26**: 304 (1976).

BUCHER, N. et al., eds., *Cell Divisions & Aging*. New York: Mss Information Corp., 1977.

BURNET, F. M., *Immunology, Aging and Cancer: Medical Aspects of Mutation and Selection*. San Francisco: Freeman, 1976. (Paperback.)

COOPER, E. H., A. BEDFORD, and T. E. Kenny, "Cell Death in Normal and Malignant Tissues." *Advan. Cancer Res.*, **21**: 59 (1975).

CRISTOFALO, V. J., and E. HOLEČKOVÁ, eds., *Cell Impairment in Aging and Development*. New York: Plenum Press, 1975.

CUTLER, R. G., ed., *Cellular Aging: Concepts and Mechanisms*. New York: S. Karger, 1976. A series.

HAYFLICK, L., "Human Cells and Aging." *Scientific American*, March 1968. (Offprint 1103.)

———, "The Cell Biology of Human Aging." *New Engl. J. Med.*, **295**: 1302 (1976).

HILL, B., L. FRANKS, and R. HOLLIDAY, "The Mechanism of Ageing." *Trends Biochem. Sci.*, **2**(4): N80 (April 1977). Two opposing views.

HOLLIDAY, R., et al., "Testing the Commitment Theory of Cellular Aging." *Science*, **198**: 366 (1977).

LAMB, M., *Biology of Ageing*. New York: Halsted Press, 1977.

LITTLEFIELD, J. W., *Variation, Senescence, and Neoplasia in Cultured Somatic Cells*. Cambridge, Mass.: Harvard Univ. Press, 1976.

ORGEL, L. E., "The Maintenance of the Accuracy of Protein Synthesis and Its Relevance to Aging." *Proc. Nat. Acad. Sci. U.S.*, **49**: 517 (1963). The "error catastrophe" theory.

ORGEL, L. E., "Ageing of Clones of Mammalian Cells." *Nature*, **234**: 5407 (1973).

RHEINWALD, J. G., and H. GREEN, "Epidermal Growth Factor and the Multiplication of Cultured Human Epidermal Keratinocytes." *Nature*, **265**: 421 (1977). EGF delays senescence.

SAUNDERS, JOHN W. JR., "Death in Embryonic Systems." *Science*, **154**: 604 (1966). The role of selective cell death in morphogenesis and development.

THORBECKE, G. J., ed., *Biology of Aging and Development* (FASEB Monographs, Vol. 3). New York: Plenum Press, 1975.

WILLIAMS, J. R., "Role of DNA Repair in Cell Inactivation, Aging, and Transformation." *Advan. Radiation Biol.*, **6**: 162 (1976).

NOTES

1. The argument is also made that specialized functions and cell division are antagonistic because both make great demands on the available energy. In other words, the metabolism of a cell can support one function but not both. Since many specialized cells are indispensable to the organism, even a temporary loss of their contribution would be fatal.

2. A limited amount of limb regeneration can be induced in mammals. No complete regeneration has yet been achieved, however, nor have the experiments been extended to humans, except for the use of small electric currents to promote the healing of broken bones. However, it is not too far-fetched to hope that one day a severed human arm or leg can be regrown.

3. Heart disease is the most common cause of death, accounting for almost 40% of the total. In the 15-34 year-old age group, cancer is the fourth most common cause of death for men (after accidents, homicide, and suicide) and the second for women (after accidents). The total number of cancer deaths is about the same for young men and young women, but primarily because of the greater incidence of other causes, cancer accounts for less than 7% of the total deaths in men of this age group but almost 20% in women.

4. A number of other viruses have been associated with human tumors, mostly via circumstantial evidence. For example, women who have had multiple sexual partners have an increased risk of cancer of the uterine cervix, thought to be due at least in part to *Herpes hominis* type II. This virus can also cause a particularly unpleasant form of venereal disease. In 1977, cervical cancer still caused more deaths among women than leukemia (over 4% of cancer deaths), in spite of the ease with which it can be identified early enough (via a yearly "Pap test") to effect a complete cure.

5. HeLa cells were named in honor of Henrietta Lacks, from whom the original cells were obtained. Ms. Lacks was a young black woman who died of her cancer in 1951, but provided researchers with the first stable, vigorous human cell line suitable for cancer research.

APPENDIX
Tools of the Cell Biologist

A-1 Microscopy 577
    Light Microscopy
    Sample Preparation for Light Microscopy
    The Electron Microscope
    Sample Preparation for the Electron Microscope
    Scanning Electron Microscopy
    Advantages and Disadvantages of the Three
      Types of Microscopy
A-2 X-Ray Diffraction 589
A-3 Centrifugation and Cell Fractionation 592
    Centrifugal Fields
    Sedimentation Velocity
    Sedimentation Equilibrium
    Cell Fractionation
A-4 Isotopes 597
    Density Labeling
    Labeling with Radioactive Isotopes
    Measuring Radioactivity
Summary 599
Study Guide 600
References 600
Notes 601

A-1 MICROSCOPY

The microscope was the first powerful tool available to cell biologists, and it remains one of the most useful. Zacharias Janssen produced the first functional instrument in the 1590s. Later, Anton van Leeuwenhoek (1632–1723) greatly improved the technique of polishing lenses, thus enabling him to describe protozoa and bacteria.

Light Microscopy. The resolution of the unaided human eye is about 0.1 mm, or 100 μm. In other words, two objects within 100 μm of each other appear to be in contact. The minimum distance, d_0, at which they are resolvable as separate entities is given, approximately, by Abbe's relationship:

$$d_0 \approx \frac{0.6 \lambda}{n \sin \alpha} \quad \text{(A-1)}$$

This equation is valid for all optical instruments including the eye. The symbol λ is the wavelength ("color") of the radiation used to form the image, n is the refractive index (a function of density) of the air or fluid between the specimen and the first lens, and α is the aperture angle, or half the angle subtended by the aperture of the first lens as viewed from the specimen (see Fig. A-1). The quantity "$n \sin \alpha$" is often called the *numerical aperture*.

Abbe's relationship makes it clear that high resolution in a microscope can only be achieved by manipulating a small number of vari-

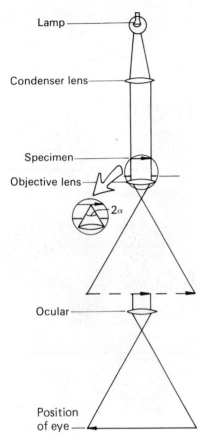

FIGURE A-1 The Light Microscope. Note the aperture angle, α. When using an oil immersion lens, the space between the specimen and the objective lens (circled) is filled with oil having a relatively high refractive index, n.

ables: the wavelength of the illuminating radiation, the refractive index, and the aperture. The aperture, of course, is limited to something less than 90° since that would have the lens and specimen in contact with one another. In fact, 85° is about the limit in good optical microscopes. Such angles require an excellent lens. In most cases the aperture is less because the edges of the lens introduce distortions and so cannot be used.

Refractive index is easy to alter, but only within narrow limits. It can be increased by using oils to fill the space between the specimen and the lens. Oils may have an n up to about 1.5, but little more. Still, 1.5 is a big improvement over air ($n = 1$), a fact accounting for the popularity of the oil immersion lenses of modern microscopes.

The wavelength of radiation is the area in which most dramatic improvement seems likely. One can, for example, use ultraviolet light instead of visible light, thus improving resolution as much as twofold. In order to do that, however, special lenses (e.g., of quartz) must be used since ordinary glass blocks much ultraviolet light. In addition, the eye cannot be used to view the image directly, for it is insensitive to ultraviolet light. And finally, absorption by the sample itself at wavelengths below about 300 nm (0.3 μm) may become a problem.

Thus, a good light microscope, with a numerical aperture of 1.4 and using light at the short end of the visible spectrum (0.4 μm), will resolve two points at about 0.17 μm separation. This is, of course, immensely better than in the unaided eye. However, while one can thus see considerable detail in most cells (which may be 10 to 20 μm in diameter), there is also a great deal that cannot be seen. The ribosomes, for instance, and the threads of chromatin in the nucleus are about 0.02 μm in diameter and quite invisible to the light microscope.

In order to make full use of resolving power that is available, special techniques have been designed to improve contrast. They include dark-field microscopy, phase contrast microscopy, and polarization microscopy.

In *dark-field microscopy*, the sample is viewed only with oblique rays. It is particularly useful for suspensions of bacteria. (It is likely that van Leeuwenhoek inadvertently used this technique in the discovery of bacteria.) Because one sees only those light rays that are scattered from objects, the images appear bright on a black background. The process is akin to seeing dust particles floating in a sunbeam.

Phase-contrast microscopy (Fig. A-2) takes advantage of the fact that different parts of a cell have different densities and hence different refractive indices. Regions where the refractive index is changing bend light rays, which is why you can sometimes see columns of hot air rising from a radiator or black roadway. These rays only blur the image in an ordinary microscope but in phase contrast they are used to form patterns of destructive interference, yielding sharp contrasts. This technique is widely used to observe living cells, which are otherwise relatively transparent and difficult to see.

Polarization microscopy is useful mainly for viewing highly ordered objects such as crystals or bundles of parallel filaments. A specimen of this type will often be *birefringent*, a property that makes it visible when placed between two crossed polarizers (Fig. A-3). You can simulate the situation by placing two Nicol's prisms or polarizing films (of the kind often used for sunglasses) at right angles to each other. No light gets

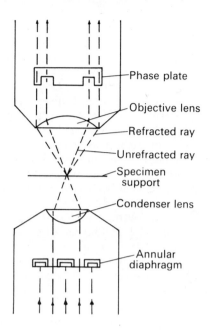
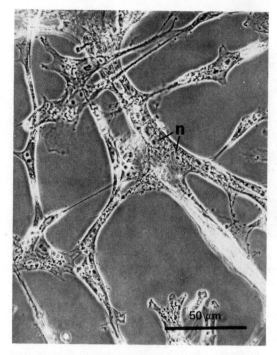

FIGURE A-2 Phase Contrast Microscopy. LEFT: Diagram of the phase contrast microscope. Rays that pass through a region of changing refractive index will be refracted and, after passing through the phase plate, will cause interference with unrefracted rays. This produces a halo effect, increasing contrast and making it easier to observe unstained, living cells. RIGHT: Phase contrast micrograph of embryonic chick myoblasts (muscle cell percursors) after four days of culture. Nuclei (n). [Micrograph courtesy of Y. Shimada, *J. Cell Biol.*, **48**: 128 (1971).]

through. But if a third polarizer is placed between the first two, light is again transmitted, with an intensity that depends on the rotational position of the third polarizer. Birefringent portions of a sample act like polarizing films and hence these portions of the sample are seen in polarizing microscopy as bright objects on a dark background.

Other optical systems are also in use, each having its special applications. The way the sample is prepared, however, is at least as important to the amount of information obtained as is the choice of optical systems. Of particular usefulness is the variety of chemical stains available for light microscopy.

Sample Preparation for Light Microscopy. The amount of information obtained from a microscope depends in large measure on how the sample is prepared.

Samples for the light microscope are commonly fixed (with alcohol, formalin, osmium salts, formaldehyde, etc.) to make their protein components insoluble and stable to subsequent procedures, then air-dried and treated with a stain having an affinity for the structure to be examined. To ensure an even penetration of the stain, the specimen is usually first embedded in paraffin (to provide support) and cut into thin sections. Then the paraffin is dissolved again with xylol before staining.

These steps are time consuming. Hence, when speed is important paraffin embedding may be replaced with rapid freezing in liquid nitrogen to produce the rigidity needed for sectioning. These *frozen sections*

FIGURE A-3 Polarization Micrograph. The sample (spindle apparatus of a fertilized sea urchin egg at two stages of mitosis) contains parallel bundles of microtubules that are birefringent and hence visible when properly placed between two crossed polarizers. [Courtesy of G. Sluder, *J. Cell Biol.*, **70**: 75 (1976).]

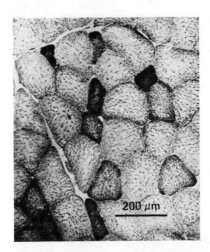

FIGURE A-4 Cytochemical Stains. This section through a monkey muscle was stained specifically for succinate dehydrogenase, a mitochondrial enzyme. Those muscle fibers (i.e., cells) having large numbers of mitochondria are stained to a much deeper extent than the surrounding fibers. [Courtesy of C. Beatty, G. Basinger, C. Dully, and R. Bocek, *J. Histochem. Cytochem.*, **14**: 590 (1966).]

can then be stained in the usual way, though they do not provide the same detail as paraffin sections.

Some of the chemical stains in common use are eosin, aniline blue, crystal violet, methylene blue, methyl green, fuchsin, Congo red, rose bengal, and so on. Generally, the stains react with (attach to) proteins. However, some are specific for nucleic acids (e.g., Feulgen's procedure for DNA, which uses a fuchsin derivative), others are specific for lipids, and still others stain starches.

In addition to stains for each of the major classes of macromolecule and a few stains for smaller substances, a large number of staining procedures have been devised to reveal the presence of specific enzymes. The utilization of these procedures is generally referred to as *histochemical* or *cytochemical staining*, an example of which is shown in Fig. A-4. The approach is usually to find a reaction that can be catalyzed by the enzyme in question and that yields a colored precipitate as a reaction product. Alternatively, the reaction might consume or produce a fluorescent substance such as NADH. This common coenzyme emits a visible fluorescence whereas its oxidized counterpart, NAD^+, does not.

Another procedure for enhancing the usefulness of light microscopy involves the application of antibodies specific for various cellular components. To permit localization of the antibodies they are either attached covalently to fluorescent dyes before use or a second layer of antibody is applied that is dye-labeled and has a specificity not for the cellular structure but for the first antibodies. These techniques, discussed in more detail in Section 9-5, are called direct and indirect *immunofluorescence microscopy*. The actual viewing is by simple *fluorescence microscopy*, using light of a wavelength that excites the fluorescent dye. The latter then emits light at a different wavelength. If a filter that blocks the exciting but not the fluorescing wavelength is placed between the sample and viewer, the location of the dye is outlined on a black background (see Fig. A-5).

FIGURE A-5 Fluorescence Microscopy. A cultured mouse cell was reacted with antitubulin antibody and then further reacted with a second layer of antibody specific for the first layer. The second layer has a covalently attached fluorescent dye. The cell is then reviewed by fluorescence microscopy where filters block out the exciting light so that only the fluorescence is seen. [Courtesy of M. Osborn and K. Weber, *Proc. Nat. Acad. Sci. U.S.*, **73**: 867 (1976).]

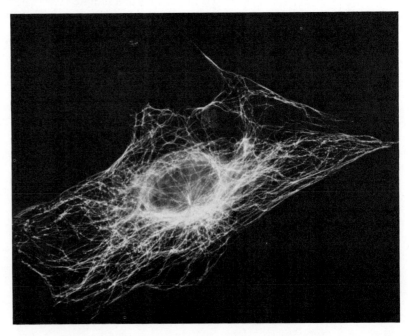

The Electron Microscope. One would like to be able to substitute a 1-Å X ray or γ ray for 4000-Å light to obtain a 4000-fold increase in resolution. In practice, there are no available lenses that can be used to focus very short wavelength electromagnetic radiation. (Medical X-ray machines do not focus the beam but use it to form shadows.) On the other hand, charged particles such as electrons respond to magnetic fields, which can thus be used to focus them. Since electrons also have associated wave properties, they can be used to form an image.

The wavelength of an electron is given approximately by

$$\lambda = \frac{12.3}{\sqrt{E}} \text{ angstroms} \quad \text{(A-2)}$$

where E is the voltage through which the electron was accelerated. Electrons at 50,000 volts have a wavelength of 0.05 Å. This value does not provide quite the improvement in resolution that one would expect from Equation A-1 because the numerical aperture of electron microscopes is much smaller than that of light microscopes (see Fig. A-6). The problem is that magnetic fields have the right properties to serve as a lens only in their center. To prevent serious spherical aberration, only a very small area in the middle of each lens is used. Although aperture angles of only a few tenths of a degree are thus common, compared to the 85° or so of light microscopes, the effective resolution of an electron microscope is still about a hundred times the wavelength of the electrons. At 50,000 volts, that provides 5-Å resolution! (See Fig. A-7)

Even higher resolution can be obtained with the electron microscope by going to greater accelerating voltages. However, as the electrons become more energetic, more and more of them pass through the sample without being significantly deflected, leading to a loss in contrast. In addition, present methodology of preparing biological samples seldom allows effective use of even 5-Å resolution. Nevertheless, high voltage electron microscopy (a million volts vs. the usual 50 to 100,000) has found some applications in viewing thick samples and, in a few cases, wet samples. Ordinarily, samples must be cut thinly and placed in a high vacuum for viewing. The greater penetration of high voltage electrons, however, permits wet, living cells to be placed in an isolation chamber and viewed directly. Although there is much interference from overlying and underlying structures when thick sections or whole cells are examined, and the detail of thin sections cannot be obtained, the use of high voltage electron microscopy to view thick specimens provides a better appreciation for the three-dimensional configuration of organelles than one can obtain by conventional electron microscopy (see Fig. A-8).

Sample Preparation for Electron Microscopy. The five most common ways of preparing samples for the electron microscope are thin sectioning, negative staining, metal shadowing, freeze-fracture, and whole mounts.

1. *Thin sectioning* uses a cutting device known as an ultramicrotome to remove slices that are only a few hundred angstroms thick. To withstand the passage of the diamond or glass knife without tearing, the specimen is first embedded in a hard plastic, such as epoxy resin, which is allowed to penetrate the sample before being polymerized. Sections are floated from the knife onto the surface of water and picked up by

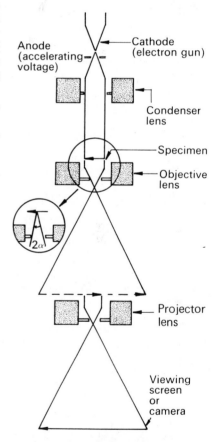

FIGURE A-6 The Electron Microscope. Compare with Fig. A-1. The entire column, from filament to screen, is maintained in a very high vacuum. Note the small aperture angle, greatly enlarged in this drawing.

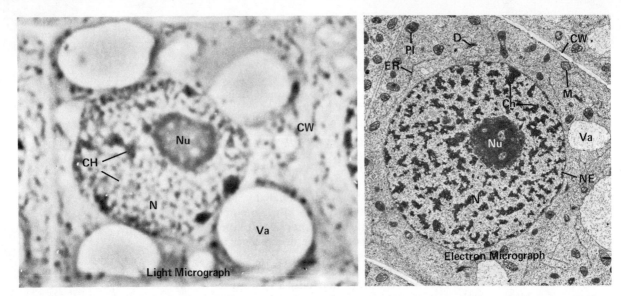

FIGURE A-7 Resolution of Optical vs. Electron Microscopes. These images were obtained from the same piece of onion root, prepared in the same way. (The tissue was fixed with gluteraldehyde and osmium tetroxide, then embedded in an epoxy resin.) The sections were cut 1.5 μm and 0.03 μm thick for light and electron microscopy, respectively. Note the difference in resolution between the two microscopes. Nuclear diameter = 1.5 μm. Nucleus (N). Nucleolus (Nu). Chromatin (CH). Nuclear envelope (NE). Mitochondria (M). Dictyosome (D). Endoplasmic reticulum (ER). Plastid (Pl). Vacuole (Va). Cell Wall (CW). [From *Cell Ultrastructure* by William A. Jensen and Roderic B. Park. © 1967 by Wadsworth Publishing Company Inc., Belmont, California 94002. Reprinted by permission of the publisher.]

touching them with a copper grid (200–300 wires/inch) that is first coated with a thin plastic or carbon membrane as a support (Fig. A-9).

Thin sectioning eliminates overlying and underlying structures that would otherwise confuse the image. However, stains must be incorporated into the tissue to improve contrast, since the structures left in these very thin sections would not otherwise scatter electrons differently enough to be seen. Various methods of fixing and staining were pioneered by George E. Palade in the 1950s. Typical stains are salts of tungsten, manganese, uranium, osmium, and so on, all of which provide electron-dense heavy metal atoms to scatter electrons.

The disadvantages of thin sectioning include: (1) the tedium of sample preparation; (2) the possibility of introducing artifacts in fixing, embedding, or staining; and (3) the difficulty of examining the three-dimensional nature of an object, which has to be done by looking at serial sections. In addition, the thin specimen is fragile and is subject to radiation damage if the beam is left on one spot too long.

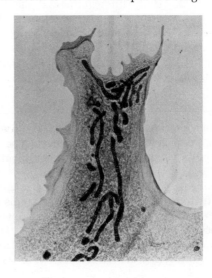

FIGURE A-8 High Voltage Electron Microscopy. Part of a flattened, dried mouse cell as seen with megavolt electron microscopy. Compare the mitochondrial configurations seen here with the thin sections commonly observed (e.g., Figs. A-7 and A-14). [Courtesy of J. Wolosewick, Univ. of Colorado.]

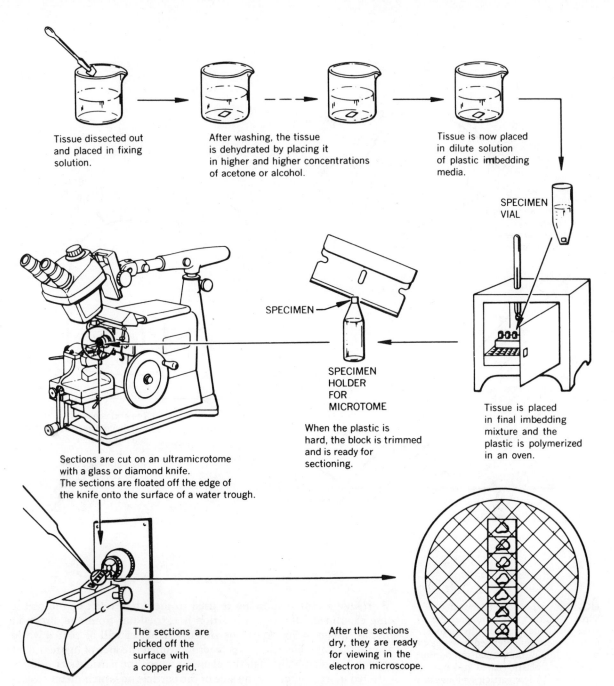

2. In *negative staining*, the specimen is embedded in a dense substance, usually phosphotungstic acid ($H_3PW_{12}O_{40}$), which is chosen for its extremely high electron density and consequent ability to scatter the beam. The portions of the specimen that exclude phosphotungstic acid transmit electrons readily, so their image can be seen. Because the stain penetrates various openings and crevices, some fine structure can often be observed, but the technique is used mostly with viruses and other particulate material. (See Fig. A-11, left, p. 585.)

FIGURE A-9 Thin Sectioning. [From *Cell Ultrastructure* by William A. Jensen and Roderic B. Park. © 1967 by Wadsworth Publishing Company, Inc., Belmont, California 94002. Reprinted by permission of the publisher.]

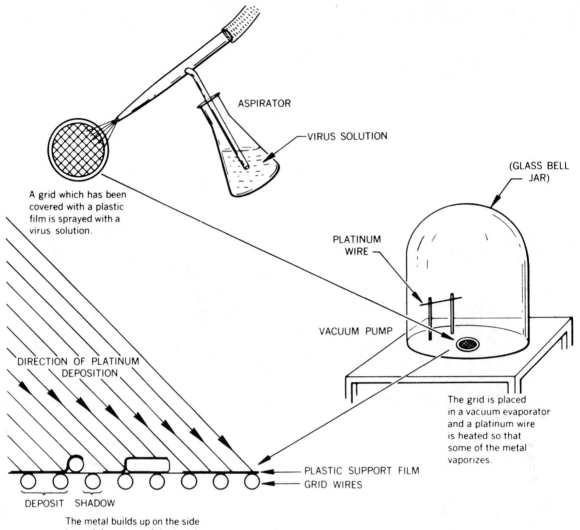

FIGURE A-10 Metal Shadowing a Virus. Another way to prepare the virus for shadowing—an alternative to the procedure described here—is to place a drop of the suspension on the surface of a plastic-coated grid and then blot. A few particles adhere to the plastic. Though gentler than the diagramed technique, blotting does not produce as even a distribution of particles. [From *Cell Ultrastructure* by William A. Jensen and Roderic B. Park. © 1967 by Wadsworth Publishing Company, Inc., Belmont California 94002. Reprinted by permission of the publisher.]

3. *Heavy metal shadowing* is used to provide a relief of the object being examined. If metal deposition is carried out from one direction only, as shown in Fig. A-10, one side of the object will be coated while the other is not. Electrons pass readily through the area of lighter metal content, less readily through the plane on which the particle sits, and are scattered most severely by the side of the particle on which metal has accumulated. As a result, the particle appears as if it is being viewed in strong light. (See Fig. A-11.)

The particle may also be rotated constantly while metal deposition is taking place so that no shadows are created. Instead, metal builds up on all sides of the particle, making it stand out above the background. This technique is often used with deoxyribonucleic acid, which by itself is a filament only about 20-Å thick. After "rotary shadowing" it has a diameter of a couple of hundred angstroms and is therefore readily visi-

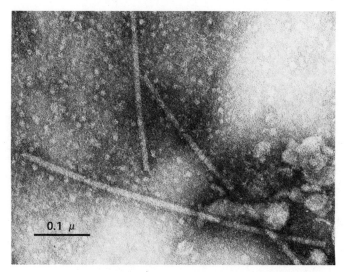
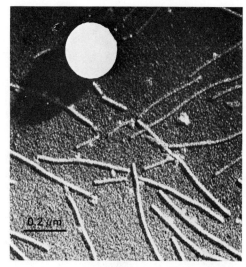

ble. The layer of metal cannot be ignored when shadowed specimens are examined, for of course it not only increases the overall size but may obscure some of the smaller surface features.

4. *Freeze-fracture* (see Fig. A-12) is carried out by rapidly freezing the sample and then sectioning it in a vacuum while it is still at −100 °C. The knife does not cut cleanly under those conditions, but tends to frac-

FIGURE A-11 A Comparison of Negatively Stained and Metal-shadowed Virus. Both preparations are of lily virus. LEFT: Negatively stained with phosphotungstic acid. RIGHT: Metal shadowed with a 20:80 platinum-palladium alloy at an angle of 15°. The round object is a polystyrene latex sphere 264 nm in diameter. Note the "shadows," representing areas where electrons were least obstructed in their passage through the specimen support. [Left photo, courtesy of A. R. Lyons and T. C. Allen, Jr., *J. Ultrastruct. Res.*, **27**: 198 (1969) © Academic Press, N.Y. Right photo, courtesy of T. C. Allen, Oregon State Univ. Similar to *Lily Yearbook*, **24**: 29 (1971).]

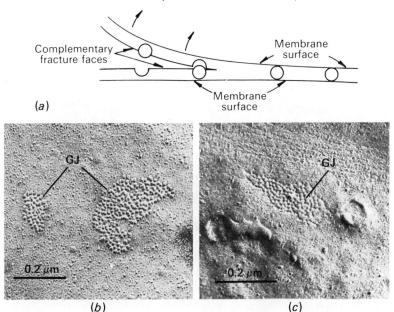

FIGURE A-12 Freeze-fracture. Some of the membrane proteins during freeze-fracture adhere to one lipid leaflet and some to the other, as shown in the diagram (*a*), and appear as pebbles or depressions when a metal replica is made and examined. The electron micrographs show complementary faces of a membrane that is participating in the formation of gap junctions between cells. Gap junctions are formed by the interaction of specialized membrane proteins from two adjacent membranes. (*b*) Fracture face of one of the membranes, showing scattered membrane proteins plus two clusters of gap junction particles (GJ) adhering to the exposed lipid leaflet. (*c*) The complementary face of a membrane at another gap junction. Again, scattered membrane proteins are seen but this time the gap junction particles have been lifted out of the exposed leaflet during freeze-fracture, leaving depressions. [Courtesy of N. Gilula and P. Satir, *J. Cell Biol.*, **51**: 869 (1971).]

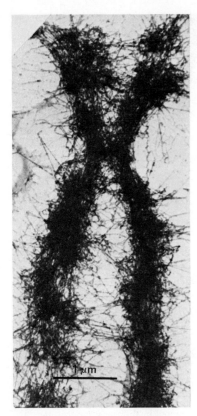

ture the specimen along lines of natural weakness, such as the middle of a membrane that runs parallel to the cut. After fracture, the sample is left in the vacuum long enough to allow some water to evaporate from the exposed surfaces, a process called *freeze etching*. The exposed face is then shadowed with metal to provide the necessary contrast, after which organic material (i.e., the specimen itself) is removed by acids to leave a metal replica for examination in the electron microscope. Much detail is retained in the replica, providing our best look at certain features of cells and our only way of seeing membrane interiors. (See also Section 6-1.)

5. Whole mounts (Fig. A-13) are often used to examine chromosomes and other relatively thick objects that can be isolated free of debris. As the name implies, the specimen is neither sectioned nor stained. Thick areas will scatter electrons more strongly than thin areas, providing enough contrast to form an image. In fact, since scattering of the beam is directly proportional to the electron density of the sample, and hence to its mass, proper calibration allows estimates of specimen mass to be obtained from the image. The weights of individual chromosomes can be measured in this way.

Cytochemical stains are also important in electron microscopy just as they are in light microscopy. The objective, of course, is to utilize an enzyme to convert a soluble substance into an insoluble, electron-dense reaction product. The location of the enzyme is thus revealed, as in Fig. A-14.

Scanning Electron Microscopy. Considerable interest has developed in applying the scanning electron microscope (SEM) to biological investigations. Although scanning instruments were available by the mid-1940s, about a decade after the appearance of commercial transmission electron

FIGURE A-13 Whole Mount. Particulate specimens like this human chromosome scatter enough electrons to be visible without stain. The tension of the fibers is due to the surface of water on which the chromosome is spread. [Courtesy of F. Lampert, *Nat. New Biol.*, **234**: 187 (1971).]

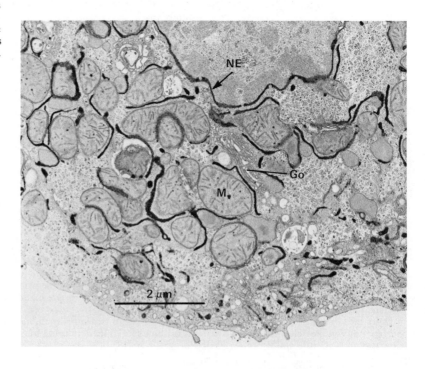

FIGURE A-14 Cytochemical Staining for the Electron Microscope. The cellular location of the enzyme glucose 6-phosphatase is revealed in this rat liver cell by virtue of its electron-dense reaction product. Note that the latter fills cisternae of the endoplasmic reticulum and nuclear envelope (NE) but spares cisternae of the golgi apparatus (Go). Mitochondrion (M). [Courtesy of P. Drochmans, J-C. Wanson, and R. Mosselmans, *J. Cell Biol.*, **66**: 1 (1975).]

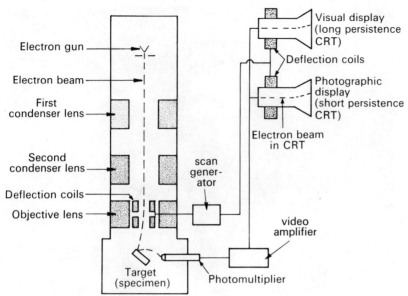

FIGURE A-15 The Scanning Electron Microscope. The beam is swept back and forth across the specimen in synchrony with a beam moving across the face of a television picture tube. Radiation (scattered or secondary electrons, etc.) from the sample is used to modulate the observation beam, producing variations in brightness that form the image.

microscopes, the early versions offered few advantages over other types of microscopes and much poorer resolution. By the early 1960s, however, significant improvements had been made and industrial applications were found. Soon thereafter, biologists also found ways to employ the new instrument.

The scanning electron microscope (see Fig. A-15) moves a thin beam of electrons back and forth across the specimen in the same way that an electron beam moves back and forth across the face of a television picture tube. In fact, a television picture tube is used to display the image, its own beam moving in synchrony with the scanning beam. As electrons strike the specimen in the microscope, they will be scattered or secondary electrons will be knocked from the sample; in either case electrons from the sample can be collected by a nearby photomultiplier tube. Scattered electrons will vary in abundance and in accessibility to the detector according to their origin. Crevices will produce fewer detectable electrons whereas projections will be highlighted. This variation is used to modulate the intensity of the beam sweeping across the picture tube, thus producing an image (Fig. A-16).

The scanning electron microscope can also be used in other ways. For example, the electron beam produces X rays after collision with various atoms, the wavelength of which is characteristic of the element hit. By analyzing emitted X rays, the elemental composition of the sample can be determined. This technique, called *energy dispersion X-ray spectroscopy*, has been used to determine the Ca^{2+} content of individual cells.

The resolution of the scanning electron microscope is much better than that of optical instruments but poorer than that of transmission electron microscopes. One obvious limitation is the beam diameter itself. The way in which samples must be prepared (to be discussed), plus the nature of the collection and display devices, add further limitations. Nevertheless, commercial instruments operate routinely at 100-Å resolution (0.01 μm) and experimental instruments with resolutions better than 5 Å—comparable to the best transmission microscopes—have been built.

APPENDIX
Tools of the Cell Biologist

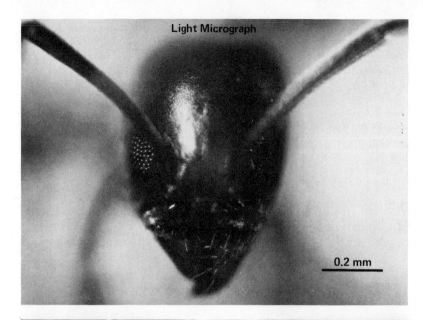

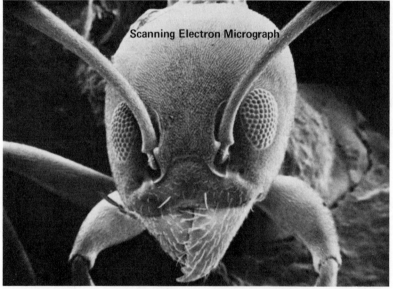

FIGURE A-16 Scanning Electron vs. Light Microscopy. A great advantage of SEM is its impressive depth of field. To demonstrate that difference an ant was photographed with a light microscope focused on one eye. The same ant was then coated with metal and examined with the scanning electron microscope. Note the depth of field of the latter instrument, making it an invaluable tool for examining surface features. [Photograph courtesy of John R. Devaney, Calif. Inst. of Technol., and Kent Cambridge Scientific, Inc.]

To prepare a specimen for the scanning electron microscope, one generally fixes or freeze-dries it in order to preserve its shape, then shadows it with a layer of heavy metal. The latter provides a conductive surface to drain off captured electrons that would otherwise cause distortions by deflecting the incoming beam. (Charge buildup can be a problem in transmission electron microscopy, too, though the thinner specimens transmit, rather than collect, most of the beam.)

Advantages and Disadvantages of the Three Types of Microscopy. Although resolution of the transmission electron microscope is clearly superior (at the present time at least) to that of the scanning electron microscope, and enormously greater than that of the light microscope, resolution is not the only criterion that affects the usefulness of an instrument. In fact, each of the three microscopes just described has applications in which it is clearly preferable to the others.

The traditional light microscope is easy to use and ignores the presence of water. All electron microscopes, on the other hand, are complicated by comparison, require more elaborate sample preparation, and with few exceptions can examine only dried specimens sitting in a vacuum chamber. (The electron beam is heavily scattered by air or water.) For these reasons, light microscopes are ordinarily the only practical way of viewing living objects.

In addition, the ability of the light microscopist to perceive color is an important asset, for variations in color—either the natural color of the specimen and its components or the color of various stains—can impart considerable information. In contrast, the transmission electron microscope is decidedly monochromatic. The ability of the scanning microscope to use several different kinds of radiation from the specimen provides some of the advantages of color but it is a weak substitute for the countless hues to which the human eye is sensitive.

The scanning electron microscope can only be used to observe surface features, for electrons that pass through the specimen are not seen. This factor is the source of both its strengths and its weakness. The light microscope is very poor at visualizing surfaces, and the transmission electron microscope can do so only indirectly by using a metal replica of the sample. Yet surface features can be of considerable interest. The scanning electron microscope reveals them with great clarity and an impressive depth of field—at least 300 times greater than that of an optical instrument (see Fig. A-16).

Where sample preparation time is critical (e.g., a pathologist's examination of a tissue while the surgeon waits by the operating table for a decision), the light microscope has no serious rival. In addition, the variety of molecular stains available for light microscopy give it a versatility unmatched by the other instruments. On the other hand, when details of intracellular structures are to be examined at the nanometer level, only the transmission electron microscope can be used. And when surface features are important, particularly when small features of varying depth are to be studied, the scanning electron microscope is the instrument of choice.

Some of these comparisons can be appreciated better by examining Fig. A-17, which shows white blood cells viewed by each of the instruments. Note for example that the lobulated, segmented nature of the granulocyte nucleus is appreciated only from the light micrograph while the electron micrograph, which was taken of a thinly sliced cell, fails to reveal nuclear shape but provides much more detail about cellular components.

A-2 X-RAY DIFFRACTION

We have seen that the working resolution of a modern electron microscope is about 5 Å and that higher resolutions are, in principle, available. At 5 Å, the shape of individual protein molecules should be readily discernible, and at 2 Å one should be able to determine the position of individual atoms within a molecule. Unfortunately, however, it isn't quite that simple.

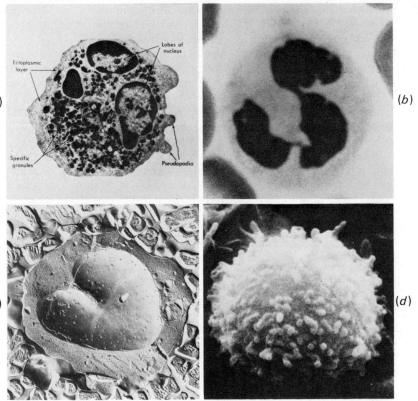

FIGURE A-17 White Blood Cells: Four Views. (a) Granular leukocyte (heterophil) from a guinea pig, viewed by transmission electron microscopy. Note that the nucleus winds in and out of the thin section, making it appear to be in three parts. (b) Comparable human granulocyte (neutrophil) as seen by light microscopy. Although cytoplasmic detail is lost, the true nuclear shape is now apparent. (c) Transmission electron micrograph of a human lymphocyte after freeze-fracture and etching (see text for discussion). A thin layer of cytoplasm surrounds the indented nucleus with its prominent pores. (d) Scanning electron micrograph of a lymphocyte. The nucleus is no longer visible, but surface features are.
[(a) Courtesy of W. Bloom and D. W. Fawcett, *A Textbook of Histology*, 10th ed., Philadelphia: W. B. Saunders, 1975; (b) courtesy of Murray L. Barr, Univ. of Western Ontario; (c) courtesy of R. Scott and V. Marchesi, *Cellular Immunology*, **3**: 301 (1972) © Academic Press, N.Y.; (d) courtesy of Aaron Polliack and Sloan Kettering Institute.]

In the first place, very high resolution electron micrographs can only be obtained from very thin specimens. But very thin specimens do not scatter electrons well. To obtain enough contrast to see an image, stains must be employed. The resolution under these conditions may be limited by the specificity and size of the molecule used as a stain. A phosphotungstic acid molecule, for instance, is considerably larger than 5 Å, so that negative staining with it cannot reveal details at the angstrom level.

In addition, the electron beam is very destructive. When it is focused on a single molecule long enough to take a picture, considerable damage is apt to result. The image will contain a correspondingly reduced amount of information.

And finally, images formed by scattered electrons can show only surface features of individual molecules, for most of the electrons are scattered near the surface. There is no immediate prospect, therefore, of using the electron microscope to determine in detail the atomic structure of biological macromolecules such as proteins.

The limitations of electron microscopy, then, are a direct result of its reliance on high-energy electrons as the incident radiation. The problem encountered when electrons are replaced by an alternate form of short-wavelength radiation—X rays, for example—is the lack of a suitable means of focusing the beam. But it isn't necessary to focus the beam to get the desired information.

If a distant point of light such as a street lamp is observed through a handkerchief or a window screen, it will appear as a cross composed of a pattern of tiny dots. With the proper knowledge of optics one could use that pattern, which is called a *diffraction pattern*, to determine the mesh and characteristics of the grid (i.e., the screen or handkerchief) that produced it.

In fact, a diffraction pattern will result when any radiation passes through a lattice or screen with openings larger than the wavelength of the radiation. A crystal of salt, for example, has a highly regular, three-dimensional array of Na^+ and Cl^- ions from which X rays will be diffracted in a predictable way. The spacing of nuclei within the lattice is readily calculated from the position and intensity of the spots in the resulting diffraction pattern.

Most proteins, because they can be obtained in a homogeneous preparation, crystallize to form regular arrays and hence produce diffraction patterns with a high informational content. The diffraction pattern formed by a good, stable crystal of protein contains the information needed to pinpoint the position of individual atoms. But as you no doubt suspect, deciphering such patterns is no easy task. The fact that a number of proteins have been successfully examined by this technique is a real tribute to the skill and resourcefulness of those who developed the needed mathematical tools and instrumentation.

Since the complexity of the task grows geometrically with the size of the protein, only small proteins (with a few exceptions) have been analyzed in detail by X-ray diffraction. However, the structure of some proteins cannot be determined in this way, either because they do not crystallize well (often the result of minor heterogeneity), because they are rapidly altered by the X-ray beam, or because it is not possible to "tag" them with heavy metal ions. (Tagging solves the "phase problem" by providing an unambiguous starting point from which to determine the relative positions of other atoms.)

Even molecules that do not crystallize can yield considerable information to the X-ray crystallographer provided only that a regular, repeating unit can be formed. Thus, DNA can be pulled into fibers in which the molecules are aligned with their long axes parallel to one another. Even though these fibers are not crystalline—there is no repeating pattern to the sequence of bases in the parallel strands—the regularity of the helix provides the basis for some order in the diffraction pattern (see Fig. A-18).

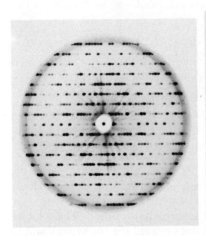 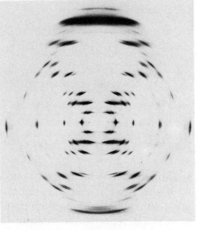

FIGURE A-18 X-Ray Diffraction. RIGHT: Diffraction pattern of a partly ordered structure, a DNA fiber. LEFT: Diffraction pattern of a highly ordered structure, a crystal of protein (adenylate kinase). Note the regular array of spots, with graded intensities. [Right photo, courtesy of M. H. F. Wilkins. Left photo, courtesy of I. Schirmer, H. Schirmer, and G. E. Schulz.]

One cannot determine the sequence of bases but, as explained in Chap. 3, the dimensions of the helix can be established.

Other noncrystalline biological systems that have been analyzed by X-ray diffraction include certain membranes (e.g., the myelin sheath of nerve cells) and the myofibrils of skeletal (striated) muscle.

For all its difficulties, X-ray diffraction is obviously a very powerful and very useful tool, capable of helping us understand the structure and hence the function of biological macromolecules. Nevertheless, X-ray diffraction can only be used to examine a limited range of biologically interesting specimens, including crystallized proteins, fibers of DNA, and the small number of other items with repeating units.

A-3 CENTRIFUGATION AND CELL FRACTIONATION

Centrifuges in their various forms are among the most versatile tools of molecular biology, for they are used not only to characterize substances but to separate them. The *analytical ultracentrifuge* provides information concerning the mass and (in a limited way) the shape of a molecule while *preparative centrifuges* permit one to use these parameters to separate molecular types.

An ultracentrifuge differs from other centrifuges only in attaining higher rotor velocities. In addition, the analytical ultracentrifuge contains an optical system, allowing one to observe changes in the solute distribution as they occur in the sample. The rotors of all ultracentrifuges spin in a vacuum in order to prevent heating from air friction. Modern commercial instruments may be operated at velocities up to about 70,000 revolutions per minute (rpm).

Centrifugal Fields. The force that any particle experiences in a spinning rotor is given as

$$F = m^*\omega^2 r \qquad \text{(A-3)}$$

where m^* is the buoyant mass of the particle (i.e., its mass less the mass of solvent it displaces), ω is the velocity of the rotor in radians per second, and r is the distance to the particle from the center of the rotor. The quantity $\omega^2 r$ is the *radial acceleration* or *centrifugal acceleration*. At 70,000 rpm, a particle 7 cm from the center of a rotor experiences an acceleration of

$$a = (70{,}000 \text{ rev/min} \times 2\pi \text{ rad/rev} \times \tfrac{1}{60} \text{min/s})^2 (7 \text{ cm})$$
$$= (7329)^2 \text{ s}^{-2} \times 7 \text{ cm} = 3.76 \times 10^8 \text{ cm/s}^2$$

The normal acceleration of the earth's gravity (g) is 980 cm/s². Hence, the particle in the ultracentrifuge experiences an acceleration that is

$$\left(\frac{3.76 \times 10^8}{980} = 384{,}000\right) \times g$$

which may be read as "384,000 g's."

Sedimentation Velocity. Any molecule or particle that is not isodense with the fluid it displaces will tend to float or sink, depending on whether it is lighter or heavier than the surrounding fluid.[1] The velocity, v, at which a particular substance moves toward the top or bottom of a liquid column will be proportional to the acceleration on it. The constant of proportionality is known as the *sedimentation coefficient, s*:

$$v = s\omega^2 r \qquad (A\text{-}4)$$

The sedimentation coefficient, in other words, is the velocity per unit acceleration. For example, the blood proteins known as "γ-globulin" have a component that sediments at a velocity of 2.6×10^{-4} cm/s (about 0.95 cm/h) at the previously computed centrifugal field ($384{,}000 \times g$). Its sedimentation coefficient is

$$s = \frac{2.6 \times 10^{-4} \text{ cm/s}}{3.8 \times 10^{8} \text{ cm/s}^2} = 7 \times 10^{-13} \text{ s}$$

The unit 10^{-13} second is known as a *svedberg* in honor of the Swedish scientist who developed the ultracentrifuge in the 1920s. Hence, we say that the above protein sediments at "7 svedbergs," or "7S" for short.

The sedimentation coefficient of a substance, or the velocity with which it sediments in a gravitational or centrifugal field, is proportional to its buoyant mass and is fastest for spherical particles. For spheres of different molecular weight, M, the sedimentation coefficient s increases as $M^{2/3}$.

Sedimentation velocity can be employed for both analytical and preparative purposes. The objective of an analytical application is usually to define either the sedimentation coefficient, which can then be related to other molecular parameters, or the number of species present (see Fig. A-19). Preparatively, the technique can be used to separate one sub-

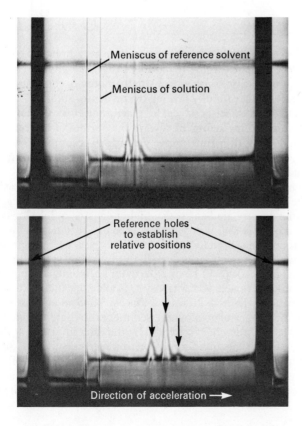

FIGURE A-19 Sedimentation Velocity. The optical system that produced these pictures is called the schlieren phaseplate system. It produces a line with vertical displacement proportional to the concentration gradient of the sedimenting species. There were three distinct species in this preparation (arrows in second picture), one of which was present in very low concentrations (leading peak).

APPENDIX
Tools of the Cell
Biologist

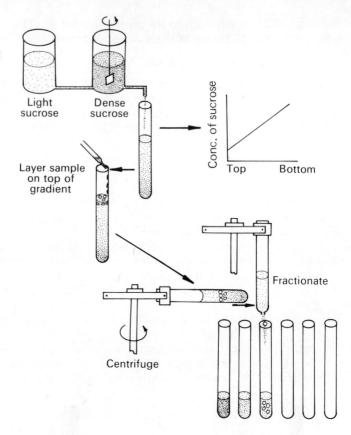

FIGURE A-20 Zone Sedimentation. A density gradient (e.g., of sucrose, as shown) is prepared in a tube. Such gradients are stable for many hours. A thin layer of solute is placed carefully on top and then the tube is spun in a centrifuge causing solute to move down the gradient in a compact zone. The solute may be recovered if the centrifuge is stopped before the zone reaches the bottom of the tube.

stance from another. The latter is most often accomplished by *zone sedimentation* or *density gradient sedimentation velocity*.

Zone sedimentation starts with a small amount of solution containing the macromolecule carefully layered over a denser solution void of macromolecule. The object is to get the macromolecules to maintain their compact band as they sediment toward the bottom of the tube (see Fig. A-20). To do that, they must move through a region that is continually increasing in density, for otherwise there would be convection and stirring in addition to sedimentation.

Sedimentation proceeds at a rate that decreases with increasing density of the supporting gradient. Thus, a molecule that diffuses ahead of its band in a density gradient sees a higher density and is slowed. Conversely, a molecule that lags behind will be in a region of lighter density and will therefore sediment faster until it catches up to the rest.

If the supporting medium is a gradient of viscosity as well as density, the banding properties will be improved, for sedimentation proceeds at a rate that varies inversely with viscosity. Molecules that diffuse ahead of the main band see both an increasing viscosity and an increasing density, with both factors acting to slow the sedimentation rate. And, of course, those that fall behind are accelerated by the drop in viscosity as well as in density of their surroundings. For this reason, solutions of sucrose and glycerol are widely used in zone sedimentation velocity experiments. A 40% solution of sucrose, for example, has a viscosity at 20 °C that is 6.2 times that of water, though it is only 1.18 times as dense. In a sucrose gradient, therefore, the sharpness of the macromolecular band will be due more to the viscosity gradient than to the density gradient.

Sedimentation Equilibrium. Most solutes of interest to us here are denser than water. Hence, in an aqueous solution they will tend to settle at a very slow rate if left undisturbed. That rate can be increased by the application of a centrifugal field, but the basic tendency to float or sink is not altered thereby. However, we do not expect any molecular-sized solute to settle out of solution in ordinary gravity, for its progress is opposed by back-diffusion due to normal thermal (Brownian) motion. Since such movements are random, they tend to cause a net migration from a region of greater to a region of lesser concentration.

Eventually, opposing movements fostered by sedimentation and diffusion will balance one another in a gravitational or centrifugal field, so that sedimentation is concentrating molecules near the bottom at the same rate as diffusion is moving them back. This condition is called *sedimentation-diffusion equilibrium,* or just *sedimentation equilibrium.* The solute, at sedimentation equilibrium, is distributed according to *Boltzmann's equation,* described in Chap. 2. In other words, the solute concentration will increase exponentially toward the bottom of the tube.

The purpose of allowing a solute to reach sedimentation equilibrium is usually to measure its molecular weight, which is related via the Boltzmann equation to the steepness of the gradient it forms. A second application of sedimentation equilibrium is the separation, for either analytical or preparative purposes, of solutes. Separation is often accomplished within an equilibrium gradient of a supporting solute as shown in Fig. A-21. The technique is called *density gradient sedimentation equilibrium.* A heavy salt, such as CsCl, is commonly used to form the gradient. Since this salt is both very dense and very soluble, solution densities of 1.7 g/ml and greater can be achieved. If a solution that is initially 1.7 g/ml is spun in a high-speed centrifuge one might, at sedimentation equilibrium, find a concentration gradient sufficient to provide a density of 1.65 at the top, where the CsCl will be less concentrated, but 1.75 g/ml at the bottom, where it is most concentrated.

Now suppose that the solution of CsCl also contains some DNA. The buoyant density of most DNA in CsCl is about 1.7 g/ml, meaning that 1.7 g of DNA displaces 1 ml of the salt solution. Although the DNA initially is isodense with its surrounding fluid, as the centrifugal field redistributes CsCl, molecules of DNA at the bottom of the tube find themselves in a medium that is more dense, while molecules of DNA at the top are left in a region of lowered density. The molecules at both extremes will tend to move toward the point in the CsCl gradient where they will be isodense with their surroundings. DNA molecules at the top sink toward that position while DNA molecules at the bottom float upward towards it. This movement to an isodense point is called *banding,* for the molecules tend to gather in a band, the width of which will decrease with the steepness of the CsCl gradient, which is a function of rotor velocity.

If more than one species of macromolecule is present—DNA isolated from two sources, for example—each may have a different buoyant density and hence band at different spots in the CsCl gradient. They will thus be separated from each other. This phenomenon can be used pre-

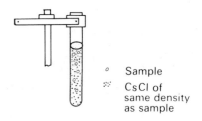
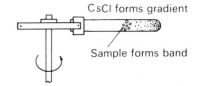
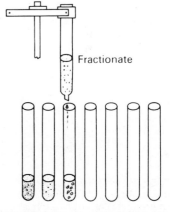

FIGURE A-21 Density Gradient Sedimentation Equilibrium. Concentrated solutions of CsCl and other heavy salts form relatively steep gradients when spun in an ultracentrifuge. Macromolecular solutes that are isodense with some level of the gradient will collect there in a band or zone that can be preserved by emptying the tube very slowly.

paratively by carefully emptying the sample tube in order to deliver each band of macromolecule to a separate container (see Fig. A-21). It can also be used analytically, to define the presence or relative amounts of the two species without necessarily preserving their separation after the centrifuge is stopped.

Cell Fractionation. The sedimentation techniques just described can be used to fractionate cells—that is, to separate various particulate components one from the other. A simple but effective procedure is outlined in Fig. A-22. In this sequence, the cell is gently broken up by an homogenizer, then subjected to centrifugations of increasing velocity. At each step larger particles form a gelatinous pellet at the bottom of the tube leaving smaller particles in the supernatant solution. By decanting the supernatant and spinning it harder, the next fraction can be brought down. Ultimately one is left with a supernatant solution having only soluble, molecular-sized components. This residual solution is the *cytosol*.

Density gradient sedimentation velocity is a refinement of the above procedure that often permits better separation of components. Sucrose gradients are sometimes used, although their osmotic strength is deleterious to some cellular organelles. To avoid that problem, high molecular weight synthetic or natural polymers (e.g., albumin) are often used to form the gradient.

Sedimentation equilibrium in density gradients is also used on occasion for cell fractionation, again using high molecular weight, low os-

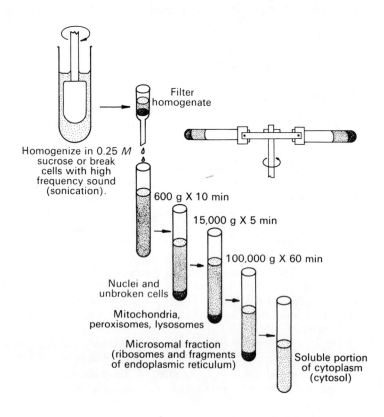

FIGURE A-22 Cell Fractionation by Centrifugation. The different sedimentation rates of various cellular components make it possible to achieve a partial separation by differential centrifugation. Nuclei and virus particles can sometimes be purified completely by such a procedure. Components of the soluble fraction can themselves be further fractionated by centrifugation, but CsCl or sucrose gradients are apt to be required, as described in the text.

motic pressure macromolecules to form the gradient. The technique of density gradient equilibrium has, however, been most widely used to separate and identify nucleic acids.

A-4 ISOTOPES

The characterization of biological systems has been made immeasurably easier by the advent of isotopes. They are used to follow the progress of an added substance through cellular processes. Their presence can be detected either because of their density or because they emit radiation. The isotopes commonly used for *density labeling* are ^{2}H (deuterium, sometimes abbreviated as D), ^{13}C, ^{15}N, and ^{18}O. Those that have been most useful as *radioactive tracers* are listed in Table A-1.

Density Labeling. As an example of density labeling, consider the effect of growing cells in the presence of ammonium salts in which ^{14}N has been replaced by ^{15}N. The cells will synthesize deoxyribonucleic acid (DNA) that is significantly more dense than usual. The difference is easily detected in a CsCl gradient equilibrium experiment. As explained in Chap. 11, this fact was used to show that the double-stranded DNA molecule unwinds as it is replicated, so that each replica contains one new and one old strand.

Labeling with Radioactive Isotopes. In order to achieve density labeling, a significant fraction of the atoms must be heavy isotopes. If the isotopes are radioactive, however, a much smaller incorporation may be detectable. Suppose, for instance, that one wishes to distinguish DNA made after viral infection of a cell from DNA made prior to infection. That can be done by adding to the cells ^{32}P (in the form of phosphate) or a tritiated precursor (i.e., one containing ^{3}H) of DNA, along with or soon after the virus. DNA made subsequent to the addition will be radioactive; that made prior to the addition will not be radioactive.

TABLE A-1. Some Commonly Used Radioactive Isotopes

Isotope	Half-life	Energy of emitted electrons
^{3}H	12.26 yr	0.0181 MeV[b]
^{14}C	5770 yr	0.156
^{24}Na[a]	15.0 h	1.39
^{32}P	14.3 days	1.71
^{35}S	87.1 days	0.167
^{36}Cl	3×10^5 yr	0.71
^{42}K	12.4 h	3.53
^{45}Ca	165 days	0.25
^{59}Fe[a]	45 days	0.46
^{131}I[a]	8 days	0.61

[a] These isotopes also emit γ rays.
[b] An MeV is a million electron volts, or the energy of an electron that has been accelerated by a million volt potential.

APPENDIX
Tools of the Cell Biologist

In a variation of the above experiment, one might determine the relative rates of DNA synthesis before and after viral infection by using *pulse labeling*. If the isotope is added to the cells for a few minutes and then followed by a "chaser"—i.e., a large quantity of the same compound, but unlabeled—competition between the common and radioactive isotopes will virtually stop uptake of the latter. (If there are a thousand nonradioactive molecules for each labeled molecule, then only 0.1% of the incorporated molecules will be labeled.) The amount of radioactivity in the cell's DNA fraction after a given pulse might be compared with the amount incorporated into the cells of a control culture given the same pulse but no virus. If the amount of radioactive DNA per cell is twice as much in the infected culture, then the rate of synthesis is probably twice as fast.

Measuring Radioactivity. The preceding experiments presume that radioactivity can be detected. There are, of course, many instruments designed to do that. The familiar Geiger counter is one, though it is not used much in biology since most isotopes employed in biology emit relatively weak β rays. (β Rays are high-energy electrons; α rays are helium nuclei; and γ rays are a high-energy, short-wavelength electromagnetic radiation.) The most useful instrument for quantitatively monitoring β

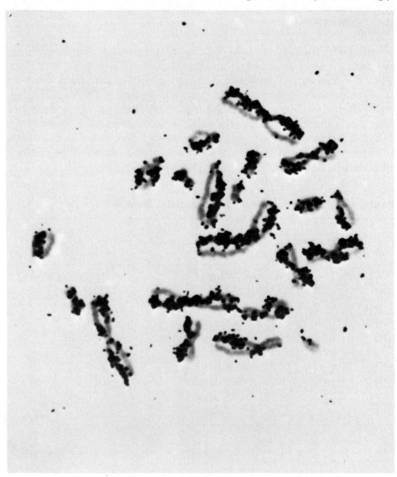

FIGURE A-23 Autoradiography. The radioactive label (^{3}H, revealed by the dense silver grains) is localized almost exclusively in only one of the two strands of each chromosome. The chromosomes themselves are visible because of a light micrograph overlay. [Courtesy of G. Marin and D. M. Prescott, *J. Cell Biol.*, **21**: 159 (1964).]

emitters is the *scintillation counter*. To use it, radioactive material is placed in a solution containing a scintillator (e.g., a naphthalene derivative), which emits photons of light when excited by β rays. When light is picked up by sensitive photomultiplier tubes, an emitted electron is "scored."

A quite different way of utilizing radioactive isotopes is in *autoradiography* (see Fig. A-23). Suppose one wished to find where protein is synthesized in a cell. To do that the cells might be pulse-labeled with $^{35}SO_4^{2-}$, or with a radioactive amino acid, and then fixed as if for microscopy. A thin photographic emulsion is layered over the slide and the "sandwich" is stored in the dark, sometimes for months. Regions of radioactivity produce silver grains in the emulsion which, after developing, can be seen with the microscope along with the cell, thus pinpointing the location of the radioactive isotopes. The technique was used in the early 1950s to show that protein synthesis occurs in the cytoplasm rather than in the nucleus, and has been used in many ways since then.

SUMMARY

A-1 There are three important types of microscopes in common use: (1) the light microscope, (2) the transmission electron microscope (TEM), and (3) the scanning electron microscope (SEM). The light microscope and TEM form images from transmitted radiation (light and electrons, respectively). The SEM scans a beam back and forth across the specimen in synchrony with the moving beam of a television picture tube. Scattered electrons or emitted radiations from the specimen are used to modulate the intensity of the picture tube's beam, and thus to form the image.

Each of the three classes of microscopes has tasks for which it is best suited. The light microscope permits observation of living cells (aided by the use of phase contrast optics), and can utilize the transparency of cells and the specificity of a host of chemical stains to observe many internal features. The transmission electron microscope can define structures at the 10-Å level, but only after they have been cut to eliminate overlying and underlying structures. Because the specimen is then only a few hundred angstroms thick, it is difficult to reconstruct the original three-dimensional relationships. Three-dimensional viewing, on the other hand, is an area in which the SEM excels, though only surface features are seen because the scattered and emitted radiations used to form the image originate mostly from the surface.

The specimen may be prepared in various ways for microscopy, each way being designed to reveal certain types of features. Shape may be preserved by chemical fixing or by freeze-drying; contrast may be improved with selective stains; and in the case of transmission electron microscopy, thin sectioning and staining (the most common technique) is augmented by negative staining, metal shadowing, or whole mounts (all used mostly with particulate specimens), and by freeze-fracture.

A-2 Any specimen that is comprised of regular, repeating units will form a diffraction pattern with radiation of the proper wavelength. That pattern can, in principle, be used to reconstruct the specimen. Fibers of DNA and the highly ordered myofibrils of muscle have been examined in this way, using X rays as the radiation. The most highly ordered specimens, however, and the ones that produce the best diffraction patterns, are crystalline. Accordingly, a number of crystalline proteins have had their complete spatial configurations determined at the angstrom level by X-ray crystallography. Although the resolution is similar to that of good electron microscopes, the information supplied is much different. Because its beam is scattered by electrons in the specimen, the electron microscope provides only a surface picture of individual molecules. Since X rays are diffracted by nuclei, molecules appear much more open to X rays, permitting their internal features to be established.

A-3 Ultracentrifuges can subject a solution to very high centrifugal fields, a factor that can sometimes be used to separate solutes from one another (preparative ultracentrifugation) or to characterize a solute by observing its behavior with an appropriate optical system (analytical ultracentrifugation).

Any solute molecule subjected to an acceleration (gravitational or centrifugal) will tend to float or sink according to whether it is lighter or heavier than the solvent it displaces. The rate at which it does so determines its sedimentation coefficient, and that, in turn, is a known function of its mass, density (buoyancy), and shape. If, however, sedimentation is continued long enough, a condition of sedimentation-diffusion equilibrium will be established. The distribution of solute molecules is then determined (for a given acceleration) by their mass and density, independent of shape.

The solute of interest is often surrounded by a *density gradient* formed by a second solute (sucrose, glycerol, CsCl, etc.). Sedimentation velocity in a density gradient starts with a thin layer of macromolecular solution at the top, and depends on the gradient of density (and sometimes viscosity) to stabilize it, causing the band to stay together as it sediments. A clean separation of two solutes with different sedimentation rates can often be achieved. During sedimentation equilibrium in a density gradient, on the other hand, the macromolecule "bands" at its isodense point regardless of its initial distribution, thus separating solutes of similar mass but of different densities.

A-4 Isotopes are used to label or "tag" molecules, allowing them to be followed through a cellular process or to be detected after isolation. Labeling may be accomplished by synthesizing (or having a cell synthesize) the molecule with a significant portion of its atoms replaced by heavy isotopes (^{2}H, ^{13}C, ^{15}N, ^{18}O, etc.). Such a molecule can then be distinguished from a normal, unlabeled molecule by its density as determined by CsCl gradient equilibrium banding (Section A-3). However, if some of the atoms are radioactive isotopes instead of heavy isotopes, the labeled molecule can be detected by its emitted radiation.

Autoradiography is another important application of radioactive isotopes. In this technique, a developed photographic emulsion, "exposed" by radiation from the specimen, is viewed along with the specimen to reveal the location of radioactive label.

STUDY GUIDE

A-1 (a) What is meant by resolution, or resolving power? What parameters affect it? (b) What are the basic operating principle and typical resolution of the light microscope? the transmission electron microscope? the scanning electron microscope? (c) What conditions might dictate a choice among the three classes of microscopes? (d) What ways are available to improve contrast in light microscopy? in electron microscopy? (e) Five types of preparation techniques for electron microscopy were discussed. What are they, and for what purposes are they employed?

A-2 (a) What is the principle on which X-ray diffraction techniques are based? (b) Instead of interpreting a complicated diffraction pattern, why can't one merely focus the X-ray beam to form an image directly? (c) What advantage does X-ray crystallography at 5-Å resolution have over electron microscopy at the same resolution?

A-3 (a) How fast would a protein having a sedimentation coefficient of 7 svedbergs (for example, a "7S" antibody) sediment through water in a gravitational field? Neglect the effects of convection and diffusion (Brownian motion). [Ans: 0.07 Å/s] (b) For what purposes are sedimentation velocity experiments used? (c) What is the purpose of using a density gradient in a sedimentation velocity experiment? How does it work? (d) What is the purpose of using a density gradient in a sedimentation equilibrium experiment? Again, how does it work? (e) Describe, without quantitative details, a common scheme for cell fractionation.

A-4 (a) What is meant by density labeling? How is the presence of label detected? (b) What is autoradiography, and for what types of problems is it used?

REFERENCES

MICROSCOPY

BREEDLOVE, J. R., Jr., and G. T. TRAMMELL, "Molecular Microscopy, Fundamental Limitations." *Science,* **170**: 1310 (1970). When molecular damage from X rays or electrons becomes limiting.

CASIDA, L., Jr., "Leeuwenhoek's Observation of Bacteria." *Science,* **192**: 1348 (1976). Dark-field microscopy.

CHANDLER, J., *X-ray Microanalysis in the Electron Microscope.* Amsterdam: Elsevier/North Holland Publ. Co., 1977.

CLARK, J. M., and S. GLAGOV, "Evaluation and Publication of Scanning Electron Micrographs." *Science,* **192**: 1360 (1976). Some of the pitfalls and artifacts.

CREWE, A. V., and J. WALL, "A Scanning Microscope with 5 Å Resolution." *J. Mol. Biol.,* **48**: 375 (1970). Uses transmitted electrons and thin sections.

DA SILVA, P., and D. BRANTON, "Membrane Splitting in Freeze-etching." *J. Cell Biol.,* **45**: 598 (1970). Proof that freeze fracture of red cell membranes splits the membrane without exposing either surface.

DAWES, CLINTON J., *Biological Techniques in Electron Mi-*

croscopy. New York: Barnes and Noble, and London: Chapman & Hall, 1971.

EVERHART, T. E., and T. L. HAYES, "The Scanning Electron Microscope." *Scientific American*, January 1972.

GLAUERT, A. M., "The High-voltage Electron Microscope in Biology." *J. Cell Biol.*, **63**: 717 (1974). A review.

GLAUERT, A. M., ed., *Practical Methods in Electron Microscopy*, Vol. 5. New York: Elsevier/North Holland, 1977. Part I on stains and Part II on X-ray microanalysis are available in separate paperbacks.

HAYAT, M. A., *Basic Electron Microscopy Techniques*. New York: Van Nostrand Reinhold, 1972. Emphasis on the preparation of biological materials.

———, *Positive Staining for Electron Microscopy*. New York: Van Nostrand Reinhold, 1975.

HAYAT, M. A., ed., *Principles and Techniques of Electron Microscopy: Biological Applications*. New York: Van Nostrand Reinhold, 1974–1976. Six volume treatise. Not for beginners.

HÜNDGEN, M., "Potential and Limitations of Enzyme Cytochemistry." *Int. Rev. Cytol.*, **48**: 281 (1977).

JAMES, J., *Light Microscopic Techniques in Biology and Medicine*. Martinus Nijhoff: The Hague, 1976.

MEEK, G. A., *Practical Electron Microscopy for Biologists*, 2d ed. New York: John Wiley, 1976.

NEEDHAM, G. H., *The Practical Use of the Microscope*. Springfield, Ill.: Chas. C. Thomas, 1977.

OSTER, GERALD, ed.,*Physical Techniques in Biological Research* (2d ed.), Vol. I, pt. A: *Optical Techniques*. New York: Academic Press, 1971. Chap. 1 is on light microscopy.

PALADE, GEORGE E., "Albert Claude and the Beginnings of Biological Electron Microscopy." *J. Cell Biol.*, **50**: 5D (1971).

PARSONS, D. F., "Structure of Wet Specimens in Electron Microscopy." *Science*, **186**: 407 (1974). High voltage EM of living cells.

ROBINSON, A. L., "Electron Microscopy: Imaging Molecules in Three Dimensions." *Science*, **192**: 360 (1976). A short review of attempts to bridge the gap between electron microscopy and X-ray diffraction.

STOLINSKI, C., and A. S. BREATHNACH, *Freeze-Fracture Replication of Biological Tissues. Techniques, Interpretation and Applications*. New York: Academic Press, 1975.

YAMADA, E., V. MIZUHIRA, K. KUROSUMI, and T. NAGANO, eds., *Recent Progress in Electron Microscopy of Cells and Tissues*. Baltimore: Univ. Park Press, 1976.

X-RAY DIFFRACTION

BRAGG, LAWRENCE, "First Stages in the X-ray Analysis of Proteins." In *Biophysics*, by W. Fuller, C. Rashbass, L. Bragg, and A. C. T. North. Menlo Park, California: W. A. Benjamin, Inc., 1969, p. 89. (Paperback.)

KARTHA, G., "Picture of Proteins by X-ray Diffraction." *Accounts Chem. Res.*, **1**: 374 (1968).

MUIRHEAD, HILARY, "X-ray Analysis of Proteins." *Sci. Progr. (Oxford)*, **53**: 17 (1965).

PERUTZ, M. F., "X-ray Analysis, Structure and Function of Enzyme Molecules." *Eur. J. Biochem.*, **8**: 455 (1969).

WILSON, H. R., *Diffraction of X-rays by Proteins, Nucleic Acids and Viruses*. New York: St. Martin's Press, Inc., 1966. An elementary account.

CENTRIFUGATION AND CELL FRACTIONATION

BIRNIE, G. D., *Subcellular Components: Preparation and Fractionation*. Baltimore: Univ. Park Press, 1972.

DE DUVE, CHRISTIAN, "Exploring Cells with a Centrifuge." *Science*, **189**: 186 (1975).

———, "The Separation and Characterization of Subcellular Particles." In *The Harvey Lectures, 1963–1964 (Ser. 59)*. New York: Academic Press, 1965, p. 49.

———, "Tissue Fractionation, Past and Present." *J. Cell Biol.*, **50**: 20D (1971).

BIOLOGICAL APPLICATION OF ISOTOPES

BASERGA, R., and W. E. KISIELESKI, "Autobiographies of Cells." *Scientific American*, August 1963. (Offprint 165.) Autoradiography.

BUDD, G. C., "Recent Developments in Light and Electron Radioautography." *Int. Rev. Cytol.*, **31**: 21 (1971).

GAHAN, P. B., ed., *Autoradiography for Biologists*. New York: Academic Press, 1972.

GUDE, WM. D., *Autoradiographic Technique*. Englewood Cliffs, N.J.: Prentice-Hall, 1968. (Paperback.)

JACOB, J., "The Practice and Application of Electron Microscope Autoradiography." *Int. Rev. Cytol.*, **30**: 91 (1971).

WOLF, GEORGE, *Isotopes in Biology*. New York: Academic Press, 1964. (Paperback.)

NOTE

1. Typical densities for the major macromolecules are 1.7 or more (i.e., 1.7 times as great as water) for nucleic acids, about 1.6 for carbohydrates, 1.4 for proteins, and around 1 for many lipids and lipoproteins. In fact, some of the latter are less dense than water and hence float during centrifugation.

Index

A

Abbe's relationship, 577
Acetabularia, gene-cytoplasm interaction, 572
Acetic acid, titration, 64
Acetylcholine, 356, 480
   and cardiac muscle, 395
   and cGMP, 168
   receptor density, 439
   and synaptic transmission, 359
   and uterine contraction, 394
Acetylcholine esterase, 359
Acetylcoenzyme A, 302
   hydrolysis of, 140
Acetylsalicylic acid (*see* Aspirin)
Acids, *defined*, 62
Acid phosphatase, 448
Acne, 470
Acriflavins, 503
Acrosomal process, 410
ACTH, synthesis of, 204
Actin, 23
   and cytokinesis, 531
   discovery and properties, 379
   filaments of, 379
      polarity, 385
   non-muscle, 395, 398
      filaments of, 395
   smooth muscle, 385
   and the spindle apparatus, 527
α-Actinin, 381, 398, 413
Actinomycin D, 553
Action potentials:
   in artificial membranes, 358
   Ca^{2+} effects, 355
   cable effect, 358
   excitatory and inhibitory, 362
   flicker fusion, 357
   in heart muscle, 355, 356
   initiation, 355
   intercellular transmission, 359
   local currents, 357
   mechanism of, 354
   in muscle, 393
   Na^+ and K^+ effects, 354
   refractory period, 354, 357
   saltatory conduction, 358
   speed of conduction, 358
   synaptic transmission, 359
   transmission by catecholamines, 362
Action spectrum, of photosynthesis, 316
Activation, energy of, 135
Active transport, *defined*, 274
   properties, 274
   tests for, 274
Actomyosin, 379
Adenine, structure, 85
Adenocarcinoma, 564
Adenyl cyclase, 166
   and hormone receptors, 438, 440
Adenylate cyclase (*see* Adenyl cyclase)
Adenylate deaminase, 386
Adenylate kinase:
   axonemal, 404
   in mitochandria, 293
   in muscle, 386
Adrenaline (*see* Epinephrine)
Aequorin, 323, 400
   and Ca^{2+} detection, 389
Afzelius, B., 402, 403
Aggregation factors, 432
Aging, theories of, 568
Albumin, 464
   synthesis of, 469
Alcohol, oxidation of, 147
Alcoholic fermentation, 143, 150
   history of, 4
Aldehyde, oxidation of, 148
Aldose, *defined*, 78
Algae:
   blue-green, 8
   chloroplasts of, 313
Alleles, *defined*, 175
Allelomorphs, *defined*, 175
Allen, R. D., 422
Allergens, 121
Allergic rhinitis, 126
Allolactose, 211
Allostery, 159
Altmann, R., 288
Ameboid locomotion, 422
Amine oxidases, in mitochondria, 293
Amino acids:
   breakdown, 303
   classification, 91
   derivatives of, 93
   diastereo isomers, 94
   isoionic pH of, 94
   optical activity, 93
   peptide bonds, 94
   residue of, 95
   sodium-coupled transport, 276
   table of, 92
   transport of, 263
   as zwitterions, 94
Aminoacyl-tRNA, 198 (*see also* RNA, transfer)
Aminoacyl-tRNA synthetase, 198
Aminoglycoside antibiotics, 201
Ammonia, ion trapping of, 257
Amniocentesis, 450
AMP, from ADP, 140 (*see also* Cyclic AMP)
Amphibia, limb regeneration, 557
Amphipathic lipids, *defined*, 77
Amphotericin B, and membrane pores, 258
Amylopectin, structure, 81
Amylose, structure, 81
Anabolism, *defined*, 143
Anaphase, meiotic, events of, 520
Anaphase, mitotic, events of, 515
Androgens, 77
   and gene control, 226
Angiotensin, and smooth muscle, 394
Ankylosing spondylitis, 434
Annulate lamellae, 458
Anomers, *defined*, 79
Antibiotics, ionophorous, 279
Antibodies:
   allotypes, 125
   anamnestic (secondary) response, 127
   cross-reacting, 122
   classes of, 125
   clonal selection, 127
   genes of, 205
   and haptens, 121

Antibodies (cont.)
  and multiple myeloma, 124
  properties, 118
  reactivity, 119
  structure, 119, 124
  synthesis, 126
Anticodon, defined, 198
Antidiuretic hormone, 258, 271
Antigens, defined, 119
  human lymphocyte, 434
Antioxidants, 504
Arginase, lifetime of, 228
Arginine phosphate, in muscle, 386
Arthritis, 434
Aryl sulfatase, 448
Ascorbic acid (see Vitamin C)
Ashley, C. C., 389
Aspirin:
  ion trapping of, 256
  and prostaglandins, 74
  structure, 256
Astbury, W. T., 96
Asters, 511, 513
  and cytokinesis, 524
Ataxia telangiectasia, 504
Atherosclerosis, 77
ATP:
  and coupled reactions, 139, 142
  and electron transport, 295
  from glycolysis, 144, 147
  in muscle, 386
  from photophosphorylation, 320
  structure of, 139
ATPase:
  Ca^{2+} activated, 391, 392
  dynein, 403
  membrane, 270
  mitochondrial, 307
  myosin, 381
  in photosynthesis, 322, 332
  of prokaryotes, 332, 335
Atropine, 361
Autoimmune disease, 435
Autophagy, 451
Autoradiography:
  technique, 598
  of DNA, 496
Autotroph, 10
Axon, 349, 372
  giant, 353
Axoneme:
  of cilia and flagella, 401
  growth of, 407
  of suctorians, 412
Axoplasm, defined, 353
Axoplasmic transport, 419
Axosome, 401, 407
Avery, O. T., 182

B

Bacteria:
  chromosomes, 206
  classification, 8, 10
  discovery, 8
  episomes, 208
  F factors, 208
  lysosomal killing, 449
  mating of, 208, 499
  and membrane lysis, 247
  phagocytosis of, 445, 446
  plasmids, 208, 236
  R factors, 208
  reproduction, 510, 538
  shape, 9
  structure, 10
  swimming pattern, 407
Bacteriophage:
  discovery, 31
  and diphtheria, 200
  life cycle, 35
  lysogenic, 35
  temperate, 35
  virulent, 35
Bacteriophage f2, 203
Bacteriophage ϕX174, 205
Bacteriophage R17, 203
Bacteriophage SP8, 193
Bacteriophage T2, genes of, 183
Bacteriophage T4, polymerase of, 215
Baltimore, David, 562
Bang, O., 560
Barr body, 219, 401
Basal body, bacterial, 11
Basal body, eukaryotic, 23
  anchors, 405
  as microtubule organizers, 114
  replication of, 406
  similarity to centrioles, 406
Basal corpuscle (see Basal body)
Base, 62
Basophils, 478, 491
Beadle, G. W., 180
Beijerinck, M. W., 31
Bence-Jones protein, 124
Berg, Howard, 407
Bernstein, Julius, 372
Binary fission, 492, 537
Bioluminescence, 323
Birefringence, of spindles, 524
Blackhead, 75, 429
Blastema, defined, 557
Blastogenesis, defined, 542
Blobel, Gunter, 463
Blood types, 433
Bohr effect, 160, 161
Boltzmann, Ludwig, 47
Bonds:
  coordinate, 59
  covalent, 48
  dative, 59
  disulfide, 93
  energy of, 48
  "high energy," 139
  hydrogen, 50
    in proteins, 96
  hydrophobic, 53
    temperature effect, 112
  ionic, 51
  isopeptide, 104, 116
  peptide:
    formation of, 199
    hydrolysis of, 137
  short-range, 54
  stability of, 46, 49
  weak, 48
Bond energy, 46
Bond stability, 46, 49
Boyer, P. D., 298
Branton, Daniel, 243
Braun, Armin C., 563
Britten, R.J., 225
Bronsted acid or base, 62
Brown, Michael, 485
Brown, Robert, 16
Brown fat, and heat production, 306
Buchner, Eduard, 4
Buffers, 64
Buller, A. J., 387
Bullough, W. S., 544
Burnet, Sir Macfarlane, 126

C

Caffeine, 173
  and adenyl cyclase, 443
  effect on muscle, 392
  as a mutagen, 503
Cairns, John, 496, 507
Calcium:
  and action potentials, 355
  and ameboid locomotion, 422
  and cell division, 542
  in cellular recognition and adhesion, 432
  and contractile rings, 531
  and cytoplasmic streaming, 418
  and the cytoskeleton, 414
  and endocytosis, 445
  and enzyme regulation, 168
  and erythrocyte shape, 416
  and gap junctions, 253
  and glycogen mobilization, 165
  and hormones, 167
  and lymphocyte activation, 435
  and microtubule assembly, 114
  and muscle contraction, 388, 391
    hormonal effects, 394
  and nerve synapses, 361
  and non-muscle contraction, 399
  regulation by mitochondria, 309
  and secretory events, 478
  and vitamin D, 265
Calcium metaphosphate, 141
Calcium pump, 392
  of muscle, 391
cAMP (see Cyclic AMP)
Cancer:
  clonal origin, 567
  is it contagious?, 565
  and immunity, 567
  the malignant cell, 558
  oncogene theory, 561

treatment by radiation and chemicals, 505
and viruses, 560
Capillaries, 447
Capsule, bacterial, 11
Carbohydrates, *defined*, 78
　caloric value, 346
　cell surface, 247
　density of, 601
　elementary composition, 71
Carbonic acid, 160
Carboxylase, *defined*, 138
Carcinogens, 502
Carcinomas, 558
Cardiac glycosides, 268
Cardiolipin, 77
　in mitochondria, 290
Caro, Lucien, 473
Casein, 480
　phosphorylated amino acids, 93
Catabolism, *defined*, 143
　stepwise, 144
Catalase, function, 106
　in peroxisomes, 336
Catalysts, properties of, 135
Catecholamines, *defined*, 163
　as neurotransmitters, 362
　receptors for, 439
　turnover, 442
Cell cycle, *defined*, 510
　checkpoints, 537
　controls, 536
　G_0 cells, 539
　G_2 arrest, 538
　typical times, 537
　variable length of G_1, 538
Cell doctrine, 4
Cell membrane (*see also* Membranes)
　eukaryotic, 14
　prokaryotic, 12
Cell plate, 534
Cell sap, 25
Cell theory, 3
Cell walls:
　eukaryotic, 25, 534
　prokaryotic, 11
Cellobiose, structure, 81, 82
Cells:
　aging of, 568
　chromaffin, 368
　culture of, 29
　cytoskeleton of, 408
　differentiation of, 550
　diversity of, 6
　division of (*see* Cytokinesis)
　equilibrium potential, 272
　eukaryotic, 13
　fractionation of, 592, 596
　fusion of, 238
　glycocalyx, 431
　HeLa, 575
　intercellular junctions, 248 (*see also* individual names)
　malignant, characteristics, 558
　mast, 478

osmotic regulation of, 270
permeability of, 254
plasmolysis of, 254
poles of, *defined*, 512
programed death, 568
prokaryotic, 7, 8
recognition of, 431
red (*see* Erythrocytes)
sensory, 364
somatic, *defined*, 177
surface features, 411
Cellulose, 11, 25
　synthesis of, 469
Centrifugal force, 592
Centrifugation, 592
　density gradient, 186, 594
　zone, 594
Centrioles, 23
　as microtubule organizers, 114
　prophase movements, 512
　replication of, 512
　similarity to basal bodies, 406
Centromeres, division of, 515
Centrosome, 24
Ceramide, 76
Cerebrosides, 76, 83
Ceruloplasm, 446
Chalones, 544
Chamberlain, C., 31
Chargaff, E., 85
Charged RNA, 198 (*see also* RNA, transfer)
Chase, M., 183
Chelate, *defined*, 60
Chemical reactions, energetics of, 58
Chemiosmotic hypothesis, 298
Chemotaxis:
　in bacteria, 408
　in eukaryotic cells, 422
Chiasmata, 177, 520
Chitin, 83
Chloramphenicol, 201, 333
Chlorophyll, 27
　bacterial, 330
　light absorption, 317
　relationship to heme, 106
　structure, 317
Chloroplasts:
　biosynthetic activities, 327
　coupling factor, 322
　development of, 312
　dimorphic, 327
　DNA of, 333
　evolutionary origin, 330
　membrane structure, 321
　membrane turnover, 484
　photophosphorylation by, 320
　pigment systems, 317
　prolamellar bodies, 312
　replication of, 332
　structure, 309
　thylakoids of, 311, 312
Chlorpromazine, 372
Cholera toxin, 441
Cholesterol, 77

excess, 491
function, 77
and membranes, 246, 248, 485
required by insects, 485
Chondriosomes, 287 (*see also* Mitochondria)
Chondroitins, 83
Chromatids, replication of, 506
　sister chromatid exchange, 507
Chromatin, 16
　attachment to nuclear membranes, 453, 512
　condensation of, 512, 539
　replication, 506
　structure of, 216
Chromatophores, 421
Chromocenter, 222
Chromomere, 222, 226
Chromoplast, 25, 309
Chromosomes, 16
　anaphase movements, 515
　autosomal, 236
　B, 236
　chromomeres of, 222
　crossover, 519
　eukaryotic, structure, 216
　and genes, 176
　heteromorphic, 178
　homologous, 177
　human, 179, 219
　interarm fibers, 509
　lampbrush, 220, 522
　mechanism of movement, 526
　number, 177
　polytene, 222
　　ecdysone effect, 226
　prokaryotic (structure of), 206
　puffing pattern, 222
　replication of, 506
　sex, 178
Chronic granulomatous disease, 449
Chylomicrons, 107
Chymotrypsin, mechanism of, 137
Cilia, 24
　growth of, 405
　metachronal waves, 405
　movement of, 403, 404
　structure of, 401
Ciliogenesis, 405
Cistron, *defined*, 181
Citric acid cycle, 302
Claude, Albert, 14
Cloning of plants and animals, 551
Cobalt-ammine, 59
Cocaine, 358
Code words, *defined*, 187 (*see also* Codon)
Codon, *defined*, 187
　assignment to amino acids, 189
　nonsense, 191
　terminator, 191, 204
　wobble hypothesis, 191, 199
Coenocyte, *defined*, 524
Coenzyme A, 302
Coenzyme Q, 295, 296

605

Colcemide, 115, 525
Colchicine, 115, 450
  and axoplasmic transport, 419
  and cell division, 524, 549
  and endocytosis, 445
  and insulin release, 476
  and lymphocyte capping, 436
  and microtubules, 412
Cole, K. S., 354, 356
Coliphages, 32
Collagen:
  assembly, 115
  and copper, 106
  cross-linking, 103
  defects in, 109
  as a glycoprotein, 108
  hydroxylated amino acids, 93
  from procollagen, 109
  registration peptides, 115
  structure of, 98
  turnover, 115
Colloids, 110
Color blindness, inheritance, 180
Comedones, 75, 429
Complement, 121, 445
Compound, coordination, 59
Concanavalin A, 437
  and lysosomal function, 450
Cone cells, 365
Conklin, E. G., 524
Contact inhibition, 540
Contractile ring, 529
Contraction:
  in non-muscle cells, 399
  in smooth muscle, 377
  in striated muscle, 382
Cooperativity, positive vs. negative, 158
Coordination compounds, 59
Copper, deficiency, 106
Corey, R. B., 96
Cork, micrograph of, 2
Cortex, defined, 397
Cortisone, and infection, 490
Coulomb's law, 52
Counter transport, 263
Cowpox, 30
Crane, R. K., 275
Creatine, in muscle, 386
Creatine kinase, 386
Creatine phosphate, 386
Crick, F. H. C., 86, 492
Crossover, 519
Cryptic mutants, 264
Cuatrecasas, P., 440
Curare, 361
Curtis, H. J., 354
Cyanide, 268
  toxicity of, 297
Cyclic AMP:
  and catabolic repression, 214
  and cell division, 542
  cytoskeletal effects, 415
  and gene control, 226
  and histone phosphorylation, 223
  and insulin, 438

  in muscle, 394
  and secretion, 479
  synthesis and breakdown, 166
  and water flux in kidney, 258
Cyclic GMP, 168
  and cell division, 542
  and insulin, 438
  in muscle, 394
  and secretion, 479
Cycloheximide, 333
Cyclosis, 418
Cystic fibrosis, 429
Cystinuria, 265
Cytochalasins, 410
  and endocytosis, 445
  and insulin release, 476
  and lymphocyte capping, 436
  and lysosomal function, 449
Cytochrome a, 297
Cytochrome a_3, 297
Cytochrome oxidase, 297
  mitochondrial synthesis, 307, 332
Cytochrome P450, 461
Cytochromes, 295, 296
  of chloroplasts, 319, 320
  of the ER, 461
  heme of, 106
Cytokinesis, defined, 510
  the contractile ring, 529
  mechanism, 528
  membrane changes, 533
  in plants, 534
Cyton, 349
Cytoplasmic streaming, 418
Cytosine, structure, 85
Cytoskeleton, 408
  and cAMP, 415
  and endocytosis, 445
  nuclear attachments, 457
  surface connections, 437
Cytosol, defined, 596

D

Dalton, defined, 70
Danielli, J. F., 241
Davidson, E. H., 225
Davson, Hugh, 241
de Duve, Christian, 14, 21
de Robertis, Eduardo, 361
Debye, P., 55
Decarboxylase, defined, 138
Dehydroascorbic acid, 148
Dehydrogenase, defined, 138
Dehydrogenation, defined, 147
Denaturation:
  of DNA, 88
  of proteins, 100
Dendrite, 349, 372
Dermatan sulfate, 83
Dermatosporaxis, 109
Desmosine, 104
Desmosomes, 249
  and α-actinin, 399

  septate, 250
Detergents, 54
Determination, 555
Dextrorotation, defined, 79
d'Herelle, Felix, 31
Diabetes, 434, 438
  and infection, 490
  therapy of, 118
Diakinesis, 520
Diastereo isomers, 94
Diauxic growth, 209
Dictyosome, 20, 467
Dictyotene, 522
Dielectric constant, 52
Diffraction patterns, 591
Diffusion, defined, 255
  coefficient of, 261
  facilitated, 259
  vs. Michaelis-Menten transport, 261
Diffusion potential, 272
Digitalis, 268, 286
Diglyceride, defined, 75
Dihydrotestosterone, receptors for, 226
Dinitrophenol, 268, 297
Dipeptide, defined, 95
1,3-Diphosphoglycerate, 147, 161
  hydrolysis of, 147
2,3-Diphosphoglycerate, 161
Diphtheria, 40
  toxin, 200
Diplonema, 520, 522
Dipoles, 50
  induced, 54, 55
Dische, Z., 151
Disulfide bond, 93
Division furrow, 524
DNA:
  amount per cell, 224
  autoradiography of, 496
  bacterial, 12, 495
    folding pattern, 206
  conservative vs. semiconservative
    replication, 493
  crossover, 502
  denaturation of, 88, 186
  discontinuous replication, 498
  effect of ionizing radiation, 503
  genetic role, 185
  highly repetitive, 224
  hybrids with RNA, 186
  length in human cells, 507
  melting of, 88
  Okazaki fragments, 498
  palindromic, 224
  of plastids and mitochondria, 333
  point mutations, 502
  polymerases, 497
  polymerase I activity, 500
  promoter sites, 193
  pyrimidine dimers, 503
  repair of, 504
  repetitive, 224
  replication of:
    errors, 501
    in eukaryotes, 506

in nucleoli, 499
proofreading, 502
rate, 507
role of RNA, 500
rolling circle model, 499
replicons, 508
from RNA, 562
satellite, 224
single copy, 224
structure, 85
transcription of (see Transcription)
transformation by, 182
unwinding proteins, 500
UV effects, 503
X-ray diffraction of, 87
DNA ligase, 499
Dolichols, 465
Donnan, Frederick, 266
Donnan equilibrium, 266
Dopamine, 372
Douglas, W. W., 477
Down's syndrome, 236
DPG (see 2,3-Diphosphoglycerate)
Drosophila, differentiation of, 556
Drugs, detoxification by liver, 461
Dubois, Raphael, 323
Dujardin, F., 395
Dulbecco, Renato, 562
DuPraw, E. J., 509
Dutrochet, R. J. H., 3
Dutton, R. W., 127
Duysens, L. N. M., 319
Dynein, 402, 403
    congenital absence, 403
    myosinlike properties, 403
    and the spindle apparatus, 527

E

E. coli (see *Escherichia coli*)
Ebashi, S., 380, 390
Eccles, J. C., 387, 429
Ecdysone, and gene control, 226
Ectoplasm, *defined*, 22, 397
Edidin, M., 238
Eggs, formation of, 521
Ehlers-Danlos syndrome, 103, 106, 109
Einstein, A., 315
Einstein unit, *defined*, 316
Elastin:
    assembly of, 116
    and copper, 106
    crosslinking, 104
Electrochemical gradient, *defined*, 274
Electron transport (see Oxidative phosphorylation)
Electron transport chain, 294
    microsomal, 462
Electronegativity, 50
Ellermann, V., 560
Elongation factors, 200
Embden, Gustav, 144

Embden-Meyerhof pathway (see Glycolysis), 144
Embryogenesis, *defined*, 550
Emerson, Robert, 318
Emerson effect, 319
Endergonic, *defined*, 56
Endocytosis, *defined*, 443
    and blood capillaries, 447
    mechanism, 443
    and membrane replacement, 481
    selectivity, 445
Endoplasmic reticulum, 18, 27, 458 ff
    and electron transport, 462
    lipid synthesis and drug metabolism, 460
    of muscle, 374
    structure and function of the rough ER, 462
    structure and function of the smooth ER, 460
    turnover of, 482
Endosymbiosis, *defined*, 335
Endothermic, *defined*, 56
Energy of activation, 135
Energy dispersion X-ray spectroscopy, 587
Energy barrier, 59
Englesberg, E., 213
Enthalpy, 56
Entropy, 53
    relation to other parameters, 56
Enzymes:
    activation energy, 135
    active site, 136
    allosteric, 159
    cascades, 165
    competitive inhibition, 155
    constitutive, 210
    control by phosphorylation, 165
    cooperativity in, 157
    discovery, 4
    inducible, 209
    interaction with substrate, 136
    maximum velocity, 152
    nomenclature, 137, 138
    noncompetitive inhibition, 156
    regulation of, 150
    repressible, 209
    specificity of, 136
    substrate inhibition, 161
    turnover number, 154
    velocity of, 152
Eosinophils, 491
Epidermal growth factor, 541
Epigenetic changes, *defined*, 550
Epinephrine, 480
    alpha vs. beta effects, 394
    and cardiac muscle, 395
    and cyclic AMP, 166
    and enzyme cascades, 165
    and glycogen, 165
    receptors, 168, 442
    structure, 163
    structure vs. binding, 439
    and uterine contraction, 394

Episomes, *defined*, 208
Equation:
    Boltzmann, 47
    Goldman, 273
    Henderson-Hasselbalch, 63
    Hill, 157, 158
    Michaelis-Menten, 152
    Nernst, 272
    van't Hoff, 267
Equatorial plane, 515
Equilibrium constants, 57
Erythrocytes:
    avian:
        and gene expression, 552
        histones of, 216
    blood types, 433
    Ca^{2+} effect, 416
    chloride-bicarbonate exchange, 271
    determinants of shape, 416
    DPG effect, 161
    freeze-fracture of, 244
    fusion of, 238, 239
    ghosts of, 269
    glycocalyx of, 431
    membranes of, 416
        asymmetry, 248
        proteins, 244
        stability, 247
    metabolism, 144
    nuclear extrusion, 470
    permeability to glycerol, 268
    phagocytosis of, 446
    reverse hemolysis, 270
    sickle cell, 188
    sodium pump of, 269
    spherical and crenated, 268
    water permeability, 258
Erythromycin, 201
Erythropoietin, 541
Escherichia coli:
    elementary composition, 72
    molecular composition, 71
Eserine, 361
Ester, *defined*, 133
Estradiol, structure, 77
Estrogens, 77
    and gene control, 226
    as mitogens, 543
Ethanol, biological production, 150
Euchromatin, *defined*, 219
Eye color, inheritance, 176
Exchange diffusion, 263
Exergonic, *defined*, 56
Exocytosis (see Secretion)
Exothermic, *defined*, 56

F

F factors, bacterial, 208
Facilitated diffusion, *defined*, 259
Fat, *defined*, 75
    caloric value, 346
Fatty acids:
    essential, 73

Fatty acids (cont.)
  mitochondrial breakdown, 303
  structure of, 73
Fawcett, D., 402
Feedback regulation, 150
Fermentation, 150
Ferredoxin, 319, 320
Ferritin, 435
Fertility factor, bacterial, 208
Fibers, stress, 398
Fibrin:
  assembly of, 116
  isopeptide bonds, 104
Fibrinogen, 116
Fibroblast growth factor, 541
Fibroblasts, and collagen secretion, 98
Fibropapillomatosis, 566
Fiers, W., 191
Filaments:
  100 Å (10 nm), 23, 399
    in melanosomes, 421
    in nerve axons, 421
    in smooth muscle, 385
  actin, 23, 379
  cytoskeletal, 408
  myosin, 23
  non-muscle myosin, 397
Filial generation, defined, 175
Filopodia, 411
Fimbriae, 11
Flagella, bacterial, 10, 407
  assembly, 112
Flagella, eukaryotic, 24
  growth of, 405
  movement of, 403, 404
  structure of, 401
Flagellin, 112, 407
Flavin adenine dinucleotide, 295
Flavin mononucleotide, 295
Flavoproteins:
  and electron transport, 295
  of the ER, 461
  reduction in the citric acid cycle, 305
Flemming, Walter, 511
Fluorescence, defined, 316
Forces, London dispersion, 55
Fox, C. F., 277
Fraenkel-Conrat, H., 184
Franklin, Rosalind, 86
Free energy, 56
  standard state, 57
Freeze-etching, 243
Freeze-fracture, 243
Frosch, P., 31
Fructose diphosphatase, 162
Frye, L. D., 238
Fucose, in glycoproteins, 107
Furan, structure, 79
Furanose, defined, 79
Fusion, heat of, 51

G

β-Galactosidase, 209
Galactoside acetylase, 209
Galactoside permease, 209
Galactosides, transport of, 264
Gamete, defined, 175
Gametogenesis, 520
Gametophyte, 523
Gamma globulin, 119
Gamow, George, 187
Gangliosides, 83
Gap junctions, 249, 363
  effect of Ca^{2+}, 253
  in electrotonic synapses, 363
  formation of, 252
  in muscle, 393
  permeability and stability, 253
Gelatin, 100, 110
Gell, Philip, 435
Generation time, 537
Genes:
  activation in eggs, 555
  autogenous regulation of, 214
  catabolite repression, 214
  and chromatin condensation, 219
  and chromosomes, 176
  crossover, 177, 519
  cytoplasmic influence, 552
  discovery of, 175
  dominant, 175
    basis for, 181
  and histones, 223
  hormonal control, 226
  induction and repression, 209
  linked, 176
  loci, 177
  Mendelian behavior, 175
  of mitochondria and chloroplasts, 333
  mutations of, 180
  and nonhistone chromosomal proteins, 223
  number of, in vertebrates, 224
  operator, 210
  recessive, 175
  regulation in eukaryotes, 216
  regulation in prokaryotes, 206
  regulator, 210
  repressor of, 211
  sex-linked, 180, 181
  single copy, 224
  transcriptional control:
    in eukaryotes, 225
    in prokaryotes, 209
  translational control, 215, 227
Genetic code, 187, 191 (see also Codons)
  table, 190
  universal nature, 191
Genome, defined, 223
Genotype, defined, 175
Germ line, defined, 205
Gibbons, Ian, 403
Gibbs free energy (see Free energy)
Gilbert, Walter, 212
Glial cells, 349
Globulins, defined, 133
Glucagon:
  action of, 168
  and glycogen, 165
Glucocorticoids, action of, 549
Glucokinase, regulation of, 151
Glucose:
  metabolism of, 143
  transport of, 262–264, 266
    in bacteria, 276
    sodium-coupled, 275
Glucose 6-phosphatase, 460
  cellular location, 586
Glucose-6-phosphate dehydrogenase, 330
  and membrane stability, 247
  and X chromosomes, 220
Glutathione, as antioxidant, 504
Glycerides, 74, 75
Glycerol:
  esters of, 74
  transport of, 266
Glycerol kinase, 266
Glycerol phosphate shuttle, of mitochondria, 308
Glycocalyx, 431
Glycogen:
  hydrolysis of, 140
  structure, 81
  synthesis and degradation, 163
Glycogen phosphorylase, 164
  subunits of, 167
Glycogen synthetase, 164
Glycolate, 339
Glycolipids, defined, 83
  in membranes, 247
Glycolysis, 143
  comparison to aerobic pathways, 305
  control of, 161
  energetics, 145, 146
  redox reactions in, 148, 149
  in white muscle, 387
Glycopeptides, in membranes, 247
Glycophorin, 244, 433
Glycoproteins, 107
  in membranes, 247
Glycosaminoglycans, defined, 82
Glycosyl transferases, and the cell surface, 432
Glyoxylate, 339
Glyoxylate cycle, 337
Glyoxysomes, 22, 337
Goldberg, N. D., 168
Goldman, D. E., 273
Goldman equation, 273
Goldstein, Joseph, 485
Golgi, Camillo, 20, 466
Golgi body (see Golgi apparatus)
Golgi apparatus, 15, 18, 20, 27, 466ff
  and cell wall growth, 534
  formation of, 467
  function of, 467
  nomenclature, 467
  and protein glycosylation, 465
  and transition vesicles, 467
Golgi complex (see Golgi apparatus)
Gorter, E., 241
Gout, 450
Gram, Christian, 10

Gram stain, 10
Grana, of chloroplasts, 310
Granule (see Vesicle)
Granulocytes, 446
Green, David E., 298
Grendel, F., 241
Griffith, F., 182
Group translocation, 276
Growth, exponential, 537
Growth factors, 540
Growth hormone, 541
Guanine, structure, 85
Gurdon, J. B., 551

H

H-2 types, 434
Hadorn, Ernst, 556
Hair, structure of, 97, 98
Hall, B. D., 186
Halobacterium halobium, 301
Hämmerling, J., 552
Hanson, J., 375, 429
Haplotype, 434
Haptens, 121
Harris, Henry, 552
Harvey, E. N., 241
Hatch, M. D., 327
Haurowitz, F., 159
Hay fever, 126
Hayflick, Leonard, 570
Heart:
    failure of, 538
    tetraploid nuclei, 538
Heat of fusion, 51
Heat of vaporization, 51
Heilbrunn, L. V., 388
HeLa cells, origin of, 575
Heme, 296
    and dative bonds, 61
    mitochondrial synthesis, 307
    structure, 106
Hemizygous, *defined*, 178
Hemocyanin, 105
Hemoglobin:
    Bohr effect, 160
    channel through, 258
    cooperativity in, 157
    DPG effect, 161
    mRNA of, 204
    oxygen binding, 157
    pH effect, 160
    polar surface, 108
    regulation of, 160
    sickle cell, 187, 236
    structure, 104
    termination mutant, 204
    variants of, 192
Hemoglobin S, 187
Hemolysis, *defined*, 268
Hemophilia, inheritance of, 180

Henderson-Hasselbalch equation, 63
Heparin, 83
Hepatocytes, and endocytosis, 446
Hershey, A. D., 183
Heterochromatin, *defined*, 219
Heterogeneous nuclear RNA, 196 (*see also* RNA)
Heterokaryon, *defined*, 238
Heterophagy, 451
Heterophil, 491
Heterotroph, 10
Heterozygous, *defined*, 175
Hexose monophosphate shunt, 329 (*see* Phosphogluconate pathway)
Hibernation, and heat production, 306
Hildebrandt, A.C., 551
Hill coefficient, 157
Hill equation, 157, 158
Histamine, 480
    release of, 479
Histiocyte, 445
Histocompatibility complex, 434
Histones:
    classification, 216
    and chromatin condensation, 539
    genes coding for, 224
    H1 and gene regulation, 223
    synthesis during the cell cycle, 511
HLA types, 434
HnRNA, 196
Hodgkin, A. L., 269, 355, 358, 429
Holley, R. W., 197
Homeostasis, *defined*, 134
Homozygous, *defined*, 175
Hooke, Robert, 1
Hormones:
    and cell division, 543
    desensitization to, 443
    effects on muscle, 394
    and gene regulation, 226
    mobile receptor hypothesis, 440
    neurosecretory, 368
    receptors for, 438, 442
    steroid, 77, 226, 543
Huebner, Robert, 561
Huie, P., 474
Humans, elementary composition, 72
Humphreys, T., 431
Hunter's disease, 450
Hurler's disease, 450
Husson, F., 243
Huxley, A. F., 358, 394, 429
Huxley, H. E., 375, 384, 429
Hyaluronic acid, 83
Hydration, of ions, 52
Hydrocarbons, oxidation of, 148
Hydrogen peroxide, 21
    and bacterial killing, 449
    production in cells, 336
Hydrogenations, *defined*, 147
Hydrolase, *defined*, 138
    lysosomal, 447
Hydrolysis, *defined*, 133, 138
    free energy of, 140
Hypercholesterolemia, 491

Hypertonic, *defined*, 268
Hypotonic, *defined*, 268

I

Ice, structure of, 51
Ig (*see* immunoglobulin)
Imaginal discs, 556
Immune globulin, *defined*, 125
Immunity:
    anamnestic (secondary) response, 127
    humoral vs. cellular, 119
    tolerance, 127
Immunoglobulin, *defined*, 125
Immunofluorescence, 397, 580
in vitro, *defined*, 28
in vivo, *defined*, 28
Inducers, embryonic, 555
Ingram, V. M., 187
Initiation factors, 203
Inosine:
    in tRNA, 198
    in muscle, 386
Inoué, S., 526
Insulin:
    action of, 95, 168, 265
    activation of, 108
    and glucose transport, 265
    as growth factor, 540
    mechanism of secretion, 474
    receptors for, 438, 439, 440
    turnover, 442
Interkinesis, 520
Interphase, 510, 520
Interactions:
    dipole-dipole, 55
    dipole-induced dipole, 55
    ion-dipole, 54
    London, 55
    van der Waals, 54
Ion trapping, 256
Ionophores, 279
    and oxidative phosphorylation, 300
    and secretion, 478
Ions:
    hydronium, 62
    sequestered, 60
Iron, oxidation and reduction, 296
Iron-sulfur proteins, 297
Isocitrate lyase, 338
Isodesmosine, 104
Isoenzymes, *defined*, 137
Isolation body, 451
Isolation envelope, 451
Isomerase, *defined*, 138
Isomers, optical, 78, 94
Isoosmotic (isosmotic), *defined*, 268
Isoprene, structure of, 74
Isotonic, *defined*, 268
Isotopes:
    labeling with, 183
    pulse labeling, 598
    table of, 597

609

Isozymes, *defined*, 137
Ivanovsky, D., 30

J

Jacob, Francois, 189, 210
Janssen, Zacharias, 577
Janssens, F. A., 177
Jenner, Edward, 30, 118
Jerne, N. K., 126
Jöbsis, F. F., 389
Johnson, R. T., 539
Junctions:
   intercellular, 248
   myoneural, 349
   neuromuscular, 349

K

K_a, 63
K_{eq}, 57
Karyokinesis, *defined*, 510
Karyotype, *defined*, 179
Katz, Bernard, 355, 359
Kendrew, J. C., 101, 104
Kennedy, E. P., 277
Keratin, 96
   isopeptide bonds, 104
Keratinocytes, 421
Keratosulfate (keratan sulfate), 83
Ketoacidosis, 337
Ketone bodies, 347
Ketose, *defined*, 78
Keynes, R. D., 269
Khorana, H. G., 189
Kinase, *defined*, 138
Kinases, protein, 166
Kinetics:
   of enzymes, 152
   Michaelis-Menten, 152
Kinetochore, 515
   as microtubule organizers, 114, 525
   and spindle assembly, 525
Kingsbury, B. F., 288
Klein, W. H., 225
Kohne, D. E., 225
Kok, Bessel, 317
Kölliker, A., 288
Kornberg, Arthur, 497
Krebs, E. G., 167
Krebs, H. A., 302
Krebs cycle (*see* Citric acid cycle)

L

Lac operon, 210
Lactate dehydrogenase, reaction of, 138
Lactation, 480
Lactoferrin, 449
Lactose, structure, 81
Lamellipodia, 422
Landsteiner, Karl, 121, 433
Langmuir, I., 240
Langmuir trough, 241
Lauffer, Max, 113
Laurence, E. B., 544
Leblond, C. P., 467
Lecithin, 76
Lectins, 437
   agglutination of cells, 534
Leder, J., 190
Leeuwenhoek, Anton van, 1, 8, 577
Lenard, Philipp, 315
Leprosy, 450
Leptonema, 517
Leucoplast (leukoplast), 25, 309
Leukemia, 558, 560
   feline, 565
Leukemia viruses, 561
Leukocytes:
   classes of, 491
   origin of PMN granules, 467
Levorotation, *defined*, 79
Lewis, G. N., 48, 59, 62
Liebig, Justus von, 4
Ligands, *defined*, 60
Lignin, 25
Lincomycin, 201
Lineweaver-Burk plot, 153
Lipase, *defined*, 75
Lipidoses, *defined*, 83
Lipids, *defined*, 72
   amphipathic, 77
   exchange proteins, 484
   self-assembly, 117
Lipmann, Fritz, 140
Lipofuscin, 451, 568
Lipomas, 558
Lipoproteins:
   high-density, 107
   low-density, 107
   and membrane composition, 248, 484
   receptors for, 485
   very low density, 106
Liposomes, 117
   and lysosomal storage disease, 450
Liver:
   drug metabolism, 460
   molecular composition, 71
   regeneration of, 539
   tetraploid nuclei, 538
Locke, Michael, 451, 474
Löffler, F., 31
Loewenstein, W. R., 252, 400
Loewi, O. H., 362
London, F., 55
London dispersion forces, 55
Luciferase, 323
Luciferin, 323
Lumisomes, 324
Lundsgaard, E., 386
Luzzati, V., 243
Lwoff, André, 211
Lymphocytes:
   activation of, 435, 542
      by lectins, 437
   and antibody production, 126
   capping of, 482
   effect of steroids, 543
   in immunity, 119
   and membrane lysis, 247
   surface receptors, 435
   T vs. B, 119
Lymphomas, 558
Lymphotoxins, 247
Lyon, Mary, 220
Lyonization, 220
Lysogeny, 35, 40
Lysosomes, 21, 27, 447
   in autophagy, 451
   and bacterial killing, 449
   defects in, 449
   and multivesicular bodies, 449
   origin of, 467
   primary, 448
   and residual bodies, 451
   secondary, 448
Lysosomal storage diseases, 450
Lysozyme, 10, 449
   structure, 136
Lysyl oxidase, 103, 104, 106

M

MacLeod, C. M., 182
Macromolecules, *defined*, 70
   density of, 601
Macromolecular assembly:
   enzymatic, 111
   forces, 112
   irreversible, 115
   self-assembly, 111
   template assembly, 111
Macronucleus, replication of, 508
Macrophage, 445, 481, 490
Macula adhaerens, 249
Malaria, 236
Malate shuttle, of mitochondria, 308
Malate synthase, 338
Maltose, structure, 81
Mammary glands, 480
Manton, I., 402
Marchesi, V. T., 244
Marek's disease, 565
Marker, K., 203
Mason, Howard S., 461
Mast cells, 126, 478
Maternal messages, 555
Matthaei, J. H., 189
McCarty, M., 182
McElroy, W. D., 323
McIntosh, J. R., 527
Meiosis, 517
   the first division, 517
   mechanism of chromosome
      movement, 526
   in plants, 523
   reduction division, 517
Melanin, 421, 429
Melanocytes, 421
Melanoma, 429

Melanosomes, 421
Membrane potential, 271 (see also Action potential)
Membrane transport, defined, 259
　active, 274
　advantages of proteins as carriers, 279
　of amino acids, 263
　competition, 261
　and counter transport, 263
　cryptic mutants, 264
　of galactosides, 264, 276
　genetics of, 264
　of glucose, 266
　of glycerol, 266
　by group translocation, 276
　hormonal effects, 265
　inherited defects in, 265
　ionophores, 280
　isolation of transport proteins, 277
　kinetics, 261
　of lactose, 264
　metabolically coupled, 266
　nonprotein carriers, 279
　periplasmic binding proteins, 278
　proton-coupled, 276
　sodium-coupled, 275
　sodium-potassium exchange pumps, 270
　sodium-potassium stimulated ATPase, 270
　sodium pump, 268
　of sugars, 262–264
Membranes:
　assembly of, 117
　asymmetry of, 247
　cellular, interconversion of, 27
　closure of holes, 531
　composition, 240
　diffusion potential, 272
　Donnan equilibrium, 266
　effect of cholesterol, 246
　effect of fatty acid composition, 246
　electron microscopy of, 242
　equilibrium potential, 272
　fluid mosaic model, 243
　fluid properties of, 238
　freeze-etching, 243
　freeze-fracture, 243
　insertion into the cell surface, 534
　mitochondrial, 290
　models of, 241
　permeability of, 254
　　to Cl^-, 271
　phase transition, 246
　plasma, 431
　pores, 243, 257, 258
　potentials, 271
　proteins of, 244
　　mobility, 248
　stability of, 247
　synthesis, 482
　transport (see Membrane transport)
　turnover, 482, 534
　unit model, 242
　X-ray diffraction of, 242

Mendel, Gregor, 174
Mendel's laws, 175
β-Mercaptoethylamine, 302
Mercury, coordination complexes, 64
Meringue, 110
Meromyosin, 381
　binding to actin, 385
　interaction with non-muscle actin, 395
Meselson, M., 493
Mesosome, 10, 13
Messenger RNA (see RNA, messenger)
Metabolism, defined, 143
　outline of, 151
　stepwise, 144
Metaphase, meiotic, events of, 520
Metaphase, mitotic, events of, 513
Metaphasic plate, 515
Metastasis, defined, 558
Meyer, A., 310
Meyerhof, Otto, 144
Micelles, 54
Michaelis constant, defined, 152, 153
　for membrane transport, 261
Michaelis-Menten equation, 152
　limitations of, 154
　and membrane transport, 261
Michaelis-Menten kinetics, 152
Microbodies (see Peroxisomes)
Microfilaments, 22 (see also Actin)
　assembly of, 410
　and cAMP, 415
　and cell division, 529
　cytochalasin effect, 410
　and cytoplasmic streaming, 418
　and endocytosis, 445, 449
Microplicae, 412
Microscopes:
　choice of, 588
　comparison to X-ray diffraction, 590
　dark-field, 578
　electron, 581
　　sample preparation, 243, 581
　fluorescence, 580
　light, 577
　　sample preparation, 579
　phase-contrast, 578
　polarization, 578
　resolution of, 577
　scanning, 586
Microsomes, 20, 185
　formation and composition, 459
Microtubules, 23
　assembly of, 113, 114
　axonemal, 401
　associated proteins, 114
　and axoplasmic transport, 419
　and cAMP, 415
　and chromosome movement, 526
　colchicine effect, 115
　cross-section, 413
　cytoplasmic distribution, 419
　and cytoplasmic streaming, 419
　cytoskeletal, 408
　and endocytosis, 445, 449
　and membrane proteins, 410

　and neuronal processes, 410
　organizing centers, 114
　regulation by Ca^{2+}, 414
　in suctorian tentacles, 414
Microvilli, 18
　contraction of, 399
　diagram, 413
　intestinal, 247
　and microfilaments, 412
　as surface redundancies, 534
Midbody, 527, 533
Midpiece, of spermatozoa, 404
Miledi, R., 359
Mishell, R. I., 127
Mitchell, Peter, 298
Mitochondria, 15, 22
　ATPase, 294
　of brown fat, 306
　and calcium regulation, 309, 391, 399
　composition of matrix, 291
　composition of membranes, 290
　cristae of, 22, 290
　　conformational changes in, 299
　discovery, 288
　distribution, 288
　DNA of, 307, 333
　evolutionary origin, 330
　factor one, 294
　genes of, 307
　half-life, 451
　heat production, 306
　inner membrane spheres, 290, 294
　intermembrane space, 289
　lipids of, 290
　matrix of, 290
　membrane turnover, 484
　and protein synthesis, 307
　replication of, 333
　respiratory assembly, 290
　size and structure, 289
　steroid and heme synthesis in, 307
　structure-function relationships, 293
　transport properties, 307
　tubulovesicular cristae, 307
Mitogen, defined, 539
Mitosis, 510
　evolution of, 517
　in lower eukaryotes, 516
　mechanism of chromosome movement, 526
Mizell, M., 564
Mobile receptor hypothesis, 440
Mollenhauer, H., 467
Mommaerts, W., 387
Monocyte, 445, 491
Monod, Jacques, 189, 210
Monoglyceride, defined, 75
Mononucleosis, 567
Mooseker, M., 399
Morell, A. G., 446
Morre, D., 467
Moscona, A. A., 431
mRNA (see RNA, messenger)
Mucopolysaccharides, defined, 82
　synthesis by the Golgi apparatus, 469

611

Mucopolysaccharidoses, *defined*, 83
Mucoproteins, 83, 107
Mueller, P., 358
Müller-Hill, B., 212
Mullerian duct inhibiting factor, 236
Multiple myeloma, 124
Multiple sclerosis, 434
Multivesicular body, 449, 470
Murexide, 389
Muscle:
 action potentials, 393
 ATP maintenance, 386
 C protein, 381
 and calcium, 388
 cardiac:
  contraction of, 393
  fibrillation, 393
  and hormones, 395
 catch, 394
 control by myosin, 391
 control by troponin, 390
 excitation-contraction coupling, 390, 392
 glycerinated, 383
 hormonal effects, 394
 initiation of contraction, 388
 innervation, 393
 of invertebrates, 378
 M line protein, 384
 molecular composition, 71
 red vs. white, 387
 rigor, 389
 slack length, 384
 sliding filament model, 382
 smooth, 377
  contraction of, 385, 392
  and hormones, 394
  proteins of, 385
  tone of, 394
 striated, 374
 tetany, 393
 tonic contraction, 393, 394
 transverse tubules, 394
Muscular dystrophy, 416
 inheritance of, 180
Mutagens, 502
Mycoplasma, 12
Myelin, 349
Myeloperoxidase, 449
Myoblast, 374
Myofibril, 374
Myofilaments, 374
Myoglobin:
 oxygen binding, 157
 polar surface, 108
 structure, 101, 103
Myokinase (*see* Adenylate kinase)
Myoneural junction, 349
Myosin, 23
 ATPase activity, 381
 and Ca^{2+} sensitivity, 391
 and cytokinesis, 531
 discovery, 378
 filaments of, 380
 interaction with non-muscle actin, 395

 non-muscle, 397
 phosphorylation of, 395
 relationship to spectrin, 416
 of smooth muscle, 385
 subunits of, 380

N

N-Acetylneuraminic acid (*see* Sialic acid)
NAD^+, 148, 149
NADH:
 and the citric acid cycle, 305
 oxidation by O_2, 150
$NADP^+$ (NADPH), 149
Nass, M., 347
Neoplasm, *defined*, 558
Nernst equation, 272
Nerve, 349
 myelinated (medullated), 349
 optic, 365
Nerve cell (*see* Neuron)
Nerve fiber, 349
Nerve growth factor, 541
Nerve impulse, *defined*, 349
Neuraminic acid (*see* Sialic acid)
Neuraminidase, 446
Neurite, 349
Neuroblastoma (micrograph), 350
 differentiation of, 564
Neurofilaments, 420
Neuromuscular junction, 349
Neurons, 349
 action potential, 353
 adrenergic vs. cholinergic, 362
 axonal microtubules, 410
 axonal regeneration, 419
 axoplasmic transport, 419
 myelinated, 349
 nodes of Ranvier, 349
 resting potential, 354
Neurosecretion, 368
Neurospora, mutations in, 180
Neurotransmitters, 362
Neutra, M., 467
Neutrophil, 445, 481, 490
 origins of specific granules, 467
Nevus, 429
Nexus, 250
Niacin, 148
Nicolson, G. L., 246
Nicotinamide, 148
Nicotinamide adenine dinucleotide (*see* NAD^+)
Nicotinamide mononucleotide, 148
Nicotinic acid, 148
Niedergerke, R., 429
Niemann-Pick disease, 450
Nirenberg, Marshall, 189, 190
Nissl substance, 361
Nitrogen, fixation of, 346
Nitrogen mustard, 502
Nitsch, J. P., 551
NMN, 148
Nonheme iron proteins, 297

Nonhistone chromosomal proteins, 223
 and gene regulation, 225
Noradrenaline (*see* Norepinephrine)
Norepinephrine, structure, 163
Nuclear body, 13
Nuclear envelope, 27
Nuclease, *defined*, 138
Nucleic acids (*see also* DNA and RNA):
 density of, 601
 elementary composition, 71
Nucleoid, 13
 structure, 207
Nucleolar organizer, 194
Nucleosomes, 218
Nucleolus, 17
 dispersal during prophase, 512
 DNA replication in, 499
 and mRNA translocation, 196
 in oocytes, 522
 structure of, 194
Nucleosides, *defined*, 84
Nucleotides, *defined*, 84
Nucleus, 16
 and annulate lamellae, 458
 bacterial, 13
 blebs of, 457
 envelope of, 14, 16, 453
  fragmentation during prophase, 512
  reassembly at telophase, 515
 pores, 14, 16, 453–455
 pore complex, 455
Nystatin, and membrane pores, 258

O

Occam, William of, 549
Occam's Razor, 526, 549
O'Connor, M., 389
Oligopeptide, *defined*, 95
Oligosaccharide, *defined*, 80
Oncogenes, 561
Oocytes, 521
Oogenesis, 521
Oogonium, 521
Ootids, 521
Operon:
 arabinose, 213
 lac:
  operator of, 213
  promoter of, 213
  repressor of, 212
 positive vs. negative control, 213
 theory of, 210
Opsonin, 445
Optical activity, *defined*, 79
Orgel, L. E., 568
Ornithine, 541
Ornithine decarboxylase, 541
 lifetime of, 228
Osmoles, *defined*, 268
Osmotic pressure, *defined*, 267
Osteomalacia, 265
Ova, formation of, 521
Ovalbumin, denaturation of, 110
Overton, E., 240, 254

Oxaloacetate, condensation with acetate, 303
Oxidases, defined, 138
  mixed function, 461
Oxidation, defined, 147
Oxidative decarboxylation, defined, 303
Oxidative phosphorylation, 294, 297
  in bacteria, 332
  chemical coupling, 297
  chemiosmotic hypothesis, 298
  conformational coupling, 298
  electrochemical coupling, 298
  uncouplers, 297, 346
Oxygen:
  as a mutagen, 503
  reduction of, 150
Oxytocin, 368, 474
  and uterine contraction, 388, 394

P

P690 (P680), 319
P700, 318
P890, 330
Pachynema, 519
Palade, George, 14, 288, 473, 582
Palindromes, defined, 224
Pancrease, β-cells, 474
  secretion by, 471
Pantothenic acid, 302
Papain, mechanism of, 137
Paramyosin, 378
Parkinson's disease, 372
Partition coefficients, 255
Pasteur, Louis, 4, 8, 30
Pasteur effect, 294
Paul, J., 223
Pauling, Linus, 96, 187
Penicillin, 10
Pentose phosphate pathway, 329 (see Phosphogluconate pathway)
Peptide bond, 94 (see also Bonds)
  stability of, 95
Peptidyl transferase, 199
Perikaryon, 349
Perinuclear cisterna, 453
Perinuclear space, 453
Periplasmic space, 278
Permeability coefficients, 255
Permeases, 259
Peroxisomes, 21, 27, 335
  evolution of, 341
  glyoxylate cycle, 337
  half-life, 451
  and photorespiration, 338
Perutz, M. F., 104, 159
Pfeffer, W., 254
pH, defined, 63
Phage (see Bacteriophage)
Phagocytosis, 443 (see also Endocytosis)
  and antibodies, 445
  and membrane replacement, 481
  and polysaccharide recognition, 446

Phagosome, defined, 443
Phenobarbital, 460, 462
Phenotype, defined, 175
Phosphatase, defined, 138
  of protein, 167
Phosphates, "high energy," 139
Phosphatidic acid, 76
Phosphatidyl choline, 76
Phosphocreatine, hydrolysis of, 140
Phosphoenolpyruvate:
  in C4 photosynthesis, 327
  hydrolysis of, 140, 147
  synthesis, 277
Phosphofructokinase, regulation of, 161, 167
Phosphogluconate pathway, 328, 329
  and membrane stability, 247
Phosphoglycerides, 74, 75, 76
Phosphorescence, defined, 316
Phosphoric acid, titration, 64
Photocytes, 324
Photoelectric effect, 315
Photophosphorylation, 320
  cyclic, 322
Photoreceptors, 365
Photorespiration, 338
Photons, defined, 315
Photosynthesis:
  C4 pathway, 327
  the Calvin pathway (C3 pathway), 325
  carbon dioxide fixation, 324
  coupling factors, 322
  Crassulacean acid metabolism, 328
  the dark reaction, 324
  efficiency, 322, 327
  evolution of, 339
  evolution of the C4 pathway, 341
  the light reaction, 314
  and photorespiration, 338
  in prokaryotes, 330
Phragmoplast, 534
Phycobilosomes, 313
Phytohemaglutinin, 437
Pigment systems, 317
Pili, 11
  and bacterial mating, 208
Pinocytosis, defined, 443 (see also Endocytosis)
  and membrane replacement, 481
pK_a, defined, 63
Planck, Max, 316
Planck's constant, 316
Plasma (blood), defined, 133
Plasma cells, 119
  and antibody production, 126
Plasma membrane (see Membranes)
Plasmalemma (see Membranes)
Plasmids, defined, 208
Plasmodesmata, 253, 535
Plasmolysis, 254
Plastids, 25, 309
Plastocyanin, 319, 320
Plastoquinones, 296, 319, 320
Pneumococcal bacteria, 182
Podophyllum, 549

Podophyllotoxin, 115
Polar body, 522
Poles (of a cell), defined, 512
Pollard, T. D., 399
Pollen, 523
Polyamines, as growth factors, 541
Polymerases (see DNA or RNA)
Polymorphonuclear cells, defined, 491
Polynucleotides, 84
  hydrolysis of, 140
Polyoma virus, 561
Polypeptide, defined, 95
  hydrolysis of, 140
Polysaccharides, defined, 80
Polysome, 200
Pores, nuclear, 14, 16, 455
Porter, K., 402
Porter, R. R., 124
Procaine, 358
Procentriole, 512
Procollagen, 109, 115
Proenzyme, defined, 108
Proinsulin, 108
Proline:
  as helix breaker, 100
  in collagen, 100
  structure, 93, 95
Prometaphase, 549
Prometheus, 549
Prophage, 35
Prophase, meiotic, 517–520
Prophase, mitotic, 511
Proplastid, 25, 309
Prostaglandins, 73, 480
  and uterine contraction, 394
Protamine, 358
Protease, defined, 109, 138
Proteins, 91, 95
  caloric value, 346
  conformation of, 95
  conformational isomers (conformers), 108
  conjugated, 104
  crystals of, 102
  denaturation of, 100, 110
  density of, 601
  effects of pH and solvents, 110
  elementary composition, 71
  glycoproteins, 107
  glycosylation, 464
  α helix, 96
  isopeptide bonds, 104
  lipoproteins, 106
  and membrane transport, 259
  multifunctional, 204
  pleated sheets, 96
  primary structure, 95
  prosthetic groups, 104
  protomers of, 103
  quaternary structure, 103
  secondary structure, 96
  self-assembly, 112
  solubility of, 110
  structural stability, 109
  subunits, 103

Proteins (*cont.*)
  synthesis in mitochondria and chloroplasts, 197
  synthesis on rough ER, 463
  tertiary structure, 100
  turnover, 227, 228
  typical size, 236
Proteoglycans (*see* Glycoproteins)
Protofilaments, of microtubules, 23
Protoelastin, 116
Protoplast, 12
Protozoa:
  diagram, 6
  discovery, 2
  micrographs of, 3, 25
Provirus, 35
Pseudopod, 422
Psoriasis, 434
Ptashne, Mark, 213
Pulse labeling, 598
Punnett square, 176
Puromycin, 204
Putrescine, 541
Pyran, structure, 79
Pyranose, 79
Pyrophosphatase, 192, 198
Pyrophosphate:
  in ATP, 139
  in DNA synthesis, 497
  hydrolysis of, 140
  in RNA polymerization, 192
Pyruvate, reduction of, 149
Pyruvate dehydrogenase complex, 302
Pyruvate kinase:
  isozymes, 137, 162
  reaction of, 147
  regulation of, 162

Q

Quantasomes, 312
Quinones, and electron transport, 296

R

R factors, 208
Racker, Efraim, 294, 301
Radiation:
  electromagnetic, 314
  "safe" doses, 505
Radon gas, 505
Rao, P. N., 539
Redox reaction, *defined*, 147
Reduction, *defined*, 147
Regeneration, of limbs, 557
Regulon, 213
Release factors, 204
Replicases, 497
Replicons, 508

Residual bodies, 451
Resistance factors, 208
Respiratory chain (*see* Electron transport chain)
Retinal, 301, 365
Retinol, and protein glycosylation, 465
Rh blood types, 433
Rheumatoid arthritis, 450
Rho factor, and DNA transcription, 193
Rhodopsin, 301, 365
Riboflavin, 295
Ribonuclease, 189
Ribonucleic acid (*see* RNA)
Ribosomal RNA (*see* RNA, ribosomal)
Ribosomes:
  70S, 202
  80S, 202
  assembly of, 203
  attachment to ER, 463
  binding sites of, 199
  during translation, 199
  half-life, 451
  of mitochondria and chloroplasts, 333
  in prokaryotes, 13
  structure, 202
Ribulose diphosphate, 325
Ribulose diphosphate carboxylase, 325
  and chloroplast genes, 332
  and photorespiration, 339
Rickets, 265
Ridgway, E. B., 389
Ringer, S., 388
RNA:
  effect of actinomycin D, 553
  genetic potential, 184
  heterogeneous nuclear (hnRNA), 196
  hybrids with DNA, 186
  maternal messages, 555
  messenger (mRNA), 185
    and 5' caps, 196
    eukaryotic, 196
    and poly A, 196
    survival of, 215
    turnover in eukaryotes, 227
  polymerases, 187, 192
  and promoter affinity, 215
  ribosomal:
    genes for, 224
    synthesis, 215
    transcription of, 194
  transfer:
    charging of, 198
    description of, 197
    genes for, 224
    initiation by, 203
    methionyl, 203
    processing of, 193
Robertson, J. D., 241
Rod cells, 365
Root cap slime, 468
Rose, B., 399
Roseman, Saul, 432
Rotenone, 300
Rous, P. F., 560
Rous sarcoma virus, 560, 561

rRNA (*see* RNA, ribosomal)
Rudin, D. O., 358

S

Sabatini, David, 463
Saccharides, *defined*, 78
Sanger, F., 96, 203
Sarcolemma, *defined*, 429
Sarcomas, 558
Sarcomeres, 374
Sarcoplasm, *defined*, 429
Sarcoplasmic reticulum, 374
  and terminal cisternae, 374
  and triads, 374
Sarcosomes, 287, 429 (*see also* Mitochondria)
Satir, Peter, 403
Schimke, R. T., 228
Schizophrenia, 372
Schleiden, M. J., 3
Schramm, G. S., 184
Schwann, Theodor, 3
Schwann cells, 349
Scurvy, signs of, 133
Sebaceous gland, 470
Sebum, *defined*, 75
Secretion:
  apocrine, 470
  calcium effect, 478
  and cyclic nucleotides, 479
  eccrine (merocrine), 471
  fate of the vesicle, 480
  holocrine, 470
  types of, 470
  vesicle transport, 474
Secretory granule, 20
Secretory vacuole, 20
Sedimentation coefficient, 593
Sedimentation equilibrium, 595
Sedimentation velocity, 592
Seeds, 523
Sell, Stewart, 435
Sendai virus, and cell fusion, 238
Serotonin, and uterine contraction, 394
Serum (blood), *defined*, 133
Sex, phenotypic development in humans, 236
Sheath, prokaryotic, 11
Sialic acid:
  and endocytosis, 446
  in glycoproteins, 107
  structure, 83
Sickle cell disease, 187, 418
Siderochromes, 281
Siderophores, 281
Sigma factor, of RNA polymerase, 192, 193
Singer, S. J., 246, 410
Skou, J. C., 269
Slack, C. R., 327
Slater, E. C., 297

Smallpox, 30
Smith, Erwin F., 560
Soap, 54, 75
Sodium pump, 268
　electrogenic, 270
　reconstitution, 278
　reversibility, 270
Sodium-potassium exchange pump, 270
Soluble RNA (see RNA, transfer)
Soma, 349
Somatic hybrid, defined, 238
Somatomedins, 541
Sörenson, S. P. L., 110
Sparsomycin, 201
Spectrin, 416
Spermatids, 520
Spermatocyte, 520
Spermatogenesis, 520
Spermatogonium, 520
Spermatozoa:
　acrosomal reaction of, 410
　defects in dynein, 403
　differentiation, 520
　function of centrioles, 406
　mitochondria of, 404
　nuclear shaping of, 516
　pattern of movement, 404
Spermidine, 541
Spermine, 541
Spherocytosis, 416
Spheroplast, 12
Sphingolipids, 74, 77
Sphingomyelin, structure, 76
Sphingosine, structure of, 74
Spiegelman, S., 186
Spindle apparatus:
　birefringence, 524
　and contractile rings, 528, 529
　during anaphase, 515
　during metaphase, 513
　mechanism of chromosome movement, 526
　pole-to-pole separation, 524
　structure and function, 523
Spontaneous generation, 8
Spores, 492, 523
Sporophyte, 523
Squid, giant axon, 353
sRNA (see RNA, transfer)
Stahl, F. W., 493
Stains, cytochemical, 580, 586
Stains, histochemical, 580, 586
Stanley, Wendell, 37, 184
Starch, structure, 81, 82
Stereoisomer, defined, 78
Steroids, 74, 77
　and gene control, 226
　mitochondrial synthesis, 307
　and smooth ER, 460
Steroid hormones, and cell division, 543
Steward, F. C., 551
Stoeckenius, Walther, 301
Stomatocytosis, 416
Streptomycin, 201
　resistance to, 201

Stress fibers, 398
Stroma, of chloroplasts, 311
Substrate, of enzymes, 136
Succinate dehydrogenase, as mitochondrial marker, 387
Succulents, photosynthesis in, 328
Sucrose:
　cell impermeability to, 254
　hydrolysis of, 140
　structure, 80
Sulfa drugs, and membrane stability, 247
Sulfatides, 83
Sulfhydryl compounds, as antioxidants, 504
Sutherland, Earl, 166
Sutton, Walter S., 176
SV40, 561
　chromosomes of, 219
Svedberg coefficient, 593
Sweat glands, 471
Synapse, defined, 349
　electrotonic, 363
　excitatory and inhibitory, 363
　and gap junctions, 363
　neuronal, 359
Synapsis, 519
Synaptonemal complex, 519
Syncytium, 374
Szent-Györgyi, Albert, 237, 379
Szent-Györgyi, Andrew, 391

T

T2, T4 (see Bacteriophage)
Tatum, E. L., 180
Tay-Sachs disease, 450
Taylor, D. L., 422
Taylor, J. Herbert, 506
Taylor, R. E., 394
TCA cycle (see Citric acid cycle)
Telophase, meiotic, 520
Telophase, mitotic, 515
Telophase neck, 533
Temin, Howard, 562
Teratocarcinoma, 564
Terminal web, 399
Terpene, defined, 74
Testicular feminization, 226
Testosterone, 77
　in fetal development, 236
Tetracycline, 201
Tetrahedron, defined, 72
Tetraploid, defined, 522
Theophylline, 173, 480
Thermiogenesis, 306
Thermodynamics:
　first law, 57
　second law, 51, 57
Thiamine, 302
Thrombin, 116
Thromboxanes, 74
Thylakoids, 311, 321
Thymine, structure, 85
Thyroglobulin, 93

Thyroxine, 93
Tight junctions, 249
Tilney, L. G., 399, 410
Tissues, defined, 248
TMV (see Viruses, tobacco mosaic)
Todaro, George, 561
Tonofilaments, 249
Townsend, C. O., 560
Toxoplasma, 450, 491
Transcriptase (see RNA, polymerase)
Transcriptase, reverse, 562
Transcription (of DNA) defined, 187
　control in eukaryotes, 225
　control in prokaryotes, 210
　rate of, 210
　and RNA polymerase, 192
　of rRNA, 194
　posttranscriptional processing, 193
　promoter sites, 193
　start and stop signals, 193
Transdetermination, 556
Transfer RNA (see RNA, transfer)
Transferrin, 281
Transformation, bacterial, 182
Transformation, malignant, 558
Translation, 198
　control of, 227
　energetics of, 200
　errors in, 200
　initiation of, 203
　rates of, 200
　termination of, 204
　termination errors, 204
Transport (see Membrane transport)
Transport proteins, 259
Tricarboxylic acid cycle (see Citric acid cycle)
Triglyceride, defined, 75
Trisomy, defined, 236
tRNA (see RNA, transfer)
Tropocollagen, 98, 115
Tropoelastin, 116
Tropomyosin, 380
　non-muscle, 398
　of smooth muscle, 385
Troponin, 380
　of smooth muscle, 385
Trypanosomes, 347
Trypsin, mechanism of, 137
Tryptophan pyrrolase, 228
Tuberculosis, 450
d-Tubocurarine, 361
Tubulin, 113
　axonemal, 401
Tumors, defined, 558
Tumorigenesis, 560
Twort, F. W., 31
Tyrian purple, 389

U

Ubiquinones, 296
UDP-glucose, 164
Ultraviolet light, as a mutagen, 503

615

Uracil, structure, 85
Urea:
  metabolic origin, 303
  synthesis of, 4

V

Vaccination, 118
Vacuolar system, 469
Vacuole, 15, 25
  digestive, 448
Valinomycin, 279
  and oxidative phosphorylation, 300
van der Waals attraction, 54, 55
van der Waals radius, 56
van Leeuwenhoek, Anton, 1, 8, 577
van't Hoff, Jacobus, 267
van't Hoff equation, 267
Vaporization, heat of, 51
Vasil, V., 551
Vasopressin, 258, 368, 474
Vegetable oil, *defined*, 73
Vesicle, *defined*, 25
Vinblastine, 115, 549
  and insulin release, 476
Vinca alkaloids, 115
Vincristine, 115
Virchow, Rudolf, 3
Viroid, 38
Viruses, 30
  animal, 31, 37
  bacterial (*see* Bacteriophage)
  and cancer, 560
  Epstein-Barr, 566
  herpes, 566, 575
  leukemia, 561
  life cycle, 40
  lysogenic, 40
  plant, 31, 37

polio, 204
polyoma, 561
Rous sarcoma, 561
SV40, 219, 561
tobacco mosaic, 30, 38
  assembly of, 112
  genes, of, 184, 185
  RNA of, 191
type C, 562
Vision, color, 365
Vision, scotopic, 365
Vitalism, 4
Vitamin A, 301, 309, 365
Vitamin B_1, 302
Vitamin B_2, 295
Vitamin C, 80, 247
  as antioxidant, 504
  deficiency of, 133
  as a mutagen, 503
  oxidation of, 148
  structure of, 148
Vitamin D, 77
  and Ca^{2+} transport, 265
Vitamin E, 247
  as antioxidant, 504
Vitamin K, 296
Voltage clamp, 356
von Liebig, Justus, 4

W

Wald, George, 41
Waller, J. P., 203
Water:
  as an acid or base, 63
  cell permeability to, 254
  in coordination compounds, 60
  formation of, 135
  intracellular, 65

  properties of, 65
  structure of, 51
Watson, J. D., 86, 492
Wax, *defined*, 74
Weismann, August, 550
Wiercinski, F. J., 388
Wilkins, M. H. F., 86
Wilson, E. B., 1, 176
Wobble hypothesis, 191, 198, 199
Wöhler, Friedrich, 4
Wood, micrograph of, 2
Wool:
  elasticity of, 98
  structure of, 99

X

X-Ray diffraction, 589
Xanthines, 173
Xenopus, cloning of, 551
Xeroderma pigmentosum, 504

Y

Yeast, fermentation in, 150
Young, J. Z., 349

Z

Zeus, 549
Zona pellucida, 522
Zonula occuludens, 249
Zwitterions, 94
Zygonema, 519
Zymogen, *defined*, 108
Zymogen granules, 473